Recent Advances in Geological Oceanography

Recent Advances in Geological Oceanography

Editors

George Kontakiotis
Assimina Antonarakou
Dmitry A. Ruban

MDPI • Basel • Beijing • Wuhan • Barcelona • Belgrade • Manchester • Tokyo • Cluj • Tianjin

Editors
George Kontakiotis
National and Kapodistrian University of Athens, School of Science, Faculty of Geology and Geoenvironment, Department of Historical Geology-Paleontology, Athens, Greece

Assimina Antonarakou
National and Kapodistrian University of Athens, School of Science, Faculty of Geology and Geoenvironment, Department of Historical Geology Paleontology, Athens, Greece

Dmitry A. Ruban
Southern Federal University, Rostov-on-Don, Russia

Editorial Office
MDPI
St. Alban-Anlage 66
4052 Basel, Switzerland

This is a reprint of articles from the Special Issue published online in the open access journal *Journal of Marine Science and Engineering* (ISSN 2077-1312) (available at: https://www.mdpi.com/journal/jmse/special_issues/cl_recent_advances_geological_oceanography).

For citation purposes, cite each article independently as indicated on the article page online and as indicated below:

LastName, A.A.; LastName, B.B.; LastName, C.C. Article Title. *Journal Name* **Year**, *Volume Number*, Page Range.

ISBN 978-3-0365-6299-5 (Hbk)
ISBN 978-3-0365-6300-8 (PDF)

Contents

About the Editors

George Kontakiotis

George Kontakiotis achieved Ph.D. degree in Paleoceanography in 2012 at the University of Athens, where he later worked as Laboratory and Teaching Staff in the fields of Marine Geology and Sedimentology. His major research contributions include developing novel approaches on the distribution and pathways of diagenesis in Mg/Ca paleothermometry. He has further worked on sedimentological and paleoceanographic reconstructions at different time scales by means of marine cores and land sections. His main research topics are summarized as follows: environmental sedimentology; marine petroleum systems; exploitation of natural energy resources; integrated bio-cyclo-tephro-stratigraphy; carbonate reservoirs; calibration–validation-application of geochemical proxies for sea surface temperature (SST) and salinity (SSS); applied environmental micropaleontology as a bio-monitoring tool; sea-level variations; ocean/climate changes.

Assimina Antonarakou

Assimina Antonarakou is a Professor of Marine Geology–Micropaleontology–Didactics on Geosciences, and President of the Faculty of Geology and Geoenvironment at the University of Athens. Her PhD Thesis dealt with Miocene cyclic sedimentary successions of the eastern Mediterranean in terms of orbital periodicities and paleoclimatic variations based on planktonic foraminiferal assemblages. Her main research topics are summarized as follows: planktonic foraminiferal eco-biostratigraphy, geobiology and paleoceanography; astronomical frequencies in paleoclimates; extreme geological events; marine environmental monitoring; ocean dynamics and sea-level changes; natural and human environmental stressors; foraminiferal trace metals and stable isotopes. She has participated in several national and international projects focused on multiproxy ecosystem responses to past/present environmental events, and she is the co-author of more than 70 peer-reviewed publications in international journals.

Dmitry A. Ruban

Dmitry A. Ruban is Associate Professor at the Southern Federal University. He received his C.Sci. (Geology and Mineralogy) degree from the Rostov State University (Russia) in 2004 and Ph.D. (Geology) degree from the University of Pretoria (South Africa) in 2009, to be followed by M.Sci. (State and Municipal Governance) degree from the Southern Federal University in 2021. His research interests are linked to Earth and environmental sciences (stratigraphy and sedimentology, palaeobiology, tectonics, physical geography and geomorphology, and environmental issues), tourism (also geotourism), and economics and management. Particularly, some of his research projects have focused on Phanerozoic mass extinction, megaclast nomenclature, and geoheritage assessment. He has published over 200 peer-reviewed articles in international journals, and he is member of the editorial board of several international and national journals.

Preface to "Recent Advances in Geological Oceanography"

Geological Oceanography: Towards Conceptual Frame

Investigations into modern oceans, seas, and their coastal zones, as well as marine ecosystems, provide valuable information for deciphering geological dynamics. For instance, studies on shores are essential for the development of sediment classifications and conceptual treatment of sedimentary differentiation. Similarly, observations of traces made on the sea-bottom by various organisms facilitate ichnological interpretations for the understanding of the deep past. Indeed, geological oceanography is not only about the present. It pays significant attention to ancient marine deposits, palaeoenvironments, and fossils, some (if not many) of which do not have modern analogues. Importantly, this discipline provides clues for solution to various practical tasks related to mineral and hydrocarbon deposits, which occur beneath modern oceans and seas or are formed in connection to vast ancient water masses. Indeed, emerging directions of geoscience research, such as geoheritage studies, are also linked to geological oceanography because significant portion of unique geological and geomorphological objects represent either ancient or modern environments.

In regard to the above, geological oceanography appears to be a vast field of research. Its conceptual frame needs regular update due to two reasons. On the one hand, numerous investigations bring a lot of new lines of evidence. On the other hand, planning research in this field requires systematic arrangement of the huge amount of knowledge to outline gaps and perspective directions. Indeed, nonsingle article or monographs can address these challenges comprehensively, but joint efforts of specialists with different research interests and representing different countries can contribute to the demanded conceptualization of geological oceanography.

The present Special Issue is comprised of around two dozen of research papers, which address various issues of geological oceanography. Although each of them focuses on a particular topic, and often a particular region, they mark the recent advances of this discipline, and, taken together, they are valuable for realizing the complexity of the conceptual frame of geological oceanography. Below, we try to overview the content of this Special Issue systematically, trying to find common ground for such diverse, but conceptually important, research. Generally, four main themes can be outlined, namely: modern marine geological environments, quaternary marine studies, palaeoenvironments and palaeoecosystems, and applied marine geology.

The themes considered in the present Special Issue, and the particular research questions raised in our 21 contributions, allow us to outline three general lessons from the recent advances in geological oceanography. First, this discipline does not only concern modern geological processes in oceans/seas and marine palaeoenvironments, taken separately. Rather, there is a kind of continuum between the two noted research domains. Second, it is geological oceanography that reveals various unexpected and unevident links and relationships in planetary dynamics, including those between remote domains of the Earth. Third, this field is not only theoretical, but also for diverse applied research. The range of possible applications is truly wide. Indeed, a lot of research questions in this discipline remain unanswered, and many are even yet to be posed. For instance, more attention should be paid to the state of Precambrian oceans, the anthropogenic factors of marine sedimentation (even on the planetary scale), and marine and underwater geoheritage. These and others research topics must be treated in-depth, and their outcomes should be put into the wide conceptual frames to achieve the maximal effect for the overall development of geological oceanography.

George Kontakiotis, Assimina Antonarakou, and Dmitry A. Ruban
Editors

Journal of
Marine Science and Engineering

MDPI

Editorial

Geological Oceanography: Towards a Conceptual Framework

George Kontakiotis [1,*], Assimina Antonarakou [1] and Dmitry A. Ruban [2]

[1] Department of Historical Geology-Paleontology, Faculty of Geology and Geoenvironment, School of Earth Sciences, National and Kapodistrian University of Athens, Panepistimiopolis, Zografou, 15784 Athens, Greece
[2] Southern Federal University, 344019 Rostov-on-Don, Russia
* Correspondence: gkontak@geol.uoa.gr

Research into modern oceans, seas, and their coastal zones, as well as marine ecosystems, provides valuable information for deciphering the geological dynamics. For instance, studies of the modern sea floor are essential for the analysis of sedimentary facies and conceptual environmental reconstruction through the integration of sequence stratigraphic and sedimentary models. Similarly, observations of traces of life from various organisms on the sea bottom facilitate ichnological interpretations of the deep past. Geological oceanography is, however, not only about the present, as it also includes marine deposits, palaeoenvironments, and fossils, some (if not many) of which do not have modern analogues. Geological oceanography also provides important frameworks for interpreting mineral and hydrocarbon deposits, which occur beneath modern oceans and seas, or are formed in connection to significant ancient water masses. Geological oceanography also includes emerging directions of geoscience research in the form of geoheritage studies, as a significant portion of unique geological and geomorphological objects represent either ancient or modern environments.

Geological oceanography is, therefore, a broad field of research. As such, any conceptual framework needs to be regularly updated due to two reasons: On the one hand, numerous investigations bring a multitude of new lines of evidence. On the other hand, planning research in this field requires systematic arrangement of the prior observations and knowledge to outline gaps and interpretive perspectives. No single article or monograph can address these challenges comprehensively, but joint efforts of specialists with different research interests and representing different countries can contribute to the ongoing conceptualization of geological oceanography.

The present Special Issue comprises twenty-one research papers addressing various issues of geological oceanography. Although each of them focuses on a particular topic and often a particular region, they mark the recent advances of this discipline, and, taken together, they are valuable for realizing the complexity of the conceptual framework of geological oceanography. Below, we introduce the content of this Special Issue systematically and outline the common aspects across the discipline. Generally, four main themes can be outlined: modern marine geological environments, Quaternary marine studies, palaeoenvironments and palaeo-ecosystems, and applied marine geology.

The first theme represents the diversity of geological processes in the modern marine environment. Petropoulos et al. [1] demonstrate how the interaction of different forces, including wave activity, landslides, and human activity, influence sediment transport and beach dynamics in a bay of the small island in the Ionian Sea. Hussian and Al-Ramadan [2] investigate the deeper environment, where sands in fans facilitate carbon burial and enhance long-term carbon cycling. Yutsis et al. [3] recognize several contourite depositional systems in the Caspian Sea, which is evidence of a complex organization of deep-marine sedimentation in the enclosed basin. Hayat et al. [4] document the spatio-temporal variations in the coccolithophore fluxes in the Eastern Mediterranean, which are partly dependent on deep-water hydrogynamics; the related processes reflect some patterns of bio-geodynamics in the deepest parts of this sea.

Citation: Kontakiotis, G.; Antonarakou, A.; Ruban, D.A. Geological Oceanography: Towards a Conceptual Framework. *J. Mar. Sci. Eng.* **2022**, *10*, 2027. https://doi.org/10.3390/jmse10122027

Received: 5 December 2022
Accepted: 15 December 2022
Published: 19 December 2022

The second theme reflects the evolution of marine environments and ecosystems throughout the Phanerozoic. Tolokonnikova and Ruban [5] demonstrate how the series of biotic and environmental crises made bryozoans vulnerable to external negative influences at the Devonian–Carboniferous transition. Ali et al. [6] document shelfal sedimentation in the Late Permian of the Salt Range of Pakistan. Garcia et al. [7] undertake a state-of-the-art approach to identify as many as ten taphofacies in the Late Jurassic coquina deposits of the Lusitanian Basin. The investigation by Wang et al. [8] offers an explanation of how deep-sea sediments with rare earth elements and yttrium found from the Pacific formed as a result of local topography (seamounts) and significant palaeoceanographical re-organizations in the Oligocene–Miocene. Ruban [9] implemented a historical approach to gather the old data for accurate outlining of the changes in the configuration of the Late Miocene Tanais Bay of the Paratethys Palaeosea. The insightful and comprehensive review by Johnson [10] focuses on the Pliocene episode of warmth when tropical cyclones became an important factor of sedimentation.

The third theme comprises Quaternary marine studies. The work by Arce-Chamorro et al. [11] traces interrelations between sea-level changes, coastline dynamics, and aeolian dune development at the very edge of the Iberian Peninsula in the Late Pleistocene–Holocene. Wang et al. [12] establish the sedimentary and geochemical archives of the global mid-Pleistocene climatic changes in the Central Pacific and demonstrate the links between the processes in this deep-ocean domain and the remote Asian landmass. Giamali et al. [13] report the Holocene ecological patterns from the North Aegean Trough, where water column stratification and upwelling influenced foraminiferal and pteropod assemblages.

The fourth theme considers the application of the marine geological process to the understanding of petroleum geology. Maravelis et al. [14] characterize the Upper Miocene possible source rocks and the related traps in the Crete Island. Janjuhah et al. [15] classify porosity in the Miocene carbonate reservoirs from Malaysia. Two contributions address the issue of overpressure in marine sediments. Li et al. [16] characterize physical, chemical, and biological mechanisms of overpressure development, with special attention given to gas hydrate systems. Dubinya et al. [17] describe various methods for prediction of overpressure zones. The other application is linked to deep-sea mineral resources. Dai et al. [18] consider seafloor massif sulphides in light of simulation experiments. Environmental aspects are also addressed in this Special Issue. Fazal et al. [19] interpret geochemical patterns of shales accumulated in the Cretaceous Hazara Basin (Pakistan) and note that the high amount of some elements in these shelfal deposits can lead to pollution of both soils and water. Khan et al. [20] demonstrate that both natural and anthropogenic factors influence the sedimentary processes in the Indus River basin and, particularly, the fan of this massive river. Finally, research in geological oceanography has importance for geotourism applications. Ruban [21] explains how finding evidence of palaeoislands of the Mesozoic Caucasian Sea facilitates the evaluation of the regional geoheritage resources and their potential for geotouristic use.

The themes considered in this Special Issue and the particular research questions raised in the 21 contributions point to the contemporary agenda of geological oceanography. Nonetheless, a number of research questions in this discipline remain unanswered. For instance, more attention should be paid to the state of Precambrian oceans, the anthropogenic factors of marine sedimentation (even on the planetary scale), and marine and underwater geoheritage.

Author Contributions: Conceptualization, G.K., A.A. and D.A.R.; methodology, G.K., A.A. and D.A.R.; software, G.K., A.A. and D.A.R.; validation, G.K., A.A. and D.A.R.; formal analysis, G.K., A.A. and D.A.R.; investigation, G.K., A.A. and D.A.R.; resources, G.K., A.A. and D.A.R.; data curation, G.K., A.A. and D.A.R.; writing—original draft preparation, G.K., A.A. and D.A.R.; writing—review and editing, G.K., A.A. and D.A.R.; visualization, G.K., A.A. and D.A.R.; supervision, G.K., A.A. and D.A.R.; project administration, G.K., A.A. and D.A.R.; funding acquisition, G.K., A.A. and D.A.R. All authors have read and agreed to the published version of the manuscript.

Funding: This research received no external funding.

Acknowledgments: We deeply thank all authors for their valuable contributions to this Special Issue, all reviewers for their thorough examination of these contributions and helpful recommendations, and the JMSE's editorial team for their outstanding support and professionalism.

Conflicts of Interest: The authors declare no conflict of interest.

References

1. Petropoulos, A.; Kapsimalis, V.; Evelpidou, N.; Karkani, A.; Giannikopoulou, K. Simulation of the Nearshore Sediment Transport Pattern and Beach Morphodynamics in the Semi-Enclosed Bay of Myrtos, Cephalonia Island, Ionian Sea. *J. Mar. Sci. Eng.* **2022**, *10*, 1015. [CrossRef]
2. Hussain, A.; Al-Ramadan, K. Organic Matter Burial in Deep-Sea Fans: A Depositional Process-Based Perspective. *J. Mar. Sci. Eng.* **2022**, *10*, 682. [CrossRef]
3. Yutsis, V.; Levchenko, O.; Putans, V. Contourite and Turbidite Features in the Middle Caspian Sea and Their Connection to Geohazards Derived from High-Resolution Seismic Data. *J. Mar. Sci. Eng.* **2022**, *10*, 990. [CrossRef]
4. Hayat, S.; Skampa, E.; Gogou, A.; Stavrakakis, S.; Parinos, C.; Triantaphyllou, M. Seasonal Variability in Present-Day Coccolithophore Fluxes in Deep Eastern Mediterranean Sea: A Multi-Year Study (2015–2017) of Coccolithophore Export in SE Ionian Sea at 4300 m Depth. *J. Mar. Sci. Eng.* **2022**, *10*, 1761. [CrossRef]
5. Tolokonnikova, Z.A.; Ruban, D.A. Bryozoan Diversity Dynamics at the Devonian–Carboniferous Transition: Evidence from Transcaucasia. *J. Mar. Sci. Eng.* **2022**, *10*, 959. [CrossRef]
6. Ali, S.K.; Janjuhah, H.T.; Shahzad, S.M.; Kontakiotis, G.; Saleem, M.H.; Khan, U.; Zarkogiannis, S.D.; Makri, P.; Antonarakou, A. Depositional Sedimentary Facies, Stratigraphic Control, Paleoecological Constraints, and Paleogeographic Reconstruction of Late Permian Chhidru Formation (Western Salt Range, Pakistan). *J. Mar. Sci. Eng.* **2021**, *9*, 1372. [CrossRef]
7. Garcia, G.G.; Garcia, A.J.V.; Henriques, M.H.P.; Marques, R.M.; Pena dos Reis, R. Taphofacies and Petrofacies Theoretical Marine Models Applied to the Coquina of the Amaral Formation (Lusitanian Basin, Portugal). *J. Mar. Sci. Eng.* **2021**, *9*, 1319. [CrossRef]
8. Wang, F.; He, G.; Deng, X.; Yang, Y.; Ren, J. Fish Teeth Sr Isotope Stratigraphy and Nd Isotope Variations: New Insights on REY Enrichments in Deep-Sea Sediments in the Pacific. *J. Mar. Sci. Eng.* **2021**, *9*, 1379. [CrossRef]
9. Ruban, D.A. The Tanais Bay of the Eastern Paratethys Sea at the Sarmatian–Maeotian Transition (Late Miocene): Widespread Desiccations and Local Uplifts in the Light of Historical Information. *J. Mar. Sci. Eng.* **2022**, *10*, 915. [CrossRef]
10. Johnson, M.E. Geological Oceanography of the Pliocene Warm Period: A Review with Predictions on the Future of Global Warming. *J. Mar. Sci. Eng.* **2021**, *9*, 1210. [CrossRef]
11. Arce-Chamorro, C.; Vidal-Romaní, J.R.; Sanjurjo-Sánchez, J. New Model of Coastal Evolution in the Ria de Vigo (NW Spain) from MIS2 to Present Day Based on the Aeolian Sedimentary Record. *J. Mar. Sci. Eng.* **2022**, *10*, 1350. [CrossRef]
12. Wang, H.; Yi, L.; Deng, X.; He, G. Geochemical and Mineral Properties of Quaternary Deep-Sea Sediments in the Central-Tropical Pacific and Its Response to the Mid-Pleistocene Transition. *J. Mar. Sci. Eng.* **2021**, *9*, 1254. [CrossRef]
13. Giamali, C.; Kontakiotis, G.; Antonarakou, A.; Koskeridou, E. Ecological Constraints of Plankton Bio-Indicators for Water Column Stratification and Productivity: A Case Study of the Holocene North Aegean Sedimentary Record. *J. Mar. Sci. Eng.* **2021**, *9*, 1249. [CrossRef]
14. Maravelis, A.G.; Kontakiotis, G.; Bellas, S.; Antonarakou, A.; Botziolis, C.; Janjuhah, H.T.; Makri, P.; Moissette, P.; Cornée, J.-J.; Pasadakis, N.; et al. Organic Geochemical Signatures of the Upper Miocene (Tortonian—Messinian) Sedimentary Succession Onshore Crete Island, Greece: Implications for Hydrocarbon Prospectivity. *J. Mar. Sci. Eng.* **2022**, *10*, 1323. [CrossRef]
15. Janjuhah, H.T.; Kontakiotis, G.; Wahid, A.; Khan, D.M.; Zarkogiannis, S.D.; Antonarakou, A. Integrated Porosity Classification and Quantification Scheme for Enhanced Carbonate Reservoir Quality: Implications from the Miocene Malaysian Carbonates. *J. Mar. Sci. Eng.* **2021**, *9*, 1410. [CrossRef]
16. Li, C.; Zhan, L.; Lu, H. Mechanisms for Overpressure Development in Marine Sediments. *J. Mar. Sci. Eng.* **2022**, *10*, 490. [CrossRef]
17. Dubinya, N.; Bayuk, I.; Hortov, A.; Myatchin, K.; Pirogova, A.; Shchuplov, P. Prediction of Overpressure Zones in Marine Sediments Using Rock-Physics and Other Approaches. *J. Mar. Sci. Eng.* **2022**, *10*, 1127. [CrossRef]
18. Dai, H.; Li, H.; Li, Y. Fragmentation Characteristics of Seafloor Massive Sulfides: A Coupled Fluid-Particle Flow Simulation. *J. Mar. Sci. Eng.* **2022**, *10*, 1306. [CrossRef]
19. Fazal, A.G.; Umar, M.; Shah, F.; Miraj, M.A.F.; Janjuhah, H.T.; Kontakiotis, G.; Jan, A.K. Geochemical Analysis of Cretaceous Shales from the Hazara Basin, Pakistan: Provenance Signatures and Paleo-Weathering Conditions. *J. Mar. Sci. Eng.* **2022**, *10*, 800. [CrossRef]
20. Khan, U.; Janjuhah, H.T.; Kontakiotis, G.; Rehman, A.; Zarkogiannis, S.D. Natural Processes and Anthropogenic Activity in the Indus River Sedimentary Environment in Pakistan: A Critical Review. *J. Mar. Sci. Eng.* **2021**, *9*, 1109. [CrossRef]
21. Ruban, D.A. Islands in the Caucasian Sea in Three Mesozoic Time Slices: Novel Dimension of Geoheritage and Geotourism. *J. Mar. Sci. Eng.* **2022**, *10*, 1300. [CrossRef]

Journal of
Marine Science and Engineering

MDPI

Review

Natural Processes and Anthropogenic Activity in the Indus River Sedimentary Environment in Pakistan: A Critical Review

Usman Khan [1,*], Hammad Tariq Janjuhah [2], George Kontakiotis [1,*], Adnanul Rehman [3] and Stergios D. Zarkogiannis [4]

1 Department of Geology and Geoenvironment, National and Kapodistrian University of Athens, Panepistimiopolis, 15784 Athens, Greece
2 Department of Geology, Shaheed Benazir Bhutto University, Sheringal 18050, Pakistan; hammad@sbbu.edu.pk
3 Department of Geography, Government College University, Faisalabad 38000, Pakistan; adnanulrehman@gmail.com
4 Department of Earth Sciences, University of Oxford, Oxford OX1 3AN, UK; stergios.zarkogiannis@earth.ox.ac.uk
* Correspondence: uskhan@geol.uoa.gr (U.K.); gkontak@geol.uoa.gr (G.K.)

Citation: Khan, U.; Janjuhah, H.T.; Kontakiotis, G.; Rehman, A.; Zarkogiannis, S.D. Natural Processes and Anthropogenic Activity in the Indus River Sedimentary Environment in Pakistan: A Critical Review. *J. Mar. Sci. Eng.* **2021**, *9*, 1109. https://doi.org/10.3390/jmse9101109

Academic Editor: Alfredo L. Aretxabaleta

Received: 17 September 2021
Accepted: 9 October 2021
Published: 12 October 2021

Abstract: The Indus River is Asia's longest river, having its origin in the Tibet Mountain northwest of Pakistan. Routed from northern Gilgit and flowing to the plains, the river passes through several provinces and is connected by numerous small and large tributaries. The river was formed tectonically due to the collusion of the Indian and Eurasian plates, which is referred to as the Indus suture Plains zone (ISPZ). The geological setting of the study area is mainly composed of igneous and metamorphic rocks. The river passed through a variety of climatic zones and areas, although the predominant climate is subtropic arid and sub arid to subequatorial. Locally and globally, anthropogenic activities such as building, dams, and water canals for irrigation purposes, mining exploration, and industries and factories all affected the physical and chemical behaviors of the sediments in various rivers. The main effect of human activities is the reworking of weathered soil smectite, a chemical weathering indicator that rises in the offshore record about 5000 years ago. This material indicates increased transport of stronger chemically weathered material, which may result from agriculture-induced erosion of older soil. However, we also see evidence for the incision of large rivers into the floodplain, which is also driving the reworking of this type of material, so the signal may be a combination of the two. Sediments undergo significant changes in form and size due to clashing with one another in the high-charge river.

Keywords: environmental sedimentology; sediment flux; sediment provenance; climate change impact; human role; depositional and erosional history; environmental controls; natural resources; aquatic environment; physical geography

1. Introduction

The Indus River is one of the most important rivers in Asia, connecting India and Pakistan (Figure 1) [1,2]. Since Alexander, the Great's expedition to the region in 325 BC, the world's biggest river has piqued academics' attention, and this is the most significant innovation of early and contemporary civilizations in the Indus valley's south Asia [3] that ever since has changed the physical geography of the area. The Indus River originates in Tibet and includes a holy peak called Kailas. The basin encompasses western Tibet, the Himalayas, and the Karakorum. While the top portion of the route passes through India, most of its characteristics (channel and drainage patterns) occur in Pakistan [4]. The river receives a massive volume of sediments as a result of water flow from various tributaries, some of which originate in the Shyok and Gilgit areas, while others originate in the Kabul area [5,6]. Sediment from the Indus River system is eroded preferentially in western Tibet and Karakorum [7,8].

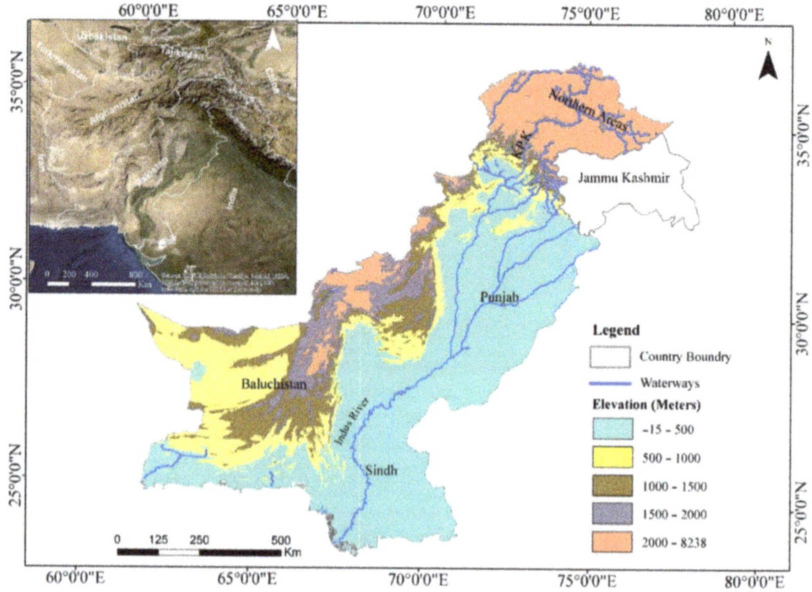

Figure 1. Location of the Indus River within Pakistan. Pakistan's topography borders with Afghanistan, India, and Iran are presented in the inset.

Integrated geotectonic and geophysical evidence suggests that the river system was formed prior to 45 million years ago because of plate tectonics (Indian and Eurasian plates) [9,10]. The Indus River system represents significant potential sources of information for determining the various uplifting and erosion activities associated with various periods in the history of the western Himalayas. More significantly, Tibet's development has been linked to the southwest monsoon [11]. The Indus River continues to follow the suture's strike through the Himalayas in northwest Pakistan, flowing continuously to the south portion of the Arabian Sea (Figure 1). The Bengal Fan is the primary feeder river; on the other hand, the Ganges and the Brahmaputra follow the Higher Himalaya along strike throughout most of the orogeny's length [12,13].

Since the dawn of human civilization, rivers such as the Indus have altered numerous fluvial landscapes [14,15]. In the late 20th century, the Indus River played a critical role in the development of modern civilization, while simultaneously raising environmental concerns [16,17]. Urbanization and related human activities pose a significant threat to the river system [18,19]. While human activities like bridge construction and artificial levee construction are necessary for human progress and welfare, they have a detrimental effect on river shape and natural features [20–22]. Prior to the 19th century, this river basin was almost deserted [23]. Following India's separation from Pakistan, many immigrants entered the neighboring rivers and settled around the Indus River [24,25]. Since then, the people have relied on the Indus River to meet their personal requirements and to profit economically. Not only are humans contaminating the river water directly or indirectly, but they are also changing the river's shape [26,27]. However, most published articles (e.g., Inam, Clift, Giosan, Tabrez, Tahir, Rabbani and Danish [4], Yang, Brown, Yu, Wescoat Jr and Ringler [26], Usman et al. [28], Hester and Doyle [29], Al-Ghanim et al. [30] or Tsering et al. [31]) focus on human effects on the river, such as environmental pollution, with just a few studies addressing changes in the physical structure of the Indus River as a result of anthropogenic activity. Moreover, several studies that dealt with human influence on various rivers around the globe have overlooked the Indus river's significance. An additional aspect of its importance relies on the fact that Pakistan is strongly dependent on the Indus River system and its tributaries for all of its water requirements. The building of irrigation and power dams has decreased the river's flow, significantly reducing the amount of freshwater available downstream. Tarbela and Mangla Dams, located upstream of the main agricultural areas, are the sites of two large hydroelectric plants powered by the river. Since Pakistan relies on Indus River Basin irrigation system for 90% of its food resources, this basin accounts for 25% of the country's gross domestic product and Pakistan may be confronted with serious food shortages because of water scarcity. Since the early 19th century, human activities, have had a major impact on the Indus River, while on the other hand the sedimentary environment had a major impact on the Indus River's course in terms of morphological, hydrogeological, and sedimentological changes.

The main objective of the present study is to assess the sedimentary environment of the Indus River system and further highlight how anthropogenic activities have affected hydrogeology and sedimentation. In this regard, water discharge from the Indus River due to the monsoon system and glacier melting due to rising temperatures are discussed in this regional case study. These factors may have a significant impact on the Indus River's water flow across the contemporary catchment areas.

2. Geological Setting

2.1. Geotectonic Framework

The Himalayan uplift encompasses and deforms a significant portion of the northern Indian continental plate (Figure 2). Prior to the collision, the Indian continental plate settled in equatorial latitudes with fine-grained sediments, particularly limestone successions, dominating the shelf and high-slope region [32,33]. When the Indian and Eurasian plates met, sedimentation rapidly changed to sandstones, resulting in the formation of new mountain ranges [34–36]. During this process, uplifting and erosion occurred. The sequence involves the development mostly of shales and sandstones, the beginning of mountain rising, and the ultimate birth of rivers along the line of collision separate between the Indian and Eurasian plates, which is known as the Indus suture Plains zone (ISPZ). Although the northwest migration of the Indian plate has slowed since its collision with Eurasia and the subcontinent, it was not halted but continued to travel north towards Asia [37–39]. The northern Indian plate edge was buried as it seems distorted and heated before rapidly returning to the surface due to erosion, while the remaining portion is driven by extensional faulting caused by the fall of the huge mountain volumes under excessive weight [40,41]. The major Himalayan range represents the surviving deformed northern Indian plate boundary, primarily intruded by granite rock bodies that elevated 22 million years ago. Over time, the main compressional force distorted plates and moved towards the southern area of the Indian plate [42]. In this instance, new ranges are emerging in the lower Himalayas [33]. Tectonic activity continued due to erosion in the valleys and adjacent mountains which uplifted to great heights (Figure 2). The most striking and abrupt example is the mountain "Nanga Parbat" which is adjacent to the Kohistan region (Indus south) [43,44]. Over the most recent geological periods, the range of the escalating Nanga Parbat has been >1 cm each year, which is considered one of the world's highest known rates [44,45] with the massif ranges moving towards northwest and southwest directions. Similarly, the plains of the province of the Sindh shelf evolved in the Cretaceous about 70 million years ago [46,47], during when India reconnected to the Seychelles Island [48,49]. The moderate rate of subsidence and sedimentation of the shelf and slope in the southeast of Karachi contrasts with the coastal line and marine geological setting in the west [33].

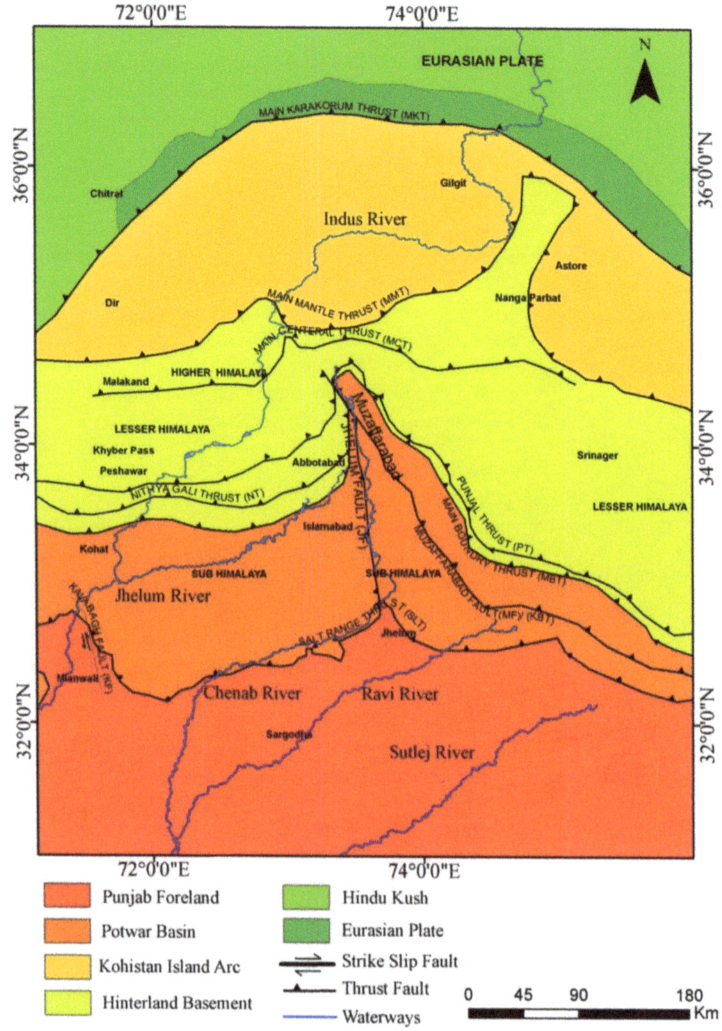

Figure 2. Tectonic map showing the main structural features in the northern area of Pakistan.

2.2. *Seasonal Controls on Hydrological Variability*

The Indus River that is passing different provinces of Pakistan covers different climatic zones (Figure 3). In the southwest region of Pakistan, Sindh and Punjab, climatic zonation consists of subtropical arid and subarid to subequatorial conditions [50]. In this regard, atmospheric circulation over the higher part of the basin mainly lies in the monsoonal system as typical of the (sub)tropic latitudes. This variation in climatic regime results in two distinct seasons. The ordinary weather season is between October and April, while the dry northeastern part sets over the plain and the vital portion of the basin. Based on spatiotemporal shifts in monsoon patterns observed during the entire monsoon season, the rainfall increases during summer months (Figure 4) [50–52]. In this period, the basin is subject to another climatic regime, i.e., the primary route of the trade winds began from the side of the Iranian highlands and are then transformed into dry and hot air conditions. The mean annual precipitation on the Indus basin varies from the source of the river to its mouth with mean annual values ranging between 500 and 125 mm, respectively. Although the calculated values by Kravtsova et al. [53] are quite different, they present a similar declining trend from 1270 mm in the mountain region to 100 mm per year. Indus River hydrological history characterized by the above seasonal constraints passes through three different altitudinal landscape zones [54,55]. The upper regions reach towards the river and belong to the high-altitude belt. With the increase in altitude, mostly forest land is encountered within the different landscapes, at a mean height of 4700 m. Hence, the river runs the zone of the savanna and thin forests, while most of the basin is situated in the plains of the Himalayas (Dessertified savanna) [56,57].

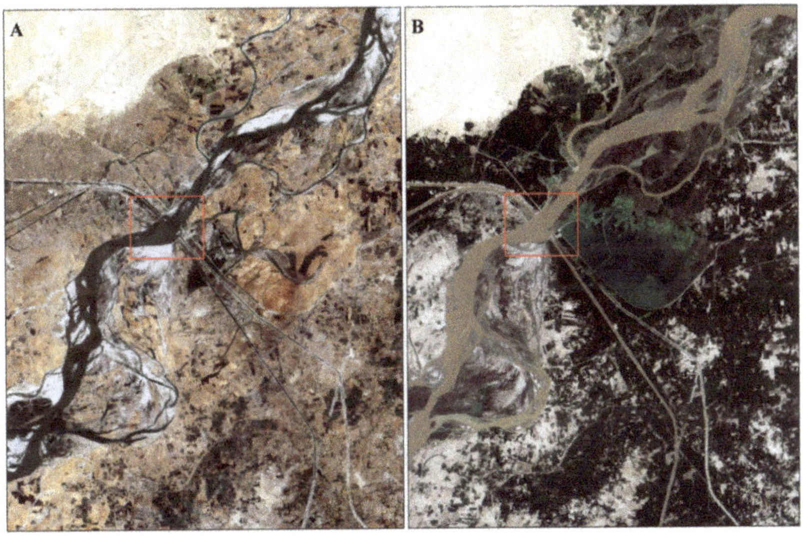

Figure 3. Satellite images from the Indus River near the Punjab and Sindh territories in opposite seasons: (**A**) typical shrinkage of the river discharge from May to November, when water is diverted for irrigation; (**B**) the swelling of the river from November to May due to high flow intensity (open access source from NASA Earth Observatory).

According to Giosan et al. [58], no significant changes in the river's route were noticed between 1940 and 1958. Even after the 1952 earthquake, the river's flow remained unchanged. Between 1959 and 2011, five sections of the river were identified where significant alterations occurred along its course [59,60]. Further back over 4000 years ago, the Harappa civilization flourished in the area that is now modern Pakistan and northern India's Indus River Valley, where they constructed complex cities, developed sewage systems that preceded those of ancient Rome, and participated in long-distance commerce with Mesopotamian towns [61]. By 1800 BCE, however, this sophisticated civilization had abandoned its metropolis in favor of smaller settlements in the Himalayan foothills [62,63]. Recent evidence suggests that climatic change forced the Harappas to relocate far from the Indus floodplains [59]. Beginning about 2500 BCE, a change in temperature and weather patterns over the Indus valley progressively reduced summer monsoonal rainfall, making cultivation near Harappa towns difficult or impossible [64,65]. Although the inconsistency of summer monsoons made cultivation challenging along the Indus, moisture and rain arrived more frequently in the foothills, since winter storms from the Mediterranean reaching the Himalayas brought rain and nourished little streams on the Pakistan side [50]. In comparison to the monsoon floods that the Harappas were used to in the Indus, it would have been a little amount of water, but it would have been dependable [66,67]. Rains in the foothills seem to have been sufficient to sustain the rural Harappas for the following century, but even those would ultimately dry up, possibly contributing to their collapse.

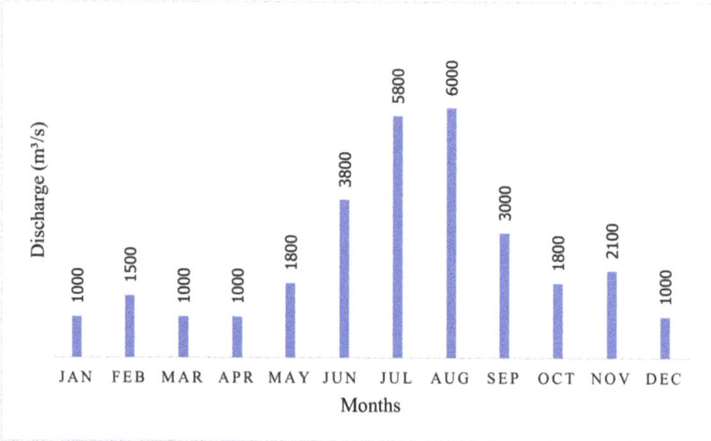

Figure 4. Mean monthly monsoon discharge with an entire annual monsoon duration during the time period of 1980–2017. The highest peak was observed from July to August due to glacier melting and monsoon rainfall. The flow rate during the post-monsoon period (September to December) increases compared to pre-monsoon period (January to May) because of a shift in temperature melting glaciers at the end of the monsoon season.

2.3. *Volumes of Sediment Eroded within the Upper Indus*

Tectonically driven rock uplift and sea-level oscillations regulate sediment production and transport from mountain source to ocean basin [68]. Climate, on the other hand, has long been acknowledged as a supporting factor. The summer monsoon climate in South Asia affects the production, transport, and deposition of sediment from the western Himalayas and Karakoram to the Indus River [69,70].

These processes interact across million-year timescales, making their separation more complicated than over interglacial–glacial timescales [71]. During the Northern Hemisphere's summer, the monsoon delivers rain far inland throughout SW Asia (May–October). The reanalysis of bibliographic data [69,72] allowed the calculation of monthly mean average riverine discharges and the results are shown in Figure 4 with evident differences between monsoonal and post-monsoonal discharges. For the Indus River, the summer monsoon controls hillside sediment production, erosion, and transport [70]. Precipitation in the western Himalaya is 60–80% monsoonal, with 4–6 rainstorms providing 40% of July precipitation and dominating the river sediment movement [73,74]. Summer monsoon precipitation varies throughout the Indus watershed, from 507 mm (76%) in Chandigarh to 385 mm (64%) in Islamabad and 178 mm (82%) in Karachi [71,75]. As a result, the Indus system may experience greater changes in erosion and transport due to monsoon variability. The Indus Fan is 4.5 × 106 km^3 of silt spread across 1.6 × 106 km^2 under the continental shelf [67]. Despite its 140 km width from the present coast to the shelf edge, the shelf only covers 46,000 km^2 [69,76]. After the tidal backwater point, the delta plain extends 12,000 km downstream, with 43,000 km^2 of advancing plains to the north, mainly in Sindh, Pakistan [71,77]. The Thar Desert presently spans 140,000 km^2. The stem Indus and its eastern tributaries cut and dig 145,000 km^2 of floodplains south of the MBT [78].

3. Study Area

3.1. *The Indus Fan*

For most of its life, the Indus fan resides along the Indian-Pakistani border's passive continental margin (Figure 5). The river setting is subtropical characterized by little rainfall (35 cm/year) [79]. The Holocene sediments (dark gray to green silts and muds with some interbedded sands) have a sedimentation rate of 45 cm/year, while accordingly the sedimentation rate during the Late Pleistocene is also considerable (40 cm/year) [80].

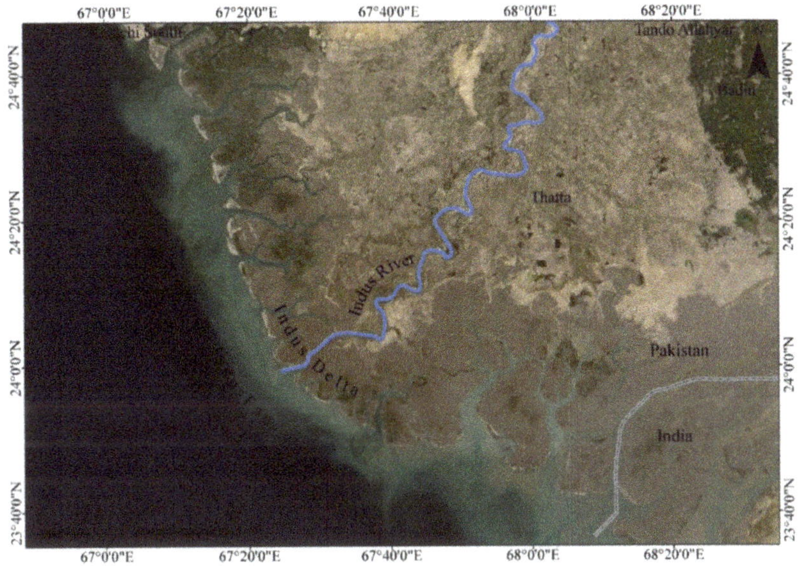

Figure 5. Location of several rivers and particularly the Indus River delta in Pakistan.

The huge migration into the Arabian Sea and the Indus fan period has been extensively debated [7,81]. The early Miocene caused rapid erosion of the upper Himalayas, resulting in the development of the Indus and Bengal fans [26,82]. Reconstructions of mass flux that moves into the Indus fan show an acceleration in the sedimentation rate during that time. The main perspective of the Indus fan shows an acceleration of sedimentation in the same period [83]. These mass flow estimates are based on in situ drilling, but the findings are variable between studies. In the foreland sequences, the primary cause of preferred sedimentation and erosion is the change in eustatic sea level. As accumulation is strongly related to changes in sea level over time, fluctuations in the breadth and depth of the foreland trough may thus not be an adequate indicator of mass flow across the Indus system [84].

3.2. *Upper Indus Basin*

The Upper Indus Basin (UIB) is located in the northernmost part of Pakistan, parallel to the Chinese border (Figure 5) [85,86]. The upstream catchment extends from approximately the Tarbela dam reservoir to the coved area, covering an area of ~200,600 km^2 [87,88]. The upper Indus basin is home to relatively some of the world's highest ranges such as Karakorum, Himalayas, and Kush Hindu Kush (KHK) with several peaks that attain a height level of >7000 m; including the second largest (K2; 8611 m) in the world [87]. Moreover, some famous glaciers occur in the (KHK). In UIB, the total cover area includes Siachen (75 km)—largest in the Karakoram and the world's second-largest glaciers outside in the Polar Regions Biafo (67.9 km), third biggest in the world Baltoro (62.1 km); Batura (59.8 km); Hisper (53.1 km) [89].

4. Environmental Conditions of the Sedimentary Indus River System

4.1. *West–East Climate and Topographic Contrasts*

The climate of Upper Indus Basin (UIB) is warm and dry in summer (around 30 °C) and cold and rainy in the winter (around 13 °C). The environmental conditions belong in the temperate zone, which is mostly located in the shadow of large mountains ranges [87,90]. Indeed, approximately 90% of the UIB falls in the Himalayan shadow zone (HSZ) [87], which results in the drastic reduction of the monsoon winds influence on north-westwards. The main annual precipitation of the UIB at the arid valley floor is 150–200 mm, whereas in the Gilgit region river valley at 4400 m altitude, the precipitation increases to almost 600 mm, and at 5500 m is 1500–2000 mm [91,92]. The climate of the UIB is quite similar to that of the eastern Himalayas [87]. Both the eastern and western parts of the Himalayas have different climatic conditions, with the eastern portion receiving more than 3000 mm more than the western part [93,94]. On the other hand, the western areas receive less than 300 mm per year, which directly and indirectly impact the river's discharge period. Moreover, the spatial and seasonal snow fluctuations vary significantly among areas within UIB. During the monsoon season, the eastern and central Himalayas get massive amounts of snowfall in the winter, that are mainly snow heaps that originate from westerly circulating winds due to northwest (Karakoram and UIB) and westerly circulations [95]. The Himalayan glacier implies the middle Karakorum holds back a large amount of snow, causing avalanches to move downhill to steep slopes. The topography of the UIB also shows great variability from the minimum elevation less than 1000 m in the flat area near the area of Tarbela and Mangla dam reservoirs to over 8000 m at a number of peaks [87,96]. Although, the LIB is mostly dependent on snow and glacier melting water for various sources.

4.2. Sedimentary Processes and Characteristics of the Indus River

The source of eroded sediments into the Indus River from the Himalayans Mountains is due to tectonics. These marine sediments represent the longest succinct archives of continental environmental history that allow reconstructions of the previous geographical patterns, altering habitats, or erosional rates at the time of deposition [60,97]. These archives may then be utilized to evaluate the impact of climatic and tectonic forces on continental environmental conditions. With the marine record, we can determine the flow limit of clastic sedimentary rock from continental sources to the marine sink. The ocean's efficiency varies from days to weeks to >105–106 years [80], while the Asian River systems maintained a relatively constant and mass flow into the ocean for the past two million years due to sediment buffering in a flood plain [76,98–100]. However, there is presently insufficient control in place to determine how efficient this buffering mechanism may be on short time periods.

4.3. Traveling Path of Sediments

Due to the intense tectonic activity, uplift of the mountain range (including the derived sediments) such as the Himalayas Koh-Hindukush, Koh-I-Safaid, Parachinar, the Waziristan agencies ranges, Sulaiman-Kirthar fold belt and Khojak Flysch sub-basins, downstream by several river systems that include Indus, Panjkora, Swat, Kabul, Kunhar, Kurram, Gambila, and Gomal Rivers. The variety of sediments which feed all the above rivers with detritus differ in their origin (igneous, sedimentary, and metamorphic rocks), source areas (even for the weathering products which present a differential flow regime from variable directions) and their age which ranges from Precambrian to the Holocene [101,102].

Following the Last Glacial Maximum (LGM), sediments were carried to the Arabian Sea by a river whose course fluctuated greatly. The deposited silt area is between 4050 and 5675 km^3. Most of the depositional volumes come from the upper alluvial plain (10 ka) with the Nanga Parbat syntaxes estimated to contribute 32–40% of the sediment flux (Himalaya), while the Karakoram Mountain contributes approximately 21–27% to the total sediment released (Table 1) [103]. However, in the mountains, sediment is slowed by land sliding, which is influenced by climate [102]. It is found that 5% of terraces in the monsoonal area of the lower and upper Himalayas had been degraded with major bedrock weathering of up to 46% of post-glacial sediments reaching the delta and since changes in the erosional patterns after the 8 ka delta are not reflected in composition shifting, extensive reworking from terraces and flood plains throughout the mid-late Holocene is implied.

Table 1. Summary of the volumes eroded along the major rivers of the Karakoram, Hindu Kush and Kohistan rock landslide.

Source	Average (km^3)	Maximum (km^3)	Minimum (km^3)	Average Contribution Deposited Mass (%)	Average (Gt)	Maximum (Gt)	Minimum (Gt)
River Terraces							
Karakoram	15.40	240.2	67.7	2.5	416	649	183
Kohistan and KHK Rockslides	165.7	248.6	82.9	2.7	447	671	224
Western-River Terraces	19.0	28.4	9.5	0.3	51	77	26
Flood plains	647.5	925.0	370.0	10.7	1748	2498	999
Rockslides N.P	43.5	56.6	30.5	0.7	117	153	82
Rain-shadow catchments	76.0	118.5	33.4	1.3	205	320	90
Upper-most-river basin	12.2	15.8	8.5	0.2	33	43	23
Grand total	1507.8	2218.1	797.5	24.9	3993	4293	1587
Deposited	6065.0	7278.0	4852.0	-	16,376	19,651	13,100
Bedrock erosion	4557.2	6480.5	3275.4	-	12,304	17,497	8844
%	75.1	89.0	67.5	-	-	-	-

4.4. Source to Sink Transport of Sediment

Erosion is caused by weathering tectonic forces and climatic changes, which transport sediments and deposit them in the open ocean. The Asian monsoon system influences Indus River and allows the exposure of the various types of the sediments under different climatic conditions. In the northwest region, the rate of erosion is mainly linked to the summer monsoon rain as well as to tectonically uplifted rocks [104,105]. Erosion and sediment transport rates increase with chemical weathering due to the climate change. Since the LGM, the sediments have been stored within the delta plain and submarine fan/shelf clinoforms system [71]. Thus, high monsoon influence reflects the high supply rate of sediments. In Table 2, we present a synopsis of the modern catchment sediment of the major five tributaries of the UIB (based on statistical Kolmogorov–Smirnov (KS) and Kuiper tests for assessing the relative similarities between the observed zircon U-Pb age). The modern river appears to be mostly supplied by the Jhelum and Sutlej rivers. However, the upper Indus basins are in general within a dry source area during the strong monsoon time [106]. The possible anthropogenic activity disruption has resulted in unusual zircon concentrations at the Indus delta [71].

Table 2. Modern catchment sediment analysis of five river tributaries.

Modern Catchment Sediment	Kuiper V	$\pm 1\sigma$	K-S D Value	$\pm 1\sigma$
Upper Indus	0.226	0.066	0.262	0.060
Jhelum	0.164	0.045	0.129	0.049
Chenab	0.067	0.049	0.068	0.050
Ravi	0.065	0.057	0.068	0.059
Beas	0.074	0.067	0.077	0.071
Sutlej	0.403	0.174	0.397	0.242

5. Anthropogenic Activity across the Indus River

In Pakistan, dams, barrages, and bridges are still being built to address the country's energy deficit [107–109]. Due to its maximum flow area, the Indus River provided potential advantages in terms of constructing dams and using for agriculture purposes (Figure 6) [107]. These kinds of anthropogenic activities are evident especially in the northwestern areas and largely affect the sedimentary environment of Indus River, including its tributaries [4,108]. The network of the river system that stretches from the Tibetan autonomous region to China and through the Himachal Pradesh and Punjab in India, is further connected with the Indus River in Pakistan. The rivers Sutlej and Beas (that originate in India, e.g., lake Rakshasa in Tibet) adopt the route for the catchment boundaries between India and China. The active geomorphological system experienced major anthropogenic influence (to both water and sediment fluxes) by the time of civil structure constructions (hydropower bridges and roads) and extensive canal irrigation network systems [109,110]. For instance, the Bhakra and Pong dams were constructed in Sutlej and Beas rivers, respectively (Figure 6), in such a way to stream precipitation to the proposed irrigated cropland in disperse patches at the urban area as well as the grassland [111]. Both Bhakra and Pong reservoirs have multi purposes such as hydro-power generation, irrigation, and regulation of high-level flows during the monsoon seasons [76]. Certain rivers' loads have risen as a result of human activity, especially in instances of land clearing and watershed disturbance; but the loads decrease as a result of dam construction but any additional human impact will be superimposed to changes caused by the current climate change in certain river basins [112,113].

Some of the most important geomorphologic activities on both a local and a global scale related to anthropogenic interventions are as follows: (i) the new land changes in terms of deforestation with the associated mass waste, slop failure and soil erosion; (ii) agriculture and evolving tillage techniques, irrigation systems and subsurface water extraction that will contribute to an increase in soil erosion, creep, siltation and subsidence at the local level; (iii) mining and its role in river channel or hill slop changes that may cause instabilities and subsidence; (iv) transportation systems, that results in further soil erosion and riverbed scouring; (v) constructions of dams or channels, diversions and channel deepening that ultimately increase coastal erosion; (vi) global temperature rises and its various impacts on the atmospheric system and precipitation intensities, cyclonic activity, desertification and an accelerated hydrological cycle; most importantly the increasing needs for aggregate extraction [114,115].

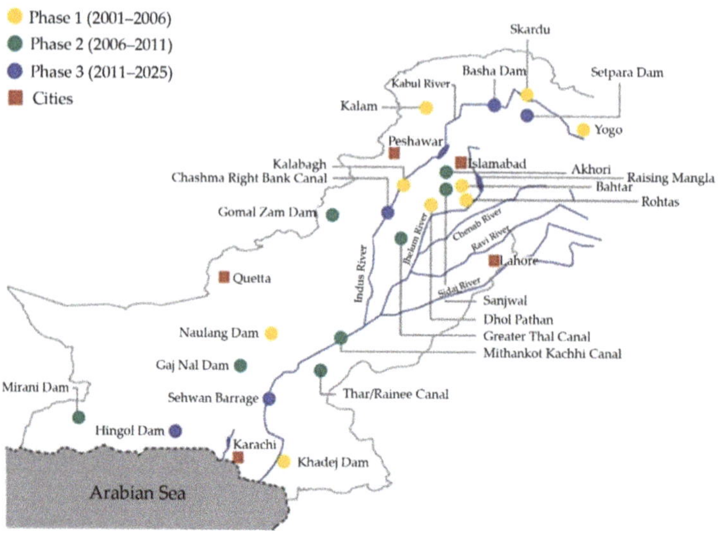

Figure 6. Location of dams constructed on different rivers in Pakistan.

5.1. Key Drivers

The river system evolution provides the evidence of two key drivers of changes in the behavior on sediments load for catchment of disturbance, which are directly or indirectly connected to human activities: (1) in terms of management land and deforestation, grazing, cultivation, irrigation and intensive monoculture, and (2) on the other hand urbanization, building constructions, sewerage and runoff as well solid waste disposal [99,102]. The industrial activity mainly lies in the sector of mining exploration, industries, and factories. Major human activity in field of civil engineering may also include roads and/or rivers engineering, canal railway lines, highways, and reservoirs (dams) construction. All of the above could be characterized as the most highlighted anthropogenic activity along different areas of the Indus River [116,117].

5.1.1. Dam Construction

Construction of dams reduces sediment load in many rivers worldwide [118]. In the study area, the construction of many barrages and irrigation channels, as well as two major dams, named Mangla and Tarbela, were built on the main Indus River near Darband (Figure 6) [119]. All these geomorphic activities which began in the 1940s were carried out for various purposes such as irrigation and water supply, flood control, and hydropower generation. The impact of these developments on the Indus River's annual discharge and sediment load is also important. In the case of the lower Indus River, annual runoff is at present-day less than 20% of what it was prior to the development of extensive irrigation systems, and the current annual sediment load is also less than 20% of what it was previously [67,117].

5.1.2. Land Clearance and Catchment Disturbance

Many rivers highly characterized by a high rate of sedimentation load caused by several factors like deforestation, intensification of agriculture, mining, land clearance, and related activities located in various areas along the Indus River system, result in increased sediment loads, as opposed to rivers where the sediment load has decreased as a result of dam construction (Figure 7A,B) [117,120]. Furthermore, while dam construction is a relatively recent phenomenon, whose impact can be documented by sediment load records dating back 40 or 50 years, there is an enhanced sensitivity of sediment loads to current environmental changes locally and globally [121,122]. We highlight the changes in the sediment loads of the world's rivers due to geochemical cycling, as well as the links to local and regional impacts and problems, such as the retreat of the delta shoreline due to reduced sediment supply and the distraction of coral reefs because of increased sediment inputs to the coastal seas [118,120].

5.1.3. Coal Mining Influence on Central Indus Basin

Mining is another human activity that has been created along the Indus River's northern and southern banks (Figure 7B) [109,123]. According to the experts, the human input of coal mining operations contaminates natural water and soil. The microbiological control, depositional environment, basin morphology, mineralogy, and hydrological conditions of the area all influenced by its function [124]. Mining water has been discovered to be a major source of toxicity due to increased concentrations of Mn, Se, Cd, and Hg elements, raising concerns about the central Indus basin's ecological status [87,125]. Rainwater has the potential to transport these potentially toxic elements (PTEs) to lower agricultural lands and, eventually, to the Central Indus Basin's main water channel and the River Jhelum [126]. The concerned organization (i.e., Ministry of Agriculture) should make proper regulation of mine tailings mandatory, as they are a potential source of toxic trace elements in the Central Indus Basin's local ecosystem. For the conservation and protection of ecological resources, regular monitoring of PTEs in water quality is recommended around active and abandoned mining sites [87,127].

Figure 7. (**A**) Desu hydropower project along the Indus River, (**B**) Pothwar coal mining in the upper Indus Basin.

5.2. Physical versus Chemical Changes in Indus River Sediments

The chemistry of the Indus River sediments attempts to constrain the provenance of the erosion and weathering that may have resulted in the characteristics of these sediments [8,128]. The grain size also determines the chemistry of the sediments [129,130]. The lithology of the study area and the weathering of silicate and carbonate rocks from the different provenances heavily influence the geochemical importance of the Indus River water analysis. Anthropogenic activities do not appear to have a significant impact on the water or sediment chemistry in these areas. Therefore, the Indus sediments indicate the important role of mechanical weathering and erosion response to the Himalayan uplift and orogeny. Isotope stability in Delta sediments during the mid-Holocene may indicate increased reworking of floodplains beginning with agriculture linked with the Indus Valley civilization [83,131]. An increase in smectite in the submarine Canyon was also observed by Li, Clift, Murray, Exnicios, Ireland and Böning [75] that may be related to enhanced reworking due to agriculture. This chemical weathering indicator reveals greater transport of stronger chemically weathered material and some of this may be caused by the erosion of old soils broken up by agriculture; although we also have evidence for the incision of the big rivers into the floodplain which is also driving reworking of this type of material, and therefore the signal may be a combination of both processes [71,75,132]. Overall, such reworking of weathered soils could be considered the primary impact of anthropogenic processes in this setting.

6. Influence of Natural and Anthropogenic Factors on Hydrogeology

Chemical water–rock interactions have a significant impact on groundwater composition because of the role played by the geosphere [133,134]. The rock type and the manner of weathering of the particulate minerals seem to be the major influences on hydrogeology in bedrock wells [135]. In Pakistan along the Indus river, the chemical composition of the groundwater in fissures and cracks is influenced by the bedrock composition, as is the chemical composition of the groundwater in Quaternary deposits [136,137]. In the absence of human intervention, water quality changes are caused by natural processes such as bedrock evapotranspiration and wind-deposited dust [138]. Water's physical and chemical composition may vary due to natural processes including soil erosion and runoff, as well as hydrological variables that cause runoff and biological activities in aquatic environments [139–142]. The Indus River's natural environment may include both dissolved and undissolved elements as a result of natural processes [125], thus this information should be taken into consideration. As lakes, rivers, streams, and groundwater are natural water sources that sustain life, they must have high-quality water [139]. Due to processes like erosion, leaching, and weathering, there are a number of contaminants present in the Indus river waters; these contaminants can only be reduced in concentration by standard water and wastewater treatment processes, so their presence in a specific water source may restrict its use [143–145]. Droughts and floods may dilute or concentrate dissolved chemicals, changing the water quality [146,147]. When the Indus River flow rates are low, temperature increases, dissolved material concentrations rise, and dissolved oxygen concentrations fall [143]. These changes have the greatest impact on water quality. Natural weathering and rainwater may leach phosphorus from rocks and minerals in the Earth's crust, contaminating surface and ground water and therefore public water systems, with fluorospar, cyolite and fluorapatite [148,149]. For instance, in Pakistan regions of with fluorine-rich volcanic rocks, Rafique et al. [150] found very high fluoride concentrations in surface water.

Heavy rainfalls transport more sediment enriched with nutrients than do light rains, although soil moisture and permeability also have a role in sediment migration [151,152]. The capacity for run-off sediment transport is determined by the relationship between dominant soil flow shear and critical rainfall stress. In coastal regions of Pakistan, large quantities of groundwater usually imply saltwater intrusion [153,154]. Increased quantities of dissolved components in groundwater, especially SO4, Cl, and Na, have been demonstrated by Naseem et al. [155] to indicate the presence of relict saltwater trapped in bedrock fissures and cracks and salty pore water in marine clay and silt deposits. For example, the Tualatin River in Oregon, the United States, has significant background amounts of phosphorus [156]; for arsenic, go to India and Bangladesh [157].

Some agricultural methods in Pakistan may increase environmental pollution, such as overgrazing grasslands, and overusing pesticides, ploughing over irrigated fields and fertilizer application [158,159]. Loss of top soil from fields is the most frequent cause of agricultural water contamination in Pakistan [160,161]. Rainwater mobilizes sediment particles and washes them into adjacent bodies of water, altering the quality of the water [162], since washed into water bodies are contaminants that adhere to soil particles, such as fertilizers, pesticides, and heavy metals. Through algal growth and oxygen depletion, these contaminants put aquatic life at danger. Increases in suspended sediment concentrations and loading rates increase turbidity, a metric for the light scattering effects of water-borne suspended particulate material [163,164]. Aside from these factors, the particle size distribution, refractive index, and water color are also important. Spring and fall have greater concentrations of suspended solids in streams than summer [165]. Agricultural pests in Pakistan are eradicated with the use of insecticides, herbicides, and fungicides. Run-off from fields or air deposition may carry these contaminants into the water, or they can be applied directly to the water. The usage of land for agriculture may lead to a variety of pollutants getting into the water, which contaminate it [166–169]. Pesticides are a class of dangerous chemicals that may be harmful to human health if used improperly [170–172]. Agricultural chemicals pollute water resources through nonpoint and point sources including field leaching, runoff, dumping, and washing. Pesticides pollute surface and groundwater. Abiotic and biotic degradation rates influence water solubility.

Urbanization is a rapidly growing land use [173,174]. Expanding impermeable surfaces like croplands and forests improve air quality and reduce greenhouse gases [175,176]. Impervious surfaces promote nonpoint (diffuse) polluted runoff harming aquatic habitats. Industrialization, sewage discharge, and other household activities damage Pakistani cities more than the rural regions [177–179]. Human activities such as industrial and household wastewater discharge, agricultural chemical usage, and land use and cover changes all impact surface water quality [180,181]. Change in land use and land cover impacts river flow and quality in Pakistan [174,182–184]. Human-induced changes in landscape pattern have significant effects on river conditions.

7. Conclusions

The Indus River is one of the largest rivers in south Asia with the maximum flow area in Pakistan. The main origin of this river is from the neighboring country India and it is a connective transboundary river between Pakistan and the Indian border. The Himalayan range is uplifted during the Indian-Asian collision. This activity resulted in the Indus River, which is known as the Indus Suture Plains Zone (ISPZ). The main route of the flow and sedimentation is to the end of the Arabian Sea (Indus fan). The flow starts from the rise in the western Tibet, northwest through the Ladakha and the Gilgit regions of Kashmir and flows southwest through Pakistan before it discharges into the Karachi (Arabian Sea). The early Holocene period was rapid sediment flux to the south Asia deltas. Indus River passes several climatic zones from subtropical arid to sub arid. The subequatorial atmospheric circulation system mainly lies in the higher part of the basin, while the monsoon is typically observed in tropic and sub tropic latitude. The dry months (October and April) in the northeastern part sets over the plain and piedmont portion of the Indus basin. The highest monsoon rainfall takes place in the summer session (May to September), while the annual rainfall drops from 500 to 125 mm from the river source to its delta. The densest anthropogenic activity takes place at the major northern part of Pakistan due to the increased needs for water storage for hydropower energy. These anthropogenic activities cause changes in the environment of the Indus River, causing increased sediment loads due to land clearance for agriculture or other facets such as land disturbance, logging activity in the forests, land mining activity and major civil engineering activity. Mining sediment significantly alters downstream sediment flux that possibly increases the effect of subsidence in the river of the deltas as well as causing scouring around engineering structures. Furthermore, coal mining influences the depositional environment, basin morphology, mineralogy, and hydrological conditions of the area.

Author Contributions: Conceptualization, U.K. and H.T.J.; methodology, U.K.; software, U.K.; validation, U.K., H.T.J. and G.K.; formal analysis, U.K.; investigation, U.K.; resources, U.K.; data curation, U.K. and H.T.J.; writing—original draft preparation, U.K.; writing—review and editing, U.K., H.T.J., G.K., A.R. and S.D.Z.; visualization, U.K.; supervision, H.T.J.; project administration, H.T.J.; funding acquisition, G.K. All authors have read and agreed to the published version of the manuscript.

Funding: This research received no external funding.

Institutional Review Board Statement: Not applicable.

Informed Consent Statement: Not applicable.

Conflicts of Interest: The authors declare no conflict of interest.

References

1. Ahmed, W.; Wu, Y.; Kidwai, S.; Li, X.; Zhang, G.; Zhang, J. Spatial and temporal variations of nutrients and chlorophyll a in the Indus River and its deltaic creeks and coastal waters (Northwest Indian Ocean, Pakistan). *J. Mar. Syst.* **2021**, *218*, 103525. [CrossRef]
2. Wang, M.; Tang, T.; Burek, P.; Havlík, P.; Krisztin, T.; Kroeze, C.; Leclère, D.; Strokal, M.; Wada, Y.; Wang, Y. Increasing nitrogen export to sea: A scenario analysis for the Indus River. *Sci. Total Environ.* **2019**, *694*, 133629. [CrossRef] [PubMed]
3. Sanyal, S. *The Ocean of Churn: How the Indian Ocean Shaped Human History*; Penguin: London, UK, 2016.
4. Inam, A.; Clift, P.D.; Giosan, L.; Tabrez, A.R.; Tahir, M.; Rabbani, M.M.; Danish, M. The geographic, geological and oceanographic setting of the Indus River. In *Large Rivers: Geomorphology and Management*; John Wiley & Sons, Ltd.: Chichester, UK, 2007; pp. 333–345.
5. Karim, A.; Veizer, J. Water balance of the Indus River Basin and moisture source in the Karakoram and western Himalayas: Implications from hydrogen and oxygen isotopes in river water. *J. Geophys. Res. Atmos.* **2002**, *107*, ACH 9-1–ACH 9-12. [CrossRef]
6. Lee, J.I.; Clift, P.D.; Layne, G.; Blum, J.; Khan, A.A. Sediment flux in the modern Indus River inferred from the trace element composition of detrital amphibole grains. *Sediment. Geol.* **2003**, *160*, 243–257. [CrossRef]
7. Clift, P.D. A brief history of the Indus River. *Geol. Soc. Lond. Spec. Publ.* **2002**, *195*, 237–258. [CrossRef]
8. Alizai, A.; Carter, A.; Clift, P.D.; VanLaningham, S.; Williams, J.C.; Kumar, R. Sediment provenance, reworking and transport processes in the Indus River by U–Pb dating of detrital zircon grains. *Glob. Planet. Chang.* **2011**, *76*, 33–55. [CrossRef]
9. Hall, R.; Cottam, M.A.; Wilson, M.E. The SE Asian gateway: History and tectonics of the Australia–Asia collision. *Geol. Soc. Lond. Spec. Publ.* **2011**, *355*, 1–6. [CrossRef]
10. Fielding, C.R.; Gupta, A. Sedimentology and stratigraphy of large river deposits: Recognition in the ancient record, and distinction from "incised valley fills". In *Large Rivers: Geomorphology and Management*; John Wiley & Sons, Ltd.: Chichester, UK, 2007; pp. 97–113.
11. Clift, P.D.; Zheng, H.; Carter, A.; Böning, P.; Jonell, T.N.; Schorr, H.; Shan, X.; Pahnke, K.; Wei, X.; Rittenour, T. Controls on erosion in the western Tarim Basin: Implications for the uplift of northwest Tibet and the Pamir. *Geosphere* **2017**, *13*, 1747–1765. [CrossRef]
12. Pickering, K.T.; Carter, A.; Andò, S.; Garzanti, E.; Limonta, M.; Vezzoli, G.; Milliken, K.L. Deciphering relationships between the Nicobar and Bengal submarine fans, Indian Ocean. *Earth Planet. Sci. Lett.* **2020**, *544*, 116329. [CrossRef]
13. Bastia, R.; Das, S.; Radhakrishna, M. Pre-and post-collisional depositional history in the upper and middle Bengal fan and evaluation of deepwater reservoir potential along the northeast Continental Margin of India. *Mar. Pet. Geol.* **2010**, *27*, 2051–2061. [CrossRef]
14. Macklin, M.G.; Lewin, J. The rivers of civilization. *Quat. Sci. Rev.* **2015**, *114*, 228–244. [CrossRef]
15. Bhattacharyya, K. *The Lower Damodar River, India: Understanding the Human Role in Changing Fluvial Environment*; Springer Science & Business Media: New York, NY, USA, 2011.
16. Angelakis, A.N.; Asano, T.; Bahri, A.; Jimenez, B.E.; Tchobanoglous, G. Water reuse: From ancient to modern times and the future. *Front. Environ. Sci.* **2018**, *6*, 26. [CrossRef]
17. Marks, R.B. *The Origins of the Modern World: A Global and Environmental Narrative from the Fifteenth to the Twenty-First Century*; Rowman & Littlefield: New York, NY, USA, 2019.
18. Eqani, S.A.M.A.S.; Kanwal, A.; Bhowmik, A.K.; Sohail, M.; Ullah, R.; Ali, S.M.; Alamdar, A.; Ali, N.; Fasola, M.; Shen, H. Spatial distribution of dust–bound trace elements in Pakistan and their implications for human exposure. *Environ. Pollut.* **2016**, *213*, 213–222. [CrossRef] [PubMed]
19. Zia, A.; Wagner, C.H. Mainstreaming early warning systems in development and planning processes: Multilevel implementation of Sendai framework in Indus and Sahel. *Int. J. Disaster Risk Sci.* **2015**, *6*, 189–199. [CrossRef]
20. Debnath, J.; Ahmed, I.; Pan, N.D. Impact of anthropogenic activities on channel characteristics: A case study of Muhuri River, Tripura, North-East India. *Arch. Appl. Sci. Res.* **2015**, *7*, 27–36.
21. Huang, M.-W.; Liao, J.-J.; Pan, Y.-W.; Cheng, M.-H. Rapid channelization and incision into soft bedrock induced by human activity—Implications from the Bachang River in Taiwan. *Eng. Geol.* **2014**, *177*, 10–24. [CrossRef]
22. Roy, S. Impact of linear transport infrastructure on fluvial connectivity across the catchments of West Bengal, India. *Geocarto Int.* **2021**, 1–26. [CrossRef]
23. Syvitski, J.P.; Kettner, A.J.; Overeem, I.; Giosan, L.; Brakenridge, G.R.; Hannon, M.; Bilham, R. Anthropocene metamorphosis of the Indus Delta and lower floodplain. *Anthropocene* **2013**, *3*, 24–35. [CrossRef]
24. Sohail, M.T.; Delin, H.; Siddiq, A.; Idrees, F.; Arshad, S. Evaluation of historic Indo-Pak relations, water resource issues and its impact on contemporary bilateral affairs. *Asia Pac. J. Multidiscip. Res.* **2015**, *3*, 48–55.
25. Petrie, C.A.; Orengo, H.A.; Green, A.S.; Walker, J.R.; Garcia, A.; Conesa, F.; Knox, J.R.; Singh, R.N. Mapping archaeology while mapping an empire: Using historical maps to reconstruct ancient settlement landscapes in modern India and Pakistan. *Geosciences* **2019**, *9*, 11. [CrossRef]
26. Yang, Y.-C.E.; Brown, C.; Yu, W.; Wescoat Jr, J.; Ringler, C. Water governance and adaptation to climate change in the Indus River Basin. *J. Hydrol.* **2014**, *519*, 2527–2537. [CrossRef]
27. Shaheen, F.; Shah, F.A. Climate Change, Economic Growth, and Cooperative Management of Indus River Basin. In Proceedings of the 2017 Agricultural & Applied Economics Association Annual Meeting, Chicago, IL, USA, 30 July–1 August 2017.

28. Usman, Q.A.; Muhammad, S.; Ali, W.; Yousaf, S.; Jadoon, I.A. Spatial distribution and provenance of heavy metal contamination in the sediments of the Indus River and its tributaries, North Pakistan: Evaluation of pollution and potential risks. *Environ. Technol. Innov.* **2021**, *21*, 101184. [CrossRef]
29. Hester, E.T.; Doyle, M.W. Human impacts to river temperature and their effects on biological processes: A quantitative synthesis 1. *JAWRA J. Am. Water Resour. Assoc.* **2011**, *47*, 571–587. [CrossRef]
30. Al-Ghanim, K.; Mahboob, S.; Seemab, S.; Sultana, S.; Sultana, T.; Al-Misned, F.; Ahmed, Z. Monitoring of trace metals in tissues of Wallago attu (lanchi) from the Indus River as an indicator of environmental pollution. *Saudi J. Biol. Sci.* **2016**, *23*, 72–78. [CrossRef]
31. Tsering, T.; Sillanpää, M.; Sillanpää, M.; Viitala, M.; Reinikainen, S.-P. Microplastics pollution in the Brahmaputra River and the Indus River of the Indian Himalaya. *Sci. Total Environ.* **2021**, *789*, 147968. [CrossRef]
32. Westerweel, J.; Roperch, P.; Licht, A.; Dupont-Nivet, G.; Win, Z.; Poblete, F.; Ruffet, G.; Swe, H.H.; Thi, M.K.; Aung, D.W. Burma Terrane part of the Trans-Tethyan arc during collision with India according to palaeomagnetic data. *Nat. Geosci.* **2019**, *12*, 863–868. [CrossRef]
33. Henderson, A.L.; Najman, Y.; Parrish, R.; BouDagher-Fadel, M.; Barford, D.; Garzanti, E.; Andò, S. Geology of the Cenozoic Indus Basin sedimentary rocks: Paleoenvironmental interpretation of sedimentation from the western Himalaya during the early phases of India-Eurasia collision. *Tectonics* **2010**, *29*. [CrossRef]
34. Wang, E. Timing of the initial collision between the Indian and Asian continents. *Sci. China Earth Sci.* **2017**, *60*, 626–634. [CrossRef]
35. Hu, X.; Wang, J.; An, W.; Garzanti, E.; Li, J. Constraining the timing of the India-Asia continental collision by the sedimentary record. *Sci. China Earth Sci.* **2017**, *60*, 603–625. [CrossRef]
36. Khan, M.; Liu, Y. Geodynamic evolution of the offshore Indus Basin Pakistan: The western Indian plate passive continental margin. *Geophys. J. Int.* **2019**, *217*, 1366–1386. [CrossRef]
37. Ali, J.R.; Aitchison, J.C. Gondwana to Asia: Plate tectonics, paleogeography and the biological connectivity of the Indian sub-continent from the Middle Jurassic through latest Eocene (166–35 Ma). *Earth-Sci. Rev.* **2008**, *88*, 145–166. [CrossRef]
38. Aitchison, J.C.; Ali, J.R.; Davis, A.M. When and where did India and Asia collide? *J. Geophys. Res. Solid Earth* **2007**, *112*, B05423. [CrossRef]
39. Chatterjee, S.; Bajpai, S. India's northward drift from Gondwana to Asia during the Late Cretaceous-Eocene. *Proc. Indian Natl. Sci. Acad.* **2016**, *82*, 479–487. [CrossRef]
40. Kohn, M.J.; Paul, S.K.; Corrie, S.L. The lower Lesser Himalayan sequence: A Paleoproterozoic arc on the northern margin of the Indian plate. *Bulletin* **2010**, *122*, 323–335. [CrossRef]
41. Chatterjee, S.; Goswami, A.; Scotese, C.R. The longest voyage: Tectonic, magmatic, and paleoclimatic evolution of the Indian plate during its northward flight from Gondwana to Asia. *Gondwana Res.* **2013**, *23*, 238–267. [CrossRef]
42. Shaikh, M.A.; Maurya, D.M.; Mukherjee, S.; Vanik, N.P.; Padmalal, A.; Chamyal, L.S. Tectonic evolution of the intra-uplift Vigodi-Gugriana-Khirasra-Netra Fault System in the seismically active Kachchh rift basin, India: Implications for the western continental margin of the Indian plate. *J. Struct. Geol.* **2020**, *140*, 104124. [CrossRef]
43. Zeitler, P.K.; Koons, P.O.; Bishop, M.P.; Chamberlain, C.P.; Craw, D.; Edwards, M.A.; Hamidullah, S.; Jan, M.Q.; Khan, M.A.; Khattak, M.U.K. Crustal reworking at Nanga Parbat, Pakistan: Metamorphic consequences of thermal-mechanical coupling facilitated by erosion. *Tectonics* **2001**, *20*, 712–728. [CrossRef]
44. Butler, R.W.H.; Casey, M.; Lloyd, G.E.; Bond, C.E.; McDade, P.; Shipton, Z.; Jones, R. Vertical stretching and crustal thickening at Nanga Parbat, Pakistan Himalaya: A model for distributed continental deformation during mountain building. *Tectonics* **2002**, *21*, 9-1–9-17. [CrossRef]
45. Pêcher, A.; Giuliani, G.; Garnier, V.; Maluski, H.; Kausar, A.; Malik, R.; Muntaz, H. Geology, geochemistry and Ar–Ar geochronology of the Nangimali ruby deposit, Nanga Parbat Himalaya (Azad Kashmir, Pakistan). *J. Asian Earth Sci.* **2002**, *21*, 265–282. [CrossRef]
46. Malkani, M.S.; Mahmood, Z. Revised stratigraphy of Pakistan. *Geol. Surv. Pak. Rec.* **2016**, *127*, 1–87.
47. Kazmi, A.H.; Abbasi, I.A. *Stratigraphy & Historical Geology of Pakistan*; Department & National Centre of Excellence in Geology: Peshawar, Pakistan, 2008.
48. Labisko, J.; Griffiths, R.A.; Chong-Seng, L.; Bunbury, N.; Maddock, S.T.; Bradfield, K.S.; Taylor, M.L.; Groombridge, J.J. Endemic, endangered and evolutionarily significant: Cryptic lineages in Seychelles' frogs (Anura: Sooglossidae). *Biol. J. Linn. Soc.* **2019**, *126*, 417–435. [CrossRef]
49. Deepak, V.; Maddock, S.T.; Williams, R.; Nagy, Z.T.; Conradie, W.; Rocha, S.; Harris, D.J.; Perera, A.; Gvoždík, V.; Doherty-Bone, T.M. Molecular phylogenetics of sub-Saharan African natricine snakes, and the biogeographic origins of the Seychelles endemic Lycognathophis seychellensis. *Mol. Phylogenet. Evol.* **2021**, *161*, 107152. [CrossRef]
50. Rajbhandari, R.; Shrestha, A.; Kulkarni, A.; Patwardhan, S.; Bajracharya, S. Projected changes in climate over the Indus river basin using a high resolution regional climate model (PRECIS). *Clim. Dyn.* **2015**, *44*, 339–357. [CrossRef]
51. Su, B.; Huang, J.; Gemmer, M.; Jian, D.; Tao, H.; Jiang, T.; Zhao, C. Statistical downscaling of CMIP5 multi-model ensemble for projected changes of climate in the Indus River Basin. *Atmos. Res.* **2016**, *178*, 138–149. [CrossRef]
52. Safdar, F.; Khokhar, M.F.; Arshad, M.; Adil, I.H. Climate change indicators and spatiotemporal shift in monsoon patterns in Pakistan. *Adv. Meteorol.* **2019**, *2019*, 8281201. [CrossRef]

53. Kravtsova, V.; Mikhailov, V.; Efremova, N. Variations of the hydrological regime, morphological structure, and landscapes of the Indus River delta (Pakistan) under the effect of large-scale water management measures. *Water Resour.* **2009**, *36*, 365–379. [CrossRef]
54. Sharma, A.; Kumar, K.; Laskar, A.; Singh, S.K.; Mehta, P. Oxygen, deuterium, and strontium isotope characteristics of the Indus River water system. *Geomorphology* **2017**, *284*, 5–16. [CrossRef]
55. Thayyen, R.; Gergan, J. Role of glaciers in watershed hydrology: A preliminary study of a "Himalayan catchment". *Cryosphere* **2010**, *4*, 115–128. [CrossRef]
56. Singh, N.; Li, J. An efficient extraction of lead metal from waste cathode ray tubes (CRTs) through mechano-thermal process by using carbon as a reducing agent. *J. Clean. Prod.* **2017**, *148*, 103–110. [CrossRef]
57. Bajpai, O.; Dutta, V.; Singh, R.; Chaudhary, L.; Pandey, J. Tree Community Assemblage and Abiotic Variables in Tropical Moist Deciduous Forest of Himalayan Terai Eco-Region. *Proc. Natl. Acad. Sci. India Sect. B Biol. Sci.* **2020**, *90*, 873–883. [CrossRef]
58. Giosan, L.; Constantinescu, S.; Clift, P.D.; Tabrez, A.R.; Danish, M.; Inam, A. Recent morphodynamics of the Indus delta shore and shelf. *Cont. Shelf Res.* **2006**, *26*, 1668–1684. [CrossRef]
59. Gaurav, K.; Sinha, R.; Panda, P. The Indus flood of 2010 in Pakistan: A perspective analysis using remote sensing data. *Nat. Hazards* **2011**, *59*, 1815–1826. [CrossRef]
60. Ijaz, M.W.; Mahar, R.B.; Ansari, K.; Siyal, A.A.; Anjum, M.N. Integrated assessment of contemporary hydro-geomorphologic evolution of the Indus River Estuary, Pakistan in context to regulated fluvial regimes. *Estuar. Coast. Shelf Sci.* **2020**, *236*, 106657. [CrossRef]
61. Valipour, M.; Krasilnikof, J.; Yannopoulos, S.; Kumar, R.; Deng, J.; Roccaro, P.; Mays, L.; Grismer, M.E.; Angelakis, A.N. The evolution of agricultural drainage from the earliest times to the present. *Sustainability* **2020**, *12*, 416. [CrossRef]
62. Ahmed, M. *Ancient Pakistan–An Archaeological History: Volume III: Harappan Civilization–The Material Culture*; Amazon: Bellevue, WA, USA, 2014.
63. Robinson, A. *The Indus: Lost Civilizations*; Reaktion Books: London, UK, 2021.
64. Uprety, D.C.; Reddy, V.; Mura, J.D. Historical Analysis of Climate Change and Agriculture. In *Climate Change and Agriculture*; Springer: Berlin/Heidelberg, Germany, 2019; pp. 7–29.
65. Miller, H.M.L. Associations and ideologies in the locations of urban craft production at Harappa, Pakistan (Indus Civilization). *Archeol. Pap. Am. Anthropol. Assoc.* **2007**, *17*, 37–51. [CrossRef]
66. Giosan, L.; Clift, P.D.; Macklin, M.G.; Fuller, D.Q.; Constantinescu, S.; Durcan, J.A.; Stevens, T.; Duller, G.A.; Tabrez, A.R.; Gangal, K. Fluvial landscapes of the Harappan civilization. *Proc. Natl. Acad. Sci. USA* **2012**, *109*, E1688–E1694. [CrossRef]
67. Winston, H.Y.; Yang, Y.-C.; Savitsky, A.; Alford, D.; Brown, C. *The Indus Basin of Pakistan: The Impacts of Climate Risks on Water and Agriculture*; World Bank Publications: Washington, WA, USA, 2013.
68. Carter, L.; Orpin, A.R.; Kuehl, S.A. From mountain source to ocean sink–the passage of sediment across an active margin, Waipaoa Sedimentary System, New Zealand. *Mar. Geol.* **2010**, *270*, 1–10. [CrossRef]
69. Berkelhammer, M.; Sinha, A.; Stott, L.; Cheng, H.; Pausata, F.S.; Yoshimura, K. An abrupt shift in the Indian monsoon 4000 years ago. *Geophys. Monogr. Ser.* **2012**, *198*, 75–87.
70. Kumar, K.A.; Thayalan, S.; Reddy, R.; Lalitha, M.; Kalaiselvi, B.; Parvathy, S.; Sujatha, K.; Hegde, R.; Singh, S.; Mishra, B.B. Geology and Geomorphology. In *The Soils of India*; Springer: Berlin/Heidelberg, Germany, 2020; pp. 57–79.
71. Clift, P.D.; Jonell, T.N. Monsoon controls on sediment generation and transport: Mass budget and provenance constraints from the Indus River catchment, delta and submarine fan over tectonic and multi-millennial timescales. *Earth-Sci. Rev.* **2021**, *2021*, 103682. [CrossRef]
72. Prat, O.P.; Nelson, B.R. Mapping the world's tropical cyclone rainfall contribution over land using the TRMM Multi-satellite Precipitation Analysis. *Water Resour. Res.* **2013**, *49*, 7236–7254. [CrossRef]
73. Wulf, H.; Bookhagen, B.; Scherler, D. Seasonal precipitation gradients and their impact on fluvial sediment flux in the Northwest Himalaya. *Geomorphology* **2010**, *118*, 13–21. [CrossRef]
74. Dimri, A.; Thayyen, R.; Kibler, K.; Stanton, A.; Jain, S.; Tullos, D.; Singh, V. A review of atmospheric and land surface processes with emphasis on flood generation in the Southern Himalayan rivers. *Sci. Total Environ.* **2016**, *556*, 98–115. [CrossRef]
75. Li, Y.; Clift, P.D.; Murray, R.W.; Exnicios, E.; Ireland, T.; Böning, P. Asian summer monsoon influence on chemical weathering and sediment provenance determined by clay mineral analysis from the Indus Submarine Canyon. *Quat. Res.* **2020**, *93*, 23–39. [CrossRef]
76. Ali, S.; Hathorne, E.C.; Frank, M. Persistent Provenance of South Asian Monsoon-Induced Silicate Weathering Over the Past 27 Million Years. *Paleoceanogr. Paleoclimatol.* **2021**, *36*, e2020PA003909. [CrossRef]
77. Vannucci, M. Indo-west Pacific mangroves. In *Mangrove Ecosystems*; Springer: Berlin/Heidelberg, Germany, 2002; pp. 123–215.
78. Garzanti, E. The Himalayan Foreland Basin from collision onset to the present: A sedimentary–petrology perspective. *Geol. Soc. Lond. Spec. Publ.* **2019**, *483*, 65–122. [CrossRef]
79. Gopal, B. Future of wetlands in tropical and subtropical Asia, especially in the face of climate change. *Aquat. Sci.* **2013**, *75*, 39–61. [CrossRef]
80. Harrison, R.; Tsiolakis, E.; Stone, B.; Lord, A.; McGeehin, J.; Mahan, S.; Chirico, P. Late Pleistocene and Holocene uplift history of Cyprus: Implications for active tectonics along the southern margin of the Anatolian microplate. *Geol. Soc. Lond. Spec. Publ.* **2013**, *372*, 561–584. [CrossRef]

81. Dong, G.; Liu, F.; Chen, F. Environmental and technological effects on ancient social evolution at different spatial scales. *Sci. China Earth Sci.* **2017**, *60*, 2067–2077. [CrossRef]
82. Clift, P.D.; Giosan, L.; Carter, A.; Garzanti, E.; Galy, V.; Tabrez, A.R.; Pringle, M.; Campbell, I.H.; France-Lanord, C.; Blusztajn, J. Monsoon control over erosion patterns in the western Himalaya: Possible feed-back into the tectonic evolution. *Geol. Soc. Lond. Spec. Publ.* **2010**, *342*, 185–218. [CrossRef]
83. Clift, P.D.; Giosan, L.; Henstock, T.J.; Tabrez, A.R. Sediment storage and reworking on the shelf and in the canyon of the Indus River-fan system since the last glacial maximum. *Basin Res.* **2014**, *26*, 183–202. [CrossRef]
84. Clift, P.D.; Shimizu, N.; Layne, G.; Blusztajn, J.; Gaedicke, C.; Schluter, H.-U.; Clark, M.; Amjad, S. Development of the Indus Fan and its significance for the erosional history of the Western Himalaya and Karakoram. *Geol. Soc. Am. Bull.* **2001**, *113*, 1039–1051. [CrossRef]
85. Mukhopadhyay, B.; Dutta, A. A stream water availability model of Upper Indus Basin based on a topologic model and global climatic datasets. *Water Resour. Manag.* **2010**, *24*, 4403–4443. [CrossRef]
86. Bocchiola, D.; Soncini, A. Water resources modeling and prospective evaluation in the Indus River under present and prospective climate change. In *Indus River Basin*; Elsevier: Amsterdam, The Netherlands, 2019; pp. 17–56.
87. Jabbar Khan, A.; Akhter, G.; Gabriel, H.F.; Shahid, M. Anthropogenic effects of coal mining on ecological resources of the Central Indus Basin, Pakistan. *Int. J. Environ. Res. Public Health* **2020**, *17*, 1255. [CrossRef]
88. Hewitt, K. Glacier change, concentration, and elevation effects in the Karakoram Himalaya, Upper Indus Basin. *Mt. Res. Dev.* **2011**, *31*, 188–200. [CrossRef]
89. Atif, I.; Mahboob, M.A.; Iqbal, J. Snow cover area change assessment in 2003 and 2013 using MODIS data of the Upper Indus Basin, Pakistan. *J. Himal. Earth Sci.* **2015**, *48*, 117.
90. Wijngaard, R.R.; Lutz, A.F.; Nepal, S.; Khanal, S.; Pradhananga, S.; Shrestha, A.B.; Immerzeel, W.W. Future changes in hydro-climatic extremes in the Upper Indus, Ganges, and Brahmaputra River basins. *PLoS ONE* **2017**, *12*, e0190224.
91. Khan, A.; Ghoraba, S.; Arnold, J.G.; Di Luzio, M. Hydrological modeling of upper Indus Basin and assessment of deltaic ecology. *Int. J. Mod. Eng. Res.* **2014**, *4*, 73–85.
92. Sharif, M.; Archer, D.; Fowler, H.; Forsythe, N. Trends in timing and magnitude of flow in the Upper Indus Basin. *Hydrol. Earth Syst. Sci.* **2013**, *17*, 1503–1516. [CrossRef]
93. Kulkarni, A.; Patwardhan, S.; Kumar, K.K.; Ashok, K.; Krishnan, R. Projected climate change in the Hindu Kush–Himalayan region by using the high-resolution regional climate model PRECIS. *Mt. Res. Dev.* **2013**, *33*, 142–151. [CrossRef]
94. Miller, J.D.; Immerzeel, W.W.; Rees, G. Climate change impacts on glacier hydrology and river discharge in the Hindu Kush–Himalayas. *Mt. Res. Dev.* **2012**, *32*, 461–467. [CrossRef]
95. Dahri, Z.H.; Ludwig, F.; Moors, E.; Ahmad, B.; Khan, A.; Kabat, P. An appraisal of precipitation distribution in the high-altitude catchments of the Indus basin. *Sci. Total Environ.* **2016**, *548*, 289–306. [CrossRef] [PubMed]
96. Eaton, D. Variability of Climate and Riverflow in Partially Glacierised Basins in the Karakoram and Western Himalaya. Ph.D. Thesis, University of Salford, Salford, UK, 2015.
97. Turkel, W. *The Archive of Place: Unearthing the Pasts of the Chilcotin Plateau*; UBC Press: Vancouver, BC, Canada, 2011; Volume 2.
98. Diderot, D. Stability of output fluxes of large rivers in South and East Asia during the last 2 million years: Implications on floodplain processes. *Basin Res.* **1999**, *11*, 293–303.
99. Walling, D. Human impact on land–ocean sediment transfer by the world's rivers. *Geomorphology* **2006**, *79*, 192–216. [CrossRef]
100. Romans, B.W.; Castelltort, S.; Covault, J.A.; Fildani, A.; Walsh, J. Environmental signal propagation in sedimentary systems across timescales. *Earth-Sci. Rev.* **2016**, *153*, 7–29. [CrossRef]
101. Ali, K.F.; De Boer, D.H. Spatially distributed erosion and sediment yield modeling in the upper Indus River basin. *Water Resour. Res.* **2010**, *46*. [CrossRef]
102. Sabir, M.A.; Shafiq-Ur-Rehman, S.; Umar, M.; Waseem, A.; Farooq, M.; Khan, A.R. The Impact of Suspended Sediment Load on Reservoir Siltation and Energy Production: A Case Study of the Indus River and Its Tributaries. *Pol. J. Environ. Stud.* **2013**, *22*, 219–225.
103. Hinderer, M. From gullies to mountain belts: A review of sediment budgets at various scales. *Sediment. Geol.* **2012**, *280*, 21–59. [CrossRef]
104. Garzanti, E.; Padoan, M.; Andò, S.; Resentini, A.; Vezzoli, G.; Lustrino, M. Weathering and relative durability of detrital minerals in equatorial climate: Sand petrology and geochemistry in the East African Rift. *J. Geol.* **2013**, *121*, 547–580. [CrossRef]
105. Robinson, R.A.; Brezina, C.A.; Parrish, R.R.; Horstwood, M.S.; Oo, N.W.; Bird, M.I.; Thein, M.; Walters, A.S.; Oliver, G.J.; Zaw, K. Large rivers and orogens: The evolution of the Yarlung Tsangpo–Irrawaddy system and the eastern Himalayan syntaxis. *Gondwana Res.* **2014**, *26*, 112–121. [CrossRef]
106. Wells, J. Effects of sea level rise on coastal sedimention and erosion. In *Climate Change*; CRC Press: Boca Raton, FL, USA, 2021; pp. 111–136.
107. Rasul, G.; Neupane, N.; Hussain, A.; Pasakhala, B. Beyond hydropower: Towards an integrated solution for water, energy and food security in South Asia. *Int. J. Water Resour. Dev.* **2021**, *37*, 466–490. [CrossRef]
108. Chen, D.; Li, X.; Saito, Y.; Liu, J.P.; Duan, Y.; Zhang, L. Recent evolution of the Irrawaddy (Ayeyarwady) Delta and the impacts of anthropogenic activities: A review and remote sensing survey. *Geomorphology* **2020**, *365*, 107231. [CrossRef]
109. Best, J. Anthropogenic stresses on the world's big rivers. *Nat. Geosci.* **2019**, *12*, 7–21. [CrossRef]

110. Scott, C.A.; Zhang, F.; Mukherji, A.; Immerzeel, W.; Mustafa, D.; Bharati, L. Water in The Hindu kush himalaya. In *The Hindu Kush Himalaya Assessment*; Springer: Berlin/Heidelberg, Germany, 2019; pp. 257–299.
111. Vercruysse, K.; Grabowski, R.C. Human impact on river planform within the context of multi-timescale river channel dynamics in a Himalayan river system. *Geomorphology* **2021**, *381*, 107659. [CrossRef]
112. Palmer, M.A.; Reidy Liermann, C.A.; Nilsson, C.; Flörke, M.; Alcamo, J.; Lake, P.S.; Bond, N. Climate change and the world's river basins: Anticipating management options. *Front. Ecol. Environ.* **2008**, *6*, 81–89. [CrossRef]
113. Gosain, A.; Rao, S.; Basuray, D. Climate change impact assessment on hydrology of Indian river basins. *Curr. Sci.* **2006**, *90*, 346–353.
114. Huang, J.; Li, Y.; Fu, C.; Chen, F.; Fu, Q.; Dai, A.; Shinoda, M.; Ma, Z.; Guo, W.; Li, Z. Dryland climate change: Recent progress and challenges. *Rev. Geophys.* **2017**, *55*, 719–778. [CrossRef]
115. Syvitski, J.P.; Kettner, A. Sediment flux and the Anthropocene. *Philos. Trans. R. Soc. A Math. Phys. Eng. Sci.* **2011**, *369*, 957–975. [CrossRef] [PubMed]
116. Jaboyedoff, M.; Michoud, C.; Derron, M.-H.; Voumard, J.; Leibundgut, G.; Sudmeier-Rieux, K.; Nadim, F.; Leroi, E. Human-induced landslides: Toward the analysis of anthropogenic changes of the slope environment. In *Landslides and Engineered Slopes. Experience, Theory and Practice*; CRC Press: Boca Raton, FL, USA, 2018; pp. 217–232.
117. Walling, D. The changing sediment loads of the world's rivers. *Ann. Wars. Univ. Life Sci.-SGGW. Land Reclam.* **2008**, *39*. [CrossRef]
118. Gupta, H.; Kao, S.-J.; Dai, M. The role of mega dams in reducing sediment fluxes: A case study of large Asian rivers. *J. Hydrol.* **2012**, *464*, 447–458. [CrossRef]
119. Nazeer, M.M. Review of World Bank Two Most Terrible Moves Resulting in All Ever Increasing, Terrible Climate Calamities over the Entire Earth Globe along with Continuous Huge Disasters in Pakistan. *Open Access Libr. J.* **2020**, *7*, 1–42.
120. Sajjad, I. *Study of Recharge Mechanism in Haripur Plain Contribution to the Area from Tarbela and Khanpur Dams. Final Report for the Period 15 June 1987–14 October 1990*; International Atomic Energy Agency: Vienna, Austria, 1991.
121. Rădoane, M.; Obreja, F.; Cristea, I.; Mihailă, D. Changes in the channel-bed level of the eastern Carpathian rivers: Climatic vs. human control over the last 50 years. *Geomorphology* **2013**, *193*, 91–111. [CrossRef]
122. Alonso, E.; Olivella, S.; Pinyol, N. A review of Beliche Dam. *Géotechnique* **2005**, *55*, 267–285. [CrossRef]
123. Ullah, M.F.; Mahmood, K.; Akram, M.S. Coal mining trends and future prospects: A case study of Eastern Salt Range, Punjab, Pakistan. *J. Himal. Earth Sci.* **2018**, *51*, 87–93.
124. Fang, Q.; Hong, H.; Algeo, T.J.; Huang, X.; Sun, A.; Churchman, G.J.; Chorover, J.; Chen, S.; Liu, Y. Microtopography-mediated hydrologic environment controls elemental migration and mineral weathering in subalpine surface soils of subtropical monsoonal China. *Geoderma* **2019**, *344*, 82–98. [CrossRef]
125. Jabeen, F.; Chaudhry, A.S. Monitoring trace metals in different tissues of Cyprinus carpio from the Indus River in Pakistan. *Environ. Monit. Assess.* **2010**, *170*, 645–656. [CrossRef] [PubMed]
126. Shukla, A.; Ali, I. Major River Systems of Jammu and Kashmir. In *The Indian Rivers*; Springer: Berlin/Heidelberg, Germany, 2018; pp. 383–411.
127. Ramadas, M.; Samantaray, A.K. Applications of remote sensing and GIS in water quality monitoring and remediation: A state-of-the-art review. In *Water Remediation*; Energy, Environment, and Sustainability Series; Springer: Singapore, 2018; pp. 225–246.
128. Ahmad, T.; Khanna, P.; Chakrapani, G.; Balakrishnan, S. Geochemical characteristics of water and sediment of the Indus river, Trans-Himalaya, India: Constraints on weathering and erosion. *J. Asian Earth Sci.* **1998**, *16*, 333–346. [CrossRef]
129. Gao, X.; Chen, S.; Long, A. Chemical speciation of 12 metals in surface sediments from the northern South China Sea under natural grain size. *Mar. Pollut. Bull.* **2008**, *56*, 786–792. [CrossRef]
130. Janjuhah, H.T.; Alansari, A. Offshore Carbonate Facies Characterization and Reservoir Quality of Miocene Rocks in the Southern Margin of South China Sea. *Acta Geol. Sin.-Engl. Ed.* **2020**, *94*, 1547–1561. [CrossRef]
131. Limmer, D.R.; Köhler, C.M.; Hillier, S.; Moreton, S.G.; Tabrez, A.R.; Clift, P.D. Chemical weathering and provenance evolution of Holocene–recent sediments from the Western Indus Shelf, Northern Arabian Sea inferred from physical and mineralogical properties. *Mar. Geol.* **2012**, *326*, 101–115. [CrossRef]
132. Huang, J.; Jiao, W.; Liu, J.; Wan, S.; Xiong, Z.; Zhang, J.; Yang, Z.; Li, A.; Li, T. Sediment distribution and dispersal in the southern South China Sea: Evidence from clay minerals and magnetic properties. *Mar. Geol.* **2021**, *439*, 106560. [CrossRef]
133. Qadir, A.; El-Rawy, M. Impact Assessment of Physiography, Subsurface Hydraulic Gradients and Lithologic Heterogeneity on the Groundwater Quality. *Iran. J. Sci. Technol. Trans. Civ. Eng.* **2021**, 1–22. [CrossRef]
134. Mukherjee, A.; Gupta, S.; Coomar, P.; Fryar, A.E.; Guillot, S.; Verma, S.; Bhattacharya, P.; Bundschuh, J.; Charlet, L. Plate tectonics influence on geogenic arsenic cycling: From primary sources to global groundwater enrichment. *Sci. Total Environ.* **2019**, *683*, 793–807. [CrossRef] [PubMed]
135. Bretschneider, L.; Hathorne, E.C.; Huang, H.; Lübbers, J.; Kochhann, K.G.; Holbourn, A.; Kuhnt, W.; Thiede, R.; Gebregiorgis, D.; Giosan, L. Provenance and weathering of clays delivered to the Bay of Bengal during the middle Miocene: Linkages to tectonics and monsoonal climate. *Paleoceanogr. Paleoclimatol.* **2021**, *36*, e2020PA003917. [CrossRef]
136. Lone, S.A.; Jeelani, G.; Mukherjee, A.; Coomar, P. Arsenic fate in upper Indus river basin (UIRB) aquifers: Controls of hydrochemical processes, provenances and water-aquifer matrix interaction. *Sci. Total Environ.* **2021**, *795*, 148734. [CrossRef] [PubMed]

137. Jeelani, G.; Lone, S.A.; Nisa, A.U.; Mukherjee, A.; Deshpande, R. Sources and processes of groundwater arsenic mobilization in upper Jhelum basin, western Himalayas. *J. Hydrol.* **2020**, *591*, 125292. [CrossRef]
138. Cook, C.G. Late Pleistocene-Early Holocene Glacial Dynamics, Asian Palaeomonsoon Variability and Landscape Change at Lake Shudu, Yunnan Province, Southwestern China. Ph.D. Thesis, University of Exeter, Exeter, UK, 2009.
139. Khatri, N.; Tyagi, S. Influences of natural and anthropogenic factors on surface and groundwater quality in rural and urban areas. *Front. Life Sci.* **2015**, *8*, 23–39. [CrossRef]
140. Zuazo, V.H.D.; Pleguezuelo, C.R.R. Soil-erosion and runoff prevention by plant covers: A review. In *Sustainable Agriculture*; Springer: Dordrecht, The Netherlands, 2009; pp. 785–811.
141. Makri, P.; Stathopoulou, E.; Hermides, D.; Kontakiotis, G.; Zarkogiannis, S.D.; Skilodimou, H.D.; Bathrellos, G.D.; Antonarakou, A.; Scoullos, M. The Environmental Impact of a Complex Hydrogeological System on Hydrocarbon-Pollutants' Natural Attenuation: The Case of the Coastal Aquifers in Eleusis, West Attica, Greece. *J. Mar. Sci. Eng.* **2020**, *8*, 1018. [CrossRef]
142. Hermides, D.; Makri, P.; Kontakiotis, G.; Antonarakou, A. Advances in the Coastal and Submarine Groundwater Processes: Controls and Environmental Impact on the Thriassion Plain and Eleusis Gulf (Attica, Greece). *J. Mar. Sci. Eng.* **2020**, *8*, 944. [CrossRef]
143. Podgorski, J.E.; Eqani, S.A.M.A.S.; Khanam, T.; Ullah, R.; Shen, H.; Berg, M. Extensive arsenic contamination in high-pH unconfined aquifers in the Indus Valley. *Sci. Adv.* **2017**, *3*, e1700935. [CrossRef]
144. Memon, A.A. Devastation of the Indus river delta. In *Impacts of Global Climate Change*; American Society of Civil Engineers: Reston, VA, USA, 2005; pp. 1–12.
145. Khan, A.; Yousafzai, A.M.; Latif, M.; Rehman, A.; Khan, Q.; Zaib, A.; Ullah, A.; Sthanadar, A.A.; Haq, I.; Aziz, A. Analysis of selected water quality parameters and heavy metals of Indus River at BekaSwabi, Khyber Pakhtunkhwa, Pakistan. *Int. J. Biosci. (IJB)* **2014**, *4*, 28–38.
146. Mosley, L.M. Drought impacts on the water quality of freshwater systems; review and integration. *Earth-Sci. Rev.* **2015**, *140*, 203–214. [CrossRef]
147. Zwolsman, J.; Van Bokhoven, A. Impact of summer droughts on water quality of the Rhine River-a preview of climate change? *Water Sci. Technol.* **2007**, *56*, 45–55. [CrossRef] [PubMed]
148. Loganathan, P.; Vigneswaran, S.; Kandasamy, J.; Naidu, R. Defluoridation of drinking water using adsorption processes. *J. Hazard. Mater.* **2013**, *248*, 1–19. [CrossRef] [PubMed]
149. Mukherjee, I.; Singh, U.K. Groundwater fluoride contamination, probable release, and containment mechanisms: A review on Indian context. *Environ. Geochem. Health* **2018**, *40*, 2259–2301. [CrossRef]
150. Rafique, T.; Naseem, S.; Usmani, T.H.; Bashir, E.; Khan, F.A.; Bhanger, M.I. Geochemical factors controlling the occurrence of high fluoride groundwater in the Nagar Parkar area, Sindh, Pakistan. *J. Hazard. Mater.* **2009**, *171*, 424–430. [CrossRef]
151. Curtosi, A.; Pelletier, E.; Vodopivez, C.L.; Mac Cormack, W.P. Polycyclic aromatic hydrocarbons in soil and surface marine sediment near Jubany Station (Antarctica). Role of permafrost as a low-permeability barrier. *Sci. Total Environ.* **2007**, *383*, 193–204. [CrossRef]
152. Zheng, Q.; Cao, C.; Zhang, M. Sedimentary features of the Permian-Triassic boundary sequence of the Meishan section in Changxing County, Zhejiang Province. *Sci. China Earth Sci.* **2013**, *56*, 956–969. [CrossRef]
153. Khan, A.; Raza, S.A.; Fatima, A.; Haider, S.W. Assessment of Groundwater Quality in Coastal Region a Case Study of Qayyumabad, Karachi, Pakistan. *Asian Rev. Environ. Earth Sci.* **2020**, *7*, 9–17. [CrossRef]
154. Hasan, M.; Shang, Y.; Metwaly, M.; Jin, W.; Khan, M.; Gao, Q. Assessment of Groundwater Resources in Coastal Areas of Pakistan for Sustainable Water Quality Management using Joint Geophysical and Geochemical Approach: A Case Study. *Sustainability* **2020**, *12*, 9730. [CrossRef]
155. Naseem, S.; Bashir, E.; Ahmed, P.; Rafique, T.; Hamza, S.; Kaleem, M. Impact of seawater intrusion on the geochemistry of groundwater of Gwadar District, Balochistan and its appraisal for drinking water quality. *Arab. J. Sci. Eng.* **2018**, *43*, 281–293. [CrossRef]
156. Labbe, J.M.; Hadley, K.S.; Schipper, A.M.; Leuven, R.S.; Gardiner, C.P. Influence of bank materials, bed sediment, and riparian vegetation on channel form along a gravel-to-sand transition reach of the Upper Tualatin River, Oregon, USA. *Geomorphology* **2011**, *125*, 374–382. [CrossRef]
157. Khalequzzaman, M.; Faruque, F.S.; Mitra, A.K. Assessment of arsenic contamination of groundwater and health problems in Bangladesh. *Int. J. Environ. Res. Public Health* **2005**, *2*, 204–213. [CrossRef] [PubMed]
158. Atapattu, S.S.; Kodituwakku, D.C. Agriculture in South Asia and its implications on downstream health and sustainability: A review. *Agric. Water Manag.* **2009**, *96*, 361–373. [CrossRef]
159. Ayub, M.A.; Usman, M.; Faiz, T.; Umair, M.; ul Haq, M.A.; Rizwan, M.; Ali, S.; ur Rehman, M.Z. Restoration of degraded soil for sustainable agriculture. In *Soil Health Restoration and Management*; Springer: Berlin/Heidelberg, Germany, 2020; pp. 31–81.
160. Iqbal, S.; Guber, A.K.; Khan, H.Z. Estimating nitrogen leaching losses after compost application in furrow irrigated soils of Pakistan using HYDRUS-2D software. *Agric. Water Manag.* **2016**, *168*, 85–95. [CrossRef]
161. Brammer, H.; Ravenscroft, P. Arsenic in groundwater: A threat to sustainable agriculture in South and South-east Asia. *Environ. Int.* **2009**, *35*, 647–654. [CrossRef]
162. Issaka, S.; Ashraf, M.A. Impact of soil erosion and degradation on water quality: A review. *Geol. Ecol. Landsc.* **2017**, *1*, 1–11. [CrossRef]

163. Jones, R.; Ricardo, G.; Negri, A. Effects of sediments on the reproductive cycle of corals. *Mar. Pollut. Bull.* **2015**, *100*, 13–33. [CrossRef]
164. Brodie, J.; Wolanski, E.; Lewis, S.; Bainbridge, Z. An assessment of residence times of land-sourced contaminants in the Great Barrier Reef lagoon and the implications for management and reef recovery. *Mar. Pollut. Bull.* **2012**, *65*, 267–279. [CrossRef]
165. Bright, C.E.; Mager, S.M. Contribution of particulate organic matter to riverine suspended material in the Glendhu Experimental Catchments, Otago, New Zealand. *J. Hydrol.* **2016**, *55*, 89–105.
166. Asghar, M.; Arshad, M.; Fiaz, M.; Suhail, A.; Sabir, A.M. A survey of rice farmers farming practices posing threats to insect biodiversity of rice crop in the Punjab, Pakistan. *Int. J. Biodivers. Conserv.* **2013**, *5*, 647–654.
167. Hashmi, M.S. Agrochemical and agricultural sustainability: A case study of Pakistan. *J. Environ. Agric. Sci.* **2016**, *9*, 1–9.
168. Shahid, M.; Ahmad, A.; Khalid, S.; Siddique, H.F.; Saeed, M.F.; Ashraf, M.R.; Sabir, M.; Niazi, N.K.; Bilal, M.; Naqvi, S.T.A. Pesticides pollution in agricultural soils of Pakistan. In *Soil Science: Agricultural and Environmental Prospectives*; Springer: Berlin/Heidelberg, Germany, 2016; pp. 199–229.
169. Abbas, A.; Ali, M.A.; Ahmad, A.; Ali, S.; Habib, A.; Amrao, L.; Rehman, A. Status of Pesticides and Their Alternatives for Crop Sustainability in Pakistan. In *Developing Sustainable Agriculture in Pakistan*; CRC Press: Boca Raton, FL, USA, 2018; pp. 485–501.
170. Yadav, I.C.; Devi, N.L. Pesticides classification and its impact on human and environment. *Environ. Sci. Eng.* **2017**, *6*, 140–158.
171. Marcelino, A.F.; Wachtel, C.C.; Ghisi, N.D.C. Are our farm workers in danger? Genetic damage in farmers exposed to pesticides. *Int. J. Environ. Res. Public Health* **2019**, *16*, 358. [CrossRef]
172. Sarwar, M. The dangers of pesticides associated with public health and preventing of the risks. *Int. J. Bioinform. Biomed. Eng.* **2015**, *1*, 130–136.
173. Hassan, Z.; Shabbir, R.; Ahmad, S.S.; Malik, A.H.; Aziz, N.; Butt, A.; Erum, S. Dynamics of land use and land cover change (LULCC) using geospatial techniques: A case study of Islamabad Pakistan. *SpringerPlus* **2016**, *5*, 1–11. [CrossRef]
174. Hussain, S.; Mubeen, M.; Ahmad, A.; Akram, W.; Hammad, H.M.; Ali, M.; Masood, N.; Amin, A.; Farid, H.U.; Sultana, S.R. Using GIS tools to detect the land use/land cover changes during forty years in Lodhran district of Pakistan. *Environ. Sci. Pollut. Res.* **2020**, *27*, 39676–39692. [CrossRef] [PubMed]
175. Li, B.; Chen, D.; Wu, S.; Zhou, S.; Wang, T.; Chen, H. Spatio-temporal assessment of urbanization impacts on ecosystem services: Case study of Nanjing City, China. *Ecol. Indic.* **2016**, *71*, 416–427. [CrossRef]
176. Don, A.; Osborne, B.; Hastings, A.; Skiba, U.; Carter, M.S.; Drewer, J.; Flessa, H.; Freibauer, A.; Hyvönen, N.; Jones, M.B. Land-use change to bioenergy production in E urope: Implications for the greenhouse gas balance and soil carbon. *GCB Bioenergy* **2012**, *4*, 372–391. [CrossRef]
177. Murtaza, G.; Ghafoor, A.; Qadir, M.; Owens, G.; Aziz, M.; Zia, M. Disposal and use of sewage on agricultural lands in Pakistan: A review. *Pedosphere* **2010**, *20*, 23–34. [CrossRef]
178. Daud, M.; Nafees, M.; Ali, S.; Rizwan, M.; Bajwa, R.A.; Shakoor, M.B.; Arshad, M.U.; Chatha, S.A.S.; Deeba, F.; Murad, W. Drinking water quality status and contamination in Pakistan. *BioMed Res. Int.* **2017**, *2017*, 7908183. [CrossRef]
179. Shahid, M.; Khalid, S.; Murtaza, B.; Anwar, H.; Shah, A.H.; Sardar, A.; Shabbir, Z.; Niazi, N.K. A critical analysis of wastewater use in agriculture and associated health risks in Pakistan. *Environ. Geochem. Health* **2020**, 1–20. [CrossRef]
180. Chotpantarat, S.; Boonkaewwan, S. Impacts of land-use changes on watershed discharge and water quality in a large intensive agricultural area in Thailand. *Hydrol. Sci. J.* **2018**, *63*, 1386–1407. [CrossRef]
181. Dai, X.; Zhou, Y.; Ma, W.; Zhou, L. Influence of spatial variation in land-use patterns and topography on water quality of the rivers inflowing to Fuxian Lake, a large deep lake in the plateau of southwestern China. *Ecol. Eng.* **2017**, *99*, 417–428. [CrossRef]
182. Butt, A.; Shabbir, R.; Ahmad, S.S.; Aziz, N. Land use change mapping and analysis using Remote Sensing and GIS: A case study of Simly watershed, Islamabad, Pakistan. *Egypt. J. Remote Sens. Space Sci.* **2015**, *18*, 251–259. [CrossRef]
183. Younis, S.M.Z.; Ammar, A. Quantification of impact of changes in land use-land cover on hydrology in the upper Indus Basin, Pakistan. *Egypt. J. Remote Sens. Space Sci.* **2018**, *21*, 255–263. [CrossRef]
184. Sohail, M.T.; Mahfooz, Y.; Azam, K.; Yen, Y.; Genfu, L.; Fahad, S. Impacts of urbanization and land cover dynamics on underground water in Islamabad, Pakistan. *Desalin. Water Treat.* **2019**, *159*, 402–411. [CrossRef]

Journal of
Marine Science and Engineering

MDPI

Review

Geological Oceanography of the Pliocene Warm Period: A Review with Predictions on the Future of Global Warming

Markes E. Johnson

Department of Geosciences, Williams College, Williamstown, MA 01267, USA; mjohnson@williams.edu;
Tel.: +1-413-597-2329

Abstract: Atmospheric carbon dioxide reached a record concentration of 419 parts per million in May 2021, 50% higher than preindustrial levels at 280 parts per million. The rise of CO_2 as a heat-trapping gas is the principal barometer tracking global warming attributed to a global average increase of 1.2 °C over the last 250 years. Ongoing global warming is expected to perturb extreme weather events such as tropical cyclones (hurricanes/typhoons), strengthened by elevated sea-surface temperatures. The melting of polar ice caps in Antarctica and Greenland also is expected to result in rising sea levels through the rest of this century. Various proxies for the estimate of long-term change in sea-surface temperatures (SSTs) are available through geological oceanography, which relies on the recovery of deep-sea cores for the study of sediments enriched in temperature-sensitive planktonic foraminifera and other algal residues. The Pliocene Warm Period occurred between ~4.5 and 3.0 million years ago, when sea level and average global temperatures were higher than today, and it is widely regarded as a predictive analog to the future impact of climate change. This work reviews some of the extensive literature on the geological oceanography of the Pliocene Warm Period together with a summary of land-based studies in paleotempestology focused on coastal boulder deposits (CBDs) and coastal outwash deposits (CODs) from the margin of the Pacific basin and parts of the North Atlantic basin. Ranging in age from the Pliocene through the Holocene, the values of such deposits serve as fixed geophysical markers, against which the micro-fossil record for the Pliocene Warm Period may be compared, as a registry of storm events from Pliocene and post-Pliocene times.

Keywords: El Niño-Southern Oscillation; coastal erosion; storm surge; paleotempestology

Citation: Johnson, M.E. Geological Oceanography of the Pliocene Warm Period: A Review with Predictions on the Future of Global Warming. *J. Mar. Sci. Eng.* **2021**, *9*, 1210. https://doi.org/10.3390/jmse9111210

Academic Editor: George Kontakiotis

Received: 11 October 2021
Accepted: 30 October 2021
Published: 2 November 2021

1. Introduction

The Iron Bridge across the Severn River in the West Midlands of England was the first of its kind, in 1779, to be erected anywhere in the world, and it is recognized as a World Heritage Site by the United Nations Educational, Scientific and Cultural Organization (UNESCO) [1]. Due to the local availability of iron ore, in addition to coal and limestone deposits necessary for the manufacture of iron, all the ingredients were present to launch the start of the industrial revolution in the British Isles. Other regions in other countries with those same natural resources were soon to follow. The atmospheric concentration of carbon dioxide fluctuated but did not rise above 280 parts per million prior to that era, based on the method of dry analysis of air inclusions preserved in ice cores recovered from deep within the glacial ice at the Vostok site in Antarctica [2]. Samples of ancient air, trapped in bubbles frozen in glacial ice at different levels within ice cores, track variations in CO_2 concentrations over a span of 800,000 years, all at or below the level characterized by the pre-industrial age. In contrast, the Keeling Curve is based on analysis of air samples for resident CO_2 measured at the Mauna Loa Observatory on the big island of Hawaii, initiated in 1958 and maintained to the present day by the National Oceanographic and Atmospheric Administration (NOAA). The upward slope of the plot deviates in a micro-rippled fashion based on the monthly mean of daily averages, but it reflects a steady rise overall in the concentration of atmospheric CO_2 on a yearly basis. As such, the curve devised by the climate scientist Charles Keeling from the Scripps Institution of Oceanography has become the principal barometer tracking global climate change linked to a specific heat-trapping gas during the expanding industrial age [3]. In May 2021, the level of atmospheric CO_2 measured on Mauna Loa peaked at 419 parts per million [4], the highest yet recorded, and with a value 50% higher than preindustrial levels. At this record level, it is calculated that humans are now adding roughly 40 billion metric tons of CO_2 to the atmosphere each year [4].

Additional CO_2 not only acts to increase the global average air temperate, but affects an increase in sea-surface temperatures (SSTs) registered around the globe, as the world's oceans absorb much of the excess heat. Measurement of ocean heat content is regarded as another way to quantify the rate of ongoing global warming. The years 2019, 2018, 2017, 2015, and 2016 rank as the five warmest years (in that order) for ocean heat values within the upper 2000 m [5]. Moreover, the ten years between 2010 and 2019 also rank as the top ten years on record [5]. Although variable on a seasonal basis in any given region, high SSTs provide the fuel to power tropical cyclones that originate in low latitudes but are capable of migrating as sea storms to much higher latitudes including boreal regions. Analysis of global data suggests that pools of warm surface water have formed more often and lasted longer during seasonal variations tracked since 1925 [6]. In turn, these data have led to the discovery that the exceedance probability of major tropical cyclones has increased over the last four decades [7]. Most big storms expire harmlessly in the open ocean, but a statistical review also indicates that ever stronger storms also have approached closer to land since 1982 [8].

Between ~4.5 and 3.0 million years ago during the Pliocene Warm Period, the average global temperature is interpreted as >2 °C higher than at present, and sea level stood no less than 16 m above today's datum [9,10]. During that interval, the pole-to-equator gradient in temperature was so flat that deposits near the Pliocene ice wedge in Canada's high Arctic on Ellesmere Island feature fossil camel bones and fossil tree remains dominated by a larch-forest habitat with mean annual temperature as much as 3.0 °C above present [11]. Hence, the Pliocene Warm Period is regarded as a predictive analog to the future impact of ongoing global warming with the connotation raised as early as 1982 by M.I. Budyko in *The Earth's Climate: Past and Future* [12] and more explicitly in a seminal paper by Zubakov and Borzenkova in 1988 [13]. Much of the discussion regarding the Pliocene Warm Period revolves around the pattern of the El Niño Southern Oscillation (ENSO) in the equatorial zone of the Pacific Ocean, where semi-cyclical events that peak for a year or more result from the slowdown and stagnation of currents that move surface water from east to west and elevate warmer SSTs. These conditions not only spawn hurricanes off the southwest coast of Mexico and increase rainfall affecting the Pacific shores of North America, but also generate comparable storms and flooding farther afield called typhoons that reach landfall in countries across the western and northwest Pacific Ocean, including the shores of mainland China [14,15].

This work offers a review of literature on the geological oceanography of the Pliocene Warm Period, including techniques by which Pliocene sea temperatures are estimated from micro-fossils preserved in deep-sea cores. Opinions differ whether continual (permanent) El Niño conditions prevailed during the Pliocene Warm Period [16–18] or were more intermittent (dampened) during the same time interval [19,20]. A parallel approach to treatment of the Pliocene Warm Period, as well as the ensuing Pleistocene and Holocene time frames, relies on analysis of coastal storm deposits based on methods in paleo-tempestology. Such methods include evidence from storm-driven over-wash deposits in former marshes and beaches, oxygen-isotope anomalies recorded in speleothems, and tree-ring variations [21,22]. The evidence summarized in this review is focused foremost on coastal boulder deposits (CBDs) and coastal outwash deposits (CODs) from the margins of the Pacific basin and parts of the North Atlantic basin associated with heightened shore erosion and rainfall [23–30]. It is problematic to distinguish between genuine storm deposits and those caused by tsunami events [31,32]. Such difficulties are avoidable where paleo-deposits may be compared directly with deposits described in the aftermath of actual hurricanes, a research format pioneered by Ball et al. (1967) [33] with respect to Hurricane Donna that struck southern Florida in 1960 and by Paris et al. (2010) [34] with respect to the tsunami that struck the Indonesian island of Sumatra in 2004.

2. Background Geological and Geographical Oceanography

An authoritative summary by Miller et al. (2020) [10] covers long-term climate change in the Pacific basin during the Cenozoic, showing a marked shift from a warmer greenhouse world 65 million years ago to the fluctuating icehouse worlds of the Pleistocene during the last 2 million years. Smoothed curves for sea-surface temperature and sea level variations are adapted from this work (Figure 1) with alignments for the Early Eocene, Middle Miocene, and Pliocene climatic optima.

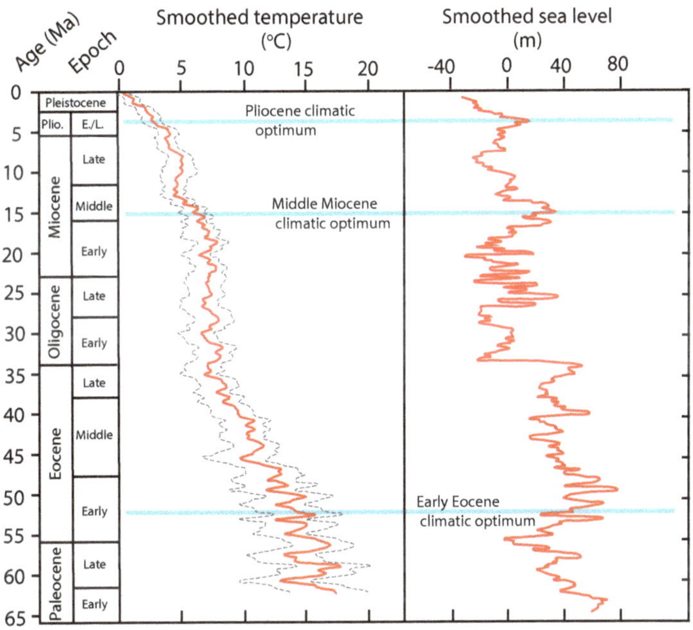

Figure 1. Geological history of sea-surface temperature and sea-level changes for the Pacific Ocean basin through the last 65 million years (modified from Miller et al., 2020) [9].

Compared to earlier intervals, the Pliocene Warm Period represents a diminishment in overall global temperature, from as much as 15 °C warmer and with a sea level 75 m higher than today, during the ice-free Early Eocene climatic optimum (Figure 1). The suitability of the Pliocene Warm Period, as the closest analog to the effects of today's rapid change in global warming, is predicated on the similarity in global geography between the Pliocene world and today's world, as well as parallel external factors such as the intensity of sunlight falling on the Earth's surface [18]. The farther back in geologic time that climatic change is gauged, the more difficult it becomes to account for major differences in paleogeography and other possible external factors related, for example, to increased volcanic activity, extent of vegetative land cover, and other variations in topography compared to the present.

The basic mechanics of oceanic circulation and cyclonic storm development are summarized in the cartoon model for a generalized ocean with broad latitudinal extent, as found today in the Pacific and Atlantic oceans (Figure 2). The Inter Tropical Convergence Zone (ITCZ) is the irregular boundary along the Earth's equator where trade winds emanating from the northeast and the southeast converge, and where warm, moist air rises upwards into the atmosphere by convection. A pair of Hadley Cells are strongly formed astride the equator, where atmospheric low pressure dominates and moisture-laden air ascends between 10 to 15 km into the troposphere. Moisture is released as rain, after which upper-air currents convey dry air to latitudes centered around 30° north and south of the equator. There, atmospheric high pressure brings the dry air back to the Earth's surface, some of which returns to the equator carried by the trade winds, and some is conveyed to higher latitudes to join westerly-directed winds located around 45° north and south of the equator.

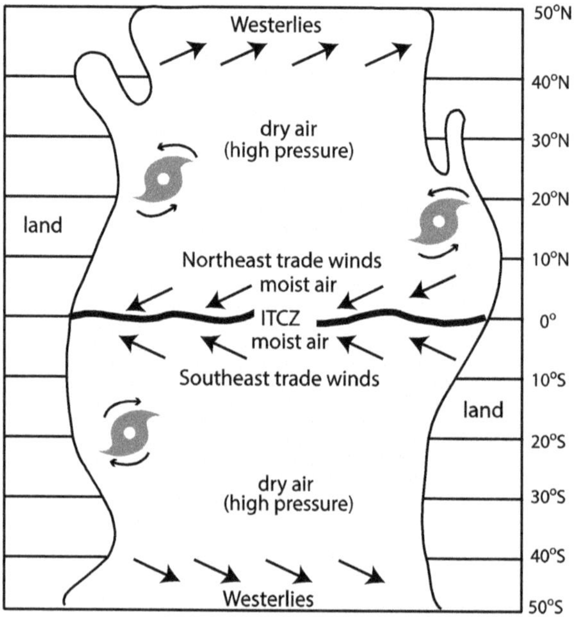

Figure 2. Cartoon model for present-day ocean-current circulation stimulated by the Westerlies and Trade winds with icons representing tropical cyclones (hurricanes/typhoons) that rotate counter-clockwise in the northern hemisphere and clockwise in the southern hemisphere (ITCZ = Inter Tropical Convergence Zone). Marginal seas are figurative, but shown as coastal embayments characteristic of Mexico's Gulf of California and China's Yellow Sea.

Under localized climatic anomalies, where super-heated tropical air warms SSTs to or above 26.6 °C through water depths reaching the upper 50 m, the stage is set for the creation of a tropical depression leading to a tropical cyclone. As warm air rises above the sea surface, cooler air from surrounding areas swirls in to take its place, and the entire rotating mass undergoes further cooling with altitude leading to condensation and cloud formation. When wind speed within such a system reaches a minimum speed of 119 km/h, it is classified as a Category 1 hurricane. By definition on the Saffir-Simson scale, a Category 5 hurricane is initiated with wind speeds of 252 km/h. In the North Atlantic Ocean today, hurricanes tend to form at a latitude around 15° north of the equator off the coast of Africa just south of the Cape Verde Islands. In the Northeast Pacific Ocean, hurricanes regularly form off the west coast of Mexico at the same latitude south of Acapulco. These disturbances may reach heights of 15,000 m, but trade winds push the storms westward as discrete entities. Tropical cyclones rotate in a counter-clockwise direction in the northern hemisphere, whereas they rotate in the opposite clockwise direction in the southern hemisphere (Figure 2). It is strictly a matter of regional terminology, but these tropical storms are called hurricanes in the Atlantic and northeastern Pacific oceans. They are called typhoons in the central and northwestern Pacific Ocean. Due to the expanse of the Pacific Ocean along the equatorial zone, there exist other peculiarities ascribed to Walker circulation that affect cycles in the El Niño Southern Oscillation.

3. Operational Definitions and Study Methods

The unique environmental attributes of the Equatorial Pacific zone during normal years entail multiple temperature gradients both in zonal (latitudinal) and meridional (longitudinal) scale, as well as below the water surface and within the overlying atmosphere (Figure 3). Along the ITCZ where the trade winds converge, there typically exists a 5 °C temperature difference between the sea surface in the warmer Western Equatorial Pool (WEP) and the cooler Eastern Equatorial Pool (EEP). The temperature gradient with depth below the WEP and EEP, or thermocline, is as much as four times as great in the west as it is in the east [17]. In addition, Walker circulation acts as a kind of zonal variation on the meridional Hadley Cell circulation, whereby warmer air in the west rises aloft and streams eastward above the equator where it descends to the EEP.

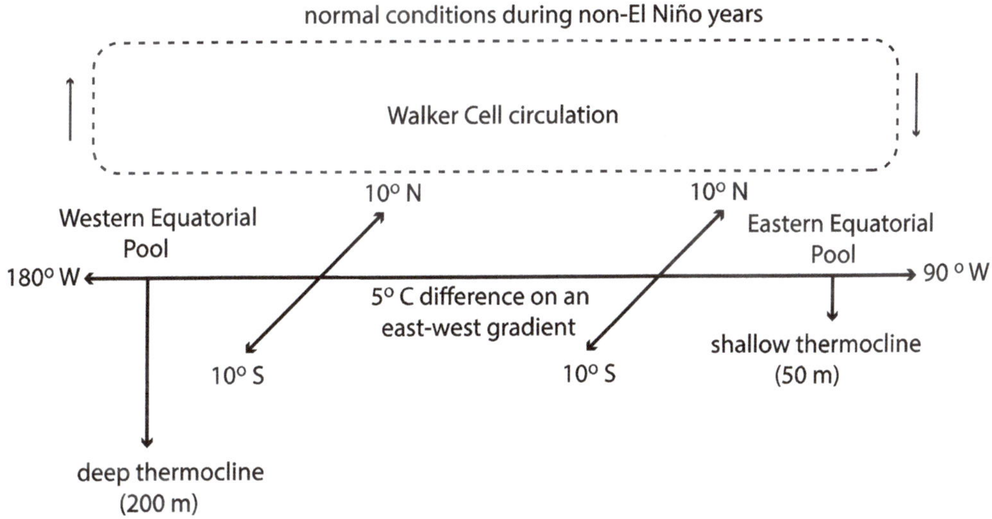

Figure 3. Sketch diagram showing the three-dimensional arrangement through the Equatorial Pacific Ocean zone under normal climatic conditions (non-El Niño years).

During El Niño years, this pattern breaks down where the cooler waters of the EEP are less insolated and more easily elevated in temperature due to the shallowness of the thermocline. Weakening of the south-flowing coastal California current and north-flowing Humboldt current that converge in the eastern Pacific deprive the EEP of a relatively cooler-water influx dependent on Ekman upwelling. As a result, the east-west asymmetry in temperature between the pools along the entire zone is ameliorated to the extent that it disappears [17]. Under these conditions, Walker circulation also dissipates. Hurricanes are more readily spawned off the coast of western Mexico, as well as typhoons in the mid-Pacific Ocean. Based on satellite surveillance over the past several decades, most of these storms are tracked to higher latitudes of the open ocean where they expire harmlessly. However, where landfall over ocean islands or against continental mainland occurs, the effects of coastal erosion due to storm waves and flooding due to attendant rainfall is devastating.

Direct observations of ENSO climate cycles underscore these variables along the Equatorial Pacific zone. For example, recurrent El Niño events were especially severe in 1972–1973, 1982–1983, and 1997–1998 [19], as well as more recently in 2014–2016. The technology involved in testing for ENSO conditions during past eras is complex and entails the sampling of temperature-sensitive foraminifera with a planktonic habit that accumulates after death as micro-fossils on the ocean floor. Retrieval of these samples is drawn from deep-ocean cores drilled into the ocean floor at hundreds of sites around the world. One of the more common methods is the application of a temperature proxy based on variations in $\delta^{18}O$ derived from fossil foraminifera of various species [10]. Other proxy methods examine variations in Mg/Ca, also from fossil foraminifera such as *Globigerinoides sacculifer* [16]. Another proxy method targets alkenones produced by certain species of haptophyte algae equilibrated to a temperature proxy based on a $U^{k'/37}$ index [35,36]. Yet another index in use is the TetraEther index (TEX_{86}) found in integrated studies applying a combination of paleotemperature proxies [35]. The complexity in all this is exacerbated by the fact that different studies typically rely on a choice of different ocean cores drawn from a wide range in east-west spread but also a variable distance north or south of the equator. Moreover, expansion of studies into the Atlantic Ocean basin adds yet more complexity based on data derived from widespread ocean cores at higher latitudes in that region [37]. Finally, modeling studies expand on predictions based on some or all of the techniques cited above [19,38].

The effectiveness of warmer pools of ocean water in triggering major storms of hurricane/typhoon intensity during the Pliocene Warm Period in a positive feed-back loop [38] may be checked by examination of coastal boulder deposits (CBDs) that result from the impact major storms generating extreme wave energy. Likewise, coastal outwash deposits (CODs) signify the impact of rainfall and erosion during landfall of major storms. Regarding CBDs, various mathematical equations have evolved in complexity that allow for the measurement of eroded shore boulders to yield estimates of wave heights necessary for dislodgement from sea cliffs [39–41]. These may be applied to geological deposits of any age, but also to deposits in the direct aftermath of a modern storm as well as a modern tsunami in their shoreward impacts. For geological examples from the past, the challenge is to differentiate between the style of deposits due to sea storms and tsunamis [30,31].

4. Results

4.1. Literature Review on the Pliocene Warm Period Based on Deep-Sea Cores

Since the earliest suggestions on the potential of the Pliocene Warm Period as an analog to the future impact of climate change [12,13], a sizable amount of literature has followed, which remains both current and controversial [19]. Table 1 summarizes details from a sample of a dozen research reports published between 2005 and 2020. This review includes notations on the kind of study techniques applied, the number of locations around the world where primary research materials were collected, and a tally on opinions expressed as to the severity of the Pliocene Warm Period as an ENSO precursor. Studies are listed in Table 1 in order of appearance in the published literature [16–20,35,36,38,42–46]. Among the earliest such reports [16], the phrase "permanent El Niño" was introduced as a Pliocene phenomenon. The latest study consulted [20] argues against the "permanent El Niño" as rather a "dampened El Niño." To an outside reviewer, such terminology is needlessly extreme, and the substitution offered by Fedorov et al. (2010) [38] for a "continual El Niño" is adapted here, as is the counter appraisal for an "intermittent El Niño" (Table 1).

Several trends are evident in the growing sophistication of this research topic over a period of 15 years. The earliest study to advocate for a "permanent" Pliocene El Niño is based on data from only two ODP sites, separated by 90° of longitude between the EEP and WEP [16]. A subsequent study increased the number of ODP sites to 85 [46] with inclusion of the Atlantic Ocean basin and an expansion in latitudinal range. Overall, the number of studies surveyed favors an interpretation of a continual El Niño state during the Pliocene against a slightly smaller number of skeptics. However, a significant number remained neutral, urging for additional studies. As debate among key participants intensified with time, later studies began to criticize the validity of techniques applied by earlier workers. The study by Tierney et al. (2019) [47] relies entirely on evidence derived from alkernones while rejecting, as flawed evidence, the Ma/Ca ratios in planktonic foraminifera. Even so, that study concludes by admitting that a Pliocene atmospheric level of CO_2 concentration higher than 400 ppm is likely to weaken Walker circulation in tandem with a meridional weakening of Hadley Cell circulation. The hard fact remains from land-based evidence, irrespective of data from deep-sea cores, that Pliocene temperatures in the high Arctic of Canada reached well above present-day conditions [11]. This kind of input suggests that other kinds of land-based evidence are relevant.

Table 1. Opinions on the Pliocene Warm Period and its severity as a precursor to future El Niño conditions. The abbreviation "alk" is for alkernones based on the $U^{k'/37}$ index. ODP refers to Ocean Drilling Program sites. A neutral opinion registered in this summary signifies a call for additional research.

Reference List	Authors	Techniques Applied	Number ODP Sites	Continual El Niño	Neutral Opinion	Intermittent El Niño
[16]	Wara et al. (2005)	$\delta^{18}O$, Mg/Ca	2	x		
[17]	Ravelo et al. (2006)	$\delta^{18}O$, Mg/Ca	4	x		
[42]	Dekens et al. (2007)	δ18O, alk.	8		x	
[36]	Haywood et al. (2007)	δ18O, Mg/Ca	2		x	
[19]	Molnar & Cane (2007)	modeling	other			x
[18]	Brierley et al. (2009)	Mg/Ca, alk.	2	x		
[35]	Dowsett & Robinson (2009)	Mg/Ca, alk.	13		x	
[38]	Fedorov et al. (2010)	modeling	other	x		
[43]	von der Heydt et al. (2011)	modling	other			x
[44]	O'Brien et al. (2014)	Mg/Ca, alk., TEX86	3	x		
[45]	Haywood et al. (2016)	δ18O, Mg/Ca, alk.	85		x	
[46]	Tierney et al. (2019)	alk.	28			x
[20]	White & Ravelo (2020)	δ18O, Mg/Ca	1			x
Total	13	5		5	4	4

4.2. *Review of Pliocene to Holocene Coastal Deposits in the Gulf of California*

In a summary of coastal boulder deposits from the Miocene and Pliocene, Ruban (2019) [48] examined 21 studies almost evenly divided between the two epochs and well distributed between the northern and southern hemispheres. All of these studies feature a mixture of boulder-size clasts exceeding 25.6 cm in diameter, although the original author's intent was not always explicit with identification of a CBD. Indeed, some are CODs attributed to a delta setting and some are tsunami deposits. Due to less time for erosion after deposition, the geological record for CBDs and CODs is potentially better when the search is widened to the last 2 million years of the Pleistocene and succeeding ten thousand years of the Holocene [49]. Deep geographic embayments (Figure 2), such as Mexico's Gulf of California, provide a fruitful place to search for and document storm-related CBDs and CODs. That the incidence rate of hurricanes reaching the Gulf of California decreased from an average return rate of 3.8 years between 1950 and 2000 to 1.3 years afterwards [50], confirms this region has become more vulnerable to hurricanes during the last decade. Table 2 summarizes storm deposits from the Pliocene [51–53], Pleistocene [27,54–56], and Holocene [23–26,57–59] preserved within the Gulf of California region.

The distance between the most northwesterly and southeasterly study localities, with documented storm deposits along the Gulf of California coastline, amounts to a spread of 900 km (Figure 3). All are well north of the Tropic of Cancer, far beyond the source of tropical cyclones that normally originate off the west coast of Mexico at a latitude of about 15° north of the equator.

A complicating factor in the formation of CBDs is due to the density of the source rocks in sea cliffs subject to erosion by storms and tsunamis. The coastal geology within the Gulf of California is highly varied with examples of igneous rocks such as granite, andesite, and rhyolite, metamorphic rocks such as gneiss, and sedimentary rocks like limestone. Among these, the local granite was found to have a density of 2.52 g/cm^3 [27], which is denser than andesite, with levels measured between 2.26 and 2.34 g/cm^3 [26], which in turn is denser than rhyolite, with a sample density measured at 2.16 g/cm^3 [24]. In turn, all these igneous rocks are denser than the local limestone on Isla del Carmen, sampled with a density of 1.86 cm/cm^3 [23]. Essentially, a surface wave that shoals against the coast must work harder to extract a boulder of granite from its source than a slab of limestone of exactly the same size with lesser weight. Even so, some of the largest and heaviest rocks observed anywhere in CBDs from the Gulf of California are represented by enormous blocks of limestone with an estimated weight of nearly 6 metric tons [23].

Table 2. Pliocene to Holocene coastal boulder deposits (CBDs) and coastal outwash deposits (CODs) among islands and shores of Mexico's Gulf of California on the Baja California peninsula. Listing is north to south in each age category.

Reference List	Authors	Age	Location	Class	Lithology	Maximum Boulder Diam. (cm)
[51]	Johnson et al. (2017)	Pliocene	Ballena Bay	COD	sandstone	-
[52]	Johnson et al. (2016)	Pliocene	Isla del Carmen	COD	andesite	64
[53]	Emhoff et al. (2012)	Pliocene	Isla Cerrolvo	COD	granite, gneiss	~35
[54]	Johnson & Ledesma- Vázquez (1999)	Pleistocene	Punta Antonio	CBD	granite, andes.	32
[55]	Ledesma-Vázquez et al. (2007)	Pleistocene	Isla Coronado	overwash	sandstone	-
[27]	Callahan et al. (2021)	Pleistocene	Isla San Diego	CBD	granite	120
[56]	Tierney et al. (2012)	Pleistocene	Isla Cerrolvo	COD	granite, gneiss	~35
[26]	Guardado-France et al. (2020)	Holocene	Isla San Luis	CBD	andesite	80
[57]	Kozlowski et al. (2018)	Holocene	Volcán Prieto	CBD	basalt	~35
[58]	Johnson et al. (2012)	Holocene	Angel Guarda	CBD	andesite	~35
[24]	Johnson et al. (2019)	Holocene	Almeja Bay	CBD	rhyolite	268
[23]	Johnson et al. (2018)	Holocene	Isla del Carmen	CBD	limestone	182
[25]	Johnson et al. (2020)	Holocene	Escondido Port	CBD	andesite	131
[59]	Backus et al. (2012)	Holocene	Isla Cerrolvo	COD	granite, gneiss	~35
Total	14 studies	0 to 4.5 my	900-km spread		mixed	

Storm deposits of Pliocene age within the Gulf of California are represented by CODs of unusual thickness [51–53]. Crude layering within CODs provides some insight regarding the repetition and overall span of time during which such deposits continued to be formed at the same locality. Among the examples from Table 2, the most intriguing is the 60-m thick Tiombó conglomerate, representing a mega-delta complex on the east coast of Isla del Carmen [52]. Boulders within the conglomerate are exclusively andesite in composition (Figure 4), the largest of which has a maximum diameter of 64 cm. For the most part, these clasts are well rounded as a consequence of travel to the coast under flood conditions from a source high in the peninsular mountains as much as 35 km to the west.

The actual thickness of the complex is unknown, because contact between the base of the conglomerate and underlying andesite in the Miocene Comondú Group is below present sea level and most likely buried in the subsurface. However, scarce fossil pecten shells from the lowest part of the exposed deposit confirm a Pliocene age. Raised marine terraces, on the margins of the complex correlated with the middle and youngest parts of the Pleistocene, suggest that the uplifted complex spans the Pliocene and may include material from the oldest part of the Pleistocene [23]. In cross-section, the Tiombó conglomerate traces an arch that extends for 2 km (Figure 5), with thick layers that form an arch (Figure 6) tilt seaward 5° and thin to a wedge shape in the landward direction. This configuration conforms to a massive delta complex that remained active through much of the Pliocene with distinct pulses of conglomerate fed to the delta front from highlands before Isla del Carmen was tectonically severed from the peninsular mainland.

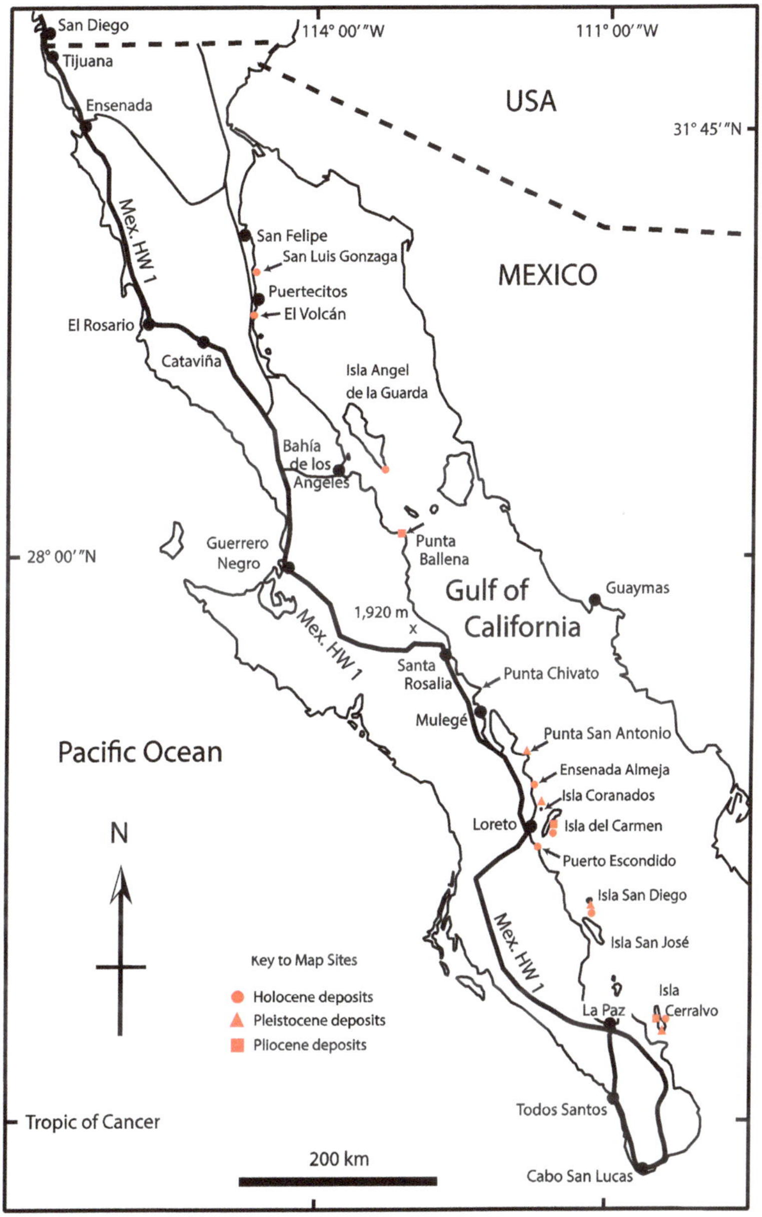

Figure 4. Mexico's Baja California and adjacent Gulf of California showing 14 localities with storm deposits of varying ages: Pliocene (square), Pleistocene (triangle), and Holocene (circle). See Table 2 for details.

Figure 5. Exposed lower part of the Pliocene Tiombó conglomerate on the east coast of Isla del Carmen, showing typical composition of well-rounded andesite boulders parted by sandstone seams. Each of several packages offset by such seams represents a major rainfall event that carried coarse material through stream beds to the coast.

Figure 6. Oblique view of the arch-shaped PlioceneTiombó conglomerate in cross-section, stretching south to north over a distance of 2 km on the east coast of Isla del Carmen.

Another Pliocene deposit, also linked with outwash dynamics dependent on storm-related rainfall, occurs far to the north of Isla del Carmen (Figure 4), where the 50-m thick Ballena fan delta covers a map area of 4 km^2 with sandstone layers dipping nearly 3° seaward in an arcuate pattern [51]. Equally far to the south of Isla del Carmen, yet another Pliocene deposit, with carbonate sands framed above and below by thick boulder deposits, occurs at Paredones Blancos on Isla Cerravo (Figure 4) [52]. In this case, coarse outwash deposits stripped from the interior of a large island are interrupted by an interval of subsidence and relative rise in sea level that filled a 0.5-km wide embayment.

Upper Pleistocene deposits, correlated with Marine Isotope Substage 5e during the last interglacial episode roughly 125,000 years ago, also are well represented in the Gulf of California (Table 2), although not nearly at the same scale as older Pliocene storm deposits. Rhodolith sand deposited in a paleolagoon on Isla Coronados (Figure 7) are interpreted as repetitive overwash events during major storms arriving from the south that topped a barrier formed by solid andesite basement rocks [55].

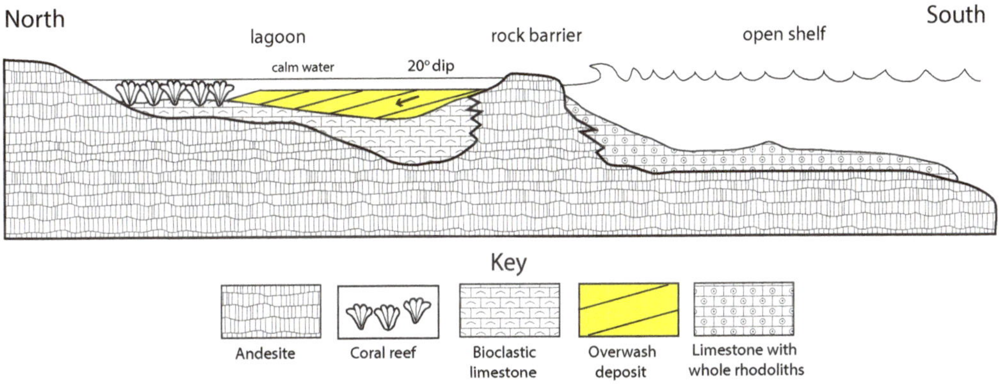

Figure 7. Cross-section through a Late Pleistocene lagoon uplifted on the south shore of Isla Coronados, showing an overwash deposit with multiple layers representing individual storm events that arrived from the south and left a succession of tilted beds off-lapped to the north [55].

The tilted beds within the lagoon amount to 6.5 m of vertical fill, but they are organized as individual beds from 40 to 60 cm in thickness dipping 20° to the north off the top of the barrier [53]. Each bed is interpreted as a separate storm event that required sufficient energy to carry carbonate sand up and over a rigid barrier from the sea floor to the south. Three other Upper Pleistocene deposits are noteworthy, located north of Isla Coronados at Punta San Antonio and farther south on Isla San Diego and Isla Cerralvo in the lower Gulf of California (Table 2). The Punta San Antonio locality features a meter-thick CBD formed by a mix of eroded granite and andesite boulders that became colonized by a distinctive intertidal biota of bivalves and encrusting red algae [54]. The deposit on Isla San Diego features a stack of crudely imbricated granite boulders that incorporate eroded coral colonies dislodged by storm waves striking the island's east shore [27]. On Isla Cerralvo, thin cobble/boulder beds buried a succession of five coral colonies that grew in place through a vertical thickness of 3 m, with each cycle initiated by a storm event that washed conglomeratic debris from the island's interior to the shore [56].

Several examples of Holocene storm beds are described from the Gulf of California, most of which are modified CBDs that form distinctive bars connected to and eroded from rocky shores under episodic storm attack. The composition and size of boulders incorporated within these deposits vary considerably (Table 2). The boulder deposit at Almeja Bay (Figure 8) is noteworthy for various reasons [24], among which is that it includes some of the largest eroded boulders on record in the gulf region. Coastal erosion contributing to bar development is implied to be ongoing at this site. Table 2 lists an individual rhyolite boulder with a diameter of 268 cm at this locality, estimated to weigh 4.3 metric tons and requiring wave shock from a 16-m high wave to dislocate it from nearby cliffs [24]. Such a huge boulder is an outlier, but an eye-witness account with storm waves filmed from the last major hurricane to strike the area in 2014 offers evidence of wave heights easily reaching 8 m [24]. This level of storm activity would be sufficient to shift most of the lesser boulders in the unconsolidated deposit.

Figure 8. Inner side of an unconsolidated Holocene deposit at Ensenada Almeja formed by rhyolite boulders eroded from an adjacent sea cliff (out of view to the right) stacked to a height of 2 m [24].

On Isla Cerralvo in the lower Gulf of California (Figure 4), Backus et al. (2012) [59] documented 39 fan deltas around the island's circumference that represent CODs formed by granite and gneiss debris washed through a network of radial drainage channels from the interior. What is most intriguing about this mid-size island, 136,000 km^2 in area, is its vulnerability to hurricanes, several of which made direct hits in recent years, including Lorena in 2019, John in 2006, Ignacio in 2003, and Marty also in 2003. Although maximum boulder size is small, around 35 cm in diameter (Table 2), the island provides a good model for the dynamic development of storm-related CODs in the past.

4.3. *Review of Pliocene to Holocene Coastal Deposits in the North Atlantic Basin*

Storm deposits attributed specifically to the Pliocene Warm Period are rare, but an unusual example from Lower Pliocene strata at Mallbusca on Santa Maria Island in the Azores is well documented [60]. Santa Maria is the only island among the archipelago's nine islands with a stratigraphic record spanning the Pliocene and parts of the Pleistocene, complete with abundant fossils. The storm sequence (Figure 9) consists of a continuous 5-m thick package distinguished by laminae composed of heavier, dark minerals dominated by pyroxene, olivine and plagioclase, alternating with lighter carbonate grains draped over a disconformity surface represented by a low rocky shore, distinguished by an intertidal to shallow-water biota including encrusting bivalves and coralline red algae. The top part of the storm bed is penetrated by a variety of trace fossils that burrowed from above as much as a meter into the sands from sometime after the storm ended. The interval in between is entirely barren of fossils.

Figure 9. Storm deposit from the Lower Pliocene represented by a singular 5-m thick sandstone body that washed over a low-relief rocky shore at Malbusca on the south coast of Santa Maria Island in the Azores [60].

The Pliocene sand body may be traced laterally over a distance of 750 m before termination against the flanks of a lava delta. The deposit is largely two-dimensional, but the size of the open bench on which it sits allows for a conservative estimate of volume at 14,500 m^3 interpreted to have been transported onshore from an adjacent sandbar during passage of a major hurricane [60].

Upper Pleistocene (Marine Isotope Substage 5e) and contemporary CBDs from Santa Maria Island also have been studied for CBDs [28] compatible with the methodology undertaken with regard to Mexico's Gulf of California [23–27]. In particular, the southeastern corner of the island at Punta do Castelo exhibits a matching set of Pleistocene and Holocene conglomerates with individual boulders that register a maximum half-meter in diameter. According to the mathematical models of Nott (2003) [39], these are estimated to have formed under wave heights between 5.3 and 5.5 m. The mathematical equation applied by Pepe et al. (2018) [41] yields a more exaggerated result for the same units. Passage of modern-day tropical cyclones in this area is not unusual, and full-fledged hurricanes have struck the Azores most recently in 1998, 2006, and 2012, when wave heights of 5.5 m or more would be expected against island shores.

A similar study was performed on boulder slumps from the Upper Pleistocene (Marine Isotope Substage 5e) at El Confital beach on Gran Canaria in the Canary Islands [29], which like the CBDs in the Azores are derived from basalt. Basalt has a higher rock density than granite, and the rock density of basalt from El Confital beach was determined to yield an average value of 2.84 g/cm^3. Six trials based on samples between 24 and 30 basalt boulders each yielded average estimates of wave heights between 3.9 and 4.8 m based on the equations of Nott (2003) [39]. Here, the mathematics from Pepe et al. (2018) [41] makes a reasonable match. Moreover, the study from El Confital also lists 26 major storms impacting Gran Canaria since 1713. Few such records provide eye-witness details on observed wave heights, but a major storm in 1966 involved wave heights between 10 and 12 m at the coast, and another storm in 1968 was reported to entail wave heights of 8 m [28]. Wave heights within that range during the late Pleistocene would have been adequate to dislodge the largest boulders at El Confital.

Elsewhere in the North Atlantic basin, huge limestone boulders stranded above the island shores of Bermuda and the Bahamas [61,62] have attracted much the same interest in storm waves during the Late Pleistocene as a possible warning about future superstorms related to global warming. Likewise, tropical cyclones that form hurricanes are not expected to migrate to high latitudes in boreal settings, but during the last decade at least four storms of hurricane intensity struck Norway's Arctic coast [30], and Holocene CBDs along that coast are of interest. The boulder beach at Støypet on Leka Island formed after glacial retreat around 10,000 years ago and is composed of unconsolidated cobbles and boulders eroded from sea cliffs exposing low-grade chromite ore with a rock density of 3.32 g/cm^3. This is the highest-density rock type yet studied for its hydrological properties in a coastal setting, and the results based on the predictive equations of Nott (2003) and Pepe et al. (2018) [39,41] are compatible in suggesting that shore erosion from wave heights between 5 and 7.5 m was possible.

5. Discussion

Studies predicting future superstorms based on CBDs from the geologic past elicit controversy related to: (1) challenges to the efficacy of mathematical equations to estimate wave heights and flow regimes based on boulder size and rock density, (2) alternate sources other than sea storms and attendant rainfall that explain similar deposits, and (3) past geographic configurations different from today's world as an influence on climate unrelated to global warming.

5.1. On the Efficacy of Mathematical Equations

Hydrodynamic equations following the approach pioneered by Nott (2003) [39] with subsequent iterations [40,41] are criticized on the basis of clifftop boulders along the Atlantic coasts of Ireland and Scotland [63], which have a storied record of study by different investigative teams [64–66]. The claim is made that such equations are flawed and should be abandoned because they yield unrealistic estimates. These concerns appear to be valid due the enormous size of limestone blocks quarried from clifftops during extreme wave events [63]. Clifftop and other boulder deposits from the Reykjanes Peninsula of Iceland [67] partly demonstrate the over performance of computational estimates in contrast to known wave heights during that region's Atlantic storms that regularly reach 15 m or more. Among 46 study sites from 10 different areas on the Reykjanes Peninsula, more than half the formulaic wave heights calculated for basalt boulders fall above the bench mark of 15 m. Twenty percent of the formulaic calculations predict wave heights of 25 m or more. The Selatangar site from this study [67] was the additional single palaeo-deposit to yield a wave-height estimate of 14 m in agreement with observations on contemporary storms.

In the Pacific Ocean basin, Super Typhoon Haiyan passed over the Philippine islands in November of 2013 as a Category 5 storm packing top wind speeds of 315 km/h with devastating consequences as one of the most powerful storms to make landfall yet recorded [68]. Studies on the volume and density of blocks dislodged from low-lying limestone cliffs on the Calicoan island in the path of the typhoon [69,70] reached the same conclusion that formulaic predictions [39–41] on wave height for large blocks far exceed the maximum wave height of 18.7 m that occurred offshore the study area [71]. The dynamics of clifftop erosion may be peculiar in this regard, involving differences between compressive hydraulic pressure due to head-on wave impact against seams and joints in a layered rock face and tensile stress on the underside of lip overhangs [72]. It is not clear that caution against use of the equations inspired by Nott (2003) [38–40] is justified when dealing with rocky shores other than limestone, whether layered or not. In any case, those formulae remain the only predictive tool available for application to CBDs from the geologic past, including those much older than the Pliocene. An approach certain to be useful in future studies of modern CBDs of all types is one that tags boulders with identification markers so they may be tracked for movement following the next major storm event having verifiable wave heights. With data on hand regarding a range in boulder size and density from different rock types susceptible to movement, it should be possible to make comparisons with boulder conglomerates that formed in the distant past.

5.2. On Mistaken or Inconclusive Classifications

Earthquakes that rupture the seabed and/or trigger the collapse of volcanic island flanks are the source of tsunamis that impact coastal zones and may result in CBDs similar to those caused by sea storms. Regarding studies in the Philippines conducted soon after Super Typhoon Haiyan [69,70], the CBD consisting of limestone blocks peeled off coastal cliffs is shown, by the timing of the study, to be clearly storm induced. Likewise, the study of coastal deposits at Banda Aceh on Sumatra in Indonesia [34] is indisputable as to its source due to a major tsunami in 2004. In some regions, such as the Canary Islands in the North Atlantic, CBDs may originate from sea storms, as argued for the Pleistocene boulder slumps at El Confital beach on Gran Canaria [29], or from a Pleistocene tsunami related to volcanic flank collapse on the same island [73]. The sedimentology of the two deposits is entirely different, although involving a wide range of boulder sizes. At El Confital, the deposit is formed by well-rounded boulders limited in vertical thickness and located adjacent to the present shore, whereas the inland deposit from the Agaete valley on the same island occurs between 41 and 188 m above sea level, and is dominated by angular volcanic clasts that decrease in size with altitude intermixed with soil and colluvium as well as broken marine shells.

Deposits on volcanic islands elsewhere in the northeast Atlantic fit the same profile of poorly sorted materials, including huge basalt blocks and mixed soil found at elevations high above present sea level, among them the megatsunami deposit on Santiago Island in the Cape Verde archipelago studied by Ramalho et al. (2015) [74]. The deposits feature enormous boulders up to 8 m in diameter and point to wave run-up heights exceeding 270 m above sea level. In this example, the tsunami appears to have been triggered by a volcanic flank collapse on the adjacent island of Fogo about 73,000 years ago. Authors studying the clifftop boulders in western Ireland [63,65] are adamant that the possibility for an origin due to offshore tsunami events at that latitude is mistaken. At the same time, others studying CBDs in the Mediterranean region, such as the Pleistocene and Holocene boulders on Malta [75] favor extreme storm events but are unable to rule out rare tsunami events also known to have occurred in that region.

Controversy likewise surrounds the 10-m thick Pliocene deposit at Caleta Hornitos in northern Chile, which includes boulders, large rock slabs, and breccia laterally traceable for nearly 2 km, first linked to a possible tsunami event [76] but reclassified as a massive debris flow triggered by an earthquake in the Andean subduction zone of coastal Chile [77]. The example is intriguing, because the tsunami connection is tied to the Eltanin impact site in the Drake Passage off the southwestern tip of South America, which represents a different non-climate related source of disruption capable of producing a significant CBD. However, the two events are correlated with different time horizons, although both within the Pliocene [77].

5.3. On the Bearing of Subtle Geographic Differences

As early proponents for a permanent Pliocene El Niño, Wara et al. (2005) [16] argued that the similar configuration of Pliocene continents and oceans placed few constraints on comparisons with today's world and the forecast of future global warming. Although subtle in scale, the open passage between North America and South America, prior to final closure by the Isthmus of Panama 3.5 million years ago [78], overlaps with the Pliocene Warm Period. A global paleogeographic reconstruction for the Pliocene Period (Figure 10) marks the location several Pliocene study sites covered in this review, and illustrates the ramifications with respect to atmospheric and oceanic circulation attendant on such and open passageway.

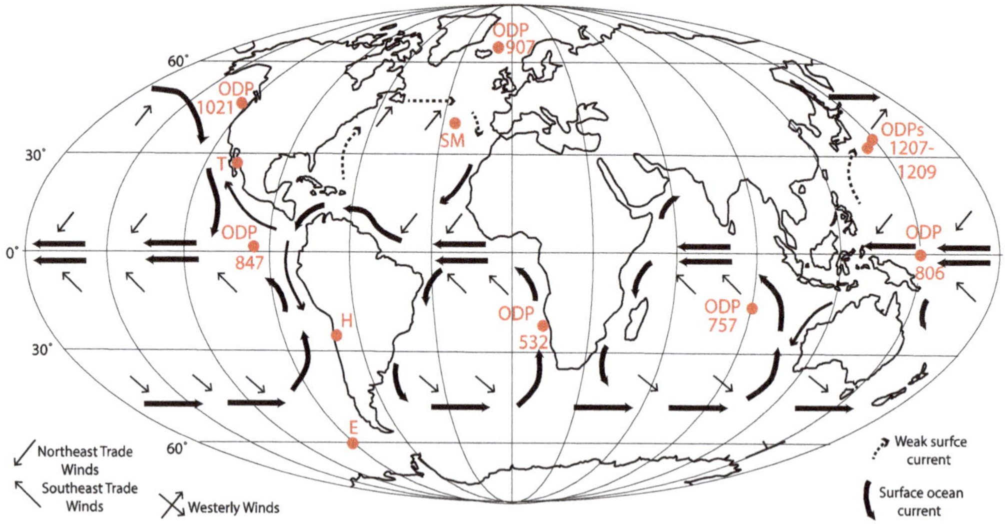

Figure 10. Global geography for the Pliocene Period, showing projected atmospheric circulation and ocean currents as well as the locations of many key study sites referenced in this review. Weakened surface currents shown by a dashed line. Map abbreviations: SM = Santa Marina Island (Azores archipelago), T = Tiombó conglomerate (Gulf of California), ODP = Ocean Drilling Project sites, H = Caletas Hornitos (northern Chile), and E = Eltanin impact site (off the southern tip of South America).

A small but significant difference is that a through-going ocean current was able to flow westward between the continents from the Atlantic Ocean into the Pacific Ocean. As consequence, the Gulf Current in the North Atlantic was almost certainly weaker compared to today [43]. Among studies probing evidence on the status of El Niño cycles during the Pliocene Warm Period (Table 2), the modeling study by von der Heydt et al. (2011) [44] is one of the few to consider the effect of an open seaway. The mathematical modeling undertaken in that study is sophisticated and difficult to evaluate. However, it concludes that weak trade winds during the Pliocene Warm Period were unlikely, and hence a permanent El Niño would have been unrealistic. A more recent study by Lam et al. (2021) [79] that relies on microfossils posits a multi-stage history in the development of the Kuroshio Current off present-day Japan. During the Pliocene Warm Period, an early stage reflects warmer SSTs and sluggish movement that evolved though a strong temperature/salinity gradient. Only after the Pliocene Warm Period at ~2.9 Ma, the Kuroshio Current is said to approximate its modern-day configuration as a major western boundary current bringing moisture to high latitudes [79]. By implication, a slow-down in the western moving Pacific equatorial currents may have influenced a weaker Kuroshio Current at the same time.

With regard to paleoecology, the local effect of a throughgoing current passing across the Caribbean into the eastern Pacific is evident due to immigration by Atlantic and Caribbean based coral and bryozoan faunas into the Gulf of California on Mexico's western coast [80,81]. This outcome implies the Pliocene presence of a north-drifting counter current along the shores of Central America and probably a similar south-drifting current off the northwest coast of South America due to a pooled bulge in surface waters that formed as a spill-over effect. The mechanics of such a bulge were pointed out by Weaver (1990) [82] as a distinct likelihood during the early Pliocene, by analogy with the present day through-flow of equatorial currents from the Pacific to Indian oceans between Indonesia and New Guinea, that stimulates the north to south poleward drift of the warm Leeuwin Current along the west coast of Australia. That anomalous current supports the 260-km long Ningaloo reef track off Western Australia in a position otherwise unsuited for coral growth.

The question is whether or not the Eastern Pacific Pool (EPP) of surface water may have been affected during the Pliocene due to reduced upwelling from the south moving California and north moving Humboldt currents, which in turn may have altered Walker Cell circulation in the atmosphere above the Pacific equator. The modeling by von der Heydt et al. (2011) [44] projects a western shift in the position of the cold tongue due to equatorial upwelling by up to 2000 km, but does not consider the possible influence of counter flow against the California and Humboldt currents or the altered strength of the Kuroshio Current.

6. Conclusions

Pertinent to the Pliocene Warm Period, as generally defined between ~4.5 and 3.0 million years ago [16,17], this review combines and attempts to balance research results not usually treated in the same discussion. Namely, the two fields of study consulted are those based on temperature-sensitive planktonic foraminifera and other organics retrieved as fossil remains from deep-sea cores, contrasted against land-based studies of non-biological accumulations preserved as coastal boulder deposits (CBDs) and coastal outwash deposits (CODs). Where alternate sources of energy such as tsunamis or extra-terrestrial impacts of bolides can be eliminated, the CBDs reflect events of hurricane intensity more often responsible for erosion on rocky shorelines resulting in coarse conglomerate. The same storms bring heavy rain with landfall, which can lead to the development of extensive CODs. A secondary goal of this review is to examine similar CBDs from the end of the Pleistocene Epoch roughly 125,000 years ago (correlative with Substage Marine Isotope 5e) and the Holocene during the last 10,000 years. The following conclusions are drawn from this process.

- The latest findings from the extensive literature on the Pliocene Warm Period argue against permanent El Niño conditions not only in the Pliocene Pacific Ocean, but more broadly worldwide. However, even the strongest proponents of a normal Pliocene El Niño suggest that elevations of CO_2 concentrations much higher than 400 ppm should induce a more persistent El Niño effect.
- Opposing studies tend to ignore the notion of a feed-back loop, whereby the steady increase in tropical disturbances of hurricane strength induce greater surface-water mixing and higher SSTs spreading beyond the equatorial zone and thereby more continuous El Niño conditions. In the opposite camp, studies that favor persistent El Niño conditions during the Pliocene Warm Period tend to ignore the notion that warming of SSTs in the equatorial Pacific may have been influenced by the open passageway between the North and South American continents irrespective of rising CO_2 levels.
- Repetitive Pliocene CODs found throughout Mexico's Gulf of California required higher rainfall in that region than today in order to achieve thicknesses of several meters between natural breaks. In turn, their presence suggests more persistent El Niño conditions as a source of local rainfall. Evidence for intense Pliocene activity is less known in the Atlantic realm but does exist as for example in the Azores, although limited overall by the availability of Pliocene island deposits.
- The end of the Pleistocene Epoch and succeeding 10,000 years of the Holocene represent a much smaller time frame than the Pliocene Warm Period, but evidence for CBDs during those intervals is even more extensive. With regard to climate-change studies, however, care must be exercised to eliminate examples due to non-climate related influences such as earthquake-induced tsunamis.

Future research on the Pliocene Warm Period is certain to provide greater insight on an interval of time during which average global temperatures and sea level exceeded what we experience today.

Funding: This research received no external funding.

Institutional Review Board Statement: Not Applicable.

Informed Consent Statement: Not Applicable.

Acknowledgments: The line drawings for this article were improved by B. Gudveig Baarli. Three anonymous readers contributed peer reviews with useful comments that led to the improvement of this contribution.

Conflicts of Interest: The author declares no conflict of interest.

References

1. UNESCO World Heritage List. Available online: https://whc.unesco.org/en/list/ (accessed on 29 October 2021).
2. Petit, J.-R.; Raynaud, D. Forty years of ice-core records of CO_2. *Nature* **2020**, *579*, 505–506. [CrossRef] [PubMed]
3. Howe, J.P. This is nature; this is un-nature: Reading the Keeling curve. *Environ. Hist.* **2015**, *20*, 286–293. [CrossRef]
4. NOAA Research News. Available online: https://research.noaa.gov/article/ArtMID/587/ArticleID/2764 (accessed on 29 October 2021).
5. Cheng, L.; Abraham, J.; Zhu, J.; Trenbereth, K.E.; Fasullo, J.; Boyer, T.; Locarnini, R.; Zhang, B.; Yu, F.; Wan, L.; et al. Record-setting ocean warmth continued in 2019. *Adv. Ocean Sci.* **2020**, *37*, 137–142. [CrossRef]
6. Oliver, E.C.J.; Donat, M.G.; Burrows, M.T.; More, P.J.; Smale, D.A.; Alexander, L.V.; Benthuysen, J.A. Longer and more frequent marine heatwaves over the century. *Nat. Commun.* **2018**, *9*, 1324. [CrossRef] [PubMed]
7. Kossin, J.P.; Knapp, K.R.; Olander, T.L.; Velden, C.S. Global increase in major tropical cyclone exceedance probability over the past four decades. *Proc. Natl. Acad. Sci. USA* **2020**, *117*, 11975–11980. [CrossRef]
8. Wang, S.; Toumi, R. Recent migration of tropical cyclones toward coasts. *Science* **2021**, *371*, 514–517. [CrossRef]
9. Dumitru, O.; Austermann, J.; Pollyak, V.J.; Fornós, J.J.; Asmerom, Y.; Ginees, J.; Ginés, A.; Onac, B.P. Constraints on global mean sea level during Pliocene warmth. *Nature* **2019**, *574*, 233–236. [CrossRef] [PubMed]
10. Miller, K.G.; Browning, J.V.; Schmelz, W.J.; Kopp, R.E.; Mountain, G.S.; Wright, J.D. Cenozoic sea-level and cryospheric evolution from deep-sea geochemical and continental margin records. *Sci. Adv.* **2020**, *6*, eaaz1346. [CrossRef]
11. Rybczynski, N.; Gosse, J.C.; Harington, C.R.; Wogeliums, R.A.; Hidy, A.J.; Buckley, M. Mid-Pliocene warm-period deposits in the high Arctic yield insight into camel evolution. *Nat. Commun.* **2013**, *4*, 1–9. [CrossRef] [PubMed]
12. Budyko, M.I. *The Earth's Climate: Past and Future*; Academic Press: New York, NY, USA, 1982; 336p.
13. Zubakov, V.A.; Borzenkova, I.I. Pliocene palaeoclimates: Past climates as possible analogues of mid-twenty-first century climate. *Palaeogeog. Palaeoclim. Palaeoecol.* **1988**, *65*, 35–49. [CrossRef]
14. Mei, W.; Xie, S.-P. Intensification of landfalling typhoons over the northwest Pacific since the late 1970s. *Nat. Geosci.* **2016**, *9*, 753–757. [CrossRef]
15. You, Z.-J. Tropical cyclone-induced hazards caused by storm surges and large waves on the coast of China. *Geosciences* **2019**, *9*, 131. [CrossRef]
16. Wara, M.W.; Ravelo, A.C.; Delaney, M.L. Permanent El Niño-like conditions during the Pliocene Warm Period. *Science* **2005**, *309*, 758–761. [CrossRef] [PubMed]
17. Ravelo, A.C.; Dekens, P.S.; McCarthy, M. Evidence for El Niño-like conditions during the Pliocene. *GSA Today* **2006**, *16*, 4–11. [CrossRef]
18. Brierley, C.M.; Fedorov, A.V.; Honghui, L.; Herbert, T.D.; Lawrence, K.T.; LaRiviere, J.P. Greatly expanded tropical warm pool and weakened Hadley circulation in the early Pliocene. *Science* **2009**, *323*, 1714–1718. [CrossRef]
19. Molnar, P.; Cane, M. Early Pliocene (pre-Ice Age) El Niño-like global climate: Which El Niño? *Geosphere* **2007**, *3*, 337–365. [CrossRef]
20. White, S.M.; Ravelo, A.C. Dampened El Ninño in the early Pliocene warm period. *Geophys. Res. Lett.* **2020**, *47*, e2019GL085504. [CrossRef]
21. Emanuel, K.; Sundararajan, R.; Williams, J. Hurricanes and global warming: Results from downscaling IPCC AR4 simulations. *Bull. Am. Meteorol. Soc.* **2008**, *89*, 347–368. [CrossRef]
22. May, S.M.; Engel, M.; Brill, D.; Squire, P.; Scheffers, A.; Kelletat, D. Coastal hazards from tropical cyclones and extratropical winter storms based on Holocene storm chronologies. In: Finkl, C.W. (ed.) *Coastal Hazards. Coast. Res. Libr.* **2013**, *6*, 557–585.
23. Johnson, M.E.; Ledesma-Vázquez, J.; Guardado-Grance, R. Coastal geomorphology of a Holocene Hurricane Deposit on a Pleistocene Marine Terrace from Isla Carmen (Baja California Sur, Mexico). *J. Mar. Sci. Eng.* **2018**, *6*, 108. [CrossRef]
24. Johnson, M.E.; Guardado-France, R.; Johnson, E.M.; Ledesma-Vázquez, J. Geomorphology of a Holocene Hurricane deposit eroded from rhyolite sea cliffs on Ensenada Almeja (Baja California Sur, Mexico). *J. Mar. Sci. Eng.* **2019**, *7*, 193. [CrossRef]
25. Johnson, M.E.; Johnson, E.M.; Guardado-France, R.; Ledesma-Vázquews, J. Holocene hurricane deposits eroded as coastal barriers from andesite sea cliffs at Puerto Escondido (Baja California Sur, Mexico). *J. Mar. Sci. Eng.* **2020**, *8*, 75. [CrossRef]
26. Guardada-France, R.; Johnson, M.E.; Ledesma-Vázquez, J.; Santa Rosa-del Rio, M.A.; Herrera-Gutiérrez, Á. Multiphase storm deposits eroded from andesite sea cliffs on Isla San Luis Gonzaga (Northern Gulf of California Mexico). *J. Mar. Sci. Eng.* **2020**, *8*, 525. [CrossRef]

27. Callahan, G.; Johnson, M.E.; Guardado-Grance, R.; Kedesna-Vázquez, J. Upper Pleistocene and Holocene storm deposits the granodiorite coast on Isla Sn Diego (Baja California Sur, Mexico). *J. Mar. Sci. Eng.* **2021**, *9*, 555. [CrossRef]
28. Ávila, S.P.; Johnson, M.E.; Rebelo, A.C.; Baptista, L.; Melo, C.S. Comparison of modern and Pleistocene (MIS 5e) coastal Boulder deposits from Santa Maria Island (Azores Archipelago, NE Atlantic Ocean). *J. Mar. Sci. Eng.* **2020**, *9*, 386. [CrossRef]
29. Galindo, I.; Johnson, M.E.; Martín-González, E.; Romero, C.; Vegas, J.; Melo, C.S.; Ávila, S.P.; Sánchez, N. Late Pleistocene boulder slumps eroded from a basalt shoreline at El Confital Beach on Gran Canaria (Canary Islands, Spain). *J. Mar. Sci. Eng.* **2020**, *9*, 138. [CrossRef]
30. Johnson, M.E. Holocene boulder beach eroded from chromite and dunite sea cliffs at Støpet on Leka Island (Northern Norway). *J. Mar. Sci. Eng.* **2020**, *8*, 644. [CrossRef]
31. Morton, R.A.; Richmond, B.M.; Jaffe, B.E.; Gelfenbaum, G. Coarse-clast ridge complexes of the Caribbean: A preliminary basis for distinguishing tsunami and storm-wave origins. *J. Sed. Res.* **2008**, *78*, 624–637. [CrossRef]
32. Lorang, M.S. A wave-competence approach to distinguish between boulder and megaclast deposits due to storm waves versus tsunamis. *Mar. Geol.* **2011**, *283*, 90–97. [CrossRef]
33. Ball, M.M.; Shinn, E.A.; Stockman, K.W. The geologic effects of Hurricane Donna in South Florida. *J. Geol.* **1967**, *75*, 583–597. [CrossRef]
34. Paris, R.; Fournier, J.; Poizot, E.; Etienne, S.; Morin, J.; Lavigne, F.; Wassmer, P. Boulder and fine sediment transport and deposition by the 2004 tsunami in Lhok Nga (western Banda Acheh, Indonesia): A coupled offshore-onshore model. *Mar. Geol.* **2010**, *268*, 43–54. [CrossRef]
35. Dowsett, H.J.; Robinson, M.M. Mid-Pliocene equatorial Pacific sea surface temperature reconstruction: A multi-proxy perspective. *Philos. Trans. R. Soc.* **2009**, *367*, 109–125. [CrossRef] [PubMed]
36. Haywood, A.M.; Valdes, P.J.; Peck, V.L. A permanent El Niño during the Pliocene? *Paleoceanography* **2007**, *22*, PA1213. [CrossRef]
37. Dowsett, H.J.; Chandler, M.A.; Robinson, M.M. Surface temperatures of the Mid-Pliocene North Atlantic Ocean: Implications for future climate. *Philos. Trans. R. Soc. A* **2009**, *367*, 69–84. [CrossRef] [PubMed]
38. Fedorov, A.V.; Brierly, C.M.; Emanuel, K. Tropical cyclones and permanent El Niño in the early Pliocene epoch. *Nature* **2010**, *463*, 1066–1071. [CrossRef] [PubMed]
39. Nott, J. Waves, coastal bolder deposits and the importance of pre-transport setting. *Earth Planet. Sci. Lett.* **2003**, *210*, 269–276. [CrossRef]
40. Nandasena, N.A.K.; Paris, R.; Tanaka, N. Reassessment of hydrodynamic equations: Minimum flow velocity to initiate boulder transport by high energy events (storms, tsunamis). *Mar. Geol.* **2011**, *281*, 70–84. [CrossRef]
41. Pepe, F.; Corradino, M.; Parrino, N.; Besio, G.; Presti, V.L.; Renda, P.; Calcagnile, L.; Quarta, G.; Sulli, A.; Antonioli, F. Boulder coastal deposits at Favignana Island rocky coast (Sicily, Italy): Litho-structural and hydrodynamic control. *Geomorphology* **2018**, *303*, 191–209. [CrossRef]
42. Dekens, P.S.; Ravelo, A.C.; McCarthy, M.D. Warm upwelling regions in the Pliocene warm period. *Paleoceanography* **2007**, *22*, PA3211. [CrossRef]
43. Schepper, S.D.; Greoeneveld, J.; Naafs, B.D.A.; Van Renterghem, C.; Hennissen, J.; Head, M.J.; Louwye, S.; Fabian, K. Northern Hemisphere glaciation during the Globally Warm early late Pliocene. *PLoS ONE* **2013**, *8*, e81508. [CrossRef]
44. Von der Heydt, A.S.; Nnafie, A.; Dijkstra, H.A. Cold tongue/Warm pool and ENSO dynamics in the Pliocene. *Clim. Past* **2011**, *7*, 903–915. [CrossRef]
45. O'Brien, C.L.; Foster, G.L.; Martínez-Boti, M.A.; Abell, R.; Rae, J.W.B.; Pancost, R.D. High sea surface temperatures in tropical warm pools during the Pliocene. *Nat. Geosci.* **2014**, *7*, 606–611. [CrossRef]
46. Haywood, A.M.; Dowsett, H.J.; Dolan, A.M. Integrating geological archives and climate models for the mid-Pliocene warm period. *Nat. Commun.* **2016**, *7*, 10646. [CrossRef] [PubMed]
47. Tierney, J.E.; Haywood, A.M.; Fend, R.; Bhattacharya, T.; Otto-Bliesner, B.L. Pliocene warmth consistent with greenhouse gas forcing. *Geophys. Res. Lett.* **2019**, *46*, 9136–9144. [CrossRef]
48. Ruban, D.A. Coastal bolder deposits of the Neogene world: A synopsis. *J. Mar. Sci. Eng.* **2019**, *7*, 446. [CrossRef]
49. Ruban, D.A. Finding coastal megaclast deposits: A virtual perspective. *J. Mar. Sci. Eng.* **2020**, *8*, 164. [CrossRef]
50. Yao, Q.; Liu, K.-b.; Wu, Y.; Aragón-Moreno, A.A.; Rodriques, E.; Cohen, M.; de Souza, A.V.; Farfán, L.M.; Antinao, J.L. A multiproxy record of hurricanes, tsunami, and post-disturbance ecosystem changes from coastal southern Baja California. *Sci. Total Environ.* **2021**, *796*, 149011. [CrossRef]
51. Johnson, M.E.; Backus, D.H.; Ledesma-Vázquez, J. Growth of the Ballena fan delta on the Gulf of California (Mexico) at the close of the Pliocene Warm Period. *Facies* **2017**, *63*, 14. [CrossRef]
52. Johnson, M.E.; Ledesma-Vázquez, J.; Backus, D.H. Tectonic decapitation of a Pliocene mega-delta on Isla del Carmen in the Gulf of California (Mexico): And a river ran through it. *J. Geol.* **2016**, *124*, 55–74. [CrossRef]
53. Emhoff, K.F.; Johnson, M.E.; Backus, D.H.; Ledesma-Vázquez, J. Pliocene stratigraphy at Paredones Blancos: Significance of a massive crushed-rhodolith deposit on Isla Cerralvo, Baja California Sur (Mexico). *J. Coast. Res.* **2012**, *28*, 234–243. [CrossRef]
54. Johnson, M.E.; Ledesma-Vázquez, J. Biological zonation on a rocky-shore boulder deposit: Upper Pleistocene Bahía San Antonio (Baja California Sur, Mexico). *Palaios* **1999**, *14*, 569–584. [CrossRef]
55. Ledesma-Vázquez, J.; Johnson, M.E.; Backus, D.H.; Mirabal-Davial, C. Coastal evolution from transgressive barrier deposit to marine terrace on Isla Coronados, Baja California Sur, Mexico. *Cienc. Mar.* **2007**, *33*, 335–351. [CrossRef]

56. Tierney, P.W.; Johnson, M.E. Stabilization role of crustose coralline algae during Late Pleistocene reef development on Isla Cerralvo, Baja California Sur (Mexico). *J. Coast. Res.* **2012**, *28*, 244–254. [CrossRef]
57. Kozlowski, J.A.; Johnson, M.E.; Ledesma-Vázquez, J.; Birgel, D.; Peckmann, J.; Schleper, C. Microbial diversity of a closed salt lagoon in the Puertecitos area, Upper Gulf of California. *Cienc. Mar.* **2018**, *44*, 71–90. [CrossRef]
58. Johnson, M.E.; Ledesma-Vázquez, J.; Backus, D.H.; González, M.R. Lagoon microbialites on Isla Angel de la Guarda and associated peninsular shores, Gulf of California (Mexico). *Sed. Geol.* **2012**, *263–264*, 76–84. [CrossRef]
59. Backus, D.H.; Johnson, M.E.; Riosmena-Rodríguez, R. Distribution, sediment source, and coastal erosion of fan-delta systems on Isla Cerralvo (Lower Gulf of California Mexico). *J. Coast. Res.* **2012**, *28*, 210–224. [CrossRef]
60. Johnson, M.E.; Uchman, A.; Costa, P.J.M.; Ramalho, R.S.; Ávila, S.P. Intense hurricane transports sand onshore: Example from the Pliocene Malbusca secion on Santa Maria Island (Azores Portugal). *Mar. Geol.* **2017**, *385*, 244–249. [CrossRef]
61. Hearty, P.J.; Tormey, B.R. Sea-level change and superstorms; geologic evidence from the last interglacial (MIS 5e) in the Bahamas and Bermuda offers ominous prospects for a warming Earth. *Mar. Geol.* **2017**, *390*, 347–365. [CrossRef]
62. Rovere, A.; Casella, E.; Harris, D.L.; Lorscheid, T.; Nandesena, N.A.K.; Dyer, B.; Sandstrom, M.R.; Stocchi, P.; D'Andrea, W.J.; Raymo, M.E. Giant boulders and last interglacial storm intensity in the North Atlantic. *Proc. Natl. Acad. Sci. USA* **2017**, *114*, 12144–12149. [CrossRef]
63. Cox, R.; Ardhuin, F.; Dias, F.; Autret, R.; Beisiegel, N.; Earlie, C.S.; Herterich, J.G.; Kennedy, A.; Paris, R.; Raby, A.; et al. Systematic review shows that work done by storm waves can be misinterpreted as tsunami- because commonly used hydrodynamic equations are flawed. *Front. Mar. Sci.* **2020**, *7*, 4. [CrossRef]
64. Hall, A.M.; Hansom, J.D.; Williams, D.M.; Jarvis, J. Distribution geomophology and lithofacies of cliff-top storm deposits: Examples from the high-energy coasts of Scotland and Ireland. *Mar. Geol.* **2006**, *232*, 131–155. [CrossRef]
65. Cox, R.; Zentner, D.B.; Kirchner, B.J.; Cook, M.S. Boulder ridges on the Aran Islands (Ireland): Recent movements caused by storm waves, not tsunamis. *J. Geol.* **2012**, *120*, 249–272. [CrossRef]
66. Erdmann, W.; Scheffers, A.M.; Kelletat, D.H. Holocene coastal sedimentation in a rocky environment: Geomophological evidence from the Aran Islands and Galway Bay (Western Ireland). *J. Coast. Res.* **2018**, *34*, 772–792. [CrossRef]
67. Etienne, S.; Paris, R. Boulder accumulations related to storms on the south coast of the Reykjanes peninsula (Iceland). *Geomorphology* **2010**, *114*, 55–70. [CrossRef]
68. Super Typhoon Haiyan. Available online: https://en.wikipedia.org/wiki/Typhoon_Haiyan (accessed on 29 October 2021).
69. Kennedy, A.B.; Mori, N.; Zhang, Y.; Yasuda, T.; Chen, S.E.; Tajima, Y.; Pecor, W.; Toride, K. Observations and modeling of coastal boulder transport and loading during Super Typhoon Haiyan. *Coast. Eng. J.* **2016**, *58*, 164004. [CrossRef]
70. Kennedy, A.B.; Mori, N.; Yasuda, T.; Shimozona, T.; Tomiczek, T.; Donahue, A.; Shimura, T.; Imai, Y. Extreme block and boulder transport along a cliffed coastline (Calicoan island Philippines) during Super Typhoon Haian. *Mar. Geol.* **2017**, *383*, 65–77. [CrossRef]
71. Mori, N.; Kato, M.; Kim, S.; Mase, H.; Shibutani, Y.; Takemi, T.; Tsuboki, K.; Yasuda, T. Local amplification of storm surge by Super Typhoon Haiyan in Leyte Gulf. *Geophys. Res. Lett.* **2014**, *41*, 5106–5113. [CrossRef]
72. Herterich, J.G.; Cox, R.; Dias, F. How does wave impact generate large boulders? Modelling hydraulic fracture of cliffs and shore platforms. *Mar. Geol.* **2018**, *399*, 34–46. [CrossRef]
73. Pérez-Torrado, F.J.; Paris, R.; Cabrera, M.C.; Schneider, J.L.; Wassmer, P.; Carracedo, J.C.; Santana, F. Tsunami deposits related to flank collapse in oceanic volcanoes: The Agaete Valley evidence, Gran Canaria, Canary Islands. *Mar. Geol.* **2006**, *227*, 135–149. [CrossRef]
74. Ramalho, R.S.; Winckler, G.; Madeira, J.; Helffrich, G.R.; Hipólito, A.; Quartau, R.; Adena, K.; Schaefer, J.M. Hazard potential of volcanic flank collapses raised by new megatsunami evidence. *Sci. Adv.* **2015**, *1*, e1500456. [CrossRef]
75. Biolchi, S.; Furlani, S.; Antonioli, F.; Baldassini, N.; Causon Deguara, J.; Devoto, S.; Di Stefano, A.; Evans, J.; Gambin, T.; Gauci, R.; et al. Boulder accumulations related to extreme wave events on the eastern coast of Malta. *Nat. Hazards Earth Syst. Sci.* **2016**, *16*, 737–756. [CrossRef]
76. Hartley, A.; Howell, J.; Mather, A.E.; Chong, G. 2001. A possible Plio-Pleistocene tsunami deposit, Hornitos, northern Chile. *Rev. Geol. Chile* **2001**, *28*, 117–125. [CrossRef]
77. Spliske, M.; Bahlburg, H.; Weiss, R. Pliocene mass failure deposits mistaken as submarine tsunami backwash sediments—An example from Hornitos, northern Chile. *Sediment. Geol.* **2014**, *305*, 69–82. [CrossRef]
78. Coates, A.G.; Jackson, J.B.; Collins, L.S.; Cronin, T.M.; Dowsett, H.J.; Bybell, L.M.; Jung, P.; Obando, J.A. Closure of the Isthmus of Panama: The near-shore marine record of Costa Rica and western Panama. *Geol. Soc. Am. Bull.* **1992**, *104*, 814–828. [CrossRef]
79. Lam, A.R.; MacLeod, K.G.; Schilling, S.H.; Leckie, R.M.; Fraass, A.J.; Patterson, M.O.; Venti, N.L. Pliocene to earlies Pleistocene (5-2.5 Ma) reconstruction of the Kuroshio Current extension reveals a dynamic current. *Paleoceanog. Paleoclimat.* **2021**, *36*, e2021PA004318. [CrossRef]
80. Foster, A.B. Environmental variation in a fossil scleractinian coral. *Lethaia* **1979**, *121*, 245–264. [CrossRef]
81. Cuffey, R.J.; Johnson, M.E. Bryozoan nodules build around andesite clasts from the upper Pliocene of Baja California: Paleoecological implications and closure of the Panama Isthmus. *Geol. Soc. Am. Spec. Pap.* **1997**, *318*, 111–117.
82. Weaver, A.J. Ocean currents and climate. *Nature* **1990**, *347*, 432. [CrossRef]

Journal of
Marine Science and Engineering

MDPI

Review

Mechanisms for Overpressure Development in Marine Sediments

Chong Li [1], Linsen Zhan [1,2] and Hailong Lu [1,*]

1 Beijing International Center for Gas Hydrate, School of Earth and Space Sciences, Peking University, Beijing 100871, China; lichong189@pku.edu.cn (C.L.); zhanlinsen@pku.edu.cn (L.Z.)
2 College of Engineering, Peking University, Beijing 100871, China
* Correspondence: hlu@pku.edu.cn

Abstract: Overpressure is widely developed in marine sediments; it is not only a critical factor related to hydrocarbon accumulation, but also a serious safety issue for oil/gas exploration and exploitation. Although the mechanisms for overpressure development in sedimentary basins have been intensively studied, some new mechanisms are proposed for overpressure development with the advancements in marine geological investigation, e.g., natural gas hydrate formation and microbial activity. In this study, the mechanisms for overpressure development are reviewed and further classified as being related to associated physical, chemical, and biological processes. The physical overpressure mechanisms include disequilibrium compaction, hydrate formation sealing, degasification, buoyancy, hydrothermal pressuring, tectonic movement, overpressure transfer, etc. The chemical overpressure mechanisms are ascribed to hydrate decomposition, diagenesis, hydrocarbon generation, etc. The biological overpressure mechanisms are mainly induced by microbial gas production and microbial plugging. In gas hydrate-bearing sediments, overpressure is a critical factor affecting the formation and distribution of gas hydrate. The mechanisms for overpressure development in marine gas hydrate systems are associated with permeability deterioration due to hydrate formation and free gas accumulation below bottom-simulating reflectors (BSR). In marine sediments, overpressure developments are generally related to a sediment layer of low permeability above and natural gas accumulation below, and overpressure is mainly developed below a sulphate–methane interface (SMI), because methane will be consumed by anaerobic oxidation above SMI.

Keywords: overpressure; disequilibrium compaction; hydrocarbon generation; natural gas hydrate; microbial activity

Citation: Li, C.; Zhan, L.; Lu, H. Mechanisms for Overpressure Development in Marine Sediments. *J. Mar. Sci. Eng.* **2022**, *10*, 490. https://doi.org/10.3390/jmse10040490

Academic Editors: George Kontakiotis, Assimina Antonarakou and Dmitry A. Ruban

Received: 25 February 2022
Accepted: 29 March 2022
Published: 1 April 2022

1. Background

Marine sediments refer to the deposits of insoluble materials from various sources that accumulate on the seafloor [1]. Based on their origins, the components of marine sediments are classified as lithogenous, biogenous, hydrogenous, and cosmogenous [2]. The lithogenous sediments originate from the weathering of land materials, volcanic eruptions and blown dusts, and their major components are quartz sands, muds and clays. Silica and calcium carbonate are the most common chemical compounds in biogenous sediments, which are formed by marine organisms such as corals, mollusks, foraminifera, etc. The hydrogenous sediments are produced by the chemical reactions and precipitation in supersaturated seawater. The cosmogenous sediments refer specifically to micrometeorites from outside the earth, and form a rare component of marine sediments [1]. Roughly 90 percent by volume of marine sediments are deposited near continents, covering approximately 25 percent of the seafloor, and the remaining 75 percent of the deep seafloor is covered by slowly accumulated pelagic sediments [1].

Diagenesis refers to all processes, chiefly chemical, by which changes in a sediment are brought about after its deposition, but before its final lithification (conversion to rock) [3].

Considering the alterations of both inorganic and organic materials, the process of diagenesis can be divided into three main phases: the pre-burial stage, shallow-burial stage and deep-burial stage [4]. During the pre-burial stage, the physicochemical processes take place in the surficial layer of sediments in the presence of oxygen [4]. However, this is negligible for the development of overpressure in sediments. In the shallow-burial stage, after deposition, the sediments are subjected to the influence of physical, chemical and biological processes. Above the sulfate–methane interface (SMI), chemical reactions play a dominant role. Oxidants such as O_2, NO_3^-, SO_4^{2-} and certain amounts of Fe^{3+} and Mn^{4+} are reduced by hydrocarbons in the sediments and methane migrating from the deep. At the same time, authigenic minerals such as carbonate and iron sulfide are formed. When the pore water is saturated with methane and other natural gases under the thermobaric conditions of the gas hydrate stability zone (GHSZ), gas hydrate will be formed. With the formation of natural gas hydrate, the porosity of hydrate-bearing sediments decreases [5]. In the deep-burial stage, the pore space of the sediments is further compressed. However, there is not much room for porosity reduction, because most of the pore space is compressed in the shallow burial stage. With the increases in burial depth and temperature, the organic matter in the sediments gradually matures and generates oil and gas. Gypsum, montmorillonite and other minerals will dehydrate in a certain temperature range, increasing the water content in sediments. On the other side, the cementation of authigenic carbonate and silica will reduce permeability. These processes have the potential to result in overpressure development in marine sediments. When the temperature exceeds 200 °C, it is considered to be in the metamorphic stage.

Pressure is an important physical property in sedimentary layers and mainly involves overburden pressure, pore pressure, and effective stress (Figure 1).

The overburden pressure (lithostatic pressure), S, is generated by the total weight of sediment matrix and pore fluid overlying a certain portion in sediments. Generally, the overburden pressure increases with depth.

$$S = g\int_0^z \rho_b(z)dz \tag{1}$$

where g, ρ_b, and z are the acceleration of gravity, bulk density, and depth, respectively.

Pore pressure (formation pressure), P, refers to the pressure exerted on the fluid at certain points in sediment pores. The difference between overburden pressure and pore pressure is known as the differential pressure, represented by ΔP, i.e., $\Delta P = S - P$.

The effective stress, σ, refers to the pressure acting on the solid sediment frame skeleton. The effective stress is equal to the difference between the overburden pressure and pore pressure, based on Terzaghi and Biot's effective stress principle [6,7]. The modified effective stress principle is that the effective stress is equal to the difference between the overburden pressure and equivalent pore pressure [7]. The equivalent pore pressure equals the product of pore pressure and effective stress coefficient. Therefore, the effective stress can be expressed as:

$$\sigma = S - \alpha P \tag{2}$$

where P is the pore pressure; S the overburden pressure; σ the vertical effective stress; and α the Biot effective stress coefficient, ranging from 0 to 1.

Hydrostatic pressure (normal pressure), P_h, is generated by the weight of the fluid in the sediment.

$$P_h = \rho_f gh \tag{3}$$

where ρ_f, g and h are fluid density, acceleration of gravity, and the height of the fluid column, respectively.

Overpressure means that the pore pressure is greater than the hydrostatic pressure in sediments. The overpressure represents the difference between the pore pressure and hydrostatic pressure. In the overpressure system, the effective stress coefficient, α, is usually set to 1. Therefore, effective stress is equal to differential pressure:

$$\sigma = S - P \tag{4}$$

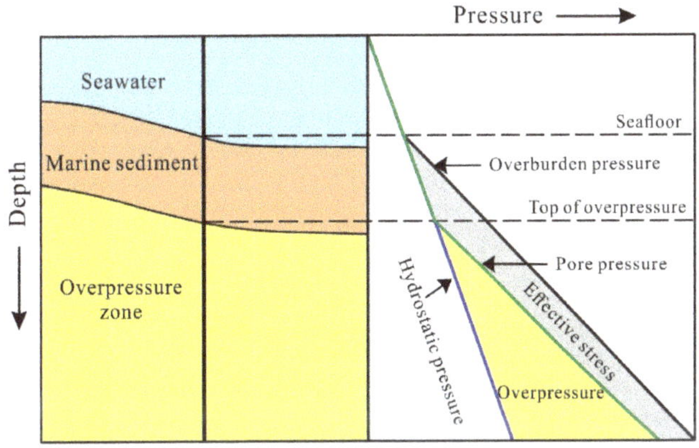

Figure 1. Pressure–depth profile in marine sediments (modified from Dutta [8]).

Sediment overpressure is found to significantly affect hydrocarbon accumulation and well drilling. Knowledge of the mechanism for overpressure development is required for pressure prediction and is critical for the understanding of hydrocarbon accumulation. It is generally believed that disequilibrium compaction (also known as under compaction) is the main cause of sediment overpressure. However, more studies suggest that the development of sediment overpressure is often the result of the interaction of several factors [9–11]. Regarding the mechanism of overpressure development in sediments, quite some studies have been carried out [8,12–15]. However, previous studies are mainly on terrestrial sedimentary basins. Marine sediments are deposited and compacted under water-saturated conditions, thus sediment pores are always hydraulically connected. The authigenic minerals (including natural gas hydrate) formed in the marine sediments are different from those in a terrestrial sedimentary basin. As a result, the mechanisms for overpressure development should be different between marine and terrestrial sedimentary basins.

Overpressure is widely developed in marine sedimentary basins [16,17]; it is not only closely related to the formation and distribution of marine oil and gas reservoirs, but is also considered to be an impetus of submarine geological hazards [18,19]. Evidence indicates that it has caused many risks and challenges to offshore drilling safety [20–23]. As an important type of marine resource [24–26], the formation and decomposition of natural gas hydrate have a significant impact on pore pressure [27,28]. For instance, in a gas hydrate occurrence area, when the temperature increases or the pressure decreases, natural gas hydrate will dissociate [29], resulting in fluid volume expansion and generation of overpressure in sediment pores [28]. In gas hydrate exploitation, the overpressure caused by hydrate decomposition will increase the risk and difficulty of drilling in hydrate reservoirs. Although natural gas hydrate is attracting more interest, the effect of gas hydrate on overpressure development has not been well discussed. Recently microbial activities have also been recognized as a possible factor for overpressure in marine sedimentary layers, but such a mechanism has not been investigated in the context of overpressure.

In this review, the mechanisms of overpressure development in marine sediments are systematically summarized, including those associated with natural gas hydrate and microbes.

2. Classification of the Mechanisms for Overpressure Development in Marine Sediments

The mechanisms for overpressure developed in sedimentary basins are commonly classified into disequilibrium compaction, tectonic movement, buoyancy, overpressure transfer, hydrothermal pressuring, hydraulic head, hydrocarbon generation, and diagenesis [9,12,30]. However, marine sediments are water-saturated, and the sediment pores are generally hydraulically connected in the longitudinal direction. As a result, overpressure caused by the difference in the height of hydraulic heads between the datum plane (ground surface or water level) and the recharge area of a sealed caprock, which is common in terrestrial basins, will not be found in marine sediments. Although osmosis is proven to be able to induce overpressure development in shales, it is not yet reported in marine sediments. In addition, the existence of gas hydrate and microbes are the mechanisms for overpressure in marine sediments, which are not found in terrestrial sedimentary basins.

In this review, the mechanisms for overpressure development in marine sediments are divided into physical, chemical, and biological overpressure (Table 1), each of which is briefly discussed in the following sections.

Table 1. The mechanisms for overpressure development in marine sediments.

Classification	Mechanism	Intrinsic Driving Force	Cases Reported	References
Overpressure development associated with physical process	Disequilibrium compaction	Deposition rate greater than drainage rate in compaction	Gulf of Mexico; Baram province, Brunei; South China Sea; Nile Delta; Offshore Mumbai, Western India	[8,9,13,16,31,32]
	Hydrate formation sealing	Permeability reduction	Mahanadi basin, India; Krishna-Godavari basin, India	[5,27,33,34]
	Degasification	Gas solubility change	Central diapir zone of the Yinggehai basin	[35,36]
	Buoyancy	Density difference between fluids	Dabis oilfield in the Malay basin	[37–39]
	Hydrothermal pressuring	Difference in thermal expansion coefficients of fluids	Gulf Coast	[40–42]
	Tectonic movement	Stress related permeability reduction	Barbados accretionary prism	[43–45]
	Overpressure transfer	Overpressured fluid migration due to permeability difference	New Jersey continental slope; Yinggehai basin; Baram province, Brunei	[9,19,46,47]
Overpressure development associated with chemical process	Hydrate decomposition	Fluid volume expansion by gas released from hydrate dissociation	Nankai Trough; Hikurangi margin; Shenhu in the northern South China Sea	[28,29,48–50]
	Diagenesis	Fluid volume expansion or permeability reduction by authigenic mineral formation	Gulf Coast; Halten Terrace offshore mid-Norway and the Gulf of Mexico	[13,51–53]
	Hydrocarbon generation	Fluid volume expansion by oil/gas generation	Gulf Coast; Amazon Fan; Krishna-Godavari basin, India	[13,54–57]
Overpressure development associated with biological activity	Microbial gas production	Generation of biogenic gas	Offshore Nile Delta, Egypt; Shenhu area in the northern South China Sea	[58–61]
	Microbial plugging	Permeability reduction	Nankai Trough	[62,63]

3. Overpressure Development Associated with Physical Processes

Overpressure development associated with physical processes refers to that caused by mechanical processes and/or differences in physical properties between sediments and pore fluids, while no new chemical product will be formed. Several physical processes can result in overpressure in marine sedimentary environments, such as disequilibrium compaction, tectonic movement, buoyancy, overpressure transfer, hydrothermal pressuring, etc. In addition, degasification from water can also cause abnormal pressure and form overpressured gas reservoirs. The formation of gas hydrate will cause the permeability to be reduced due to the occupation of gas hydrate in sediment pores, leading overpressure from gas accumulation below the hydrate layer.

3.1. Disequilibrium Compaction

Disequilibrium compaction, also known as under compaction, is generally developed in low-permeability sediment, where the deposition rate is greater than the gravity dehydration. Due to the low permeability of sediment and lower compaction rate than the overlying sediment, the pore fluid will be trapped in the sediment. Disequilibrium compaction also occurs when sediments with different particle sizes are interbedded, for example in turbidites. In such cases, the pore fluid will sustain the partial weight of overlying sediments and fluids, and the pore pressure is then greater than hydrostatic pressure, resulting in overpressure [8] (Figure 2). Overpressure caused by disequilibrium compaction mainly occurs in the shallow-burial diagenetic stage.

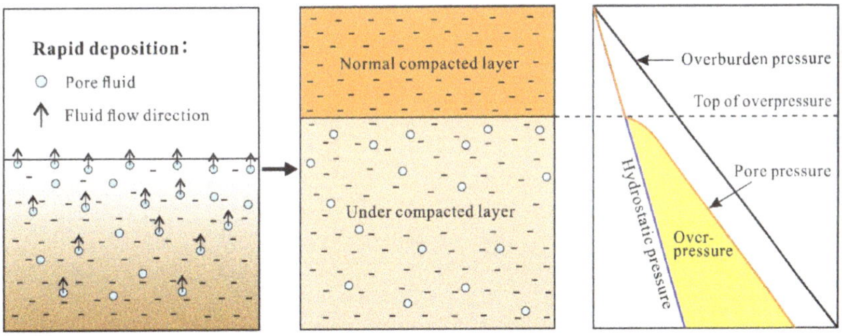

Figure 2. Schematic diagram for overpressure development associated with disequilibrium compaction.

The development of overpressure in many sedimentary basins is related to disequilibrium compaction [64]. For example, about 75% of the overpressure in the Pleistocene basin along the Gulf of Mexico originates from disequilibrium compaction [16]. The overpressures of 54 areas in the Brunei Baram basin are mainly caused by the disequilibrium compaction in the prodelta shales, and for the rest the overpressure is transferred by fluid expansion which is developed in the delta sequence on the continental shelf [9]. The gradient of the pore overpressure up to 12 MPa/km is caused by disequilibrium compaction in the sediments of about 240–370 mbsf (meters below the seafloor) at Site 1144 in the north of the South China Sea [31]. Based on the one-dimensional mathematical model of Mann and Mackenzie [65], the fitting results of the actual pressure data show that the pressure due to under compaction is with a similar gradient to the overburden pressure observed in the Nile Delta [14]. Likewise, the prevalent disequilibrium compaction directly led to overpressure with a near-lithostatic gradient (22 MPa/km) in the underlying Paleocene shales in western India [32].

Before the mid-1990s, disequilibrium compaction had been considered the dominant factor for overpressure in many sedimentary basins. However, other mechanisms for overpressure development have attracted more attention in recent years [13].

3.2. Hydrate Formation Sealing

The gas trapped in gas hydrate is usually of biogenic or thermogenic origin [66–69]. Water and gas molecules can form gas hydrate in a certain range of temperatures and pressures after the gas migrates upward through sediment pores or fractures to the gas hydrate stability zone (GHSZ) [70]. Pore pressure should be increased because the density of methane hydrate (~0.93 g/cm^3) is less than water. However, generally, hydrate formation does not increase pore pressure significantly, because the formation rate of natural gas hydrate is very low due to limited CH_4 availability. With the growth of gas hydrate, sediment permeability will be decreased by hydrate occupation of pore space. Experimental and numerical studies on methane or carbon dioxide hydrate [5,27,71–74] suggest that the permeability of the sediment decreases with increasing hydrate saturation. Konno et al. [75] found that the absolute permeability of clay-rich sediments and silty sediments without gas hydrate are in the tens mD, and the absolute permeability of sandy sediments in the methane hydrate reservoir can reach 1.5 D in the eastern Nankai Trough, Japan. Further study shows that in the gas hydrate-bearing layer of sandy sediments, the effective permeability of water is only 47 mD when the hydrate saturation is 70%, while the absolute permeability is about 840 mD after hydrate decomposition [75]. As a result, the formation of gas hydrate will significantly decrease permeability and hinder the upward migration of fluid [76]. When the permeability of the gas hydrate-bearing layer is too low for free gas to pass through, the gas from deep strata will be sealed by the less permeable hydrate-containing sediments, below which gas accumulates to develop overpressure.

The large difference in wave impedance between the gas hydrate-bearing layer and the underlying gas-bearing layer often generates a strong reflection called BSR (bottom-simulating reflector) on the seismic profile. BSR is approximately parallel to the seafloor reflection and has opposite polarity [77–79]. As the result, BSR is usually regarded as the bottom of the gas hydrate stability zone. The deepest BSR found to date is about 500 mbsf in Black Ridge. In the sediments with hydrate occurrence, overpressure usually develops below BSR, associated with free gas accumulation. However, when a layer of mixed hydrate and free gas appears beneath a gas hydrate-bearing layer, overpressure can also develop. Overpressure under BSR is caused by hydrate formations [33,34].

3.3. Degasification

Gas solubility refers to the volume of gas dissolved in one volume of water at a certain temperature and standard atmospheric pressure (101 kPa). In sediments, CH_4, H_2S, CO_2, and other gases are more or less dissolved in water. The gas solubility is affected by the combined effects of temperature, pressure, and salinity. Methane solubility generally decreases first and then increases with the temperature, increases at higher pressure, and decreases at higher salinity [35,80]. Wang et al. [81] found that methane solubility actually increases first and then decreases with increases in salinity. However, methane solubility at liquid–hydrate equilibrium in the gas hydrate occurrence area shows a monotonically increasing trend with temperature and pressure [68,82–85]. Assuming that the seawater depth is 2 km, the seafloor temperature is 3 °C and the geothermal gradient is 30 °C/km, Figure 3 depicts the changing trend of methane solubility in marine sediments [35,83]. The pressure in the liquid phase is equal to the hydrostatic pressure (10 MPa/km). The black dashed lines represent the base of the methane hydrate layer in marine sediments. The purple, red and blue lines represent methane solubility in pure water, seawater with 3.5 wt% NaCl and seawater with 7.0 wt% NaCl respectively. In marine sediments, when the temperature is above 20 °C, methane solubility changes insignificantly with pressure, and even decreases slightly with pressure/depth increases (Figure 3). When the temperature is below 20 °C, methane solubility decreases significantly with the decrease in temperature and burial depth. As a result, the methane dissolved in the fluid migrating upward will be partly exsolved, and overpressure will be developed when it encounters a low permeability cap, such as a hydrate-containing sediment layer.

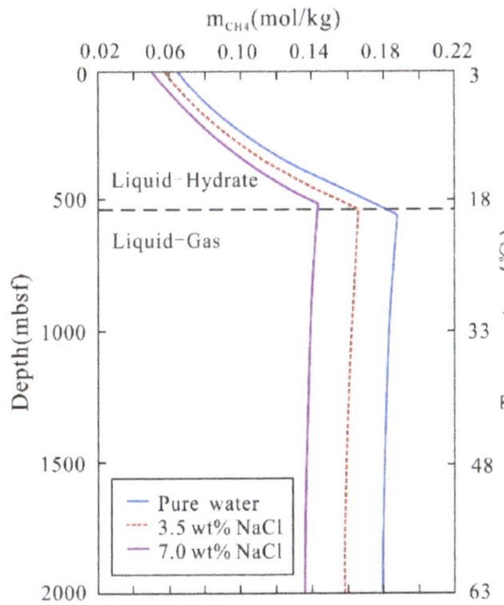

Figure 3. Trend of methane solubility in seawater with temperature and water depth (modified from Duan et al. [35] and Rui and Duan [83]).

"Aqueous phase exsolution type" in the isolated overpressure system is a gas accumulation model in the central diapir zone in the Yinggehai Basin [36]. Degasification may release a great amount of gas. However, the relationship between gas dissolution and overpressure generation seems to be two-way, rather than just promotion. Degasification will increase the amount of free gas, expand the fluid volume, increase the pore pressure, and result in overpressure in reservoir pores. In turn, overpressure will enhance gas solubility and reduce the rate of degasification. Although their relationship is complex, degasification can cause local overpressure. The contribution of degasification to basin overpressure needs to be further studied and understood.

3.4. *Buoyancy*

Buoyancy, due to the difference in the density of fluids in sediment pores, has the potential to cause overpressure. The buoyancy phenomenon is common in oil–water, gas–water, or oil–gas–water reservoirs. Buoyancy will occur when the water is replaced by hydrocarbons in sediment pores, because the density of oil or gas is less than that of water, leading to the development of pore pressure. The main factors associated with overpressure are the density difference between pore water and oil (or gas), and the positions of oil and gas in sediment column [14,37,38]. In gas hydrate-bearing sediment, buoyancy can also cause overpressure in the free gas layer below the low-permeability hydrate layer, as the gas density is much less than the density of pore water in sediment layers (Figure 4).

Sahagian and Proussevitch [86] suggested that the buoyancy effect of gas bubbles will transport the hydrostatic pressure upward from the lower portion to the top in a reservoir via incompressible fluid (e.g., water), resulting in the increase in pressure. However, the effect of the compressibility of the actual fluid, gas content, sediment permeability, gas solubility, and the changing properties of gas with temperature and pressure on sediment pressure need further quantitative studies [14]. Buoyancy occurs in oil and gas reservoirs; however, the overpressure originating from buoyancy varies greatly in different oil and gas reservoirs. This mechanism plays an obvious role in the pressure field in Dabis oilfield in the southeast of the Malay Basin, whereas only insignificant overpressure is generated in some hydrocarbon traps in the North Sea Basin [39]. Overall, overpressure related to buoyancy mainly arises from the difference in density between different fluids (oil, gas, water, etc.), generally occurring in oil/gas reservoirs or under BSR [87].

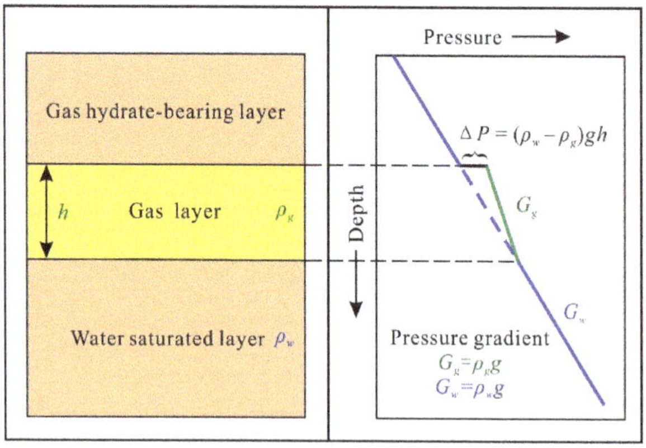

Figure 4. Schematic diagram to show buoyancy related overpressure development in gas hydrate occurrence area (modified from Zhang [38]).

3.5. Hydrothermal Pressuring

Hydrothermal pressuring was first proposed by Barker [40], based on the overpressure isolations observed on the northern coast of the Gulf of Mexico. The fluid in the sediment will expand when it is heated. Hydrothermal pressuring is produced because the thermal expansion coefficient of pore fluid is greater than that of the sediment matrix. The increasing temperature will lead to the volume expansion of pore fluid when it is buried and isolated, thereby creating overpressure (Figure 5) [8]. The pressure generated by hydrothermal pressuring depends on the burial depth and geothermal gradient [40,88,89]. There is no transition zone on the pressure curve since hydrothermal pressuring can only occur in isolation [89].

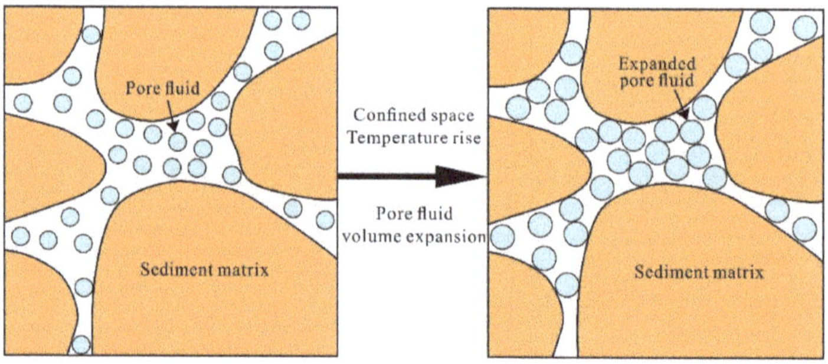

Figure 5. Schematic diagram for overpressure development by hydrothermal pressuring.

The hydrothermal pressuring mechanism is still a controversial issue. Some scholars believe that it can play an important role for effective closure [88,90], however, other scholars pointed out that unless it is completely closed, the fluid volume expansion due to hydrothermal expansion is too small to produce significant overpressure [14]. Numerical simulations reveal that the overpressure generated by hydrothermal mechanisms is very small, even in a completely closed system [91]. In addition, this overpressure development requires a rapid temperature rise [89]. In marine sediments, these two conditions are difficult to meet simultaneously. The pore water cannot be completely isolated because of the actual non-zero permeability in the sediment. Hydrothermal pressuring is supposed to play a role in overpressure generation, but it might not be the main mechanism. Overpressure of this type may occur in some special geological conditions, such as upon magma intrusion and thermal fluid arching [39]. Magara [41]'s study on the overpressure along the coast of

Gulf of Mexico shows that, in the overpressure zone, the average increasing rate of pore pressure is about 1.4 psi/ft after isolation, which is apparently higher than the overburden pressure gradient (usually less than 1 psi/ft). Because relative isolation and temperature rises in a system are the prerequisite for the overpressure development on this mechanism, hydrothermal pressurization mainly occurs in the deep-burial stage.

3.6. Tectonic Movement

Tectonic movement will change the transverse and/or longitudinal stress of sediments. For instance, tectonic uplift and subsidence will change the overburden pressure, meanwhile, compression and shear deformation will destroy the structure of original strata, resulting in the rearrangement of rock skeleton stress and abnormal pressure [43,92]. Nevertheless, tectonic stress directly acts on sediment particles rather than the pore fluid. Luo [44] suggested that the tectonic effect on pore pressure can be significant through sediment compaction by reducing sediment pore volume and causing permeability and porosity decreases. When sediments are consolidated to some extent, tectonic movement might create overpressure. In such a condition, tectonic overpressure mainly occurs in deep-burial stage.

Tectonic compression and shear stress can produce strong tectonic overpressure. The pressure peak usually appears at the position deformed most intensely and gradually attenuates away from it (Figure 6). In provinces such as the Orinoco Delta, Venezuela, Trinidad, Sumatra, and California, significant overpressure can be induced by tectonic compression and wrench faults [8]. The quantitative study of tectonic overpressure also reveals that the additional pressure generated by tectonic movement in the lithosphere may have the same order of magnitude as the lithostatic pressure. The abnormally high fluid pressure (>90% of lithostatic pressure), which is calculated from the logging data collected during ODP Leg 156 below the thrusts in the Barbados accretionary prism, is presumably due to the increase in overburden and lateral tectonic loading [45].

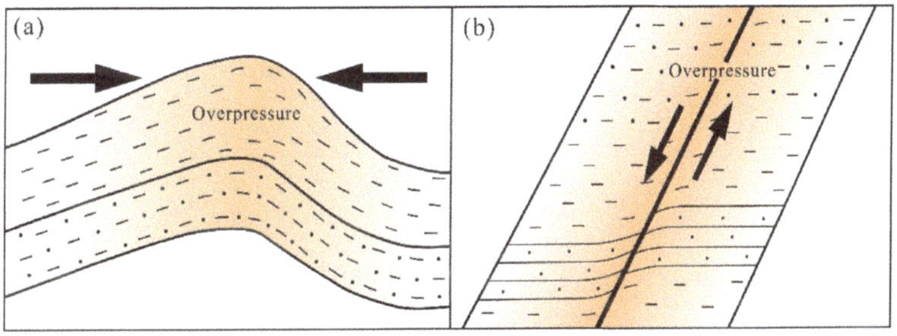

Figure 6. Schematic diagram of tectonic-related overpressure development: (**a**) compression, (**b**) shear.

3.7. Overpressure Transfer

The distribution of overpressure is dynamic in sediment sections. The overpressure developed in deep sediments can be transferred to a shallow portion through fluid flow, and it will be redistributed in an isolated and closed reservoir to maintain hydrostatic balance [9]. Overpressure transfer usually involves fluid expansion. Faults and/or inclined sand bodies which are inter-bedded with clay-rich sediment may behave as fluid migration channels. Driven by abnormally high pressure, deep overpressure fluid migrates from the lower sediment section and releases at the top of the faults or sand layer due to permeability differences. These activities cause the pore pressure around the channel bottom to be lower than that in the sediments of the overpressure interval at about the same depth. At the top of the fluid migration channel, the opposite is the case [8,12,19]. This process is referred to as the lateral migration of overpressure by Yardley and Swarbrick [93]. Overpressure may also be transferred vertically through the fractures or faults in the cap layer if one overpressure isolation keeps hydraulic connection with another isolation with lower pressure [10,94,95].

Based on the actual seismic and logging data on the continental slope of New Jersey, Dugan and Flemings [46] created a deposition–compaction model. In the model, the flat and permeable silt strata with permeability of 3×10^{-16} m^2 is overloaded by low-permeability (1×10^{-18} m^2) silt and clay. The distribution of pressure and fluid fields on the slope profile is simulated after the lateral differential deposition and compaction for 1 Ma. In this process, the excess fluid in the low-permeability formation transfers to the permeable silt layers. The flow model successfully predicts the overpressure observed at Site 1073 at the downdip part of the slope. If the covering layer is thin at the lower part of the slope, the pore pressure will be close to the overburden pressure, and the transfer of pressure may trigger slope failure [19]. The activities of the strong lateral and vertical overpressure fluid in the Yinggehai basin reveal the interaction between the fault system and overpressure transfer. Tensile faults first induce the episodic breakthrough of overpressure fluid, then overpressured fluid flows through the main faults or fractures, and finally the overpressured fluid is released along the transport corridor. In parallel, the overpressure fluid event will open new faults or activate old faults, which will become an incentive for further overpressure fluid breakthroughs [47]. Overpressure in the inner-shelf deltaic sequences in Baram basin in Brunei is transferred from the prodelta shales through faults as well [9]. Diapir/mud volcanoes and gas chimneys are also the main migration channels for the deep overpressured fluid in both Shenhu area in China and offshore Nile Delta in Egypt [49,60,61]. Therefore, the fault system, inclined sandstone, mud diapir, and gas chimneys are all important channels for overpressure transfer [96].

It should be noted that engineering accidents in the drilling process may cause the quality of well cementation to deteriorate, and it will artificially establish hydrodynamic communication between the upper and lower formations that were originally isolated from each other. As a result, such wells can become a special channel for overpressure transfer as well [15]. As long as the condition of isolation is present and the connection between them is established, the transfer of overpressure could occur.

4. Overpressure Development Associated with Chemical Process

Overpressure development associated with chemical processes refers to overpressure that is related to chemical reactions, by which fluid volume expansion and large-scale or local overpressure will result due to the formation of new material in sediment pores. In marine sedimentary basins, chemical reactions including kerogen maturation, oil cracking, smectite dehydration, gypsum dehydration, hydrate decomposition, etc. can lead to fluid expansion. In addition, chemical cementation in diagenetic processes can induce overpressure because the cementation may reduce sediment permeability. Based on the direct mechanism that results in the development of overpressure, chemical reaction-related overpressures are divided into hydrate decomposition, diagenesis and hydrocarbon generation.

4.1. Hydrate Decomposition

Natural gas hydrate usually occurs in the sediments within a certain range of temperatures and pressures. When temperature or pressure changes exceed the stability regime of gas hydrate, the hydrate will decompose to free gas and water. As the released gas and water from hydrate decomposition is sealed by the overlying hydrate-bearing sediment layer with low permeability, overpressure will develop [33].

The water and gas from hydrate decomposition will lead to fluid volume expansion and overpressure [25,97]. Xu [28] found that the overpressure caused by hydrate decomposition in marine sediments can be up to tens of megapascals (MPas) in confined pore spaces. It is estimated that about 164 units volume of methane can be released from the decomposition of one unit volume of gas hydrate under standard conditions [26]. According to the ideal gas equation of state, the decomposition of gas hydrate per unit volume will generate about 2.36, 1.95, 1.52 units volume of gas respectively around the BSRs at the sites of ODP 204, NGHP expedition 01, and IODP expedition 311. The shallower the burial depth, the larger the volume of released gas and the greater the overpressure development. As a result, hydrate decomposition is an important factor for overpressure development.

Due to the thermodynamic and kinetic principles of natural gas hydrate, hydrate decomposition is most likely to occur in the three-phase zone in which gas hydrate, water and free gas coexist under equilibrium conditions, and the place with such conditions is around BSR [85,98]. Tectonic uplift or sea level fall, continuous fast sedimentation, and hydrothermal fluid rise have the potential to induce hydrate decomposition [28,99].

Tectonic uplift or sea level fall will decrease the pressure of the gas hydrate stability zone, and consequently the gas hydrate near BSR will be outside its stable regime and begin to decompose, releasing free gas into the sediment pores. As the less permeable upper hydrate-bearing layer is still in the regime of hydrate stability, the free gas released will be sealed and overpressure will develop (Figure 7a). The formation and decomposition of gas hydrate are dynamic processes controlled by temperature and pressure. To maintain the dynamic balance, the bottom of the hydrate stability zone will move upward. The three-phase medium, consisting of gas hydrate, free gas, and water, will eventually reach equilibrium in the rebuilt stability zone [85]. After reaching the new balance, a new BSR (BSR2) above the original BSR (BSR1) will appear when only part of the hydrate above the old BSR is decomposed, and double BSRs will be observed on the seismic profile [29,100]. The magnitude of overpressure generated by tectonic uplift or sea level fall is related to the rate and duration of the movement. However, it should be noted that the overpressure developed by hydrate decomposition is with remarkable impact only in the relatively shallow hydrate stability zone [28].

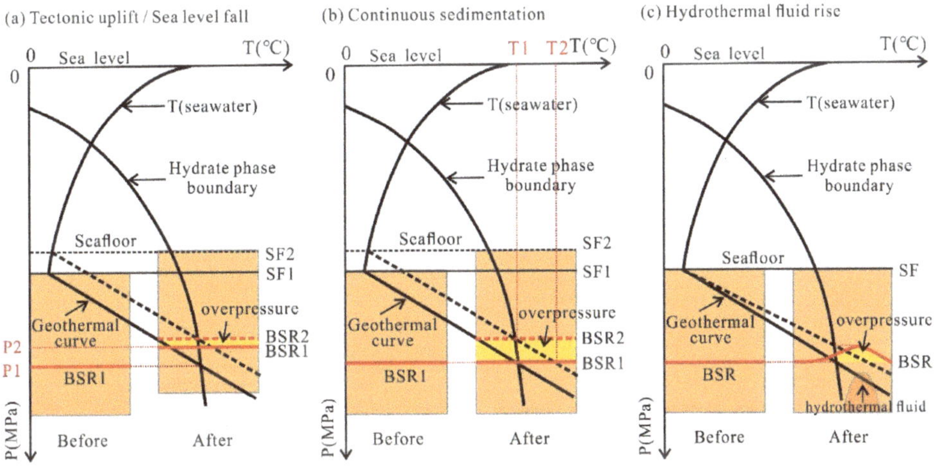

Figure 7. Overpressure development associated with hydrate decomposition due to tectonic uplift/Sea level fall, continuous sedimentation, hydrothermal fluid rise (modified from Foucher et al. [29]). T, temperature; P, pressure; SF, seafloor; BSR, bottom-simulating reflector.

The continuous sedimentation will constantly thicken the sediment section and increase the temperature of the hydrate stability zone. With the thickening of the sedimentary layer, the gas hydrate near the BSR moves downward relative to the seafloor. When it reaches the phase equilibrium boundary, the gas hydrate begins to decompose (Figure 7b), so the continuous sedimentation will lead to the upward migration of the bottom of the hydrate zone. The magnitude of the resultant overpressure is related to the sedimentation rate [28]. However, if the sedimentation rate is so fast that the gas produced by hydrate decomposition has no time to migrate, a new BSR will not form, and the original BSR may become weakened or even disappear [99].

The thermal effect of the hydrothermal fluid, which is migrated from deep in the basin through faults, mud diapirs, and mud volcanoes, etc., could also cause hydrate decomposition. Site SH5 in the northern South China Sea is a typical case. Although the BSR is clear at site SH5, the drilling results confirm that there is no gas hydrate. That is because the deep hydrothermal fluid invades along the mud diapir and fault channel, increasing the temperature of the hydrate stability zone and leading to the decomposition of the gas hydrate [49,50]. However, if the mud diapir does not penetrate the hydrate stability zone and the thermal effect of mud diapir is not strong enough to break the gas hydrate-bearing layer, the thermal effect may only result in the decomposition of gas hydrate near the BSR. Then, a large amount of gas will be released and accumulate at the top of the mud diapir, and overpressure will develop. Meanwhile, the interface between the gas hydrate-bearing layer and the gas-bearing layer will change due to the local decomposition of gas hydrate. When a new equilibrium state is reached, the BSR above the mud diapir may subsequently bend toward the seafloor (Figure 7c).

As discussed above, hydrate decomposition mainly occurs in a mixed layer of mixed hydrate and natural gas and in hydrate-bearing sediments above the BSR.

4.2. Diagenesis

After the deposition and burial of sediments, dissolved oxidants such as O_2, NO_3^-, SO_4^{2-}, Fe^{3+} and Mn^{4+} may be present in surficial marine sediments. With the involvement of microbes, these oxidants react with the organic matter and migrating hydrocarbons from the deep [101]. As a result, overpressure does not easily develop in sediments above SMI due to the consumption of methane and other hydrocarbons. Below SMI, dehydration of gypsum or montmorillonite and cementation of calcite or quartz can lead to overpressure development.

The reduction in the porosity of clay-rich sediments during diagenesis may be accompanied by slow chemical changes, such as the transformation of smectite into illite [102]. Smectite or swelling clay has been commonly identified. At 65–120 °C, smectite begins to dehydrate and is transformed to illite by the catalysis of potassium feldspar. The bound water in the interlayer of smectite is released into the pore space, resulting in the increase of pore pressure and the decrease of effective stress [8], so smectite dehydration can cause pore fluid expansion and weaken the sediment skeleton. This process causes the pressure originally borne by the rock skeleton to partially transfer to the pore fluid, and such load transfer induced by illitization will also lead to an increase in pore pressure [103].

Because the transformation of gypsum to anhydrite will decrease its volume by about 39%, it is considered to be a crucial factor for overpressure development in evaporite layers. The reaction is mainly controlled by the temperature and pressure of pore fluid and fluid activities in pores. In the temperature range of 40–60 °C, great overpressure could be generated at depth of about 1.0 km [51]. Since this process usually occurs during shallow burial, it will not be a factor for overpressure in deep sediments [14].

It is suggested that the dissolution and cementation of quartz in sandstone and mudstone can also induce overpressure [52,104–106], because dissolution and cementation can significantly reduce the porosity and permeability of sediments. When the fluid is trapped in sediments due to extremely low permeability, overpressure will be generated. As indicated by carbonate concretions, calcite cementation may take place at the beginning of shallow-burial stage, however, generally during a later period in deep-burial stage [4]. Zhao et al. [13] suggested that quartz cementation could locally affect overpressure generation; however, Wangen [107] pointed out that pore space cementation is the most likely mechanism of overpressure development in deep reservoirs.

Shaw and Primmer [52] took illitization and calcite cementation as having the same importance as disequilibrium compaction for the overpressure development in the Tertiary strata off the coast of Texas. The models on fluid flow and pressure development reveal that diagenesis contributes about 25–80% to the overpressure in the Halten Terrace, offshore mid-Norway, and the Gulf of Mexico [53].

Diagenesis can generate overpressure, and in turn overpressure also affects diagenesis. Overpressure may inhibit pressolution and quartz generation in the late diagenesis process [108], restraining the transformation of clay minerals and carbonate cement [109].

4.3. Hydrocarbon Generation

With increasing burial depth and temperature, hydrocarbons can be generated from organic matter in sediments by either biological or chemical processes. Hydrocarbon generation can be divided into four stages. In the first stage, biochemical gas is generated, being part of the shallow-burial diagenetic stage. In this stage, generally the temperature is 0–80 °C, the depth less than 1500–2000 mbsf, and the Ro (vitrinite reflectance) $\leq$0.5%. It will be discussed in detail in the section concerning microbial gas production. The second is the thermo-catalytic oil/gas generation stage, in which the temperature is 80–150 °C, the depth greater than 1500–2000 mbsf, and Ro 0.5–1.3%. The third is the thermo-cracking condensate gas generation stage, with a temperature of 150–200 °C, a depth of more than 3500–4000 mbsf and Ro 1.3–2.0%. The fourth is the deep pyrometric gas generation stage, with a temperature of 200–300 °C, a depth of more than 5000–6000 mbsf and Ro > 2.0%, which belongs to the scope of metamorphism [110].

Within the hydrocarbon generation threshold, either maturated kerogen decomposition or hydrocarbon cracking will generate certain amounts of oil and gas, and such extra fluid in sediment pores will cause the expansion of fluid volume and result in overpressure in a closed system [54]. Although overpressure generation related to oil/gas generation is primarily dependent on kerogen type, organic matter abundance, temperature history, and rock permeability [14], thermal effect is the controlling factor. It should be noted that only when the thick sediments contain a large amount of organic matter and the evolution of kerogen has reached the massive oil and gas generation stage will hydrocarbon generation play a role. It means that the overpressure may only occur in mature enough or overmature, organic, rich source rocks [39]. Hydrocarbon generation can continuously generate pressure in the course of sedimentation and burial, mainly in the deep-burial stage. As a result, overpressure can be sustained over a long period, except that the source rocks are greatly lifted to a much shallower location where the condition is not favorable for oil/gas generation [13].

Hydrocarbon generation is considered to be the main mechanism for overpressure development in many basins and has attracted more and more attention. It has been found that, to a certain degree, hydrocarbon generation contributed to overpressure development in the Beihai basin, Gulf of Mexico basin, Mahakam Delta, Niger Delta, Malay basin, and Carnarvon basin in Western Australia [13], and Hunt et al. [55] suggested that gas generation is the main cause of overpressure in the Gulf of Mexico. It was overpressure, induced by hydrocarbon generation, that induced the detachment of Cretaceous rocks in the Amazon sedimentary fan [57]. Logging data in the east of the Krishna-Godavari basin off India reveals that the abnormally high pore pressure might be caused by gas generation [56].

5. Overpressure Development Associated with Biological Activity

Biological overpressure development mainly refers to the involvement of microbiological activities in overpressure generation, including microbial gas production and microbial plugging.

Microbial gas production is essentially similar to hydrocarbon generation in the chemical development of overpressure, which can result in the volume expansion of pore fluid as well. Different from thermogenic gas, the gas produced by biological processes is usually called biogenic gas or biogas. Biogenic gas and thermogenic gas are different in molecular and isotopic composition, the former being predominantly composed of methane with carbon isotope ($\delta^{13}C_1$) lighter than -60‰ [58] and hydrogen isotope between -158‰ and -368‰ [59]. The heavy hydrocarbons in biogenic gas are generally less than 2%, or else the gas must be mixed with thermogenic gas [58].

The high levels of molecular biopolymer produced or $CaCO_3$ precipitation induced by microbes have the potential to block sediment pore space or cement the sediments, leading to a reduction in sediment permeability and deterioration of pore fluid migration. These processes will cause microbial plugging and are with the potential to induce local overpressure.

5.1. Microbial Gas Production

Microbial gas production generally occurs in the shallow-burial stage. Except for the participation of microbes during gas production, microbial gas production has the same effect as hydrocarbon generation. In a reductive environment, the organic matter in the shallow sediments can generate methane-rich gas by microbial activity, and the gas generated can accumulate to form a biogas reservoir [59]. In rapidly accumulated marine sediments, microbes commonly employ aerobic respiration in the beginning, and then sulfate reduction becomes the main form of respiration after the oxygen is consumed. The generation and accumulation of methane will play a dominant role when the sulfate in the pore water is exhausted [58,101]. Microbial activities are restricted by the surrounding conditions, such as the temperature, nutrients, pressure, salinity, and acidity of the water. Temperature will be the most important factor if sufficient nutrients are given to the microbes. The suitable temperature is 0–80 °C; outside this range, the microbial activity decreases sharply or even disappears [59]. Therefore, most biogenic gases occur in shallow immature sediments with organic matter.

Biogas reservoirs are widely distributed and generally under-pressurized. That is because biogas reservoirs are usually shallowly buried and the cap sediment layers are poorly consolidated. Therefore, the gas can easily break through the sediments [59,111]. However, a few studies reveal that biogas can also contribute to overpressure. Helal et al. [60] suggest that the origin of the gas chimney in the Baltim area is related to the overpressure in deeper areas. The overpressure was generated by fast burial during the mega deposition of Post-Messinian succession. The hydrocarbon source of the Post-Messinian succession is only biogas, although Pre-Messinian oil and gas are thermogenic. Similarly, most of the methane gases in hydrate-bearing sediments are biogenic in the Shenhu area in the northern South China Sea [112], while gas chimneys and a large number of small-scale fault systems provide vertical and lateral migration channels for gas-bearing fluids [61].

In marine sediments, pore spaces are always water-saturated due to the existence of overlying seawater. The cracks do not develop easily in soft sediments [113], which are conducive to biogas accumulation. At certain pressures and temperatures, biogenic methane and water can combine to form natural gas hydrate when pore water is over saturated by methane [58]. With hydrate formation sealing, microbial gas production could also result in overpressure in the hydrate stability zone.

5.2. Microbial Plugging

The basis of microbial oil recovery technology is that microbial metabolism can produce acid (formic acid, propionic acid, sulfuric acid), organic solvent (alcohol, ester, etc.), biosurfactant (alkanoic acid, lipopeptide, glycolipid, etc.), biopolymers (polymer polysaccharide), and gas (carbon dioxide, hydrogen, methane, etc.), and such products can dissolve and emulsify crude oil, reduce oil viscosity, increase pore pressure, and improve crude oil fluidity [62,114]. The plugging effect of the microbes can be achieved in two ways, by occupying the pore space of sediments via the growth and reproduction of the microbes on the surface of the sediment particles, and by producing biopolymers by metabolism to block the fluid channels. Among them, the biopolymers have a stronger plugging effect [114].

Experimental studies found that microbial plugging can remarkably decrease the permeability of sediments. The biofilm formed during microbial growth can reduce the average permeability of the porous media to 12% [115]. Cheng et al. [116] also reported that microbial growth can reduce permeability by more than 80%, although it may form a relatively stable plugging only when the sediment is with low permeability (less than

350×10^{-3} μm^2). Microbial plugging technology can also help consolidate the soft sediments at the seafloor. Microbes are applied to induce $CaCO_3$ precipitation to cement porous media at high pH and to reduce geological disasters such as submarine landslide and sand production during oil and gas exploitation. Currently, the primary flora studied in the field of biogeotechnics are *Sporosarcina pasteurii* and *Sporosarcina aquimarina* [117,118]. *Sporosarcina pasteurii* is an alkalophilic bacterium with high urease activity isolated from the surface soil, while *Sporosarcina aquimarina* from the estuaries of South Korea is a urease-producing bacterium with high salt tolerance that can survive in seawater. Both of them have the characteristics of producing urease to hydrolyze urea into ammonia and carbon dioxide. Ammonia increases the pH value of the environment, and then induces $CaCO_3$ precipitation [63,119]. Recently, Hata et al. [63] obtained a new urease-producing microbe, *Sporosarcina newyorkensis*, from the pressure cores of the methane hydrate-bearing layer in the Nankai Trough, Japan. The triaxial experimental results demonstrate that this native microorganism can promote $CaCO_3$ precipitation through urea hydrolysis, improving the strength of the seabed sediments, and then benefit the safe production of natural gas hydrate.

According to the experimental results and field application of microbial plugging technology, microbial plugging has great potential in overpressure development in marine environments. The growth and metabolism of some flora in marine sediments may effectively reduce permeability and prevent the migration of pore fluid under certain conditions. Because microbial activity is affected by temperature, salinity, pH, pressure, permeability, and crude oil properties [62,120], the overpressure caused by microbial plugging might only have local impact and probably occurs in the shallow-burial stage. The development conditions, scope, and magnitude of such kinds of overpressure need to be further studied.

After the introduction of the mechanisms of overpressure development in marine sediments, each overpressure mechanism, which is associated with either physical, chemical, or biological processes, is located in certain stage after deposition, as shown in Figure 8.

Overpressure development is often associated with not only a single factor, but several factors can be involved. For example, overpressure in many sediments has been proven to be caused by hydrocarbon generation and clay mineral diagenesis because they occur in close temperature and depth conditions [13]. For another instance, tectonic movement can generate physical overpressure by changing the overburden pressure; meanwhile, it may produce chemical overpressure from hydrate decomposition by changing the temperature and pressure in the gas hydrate occurrence area. Previous studies suggest that, except for disequilibrium compaction and hydrocarbon generation, other mechanisms can only cause insignificant and local overpressure [107,121]. The overpressure developments associated with chemical and biological processes are closely related with temperature, sedimentary history and rate. Therefore, these factors should also be taken into account [30].

To determine the factors of overpressure development, geophysical and geochemical results, regional geological history, structural evolution, and pressure and temperature conditions should be comprehensively considered.

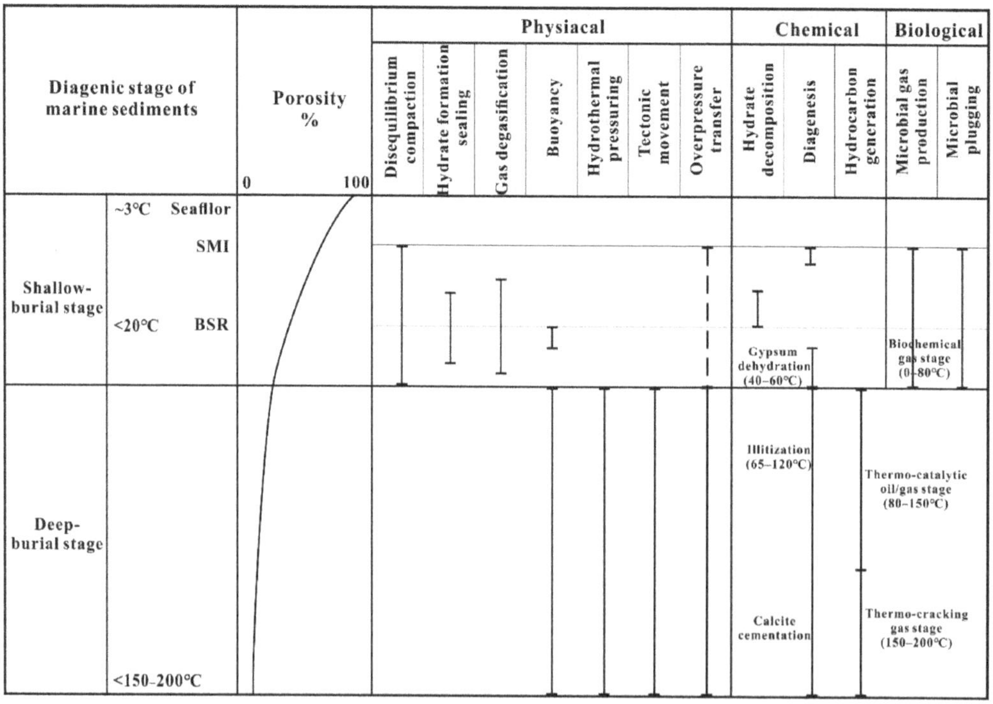

Figure 8. The appearance of each overpressure mechanism in the process after deposition of marine sediments. The solid lines indicate the most likely stage where overpressure appears, and the dotted lines indicate the stage where overpressure is likely to appear. SMI, sulphate methane interface; BSR, bottom-simulating reflector. Not to Scale.

6. Conclusions

As discussed in the previous sections, overpressure development can occur in a certain stage in the marine sediments below SMI. Overpressure developments in marine sediments are classified into physical, chemical, and biological mechanisms. The factors related to physical overpressure development include disequilibrium compaction, hydrate formation sealing, degasification, buoyancy, hydrothermal pressuring, tectonic movement and overpressure transfer; the chemical overpressure mechanisms are generally associated with hydrate decomposition, diagenesis and hydrocarbon generation; and biological overpressure mechanisms are induced mainly by microbial gas production and microbial plugging.

Regarding the physical overpressure mechanisms, (1) disequilibrium compaction is an important factor for overpressure development in most cases because the deposition rate is greater than the drainage rate, which greatly reduces the permeability of the overlying sediments; (2) the overpressure developed below the hydrate layer is caused by gas accumulation below the hydrate-bearing sediment, which has low permeability due to hydrate formation; (3) degasification may yield a great quantity of free gas with the gas solubility decreasing; (4) buoyancy-associated overpressure usually occurs in oil and gas reservoirs due to the different densities of oil, gas, and water. It can also be developed below gas hydrate-bearing layers in marine environments; (5) hydrothermal pressuring is generated due to the difference in the thermal expansion coefficient between the pore fluid and sediment matrix; (6) tectonic movement-related overpressure is mainly caused by the stress change resulting from tectonic compression and shear activities, resulting in strong local overpressure depending on the degree of compaction; (7) overpressure transfer occurs through overpressured fluid activity, which is closely related to fluid channels such as permeable sandstone, fault systems, mud diapirs, and gas chimneys.

About the chemical overpressure mechanisms, (1) a large amount of free gas produced by hydrate decomposition may be sealed by a low-permeability gas hydrate-bearing layer, and they will accumulate to generate overpressure near the BSR. Tectonic uplift or sea-level fall, continuous sedimentation, and hydrothermal fluid rise will induce hydrate decomposition. (2) During the process of diagenesis, illitization and gypsum dehydration will generate a lot of excess pore fluid, whereas cementation will reduce the porosity and permeability of the sediments. Therefore, diagenesis may cause fluid volume expansion and restrain the migration of fluid, forming overpressure. (3) Overpressure associated with hydrocarbon generation mainly occurs in sufficiently mature or overmatured source rocks. Hydrocarbon generation can produce strong and large-scale overpressure, and its effect on overpressure has been widely recognized.

In marine sediments, (1) microbial gas production, accompanied by hydrate formation sealing, can induce overpressure; (2) under certain conditions the growth, reproduction, and metabolism of microbes can cause biological plugging, significantly reducing permeability and leading to overpressure development.

Author Contributions: H.L. conceptualized the study; C.L. wrote the original draft; L.Z. refined the draft text and all authors contributed to the editing and final reviews of the manuscript. All authors have read and agreed to the published version of the manuscript.

Funding: This research received funding from the China Geological Survey (Grant No. DD20221703).

Institutional Review Board Statement: Not applicable.

Informed Consent Statement: Not applicable.

Data Availability Statement: Not applicable.

Acknowledgments: We highly appreciate the two anonymous reviewers for their comments and suggestions, which benefited us greatly in revising this manuscript.

Conflicts of Interest: The authors declare no conflict of interest.

References

1. Balasubramanian, A. *Marine Sediments*; University of Mysore: Mysore, India, 2017.
2. Chester, R. *Marine Sediments*; Springer: Dordrecht, The Netherlands, 1990; pp. 441–467.
3. Encyclopaedia Britannica. *Diagenesis. The Editors of Encyclopaedia Britannica*; Encyclopedia Britannica: Scotland, UK, 2017.
4. Singer, A.; Müller, G. Chapter 3 Diagenesis in Agrillaceous Sediments. *Dev. Sedimentol.* **1979**, *25*, 115–212.
5. Wang, L.; Gu, L.; Lu, H. Sediment permeability change on natural gas hydrate dissociation induced by depressurization—ScienceDirect. *China Geol.* **2020**, *3*, 221–229.
6. Terzaghi, K. *Theoretical Soil Mechanics*; John Wiley and Sons: Hoboken, NJ, USA, 1943.
7. Biot, M.A. General Theory of Three-Dimensional Consolidation. *J. Appl. Phys.* **1941**, *12*, 155–164. [CrossRef]
8. Dutta, N.C. Geopressure prediction using seismic data; current status and the road ahead. *Geophysics* **2002**, *67*, 2012–2041. [CrossRef]
9. Tingay, M.R.P.; Hillis, R.R.; Swarbrick, R.E.; Morley, C.K.; Damit, A.R. Origin of overpressure and pore-pressure prediction in the Baram province, Brunei. *AAPG Bull.* **2009**, *93*, 51–74. [CrossRef]
10. Li, C.; Zhang, L.; Luo, X.; Lei, Y.; Yu, L.; Cheng, M.; Wang, Y.; Wang, Z. Overpressure generation by disequilibrium compaction or hydrocarbon generation in the Paleocene Shahejie Formation in the Chezhen Depression: Insights from logging responses and basin modeling. *Mar. Pet. Geol.* **2021**, *133*, 105258. [CrossRef]
11. Duan, M.D.; Ye, J.R.; Wu, J.F.; Tian, Y.H.; Cui, Y. Prediction of pressure distribution and Formation Mechanism in low exploration area: A case study of the Xihu depression in East China Sea Shelf Basin. *Bull. Geol. Sci. Technol.* **2020**, *39*, 129–139.
12. Bowers, G.L. Detecting high overpressure. *Lead. Edge* **2002**, *21*, 174–177. [CrossRef]
13. Zhao, J.Z.; Li, J.; Xu, Z.Y. Advances in the origin of overpressures in sedimentary basins. *Acta Pet. Sin.* **2017**, *38*, 973–998. [CrossRef]
14. Osborne, M.J.; Swarbrick, R.E. Mechanisms for generating overpressure in sedimentary basins: A Reevaluation. *AAPG Bull.* **1997**, *81*, 1023–1041.
15. Luo, X.R.; Yang, J.H.; Wang, Z.F. The Overpressure Mechanisim in Aquifers and Pressure Prediction in Basins. *Geol. Rev.* **2000**, *46*, 22–31.

16. Hart, B.S.; Flemings, P.B.; Deshpande, A. Porosity and pressure: Role of compaction disequilibrium in the development of geopressures in a Gulf Coast Pleistocene basin. *Geology* **1995**, *23*, 45–48. [CrossRef]
17. Wan, Z.F.; Xia, B.; Lin, G.; Li, J.T.; Liu, B.M. Hydrocarbon Accumulation Model for Overpressure Basin: An Example from the Yinggehai Basin. *MARINE Geol. Quat. Geol.* **2010**, *30*, 91–97.
18. Liu, T.; Lu, Y.; Zhou, L.; Yang, X.; Guo, L. Experiment and Analysis of Submarine Landslide Model Caused by Elevated Pore Pressure. *J. Mar. Sci. Eng.* **2019**, *7*, 146. [CrossRef]
19. Flemings, P.B.; Stump, B.B.; Finkbeiner, T.; Zoback, M. Flow focusing in overpressured sandstones: Theory, observations, and applications. *Am. J. Sci.* **2002**, *302*, 827. [CrossRef]
20. Zhang, Y.; Wang, W.; Zhang, P.; Li, G.; Tian, S.; Lu, J.; Zhang, B. A Solution to Sand Production from Natural Gas Hydrate Deposits with Radial Wells: Combined Gravel Packing and Sand Screen. *J. Mar. Sci. Eng.* **2022**, *10*, 71. [CrossRef]
21. Alberty, M. Cost analysis of SWF preventative, remedial measures in deepwater drilling. *Offshore* **2000**, *60*, 58–64.
22. Hardage, B.A.; Remington, R. Gas hydrate-a source of shallow water flow? *Lead. Edge* **2006**, *25*, 634–635. [CrossRef]
23. Ostermeier, R.M.; Pelletier, J.H.; Winker, C.D.; Nicholson, J.W.; Rambow, F.H.; Cowan, K.M. Dealing with shallow-water flow in the deepwater Gulf of Mexico. In Proceedings of the Offshore Technology Conference One Petro, Houston, TX, USA, 1 May 2000.
24. Kvenvolden, K.A. Gas hydrates—Geological perspective and global change. *Rev. Geophys.* **1993**, *31*, 173–187. [CrossRef]
25. Makogon, Y.F.; Holditch, S.A.; Makogon, T.Y. Natural gas-hydrates—A potential energy source for the 21st Century. *J. Pet. Sci. Eng.* **2007**, *56*, 14–31. [CrossRef]
26. Boswell, R.; Collett, T.S. Current perspectives on gas hydrate resources. *Energy Environ. Sci.* **2011**, *4*, 1206–1215. [CrossRef]
27. Delli, M.L.; Grozic, J.L.H. Experimental determination of permeability of porous media in the presence of gas hydrates. *J. Pet. Sci. Eng.* **2014**, *120*, 1–9. [CrossRef]
28. Xu, W. Excess pore pressure resulting from methane hydrate dissociation in marine sediments: A theoretical approach. *J. Geophys. Res. Solid Earth* **2006**, *111*, B01104. [CrossRef]
29. Foucher, J.-P.; Nouzé, H.; Henry, P. Observation and tentative interpretation of a double BSR on the Nankai slope. *Mar. Geol.* **2002**, *187*, 161–175. [CrossRef]
30. Dasgupta, T.; Mukherjee, S. Pore Pressure Determination Methods. *Adv. Oil Gas Explor. Prod.* **2020**, 19–30.
31. Xie, Y.; Wu, T.; Sun, J.; Zhang, H.; Wang, J.; Gao, J.; Chen, C. Sediment compaction and pore pressure prediction in deepwater basin of the South China Sea: Estimation from ODP and IODP drilling well data. *J. Ocean. Univ. China* **2018**, *17*, 25–34. [CrossRef]
32. Sen, S.; Kundan, A.; Kumar, M. Modeling Pore Pressure, Fracture Pressure and Collapse Pressure Gradients in Offshore Panna, Western India: Implications for Drilling and Wellbore Stability. *Nat. Resour. Res.* **2020**, *29*, 2717–2734. [CrossRef]
33. Singha, D.K.; Chatterjee, R.; Sen, M.K.; Sain, K. Pore pressure prediction in gas-hydrate bearing sediments of Krishna–Godavari basin, India. *Mar. Geol.* **2014**, *357*, 1–11. [CrossRef]
34. Singha, D.K.; Shukla, P.K.; Chatterjee, R.; Sain, K. Multi-channel 2D seismic constraints on pore pressure- and vertical stress-related gas hydrate in the deep offshore of the Mahanadi Basin, India. *J. Asian Earth Sci.* **2019**, *180*, 103882. [CrossRef]
35. Duan, Z.; Møller, N.; Greenberg, J.; Weare, J.H. The prediction of methane solubility in natural waters to high ionic strength from 0 to 250 °C and from 0 to 1600 bar. *Geochim. Cosmochim. Acta* **1992**, *56*, 1451–1460. [CrossRef]
36. Xie, Y.; Li, X.; Tong, C.; Liu, P.; Wu, H.; Huang, Z. High temperature and high pressure gas enrichment condition, distribution law and accumulation model in central diapir zone of Yinggehai basin. *China Offshore Oil Gas.* **2015**, *27*, 1–12.
37. Herkommer, M.A. How gas buoyancy creates shallow-zone geopressures. *World Oil.* **1993**, *214*, 63.
38. Zhang, J. Pore pressure prediction from well logs: Methods, modifications, and new approaches. *Earth-Sci. Rev.* **2011**, *108*, 50–63. [CrossRef]
39. Gao, G.; Huang, Z.L.; Wang, Z.F.; Quan, Y. Study on the mechanisms of the formation of formation abnormal high-pressure. *J. Xi'Shiyou Univ. (Nat. Sci. Ed.)* **2005**, *20*, 1–7.
40. Barker, C. Aquathermal Pressuring; Role of Temperature in Development of Abnormal-Pressure Zones. *AAPG Bull.* **1972**, *56*, 2068–2071.
41. Magara, K. Importance of Aquathermal Pressuring Effect in Gulf Coast: GEOLOGIC NOTE. *AAPG Bull.* **1975**, *59*, 2037–2045.
42. Sharp, J.M., Jr.; Kort, M.H. Numerical Model of Shale Compaction, Aquathermal Pressuring, and Hydraulic Fracturing: ABSTRACT. *AAPG Bull.* **1982**, *67*, 547.
43. Dai, J.X.; Wang, T.B.; Song, Y.; Zhang, H.N.; Xu, Y.C.; Zhang, Q.M. *Formation Conditions and Distribution Law of Large and Medium-Sized Natural Gas Fields in China*; Geology Press: Beijing, China, 1997.
44. Luo, X.R. Quantitative Analysis on Overpressuring Mechanism Resulted from Tectonic Stress. *Chin. J. Geophys.* **2004**, *47*, 1086–1093. [CrossRef]
45. Moore, J.C.; Shipley, T.H.; Goldberg, D.; Ogawa, Y.; Filice, F.; Fisher, A.; Jurado, M.J.; Moore, G.F.; Rabaute, A.; Yin, H. Abnormal fluid pressures and fault-zone dilation in the Barbados accretionary prism: Evidence from logging while drilling. *Geology* **1995**, *23*, 605. [CrossRef]
46. Dugan, B.; Flemings, P.B. Overpressure and Fluid Flow in the New Jersey Continental Slope: Implications for Slope Failure and Cold Seeps. *Science* **2000**, *289*, 288–291. [CrossRef]
47. Yin, X.L.; Li, S.T.; Yang, J.H.; Zhang, Q.M. Correlations between Overpressure Fluid Activity and Fault System in Yinggehai Basin. *Acta Geosci. Sin.* **2002**, 23, 141–146.

48. Han, S.; Bangs, N.L.; Hornbach, M.J.; Pecher, I.A.; Tobin, H.J.; Silver, E.A. The many double BSRs across the northern Hikurangi margin and their implications for subduction processes. *Earth Planet. Sci. Lett.* **2021**, *558*, 116743. [CrossRef]
49. Yang, X.; Zhong, S.; Wan, Z. The Thermodynamics of Mud Diapir/Volcano Fluid and Its Influence on Gas Hydrate Occurrence. *Mar. Geol. Front.* **2018**, *34*, 15–23.
50. Wang, L.; Sha, Z.; Liang, J.; Lu, J. Analysis of Gas Hydrate Absence Induced by the Late-stage Diapir Domination in the Borehole SH5 of Shenhu Area. *Geoscience* **2010**, *24*, 450–456.
51. Jowett, E.C.; Cathles, L.M.; Davis, B.W. Predicting depths of gypsum dehydration in evaporitic sedimentary basins. *AAPG Bull.* **1993**, *77*, 402–413.
52. Shaw, H.F.; Primmer, T.J. Diagenesis in shales from a partly overpressured sequence in the Gulf Coast, Texas, USA. *Mar. Pet. Geol.* **1989**, *6*, 121–128. [CrossRef]
53. Helset, H.M.; Lander, R.H.; Matthews, J.C.; Reemst, P.; Bonnell, L.M.; Frette, I. The role of diagenesis in the formation of fluid overpressures in clastic rocks. *Nor. Pet. Soc. Spec. Publ.* **2002**, *11*, 37–50.
54. Meissner, F.F. Petroleum Geology of the Bakken Formation Williston Basin, North Dakota and Montana. In Proceedings of the Montana Geological Society: Twenty-Fourth Annual Conference: 1978 Williston Basin Symposium: The Economic Geology of Williston Basin, Denver, CO, USA, 24–27 September 1978.
55. Hunt, J.M.; Whelan, J.K.; Eglinton, L.B.; Cathles, L.M.I. Gas generation—A major cause of deep Gulf Coast overpressures. *Oil Gas. J.* **1994**, *92*, 59–63.
56. Dasgupta, S.; Chatterjee, R.; Mohanty, S.P. Magnitude, mechanisms, and prediction of abnormal pore pressure using well data in the Krishna-Godavari Basin, east coast of India. *AAPG Bull.* **2016**, *100*, 1833–1855. [CrossRef]
57. Cobbold, P.R.; Mourgues, R.; Boyd, K. Mechanism of thin-skinned detachment in the Amazon Fan: Assessing the importance of fluid overpressure and hydrocarbon generation. *Mar. Pet. Geol.* **2004**, *21*, 1013–1025. [CrossRef]
58. Rice, D.D.; Claypool, G.E. Generation, Accumulation, and Resource Potential of Biogenic Gas. *AAPG Bull.* **1981**, *65*, 5–25.
59. Li, X.; Zhang, S.; Zhu, G.; Liang, Y. Types and Research Direction of Biogenic Gas in China. *Nat. Gas Geosci.* **2005**, *16*, 477–484.
60. Helal, E.N.; Lala, A.; Ahmed, S.; Amr, T. The impact of gas chimneys on the reservoir characteristics, offshore Nile Delta, Egypt. *Arab J. Geosci* **2015**, *8*, 7929–7939. [CrossRef]
61. Su, M.; Yang, R.; Wu, N.; Wang, H.; Liang, J.; Sha, Z.; Cong, X.; Qiao, S. Structural Characteristics in the Shenhu Area, Northern Continental Slope of South China Sea, and Their Influences on Gas Hydrate. *Acta Geol Sin. Engl.* **2014**, *88*, 318–326.
62. Dou, Q.L.; Chen, J.F.; Wang, J.; Zhang, D.W. Advances in researches and outlook for microbial enhanced oil recovery. *Nat. Gas. Geosci.* **2004**, *15*, 559–563.
63. Hata, T.; Saracho, A.C.; Haigh, S.K.; Yoneda, J.; Yamamoto, K. Microbial-induced carbonate precipitation applicability with the Methane Hydrate-bearing layer microbe. *J. Nat. Gas. Sci. Eng.* **2020**, *81*, 103490. [CrossRef]
64. Li, C.; Luo, X.; Zhang, L.; Wang, B.; Lei, Y. Overpressure Generation Mechanisms and Its Distribution in the Paleocene Shahejie Formation in the Linnan Sag, Huimin Depression, Eastern China. *Energies* **2019**, *12*, 3183. [CrossRef]
65. Mann, D.M.; Mackenzie, A.S. Prediction of pore fluid pressures in sedimentary basins. *Mar. Pet. Geol.* **1990**, *7*, 55–65. [CrossRef]
66. Kida, M.; Khlystov, O.; Zemskaya, T.; Takahashi, N.; Minami, H.; Sakagami, H.; Krylov, A.; Hachikubo, A.; Yamashita, S.; Shoji, H. Coexistence of structure I and II gas hydrates in Lake Baikal suggesting gas sources from microbial and thermogenic origin. *Geophys. Res. Lett.* **2006**, *33*, L24603. [CrossRef]
67. Pohlman, J.W.; Canuel, E.A.; Chapman, N.R.; Spence, G.D.; Coffin, R.B. The origin of thermogenic gas hydrates on the northern Cascadia Margin as inferred from isotopic (13C/12C and D/H) and molecular composition of hydrate and vent gas. *Org. Geochem.* **2005**, *36*, 703–716. [CrossRef]
68. Davie, M.K.; Zatsepina, O.Y.; Buffett, B. Methane solubility in marine hydrate environments. *Mar. Geol.* **2004**, *203*, 177–184. [CrossRef]
69. Lu, H.; Lorenson, T.D.; Moudrakovski, I.L.; Ripmeester, J.A.; Collett, T.S.; Hunter, R.B.; Ratcliffe, C.I. The characteristics of gas hydrates recovered from the Mount Elbert gas hydrate stratigraphic test well, Alaska North Slope. *Mar. Pet. Geol.* **2011**, *28*, 411–418. [CrossRef]
70. Kida, M.; Suzuki, K.; Kawaraura, T.; Oyama, H.; Nagao, J.; Ebinuma, T.; Narita, H.; Suzuki, H.; Sakagami, H.; Takahashi, N. Characteristics of Natural Gas Hydrates Occurring in Pore-Spaces of Marine Sediments Collected from the Eastern Nankai Trough, off Japan. *Energy Fuel* **2009**, *23*, 5580–5586. [CrossRef]
71. Kumar, A.; Maini, B.; Bishnoi, P.R.; Clarke, M.; Zatsepina, O.; Srinivasan, S. Experimental determination of permeability in the presence of hydrates and its effect on the dissociation characteristics of gas hydrates in porous media. *J. Pet. Sci. Eng.* **2010**, *70*, 114–122. [CrossRef]
72. Song, Y.C.; Huang, X.; Liu, Y.; Yang, M.J. Experimental study of permeability of porous medium containing methane hydrate. *J. Therm. Sci. Technol.* **2010**, *9*, 51–57.
73. Yamawaki, M.; Takeyama, N.; Katsumura, Y. Numerical study on permeability hysteresis during hydrate dissociation in hot water injection. *Jpn. J. Appl. Phys.* **2008**, *47*, 1104–1109. [CrossRef]
74. Li, C.-H.; Zhao, Q.; Xu, H.-J.; Feng, K.; Liu, X.-W. Relation between relative permeability and hydrate saturation in Shenhu area, South China Sea. *Appl. Geophys.* **2014**, *11*, 207–214. [CrossRef]
75. Konno, Y.; Yoneda, J.; Egawa, K.; Ito, T.; Jin, Y.; Kida, M.; Suzuki, K.; Fujii, T.; Nagao, J. Permeability of sediment cores from methane hydrate deposit in the Eastern Nankai Trough. *Mar. Pet. Geol.* **2015**, 487–495. [CrossRef]

76. Katoh, A.; Nakayama, K.; Baba, K.; Uchida, T. Model simulation for generation and migration of methane hydrate. *Energy Explor. Exploit.* **2000**, *18*, 401–421. [CrossRef]
77. Posewang, J.; Mienert, J. The enigma of double BSRs: Indicators for changes in the hydrate stability field? *Geo-Mar. Lett.* **1999**, *19*, 157–163. [CrossRef]
78. Andreassen, K.; Hart, P.E.; Grantz, A. Seismic studies of a bottom simulating reflection related to gas hydrate beneath the continental margin of the Beaufort Sea. *J. Geophys. Res. Solid Earth* **1995**, *100*, 12659–12673. [CrossRef]
79. Zhang, W.; Liang, J.Q.; Su, P.B.; Wang, L.F.; Lin, L.; Huang, W.A.; Wei, J.G.; Liang, J. Research progress and prospect of relationship between double bottom simulating reflector and the accumulation of gas hydrates. *Geol. China* **2020**, *47*, 29–42.
80. Duan, Z.H.; Wei, Q. Model for the Calculation of the Solubility of CH_4, H_2S and CO_2 in Aqueous Solutions. *Acta Geol. Sin. Engl.* **2011**, *85*, 1079–1093.
81. Wang, M.Z.; Niu, Y.B.; Wang, B.Y. Discussion about Relationship between Water′s Salinity and Methane′s Solubility. *Coal Technol.* **2016**, *35*, 178–180.
82. Lu, W.; Chou, I.M.; Burruss, R.C. Determination of methane concentrations in water in equilibrium with sI methane hydrate in the absence of a vapor phase by in situ Raman spectroscopy. *Geochim. Cosmochim. Acta* **2008**, *72*, 412–422. [CrossRef]
83. Rui, S.; Duan, Z. An accurate model to predict the thermodynamic stability of methane hydrate and methane solubility in marine environments. *Chem. Geol.* **2007**, *244*, 248–262.
84. Tsimpanogiannis, I.N.; Economou, I.G.; Stubos, A.K. Methane solubility in aqueous solutions under two-phase (H–Lw) hydrate equilibrium conditions. *Fluid Phase Equilibria* **2014**, *371*, 106–120. [CrossRef]
85. Xu, W. Modeling dynamic marine gas hydrate systems. *Am. Miner.* **2004**, *89*, 1271–1279. [CrossRef]
86. Sahagian, D.L.; Proussevitch, A.A. Bubbles in volcanic systems. *Nature* **1992**, *359*, 485. [CrossRef]
87. Wan, Z.F.; Xia, B.; He, J.X.; Liu, B.M. Analogy analysis on petroleum geological characteristics of the foreland basins between western China and central Asia. *Nat. Gas Geosci.* **2007**, *18*, 219–223.
88. Guo, Z.F.; Zhen, L.; Peng, L.; Liu, W.C. Experimental analysis of aquathermal pressuring under high temperature conditions and its geological implications. *Pet. Geol. Exp.* **2016**, *38*, 836–841.
89. Daines, S.R. Aquathermal pressuring and geopressure evaluation. *AAPG Bull.* **1982**, *66*, 931–939.
90. Sharp, J.M., Jr. Permeability controls on aquathermal pressuring. *AAPG Bull.* **1983**, *67*, 2057–2061.
91. Luo, X.; Vasseur, G. Contributions of compaction and aquathermal pressuring to geopressure and the influence of environmental conditions. *AAPG Bull.* **1992**, *76*, 1550–1559.
92. Shi, X.; Cheng, Y.F.; Mei, W. Method for Formation Pore Pressure Prediction Based on Logging Data. *J. Oil Gas Technol.* **2012**, *34*, 94–98.
93. Yardley, G.S.; Swarbrick, R.E. Lateral transfer: A source of additional overpressure? *Mar. Pet. Geol.* **2000**, *17*, 523–537. [CrossRef]
94. Tingay, M.; Hillis, R.R.; Swarbrick, R.E.; Morley, C.K.; Damit, A.R. 'Vertically transferred' overpressures in Brunei: Evidence for a new mechanism for the formation of high-magnitude overpressure. *Geology* **2007**, *35*, 1023–1026. [CrossRef]
95. Finkbeiner, T.; Zoback, M.; Flemings, P.; Stump, B. Stress, pore pressure, and dynamically constrained hydrocarbon columns in the South Eugene Island 330 Field, northern Gulf of Mexico. *AAPG Bull.* **2001**, *85*, 1007–1031.
96. Ma, G.; Zhan, L.; Lu, H.; Hou, G. Structures in Shallow Marine Sediments Associated with Gas and Fluid Migration. *J. Mar. Sci. Eng.* **2021**, *9*, 396. [CrossRef]
97. Mienert, J.; Posewang, J. Evidence of shallow- and deep-water gas hydrate destabilizations in North Atlantic polar continental margin sediments. *Geo-Mar. Lett.* **1999**, *19*, 143–149. [CrossRef]
98. Xu, W. Phase balance and dynamic equilibrium during formation and dissociation of methane gas hydrate. In Proceedings of the Fourth International Conference on Gas Hydrates, Yokohama, Japan, 19–23 May 2002.
99. Ashi, J.; Tokuyama, H.; Taira, A. Distribution of methane hydrate BSRs and its implication for the prism growth in the Nankai Trough. *Mar. Geol.* **2002**, *187*, 177–191. [CrossRef]
100. Nathan, L.B.B.; Musgrave, R.J.; Tréhu, A.M. Upward shifts in the southern Hydrate Ridge gas hydrate stability zone following postglacial warming, offshore Oregon. *J. Geophys. Res. Solid Earth* **2005**, *110*, B03102.
101. Yang, H.; Yu, S.; Lu, H. Iron-coupled anaerobic oxidation of methane in marine sediments: A review. *J. Mar. Sci. Eng.* **2021**, *9*, 875. [CrossRef]
102. Weaver, C.E. Potassium, illite and the ocean. *Geochim. Cosmochim. Acta* **1967**, *31*, 2181–2196. [CrossRef]
103. Lahann, R.W.; Swarbrick, R.E. Overpressure generation by load transfer following shale framework weakening due to smectite diagenesis. *Geofluids* **2011**, *11*, 362–375. [CrossRef]
104. Bols, H.; Hermanrud, C.; Teige, G. Origin of overpressures in shales: Constraints from basin modeling. *AAPG Bull.* **2004**, *88*, 193–211.
105. Vidar, S.; Ivar, B. Identifying time, temperature, and mineralogical effects on chemical compaction in shales by rock physics relations. *Lead. Edge* **2008**, *27*, 750–756.
106. Chilingar, G.V.; Robertsonjr, J.; Riekeiii, H. Chapter 2 Origin of abnormal formation pressures. *Dev. Pet. Sci.* **2002**, *50*, 21–67.
107. Wangen, M. A quantitative comparison of some mechanisms generating overpressure in sedimentary basins. *Tectonophysics* **2001**, *334*, 211–234. [CrossRef]
108. Osborne, M.J.; Swarbrick, R.E. How Overpressure and Diagenesis Interact in Sedimentary Basins—Consequences for Porosity Preservation in HPHT Reservoir Sandstones. *Indones. Pet. Assoc.* **1997**, 947–954.

109. Ma, Y.; Huang, Y.; Yao, G.; Tao, C.; Pan, S. Effecting of Overpressure on Diagenesis of Huangliu Formation in DX Area, Yinggehai Basin. *Geol. Sci. Technol. Inf.* **2015**, *34*, 7–14.
110. Tissot, B.P.; Welte, D.H. *From Kerogen to Petroleum. in Petroleum Formation and Occurrence*; Springer: Berlin/Heidelberg, Germany, 1984; pp. 160–198.
111. Shurr, G.W.; Ridgley, J.L. Unconventional shallow biogenic gas systems. *AAPG Bull.* **2002**, *86*, 1939–1969.
112. Wu, N.; Zhang, H.; Yang, S.; Zhang, G.; Liang, J.; Lu, J.A.; Su, X.; Schultheiss, P.; Holland, M.; Zhu, Y. Gas Hydrate System of Shenhu Area, Northern South China Sea: Geochemical Results. *J. Geol. Res.* **2011**, *2011*, 1–10. [CrossRef]
113. Dai, C.-X.; Zhang, Q.-F.; He, S.-H.; Zhang, A.; Shan, H.-F.; Xia, T.-D. Variation in Micro-Pores during Dynamic Consolidation and Compression of Soft Marine Soil. *J. Mar. Sci. Eng.* **2021**, *9*, 750. [CrossRef]
114. Wang, W. Laboratory research and field trials of microbial oil recovery technique. *Oil Drill. Prod. Technol.* **2012**, *34*, 107–113.
115. Yang, J.; Ye, S.J.; Wu, J.C. Study on the influence of bioclogging on permeability of saturated porous media by experiments and models. *Environ. Sci.* **2011**, *32*, 1364–1371.
116. Cheng, H.Y.; Wang, L.; Zhang, J.; Zhang, R.Y. Impacts of in situ microorganism growth and multiplication on permeability in porous medium. *Spec. Oil Gas Reserv.* **2010**, *17*, 98–101.
117. Bang, S.S.; Galinat, J.K.; Ramakrishnan, V. Calcite precipitation induced by polyurethane-immobilized Bacillus pasteurii. *Enzyme Microb. Technol.* **2001**, *28*, 404–409. [CrossRef]
118. Dejong, J.T.; Soga, K.; Kavazanjian, E.; Burns, S.; Paassen, L.A.V.; Qabany, A.A.; Aydilek, A.; Bang, S.S.; Burbank, M.; Caslake, L.F.; et al. Biogeochemical processes and geotechnical applications: Progress, opportunities and challenges. *Geotechnique* **2013**, *63*, 287–301. [CrossRef]
119. Wei, Z.; Xu, T.; Shang, S.; Tian, H.; Cao, Y.; Wang, J.; Shi, Z.; Liu, X. Laboratory Experimental Study on the Formation of Authigenic Carbonates Induced by Microbes in Marine Sediments. *J. Mar. Sci. Eng.* **2021**, *9*, 479. [CrossRef]
120. Bao, M.T.; Yuan, S.W.; Li, X.M.; Song, Z.Y.; Guo, L.Y. Effects of permeability of porous medium on the growth and metabolism of microorganism in reservoir. *J. Shenzhen Univ. Sci. Eng.* **2011**, *28*, 35–40.
121. Swarbrick, R.E.; Osborne, M.J.; Yardley, G.S. Comparison of overpressure magnitude resulting from the main generating mechanisms. *AAPG Mem.* **2002**, *76*, 1–12.

Journal of
Marine Science and Engineering

MDPI

Review

Organic Matter Burial in Deep-Sea Fans: A Depositional Process-Based Perspective

Arif Hussain * and Khalid Al-Ramadan

Department of Geosciences, College of Petroleum Engineering and Geosciences (CPG), King Fahd University of Petroleum and Minerals, Dhahran 31261, Saudi Arabia; ramadank@kfupm.edu.sa
* Correspondence: arif.hussain@kfupm.edu.sa

Abstract: Organic matter burial in the deep-sea fan sediments is an important component of the long-term carbon cycle. Although there is increasing recognition of the importance of organic matter in deep-sea sediments, a major focus has been on mudstones, commonly interpreted as the background sediments, deposited by pelagic or hemipelagic vertical suspension fallout in low-energy fan environments. Emerging evidence suggests that relatively coarse-grained sediment gravity flow deposits (e.g., turbidites and hybrid event beds) can also store a significant quantity of organic carbon, implying that a wide range of depositional processes can result in the concentration and enrichment of organic matter in submarine fans. However, the role of these processes on carbon burial is still not fully understood. This review aims to discuss the impact of three widely documented deep-sea depositional mechanisms/processes, namely vertical suspension settling, grain-by-grain (incremental aggradation), and the en-masse deposition on distribution, burial, and preservation of organic matter in deep-marine deposits. Organic matter accumulated from slowly settling suspension in mud caps (Te or H5 divisions of turbidites and hybrid beds, respectively) is prone to higher oxidation compared to the carbon buried in sandy components of turbidity currents (Ta-Tc units) and hybrid beds (H2/H3 divisions). The burial of organic matter in sandy parts of the deposits has important implications for understanding the fundamental physical processes that control carbon accumulation and preservation in deep-marine rock record.

Keywords: organic carbon; deep-sea fans; carbon burial; turbidity current; suspension settling; hybrid event bed; mudstone; linked debrite

Citation: Hussain, A.; Al-Ramadan, K. Organic Matter Burial in Deep-Sea Fans: A Depositional Process-Based Perspective. *J. Mar. Sci. Eng.* **2022**, *10*, 682. https://doi.org/10.3390/jmse10050682

Academic Editors: George Kontakiotis, Assimina Antonarakou and Dmitry A. Ruban

Received: 10 April 2022
Accepted: 30 April 2022
Published: 17 May 2022

1. Introduction

Organic matter (hereafter OM) deposition and burial in deep-sea fans is an important component of the long-term carbon cycle [1–13] and represents the second largest global sink (~170 megatonnes of carbon) of atmospheric carbon dioxide after the combined processes of silicate weathering and carbonate precipitation [2]. The amount of carbon buried in these sediments can be stored over geologic timescales, thereby removing atmospheric carbon dioxide and regulating long-term climatic trends [14–18].

Off-shelf transport and ultimate burial of large volumes of terrestrial OM in different parts of the deep-sea fans is commonly attributed to a range of sediment gravity flow and suspension settling processes [1–13]. Traditionally, sediment gravity flows have been subdivided into two end-member flow states, namely debris flows and turbidity currents. Debris flows are characterized by laminar rheology with particle support by a combination of matrix strength, buoyancy and grain-to-grain interaction [19–22], whereas turbidity currents are turbid flows of variable concentration but with significant particle support by turbulence [23–27]. Several earlier flow classifications schemes have been proposed to capture the spectrum extending from cohesive debris flows through to dilute turbidity currents [27–30].

Whilst most of the flow classification schemes envisage discrete flow states, albeit encompassing a spectrum (defined by the degree of turbulence and type of particle support mechanism), the definition of intervening flow types that exhibit mixed behavior (part turbidity current and part debris flow in the same event) has proven less satisfactory. This is despite the fact that mixed flow deposits are widespread in many deep-water basins [6,31–33]. The mixed flow deposits were initially reported from the Aberystwyth Grits on the Welsh coast and were termed 'slurry' beds [34]. Subsequently they were encountered in the Cretaceous Britannia system in the North Sea [35–37]. These deposits with alternating clay-poor and clay-rich sandstones in the Britannia system were attributed to episodic gelation of mud [36] in near-bed layers (i.e., neither laminar nor turbulent but somewhere in between) or cohesion-dominated traction carpets [35]. A recurring facies association in the fringes of several sand-prone, syn-rift Jurassic submarine fans in the North Sea Basins was identified by Haughton et al [31]. These authors documented that sandstone beds displaying full or partial Bouma sequences were surprisingly rare in these systems [31]. Instead, the event beds commonly comprise a basal, dewatered and/or structureless sandstone (H1), capped by a muddy sandstone or sandy mudstone with scattered mudstone clasts and carbonaceous matter, and complex internal fabrics including sheared sand injections (H3 division). These beds were interpreted as 'linked debrites' formed by distal flow transformations [31], prefixed 'linked' to emphasize that the debrites were accumulated as part of the same event that emplaced the sand, as opposed to a separate or standalone debris flow. The widespread recognition of analogous deposits in deep-water fan successions has demonstrated that schemes that place deposits into separate end-member flow states (turbidites and debrites) fail to capture the variability observed in deep-water deposits and a single flow can have zones with very different rheology [31–54]. The flow deposits showing evidence of deposition from turbulent, transitional and laminar flow, all as part of the same event, have been termed "hybrid event beds (HEBs)" [32]. An alternative classification scheme (Figure 1) was thus proposed that distinguishes the deposits of flows that were either cohesive (debris flows, mudflows), non-cohesive (low- and high-density turbidity currents, cohesionless debris flows) or show evidence for mixed behavior at a point (hybrid flows and other types of linked event bed, e.g., debrite–turbidite couplets commonly forming megabeds) [32].

The fine-grained muddy material including OM carried by the sediment gravity currents disperses above and travels beyond the final deposit as a low-density wake and mixes with suspended material in the water column, and slowly accumulates from suspension settling [55]. Periods representing quiescence between the sediment gravity flow events are also characterized by the deposition of background mudstones from suspension settling [56–58]; however, these mudstones are not part of the sediment gravity flow events and hence their role in OM burial and preservation is not discussed here. Traditionally, deep-marine mudstones have been commonly interpreted as the background sediments, deposited by pelagic or hemipelagic vertical suspension fallout in low-energy fan environments [51,56,57], but recent scientific evidence suggest that these deposits may be formed by a range of depositional processes including relatively energetic turbidity currents and transitional flows [57–62]. The observation of the drilled cores in the Laingsburg depocentre, Karoo Basin, South Africa has allowed us to document the variability in depositional processes and stacking patterns between slope and basin-floor mudstones within the same basin margin succession [63]. Basin-floor mudstones exhibited a repeated and predictable alternation of bedsets and massive packages and were interpreted to be deposited by a wide range of low-density turbidity currents (waning, waxing, waxing-to-waning, multi-pulsed), and occasional transitional flows. The slope mudstones were characterized by a higher proportion of low-density turbidites compared to debrites with a less predictable facies stacking pattern, and a higher degree of bioturbation that may affect OM preservation. Hence, understanding facies and related depositional process variability within deep-water mudstone successions is critical to constrain preservation of OM in these rocks.

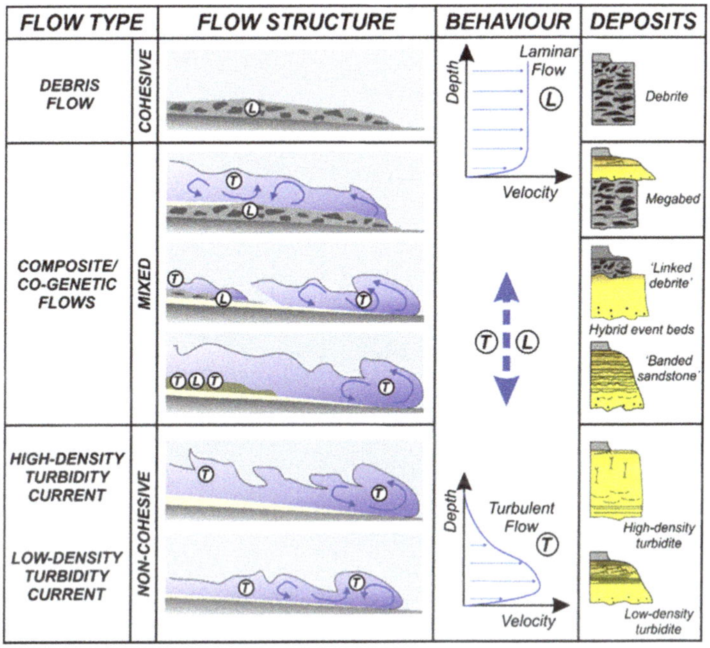

Figure 1. Haughton et al. (2009) classification scheme for event beds emplaced by a wide variety of subaqueous sediment gravity flows emphasizing that many of the deposits are composite or transitional in character, lying between end-member non-cohesive (turbulent) and cohesive (laminar) flow states. Reprinted/adapted with permission from Ref. [32], 2009, Elsevier.

Particulate OM in deep-sea fans comprises heterogeneous mixtures of organic particles from different sources [64–66]. It includes fragments of terrestrial plants such as leaves, wood, spores and pollen grains [1–13,67], remains of marine algae/bacteria referred to as modern carbon [2,68], and a portion of recycled and aged carbon stored in soils and sediments for thousands of years, and referred to as old carbon [2,68]. The presence of these different OM varieties leads to a continuum of age and reactivity of OM in deep-marine sedimentary systems [66]. Although the presence of dissolved organic carbon cannot be ruled out, here we only focus on particulate OM because it represents the fraction that is ultimately buried and has the potential to enter the long-term carbon cycle.

Previous work [1–13,68–70] in deep-marine fan systems has shown that sediment-gravity flows are highly efficient at transporting these different types of OM from the continental shelf into the deep-marine settings, where it becomes preferentially concentrated in specific sub-environments such as channels–levees, lobes and basin floor sheets. During downslope transport, only a small proportion of the carbon survives oxidation (between 50–80% of the organic carbon is remineralized) and ultimately becomes sequestered in deep-sea sediments [71]. Despite increasing recognition of the importance of OM in deep-sea sediments, the focus has been on mudstones (turbidite (Te) and hybrid bed (H5) mud caps and hemipelagites) with very limited emphasis on sand-rich facies (Tb/Tc parts of turbidites) and muddy sandstones (H3 divisions of HEBs). Recent evidence [1,10,11] highlights that sandy divisions in turbidites and muddy H3 divisions of HEBs can also store significant concentrations of land-derived organic carbon to the extent that they could act as conventional source rocks and unconventional reservoirs [1,8]. There is also growing evidence that mud can offer enhanced physical protection to the OM and that the surrounding mineral matrix may regulate long-term global preservation of organic carbon [11,68]. Thus, mechanisms that can sort clay and OM together during transport and deposition are potentially going to lead to enhanced OM burial in the stratigraphic record.

A better mechanistic understanding of OM distribution, the thresholds of its segregation between various deep-water sedimentary facies and of constraining the impact of a range of deep-water depositional mechanisms (suspension settling, incremental aggradation, en-masse deposition) on carbon burial and preservation is important to develop models that aim to forecast the fate and dispersal of organic carbon in deep-sea environments. Furthermore, understanding the nature and occurrence of OM burial in ancient deep-marine deposits is critical to understand the mechanisms and efficiency by which OM has been preserved throughout geologic time, and in turn, better constrain models describing global carbon budgets. The aim of this review is to constrain the impact of different deep-sea depositional mechanisms on the distribution, burial and preservation of OM in the stratigraphic record. This study discusses OM distribution data from the literature on both ancient and modern submarine fans in order to assess patterns of OM distribution and preservation in different deep-water event beds and related mudstones. The key questions discussed include: (i) How is terrestrial OM distributed in different types of sediment gravity flow event deposits (turbidite, hybrid event beds) and mudstones; (ii) what are the controls on OM distribution, burial and preservation in the deep-water rock record; and (iii) what are its implications for the development of improved process-based depositional models for carbon burial. Significantly, the depositional mechanisms discussed here define the common facies associations of OM and thus help illustrate the fundamental physical processes that govern its accumulation and preservation in the rock record. Understanding carbon burial in the context of deep-sea depositional processes also has important implications for understanding the source rock and unconventional reservoir potential of submarine fan deposits.

In the following sections of this study, a comprehensive but non-exhaustive scientific review on OM distribution and burial in different deep-marine event deposits is provided. Where required, the study also combines scientific literature data with some analytical data (acquired by the primary author) to better constrain the distribution of OM in studied deposits. The applicability of the concepts inferred from the literature for terrestrial OM burial elsewhere is touched upon, and wider challenges associated with trying to determine the distribution of OM in different deep-water event beds are considered, with the last part synthesizing the overall conclusions of this study. Furthermore, in this paper, we only focus on terrestrial OM fractionation and burial by three widely documented deep-marine depositional processes, i.e., turbidity currents, hybrid and transitional flows, and suspension settling, without providing exhaustive reviews on different deep-water flow processes and their deposits characteristics. Detailed reviews on different deep-water deposits and related depositional processes including those that are not discussed in the present study (e.g., bottom currents) are provided by [32,33,35,55].

2. Organic Matter Deposition and Burial in Submarine Fans

2.1. Suspension Settling

Suspension settling is an important deep-marine depositional mechanism by which fine-grained materials including clay, silt and OM fall slowly to the seafloor [49,51,56,57]. Muds can be deposited from a low-density suspension and left on top of the sandy component of the turbidites as Te division (Figure 2a) or as H5 division in hybrid event beds (Figure 2b); see [23,32], respectively. Background mudstone deposits are also accumulated from the water column via slow vertical settling and are often vertically stacked with sediment gravity flow deposits in cores and outcrops, but they are not considered as part of flow events (Figure 2c). Muds are commonly interpreted to be deposited in low-energy (distal) fan environments under oxic to anoxic conditions. In proximal energetic fan sectors muds may either be very thin or not expressed at all [11,57]. Organic carbon carried by sediment gravity flows is also often sorted with mud in the mudstones because of their platy particle shape, which facilitates slower settling than denser, relatively rounded, coarser clastic grains (such as quartz and feldspar) with which they are transported [72,73]. Conventional models argued that for preservation of OM bottom water anoxia is prerequisite [71]. However, emerging evidence suggests that low-energy conditions and persistent bottom-water anoxia are not necessary pre-requisites for enhanced OM preservation in these units [8,11,60,74]. The important factor to sequester OM is that at least some OM is rapidly buried in the sediments without being completely oxidized (e.g., [75–78]). This is commonly accomplished if large volumes of OM are supplied to the regions of high primary productivity [79] or when OM is rapidly buried and preserved due to high frequency recurrence of sediment gravity flows events [74] or clay-forced turbulence dampening [11]. It was noted by [74] that during deposition of the Whitby Mudstone Formation, large volumes of organic carbon were delivered episodically to the seafloor and the frequency of depositional events was sufficiently high to minimize oxidation of the OM. However, a limited number of studies have documented the impact of rapid deposition on enhanced preservation of OM in mudstones and transitional flow processes in fine-grained deep-water successions.

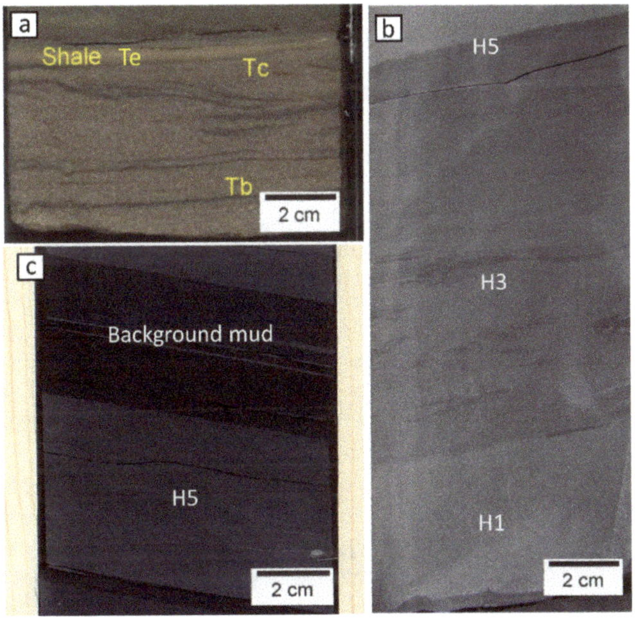

Figure 2. Core photographs showing (**a**) a low-density turbidite with organic-rich (average TOC = 8%) lower sandy divisions (Tb and Tc) and overlain by relatively organic-lean (average TOC = 2%) Te mudstone/shale from a Miocene deep-water succession of Kutei Basin, Indonesia [1]. Reprinted/adapted with permission from Ref. [1], 2006, AAPG; (**b**) Organic-rich H3 division (with 1.24 wt% TOC) of a hybrid event bed overlain by co-genetic H5 mudstone (with 0.6% TOC) from the Ross Sandstone Formation; and (**c**) vertically stacked H5 (0.7 wt% TOC) and background mudstone (0.6 wt% TOC) from the Ross Sandstone Formation, western Ireland.

Deposition of OM via suspension settling could be sufficiently slow and OM is prone to increased bioturbation and oxic degradation [80]. This is evident from the lower TOC content of the deep-marine mudstones compared to their co-genetic or co-eval turbidity current and transitional flow facies in many examples [1,3–5,11]. The average TOC content in sand-rich, parallel laminated Tb sandstones (average TOC = 8%) is significantly higher than co-genetic laminated Te mudstone facies (average TOC up to 2%) in Miocene turbidites from the Kutei Basin, Indonesia (Figure 2a) [1]. Similar TOC trends have been reported in the Ogooué and Congo Fan systems where sandy turbidite facies have significantly higher OM content than their pelagic/hemipelagic muddy counterparts [3–5]. Recently, [11] documented that H5 mudstone division has lower TOC contents (0.6% TOC) than underlying muddy sandstone or H3 division (1.24% TOC) in HEBs from the Pennsylvanian Ross Sandstone Formation, western Ireland (Figure 2b). Higher OM content in deep-water mudstones has been commonly linked to bottom water anoxia ([79,81,82] and references therein). In anoxic conditions, the slowly settling carbon is relatively less prone to oxidation and bacterial degradation. Despite the fact that many deep-marine anoxic mudstones are organic-rich, in some cases these mudstones could be significantly thin and constitute only a minor portion (small percentage) of the entire stratigraphic succession, thereby not sequestrating significant quantities of OM. For example, deep-water mudstones constitute only a small proportion (up to 10%) by thickness compared to HEBs (27%) and turbidites (31%) of the Pennsylvanian Ross Sandstone Formation, western Ireland, an ancient tropical deep-sea fan system [11,51].

2.2. Turbidity Currents

Turbidity currents are turbulent suspensions of sand and mud (with occasional gravel), which are propelled by the downslope component of gravity acting on the excess density. It is one of the most important mechanisms by which OM is transferred from shelf settings to the continental slope and deep-marine fans and basin plain. The deposits of turbidity currents are referred to as 'turbidites' [83]. Refer to Figure 1 for the texture and structure of different turbidity current deposits.

Initial turbidity current models advocated for efficient segregation of the sand and mud in the flows, with the mud deposited as mudstone caps (Figure 3a) [23,25]. Organic carbon carried in largely turbulent flows is often sorted with mud in the mudstone caps (Figure 3a). As discussed in earlier section, OM buried in mudstone caps is prone to oxidation, implying that in many cases the preservation potential of OM in mudstone caps could be relatively low. Prevailing turbidity current models suggest that sandy fractions (Ta-Tc divisions) in turbidites are commonly organically-lean [23,25]. The lower portions of even finer-grained turbidites (Ta divisions) such as those in the Ross Sandstone Formation are organically-lean, although mud-chips and associated OM may preferentially segregate in the lee side of the ripple-laminated divisions sandstones as shown in Figure 3d in [11]. Nevertheless, a limited number of studies have described organic-rich layers in sandy deep-sea turbidites [1,10,84–86], in some cases even within sands of units Ta and Tb (Figure 3b,c), depending on the density and size of organic debris [1,12]. Fine- to very fine-grained Miocene turbidite sandstones in Kutei Basin, Indonesia contain significant quantities of organic carbon (average TOC in Tb division = 8%) in the form of terrestrial leaves and coaly fragments (Figure 3b,c) [1]. The concentration of leaf debris within turbidite sandstones was a function of leaf density, as leaf debris is of slightly higher density and larger size compared to woody fragments, which is why it may have been carried as bedload rather than its common transportation in suspension load by the gravity currents [1]. Woody material has been reported in turbidities from multiple locations such as Gres d'Annot, southeast France and the Marnoso-Arenacea Formation, Italy [12,13], as well as sands in the Bengal Fan in the Indian Ocean [86] and in the Bute Inlet Fjords [10]. Other examples of plant material transported into deep-water marine settings via hyperpycnal flows (direct river discharge) are known from Eocene rocks exposed within the Central Basin of Spitsbergen, Norway [87] and sedimentary basins in Argentina and Trinidad [84]. These studies suggested that one criterion for recognizing hyperpycnal flow turbidites is the presence of continental plant material within the deposits. In many of these examples where organically-rich sandy components have been reported, the sandstones are very fine- to fine-grained with a likelihood of clastic grains being hydrodynamically equivalent to the organic debris, implying that perhaps documenting variation in type, density and shape parameters of the organic material is important to predict OM type zonation in submarine fans [12,13].

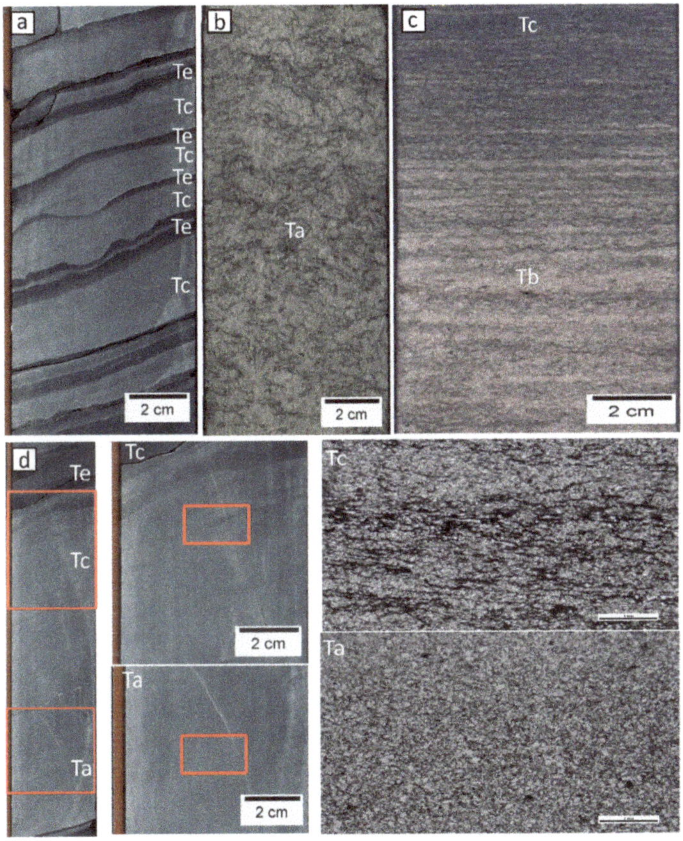

Figure 3. Core photographs showing (**a**) a stack of low-density turbidites with efficient segregation of sand and mud components in the Ross Sandstone Formation, western Ireland; (**b**,**c**) organic-rich Ta (**b**) and Tb-Tc (**c**) divisions in sandy turbidites from the Miocene deep-water succession of Kutei Basin, Indonesia [1]. Reprinted/adapted with permission from Ref. [1], 2006, AAPG; and (**d**) organic-lean sandy Ta and Tc divisions in a turbidite from the Ross Sandstone Formation, western Ireland. In ripple-laminated sandstones, mud chips and minor OM is occasionally fractionated in the lee side of the ripples, as seen in the photomicrograph Tc. Red rectangles on the core photographs show the location of plug samples.

Although a large number of studies [1–6,12,13,62,66,67,88,89] have attempted to characterize and quantify OM contents in both ancient and modern turbidites, they have focused and analyzed the finer grained upper tiers (Td/Te) of the event beds, where OM is commonly thought to be fractionated. These studies may reflect a sampling bias towards more fine-grained lithologies in otherwise sand-rich turbidity current deposits. Furthermore, turbidites commonly dominate in relatively channelized sectors of the fan systems, where these muddy tiers (Td/Te divisions) either constitute a very small part of the turbidity current events, or sometimes are not at all expressed due to fine sediment bypass in the downcurrent direction [90,91]. Where mudstones are present, their accumulation from slow settling suspensions makes them prone to bioturbation and degradation. Moreover, in channelized/proximal settings, the channel edges collapse regularly, supplying turbid flows and redistributing the sediment and OM. This reflects that OM within turbidites may be more prone to oxidation and remineralization due to several cycles of re-suspension [66].

2.3. Hybrid Sediment Gravity Flows

Hybrid and related transitional flows form an enigmatic yet common class of sediment gravity flows and exhibit components of both turbidity currents and debris flows as part of the same event [32]. The resulting deposits typically contain a basal clay-poor turbidite sandstone and an upper matrix-supported linked debrite (Figure 4a–c). High matrix and OM content is one of the key characteristics of HEBs and is commonly linked to sudden changes in original flow behavior. Hybrid flow deposits significantly diverge from classic turbidite models [23–25], and an idealized HEB is comprised of up to five vertically stacked divisions, including a basal, clean, and often graded 'turbidite' sandstone (H1), a banded sandstone (H2) with alternating darker mud-rich and paler mud-poor sandstone bands, a mud-rich sandstone (H3) forming a 'linked debrite' typically with mudstone clasts, mud chips, dispersed mud, and OM. Sometimes H3 divisions can be subdivided into two parts, H3a and H3b, on the basis of differences in textural, mud and OM contents (Figure 4b). For more details on the H3 division sub-structure refer to [53]. A thin well-structured parallel and/or ripple laminated sandstone/siltstone (H4) and a mudstone (H5) couplet caps the event bed. Sometimes the H5 mudstone is capped by a background/hemipelagic mudstone, representing periods of quiescence between the sediment gravity flow events. Hybrid beds are thus complex tiered deposits with different parts of the bed emplaced by distinct depositional mechanisms (incremental aggradation, en-masse deposition, and suspension settling).

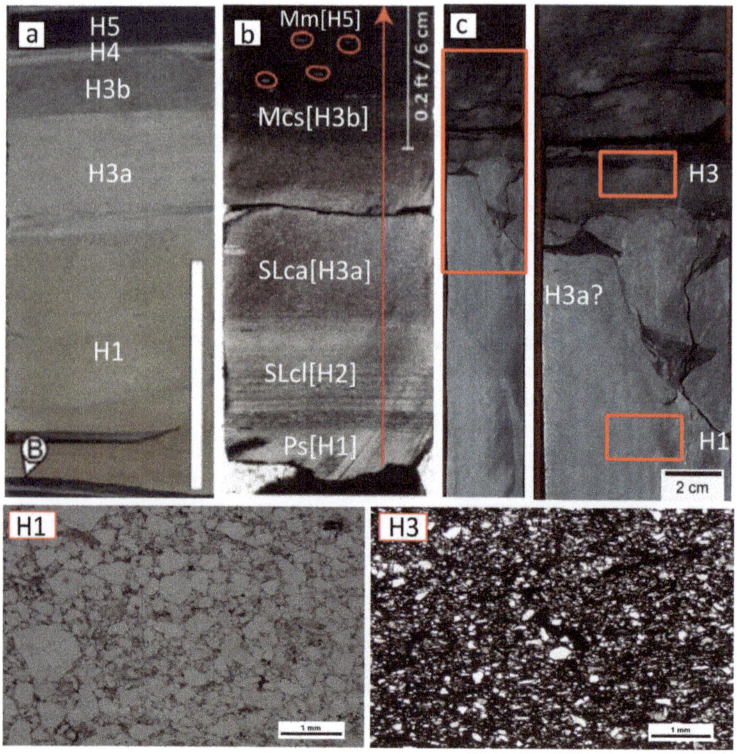

Figure 4. Core photographs showing; (**a**) a type IV hybrid event bed with organic-rich H3b and H5 divisions from the Palaeogene Gulf of Mexico [48]. Reprinted/adapted with permission from Ref. [48], 2012, GSA; (**b**) a mixed siliciclastic–carbonate hybrid event bed with organic-rich muddy

H3b and H5 divisions from the Wolfcamp A Formation in the Permian Basin, USA [8]. Reprinted/adapted with permission from Ref. [9], 2019, AAPG; and (c) a hybrid event bed with organically-lean H1 and organic-rich H3 divisions from the Ross Sandstone Formation, western Ireland. Organic matter is preferentially segregated in H3 divisions in hybrid event beds. Red rectangles on the core photographs show the location of plug samples.

Vertical TOC profiles for a range of HEBs show that H1 divisions are generally organically-lean (average TOC << 1%) and most of the OM is hydrodynamically segregated into H3a and H3b sub-divisions (average TOC > 2%) (Figure 4c) [11]. The co-genetic H5 divisions have either equal or lower TOC contents than their H3 counterparts (refer to Figure 2b). The flows preferentially fractionate OM (on the basis of shape and density) into linked debris flows (top and rear of the flows) as revealed by the occurrence of H3 divisions with significant TOC contents (Figure 4c). A positive correlation between OM and mud content exists in HEBs and was interpreted as a function of both hydraulic segregation and enhanced preservation of OM [11]. Clay flocs were also implicated for encapsulation of the significant quantities of the organic particles at the onset of flocculation and during transport, thereby providing a long-term 'physical protection' mechanism [11]. Similar vertical TOC trends have been reported from mixed siliciclastic–carbonate hybrid event beds in the Permian Wolfcamp A succession (Figure 4b) in the Delaware Basin, USA [8], in which the best unconventional reservoir facies are the clay- and organic-rich and carbonate-depleted uppermost parts of HEBs (their SLca and Mcs facies are equivalent to the H3a and H3b parts described by [11] and others). Thus, predicting areas where thick H3 and particularly H3b divisions are developed can be a part of an effective hydrocarbon exploration and development strategy in the deep-water basins [8]. Although OM is dominantly particulate in nature, the chemical sorption of the organic molecules on clay surfaces cannot be completely ruled out due to the presence of abundant clay. Organic carbon protection by clay minerals is a well-known process in sedimentary rocks [68]. Thus, mud-driven flow transformations in sediment gravity currents provide a particularly effective mechanism for the burial of OM in deep-sea fan systems. HEBs are usually dominant in distal lobes [31,32], and thus abundant linked debrites in the distal parts of many deep-water systems are prone to very limited erosion or reworking by subsequent events. In short, high mud and TOC contents and en-masse deposition all favor enhanced OM burial and preservation in HEBs [11].

The widespread occurrence of beds with linked debrites indicates that these are important components of many deep-water successions [31–33,38–40,42], including those basins actively explored for hydrocarbons [31,32,49,92,93]. In some instances entire or parts of the submarine fans (such as distal lobes) are dominated by HEBs. The Pennsylvanian Ross Sandstone Formation, western Ireland, is a case in point where the basal sandy basin-floor sheet system is dominated (>70%) by HEBs with minimal mudstones [11]. This implies that hybrid sediment gravity flows could be important vectors for transferring terrestrial OM to deep-sea floor repositories where it can be efficiently buried in muddy sandstone and/or sandy mudstone divisions over geological time scales.

3. Discussion

Sediment gravity flows have been reported as important vectors for transferring terrestrial OM to deep-sea floor repositories and emerging evidence suggests that deep-marine event beds may incorporate significant TOC [1,6,8]. Inferring the extent of hydraulic fractionation of OM and related components within sediment gravity flows and mapping where this ends up in the deposit is critical for understanding the ultimate fate of OM exported to deep-water sinks and how these components are partitioned across the deep-sea sediment routing system between canyons, channel-levees and lobes [2,3].

Mudstones are accumulated from vertical suspension settling as: (i) organic-rich mud caps (Td/Te units) in turbidites [4,10]; (ii) H5 mudstone caps in hybrid event beds [32,94]; and (iii) background mudstones, albeit not a focus of this study (Figure 5a). In all three cases the deposition of OM could be sufficiently slow and when the sedimentation rates are lower, the diffusion of oxygen from the bottom waters is more efficient and ultimately, even relatively resistant terrestrial OM is prone to significant remineralization [66]. By contrast, OM buried deeper within sand-rich Ta-Tc divisions of turbidites and H3 divisions of HEBs is less prone to oxidation because the sedimentation rate is sufficiently high (Figure 5b,c). However, a majority of the studies on turbidites have documented the mud caps as significant OM sinks [1–6,12,13,62,66,67,88,89] rather than sand-rich divisions. Studies on carbon burial by hybrid gravity flows are still largely under-investigated, but preliminary studies suggest that OM is better preserved in HEBs compared to turbidites and mudstones (Figure 5), due to clay-forced rapid deposition, encapsulation of OM particles in clay flocs at the onset of flocculation and during transport, and subsequent deep-burial beneath the sea floor [11]. In HEBs, the sandstone/mudstone couplets (H4/H5) divisions may further hinder downward diffusion and/or percolation of oxidants into the sediment pore waters [74], thereby further enhancing the OM preservation in H3 divisions. The proportions of different event bed types and mudstones—hybrid event bed versus turbidite versus mudstones—making up most of the deep-water fan successions is thus an important metric to quantify and map OM burial and preservation in deep-sea settings.

Several other factors also dictate OM burial efficiency in submarine fans (e.g., sedimentation rate, sea level, oxygen level, depositional mechanism). The exceptionally high burial efficiency of OM on the Bengal fan, which is close to 100%, is caused by rapid accumulation rates combined with low oxygen concentration in the waters of the Bay of Bengal [2]. Furthermore, relative sea level changes linked to glacial–interglacial cycles greatly affect the sediment transfer and specifically OM sequestration over the geological times scales in the deep-sea fan systems [95,96]. In the Amazon deep-sea fan OM accumulation is controlled by glacio-eustatic sea-level oscillations [96]. Interglacial sea-level high-stand sediments are dominated by marine OM (100%), whereas during glacial sea-level low-stands terrestrial OM (60–90%) is transported beyond the continental shelf through the Amazon canyon and deposited directly onto the Amazon deep-sea fan [95].

This review article also shows that the depositional mechanism/process can also have a profound control on the fate of OM distribution, burial and preservation. It was highlighted by Kvale et al. [8] that the depositional mechanism is an important factor (perhaps more so than water column anoxia) in preserving high TOC contents within the HEB-dominated fan fringe settings in the mixed carbonate–clastic Wolfcamp Fm, in the Delaware Basin. Recently, Baudin et al. [66] showed that it is possible to estimate the efficiency of the different depositional environments and to some extent depositional mechanism in trapping OM fluxes by calculating the percentage of the OM burial in each sector of the fan system. They documented that lobes and levees are significant OM sinks and accumulate about 45% and 52% of the total OM burial in Congo Fan, respectively, whereas the canyon and channel accumulate the remaining 3%. The lobes are more effective than levees because they accumulate about four times more OM compared to levees at the same amount of surface area (inferred from Figure 6 in [66]). Lobes have been referred to as 'mega-sinks' for terrestrial OM by other researchers [9]. Grain-size measurements in the cores indicate that most (up to 80–90%) of the Congo Fan sediment is made of clay and silt [3]. A large proportion of silt and finer clay grains in the distal lobes in the Mississippi Fan has been attributed to diminished turbulence once clay and silt content cross a certain threshold within sediment gravity flows as they expand from confined to unconfined settings downstream of the channel [50]. The role of these muddy components in enhanced physical protection of the OM and regulation of long-term global preservation of organic carbon is very well established [68]. It is highly likely that many of these preservation mechanisms (rapid deposition, bottom water anoxia, lack of bioturbation, etc.) may operate together to facilitate long-term OM preservation in the deep-water rock record.

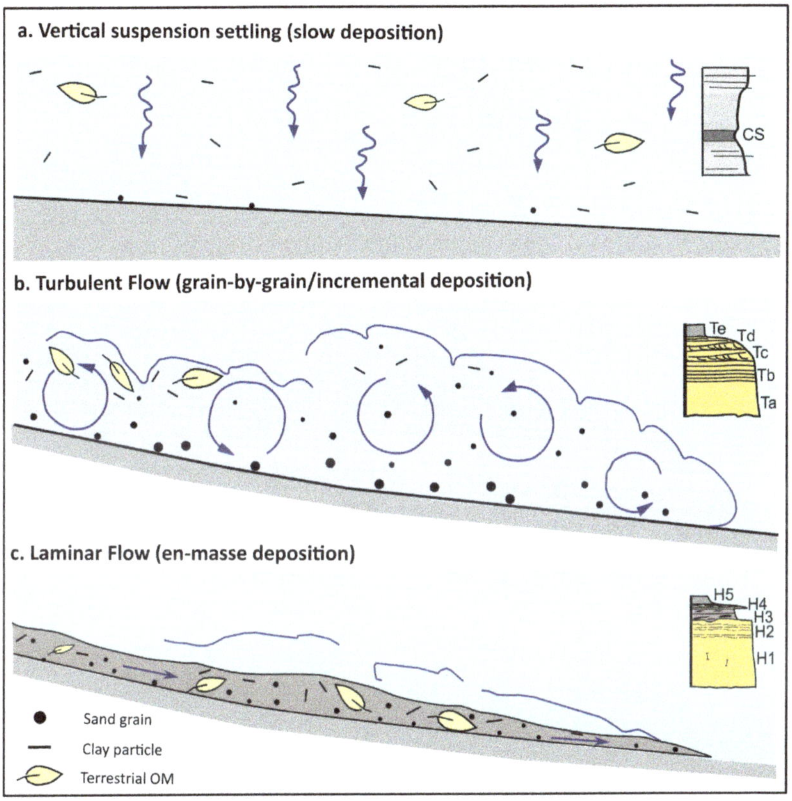

Figure 5. Conceptual models accounting for three dominant mechanisms responsible for OM deposition and segregation in the deep sea. (**a**) In bed capping (Te or H5 divisions) and background mudstones OM is accumulated from slowly settling suspensions and these flows are prone to higher OM oxidation and degradation. CS = condensed section; (**b**) turbidity currents where grain-by-grain deposition takes place, OM is often segregated towards the tops and/or tail regions of the flows (namely Td and Te divisions), although some OM could be deposited as bed load in Ta-Tc parts; and (**c**) hybrid event beds OM is also locked in lower tiers (below H5 divisions, which are equivalent to Te mud caps in turbidites), particularly in H3 divisions emplaced by turbulence-dampened laminar flows, thereby providing higher post-burial protection.

4. Conclusions

Sediment gravity flows are important agents for transferring terrestrial OM to deep-sea fan repositories and emerging evidence shows that relatively sandy deep-marine event beds may also incorporate significant quantities of OM, implying that global carbon burial budgets for the deep-sea fans may have been significantly underestimated. Organic matter accumulated from slowly settling suspension in mud caps (Te or H5 divisions in turbidites and hybrid event beds, respectively) is prone to oxidation and bioturbation compared to the OM buried in sandy components of turbidity currents (Ta-Tc) and HEBs (H2/H3 divisions). Organic matter burial in sandy divisions of turbidites is still not a very widely recognized phenomenon and most of the models still favor OM accumulation in mudstone caps. In hybrid beds, OM is preferentially segregated into mud-rich linked debrite but still sandy H3 divisions. It is also thought that in HEBs clay flocs within laminar flows can also encapsulate significant quantities of the OM during transport, thereby providing a long-term 'physical protection' mechanism. Thus, HEBs can efficiently sequester large quantities of OM because rapid clay-forced deposition and physical protection of OM in cohesive linked debrites (excluding bed tops with intense scavenging) provide a useful mechanism for long-term OM burial in submarine fans. The present study shows that the flow depositional process may exert a strong control on OM distribution, burial and preservation.

Author Contributions: Conceptualization, A.H. and K.A.-R.; methodology, A.H.; validation, A.H. and K.A.-R.; investigation, A.H.; resources, K.A.-R.; data curation, A.H.; writing—original draft preparation, A.H.; writing—review and editing, K.A.-R.; visualization, A.H.; supervision, K.A.-R.; project administration, A.H. and K.A.-R.; funding acquisition, K.A.-R. All authors have read and agreed to the published version of the manuscript.

Funding: This review was compiled during A.H.'s ongoing Postdoctoral Fellowship at the College of Petroleum Engineering and Geosciences (CPG), King Fahd University of Petroleum and Minerals (KFUPM). Some core and thin section photographs were obtained from A.H.'s PhD thesis at the University College Dublin, Ireland, funded in part by a research grant from Science Foundation Ireland (SFI) under grant number 13/RC/2092 and co-funded under the European Regional Development Fund and by PIPCO RSG and its member companies. The funding of the behind-outcrop coring program by Equinor ASA is also gratefully acknowledged.

Institutional Review Board Statement: Not applicable.

Informed Consent Statement: Not applicable.

Data Availability Statement: Not applicable.

Acknowledgments: Peter Haughton, Patrick Shannon, and Emma Morris are thanked for their mentorship and supervision during the primary author's PhD.

Conflicts of Interest: The authors declare no conflict of interest.

References

1. Saller, A.; Lin, R.; Dunham, J. Leaves in turbidite sands: The main source of oil and gas in the deep-water Kutei Basin, Indonesia. *AAPG Bull.* **2006**, *90*, 1585–1608. [CrossRef]
2. Galy, V.; France-Lanord, C.; Beyssac, O.; Faure, P.; Kudrass, H.; Palhol, F. Efficient organic carbon burial in the Bengal fan sustained by the Himalayan erosional system. *Nature* **2007**, *450*, 407–411. [CrossRef] [PubMed]
3. Baudin, F.; Disnar, J.R.; Martinez, P.; Dennielou, B. Distribution of the organic matter in the channel-levees systems of the Congo mud-rich deep-sea fan (West Africa). Implication for deep offshore petroleum source rocks and global carbon cycle. *Mar. Pet. Geol.* **2010**, *27*, 995–1010. [CrossRef]
4. Baudin, F.; Stetten, E.; Schnyder, J.; Charlier, K.; Martinez, P.; Dennielou, B.; Droz, L. Origin and distribution of the organic matter in the distal lobe of the Congo deep-sea fan-a rock-Eval survey. *Deep Sea Res. Part II* **2017**, *142*, 75–90. [CrossRef]
5. Biscara, L.; Mulder, T.; Martinez, P.; Baudin, F.; Etcheber, H.; Jouanneau, J.M.; Garlan, T. Transport of terrestrial organic matter in the Ogooue deep sea turbidite system (Gabon). *Mar. Pet. Geol.* **2011**, *28*, 1061–1072. [CrossRef]
6. Talling, P.J. Hybrid submarine flows comprising turbidity current and cohesive debris flow: Deposits, theoretical and experimental analyses, and generalized models. *Geosph.* **2013**, *9*, 460–488. [CrossRef]
7. Schnyder, J.; Stetten, E.; Baudin, F.; Pruski, A.M.; Martinez, P. Palynofacies reveal fresh terrestrial organic matter inputs in the terminal lobes of the Congo deep-sea fan. *Deep Sea Res. Part II* **2017**, *142*, 91–108. [CrossRef]
8. Kvale, E.P.; Bowie, C.M.; Flenthrope, C.; Mace, C.; Parrish, J.M.; Price, B.; Anderson, S.; DiMichele, W.A. Facies variability within a mixed carbonate-siliciclastic seafloor fan (upper Wolfcamp Fm., Permian, Delaware Basin, New Mexico). *AAPG Bull.* **2019**, *104*, 525–563. [CrossRef]
9. Rabouille, C.; Dennielou, B.; Baudin, F.; Raimonet, M.; Droz, L.; Khripounoff, A.; Martinez, P.; Mejanelle, L.; Michalopoulos, P.; Pastor, L.; et al. Carbon and silica megasink in deep-sea sediments of the Congo terminal lobes. *Quat. Sci. Rev.* **2019**, *222*, 105854. [CrossRef]
10. Hage, S.; Galy, V.; Cartigny, M.; Acikalin, S.; Clare, M.A.; Grocke, D.; Hilton, R.; Hunt, J.; Lintern, D.; Mcghee, C.; et al. Efficient preservation of young terrestrial organic carbon in sandy turbidity-current deposits. *Geology* **2020**, *48*, 882–887. [CrossRef]
11. Hussain, A.; Haughton, P.; Shannon, P.; Morris, E.; Pierce, C.; Omma, J. Mud-forced turbulence dampening facilitates rapid burial and enhanced preservation of terrestrial organic matter in deep-sea environments. *Mar. Pet. Geol.* **2021**, *130*, 105101. [CrossRef]
12. McArthur, A.; Kneller, B.; Wakefield, M.; Souza, P.; Kuchle, J. Palynofacie classification of the depositional elements of confined turbidite systems: Examples from the Gres d'Annot, SE France. *Mar. Pet. Geol.* **2016**, *77*, 1254–1273. [CrossRef]
13. McArthur, A.; Gamberi, F.; Kneller, B.; Wakefield, M.; Souza, P.; Kuchle, J. Palynofacies classification of submarine fan depositional environments: Outcrop examples from the Marnoso-Arenacea Formation. Italy. *Mar. Pet. Geol.* **2017**, *88*, 181–199. [CrossRef]

14. Berner, R. Burial of organic carbon and pyrite sulfur in the modern ocean: Its geochemical and environmental significance. *Am. J. Sci.* **1982**, *282*, 451–473. [CrossRef]
15. Berner, R.A. Atmospheric carbon dioxide levels over Phanerozoic time. *Science* **1990**, *249*, 1382–1386. [CrossRef] [PubMed]
16. France-Lanord, C.; Derry, L. Organic carbon burial forcing of the carbon cycle from Himalayan erosion. *Nature* **1997**, *390*, 65–67. [CrossRef]
17. Hayes, J.M.; Strauss, H.; Kaufman, A. The abundance of 13C in marine organic matter and isotopic fractionation in the global biogeochemical cycle of carbon during the past 800 Ma. *Chem. Geol.* **1999**, *161*, 103–125. [CrossRef]
18. Hayes, J.; Waldbauer, J. The carbon cycle and associated redox processes through time: Philosophical Transections B. *R. Soc. Lond. Biol. Sci.* **2006**, *361*, 931–950. [CrossRef]
19. Abbate, E.; Bortolotti, V.; Passerini, P. Olistostromes and olistoliths. *Sediment. Geol.* **1970**, *4*, 521–557. [CrossRef]
20. Carter, R. A discussion and classification of subaqueous mass-transport with particular application to gravity flow, slurry flow and fluxo-turbidites. *Earth Sci. Rev.* **1975**, *11*, 145–177. [CrossRef]
21. Middleton, G.; Hampton, M. Subaqueous sediment transport and deposition by sediment gravity waves. In *Marine Sediment Transport and Environmental Management*; Stanley, D.J., Swift, D.J., Eds.; Wiley: New York, NY, USA, 1976; pp. 197–218.
22. Marr, J.G.; Shanmugam, G.; Parker, G. Experiments on subaqueous sandy gravity flows: The role of clay and water content in flow dynamics and depositional structures. *Bull. Geol. Soc. Am.* **2001**, *113*, 1377–1386. [CrossRef]
23. Bouma, A.H. Turbidites. *Dev. Sedimentol.* **1962**, *3*, 247–256. [CrossRef]
24. Stow, D.A.; Shanmugam, G. Sequence of structures in fine-grained turbidites: Comparison of recent deep-sea and ancient flysch sediments. *Sediment. Geol.* **1980**, *25*, 23–42. [CrossRef]
25. Lowe, D. Sediment gravity flows: II Depositional models with special reference to the deposits of high-density turbidity currents. *J. Sediment. Res.* **1982**, *52*, 279–297.
26. Kneller, B.; Buckee, C. The structure and fluid mechanics of turbidity currents: A review of some recent studies and their geological implications. *Sedimentology* **2000**, *47*, 62–94. [CrossRef]
27. Mulder, T.; Alexander, J. The physical character of subaqueous sedimentary density flows and their deposits. *Sedimentology* **2001**, *48*, 269–299. [CrossRef]
28. Pickering, K.; Hiscott, R.; Hein, F. *Deep Marine Environments: Clastic Sedimentation and Tectonics*; Unwin Hyman Ltd.: London, UK, 1989; p. 416.
29. Nemec, W. Aspects of sediment movement on steep delta slopes. In *Coarse Grained Deltas*; Colella, A., Prior, D., Eds.; Special Publication of the International Association of Sedimentologists; International Association of Sedimentologists: Oxford, UK, 1990.
30. Stow, D.V.; Johansson, M. Deep-water massive sands: Nature, origin and hydrocarbon implications. *Mar. Pet. Geol.* **2000**, *17*, 145–174. [CrossRef]
31. Haughton, P.; Barker, S.; McCaffrey, W. 'Linked' debrites in sand-rich turbidite systems–origin and significance. *Sedimentology* **2003**, *50*, 459–482. [CrossRef]
32. Haughton, P.; Davis, C.; McCaffrey, W.; Barker, S. Hybrid sediment gravity flow deposits—Classification, origin and significance. *Mar. Pet. Geol.* **2009**, *26*, 1900–1918. [CrossRef]
33. Talling, P.; Amy, L.; Wynn, R.; Peakall, J.; Robinson, M. Beds comprising debrite sandwiched within co-genetic turbidite: Origin and widespread occurrence in distal depositional environments. *Sedimentology* **2004**, *51*, 163–194. [CrossRef]
34. Wood, A.; Smith, A. The sedimentation and sedimentary history of the Aberystwyth Grits (Upper Llandoverian). *J. Geol. Soc. Lond.* **1958**, *114*, 163–195. [CrossRef]
35. Lowe, D.; Guy, M. Slurry-flow deposits in the Britannia Formation (Lower Cretaceous), North Sea: A new perspective on the turbidity current and debris flow problem. *Sedimentology* **2000**, *47*, 31–70. [CrossRef]
36. Blackbourn, G.; Thomason, M. Britannia Field, UK North Sea: Petrographic constraints on Lower Cretaceous provenance, facies, and the origin of slurry-flow deposits. *Pet. Geosci.* **2000**, *6*, 329–343. [CrossRef]
37. Barker, S.; Haughton, P.; McCaffrey, W.; Archer, S.; Hakes, B. Development of rheological heterogeneity in clay-rich high-density turbidity currents: Aptian Britannia Sandstone Member, UK continental shelf. *J. Sediment. Res.* **2008**, *78*, 45–68. [CrossRef]
38. Sylvester, Z.; Lowe, D. Textural trends in turbidites and slurry beds from the Oligocene flysch of the East Carpathians, Romania. *Sedimentology* **2004**, *51*, 945–972. [CrossRef]
39. Talling, P.; Wynn, R.; Masson, D.; Frenz, M.; Cronin, B.; Schiebel, R.; Akhmetzhanov, A.; Dallmeier, T.; Benetti, S.; Weaver, P. Onset of submarine debris flow deposition far from original giant landslide. *Nature* **2007**, *450*, 541–544. [CrossRef]
40. Talling, P.; Masson, D.; Sumner, E.; Malgesini, G. Subaqueous sediment density flows: Depositional processes and deposit types. *Sedimentology* **2012**, *59*, 1937–2003. [CrossRef]
41. Ito, M. Downfan transformation from turbidity currents to debris flows at a channel-to-lobe transitional zone: The lower Pleistocene Otadai Formation, Boso Peninsula, Japan. *J. Sediment. Res.* **2008**, *78*, 668–682. [CrossRef]
42. Hodgson, D. Distribution and origin of hybrid beds in sand-rich submarine fans of the Tanqua depocentre, Karoo Basin, South Africa. *Mar. Pet. Geol.* **2009**, *26*, 1940–1956. [CrossRef]
43. Kane, I.; Pontén, A. Submarine transitional flow deposits in the Paleogene Gulf of Mexico. *Geology* **2012**, *40*, 1119–1122. [CrossRef]
44. Patacci, M.; Haughton, P.; McCaffrey, W. Rheological complexity in sediment gravity flows forced to decelerate against a confining slope, Braux, SE France. *J. Sediment. Res.* **2014**, *84*, 270–277. [CrossRef]

45. Fonnesu, M.; Haughton, P.; Felletti, F.; McCaffrey, W. Short length-scale variability of hybrid event beds and its applied significance. *Mar. Pet. Geol.* **2015**, *67*, 583–603. [CrossRef]
46. Fonnesu, M.; Patacci, M.; Haughton, P.; Felletti, F.; McCaffrey, W. Hybrid event beds generated by local substrate delamination on a confined-basin floor. *J. Sediment. Res.* **2016**, *86*, 929–943. [CrossRef]
47. Fonnesu, M.; Felletti, F.; Haughton, P.; Patacci, M.; McCaffrey, W. Hybrid event bed character and distribution linked to turbidite sub-environments: The North Apennine Gottero Sandstone (north-west Italy). *Sedimentology* **2018**, *65*, 151–190. [CrossRef]
48. Kane, I.; Pontén, A.; Vangdal, B.; Eggenhuisen, J.; Hodgson, D.; Spychala, Y. The stratigraphic record and processes of turbidity current transformation across deep-marine lobes. *Sedimentology* **2017**, *64*, 1236–1273. [CrossRef]
49. Southern, S.; Kane, I.; Warchoł, M.; Porten, K.; McCaffrey, W. Hybrid event beds dominated by transitional-flow facies: Character, distribution and significance in the Maastrichtian Springar Formation, north-west Vøring Basin, Norwegian Sea. *Sedimentology* **2017**, *64*, 747–776. [CrossRef]
50. Fildani, A.; Clark, J.; Covault, J.; Power, B.; Romans, B.; Aiello, I. Muddy sand and sandy mud on the distal Mississippi fan: Implications for lobe depositional processes. *Geosphere* **2018**, *14*, 1051–1066. [CrossRef]
51. Pierce, C.; Haughton, P.; Shannon, P.; Pulham, A.; Barker, S.; Martinsen, O. Variable character and diverse origin of hybrid event beds in a sandy submarine fan system, Pennsylvanian Ross Sandstone Formation, western Ireland. *Sedimentology* **2018**, *65*, 952–992. [CrossRef]
52. Angus, K.; Arnott, W.C.; Terlaky, V. Lateral and vertical juxtaposition of matrix-rich and matrix-poor lithologies caused by particle settling in mixed mud–sand deep-marine sediment suspensions. *Sedimentology* **2019**, *66*, 940–962. [CrossRef]
53. Hussain, A.; Haughton, P.; Shannon, P.; Turner, J.; Pierce, C.; Obradors-Latre, A.; Barker, S.; Martinsen, O. High-resolution X-ray fluorescence profiling of hybrid event beds: Implications for sediment gravity flow behaviour and deposit structure. *Sedimentology* **2020**, *67*, 2850–2882. [CrossRef]
54. Brooks, H.; Ito, M.; Zuchuat, V.; Peakall, J.; Hodgson, D. Channel-lobe transition zone development in tectonically active settings: Implications for hybrid bed development. *Depos. Rec.* **2022**, 1–40. [CrossRef]
55. Stow, D.; Smillie, Z. Distinguishing between Deep-Water Sediment Facies: Turbidites, Contourites and Hemipelagites. *Geosciences* **2020**, *10*, 68. [CrossRef]
56. Wynn, R.; Masson, D.; Stow, D.; Weaver, P. The Northwest African slope apron: A modern analogue for deep water systems with complex seafloor topography. *Mar. Pet. Geol.* **2000**, *17*, 253–265. [CrossRef]
57. Boulesteix, K.; Poyatos-More, M.; Flint, S.; Taylor, K.; Hodgson, D.; Hasiotis, S. Transport and deposition of mud in deep-water environments: Processes and stratigraphic implications. *Sedimentology* **2019**, *66*, 2894–2925. [CrossRef]
58. Stow, D. Fine-grained sediments in deep water: An overview of processes and facies models. *GeoMarine Lett.* **1985**, *5*, 17–23. [CrossRef]
59. Plint, G. Mud dispersal across a Cretaceous prodelta: Storm-generated, wave-enhanced sediment gravity flows inferred from mudstone microtexture and Microfacies. *Sedimentology* **2014**, *61*, 609–647. [CrossRef]
60. Schieber, J.; Southard, J.; Thaisen, K. Accretion of mudstone beds from migrating floccule ripples. *Science* **2007**, *318*, 1760–1763. [CrossRef]
61. Schieber, J.; Southard, J. Bedload transport of mud by floccule ripples-direct observation of ripple migration processes and their implications. *Geology* **2009**, *37*, 483–486. [CrossRef]
62. Könitzer, S.; Davies, S.; Stephenson, H.; Leng, M. Depositional controls on mudstone lithofacies in a basinal setting–implications for the delivery of sedimentary organic matter. *J. Sediment. Res.* **2014**, *84*, 198–214. [CrossRef]
63. Boulesteix, K.; Poyatos-Moré, M.; Flint, S.; Hodgson, D.; Taylor, K.; Brunt, R. Sedimentological and stratigraphic criteria to distinguish between basin-floor and slope mudstones: Implications for the delivery of mud to deep-water environments. *Depos. Rec.* **2022**. [CrossRef]
64. Hedges, J.; Keil, R. Sedimentary organic matter preservation: An assessment and speculative synthesis. *Mar. Chem.* **1995**, *49*, 81–115. [CrossRef]
65. Blair, N.; Leithold, E.; Brackley, H.; Trustrum, N.; Page, M.; Childress, L. Terrestrial sources and export of particulate organic carbon in the Waipaoa sedimentary system: Problems, progress and processes. *Mar. Geol.* **2010**, *270*, 108–118. [CrossRef]
66. Baudin, F.; Rabouille, C.; Dennielou, B. Routing of terrestrial organic matter from the Congo River to the ultimate sink in the abyss: A mass balance approach. *Geol. Belg.* **2020**, *23*, 41–52. [CrossRef]
67. Stetten, E.; Baudin, F.; Reyss, J.; Martinez, P.; Charlier, K.; Schnyder, J.; Rabouille, C.; Dennielou, B.; Coston-Guarini, J.; Pruski, A. Organic matter characterization and distribution in sediments of the terminal lobes of the Congo deep-sea fan: Evidence for the direct influence of the Congo River. *Mar. Geol.* **2015**, *369*, 182–195. [CrossRef]
68. Hemingway, J.; Rothman, D.; Grant, K.; Rosengard, S.; Eglinton, T.; Derry, L.; Galy, V. Mineral protection regulates long-term global preservation of natural organic carbon. *Nature* **2019**, *570*, 228–231. [CrossRef]
69. Keil, R.G.; Tsamakis, E.; Wolf, N.; Hedges, J.; Goñi, M. Relationships between organic carbon preservation and mineral surface area in Amazon Fan sediments (Holes 920 932A and 942A). *Proc. ODP Sci. Results* **1997**, *155*, 531–538.
70. Masiello, C.A. Quick burial at sea. *Nature* **2007**, *450*, 360–361. [CrossRef]
71. Burdige, D. Burial of terrestrial organic matter in marine sediments: A reassessment. *Glob. Biogeochem. Cycles* **2005**, *19*, 1–7. [CrossRef]

72. Stanley, D. Turbidity current transport of organic-rich sediments: Alpine and Mediterranean examples. *Mar. Geol.* **1986**, *70*, 85–101. [CrossRef]
73. Pyles, D.; Straub, K.; Stammer, J. Spatial variations in the composition of turbidites due to hydrodynamic fractionation. *Geophys. Res. Lett.* **2013**, *40*, 3919–3923. [CrossRef]
74. Ghadeer, S.; Macquaker, J. The role of event beds in the preservation of organic carbon in fine-grained sediments: Analyses of the sedimentological processes operating during deposition of the Whitby Mudstone Formation (Toarcian, Lower Jurassic) preserved in northeast England. *Mar. Pet. Geol.* **2012**, *35*, 309–320. [CrossRef]
75. Nittrouer, C.; Sternberg, R. The formation of sedimentary strata in anallochthonous shelf. In *Developments in Sedimentology*; Elsevier: Amsterdam, The Netherlands, 1981; Volume 32, pp. 201–232.
76. Bentley, S.; Nittrouer, C.; Sommerfield, C. Development of sedimentary strata in Eckernförde Bay, southwestern Baltic Sea. *Geo-Mar. Lett.* **1996**, *16*, 148–154. [CrossRef]
77. Bentley, S.; Nittrouer, C. Physical and biological influences on the formation of sedimentary fabric in an oxygen-restricted depositional environment: Eckemforde Bay, Southwestern Baltic Sea. *Palaios* **1999**, *14*, 585–600. [CrossRef]
78. Bohacs, K.; Norton, I.; Gilbert, D.; Neal, J.; Kennedy, M.; Borkowski, W.; Rottman, M.; Burke, T. The Accumulation of Organic-Matter-Rich Rocks within an Earth Systems Framework: The integrated roles of Plate Tectonics, Atmosphere, Ocean, and Biota through the Phanerozoic. *Princ. Geol. Anal.* **2020**, 721–744. [CrossRef]
79. Pedersen, T.; Calvert, S. Anoxia vs. productivity: What controls the formation of organic-carbon-rich sediments and sedimentary rocks? *AAPG Bull.* **1990**, *74*, 454–466.
80. Pickering, K.; Pouderoux, H.; McNeill, L.; Backman, J.; Chemale, C.; Kutterolf, S.; Milliken, K.; Mukoyoshi, H.; Henstock, T.; Stevens, D.; et al. Sedimentology, stratigraphy and architecture of the Nicobar Fan (Bengal–Nicobar Fan System), Indian Ocean: Results from International Ocean Discovery Program Expedition 362. *Sedimentology* **2020**, *67*, 2248–2281. [CrossRef]
81. Collinson, J.; Martinsen, O.; Bakken, B.; Kloster, A. Early fill of the Western Irish Namurian Basin: A complex relationship between turbidites and deltas. *Basin Res.* **1991**, *3*, 223–242. [CrossRef]
82. Liang, C.; Jiang, Z.; Cao, Y.; Wu, M.; Guo, L.; Zhang, C. Deep-water depositional mechanisms and significance for unconventional hydrocarbon exploration: A case study from the lower Silurian Longmaxi shale in the southeastern Sichuan Basin. *AAPG Bull.* **2016**, *100*, 773–794. [CrossRef]
83. Kuenen, P. Sole markings of graded graywacke beds. *J. Geol.* **1957**, *65*, 231–258. [CrossRef]
84. Zavala, C.; Arcuri, M. Intrabasinal and extrabasinal turbidites: Origin and distinctive characteristics. *Sediment. Geol.* **2016**, *337*, 36–54. [CrossRef]
85. Leithold, E.; Blair, N.; Wegmann, K. Source-to-sink sedimentary systems and global carbon burial: A river runs through it. *Earth Sci. Reviws* **2016**, *153*, 30–42. [CrossRef]
86. Lee, H.; Galy, V.; Feng, X.; Ponton, C.; Galy, A.; France-Lanord, C.; Feakins, S. Sustained wood burial in the Bengal Fan over the last 19 my. *Proc. Natl. Acad. Sci. USA* **2019**, *116*, 22518–22525. [CrossRef] [PubMed]
87. Plink-Bjorklund, P.; Steel, R. Initiation of turbidity currents: Evidence for hyperpycnal flow turbidites in Eocene Central Basin of Spitsbergen. *Sediment. Geol.* **2004**, *165*, 29–52. [CrossRef]
88. Jansen, J.; Vanweering, T.; Gieles, R.; Van, J. Middle and Late Quaternary oceanography and climatology of the Zaire-Congo fan and adjacent eastern Angola Basin. *Netherland J. Sea Res.* **1984**, *17*, 201–249. [CrossRef]
89. Huc, A.; Bertrand, P.; Stow, D.; Gayet, J.; Vandenbroucke, M. Organic sedimentation in deep offshore settings: The Quaternary sediments approach. *Mar. Pet. Geol.* **2001**, *18*, 513–517. [CrossRef]
90. Sumner, E.; Talling, P.; Amy, L.; Wynn, R.; Stevenson, C.; Frenz, M. Facies architecture of individual basin-plain turbidites: Comparison with existing models and implications for flow processes. *Sedimentology* **2012**, *59*, 1850–1887. [CrossRef]
91. Stevenson, C.; Jackson, C.; Hodgson, D.; Hubbard, S.; Eggenhuisen, J. Deep-Water Sediment Bypass . *J. Sediment. Res.* **2015**, *85*, 1058–1081. [CrossRef]
92. Davis, C.; Haughton, P.; McCaffrey, W.; Scott, E.; Hogg, N.; Kitching, D. Character and distribution of hybrid sediment gravity flow deposits from the outer Forties Fan, Palaeocene central North Sea, UKCS. *Mar. Pet. Geol.* **2009**, *26*, 1919–1939. [CrossRef]
93. Porten, K.; Kane, I.; Warchol, M.; Southern, S. A sedimentological process-based approach to depositional reservoir quality of deep-marine sandstones: An example from the Springar Formation, north-western Vøring Basin, Norwegian Sea. *J. Sediment. Res.* **2016**, *86*, 1269–1286. [CrossRef]
94. Randazzo, V.; Le Goff, J.; Di Stefano, P.; Reijmer, J.; Todaro, S.; Cacciatore, M. Carbonate slope resedimentation in a tectonically active setting (western Sicily, Cretaceous Escarpment, Italy). *Sedimentology* **2020**, *67*, 2360–2391. [CrossRef]
95. Schlünz, B.; Schneider, R.; Müller, P.; Showers, W.; Wefer, G. Terrestrial organic carbon accumulation on the Amazon deep sea fan during the last glacial sea level low stand. *Chem. Geol.* **1999**, *159*, 263–281. [CrossRef]
96. Cartapanis, O.; Bianchi, D.; Jaccard, S.; Galbraith, E. Global pulses of organic carbon burial in deep-sea sediments during glacial maxima. *Nat. Commun.* **2016**, *7*, 10796. [CrossRef] [PubMed]

Journal of
Marine Science and Engineering

MDPI

Article

Ecological Constraints of Plankton Bio-Indicators for Water Column Stratification and Productivity: A Case Study of the Holocene North Aegean Sedimentary Record

Christina Giamali [1,*], George Kontakiotis [2], Assimina Antonarakou [2] and Efterpi Koskeridou [2]

1 The Goulandris Natural History Museum, Levidou 13, 14562 Kifissia, Greece
2 Department of Historical Geology and Palaeontology, Faculty of Geology and Geoenvironment, National and Kapodistrian University of Athens, Panepistimiopolis, 15784 Athens, Greece; gkontak@geol.uoa.gr (G.K.); aantonar@geol.uoa.gr (A.A.); ekosker@geol.uoa.gr (E.K.)
* Correspondence: ch.giamali@gnhm.gr

Abstract: This study presents novel findings on the drivers of the calcitic planktonic foraminiferal and aragonitic pteropod Holocene assemblages of the North Aegean Trough (northeastern Mediterranean), an area recording the interaction between dynamic water masses as they exchange between the northern and southern Mediterranean sub-basins. Both of these groups of microorganisms are the major producers of calcium carbonate in the ocean, and are particularly sensitive to climate and oceanographic changes over the late Quaternary. Downcore micropaleontological data from the gravity core AEX-15, supplemented with multivariate statistical Q-mode cluster and principal component analyses (PCA) results, provide significant insights on the water column dynamics during the Holocene. Focusing on the last ~10 calibrated thousands of years before the present day (ka cal BP), our integrated study reveals that primary productivity is the dominant factor controlling the planktonic foraminifera distribution in the North Aegean Sea, whereas water column stratification, and particularly the intensity of the oxygen minimum zone, seems to be the major driver on the pteropod distribution. Besides productivity and thermal stratification, which show the highest explanatory power for planktonic foraminifera and pteropod communities, respectively, though they affect both groups to a different extent, upwelling seems to further affect both faunal groups. Overall, our findings are consistent with those derived by published late Quaternary eastern Mediterranean records, highlighting in parallel a useful additional dimension on planktonic foraminiferal and pteropod ecology, which is inextricably linked with the factors of primary productivity and vertical stratification of the warm Holocene water column.

Keywords: planktonic foraminifera; pteropods; paleoecology; multivariate statistical analysis; primary productivity; water column stratification; upwelling; Holocene climatic variability; hydrological changes; eastern Mediterranean

Citation: Giamali, C.; Kontakiotis, G.; Antonarakou, A.; Koskeridou, E. Ecological Constraints of Plankton Bio-Indicators for Water Column Stratification and Productivity: A Case Study of the Holocene North Aegean Sedimentary Record. *J. Mar. Sci. Eng.* **2021**, *9*, 1249. https://doi.org/10.3390/jmse9111249

Academic Editor: Marco Uttieri

Received: 15 October 2021
Accepted: 6 November 2021
Published: 11 November 2021

1. Introduction

The Aegean Sea is a physical laboratory to investigate climatic oscillations at both a global and local scale due to its intermediate position between the higher- (i.e., North Atlantic-influenced) and lower-latitude (i.e., monsoonally-influenced) climate systems [1–3], high sedimentation rate marine records compared to the open Mediterranean Sea [4–6], and its latitudinal and land-locked configuration [7]. The recent Holocene subdivision (Greenlandian, Northgrippian, and Meghalayan) confirms the traditional understanding of an evolution from wetter (Greenlandian) to gradually drier (Northgrippian and Meghalayan) climatic conditions. Moreover, Holocene sediments of the Aegean Sea include the most recent sapropel S1 that was deposited during the Holocene Climatic Optimum (10.8–6.1 ka cal BP; [8,9]) under reduced oxygen and productive conditions [10–18]. Therefore, the sedimentary archive corresponding to this time interval can be considered as a natural "experiment" in order to investigate the plankton response to such severe paleoenvironmental changes. In particular, the basin's limited communication with the open ocean implies that any paleoceanographic signals will be recorded in an amplified fashion, and, therefore, this heightened sensitivity to the effects of climate variability further underlies the prominent role of such marginal basins in the understanding of the global climatic evolution.

Plankton indicators have long been used as a biomonitoring tool [19,20] and/or for investigating modern [21–25] and past [26–31] environments because of their rapid response to environmental changes. Two distinct communities of plankton bio-indicators are used in this study: planktonic foraminifera and pteropods. Planktonic foraminifera are excellent bio-eco-stratigraphic [32–35], paleoceanographic [31,36–40], and paleoclimatic [41–46] indicators, due to their wide distribution in the world oceans, the good fossilization potential for their calcitic shells, their long geological record, their upper water column habitat, and their preservation in the deep-sea sedimentary record [47,48]. Pteropods are holoplanktonic aragonitic molluscs, widespread and abundant in the world's oceans, [49–51] which play an important role in the direct export of organic carbon (12% of the carbon flux worldwide [52]) to the deep ocean [25]. Several studies (e.g., [28,29,53–59]) have shown an enhanced sensitivity of late Quaternary pteropod assemblages to environmental changes, particularly of temperature, oxygen concentration, and salinity, making them valuable for paleoenvironmental and paleoclimatic reconstructions.

The objective of the present work is to determine the plankton assemblages of the North Aegean Sea during the last 10 ka BP, and further assess the environmental factors that control their distributional pattern, based on marine sediments retrieved by a gravity core (AEX-15). In order to advance the understanding of the temporal variability of observed climatic and oceanographic changes and elucidate forcing mechanisms, this study aims at achieving high-resolution sediment core at a sub-centennial to centennial scale. The faunal distribution pattern of both planktonic foraminiferal and pteropod species identified in AEX-15 core, supplemented by the ecological interpretation obtained through multivariate statistical analyses, provide significant insights on the unraveling of the paleoclimatic history, the determination of the surface conditions, and the orbital configuration of the climatic changes recorded on Holocene sediments.

2. Oceanographic and Geologic Setting

The Aegean Sea is in the northern sector of the eastern Mediterranean, between the Turkish coastline to the east, the Greek mainland to the north and west, and bounded on the south by the island of Crete and the Cretan Arc. It is connected to the Black Sea through the Straits of Bosphorus and Dardanelles, and to the Levantine Sea through several larger and deeper straits between the Peloponnese, the islands of Crete and Rhodes, and south-western Turkey (Figure 1a). It is separated into two major sub-basins with different climatic conditions: the "north" and the "south" Aegean Sea. The north is more humid than the semiarid south basin [60]. The studied core comes from the North Aegean Trough (NAT; Figure 1a), a northeast–southwest (NE–SW) elongated depression with a depth range between 800 and 1590 m that includes several interconnected sub-basin depressions, separated from each other by 100- to 350-m-deep intervening shoals and associated islands [5,61].

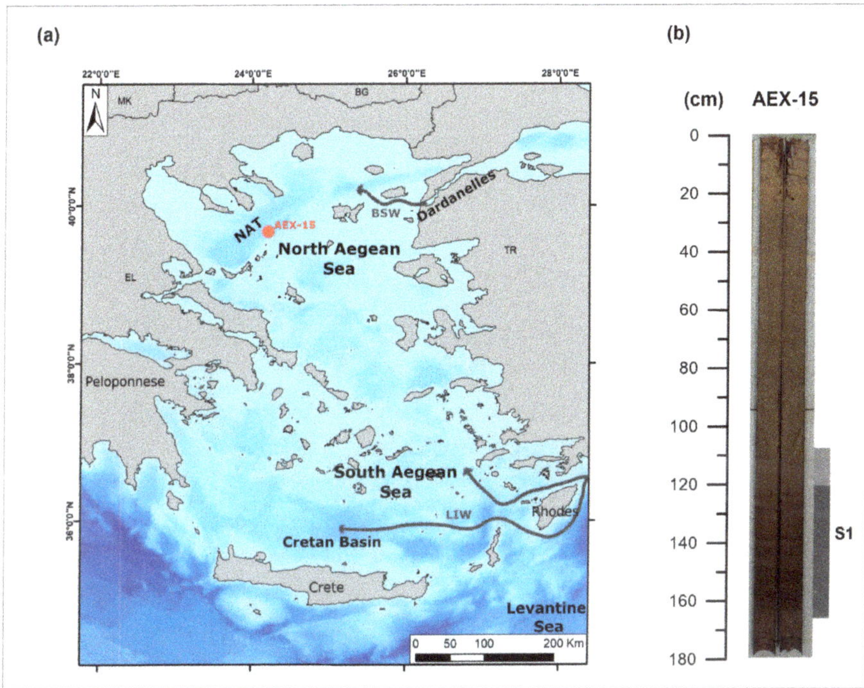

Figure 1. (**a**) Map of the Aegean Sea; inset shows location of the north Aegean sediment core AEX-15 (39°39.900′ N, 24°15.369′ E; 1242 m water depth, length 178 cm), gray arrows show the Black Sea Water (BSW) inflow and the Levantine Intermediate Water (LIW) that enters the Aegean Sea from the Levantine Sea; (**b**) lithological log of core AEX-15.

The hydrography of the Aegean Sea is characterized, in general, by a cyclonic surface water circulation, although the most active dynamic features are the mesoscale permanent and/or recurrent cyclonic and anticyclonic eddies [60], which form as a consequence of the Etesian winds [62]. In the study area, the water column structure comprises three major water masses: the superficial nutrient-rich and less saline Black Sea Water (BSW; 9–22 °C and 22–23 psu), the warmer and saline Levantine Intermediate Water (LIW; 16–25 °C and 39.2–39.5 psu), and the North Aegean Deep Water (NADW; 13–14 °C and 39.1–39.2 psu) [63,64]. The surface and intermediate waters follow the general counterclockwise circulation of the Aegean Sea and progressively mix as they flow southward along the east coast of Greece [65]. The main characteristic of the north Aegean is the fluvial freshwater inputs that discharge from the Black Sea and the river runoff from the Greek and Turkish mainland, which dilute the highly saline waters from the Levantine and south–central Aegean. The lateral and diapycnal mixing of warm subsurface intermediate waters with low salinity surficial waters leads to strong near-surface thermal stratification, which is a dominant feature of the north Aegean water column during the summer. During the winter, upwelling nutrient-rich waters seem to create favorable conditions for primary production [66]. Such seasonal upwelling pathways are very important for the low-chlorophyll and phosphorus-controlled oceanic system of the eastern Mediterranean basin [67–69], highlighting the importance of small-scale hydrographic dynamics in controlling primary productivity of this mesotrophic to oligotrophic setting.

3. Materials and Methods

3.1. Material

The study core was recovered from the NAT (39°39.900′ N, 24°15.369′ E; 1242 m water depth, length 178 cm), with R/V Aegaeo in October 2013 (Figure 1a,b). The planktonic foraminiferal and pteropod assemblages derived from 115 downcore samples, were picked from 10 cm^3 of wet sediment after washing through a 125 μm mesh sieve and cleaning using the HyPerCal protocol of [70]. All shells (300 individuals for each group) were handpicked, identified, and counted in each sample and then converted into percentages (Table S1; Supplementary Material). Detailed micropaleontological procedures, along with the time stratigraphic framework of the analyzed core are described in [30]. Based on the published age model of the study core, which is based on a combination of accelerator mass spectrometry (^{14}C AMS) radiocarbon date measurements, additional control points, and bioevents of planktonic foraminifera, its sediments cover the last 10.1 ka BP with an average sedimentation rate of 17.32 cm/ka.

3.2. Multivariate Statistical Analyses

The number of planktonic foraminifera and pteropods (Euthecosomata) counted is statistically reliable for both paleoceanographic and paleoclimatic reconstructions [71]. To determine the overall statistical similarity between samples, Q-mode cluster analysis was used, following the algorithms of [72] using the correlation coefficient matrix. In the case of planktonic foraminifera, the totality of the samples was used, whereas for pteropods, cluster analysis was carried out in 105 out of 115 samples. This was due to the lack of pteropod fauna in samples 1.5, 47, 51, 57, 59, 107, 122.5, 141.5, 142.5, and 143.5 cm. For paleoenvironmental reconstructions, multivariate statistical analyses have been performed on the data set, after exclusion of rare species (<3%) and grouping of species that have a discontinuous, scattered distribution at generic level. The results of cluster analysis were reported as Morisita similarity and arranged in two-dimensional hierarchical dendrograms, wherein locations were presented along the Y-axis while similarity level was plotted on the X-axis. Q-mode cluster analysis was performed in both faunal groups (planktonic foraminifera and pteropods) in order to investigate the differences or correlations among the biotopes identified within plankton communities.

Furthermore, principal component analysis (PCA) was applied to reduce the dimensionality of a multivariate data set to a few principal factors that determine the distributions of species. Raw data were processed using PAST (2.17) multivariate statistical software package of [73]. The factors obtained were rotated using a varimax-normalized algorithm, which allows more straightforward interpretation of the loadings of the principal components and maximization of the variances explained by the factors extracted. The resulting factor scores show the contribution of each factor in every sample and, therefore, the downcore contribution of each factor. The total number of factors was defined by minimizing the remaining "random" variability, and by the possibility to relate the factors to modern hydrographic conditions, and planktonic foraminiferal and pteropod ecology.

4. Results

Both faunal groups are abundant and well preserved in the samples of the studied core. In particular, 17 species of planktonic foraminifera lumped in 12 groups and 11 species of pteropods were identified (Table S1; Supplementary Materials). The most abundant planktonic foraminiferal species are *Globigerina bulloides*, *Turborotalita quinqueloba*, and *Globigerinoides ruber alba*, whereas the species *Globorotalia inflata* and *Neogloboquadrina pachyderma* present sporadic peaks. Pteropod species in abundance with a continuous distributional pattern are *Heliconoides inflatus* and *Boasia chierchiae*, whereas the rest of them present a scattered distributional pattern. More details on the distributional patterns of both planktonic groups are given in [30].

4.1. Q-Mode Cluster Analysis

The Q-mode hierarchical cluster analysis represents the grouping of samples based on the downcore abundance of the species higher than at least 3% in one sample. Three distinct assemblages of planktonic foraminifera and pteropods were identified by Q-mode cluster analysis in the studied core, reflecting different biotopes during the Holocene. Each assemblage is characterized by the dominant species and named accordingly after it.

4.1.1. Planktonic Foraminifera

Cluster I (*T. quinqueloba* assemblage): it contains 47 samples (samples 61 cm, 97 cm, and samples between 120 and 166 cm; Figure 2). Apart from samples 61 and 97, the remaining 45 samples correspond to the most recent sapropel S1. This assemblage is dominated by *T. quinqueloba* (up to 65%), an eutrophic species preferring low salinities [74]. In this assemblage, *G. bulloides* also occurs with high percentages (up to 50%), as well as the warm oligotrophic *Globigerinoides ruber rosea* (25%) and SPRUDTS group.

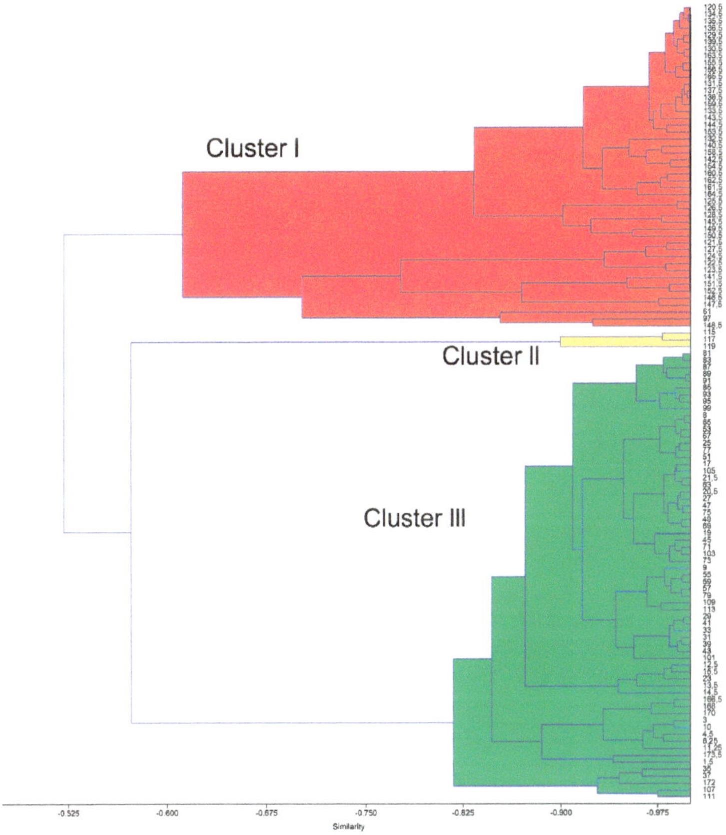

Figure 2. Dendrogram resulting from Q−mode cluster analysis on planktonic foraminifera and the clusters identified in core AEX−15: *T. quinqueloba* assemblage with pink color (Cluster I), *G. inflata* assemblage with yellow color (Cluster II), and *G. ruber alba* assemblage with green color (Cluster III).

Cluster II (*G. inflata* assemblage): this assemblage groups three samples (115 cm, 117 cm, and 119 cm; Figure 2), which belong to the oxidized part of sapropel S1. The dominant component of this assemblage is *G. inflata* (up to 32%), which is a temperate species indicative of a homogenous water column that prefers to live at the bottom of thermocline depths [47]. Additional components of this assemblage are the species *N. pachyderma*,

G. ruber alba, *G. ruber rosea*, *T. quinqueloba*, and *G. bulloides*. *Neogloboquadrina pachyderma* thrives in cool water close to or below the thermocline [75]. Its abundance is related to the development of the deep chlorophyll maximum (DCM) [75].

Cluster III (*G. ruber alba* assemblage): it includes 65 samples (from 113 to 1.5 cm, and between 166.5 and 173.5 cm; Figure 2) that correspond to the post-sapropel S1 interval, apart from the samples 166.5–173.5 cm, which derive from the pre-sapropel interval. This assemblage is characterized by the predominance of the warm, oligotrophic *G. ruber alba* (average 40%). Next in abundance are the eutrophic *N. pachyderma* (17%), *G. bulloides* (16%), and *T. quinqueloba* (13%). *Globigerinoides ruber rosea* and *Globoturborotalita rubescens* are also present but with significantly lower percentages (average 13% and 5%, respectively).

4.1.2. Pteropods

Cluster I (*Diacria trispinosa* assemblage): it contains six samples (samples 6.25 cm, 165.5 cm, and from 168 cm to 173.5 cm; Figure 3) that mainly correspond to the pre-sapropel interval. This assemblage is characterized by the high relative abundance of the mesopelagic *D. trispinosa* (up to 60%) along with the epipelagic *Creseis acicula* (up to 50%). Additional components are the species *B. chierchiae* (25%), *Clio pyramidata* (20%), and *H. inflatus* (15%).

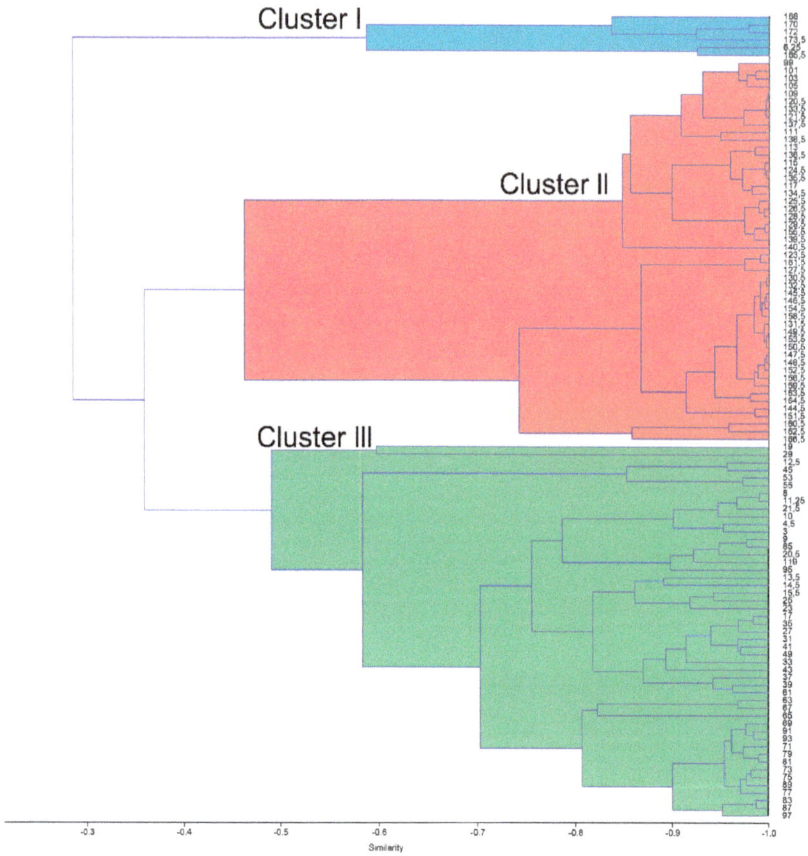

Figure 3. Dendrogram resulting from Q−mode cluster analysis in pteropods and the clusters identified in core AEX−15: *D. trispinosa* assemblage with blue color (Cluster I), *H. inflatus* assemblage with pink color (Cluster II), and *L. trochiformis* assemblage with green color (Cluster III).

Cluster II (*H. inflatus* assemblage): this cluster includes 50 samples (from 99 to 166.5 cm) the majority of which correspond to the sapropel S1 interval (exceptions are samples 99–105 cm and 166.5 cm) (Figure 3). Dominant component of the fauna is the warm oligotrophic *H. inflatus* (72%), followed by the epipelagic *B. chierchiae* (66%). The rest of the pteropod species in these samples present low relative abundances (e.g., *Cavolinia* spp. ~10%, *C. pyramidata* ~6%).

Cluster III (*Limacina trochiformis* assemblage): it groups the rest of the samples (49; Figure 3) corresponding to the post-sapropel S1 interval (3 cm–4.5 cm, 8 cm–97 cm), with the exception of the sample 119 cm, which corresponds to the oxidized part of S1. This assemblage is dominated by the typical upwelling species *L. trochiformis* (76%). Additional components of the fauna are the *B. chierchiae* (29%), *H. inflatus* (28%), *C. pyramidata* (26%), *Styliola subula* (28%), *C. acicula* (20%), and *Cavolinia* spp. (17%).

4.2. Principal Component Analysis

Principal component analysis is a commonly used method for multivariate statistical analysis, mainly because of its simple algebra and direct interpretation. The application of this statistical analysis yielded a three-factor model for both planktonic foraminiferal and pteropod communities. The interpretation of the three components in each case was based on the screen plots of eigen values, and the factor loadings of the planktonic foraminiferal and pteropod species, respectively. The distinguished factors account for 90.94% and 79.46%, of the total variance for the planktonic foraminiferal and pteropod, respectively (Tables 1 and 2), with their factor loadings showing the contribution of each factor in every sample and, therefore, the downcore contribution of each factor (Tables 3 and 4, Figure 4). In the case of a bipolar factor, which has extremes of positive and negative loadings, high positive factor scores are related to the positive pole and high negative scores to the negative pole, respectively.

Table 1. Principle component analysis (PCA) factors based on planktonic foraminifera and their percentages of the total variability for core AEX-15.

PCA Factors	Eigenvalue	% Variance	Cumulative % of the Total Variance
1	622.458	67.949	67.949
2	142.717	15.579	83.528
3	67.9386	7.416	90.944
4	27.039	2.952	93.896
5	22.7885	2.488	96.384
6	17.852	1.949	98.333
7	6.21708	0.679	99.011
8	3.39922	0.371	99.382
9	2.74832	0.300	99.682
10	1.76144	0.192	99.875
11	0.784001	0.086	99.960
12	0.342134	0.037	99.998
13	0.0172397	0.002	99.999

Table 2. Principle component analysis (PCA) factors based on pteropods and their percentages of the total variability for core AEX-15.

PCA Factors	Eigenvalue	% Variance	Cumulative % of the Total Variance
1	494.862	37.764	37.764
2	362.476	27.662	65.426
3	183.956	14.038	79.464
4	95.0715	7.255	86.719
5	78.8895	6.020	92.739

Table 2. *Cont.*

PCA Factors	Eigenvalue	% Variance	Cumulative % of the Total Variance
6	35.6337	2.719	95.459
7	29.4056	2.244	97.703
8	24.0161	1.833	99.536
9	3.7509	0.286	99.822
10	2.33109	0.178	99.999

Table 3. Ranking of the planktonic foraminifera species and their factor loadings along the PCA factors. Bold data indicate the most important factor loadings in each factor.

Species	Factor 1	Factor 2	Factor 3
Orbulina universa	−0.084	−0.170	−0.137
Globigerinoides ruber alba	**0.670**	**0.569**	0.256
Globigerinoides ruber rosea	0.077	−0.209	**−0.515**
Globigerinoides sacculifer	0.027	0.033	−0.027
Globorotalia inflata	0.002	−0.044	−0.183
Globigerina bulloides	−0.024	**−0.536**	**0.734**
Globoturborotalita rubescens	0.011	−0.158	−0.155
Globigerinella siphonifera gr.	0.004	0.034	−0.023
Neogloboquadrina pachyderma	0.050	−0.004	−0.155
Neogloboquadrina dutertrei	−0.003	−0.006	−0.001
Turborotalita quinqueloba	**−0.731**	0.536	0.159
Globigerinita glutinata	0.008	−0.005	0.063

Table 4. Ranking of the pteropod species and their factor loadings along the PCA factors. Bold data indicate the most important factor loadings in each factor.

Species	Factor 1	Factor 2	Factor 3
Heliconoides inflatus	**0.866**	**0.380**	0.162
Limacina retroversa	−0.006	0.006	−0.011
Limacina trochiformis	**−0.364**	0.300	**0.733**
Boasia chierchiae	0.218	**−0.842**	0.244
Creseis acicula	−0.133	0.071	**−0.448**
Styliola subula	−0.101	0.062	−0.017
Clio cuspidata	−0.015	0.010	0.028
Clio pyramidata	−0.190	0.166	−0.088
Diacria trispinosa	−0.011	0.035	−0.394
Cavolinia spp.	−0.079	−0.137	0.109

4.2.1. Planktonic Foraminifera

The first varimax factor (PCA1) accounts for 67.95% of the total variance (Table 1) and exhibits a bipolar character with the negative pole to be dominated by the eutrophic species *T. quinqueloba* and the positive pole by the oligotrophic *G. ruber alba* (Table 3) [76–78]. Therefore, this factor is interpreted as a productivity indicator (Figure 4a). The second factor (PCA2) explains 15.58% of the total variance (Table 1). It is characterized by positive values of the surface dwellers *G. ruber alba* and *T. quinqueloba* [78]. Negative loadings are dominated mainly by the species *G. bulloides*, which is highly dependent on enhanced food levels, upwelling, strong seasonal mixing, or freshwater inputs [79–82]. Thus, the PCA2 factor is referred to as the stratification factor (Figure 4b). The third varimax factor (PCA3) accounts for the 7.41% of the total variance (Table 1) and is interpreted as an upwelling indicator (Figure 4c) since its main positive representative species, *G. bulloides* (Table 3), is strongly associated with the seasonal upwelling [83–86]. On the other hand, the negative loadings are represented mainly by the species *G. ruber rosea*, a warm subtropical species, whose abundance is mainly controlled by the thermal stratification of the upper water column [21,22,78].

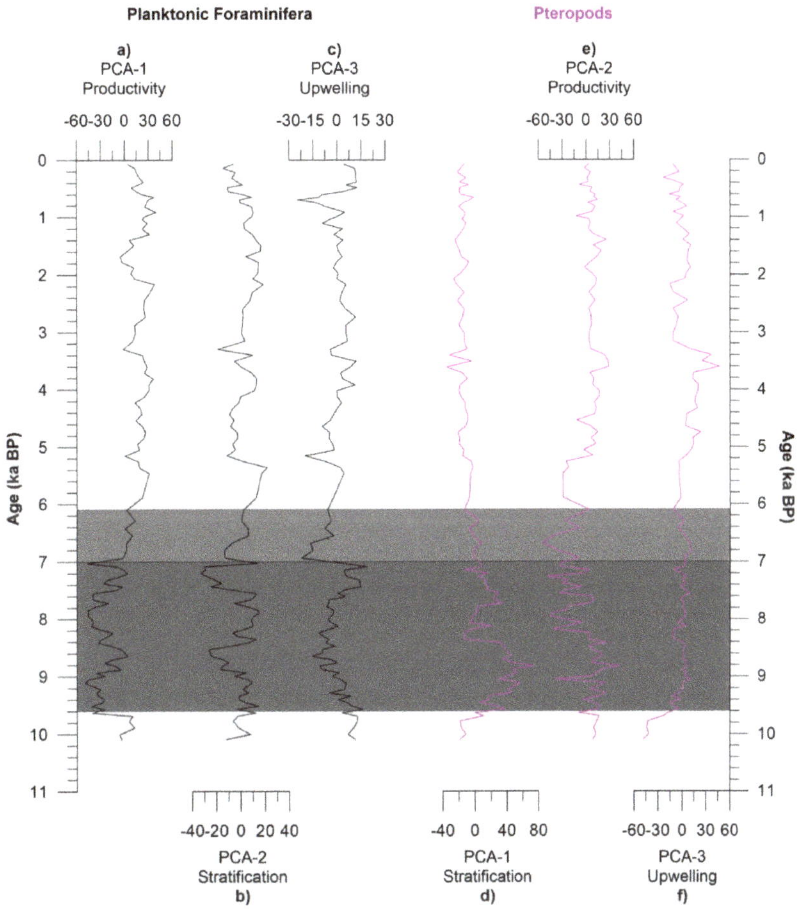

Figure 4. Environmental factors controlling planktonic foraminifera and pteropod distribution resulting from the PCA. (**a**) PCA1 of planktonic foraminifera was interpreted as a productivity factor; (**b**) PCA2 of planktonic foraminifera was interpreted as a stratification factor; (**c**) PCA3 of planktonic foraminifera was interpreted as an upwelling factor; (**d**) PCA1 of pteropods was interpreted as a stratification factor; (**e**) PCA2 of pteropods was interpreted as a productivity factor; (**f**) PCA3 of pteropods was interpreted as an upwelling factor. Gray and light gray bands correspond to the sapropel S1 and its oxidized part, respectively.

4.2.2. Pteropods

The first varimax factor (PCA1) explains 37.76% of the total variance (Table 2), with the positive loadings expressed by the mesopelagic *H. inflatus*, tolerant to a low oxygen concentration of the oxygen minimum zone (OMZ), [57] and the epipelagic *B. chierchiae* (Table 4). In contrast, negative loadings are expressed mainly by species indicative of a well-ventilated water column (*L. trochiformis* and *C. pyramidata*) [57,87]. Thus, PCA1 can be explained as a stratification factor (Figure 4d). The second factor (PCA2) explains the 27.66% of the total variance (Table 2). Its positive pole is represented mainly by the mesopelagic oligotrophic *H. inflatus* and the negative pole by the epipelagic *B. chierchiae*, which proliferates in nutrient-enriched waters (Table 4) [29,57]. Therefore, PCA2 is interpreted as a productivity factor (Figure 4e). The third factor (PCA3) describes 14.04% of the total variance (Table 2). The main representative of the positive pole, *L. trochiformis* (Table 4), is characterized as a typical upwelling species [87,88]; thus, PCA3 can be interpreted as an upwelling factor (Figure 4f).

5. Discussion

5.1. Biotopes of the North Aegean

The main assemblages identified in core AEX-15 through the cluster analysis, represent the sapropel S1 or the pre- and/or post-sapropel interval. Based on the ecological characteristics of the species participating in each assemblage, a unification of some of them into biotopes was made. Thus, each biotope is indicative of paleoenvironmental processes that affect both faunal groups.

More explicitly, the assemblage of *T. quinqueloba,* which corresponds to the sapropel S1 interval in the north Aegean, represents a biotope with low salinity and increased food availability in the surface waters [6,74,89,90]. This is explained by the larger river inflows draining the surrounding land [5,89] and supplying nutrients that fuel the productivity. In addition, the warm oligotrophic *G. ruber rosea*, and the SPRUDTS group that are also abundant in this assemblage, suggest enhanced thermal stratification of the upper water column [47,79,82]. This is further reinforced by the increased percentages of *H. inflatus* (*H. inflatus* assemblage) and the reduction/absence of mesopelagic species typical of a well-ventilated water column (e.g., *Clio cuspidata*, *C. pyramidata*, *D. trispinosa*) in pteropod's fauna. Although *Heliconoides inflatus* is a mesopelagic species, its presence in this biotope is explained by its habitat. This species adopts a variable depth habitat during its growth stages and is also tolerant to low oxygen concentration levels [57]. The presence of low salinity surface waters, as inferred from the planktonic foraminifera, is also enhanced by the high relative abundance of the epipelagic pteropod *B. chierchiae.* This species is known to proliferate in low, as well as high, salinity shallow waters [29,91,92]. Thus, both plankton assemblages (*T. quinqueloba* and *H. inflatus*) that correspond to the sapropel S1 represent a biotope with a strongly stratified water column characterized by intense OMZ, with low salinity and high fertility of the surficial waters.

The next biotope includes the *G. ruber alba* and *L. trochiformis* assemblages that correspond to the post-sapropel interval (Northgrippian and Meghalayan stages). In this biotope, oligotrophic (*G. ruber alba*, *G. ruber rosea*, and *G. rubescens*) and eutrophic (*G. bulloides*, *T. quinqueloba*, and *N. pachyderma*) species simultaneously occur, indicating a seasonal contrast in primary productivity probably related to the food availability. Seasonality is also reinforced by the continuous presence of *Globigerinita glutinata* since this species can survive, both in the oligotrophic surface and in more eutrophic waters, by changing its diet from diatoms to chrysophytes [47]. Moreover, the findings suggest a seasonal contrast in temperature as warm-characteristic species (*G. ruber alba*, *G. ruber rosea*, and *G. rubescens*) coexist with cold-indicative species (*T. quinqueloba* and *N. pachyderma*). On the other hand, the occurrence of mesopelagic pteropods along with epipelagic, suggests a well-ventilated water column [93]. The high relative abundance of *L. trochiformis* is indicative of local upwellings [87], whereas its reduction at ~3.0 ka BP, along with the increased relative abundance of *S. subula*, is suggestive of an increase in sea surface salinity [56,57,94]. Thus, our findings suggest a high seasonal biotope indicative of a well-ventilated and high saline water column as a result of the reduction in freshwater inputs. This interpretation is in accordance with clay mineral data derived from north Aegean sediments [5,95] that point to particularly dry conditions in the north Aegean catchment as well as the reduction in the surface water Black Sea outflow at ~4.5 ka BP [10,96].

The biotope characteristic of the *G. inflata* assemblage corresponds to the oxidized part of S1. The occurrence of the deep mixing species *G. inflata* and *N. pachyderma* in this assemblage reflects the prevalence of horizontally and/or vertically homogenized conditions in the water column [78,79,97]. Moreover, the biotope characteristic of the *D. trispinosa* assemblage corresponds to the pre-sapropel interval. *Diacria trispinosa* is a species typical of Atlantic-influenced waters [94,98], reflecting the influence of Atlantic-modified waters into the eastern Mediterranean basin prior to the S1 deposition. The epipelagic euryhaline species *C. acicula* is also precent in this assemblage. Its distribution pattern in the Holocene record of the Aegean sea was described for the first time in this gravity core (AEX-15; [30]), and is negatively affected by the water column stratification. Similar observations are recorded from the western continental shelf of India [57], where it appears to be related to the conspicuous differences in water column stratification linked with the vertical salinity gradient. Thus, our findings reflect a biozone with homogenized conditions in the water column influenced by Atlantic waters.

5.2. Factors Controlling Planktonic Fauna Distribution

Both faunal groups seem to be reliable indicators for past paleoenvironmental changes in this setting. The differences relay the extent to which each factor affects the fauna. In particular, the productivity factor is expressed as being more intensified in planktonic foraminifera fauna (68%) than in pteropods (28%) (Tables 1 and 2). In the case of planktonic foraminifera, this can be explained by the species *T. quinqueloba* and *G. bulloides* that render the role of productivity as the main factor, especially in the more humid and eutrophicated north Aegean region [6,60,89]. In contrast, the pteropod distribution pattern does not respond in the same way as planktonic foraminifera to changes in productivity. It is known that nutrient concentration is not a limited factor for their distribution [99], though our data suggest that fluctuations in nutrients and salinity due to the increased freshwater inputs, especially during the sapropel S1 deposition, favor the flourishment of the species *B. chierchiae* [29]. The stratification factor, on the contrary, affects the pteropod fauna (38%) more than the planktonic foraminifera (16%) (Tables 1 and 2). Our data suggest that the pteropod distribution pattern is controlled primarily by the oxygen concentration of the water column and, thus, the extent and intensity of the OMZ [29,56], which is related to the stratification of the entire water column. OMZ alterations are climatically controlled, with humid and milder climates favoring a strong and well developed OMZ, whereas cooler and/or arid periods are characterized by a more aerated OMZ [93,100]. In particular, mesopelagic species that are affected negatively by low oxygen concentration and intensified OMZ, are absent or present with very low percentages during the stratified conditions of the sapropel S1 layer. In contrast, such intense changes in species abundances of planktonic foraminifera, are not observed.

Finally, the third factor affecting equally both faunal groups' distribution in the north Aegean Sea seems to be upwelling. Upwellings control food availability and foraminifera reproductive cycles [47,101]. The participation of this factor in Holocene records of the north Aegean is explained by the presence of cyclonic and anticyclonic eddies and, thus, to the development of down and upwellings of the organic matter [60,89].

6. Conclusions

Plankton foraminiferal and pteropod faunas have been studied in core AEX-15, from the North Aegean Trough (northeastern Mediterranean), to evaluate their relationship with surface water dynamics, and to decipher the control of the Holocene productivity and stratification changes coupled with the impact of the regional eastern Mediterranean climate regime. The ecological interpretation of both of these planktonic groups, integrated with the multivariate statistical analysis data, has provided an accurate interpretation of the changes in surface water properties prior, during, and after the deposition of the most recent sapropel S1. The Q-mode cluster and principal component analyses performed in the planktonic fauna of core AEX-15 revealed three distinct assemblages in the north Aegean Sea that can be summarized into relevant biotopes, each one representative of different hydrological and climatic conditions. These biotopes correspond to the sapropel S1 and its oxidized part, pre- and, post-sapropel intervals. The PCA performed in both fauna groups, suggested that primary productivity, stratification of the water column, and upwelling are the main drivers of the planktonic foraminifera distribution. The same factors are recognized for the pteropod fauna but with a different order, indicating stratification as the primary factor controlling their distribution. Overall, the observed high-resolution water column variability allows a detailed palaeoecological reconstruction for the Holocene archive in the Aegean Sea and, furthermore, provides a notable contribution to paleoclimatic studies, facilitating intercorrelations between similar oceanographic basins and offering a better comprehension of the paleoceanographic history of these basins.

Supplementary Materials: The following are available online at https://www.mdpi.com/2077-1312/9/11/1249/s1. Table S1: Raw data of the percentages of planktonic foraminifera and pteropod species in downcore samples.

Author Contributions: Conceptualization, C.G. and G.K.; methodology, C.G., G.K. and A.A.; software, C.G.; validation, C.G., G.K. and A.A.; formal analysis, C.G.; investigation, C.G., G.K. and A.A.; resources, C.G., G.K. and A.A.; data curation, C.G. and G.K.; writing—original draft preparation, C.G.; writing—review and editing, C.G., G.K. and A.A.; visualization, G.K.; supervision, A.A. and E.K.; project administration, C.G.; funding acquisition, G.K. All authors have read and agreed to the published version of the manuscript.

Funding: This research received no external funding.

Institutional Review Board Statement: Not applicable.

Informed Consent Statement: Not applicable.

Data Availability Statement: The data presented in this study are available in Supplementary material [Table S1].

Acknowledgments: We wish to thank the Hellenic Survey of Geology and Mineral Exploration (H.S.G.M.E.) and especially Chr. Ioakim, for permitting the use of the material retrieved in the frame of "YPOTHER" project. This project was supported by the National Strategic Development Program EE (NSRF 2007–2013) and was realized at the H.S.G.M.E. in collaboration with the Institute of Oceanography of the Hellenic Centre for Marine Research (HCMR). We also thank Leonidas Giannatos for his assistance during the preparation of Figures 2 and 3.

Conflicts of Interest: The authors declare no conflict of interest.

References

1. Zervakis, V.; Georgopoulos, D.; Karageorgis, A.P.; Theocharis, A. On the response of the Aegean Sea to climatic variability: A review. *Int. J. Climatol.* **2004**, *24*, 1845–1858. [CrossRef]
2. Giorgi, F.; Lionello, P. Climate change projections for the Mediterranean region. *Global Planet. Chang.* **2008**, *63*, 90–104. [CrossRef]
3. Rohling, E.J.; Grant, K.; Bolshaw, M.; Roberts, A.; Siddall, M.; Hemleben, C.; Kucera, M. Antarctic temperature and global sea level closely coupled over the past five glacial cycles. *Nat. Geosci.* **2009**, *2*, 500–504. [CrossRef]
4. Aksu, A.E.; Yaşar, D.; Mudie, P.J. Paleoclimatic and paleoceanographic conditions leading to development of sapropel layer S1 in the Aegean Sea. *Palaeogeogr. Palaeoclim. Palaeoecol.* **1995**, *116*, 71–101. [CrossRef]
5. Rousakis, G.; Karageorgis, A.; Conispoliatis, N.; Lykousis, V. Last glacial-Holocene sediment sequences in N. Aegean basins: Structure, accumulation rates and clay mineral distribution. *Geo-Mar. Lett.* **2004**, *24*, 97–111. [CrossRef]
6. Kontakiotis, G. Late Quaternary paleoenvironmental reconstruction and paleoclimatic implications of the Aegean Sea (eastern Mediterranean) based on paleoceanographic indexes and stable isotopes. *Quatern. Int.* **2016**, *401*, 28–42. [CrossRef]
7. Poulos, S.E. The Mediterranean and Black Sea Marine System: An overview of its physico-geographic and oceanographic characteristics. *Earth-Sci. Rev.* **2020**, *200*, 103004. [CrossRef]
8. De Lange, G.J.; Thomson, J.; Reitz, A.; Slomp, C.P.; Principato, M.S.; Erba, E.; Corselli, C. Synchronous basin-wide formation and redox-controlled preservation of a Mediterranean sapropel. *Nat. Geosci.* **2008**, *1*, 606–610. [CrossRef]
9. Triantaphyllou, M.V.; Gogou, A.; Dimiza, M.D.; Kostopoulou, S.; Parinos, C.; Roussakis, G.; Geraga, M.; Bouloubassi, I.; Fleitmann, D.; Zervakis, V. Holocene climatic optimum centennial-scale paleoceanography in the NE Aegean (Mediterranean Sea). *Geo-Mar. Lett.* **2016**, *36*, 51–66. [CrossRef]
10. Kuhnt, T.; Schmiedl, G.; Ehrmann, W.; Hamann, Y.; Hemleben, C. Deep-sea ecosystem variability of the Aegean Sea during the past 22 kyr as revealed by Benthic Foraminifera. *Mar. Micropaleontol.* **2007**, *64*, 141–162. [CrossRef]
11. Schmiedl, G.; Kuhnt, T.; Ehrmann, W.; Emeis, K.-C.; Hamann, Y.; Kotthoff, U.; Dulski, P.; Pross, J. Climatic forcing of eastern Mediterranean deep-water formation and benthic ecosystems during the past 22,000 years. *Quat. Sci. Rev.* **2010**, *29*, 3006–3020. [CrossRef]
12. Louvari, M.A.; Drinia, H.; Kontakiotis, G.; Di Bella, L.; Antonarakou, A.; Anastasakis, G. Impact of latest-glacial to Holocene sea-level oscillations on central Aegean shelf ecosystems: A benthic foraminiferal palaeoenvironmental assessment of South Evoikos Gulf, Greece. *J. Mar. Syst.* **2019**, *199*, 103181. [CrossRef]

13. Vadsaria, T.; Ramstein, G.; Dutay, J.C.; Li, L.; Ayache, M.; Richon, C. Simulating the occurrence of the last sapropel event (S1): Mediterranean basin ocean dynamics simulations using Nd isotopic composition modeling. *Paleoceanogr. Paleocl.* **2019**, *34*, 237–251. [CrossRef]
14. Le Houedec, S.; Mojtahid, M.; Bicchi, E.; de Lange, G.J.; Hennekam, R. Suborbital Hydrological Variability Inferred From Coupled Benthic and Planktic Foraminiferal-Based Proxies in the Southeastern Mediterranean During the Last 19 ka. *Paleoceanogr. Paleocl.* **2020**, *35*, e2019PA003827. [CrossRef]
15. Rohling, E.J.; Marino, G.; Grant, K.M. Mediterranean climate and oceanography, and the periodic development of anoxic events (sapropels). *Earth-Sci. Rev.* **2015**, *143*, 62–97. [CrossRef]
16. Zirks, E.; Krom, M.D.; Zhu, D.; Schmiedl, G.; Goodman-Tchernov, B.N. Evidence for the Presence of Oxygen-Depleted Sapropel Intermediate Water across the Eastern Mediterranean during Sapropel S1. *ACS Earth Space Chem.* **2019**, *3*, 2287–2297. [CrossRef]
17. Tachikawa, K.; Vidal, L.; Cornuault, M.; Garcia, M.; Pothin, A.; Sonzogni, C.; Bard, E.; Menot, G.; Revel, M. Eastern Mediterranean Sea circulation inferred from the conditions of S1 sapropel deposition. *Clim. Past* **2015**, *11*, 855–867. [CrossRef]
18. Grimm, R.; Maier-Reimer, E.; Mikolajewicz, U.; Schmiedl, G.; Müller-Navarra, K.; Adloff, F.; Grant, K.M.; Ziegler, M.; Lourens, L.J.; Emeis, K.-C. Late glacial initiation of Holocene eastern Mediterranean sapropel formation. *Nat. Commun.* **2015**, *6*, 7099. [CrossRef]
19. Antonarakou, A.; Kontakiotis, G.; Zarkogiannis, S.; Mortyn, P.G.; Drinia, H.; Koskeridou, E.; Anastasakis, G. Planktonic foraminiferal abnormalities in coastal and open marine eastern Mediterranean environments: A natural stress monitoring approach in recent and early Holocene marine systems. *J. Mar. Syst.* **2018**, *181*, 63–78. [CrossRef]
20. Pallacks, S.; Ziveri, P.; Martrat, B.; Mortyn, P.G.; Grelaud, M.; Schiebel, R.; Incarbona, A.; Garcia-Orellana, J.; Anglada-Ortiz, G. Planktic foraminiferal changes in the western Mediterranean Anthropocene. *Global Planet. Change* **2021**, *204*, 103549. [CrossRef]
21. Avnaim-Katav, S.; Herut, B.; Rahav, E.; Katz, T.; Weinstein, Y.; Alkalay, R.; Berman-Frank, I.; Zlatkin, O.; Almogi-Labin, A. Sediment trap and deep sea coretop sediments as tracers of recent changes in planktonic foraminifera assemblages in the southeastern ultra-oligotrophic Levantine Basin. *Deep Sea Res. Part II Top. Stud. Oceanogr.* **2020**, *171*, 104669. [CrossRef]
22. Zarkogiannis, S.; Kontakiotis, G.; Antonarakou, A. Recent planktonic foraminifera population and size response to Eastern Mediterranean hudrography. *Rev. De Micropaleontol.* **2020**, *69*, 100450. [CrossRef]
23. Kontakiotis, G.; Efstathiou, E.; Zarkogiannis, S.D.; Besiou, E.; Antonarakou, A. Latitudinal Differentiation among Modern Planktonic Foraminiferal Populations of Central Mediterranean: Species–Specific Distribution Patterns and Size Variability. *J. Mar. Sci. Eng.* **2021**, *9*, 551. [CrossRef]
24. Ofstad, S.; Zamelczyk, K.; Kimoto, K.; Chierici, M.; Fransson, A.; Rasmussen, T.L. Shell density of planktonic foraminifera and pteropod species *Limacina helicina* in the Barents Sea: Relation to ontogeny and water chemistry. *PLoS ONE* **2021**, *16*, e0249178. [CrossRef]
25. Manno, C.; Bednaršek, N.; Tarling, G.A.; Peck, V.L.; Comeau, S.; Adhikari, D.; Bakker, D.C.E.; Bauerfeind, E.; Bergan, A.J.; Berning, M.I.; et al. Shelled pteropods in peril: Assessing vulnerability in a high CO_2 ocean. *Earth-Sci. Rev.* **2017**, *169*, 132–145. [CrossRef]
26. Casford, J.S.L.; Rohling, E.J.; Abu-Zied, R.H.; Fontanier, C.; Jorissen, F.J.; Leng, M.J.; Schmiedl, G.; Thomson, J. A dynamic concept for eastern Mediterranean circulation and oxygenation during sapropel formation. *Palaeogeogr. Palaeoclim. Palaeoecol.* **2003**, *190*, 103–119. [CrossRef]
27. Kontakiotis, G.; Antonarakou, A.; Zachariasse, W.J. Late Quaternary palaeoenvironmental changes in the Aegean Sea: Interrelations and interactions between north and south Aegean Sea. *Bull. Geol. Soc. Gr.* **2013**, *47*, 167–177. [CrossRef]
28. Wall-Palmer, D.; Smart, C.W.; Hart, M.B.; Leng, M.J.; Borghini, M.; Manini, E.; Aliani, S.; Conversi, A. Late Pleistocene pteropods, heteropods and planktonic foraminifera from the Caribbean Sea, Mediterranean Sea and Indian Ocean. *Micropaleontology* **2014**, *60*, 557–578.
29. Giamali, C.; Kontakiotis, G.; Koskeridou, E.; Ioakim, C.; Antonarakou, A. Key Environmental Factors Controlling Planktonic Foraminiferal and Pteropod Community's Response to Late Quaternary Hydroclimate Changes in the South Aegean Sea (Eastern Mediterranean). *J. Mar. Sci. Eng.* **2020**, *8*, 709. [CrossRef]
30. Giamali, C.; Koskeridou, E.; Antonarakou, A.; Ioakim, C.; Kontakiotis, G.; Karageorgis, A.P.; Roussakis, G.; Karakitsios, V. Multiproxy ecosystem response of abrupt Holocene climatic changes in the northeastern Mediterranean sedimentary archive and hydrologic regime. *Quat. Res.* **2019**, *92*, 665–685. [CrossRef]
31. Checa, H.; Margaritelli, G.; Pena, L.D.; Frigola, J.; Cacho, I.; Rettori, R.; Lirer, F. High resolution paleo-environmental changes during the Sapropel 1 in the North Ionian Sea, central Mediterranean. *Holocene* **2020**, *30*, 1504–1515. [CrossRef]
32. Casford, J.S.L.; Abu-Zied, R.; Rohling, E.J.; Cooke, S.; Fontanier, C.; Leng, M.; Millard, A.; Thomson, J. A stratigraphically controlled multiproxy chronostratigraphy for the eastern Mediterranean. *Paleoceanography* **2007**, *22*, PA4215. [CrossRef]
33. Siani, G.; Paterne, M.; Colin, C. Late glacial to Holocene planktic foraminifera bioevents and climatic record in the South Adriatic Sea. *J. Quat. Sci.* **2010**, *25*, 808–821. [CrossRef]
34. Drinia, H.; Antonarakou, A.; Tsourou, T.; Kontakiotis, G.; Psychogiou, M.; Anastasakis, G. Foraminifera eco-biostratigraphy of the southern Evoikos outer shelf, central Aegean Sea, during MIS 5 to present. *Cont. Shelf Res.* **2016**, *126*, 36–49. [CrossRef]
35. Antonarakou, A.; Kontakiotis, G.; Karageorgis, A.P.; Besiou, E.; Zarkogiannis, S.; Drinia, H.; Mortyn, G.P.; Tripsanas, E. Eco-biostratigraphic advances on late Quaternary geochronology and palaeoclimate: The marginal Gulf of Mexico analogue. *Geol. Q.* **2019**, *63*, 178–191. [CrossRef]

36. Zarkogiannis, S.D.; Antonarakou, A.; Tripati, A.; Kontakiotis, G.; Mortyn, P.G.; Drinia, H.; Greaves, M. Influence of surface ocean density on planktonic foraminifera calcification. *Sci. Rep.* **2019**, *9*, 533. [CrossRef]
37. Mojtahid, M.; Manceau, R.; Schiebel, R.; Hennekam, R.; de Lange, G.J. Thirteen thousand years of southeastern Mediterranean climate variability inferred from an integrative planktic foraminiferal-based approach. *Paleoceanography* **2015**, *30*, 402–422. [CrossRef]
38. Zachariasse, W.J.; Kontakiotis, G.; Lourens, L.J.; Antonarakou, A. The Messinian of Agios Myron (Crete, Greece): A key to better understanding of diatomite formation on Gavdos (south of Crete). *Palaeogeogr. Palaeoclim. Palaeoecol.* **2021**, *581*, 110633. [CrossRef]
39. Antonarakou, A.; Kontakiotis, G.; Mortyn, P.G.; Drinia, H.; Sprovieri, M.; Besiou, E.; Tripsanas, E. Biotic and geochemical ($\delta^{18}O$, $\delta^{13}C$, Mg/Ca, Ba/Ca) responses of *Globigerinoides ruber* morphotypes to upper water column variations during the last deglaciation, Gulf of Mexico. *Geochim. Cosmochim. Acta* **2015**, *170*, 69–93. [CrossRef]
40. Vasiliev, I.; Karakitsios, V.; Bouloubassi, I.; Agiadi, K.; Kontakiotis, G.; Antonarakou, A.; Triantaphyllou, M.; Gogou, A.; Kafousia, N.; de Rafélis, M.; et al. Large Sea Surface Temperature, Salinity, and Productivity-Preservation Changes Preceding the Onset of the Messinian Salinity Crisis in the Eastern Mediterranean Sea. *Paleoceanogr. Paleocl.* **2019**, *34*, 182–202. [CrossRef]
41. Kontakiotis, G.; Karakitsios, V.; Mortyn, P.G.; Antonarakou, A.; Drinia, H.; Anastasakis, G.; Agiadi, K.; Kafousia, N.; De Rafelis, M. New insights into the early Pliocene hydrographic dynamics and their relationship to the climatic evolution of the Mediterranean Sea. *Palaeogeogr. Palaeoclim. Palaeoecol.* **2016**, *459*, 348–364. [CrossRef]
42. Kontakiotis, G.; Besiou, E.; Antonarakou, A.; Zarkogiannis, S.D.; Kostis, A.; Mortyn, P.G.; Moissette, P.; Cornée, J.J.; Schulbert, C.; Drinia, H.; et al. Decoding sea surface and paleoclimate conditions in the eastern Mediterranean over the Tortonian-Messinian Transition. *Palaeogeogr. Palaeoclim. Palaeoecol.* **2019**, *534*, 109312. [CrossRef]
43. Margaritelli, G.; Cisneros, M.; Cacho, I.; Capotondi, L.; Vallefuoco, M.; Rettori, R.; Lirer, F. Climatic variability over the last 3000 years in the central-western Mediterranean Sea (Menorca Basin) detected by planktonic foraminifera and stable isotope records. *Global Planet. Chang.* **2018**, *169*, 179–187. [CrossRef]
44. Margaritelli, G.; Cacho, I.; Català, A.; Barra, M.; Bellucci, L.G.; Lubritto, C.; Rettori, R.; Lirer, F. Persistent warm Mediterranean surface waters during the Roman period. *Sci. Rep.* **2020**, *10*, 10431. [CrossRef]
45. Margaritelli, G.; Vallefuoco, M.; Di Rita, F.; Capotondi, L.; Bellucci, L.G.; Insinga, D.D.; Petrosino, P.; Bonomo, S.; Cacho, I.; Cascella, A. Marine response to climate changes during the last five millennia in the central Mediterranean Sea. *Global Planet. Chang.* **2016**, *142*, 53–72. [CrossRef]
46. Thirumalai, K.; Richey, J.N.; Quinn, T.M. Holocene evolution of sea-surface temperature and salinity in the Gulf of Mexico. *Paleoceanogr. Paleocl.* **2021**, *36*, e2021PA004221. [CrossRef]
47. Hemleben, C.; Spindler, M.; Anderson, O. *Modern Planktic Foraminifera*; Springer-Verlag: New York, NY, USA, 1989; Volume 22.
48. Kucera, M. Chapter six planktonic foraminifera as tracers of past oceanic environments. *Dev. Mar. Geol.* **2007**, *1*, 213–262.
49. Lalli, C.M.; Gilmer, R.W. *Pelagic Snails: The Biology of Holoplanktonic Gastropod Mollusks*; Stanford University Press: Redwood City, CA, USA, 1989.
50. Pierrot-Bultsa, A.C.; Peijnenburga, K.T.C.A. Pteropods. In *Encyclopedia of Marine Geosciences*; Harff, J., Meschede, M., Petersen, S., Thiede, J., Eds.; Springer: Dordrecht, The Netherlands, 2015. [CrossRef]
51. Hunt, B.; Pakhomov, E.; Hosie, G.W.; Siegel, V.; Ward, P.; Bernard, K. Pteropods in Southern Ocean ecosystems. *Prog. Oceanogr.* **2008**, *78*, 193–221. [CrossRef]
52. Berner, R.A.; Honjo, S. Pelagic sedimentation of aragonite: Its geochemical significance. *Science* **1981**, *211*, 940–942. [CrossRef]
53. Herman, Y. Evidence of climatic changes in Red Sea cores. In *Proceedings of the VII Congress International Association for Quaternary Research*; University of Utah Press: Salt Lake City, UT, USA, 1968; pp. 325–348.
54. Herman, Y. Vertical and horizontal distribution of pteropods in Quaternary sequences. In *The Micropalaeontology of Oceans*; Funnell, B.M., Reidel, W.R., Eds.; Cambridge University Press: Cambridge, UK, 1971; pp. 463–486.
55. Chen, C. *Pteropods in the Hot Brine Sediments of the Red Sea*; Springer: Berlin/Heidelberg, Germany, 1969. [CrossRef]
56. Almogi-Labin, A.; Edelman-Furstenberg, Y.; Hemleben, C. Variations in the biodiversity of thecosomatous pteropods during the Late Quaternary as a response to environmental changes in the Gulf of Aden-Red Sea-Gulf of Aqaba ecosystem. In *The Improbable Gulf. Environment, Biodiversity and Preservation*; Por, D.e.A.-E., Ed.; The Hebrew University Magnes Press: Jerusalem, Israel, 2008; pp. 31–48.
57. Singh, A.; Nisha, N.R.; Joydas, T.; Joydas, V. Distribution patterns of Recent pteropods in surface sediments of the western continental shelf of India. *J. Micropalaeontol.* **2005**, *24*, 39. [CrossRef]
58. Wall-Palmer, D.; Smart, C.W.; Hart, M.B. In-life pteropod shell dissolution as an indicator of past ocean carbonate saturation. *Quat. Sci. Rev.* **2013**, *81*, 29–34. [CrossRef]
59. Wall-Palmer, D.; Jones, M.T.; Hart, M.B.; Fisher, J.K.; Smart, C.W.; Hembury, D.J.; Palmer, M.R.; Fones, G.R. Explosive volcanism as a cause for mass mortality of pteropods. *Mar. Geol.* **2011**, *282*, 231–239. [CrossRef]
60. Lykousis, V.; Chronis, G.; Tselepides, A.; Price, N.B.; Theocharis, A.; Siokou-Frangou, I.; Van Wambeke, F.; Danovaro, R.; Stavrakakis, S.; Duineveld, G.; et al. Major outputs of the recent multidisciplinary biogeochemical researches undertaken in the Aegean Sea. *J. Mar. Syst.* **2002**, *33*, 313–334. [CrossRef]
61. Papanikolaou, D.; Alexandri, M.; Nomikou, P.; Ballas, D. Morphotectonic structure of the western part of the North Aegean Basin based on swath bathymetry. *Mar. Geol.* **2002**, *190*, 465–492. [CrossRef]

62. Poulos, S.E.; Drakopoulos, P.G.; Collins, M.B. Seasonal variability in sea surface oceanographic conditions in the Aegean Sea (Eastern Mediterranean): An overview. *J. Mar. Syst.* **1997**, *13*, 225–244. [CrossRef]
63. Theocharis, A.; Georgopoulos, D. Dense water formation over the Samothraki and Limnos Plateaux in the north Aegean Sea (Eastern Mediterranean Sea). *Cont. Shelf Res.* **1993**, *13*, 919–939. [CrossRef]
64. Velaoras, D.; Lascaratos, A. Deep water mass characteristics and interannual variability in the North and Central Aegean Sea. *J. Mar. Syst.* **2005**, *53*, 59–85. [CrossRef]
65. Karageorgis, A.P.; Kaberi, H.G.; Tengberg, A.; Zervakis, V.; Hall, P.O.J.; Anagnostou, C.L. Comparison of particulate matter distribution, in relation to hydrography, in the mesotrophic Skagerrak and the oligotrophic northeastern Aegean Sea. *Cont. Shelf Res.* **2003**, *23*, 1787–1809. [CrossRef]
66. Androulidakis, Y.S.; Krestenitis, Y.N.; Psarra, S. Coastal upwelling over the North Aegean Sea: Observations and simulations. *Cont. Shelf Res.* **2017**, *149*, 32–51. [CrossRef]
67. Krom, M.D.; Kress, N.; Brenner, S.; Gordon, L.I. Phosphorus limitation of primary productivity in the eastern Mediterranean Sea. *Limnol. Oceanogr.* **1991**, *36*, 424–432. [CrossRef]
68. Stergiou, K.I.; Christou, E.D.; Georgopoulos, D.; Zenetos, A.; Souvermezoglou, C. The Hellenic Seas: Physics, chemistry, biology and fisheries. *Oceanogr. Mar. Biol.* **1997**, *35*, 415–538.
69. Kontakiotis, G.; Antonarakou, A.; Mortyn, P.G.; Drinia, H.; Anastasakis, G.; Zarkogiannis, S.; Möbius, J. Morphological recognition of *Globigerinoides ruber* morphotypes and their susceptibility to diagenetic alteration in the eastern Mediterranean Sea. *J. Mar. Syst.* **2017**, *174*, 12–24. [CrossRef]
70. Zarkogiannis, S.D.; Kontakiotis, G.; Gkaniatsa, G.; Kuppili, V.S.C.; Marathe, S.; Wanelik, K.; Lianou, V.; Besiou, E.; Makri, P.; Antonarakou, A. An Improved Cleaning Protocol for Foraminiferal Calcite from Unconsolidated Core Sediments: HyPerCal—A New Practice for Micropaleontological and Paleoclimatic Proxies. *J. Mar. Sci. Eng.* **2020**, *8*, 998. [CrossRef]
71. Patterson, R.T.; Fishbein, E. Re-examination of the statistical methods used to determine the number of point counts needed for micropaleontological quantitative research. *J. Paleontol.* **1989**, *63*, 245–248. [CrossRef]
72. Davis, J.C.; Sampson, R.J. *Statistics and Data Analysis in Geology*; Wiley New York: New York, NY, USA, 1986; Volume 646.
73. Hammer, Ø.; Harper, D.A.T.; Ryan, P.D. PAST: Paleontological Statistics software package for education and data analysis. *Palaeontol. Electron.* **2001**, *4*, 9.
74. Rohling, E.; Jorissen, F.; Stigter, H. 200 Year interruption of Holocene sapropel formation in the Adriatic Sea. *J. Micropalaeontol.* **1997**, *16*, 97–108. [CrossRef]
75. Fairbanks, R.; Wiebe, P. Foraminifera and Chlorophyll Maximum: Vertical Distribution, Seasonal Succession, and Paleoceanographic Significance. *Science* **1980**, *209*, 1524–1526. [CrossRef]
76. Lourens, L.; Hilgen, F.; Gudjonsson, L.; Zachariasse, W.J. Late Pliocene to early Pleistocene astonomically forced sea surface productivity and temperature variations in the Mediterranean. *Mar. Micropaleontol.* **1992**, *19*, 49–78. [CrossRef]
77. Thunell, R. Distribution of recent planktonic foraminifera in surface sediments of the Mediterranean Sea. *Mar. Micropaleontol.* **1978**, *3*, 147–173. [CrossRef]
78. Pujol, C.; Vergnaud-Grazzini, C. Distribution patterns of live planktic foraminifers as related to regional hydrology and productive systems of the Mediterranean Sea. *Mar. Micropaleontol.* **1995**, *25*, 187–217. [CrossRef]
79. Rohling, E.J.; Jorissen, F.; Grazzini, C.V.; Zachariasse, W.J. Northern Levantine and Adriatic Quaternary planktic foraminifera; Reconstruction of paleoenvironmental gradients. *Mar. Micropaleontol.* **1993**, *21*, 191–218. [CrossRef]
80. Jonkers, L.; Kučera, M. Global analysis of seasonality in the shell flux of extant planktonic Foraminifera. *Biogeosciences* **2015**, *12*, 2207–2226. [CrossRef]
81. Weinkauf, M.F.G.; Kunze, J.G.; Waniek, J.J.; Kučera, M. Seasonal variation in shell calcification of planktonic foraminifera in the NE Atlantic reveals species-specific response to temperature, productivity, and optimum growth conditions. *PLoS ONE* **2016**, *11*, e0148363. [CrossRef] [PubMed]
82. Schiebel, R.; Hemleben, C. *Planktic Foraminifers in the Modern Ocean*; Springer: Berlin/Heidelberg, Germany, 2017.
83. Naidu, P.D. Distribution patterns of Recent planktonic foraminifera in surface sediments of the western continental margin of India. *Mar. Geol.* **1993**, *110*, 403–418. [CrossRef]
84. Peterson, L.C.; Overpeck, J.T.; Kipp, N.G.; Imbrie, J. A high-resolution Late Quaternary upwelling record from the anoxic Cariaco Basin, Venezuela. *Paleoceanography* **1991**, *6*, 99–119. [CrossRef]
85. Prell, W.L.; Curry, W.B. Faunal and isotopic indices of monsoonal upwelling-western arabian sea. *Oceanol. Acta* **1981**, *4*, 91–98.
86. Cullen, J.L.; Prell, W.L. Planktonic foraminifera of the northern Indian Ocean: Distribution and preservation in surface sediments. *Mar. Micropaleontol.* **1984**, *9*, 1–52. [CrossRef]
87. Bé, A.W.H.; Gilmer, R.W. A zoogeographic and taxonomic review of euthecosomatous Pteropoda. *Oceanic Micropaleontol.* **1977**, *1*, 733–808.
88. Almogi-Labin, A.; Hemleben, C.; Deuser, W.G. Seasonal variation in the flux of euthecosomatous pteropods collected in a deep sediment trap in the Sargasso Sea. *Deep Sea Res. Part I Oceanogr. Res. Pap.* **1988**, *35*, 441–464. [CrossRef]
89. Geraga, M.; Ioakim, C.; Lykousis, V.; Tsaila-Monopolis, S.; Mylona, G. The high-resolution palaeoclimatic and palaeoceanographic history of the last 24,000years in the central Aegean Sea, Greece. *Palaeogeogr. Palaeoclim. Palaeoecol.* **2010**, *287*, 101–115. [CrossRef]

90. Zachariasse, W.; Jorissen, F.; Perissoratis, C.; Rohling, E.; Tsapralis, V. Late quaternary foraminiferal changes and the nature of sapropel S1 in Skopelos Basin. In Proceedings of the 5th Hellenic Symposium of Oceanography and Fisheries, Kavala, Greece, 15–18 April 1997; pp. 391–394.
91. Almogi-Labin, A.; Hemleben, C.; Meischner, D.; Erlenkeuser, H. Paleoenvironmental events during the last 13,000 years in the central Red Sea as recorded by pteropoda. *Paleoceanography* **1991**, *6*, 83–98. [CrossRef]
92. Rottman, M.L. Net tow and surface sediment distributions of pteropods in the South China Sea region: Comparison and oceanographic implications. *Mar. Micropaleontol.* **1980**, *5*, 71–110. [CrossRef]
93. Sijinkumar, A.V.; Bejugam, N.; Guptha, M.V.S. Late Quaternary record of pteropod preservation from the Andaman Sea. *Mar. Geol.* **2010**, *275*, 221–229. [CrossRef]
94. Buccheri, G.; Capretto, G.; Di Donato, V.; Esposito, P.; Ferruzza, G.; Pescatore, T.; Russo Ermolli, E.; Senatore, M.R.; Sprovieri, M.; Bertoldo, M.; et al. A high resolution record of the last deglaciation in the southern Tyrrhenian Sea: Environmental and climatic evolution. *Mar. Geol.* **2002**, *186*, 447–470. [CrossRef]
95. Ehrmann, W.; Schmiedl, G.; Hamann, Y.; Kuhnt, T.; Hemleben, C.; Siebel, W. Clay minerals in late glacial and Holocene sediments of the northern and southern Aegean Sea. *Palaeogeogr. Palaeoclim. Palaeoecol.* **2007**, *249*, 36–57. [CrossRef]
96. Sperling, M.; Schmiedl, G.; Hemleben, C.; Emeis, K.C.; Erlenkeuser, H.; Grootes, P.M. Black Sea impact on the formation of eastern Mediterranean sapropel S1? Evidence from the Marmara Sea. *Palaeogeogr. Palaeoclim. Palaeoecol.* **2003**, *190*, 9–21. [CrossRef]
97. Casford, J.S.L.; Rohling, E.J.; Abu-Zied, R.; Cooke, S.; Fontanier, C.; Leng, M.; Lykousis, V. Circulation changes and nutrient concentrations in the late Quaternary Aegean Sea: A nonsteady state concept for sapropel formation. *Paleoceanography* **2002**, *17*, 1024–1034. [CrossRef]
98. Rampal, J. Biodiversité et biogéographie chez les Cavoliniidae (Mollusca, Gastropoda, Opisthobranchia, Euthecosomata). Régions faunistiques marines. *Zoosystema* **2002**, *24*, 209–258.
99. Johnson, R.; Manno, C.; Ziveri, P. Spring distribution of shelled pteropods across the Mediterranean Sea. *Biogeosci. Discuss.* **2020**, *2020*, 1–23. [CrossRef]
100. Almogi-Labin, A.; Hemleben, C.; Meischner, D. Carbonate preservation and climatic changes in the central Red Sea during the last 380 kyr as recorded by pteropods. *Mar. Micropaleontol.* **1998**, *33*, 87–107. [CrossRef]
101. Thunell, R.; Reynolds, L.S. Planktonic Foraminifera faunal and stable isotopic indices of upwelling: A sediment trap study in the San Pedro Basin, Southern California Bight. In *Upwelling Systems: Evolution Since the Early Miocene*; Summerhayes, C.P., Prell, W.L., Emeis, K.C., Eds.; Geological Society London: London, UK, 1992; Volume 64, pp. 77–91.

Journal of
Marine Science and Engineering

MDPI

Article

Geochemical and Mineral Properties of Quaternary Deep-Sea Sediments in the Central-Tropical Pacific and Its Response to the Mid-Pleistocene Transition

Haifeng Wang [1,2], Liang Yi [3,*], Xiguang Deng [1,2] and Gaowen He [1,2]

[1] Southern Marine Science and Engineering Guangdong Laboratory (Guangzhou), Guangzhou 511458, China; wanghaifeng@mail.cgs.gov.cn (H.W.); dxiguang@mail.cgs.gov.cn (X.D.); hgaowen@mail.cgs.gov.cn (G.H.)
[2] Key Laboratory of Marine Resources, Ministry of Nature Resources, Guangzhou Marine Geological Survey, China Geological Survey, Guangzhou 510075, China
[3] State Key Laboratory of Marine Geology, Tongji University, Shanghai 200092, China
* Correspondence: yiliang@tongji.edu.cn

Abstract: Global climate and oceanic water masses have undergone profound changes during the middle Pleistocene transition; however, due to a lack of foraminiferal fossils, the nonfossiliferous pelagic deposits were less detected in previous reports. In this work, a gravity core from the Kamehameha Basin in the Central Pacific was studied in terms of magnetostratigraphy, clay mineral and geochemical elements. The main results are: (1) nine magnetozones are recognized in the core, which can be correlated to the geomagnetic polarity timescale from chrons C2n to C1n; (2) smectite is the dominant clay mineral, and the others are illite, chlorite and kaolinite; and (3) the sediments are mainly composed of Al_2O_3, Fe_2O_3, MnO, Na_2O and TiO_2. Based on these results, a geochronological framework for the study area was established, and the depositional rates are estimated as 3–7 m/Myr in the Quaternary, showing an increase during the middle Pleistocene transition. By comparing the findings to various paleoenvironmental processes, it is inferred that the increased sedimentation in the Kamehameha Basin may have resulted from the induced weathering processes and the strengthened aeolian inputs from inner Asia. Moreover, regional circulation related to bottom water evolution has experienced a rapid reorganization across the middle Pleistocene transition. All these findings illustrate the potential of deep-sea sediments in the central tropical Pacific in revealing some key features in paleoclimatology and paleoceanography, which are worthy of further investigation in the future.

Keywords: magnetostratigraphy; clay minerals; geochemical properties; deep-sea sediments; mid-Pleistocene transition

Citation: Wang, H.; Yi, L.; Deng, X.; He, G. Geochemical and Mineral Properties of Quaternary Deep-Sea Sediments in the Central-Tropical Pacific and Its Response to the Mid-Pleistocene Transition. *J. Mar. Sci. Eng.* **2021**, *9*, 1254. https://doi.org/10.3390/jmse9111254

Academic Editors: Dmitry A. Ruban, George Kontakiotis and Assimina Antonarakou

Received: 5 October 2021
Accepted: 5 November 2021
Published: 12 November 2021

1. Introduction

The past climate was dominated by a symmetric 41 kyr cycle in the early Pleistocene, and by an asymmetric 100 kyr cycle since the middle Pleistocene [1], and this change in dominant cycles is the so-called Middle Pleistocene transition (MPT, ~1.45–0.9 Ma). Over the past decades, many efforts have been made to reveal the characteristics and mechanism of the MPT [2–4], and a major shift at ~900 ka was commonly found in paleoclimatic records, such as the LR04 $\delta^{18}O$ isotopic stack [5], ice-rafted debris into the Norwegian Sea [6], North Atlantic sea surface temperatures (SSTs) at DSDP Site 607 [7] and Eastern Tropical Atlantic SSTs at ODP Site 1077 off the west coast of Africa [8]. However, deep-sea sedimentary processes below the calcite compensation depth (CCD) responding to the MPT are not yet well documented.

The pelagic sediments in the tropical Pacific may have fully recorded changes in paleoceanography, provenance and global climate, and their characteristics can be employed to reveal such information [9].

The Kamehameha Basin connects the Central Pacific and the Eastern Pacific (Figure 1). The Lower Circumpolar deep water (LCDW) and the Antarctic bottom water (AABW) flowed into the open Eastern Pacific Basin through the Horizon and Clarion Passages [10]. This region is an area with significant interactions between various systems: (1) the south is the high-productivity equatorial Pacific belt, and the north is controlled by the North Pacific Gyre [11]; (2) the sediments contain biogenic debris and aeolian dust; (3) volcanogenic materials are evident from the Hawaii hotspot [12]; and (4) fresh volcanic ashes are commonly found. Thus, sedimentary properties of the deep-sea sediments in the Kamehameha Basin may record the interactions between various climatic and oceanic systems.

However, such studies on the nonfossiliferous sediments were less studied in previous research due to age uncertainties [11,13], limiting our understanding of how deep ocean was affected during global changes. In this work, we studied a gravity core PC15 collected from the Kamehameha Basin. Combining the results of magnetostratigraphy, clay minerals and geochemistry, the properties of the deep-sea sediments were well studied in the Quaternary. By comparing the results of the core with various paleoenvironmental proxies during the MPT, we attempt to study (1) the sedimentary processes, (2) changes in the provenance, and (3) bottom water evolution in the study area.

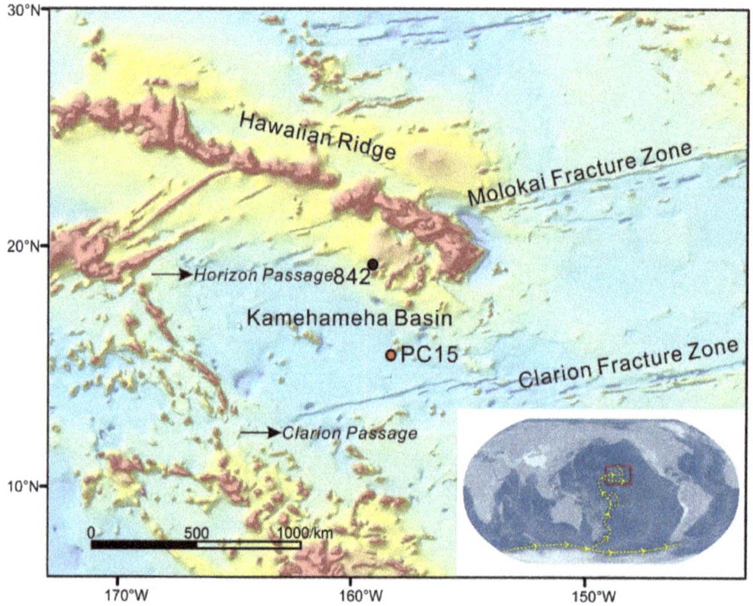

Figure 1. Schematic map of the central Pacific region showing the study sites and other referenced sites. The yellow dashed line in the bottom right insert shows the paths of the Lower Circumpolar deep water (LCDW) and the Antarctic Bottom Water (AABW). The base data are from ref. [14] and generated using the open and free software DIVA–GIS 7.5 (http://www.diva-gis.org/, accessed on 10 September 2021).

2. Materials and Methods

2.1. The Studied Core

Core PC15, located between the Horizon Passage and the Clarion Passage (158.25° W, 15.53° N, 5525 m water depth), was collected in 2013 by the Guangzhou Marine Geological Survey with a length of 5.65 m (Figure 1). The oceanographic setting of the studied area is dominated by the LCDW and the AABW [11,15]. Since the locality is generally below the CCD (~4500 m [16]), the core mainly contains homogenous carbonate-free muds.

According to the lithological description (Figure 2), core PC15 can be divided into two units around 0.92 m: the upper Unit I contains yellow muds (10YR7/4) with black laminates with volcanic ashes, and the lower Unit II contains clay in pale brown (10YR8/2) and clay with zeolite in dark brown (10YR3/3).

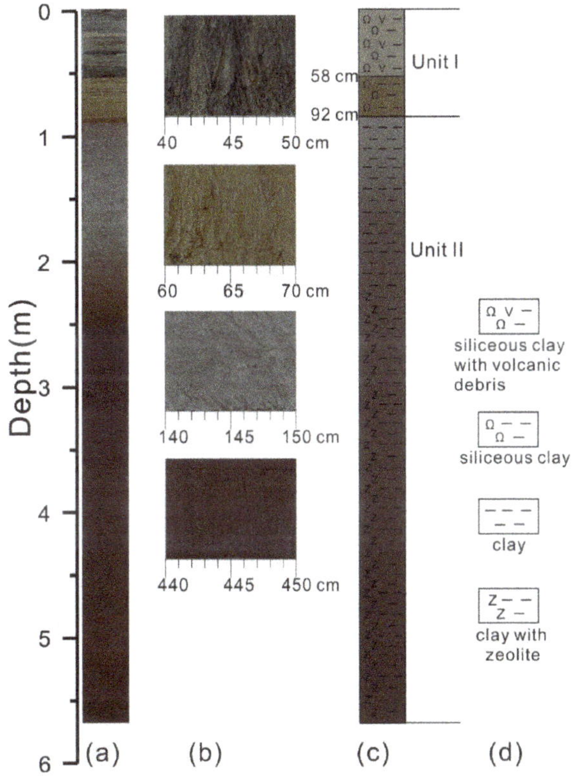

Figure 2. Photography (**a**,**b**) and lithological changes (**c**) of core PC15; (**d**) lithological legends.

2.2. Demagnetization

Paleomagnetic samples were continuously collected using nonmagnetic plastic cubic boxes (2 cm × 2 cm × 2 cm). In total, 282 samples were obtained and subject to stepwise alternating field (AF) demagnetization up to a peak field of 90 mT (13 steps). The natural remanent magnetization (NRM) was measured using a three-axis cryogenic magnetometer (2G Enterprise Model 755-4K, USA) installed in magnetically shielded room (residual fields <300 nT) at the State Key Laboratory of Marine Geology, Tongji University. Characteristic remanent magnetization (ChRM) directions were determined using principal component analysis [17] implemented by the PuffinPlot package [18], with at least four consecutive demagnetization steps.

2.3. Clay Minerals Analysis

A total of 28 samples were taken for clay mineral analysis by X-ray diffraction (XRD) at a sampling interval of 20 cm continuously from the top to the bottom of core PC15. The XRD clay mineral study was carried out on the <2 μm fraction, which was separated by conventional Stokes' settling after the removal of carbonate and organic matter by acetic acid (15%) and hydrogen peroxide (10%), respectively. Clay minerals were then identified by XRD using a (Rigaku) D/Max 2500PC 18 kW powder diffractometer (SYM125) in the Guangzhou Marine Geological Survey, with an X-ray wavelength λ = 1.5418Å

(CuKα) and a scanning rate of 2°(2θ)/min at 0.02°(2θ). Each sample was measured 3 times under conditions of air-drying, ethylene glycol solvation, and heating at 490 °C for 2 h at atmospheric pressure [19]. Clay minerals were identified according to the position of the (001) series of basal reflections observed on the XRD diagrams [19,20].

2.4. Element Analysis

For trace element content analysis, 28 bulk sediments were completely dissolved by HF and HNO_3 solutions. A sample of about 50 mg was weighed and transferred into a pre-cleaned Teflon beaker followed by the addition of ultra-pure 1 mL of HF and 1 mL of HNO_3 solution. The beakers were then placed in steel cans and subjected to high temperature (185 °C) and high pressure. After 36 h, the solution was dried on a hotplate. The residues were fully digested using a mixture of concentrated 2 mL HNO_3 and 3 mL Milli-Q water. Thereafter, the beakers were placed into steel cans at 120 °C for 5 h. After cooling, the solution was diluted to 20 mL with Milli-Q water. Major elements were analyzed with ICP-OES (Optima 8300, PerkinElmer, MA, USA), and trace elements were measured via ICP-MS (X Series2, Thermo Fisher Scientific, MA, USA). All the samples were analyzed at Guangzhou Marine Geological Survey. Certified reference material (GBW07315) was used for quality control. Precision and accuracy were both better than 5%.

3. Results

3.1. Magnetostratigraphic Results

In general, the remanence gradually decreases subject to AF demagnetization. About 50% of the NRMs was removed at ~25 mT, and up to a peak field of 80 mT, more than 90% of the remanence was removed. The normal inclinations are from 5° to 40°, and the reverse ones from −8° to −45°. The characteristics of AF stepwise of the representative samples were displayed in the form of orthogonal diagrams (Figure 3).

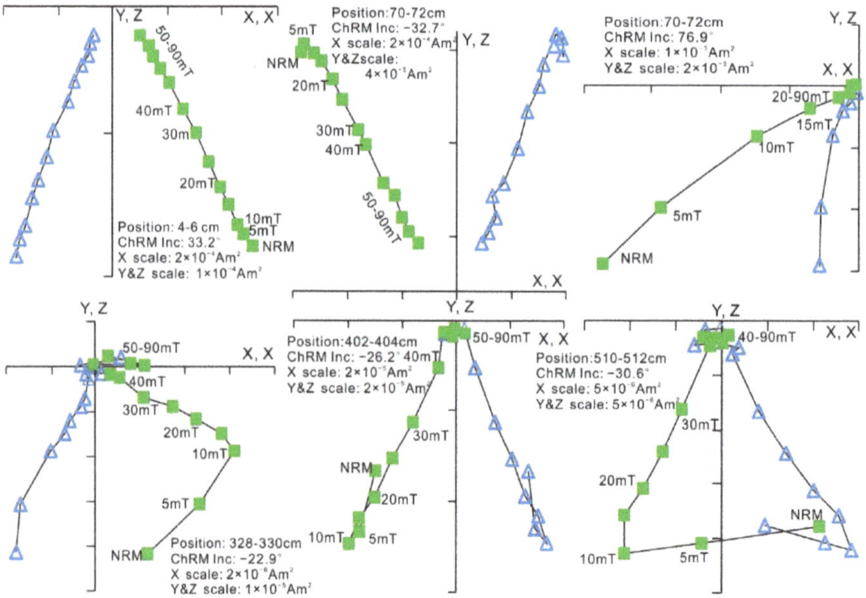

Figure 3. Orthogonal diagrams of the stepwise AF demagnetization of representative samples. Open triangles represent the horizontal planes, while solid squares represent vertical planes.

According to the obtained ChRMs, five normal (n1–n5) and four reverse magnetozones (r1–r4) were recognized. Considering the homogeneous sedimentary lithology and taking the results at ODP Site 842B as a reference [13], the magnetozones can be correlated to the geomagnetic polarity time scale (GPTS) [21]. As particularly shown in Figure 4, the normal magnetozones n1, n2, n3, n4 and n5 are correlate to Chron C1n (Brunhes, 0–0.781 Ma), C1r.1n subchron (Jaramillo, 0.988–1.072 Ma), Cobb Mountain subchron (Cobb Mtn., 186–1.221 Ma), the Gilsa excursion (~1.68 Ma) and Chron C2n (Olduvia, 1.778–1.945 Ma), respectively. Subsequently, reversal magentozones can be correlated to the reversed subchrons belonging to the Matuyama Chron. Based on these correlations (Table 1), the sediment accumulation rates (SAR) of core PC15 are estimated as 3–6 m/Myr.

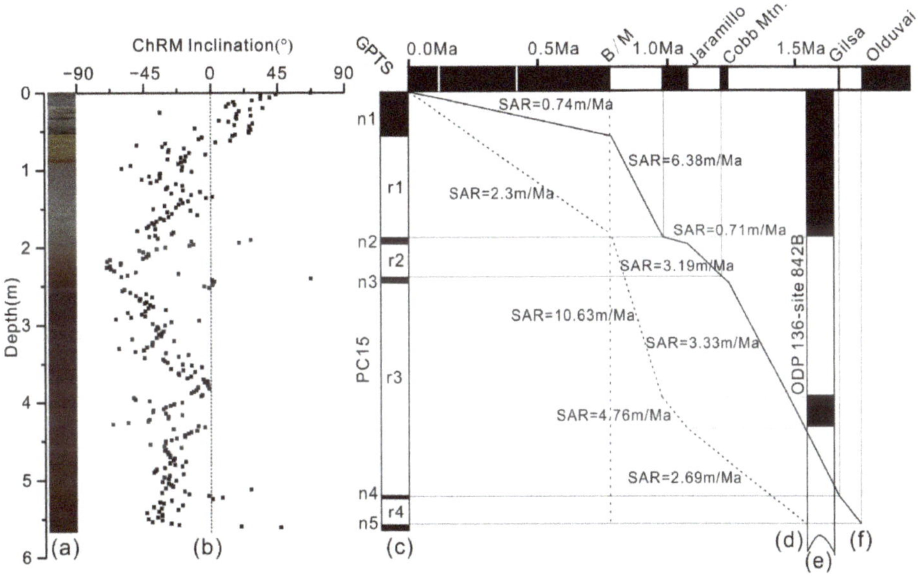

Figure 4. Magnetostratigraphy of the studied core and their correlations to the geomagnetic polarity time scale (GPTS). (**a**) Image of Core PC15; (**b**) ChRM inclination; (**c**) recognized magnetozones (n1–n5 and r1–r4) of core PC15; (**d**) sedimentation rates of ODP 842B site (dash line); (**e**) magnetostratigraphic results of core ODP 136-site 842B [13]; (**f**) sedimentation rates of core PC15 (solid line). See data in Table 1.

Table 1. Correlation of magnetozones of core PC15 to the Geomagnetic Polarity Time Scale.

Geomagnetic Polarity Time Scale	Age [21] (Ma)	PC15 Depth (m)	PC15 SARs (m/Ma)	842 [22] Depth (m)	842 [22] SARs (m/Ma)
Surface	0	0	-	0	-
C1n (bottom)	0.781	0.58	0.74	1.8	2.3
C1r.1n (top)	0.988	1.9	6.38	4	10.63
C1r.1n (bottom)	1.072	1.96	0.71	4.4	4.76
Cobb Mountain (top)	1.21	2.4	3.19	-	-
Cobb Mountain (bottom)	1.24	2.5	3.33	-	-
C2n(top)	1.778	5.6	5.76	6.3	2.69

Since the sediments of core PC15 deposited below the CCD, most calcareous foraminifera and coccolith have been completely dissolved, and only some pieces of radiolarian fossils are observed. In the topmost 50 cm, diatom fossils were well-preserved, while in the lower, no diatoms were found. The identified fossils include *Azpeitia nodulifera*, *Hemidiscus cuneiformis*, *Azpeitia Africana*, *Nitzschia marina*, *Thalassiosira oestrupii* and *Coscinodiscus excentricus*, supporting the reliability of paleomagnetic correlation (Table 1).

3.2. Clay Mineral Changes

Clay minerals in core PC15 are smectite (46–89%) in dominance, with illite (~38%), chlorite (4–11%) and kaolinite (4–8%). High percentages of smectite in clay minerals usually demonstrates that the sediments are closely related to volcano glasses, while illite and chlorite are from physical weathering and kaolinite can indicate warm and humid climates, which is strongly controlled by continental hydrolysis [23–27]. The assemblage of clay minerals of Unit II is close to those of surface sediments in West Philippine Basin [24–26], while Unit I is close to those of sediments of Mariana Trough [26] and Luzon Island [27], as displayed in Figure 5.

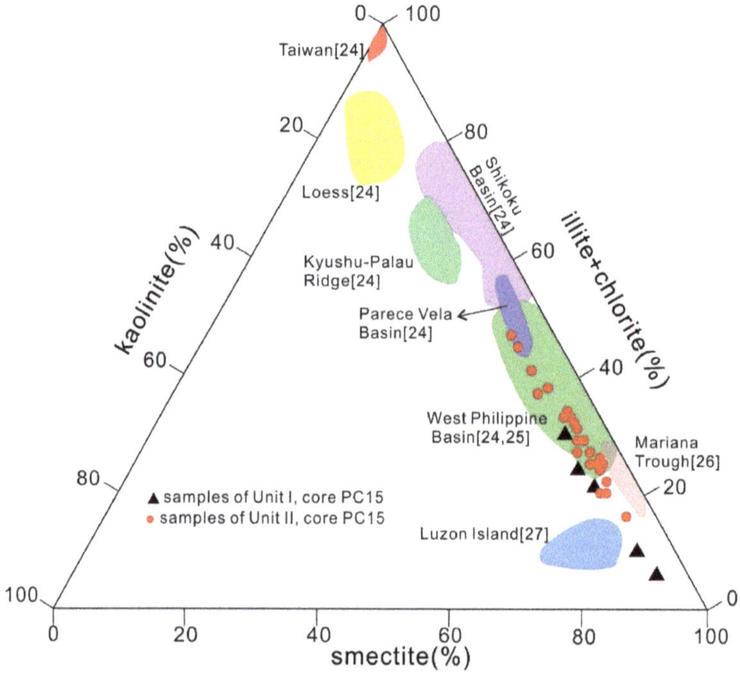

Figure 5. Triangular plot of clay minerals' composition.

Illite is the dominant clay mineral in Chinese loesses, up to 65%, and chlorite, ~20%. Kaolinite was commonly found in loesses, about 10%, while smectite was only ~5% [22]. In the West Philippine Basin, changes in clay minerals are large: smectite is high with content of 60–80%, and illite and chlorite range from 16% to 37% [24,25]. In the sediments of the Mariana Trough, more smectite was reported due to regional hydrothermal and volcanic activities [26].

Although the Kamehameha Basin is distant from the Philippine Sea, the provenance between them is similar [28,29]. For example, samples from Unit II of core PC15 displace in the area of the triangular plot similar to ones of the West Philippine Basin and the Mariana Trough, while samples from Unit I are close to samples from the Mariana Trough and the Luzon Islands, which can be correlated to the increased contribution of hydrothermal materials and volcanic debris [13]. Thus, the comparison between clay minerals indicates that aeolian deposits in the sediments generally decrease along with the distance away from the Chinese loess plateau, and hydrothermal materials and volcanic debris significantly increase. For stratigraphic changes, smectite content in the top 0.92 m (Unit I) is higher than the lower part (Unit II), illite reduces from ~30% to a much lower level and changes in chlorite and kaolinite are small (Figure 6).

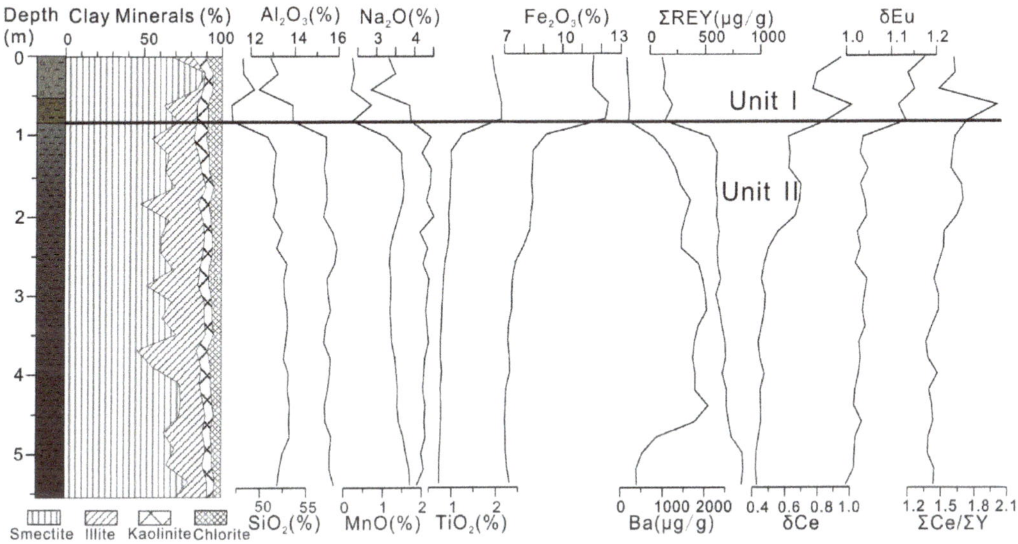

Figure 6. Stratigraphic changes in clay minerals, and major, trace and rare-earth elements in core PC15.

3.3. Geochemical Properties

Following equations of $\Sigma Ce = La + Ce + Pr + Nd + Sm + Eu$; $\Sigma Y = Gd + Tb + Dy + Ho + Er + Tm + Yb + Lu + Y$; $\delta Ce = 2Ce_N/(La_N + Pr_N)$ and $\delta Eu = 2Eu_N/(Sm_N + Gd_N)$, and normalized by the North American Shale Composition (NASC) [30], geochemical properties of core PC-15 are displayed. The difference between the major elements of SiO_2, Al_2O_3, MnO and Na_2O is minor, while TiO_2, Fe_2O_3, $\Sigma Ce/\Sigma Y$, δCe and δEu change in an opposite pattern (Figure 6). Their down-core variation can be divided into two groups: (1) below 0.92 m, all elements change little; and (2) above 0.92 m, SiO_2, Al_2O_3, MnO and Na_2O gradually decrease, and TiO_2, Fe_2O_3, $\Sigma Ce/\Sigma Y$, δCe and δEu increase. The detail geochemical properties are listed in Supplementary Tables S1 and S2.

The REE data of core PC15 can be clustered into two parts (Figure 7): (1) in the lower part (Unit II, below 0.92 m), ΣREY range from 536 to 860 μg/g, with a distinct Ce negative anomaly; and (2) in Unit I (above 0.92 m), ΣREY is less than 200 μg/g, with a slightly positive anomaly in element Eu.

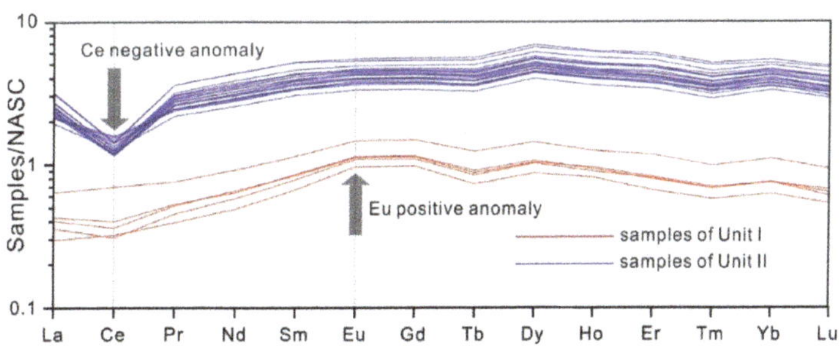

Figure 7. The NASC-normalized REE patterns of sediments from core PC15.

4. Discussion

4.1. Sedimentary Changes during the MPT

Within the MPT, SARs of core PC15 and ODP Site 842 [13] significantly increased, from 3 m/Ma to 6.38 m/Ma in core PC15, and from 4.76 m/Ma to 10.63 m/Ma at ODP

Site 842 [13]. This increase in SARs is not only evident in the Kamehameha Basin, but also in the southern slope of Mariana Trench [14], the Western Pacific and the Philippian Sea [31,32] (Figure 8).

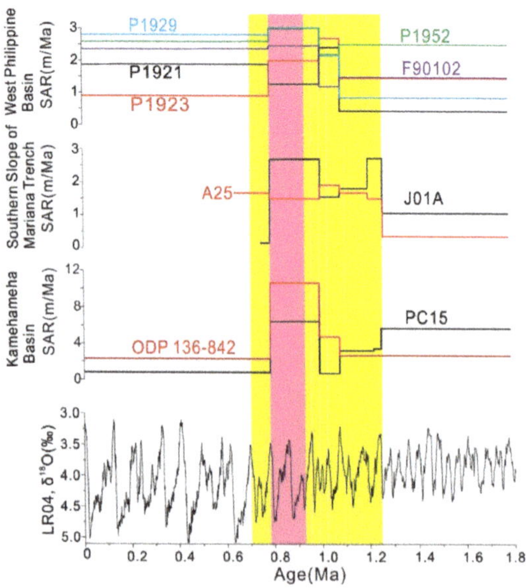

Figure 8. Changes in SARs of nine cores from the Kamehameha basin, the southern slope of Mariana Trench and the West Philippine Basin. Benthic $\delta^{18}O$ stack LR04 [6] indicates global climate changes from 41 kyr to 100 kyr world. The Mid-Pleistocene Transition is highlighted by yellow shades, the 900 kyr event with pink shades. ODP Site 842 is from Ref. [13]. Sites A25 and J01A are from Ref. [14]. Sites P1921, P1923, P1929 and P1952 are from Ref. [31]. Site F90102 is from Ref. [32].

During the MPT, there is a major shift in many geological records, such as the LR04 $\delta^{18}O$ stack [5], ice-rafted debris in the Norwegian Sea [6] and changes in SST in the Eastern Equatorial Pacific of ODP 849 Site [33]. Similarly, the mean grain size of loess sequences in the Chinese Loess Plateau increased significantly, inferring a strengthened aridity [34]. Therefore, the increased SAR from the early to middle Pleistocene in the Western Pacific can be explained as follows. The long-term average ice volume gradually increased, about 50 m down in sea-level equivalent [35]. The decreased sea levels would cause more exposure of bedrocks in glacial intervals, and induce weathering processes [34,36]. As a result, the fine-grained weathering particles were carried by regional currents and the Westerlies jet to the study area. Additionally, since biological productivity in the equatorial Pacific Ocean with high nutrients and low chlorophyll was limited by the micronutrient Fe, dust delivery may have induced marine primary productivity by aeolian Fe inputs [37], thus increasing the SAR in the study area.

In addition, since volcanic debris and hydrothermal materials, namely volcanic glasses and smectite, are remarkably higher in Unit I of core PC15, the low SARs in the Kamehameha Basin [13] after the MPT may suggest a substantial reduction in aeolian dusts and biological materials (mainly diatom).

4.2. Bottom-Water Evolution

It has been suggested that Ce can be oxidized to Ce^{4+} hydroxide in an oxidized environment due to hydroxide precipitation [38], and negative Ce anomalies are usually observed in the present oxidized deep seawater [38]. On the contrary, deep-sea ferromanganese nodules and crusts usually exhibit positive Ce anomalies, likely reflecting preferential removal of Ce^{4+} from sea water. Hence, Ce anomaly is a good indicator of sedimentary redox conditions. Moreover, element europium (Eu) is usually in Eu^{2+} in a reducing condition [39]. In high-productivity sea waters, such as the equatorial Pacific and Atlantic water masses, barites ($BaSO_4$) can indicate the upper marine productivity [40–43].

In this study, changes in δCe of core PC15 increased rapidly during 0.9–0.7 Ma (Figure 9), inferring that the bottom current changed from oxic to anoxic conditions. The increased δEu since 0.7 Ma, together with Ba and P components, indicates the ascending influence of marine production in the upper ocean onto the redox condition of bottom water. In addition, the positive Eu anomaly was observed in the top 0.92 m of core PC15, inferring that volcanic materials are an important component in the sediments.

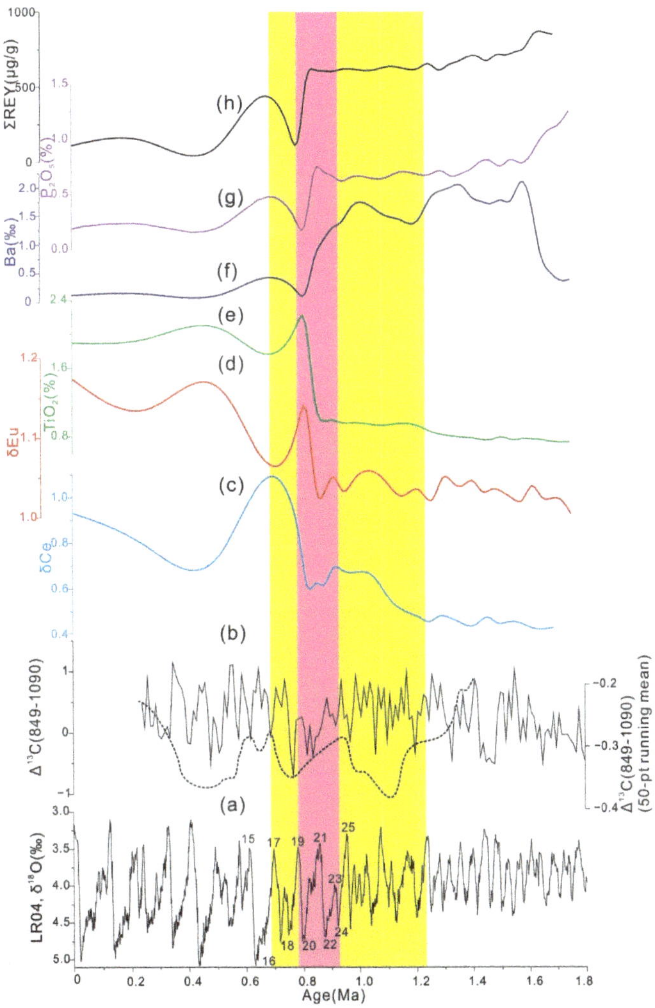

Figure 9. Comparison of selected climatic and environmental proxies of sediments from core PC15. (**a**) LR04 marine foraminifer oxygen isotope stack [5]. Marine isotopic stages (15–25) are labeled. (**b**) Inter-basinal carbon isotope gradient (solid line) between Pacific Site 849 and South Atlantic Site 1090 [44], with a 50-point running average (dash line). (**c–h**) δCe, δEu, contents of TiO_2 (%), Ba (‰), P_2O_5 (%) and REY (μg/g) of core PC15. The Mid-Pleistocene Transition is highlighted by yellow shades and labeled, the 900 ka interval is highlighted by pink shades.

Benthic foraminifer $\delta^{13}C$ records is a well-documented proxy in tracing ventilation history of intermediate and deep-water masses in the Pacific and Atlantic [44]. ODP Site 849 was located in the Eastern Pacific, which was dominated by the Pacific deep water (PDW) [45], and ODP Site 1090 in the South Atlantic was used to track the LCDW [46]. The inter-basinal carbon isotope gradient ($\Delta\delta^{13}C$) between Pacific Site 849 and South Atlantic Site 1090 was used to infer the ventilation history of the deep basin of the Pacific. As shown, the $\Delta\delta^{13}C$ (50 pt running mean) has the same trend with the δCe curve. In specific, during 900–800 ka, these water mass proxies show an evident downward trend, indicating an increased ventilation of the PDW. Around 700 ka, the curves show an upward trend, agreeing with the reduction trend in deep-ocean circulation [1]. It is noted that the provenance had undergone profound changes, and at the same time, a rapid increase in δEu and titanium (TiO_2) indicates the induced input of hydrothermal materials from the Hawaii hotspot, and the increased hydrothermal materials may similarly contribute to the reduced seawater in the study area. Therefore, we confirmed that bottom water experienced a rapid reorganization in the central Pacific during the MPT.

5. Conclusions

Based on magnetostratigraphy of the sediments of core PC15, we established a chronological framework for deep-sea sediments in the Kamehameha Basin in the Quaternary. The SARs are estimated as 3–7 m/Myr and show an evident increase around the MPT, comparable with those in the West Philippine Sea and the Mariana Trench. Comparing with various paleoenvironmental proxies, the SAR shift around the MPT likely resulted from the induced weathering processes around the study area and the strengthened aeolian inputs from inner Asia, which are all correlated to the enhanced glaciation in the Northern Hemisphere. Moreover, changes in regional bottom water during the MPT were studied, showing that the Kamehameha Basin was more oxic during 900–800 ka, and then turned to a relatively anoxic condition since ~700 ka. This evidence suggests a complex reorganization in regional bottom water across the MPT. Therefore, we propose that the deep-sea sediments in the Kamehameha Basin in the central tropical Pacific record some key features in global climate changes and are worthy of further investigation in future.

Supplementary Materials: The following are available online at https://www.mdpi.com/2077-1312/9/11/1254/s1, Table S1: Major and trace element concentrations of different depth in the sediment core PC15, Table S2: Rare earth element concentrations (μg/g) and characteristics parameters of different depth in the sediment core PC15.

Author Contributions: Conceptualization and methodology, H.W. and L.Y.; sample collection, H.W. and X.D.; formal analysis, H.W., L.Y. and G.H.; original draft preparation, H.W. and L.Y. All authors have read and agreed to the published version of the manuscript.

Funding: This research was funded by the Key Special Project for Introduced Talents Team of Southern Marine Science and Engineer Guangdong Laboratory (Guangzhou), grant number GML2019 ZD0106; the National Natural Science Foundation of China (42177422 and 41803062); and the COMRA Project from China Ocean Mineral Resources R&D Association, grant numbers DY135-C1-1-07 and DY135-N1-1-01.

Institutional Review Board Statement: Not applicable.

Informed Consent Statement: Not applicable.

Data Availability Statement: Data are available on request from the author H.W. (wanghaifeng@mail.cgs.gov.cn).

Acknowledgments: We are grateful to all of the onboard crew members of DY29 during the 2013 scientific expedition for collecting samples. We thank Peixin Lai and Qing Chen for analyzing clay minerals, and Piaoer Fu and Yinan Deng for their help with geochemical measurements.

Conflicts of Interest: The authors declare no conflict of interest.

References

1. Clark, P.U.; Archer, D.W.; Pollard, D.; Blum, J.D.; Rial, J.A.; Brovkin, V.; Mix, A.C.; Pisias, N.G.; Roy, M. The middle Pleistocene transition: Characteristics, mechanisms, and implications for long-term changes in atmospheric pCO_2. *Quat. Sci. Rev.* **2006**, *25*, 3150–3184. [CrossRef]
2. Korff, L.; von Dobeneck, T.; Frederichs, T.; Kasten, S.; Kuhn, G.; Gersonde, R.; Diekmann, B. Cyclic magnetite dissolution in Pleistocene sediments of the abyssal northwest Pacific Ocean: Evidence for glacial oxygen depletion and carbon trapping. *Paleoceanography* **2016**, *31*, 600–624. [CrossRef]
3. Lear, C.H.; Billups, K.; Rickaby, R.E.; Diester-Haass, L.; Mawbey, E.M.; Sosdian, S.M. Breathing more deeply: Deep ocean carbon storage during the mid-Pleistocene climate transition. *Geology* **2016**, *44*, 1035–1038. [CrossRef]
4. Shi, F.; Yin, Q.; Nikolova, I.; Berger, A.; Ramstein, G.; Guo, Z. Impacts of extremely asymmetrical polar ice sheets on the East Asian summer monsoon during the MIS-13 interglacial. *Quat. Sci. Rev.* **2020**, *230*, 106164. [CrossRef]
5. Lisiecki, L.E.; Raymo, M.E. A Pliocene-Pleistocene stack of 57 globally distributed benthic $\delta^{18}O$ records. *Paleoceanography* **2005**, *20*, PA103. [CrossRef]
6. Jansen, E.; Fronval, T.; Rack, F.; Channell, J.E.T. Pliocene-Pleistocene ice rafting history and cyclicity in the Nordic Seas during the last 3.5 Myr. *Paleoceanography* **2000**, *15*, 709–721. [CrossRef]
7. Ruddiman, W.F.; E Raymo, M.; Martinson, D.G.; Clement, B.; Backman, J. Pleistocene evolution: Northern hemisphere ice sheets and North Atlantic Ocean. *Paleoceanography* **1989**, *4*, 353–412. [CrossRef]
8. Schefuß, E.; Damsté, J.S.S.; Jansen, J.H.F. Forcing of tropical Atlantic sea surface temperatures during the mid-Pleistocene transition. *Paleoceanography* **2004**, *19*, PA4029. [CrossRef]
9. Hu, B.; Yi, L.; Zhao, J.; Guo, W.; Ding, X.; Wang, F.; Chen, W. Magnetostratigraphy of core XT06 and quaternary sedimentary dynamics of the deep-sea deposits in the West Philippian Basin. *Mar. Geol. Quat. Geol.* **2021**, *41*, 61–74. [CrossRef]
10. Kawabe, M.; Fujio, S. Pacific ocean circulation based on observation. *J. Oceanogr.* **2010**, *66*, 389–403. [CrossRef]
11. Shimono, T.; Yamazaki, T. Environmental rock-magnetism of Cenozoic red clay in the South Pacific Gyre. *Geochem. Geophys. Geosystems* **2016**, *17*, 1296–1311. [CrossRef]
12. Dziewonski, A.; Wilkens, R.; Firth, J.; Baker, D.J.; Briden, J.C.; Carson, B.; Collins, J.A.; De Carlo, E.H.; Duennebier, F.K.; Dürbaum, H.J.; et al. Site 842. In *Proceedings of the Ocean Drilling Program*; Winkler, W., Ed.; Initial Reports, 136; Texas A&M University: College Station, TX, USA, 1992; pp. 37–63.
13. Yi, L.; Xu, D.; Jiang, X.; Ma, X.; Ge, Q.; Deng, X.; Wang, H.; Deng, C. Magnetostratigraphy and Authigenic 10 Be/9 Be Dating of Plio-Pleistocene Abyssal Surficial Sediments on the Southern Slope of Mariana Trench and Sedimentary Processes During the Mid-Pleistocene Transition. *J. Geophys. Res. Ocean.* **2020**, *125*, e2020JC016250. [CrossRef]
14. Smith, W.H.F.; Sandwell, D.T. Global seafloor topography from satellite altimetry and ship depth soundings. *Science* **1997**, *277*, 1957–1962. Available online: https://topex.ucsd.edu/WWW_html/mar_topo.html (accessed on 1 November 2021). [CrossRef]
15. Chunhui, X.; Yonghong, W.; Jiwei, T.; Xuchen, W.; Yu, X. Mineral composition and geochemical characteristics of sinking particles in the Challenger Deep, Mariana Trench: Implications for provenance and sedimentary environment. *Deep. Sea Res. Part I Oceanogr. Res. Pap.* **2019**, *157*, 103211. [CrossRef]
16. Lyle, M.; Wilson, P. Leg 199 synthesis: Evolution of the equatorial Pacific in the early Cenozoic. In *Proceedings of the Ocean Drilling Program, Scientific Results*; Wilson, P.A., Lyle, M., Firth, J.V., Eds.; Texas A&M University: College Station, TX, USA, 2006; pp. 1–39.
17. Kirschvink, J.L. The least-squares line and plane and the analysis of palaeomagnetic data. *Geophys. J. Int.* **1980**, *62*, 699–718. [CrossRef]
18. Lurcock, P.; Wilson, G.S. PuffinPlot: A versatile, user-friendly program for paleomagnetic analysis. *Geochem. Geophys. Geosystems* **2012**, *13*, Q06Z45. [CrossRef]
19. Zhao, S.; Cai, F.; Liu, Z.; Cao, C.; Qi, H. Disturbed climate changes preserved in terrigenous sediments associated with anthropagenic activities during the last century in the Taiwan Strait, East Asia. *Mar. Geol.* **2021**, *437*, 106499. [CrossRef]
20. Biscaye, P.E. Mineralogy and sedimentation of recent deep sea clay in the Atlantic Ocean and adjacent seas and oceans. *Geol. Soc. Am. Bull.* **1965**, *76*, 803–832. [CrossRef]
21. Hilgen, F.J.; Lourens, L.J.; Van Dam, J.A. *The Neogene Period, The Geological Time Scale*; BV Elsevier Press: Cambridge, MA, USA, 2012; pp. 923–978.
22. Channell, J.; Singer, B.; Jicha, B. Timing of Quaternary geomagnetic reversals and excursions in volcanic and sedimentary archives. *Quat. Sci. Rev.* **2020**, *228*, 106114. [CrossRef]
23. Chamley, H. *Clay Sedimentology*; Springer: Berlin, Germany, 1989; pp. 259–289.
24. Huang, J.; Wan, S.; Zhang, G.; Xu, Z.; Liu, H.; Dong, J.; Li, A.; Li, T. Impact of seafloor topography on distribution of clay minerals in the east Philippines Sea. *Mar. Geol. Quat. Geol.* **2017**, *37*, 77–85. [CrossRef]
25. Ding, X.; Hu, B.; Xu, F.; Guo, J.; Cui, R.; Yi, L. Evolution of clay minerals assemblages since Late Pliocene and its paleoenvironmental implications: Evidence from Core XT4 of the Philippine Sea Basin. *Mar. Geol. Quat. Geol.* **2020**, *41*, 42–51. [CrossRef]
26. Zhang, D. Clay Mineral composition and distribution in the Mariana Trough. *J. Oceanogr. Huanghai Bohai Seas* **1994**, *12*, 32–39.
27. Liu, Z.; Zhao, Y.; Colin, C.; Siringan, F.P.; Wu, Q. Chemical weathering in Luzon, Philippines from clay mineralogy and major-element geochemistry of river sediments. *Appl. Geochem.* **2009**, *24*, 2195–2205. [CrossRef]
28. Yasukawa, K.; Ohta, J.; Miyazaki, T.; Vaglarov, B.S.; Chang, Q.; Ueki, K.; Toyama, C.; Kimura, J.; Tanaka, E.; Nakamura, K.; et al. Statistic and Isotopic Characterization of Deep-Sea Sediments in the Western North Pacific Ocean: Implications for Genesis of the Sediment Extremely Enriched in Rare Earth Elements. *Geochem. Geophys. Geosyst.* **2019**, *20*, 3402–3430. [CrossRef]

29. Wang, F.; He, G.; Lai, P. Nd isotopes of the pelagic sediments and clay fractions from the Pacific Ocean and their provenance significance. *Geotecton. Metallog.* **2019**, *43*, 292–301. [CrossRef]
30. Haskin, A.M.; Haskin, A.L. Rare earths in European shales: A redetermination. *Science* **1966**, *154*, 507–509. [CrossRef]
31. Yao, H.; Wang, F.; Wang, H.; Yu, M.; Ren, J.; He, G.; Chen, W.; Yi, L. Pleistocene magnetostratigraphy of four cores in the West Philippian Basin and regional sedimentary shift during the Mid-Pleistocene transition. *Geol. J.* **2021**, *56*, 2919–2929. [CrossRef]
32. Meng, Q.; Li, A.; Jiang, F.; Xu, Z. A geomagnetic paleointensity record over the last 2 Ma from the east Philippine Sea. *Oceanol. Limnol. Sin.* **2010**, *41*, 606–613. [CrossRef]
33. McClymont, E.L.; Rossel-Mele, A. Links between the onset of modern Walker circulation and the mid- Pleistocene climate transition. *Geology* **2005**, *33*, 386–392. [CrossRef]
34. Sun, Y.; Clemens, S.C.; An, Z.; Yu, Z. Astronomical timescale and paleoclimatic implication of stacked 3.6–Myr monsoon records from Chinese Loess Plateau. *Quat. Sci. Rev.* **2006**, *25*, 33–48. [CrossRef]
35. Dwyer, G.S.; Cronin, T.M.; Baker, P.A.; Raymo, M.E.; Buzas, J.S.; Correge, T. North Atlantic deepwater temperature change during late Pliocene and late Quaternary climatic cycles. *Science* **1995**, *270*, 1347–1351. [CrossRef]
36. Clark, P.U.; Pollard, D. Origin of the Middle Pleistocene transition by ice sheet erosion of regolith. *Paleoceanography* **1998**, *13*, 1–9. [CrossRef]
37. Ziegler, C.L.; Murray, R.W.; Plank, T.; Hemming, R.S. Source of Fe to the equatorial Pacific Ocean from the Holocene to Miocene. *Earth Planet. Sci. Lett.* **2008**, *270*, 258–270. [CrossRef]
38. Goldberg, E.D. Chemical and Mineralogical aspects of deep-sea sediments. *Phys. Chem. Earth* **1961**, *4*, 281–302. [CrossRef]
39. Elderfield, H.; Greaves, M.J. The rare earth elements in seawater. *Nature* **1982**, *296*, 214–219. [CrossRef]
40. Dymond, J.; Suess, E.; Lyle, M. Barium in deep-sea sediment: A geochemical proxy for paleoproductivity. *Paleoceanography* **1992**, *7*, 163–181. [CrossRef]
41. Griffith, E.M.; Paytan, A. Barite in the ocean–occurrence, geochemistry and palaeoceanographic applications. *Sedimentology* **2012**, *59*, 1817–1835. [CrossRef]
42. Paytan, A.; Kasten, M.; Chavez, F.P. Glacial to interglacial fluctuations in productivity in the equatorial Pacific as indicated by marine barite. *Science* **1996**, *277*, 1355–1357. [CrossRef]
43. Guichard, F.; Church, T.M.; Treuil, M.; Jaffrezic, H. Rare earths in barites: Distribution and effects on aqueous partitioning. *Geochim. Cosmochim. Acta* **1979**, *43*, 983–997. [CrossRef]
44. Hodell, D.A.; Venz-Curtis, K.A. Late Neogene history of deepwater ventilation in the Southern Ocean. *Geochem. Geophys. Geosystems* **2006**, *7*, Q09001. [CrossRef]
45. Mix, A.C.; Pisias, N.G.; Rugh, W.; JWilson Morey, A.; Hagelberg, T.K. Benthic foraminifer stable isotope record from site 849 (0–5 Ma): Local and global climate changes. In *Proceedings of the Ocean Drilling Program, 138 Scientific Results*; Stewart, S.K., Ed.; Texas A&M University: College Station, TX, USA, 1995; pp. 371–412.
46. Venz, K.A.; Hodell, D.A. New evidence for changes in Plio- Pleistocene deep water circulation from Southern Ocean ODP Leg 177 Site 1090. *Palaeogeogr. Palaeoclimatol. Palaeoecol.* **2002**, *182*, 197–220. [CrossRef]

Journal of
Marine Science and Engineering

MDPI

Article

Taphofacies and Petrofacies Theoretical Marine Models Applied to the Coquina of the Amaral Formation (Lusitanian Basin, Portugal)

Gustavo Gonçalves Garcia [1], Antônio Jorge Vasconcellos Garcia [1,2], Maria Helena Paiva Henriques [1,*], Rafael Mendes Marques [1] and Rui Pena dos Reis [1]

[1] Geosciences Center, Department of Earth Sciences, University of Coimbra, Rua Sílvio Lima, 3030-790 Coimbra, Portugal; gustavogarcia@dct.uc.pt (G.G.G.); ajvgarcia@academico.ufs.br (A.J.V.G.); uc2017257788@student.uc.pt (R.M.M.); penareis@dct.uc.pt (R.P.d.R.)

[2] Progeologia Laboratory/Nucleus of Competence in Oil, Gas and Biofuels, Federal Universiy of Sergipe, Avenida Marechal Rondon S/n, São Cristóvão 49100-000, Sergipe, Brazil

* Correspondence: hhenriq@dct.uc.pt; Tel.: +351-919-721-820

Abstract: The Amaral Formation has a wide geographic distribution within the Lusitanian Basin, at the western Iberian Margin (Portugal). The different depositional contexts for this unit enabled the distinction of three sectors: lagoon, lagoon-barrier, and marine-distal. The integration of the evolutionary taphonomic analysis of its fossil assemblages with the analysis of multiscale properties through the CAMURES methodology (Multiscale Reservoir Characterization) allowed the application of a methodology for the classification of coquina which was previously developed for the Morro do Chaves Formation (Sergipe–Alagoas Basin, Brazil). Here, it was adapted according to the complexity of the Amaral Formation deposits. The classification of ten taphofacies, in association with four lithofacies, allowed the definition of 84 petrofacies, based on the nature of the sedimentary and taphonomic processes. The relationship between the structural context, the systems tracts, the diversity of the fossil record, the classification of taphofacies and petrofacies, and the understanding of vertical and lateral variations of the sediments' deposition within the unit support the construction of geological and theoretical models for coquina deposits. These models will allow for prediction of the spatial distribution of facies in other coquina analogous hydrocarbon reservoirs, as well as specifying the delimitation of reservoir zones for 3D geocellular modeling and flow simulation of hydrocarbon-producing reservoirs, thus improving predictive analyses.

Keywords: coquina taphofacies/petrofacies; Amaral Formation; Lusitanian Basin (Portugal); Upper Jurassic; carbonate reservoirs; theoretical models; paleoenvironmental models; paleogeographic reconstructions

Citation: Garcia, G.G.; Garcia, A.J.V.; Henriques, M.H.P.; Marques, R.M.; Pena dos Reis, R. Taphofacies and Petrofacies Theoretical Marine Models Applied to the Coquina of the Amaral Formation (Lusitanian Basin, Portugal). *J. Mar. Sci. Eng.* **2021**, *9*, 1319. https://doi.org/10.3390/jmse9121319

Academic Editors: George Kontakiotis, Assimina Antonarakou and Dmitry A. Ruban

Received: 20 October 2021
Accepted: 18 November 2021
Published: 23 November 2021

1. Introduction

The need for increasing accuracy and effective predictive models is an important challenge faced by the oil industry, which seeks to understand the distribution of permoporous spaces in hydrocarbon reservoirs. In this context, multidisciplinary/multiscale studies on carbonate rocks, especially those formed by concentrations of fossils (coquina), have been developed, aiming for a better understanding on pre-salt reservoirs in the Brazilian Atlantic Border Basins, namely, in the Santos and Campos Basins [1,2].

The integrated analysis involving the characterization of petrofacies and the taphonomic approach based on the evolutionary taphonomy concepts and methods [3,4], with the application of the Multiscale Characterization Methodology of Reservoirs—CAMURES [2,5], represents a new pathway for the development of predictive modeling of reservoir quality. By improving the understanding of the lateral and vertical facies variations of the coquina deposits, this integration allows us to expand the knowledge about the genesis of paleoenvironmental models, as well as the 3D geocellular models of coquina facies, which underlie the fluid flow simulation in hydrocarbon reservoirs [6–9].

The recognition of taphonomic alteration mechanisms in the fossil assemblages of the Amaral Formation (Kimmeridgian–Tithonian, Lusitanian Basin, Portugal), such as biodegradation, disarticulation, dispersion, reorientation, abrasion, dissolution, cementation, neomorphism, and recrystallization, with the resulting textural patterns (degree of fragmentation of skeletal remains, matrix type, and pore types) enabled the characterization of different taphofacies, associated with different petrofacies [10–13]. The approach applied in this work for fossil assemblages and other nonbioclastic elements refers to a classification system for coquina reservoirs based on bioestratinomic and fossil-diagenetic processes, and not only on the resulting products [13].

The Amaral Formation coquina deposits display high diversity in terms of fossil record and a very complex genesis. The application of the methodology previously developed for the classification of taphofacies and petrofacies of the Morro do Chaves Formation [13] to the Amaral Formation required an adjustment in order to incorporate the interaction between the taphonomic mechanisms, the tectonic–structural context of the basin, and the diversity of the fossil record and of the depositional environments inherent in these coquina deposits. In addition to the validation of the classification system and theoretical models for coquina deposits, the wide paleogeographic distribution of the Amaral Formation throughout the Lusitanian Basin and the large number of outcrops and control points allowed for the extrapolation of the analysis of the lateral variations of coquina deposits. The Amaral Formation represents an important cyclical depositional context, which includes coquina with high diversity, which are interspersed with fine silty-clay deposits from deeper areas, and with sandy deposits with fluvial origins. The possibility of analyzing the vertical and lateral variations, from the microscopic scale to the basin-wide scale, makes the Amaral Formation an essential reference on the extrapolation of theoretical models to coquina reservoir analogues.

The basin's context plays an important role in the application and refinement of the analogous model over the entire extension of the studied deposits. Considering the different sectors and corresponding evolutionary particularities, factors such as tectonism, uplifts, and tilting controlled the differentiation in substrate inclination and in facies distribution. Transgressive and regressive events also play an important role in these carbonate and mixed-deposition systems.

2. Geological Setting

The sedimentation in the Lusitanian Basin was conditioned by its paleogeographic configuration represented by a NNE–SSW gulf, open to the south, where sediments from its steep east and west margins were drained, as well as from the north, through a distal fluviodelta system [14,15]. During the deposition of the Amaral Formation sediments, the paleoenvironmental evolution, somewhat carbonated, whether bioconstructed or not, was strongly affected by the substrate's positioning related to tectonism, as well as by the saliferous diapirs that generated intrabasin high bottoms, which favor concentrations of carbonate sediments. Sedimentation in the initial rift phase occurred in association with a transgressive event, changing, in the Late Kimmeridgian and Early Tithonian, to a regressive context. Then, the conditions for carbonate sedimentation in shallow waters were established in the basin, which corresponds to Sequences I 2 and I 3 of [13] of the Amaral Formation. The marine sedimentation conditions remained shallow, represented by an alternation of muddy, sandy, and carbonate sediments, configuring a lake into a deep marine environment with an expressive fossiliferous record [16]; towards the south, it gradually became deeper, drawing a large marine gulf.

During the Kimmeridgian in the proximal sectors (Boa Viagem and Alcobaça Formations), the sedimentation was predominantly continental; the distal sectors (Abadia Formation) record turbiditic fans. The deposition of the carbonate sediments from the Amaral Formation [17] and their relationship with other Upper Jurassic units is referred to as the upper Kimmeridgian–lower Tithonian boundary, framed in a tectonic context of sub-basin formation, which marks the post-climax of the rift [18,19] (Figure 1).

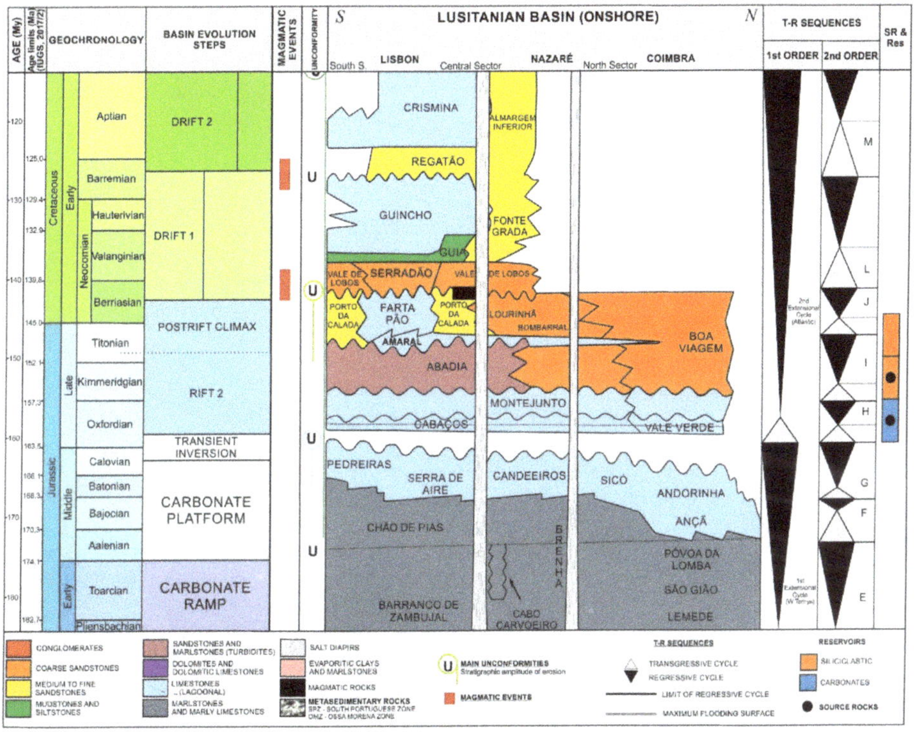

Figure 1. Stratigraphic framework of the Lusitanian Basin for the Pliensbachian–Aptian, i.e., discarding the underlying (RIFT 1 and marine invasion) and the overlying (DRIFT 3, inversion and uplift) intervals (modified after Pena dos Reis et al. [20]).

The carbonates of the Amaral Formation are composed of thin shallow marine deposits which are associated with a transgressive event. These carbonates cover the deep-sea sediments of the Abadia Formation, and, therefore, they can be considered as indicative of a significant decrease in bathymetry [21]. Pena dos Reis et al. [22] refer to this interval as facies of oolitic barriers and locally bioconstructed coquinoid reefs, represented by biofacies rich in fragments of echinoids, bivalves, stromatoporoids, spongy spicules, and microfauna (foraminifera, algae, and ostracods), and related to inter to infralittoral environments with clear marine influence [23]. Choffat [24] and Seifert [25] classified the Amaral Formation lithotypes as intraclastic limestones, dolomitic limestones, oolitic limestones, and conglomeratic limestones with intercalations of reef limestones.

Wilson [26] and Teixeira and Gonçalves [27] argued that the reef units which occurred from Montejunto to Tejo during the Kimmeridgian, whether of biostroma type or of bioherm type, correspond to lateral variations of the Amaral Formation, but with indefinite ages ("Corálico do Amaral", Ota Limestones, Monte Redondo Limestones, and Atouguia Limestones).

More recently, Leinfelder et al. [28] recognized the fossil diversity of the Amaral Formation reefs, as well as the paleoenvironmental implications of the corresponding distribution within the context of a mixed carbonated platform. These were developed during the Late Jurassic in the central-west region of Portugal, namely during the Kimmeridgian and Tithonian transgressive events, despite the limitations on using biostratigraphic tools (based on macro and microfossils) in the establishment of lateral correlations in reef deposits. In this sense, sequence stratigraphy seems to be the only method allowing the recognition of the depositional models for this interval in the Lusitanian Basin [29–34]. However, the sequence stratigraphy does not consider the different source areas and the basin structure, which can lead to distinct paleoenvironmental interpretations, such as the Ota Limestones and the Amaral Formation [28].

Fürsich et al. [35] describe and interpret, from the taphonomic and paleoecological points of views, the Upper Jurassic bivalve assemblages of these units from the central part of the Lusitanian Basin, outcropping between the regions of Torres Vedras and Nazaré. The authors consider that the development of bivalve concentrations is associated with phases of maximum flooding associated with coastal environments, with controlled salinity, being less common in open platform configurations. However, they also admit that, although rarely, they can also occur within a transgressive system. This happens when the remobilization of sediments is not as expressive as it is during the late transgression. Finally, the authors claim that several of the described concentrations are not restricted to the Lusitanian Basin, but also occur in other places in the epicontinental seas located in the north of the Neo-Tethys Ocean.

3. Materials and Methods

The Amaral Formation is recognizable throughout the Lusitanian Basin. Due to its wide geographical distribution and based on the depositional contexts interpreted in this work, the study area was divided into three sectors, namely, lagoon, lagoon-barrier, and marine-distal (Figure 2a,b). The lagoon sector covers the region between Leiria and Figueira da Foz, on the current coastal strip, extending to the latitude of Coimbra to the east. In this region, the Amaral Formation was analyzed through two stratigraphic sections: Cumieira (Pombal) and Abiul (Pombal). The lithofaciological and taphonomic analyses were carried out based on macroscopic descriptions of 58 samples: 53 from the Cumieira Section and five from the Abiul Section. Fifteen outcrops were also analyzed for lateral and vertical control of the units, and 41 samples were collected (Figure 2c,d).

The lagoon-barrier sector covers the region located between Lourinhã and Leiria, on the coastal strip, extending to the latitude of Batalha to the east. In this region, the Amaral Formation was characterized through the study of two stratigraphic sections: Praia do Salgado (Famalicão) and Praia da Consolação (Atouguia da Baleia). The lithofaciological and taphonomic analyses were performed based on macroscopic descriptions of 102 samples, 61 from the Praia do Salgado Section and 41 from the Praia da Consolação Section. Two other outcrops were also analyzed for lateral and vertical control of the units and nine samples were collected (Figure 2e,f).

The marine-distal sector covers the region located between Cascais and Lourinhã, on the coastal strip, extending to the latitude of Ota to the east. In this region, the Amaral Formation was characterized through the study of one section: Pedralvo (Arruda dos Vinhos). The lithofaciological and taphonomic analyses of this section were carried out based on macroscopic descriptions of nine samples. Eight other outcrops were also analyzed for lateral and vertical control of the units, and eight samples were collected (Figure 2g).

The methodological approach presented in this work for the coquina of the Amaral Formation follows the methods applied by Garcia et al. [13] for the coquina of the Morro do Chaves Formation (Sergipe Alagos Basin, Brazil). However, it was necessary to adjust them for the characterization of petrofacies, considering the existence of other variables, namely, the higher diversity of fossil assemblages, as well as the influence of local tectonic constraints, and also the impacts of transgressive events when framed within a regressive context.

The classification of taphofacies and petrofacies proposed by Garcia et al. [13] is based on concepts and methods of evolutionary taphonomy [36]; it considers the description and identification of all taphonomic mechanisms, i.e., biostratinomic and fossil diagenetic. The principles grounding the CAMURES methodology [2,5] improve the understanding of taphonomic processes over time and space, thus contributing to deepen the knowledge regarding the porosity–permeability distribution in a lithostratigraphic unit.

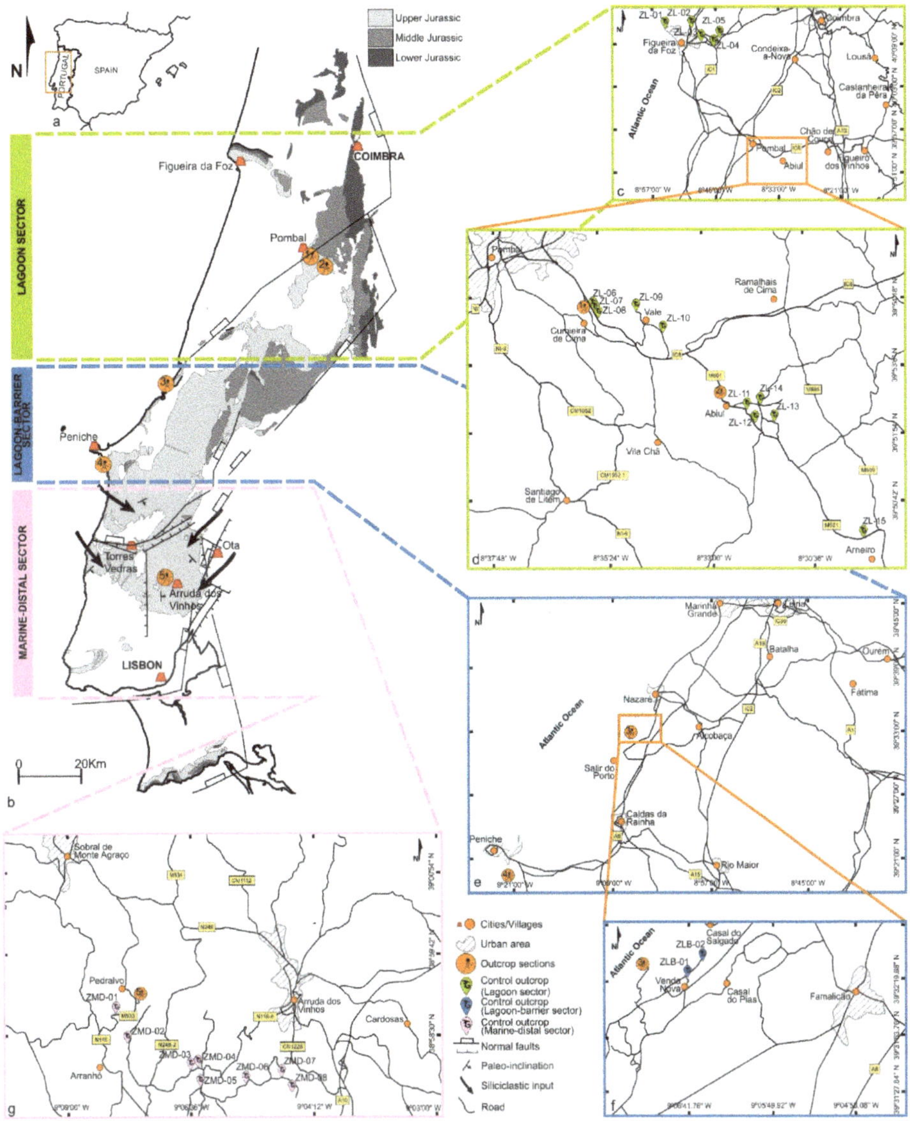

Figure 2. (**a**) Geographical location of the Lusitanian Basin; (**b**) Jurassic geological map of the Lusitanian Basin; (**c**) Location of the lagoon sector showing the Cumieira Section, the Abiul Section, and five control outcrops; (**d**) Location of the Cumieira Section, the Abiul Section, and ten control outcrops in detail; (**e**) Location of the lagoon-barrier sector showing the Praia do Salgado Section and the Praia da Consolação Section; (**f**) Location of the lagoon-barrier sector showing the Praia do Salgado Section and two control outcrops in detail; (**g**) Location of the marine-distal sector showing the Pedralvo Section and eight control outcrops.

The CAMURES methodology [2,5] consists of the integration of tools and data obtained at different observation scales (ranging from facies interpretation of rock samples to seismic interpretation of the basin) and in the representation of the heterogeneities existing in each of them. Thereby, an ideal geological model is obtained, both from the point of view of its genetic conception (conceptual geological model) and from the point of view of 3D geocellular modeling of properties. The CAMURES methodology assumes the recognition of the genetic processes, associated with their corresponding products/properties. The processes are classified in different scales of observation in a descriptive way. The analysis of processes and not just products/properties allow the transfer of information between scales. Thereby, the positioning of the petrofacies in the systems tracts and depositional sequences increases the accuracy of 3D geocellular facies models of the reservoir on the seismic scale.

In this context, a taphofacies/petrofacies definition is proposed for the hybrid coquina deposits of the Amaral Formation, based on the types and intensities of the identified sedimentary and taphonomic processes (Figure 3). Environmental energy is the primary control for the origin and distribution of petrofacies located in different "petrofacies clusters" (A, B and C) as shown in Figure 3. These clusters are individualized according to the nature of the interstitial material, being cluster (A) muddy carbonate matrix; cluster (B) terrigenous, coarse-grained siliciclastic material, and cluster (C) dominance of fragments of broken carbonate bioclasts.

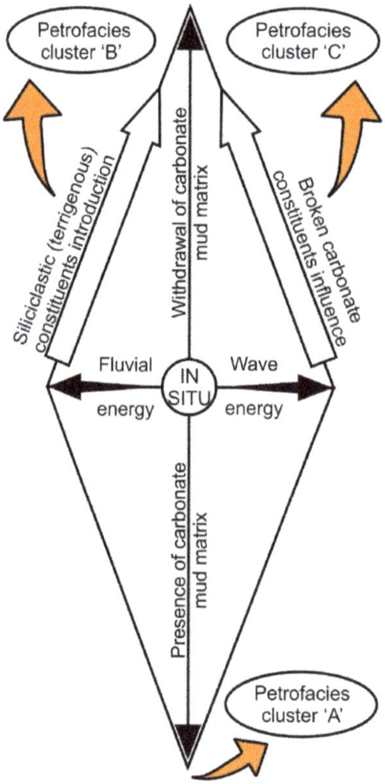

Figure 3. Environmental energy factors controlling the distribution of petrofacies characterized in this work which define the main petrofacies clusters (A, B and C) (according to Garcia et al. [13]).

The petrofacies designation used in this article follows the alphanumeric classification, designed with the goal of identifying types and intensities of processes and products associated with coquina [13]. The alphanumeric classification presented here aims to facilitate the understanding of facies types according to their petrographic and taphonomic characteristics. This proposal seeks to validate and emphasize the different recognized taphonomic processes and their relationship with depositional environments (Table S1). The facies distribution, according to the depositional environment and considering the origin and energy intensity (fluvial and wave) of the remobilization, is represented in Figure 4.

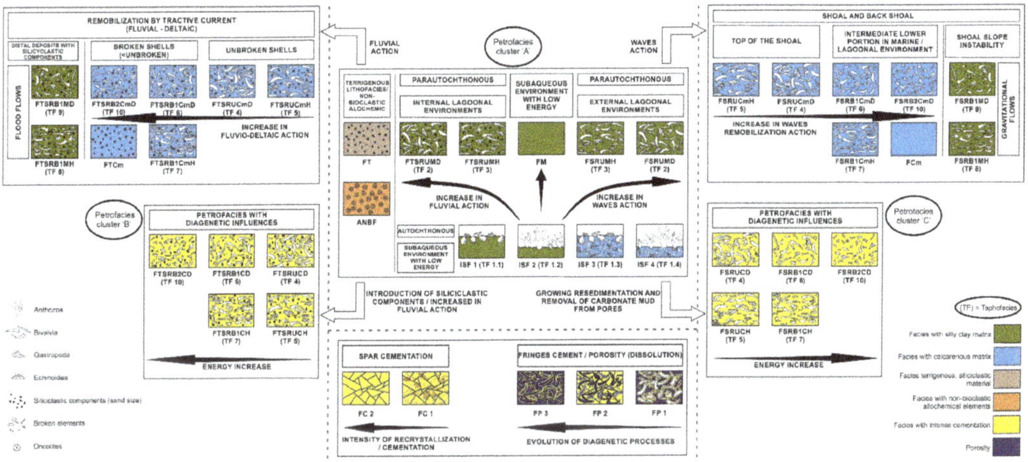

Figure 4. Petrofacies characterization and petrofacies clusters, associated with depositional environments, degree of remobilization, and fossil-diagenetic intensity (modified after Garcia et al. [13]).

The coquina petrofacies can be differentiated into two groups: coquina composed almost exclusively of representatives of a single taxon (e.g., bivalve coquina), and coquina made up of representatives of different taxa. In addition, coquina may consist of organically bounded elements (e.g., corals) and not organically bounded elements (e.g., bivalve). This faciologic/compositional distinction of coquina led to the creation of a terminology in addition to the alphanumeric classification previously described [13]. Letters are then used which resulted from the abbreviation of the type of organic relationship that exists between elements. This characteristic is expressed as described below, respectively: nob—not organically bounded elements (e.g., bivalve, gastropods); ob—organically bounded elements (e.g., coral reefs). Organically bounded elements are considered unbroken (U) only in the in situ facies (ISF), while in other contexts, they are always recognized as being broken (B, 1 or 2).

4. Results

The results for the macroscopic and taphonomic analyses of the Amaral Formation, framed in a multiscalar perspective, allowed the establishment of a stratigraphic evolutionary model for the Amaral Formation, based on the recognition of different facies according to their compositional, textural, and taphonomic properties. Thus, the analysis of different lithological and taphonomic aspects of the Amaral Formation sedimentary record resulted in the establishment of 10 taphofacies (1 to 10, related to deposits with fossils) that, in association with four lithofacies (1, 2, 4 and 5, related to deposits without fossils, adapted from Garcia et al. [13]), allowed the definition of 84 petrofacies (Figure 4 and Table S2). The petrofacies characterization and petrofacies clusters, associated with depositional environments, degree of remobilization, and fossil-diagenetic intensity described in this work (Figure 4), illustrate in a synthetic way the characteristics of this type of deposit, not distinguishing, for example, if the elements are organically bounded or not organically bounded. Likewise, in the lithofacies and taphofacies classification of the coquina from the Amaral Formation, there are petrofacies that occur in more than one type of environment (eight in total; Table S2). For the taphofacies definition, the different mechanisms of taphonomic alterations that had acted on the fossil elements recorded in the coquina from the Amaral Formation, which are described below, were taken into consideration.

4.1. Mechanisms of Taphonomic Change in the Coquina of the Amaral Formation

The high concentration and diversity of bioclasts and other allochemical constituents recorded in the Amaral Formation are the result of different long-lasting taphonomic mechanisms and of the conditions that controlled the different paleoenvironmental contexts to which these deposits are related. Taphonomic mechanisms recognized in the recorded elements and assemblages of the Amaral Formation include cementation (permineralization, formation of concretions), neomorphism (recrystallization), replacement, abrasion, dissolution, mechanical distortion, disarticulation, reorientation (dispersion and regrouping), and taphonomic removal or remobilization (accumulation and resedimentation). Taphonomic analysis of hand samples, together with fossil separation techniques, allowed the identification of the main taphonomic processes, which generally occurred in the following order: disarticulation, mechanical distortion and abrasion, reorientation (dispersion and regrouping), cementation (permineralization, formation of concretions) and neomorphism (recrystallization), and taphonomic removal or remobilization (accumulation and resedimentation).

4.1.1. Disarticulation

The disarticulation process is identified essentially in elements that are not organically bounded. This is one of the first processes to occur and is the most common, especially in environments subject to intense wave action. The elements recognized in the Amaral Formation present mainly two states of disarticulation: with the mollusca valves and/or echinoderm plates joined, but open, and slightly fragmented; or with all the elements disarticulated (spines and plates of echinoderms, mollusca valves, among others) (Figure 5a–d). The TF 1 taphofacies, usually related to muddy facies, is the only example in which the elements are complete.

4.1.2. Mechanical Distortion and Abrasion

The processes of mechanical distortion and abrasion occur simultaneously and result from the action of waves (normal or storm) and river flows. The degree of taphonomic alteration of these processes is defined by the duration of its activity and intensity. The mechanical distortion corresponds mainly to change in the size of elements by fragmentation, usually before abrasion [37]. The allochemical components of carbonate rocks (e.g., bivalves, gastropods, corals, among others) have naturally rounded surfaces. However, the abrasion process increases the roundness of these surfaces (Figure 5e,f). In the Amaral Formation deposits, elements are commonly recognized in two stages of fragmentation: totally or partially fragmented. Such situation is found in almost all petrofacies (Figure 5g,h), except for the TF 1 taphofacies. In turn, the mechanical distortion mechanism can be recognized for elements not organically bounded in TF 4.3, TF 5.3, TF 6, TF 7, TF 8.1, TF 8.2, TF 9.1, TF 9.2, and TF 10 taphofacies, and for organically bounded elements in TF 2.2, TF 3.2, TF 4.2, TF 4.3, TF 5.2, TF 5.3, TF 6, TF 7, TF 8, TF 9, and TF 10 taphofacies. However, it is not recognized in TF 2.1, TF 3.1, TF 4.1, and TF 5.1 taphofacies.

4.1.3. Reorientation (Dispersion and Regrouping)

The elements (organically bounded or not) may occur scattered in the deposits or display some pattern of regroupment. In both cases, these elements were subjected to different types of displacements with different origins: waves (normal or storm), fluvial (normal or flood discharges), induced currents, or gravitational flows. These agents tend to organize the elements in two ways: aligning them with a preferential orientation, imbricated or not (easier to recognize among not organically bounded elements), or dispersing them in a chaotic way. This process is, therefore, responsible for two aspects recognized in the described petrofacies: dispersion and regrouping. They were recognized in almost all taphofacies (Figure 6a,b) and are partially absent only in TF 1 and TF 2 taphofacies.

Figure 5. Photographs showing the results of different mechanisms of taphonomic alteration in coquina deposits, namely, (**a**) disarticulated elements with orientation; (**b**) unbroken elements with low fragmentation (partial); (**c**) unbroken and broken elements disarticulated with preferential orientation; (**d**) elements with chaotic orientation; (**e**,**f**) abrasion intense processes; (**g**) fragmentation intense processes; (**h**) fragmentation intense processes and important contribution of terrigenous constituents.

Figure 6. Photographs showing the results of different mechanisms of taphonomic alteration in coquina deposits, namely, (**a**) elements with orientation and/or imbrication; (**b**) elements with chaotic orientation; (**c**,**d**) micritization; (**e**) unbroken elements with joined valves, fragmented, and recrystallized; (**f**) recrystallization; (**g**,**h**) formation of concretions.

4.1.4. Cementation (Permineralization, Formation of Concretions) and Neomorphism (Recrystallization)

Most of the allochemical constituents recognized in the Amaral Formation display a large size. In this context, the recognition of diagenetic processes can be performed through macroscopic analysis carried out in the field and in the laboratory. Diagenetic processes can be recognized in all taphofacies which, in this case, refer to neomorphism processes, namely micritization (Figure 6c,d) and recrystallization (Figure 6e,f). The first is recognizable in TF 1.1, TF 1.2, TF 2, TF 3, TF 8, and TF 9 taphofacies; the second, on the other hand, is particularly well represented in taphofacies that show complete and large not organically bounded elements, namely in TF 1, TF 2, TF 3, TF 4, and TF 5 taphofacies. The formation of concretions has been recognized in taphofacies related to low-energy environments. These occur due to an increase in the alkaline concentration around the elements and are usually related to the first stages of diagenesis; they occur in TF 4.3, TF 5.3, TF 6.3, TF 7.3, TF 8, and TF 9.3 taphofacies (Figure 6g,h).

4.1.5. Taphonomic Removal or Remobilization (Accumulation and Resedimentation)

Almost all the taphofacies described in this work present resedimented allochemical constituents. These are elements that correspond to skeletal remains that were subject to transport before the final burial in the sediments (Figure 7a,b); no evidence of taphonomic re-elaboration has been detected. The assemblages corresponding to lagoon and lagoon-barrier paleoenvironments are accumulated, which is attested by the occurrence of in situ elements of the Anthozoa Class in muddy facies (TF 1.2; Figure 7c–e)

Figure 7. *Cont.*

Figure 7. Photographs showing the results of different mechanisms of taphonomic alteration in coquina deposits, namely, (**a**) resedimented elements; (**b**) resedimented element with an excellent state of preservation; (**c–e**) accumulated Anthozoa class elements.

4.2. Definition of Taphofacies and Petrofacies of the Amaral Formation

For the Amaral Formation, and based on the identification of different lithological, taphonomic, and structural aspects, four lithofacies and 10 taphofacies were identified and characterized. Taphofacies 1 to 10 are related to a mixed clastic–carbonate platform and represent the biostratinomic and fossil-diagenetic processes that gave rise to the recorded elements (Table S2). Moreover, their spatial distribution reflects different sub-environments, and corresponding energy levels, across different platforms' contexts. Lithofacies 1, 2, 4 and 5 correspond to deposits without fossils. Lithofacies 3 (sabkha) was classified for the Morro do Chaves Formation, but it was not identified for the coquina deposits of the Amaral Formation.

The integrated analysis of taphonomic and sedimentological data enabled the definition of 84 petrofacies. Among them, 53 display direct correlation with specific taphofacies (with bioclasts), three refer to lithofacies (without bioclasts), and 28 showing intense cementation, some of them with almost total replacement of the original constituents. The taphofacies and petrofacies' distribution, in association with the paleoenvironmental settings of the coquina record, lead to the differentiation of several depositional environments, in the context of a mixed platform.

4.3. Stratigraphic Evolution Model for the Depositional Systems of the Coquina from the Amaral Formation

The deposits of the Amaral Formation display important lateral variations. These are related not only to the dynamics inherent to each of the sub-environments where they are located, but also to the paleogeographic configurations recognized for the interval between the Kimmeridgian and the Tithonian. Due to these particularities, in this work, five distinct sequences widespread in three depositional environments were identified, namely, lagoon sector, lagoon-barrier sector, and marine-distal sector (Figure 2a).

4.3.1. Lagoon Sector

The lagoon sector is widely represented in most parts of the inland and the protected region of the large marine gulf, in which the Lusitanian Basin was formed at the time of deposition for the Amaral Formation sediments. The lagoon deposits surround the body of water; the occurrence of reef bodies or even barriers made up of bioclastic or oncolitic fragments protects it from the direct action of the waves. This depositional environment is widely expressed in the northern part, narrowing to the south with overlapping barrier islands and closer proximity to the uplifted basin margins (Berlengas and Ota). The Cumieira (Pombal) and Abiul (Pombal) sections and 15 outcrops for lateral and vertical control of facies represent this lagoon sector of the basin (Figure 2b,c).

Taking into consideration paleoecological, taphonomic, sedimentological, and stratigraphic criteria, five distinct sequences were identified for this sector, reflecting different depositional environments.

Lagoon-Barrier Sequence (Internal Lagoon Context)

This first sequence is recorded only at the Pombal Section. It shows amalgamated layers with external lenticular geometry with lateral continuity. It consists of mudstones/wackstones, displaying plane-parallel and crossed stratifications truncated by waves, with ripple marks on the upper surface of some layers. Representatives of the Bivalvia and Gastropoda classes were recognized, supported by carbonate silty-clay matrix. At the bottom of this interval, oncolites up to 0.1 mm in diameter are also recorded. Fragments of carbonized fossil wood (between 0.1 and 1.1 mm) are frequent along the whole sequence. FM and FSRUnobMD petrofacies were recognized in this sequence of the lagoon sector.

This interval corresponds to the most basal sequence of the Amaral Formation and is interpreted as representative of an internal lagoon environment with low energy, subject to normal waves, and far from the influence of the fluvial action, which can be attested by the large number of carbonized plant fragments transported to the lagoon. In the lagoon-barrier sector it is possible to laterally follow this sequence, namely, at the Praia do Salgado Section.

High-Energy Transitional Sequence

The second sequence is observed only at the Pombal Section. It presents external lenticular geometry with lateral continuity. The layers are composed of mudstones/wackstones (shales). Representatives of the Bivalvia and Gastropoda classes were recognized, supported by a silty-clay matrix with carbonate sand. They show ripple marks, plane-parallel stratification, and truncated stratification by waves. FM and FSRBnob1MD petrofacies were recognized in this sequence of the lagoon sector.

This second sequence of the Amaral Formation is interpreted as characteristic of a transitional environment, with moderate to high energy, subject to the frequent action of normal waves. In the lagoon-barrier sector it is possible to laterally follow this sequence at the Praia do Salgado and Praia da Consolação sections.

Low-Energy Transitional Sequence

The third sequence is observed only at the Pombal Section. It is composed of some amalgamated layers, with the thicker ones showing external lenticular geometry with good lateral continuity. The layers are made up of mudstones/wackstones. In general, they present wavy and crossed stratifications truncated by waves and ripple marks on the upper surface of some layers. Representatives of the Bivalvia and Gastropoda classes were recognized, supported by a silty-clay to very fine and carbonate sandy matrix. In the middle part of the sequence, some layers record rare fragments of carbonized fossil wood. FM, FCm, FSRBnob1CmD, FSRBnob1MD, and FSRBnob2MD petrofacies were recognized in this sequence of the lagoon sector.

The third sequence of the Amaral Formation is interpreted as representative of low-energy depositional context, subject to the continuous action of normal waves. In the lagoon-barrier sector it is possible to laterally follow this sequence at the Praia do Salgado and Praia da Consolação sections.

Lagoon-Barrier Sequence

The fourth depositional sequence of the Amaral Formation is represented at the Cumieira and Abiul sections. Amalgamated layers show external lenticular geometry and thicker layers present good lateral continuity. At the base, the beds are made up of grainstone, containing representatives of the Bivalvia class in sandy matrix with grain-supported biofactory. Oncolites (0.2 to 2.5 cm) are also recorded. The upper layers are made up of mudstones/wackstones, containing representatives of the Bivalvia and Gastropoda classes, in addition to rare oncolites, supported by a very fine, carbonate silty-clay matrix. They present wavy and crossed stratifications truncated by waves, with some layers displaying ripple marks at the upper surface. At the top of this sequence, Echinoidea spicules (0.8 to 1.3 cm) and coral fragments (3.5 to 5.0 cm) occur. FM, FCm, FSRBnob1CD, FSRBnob1MD,

FSRUnobCmD, FSRBnob2MD, FSRUnobBob1CmD, and FSRBnob1Bob1CmD petrofacies were recognized in this sequence of the lagoon sector.

The occurrence of Echinoidea and coral fragments suggests a depositional context of fast sea level rise. This sequence is laterally recognized in all sectors except at the Praia do Salgado Section.

Lagoon Alluvial-Deltaic Sequence

The top of the stratigraphic record is identified at the Cumieira and Abiul sections and displays limited exposure. It is represented, from the bottom to the top, by a layer of medium to fine sandstone, followed by amalgamated layers of grainstones, packstones, and wackstones, whose main component is oncolites with dimensions ranging from 0.1 to 1.0 cm, and, finally, a sandstone layer, containing dispersed bivalves, which marks the starting of a new regressive cycle. FCm and FT petrofacies were recognized in this sequence of the lagoon sector.

The top of the stratigraphic sequence described in the lagoon sector clearly shows the transition between the carbonates of the Amaral Formation and the sandstones of the Lourinhã Formation. This fifth sequence records the establishment of a lagoon environment with fluvio-delta input. This sequence is also laterally recognized at the Praia do Salgado Section in the lagoon-barrier sector.

4.3.2. Lagoon-Barrier Sector

The lagoon-barrier sector corresponds to the narrowing, towards the south, of the lagoon located to the north of the basin. It is characterized by the occurrence of expressive reef bodies, as well as barrier islands, composed of bioclastic or oncolitic fragments. It is noteworthy, in this sector, to mention the occurrence of coarser continental sediments, coming from the uplifted margins of the basin (Berlengas to the west, and Ota to the east). In this region, the Amaral Formation was analyzed based on the study of the Praia do Salgado (Famalicão) and Praia da Consolação (Atouguia da Baleia) sections and two outcrops for lateral and vertical control of facies (Figure 2d,e).

In this context, and based on paleoecological, taphonomic, sedimentological, and stratigraphic criteria, five distinct sequences reflecting different depositional environments were recognized in this sector.

Lagoon-Barrier Sequence (Regressive Context)

This sequence is observed only at the Praia do Salgado Section. It consists of layers with amalgamated external lenticular geometries of wide lateral continuity, composed of grainstones, packstones, and mudstones. They present plane-parallel and crossed stratifications. In this sector, the top of the sequence is represented by a scraping surface, where dinosaur footprints are found. Representatives of the Anthozoa, Bivalvia, Gastropoda, and Echinoidea classes were also recognized, predominantly supported by a medium to fine, carbonate sand matrix. Locally, at the base of some layers, oncolitic levels occur.

FM, FSRBnob1Bob2MD, FSRBnob1CmD, FSRBnob1Bob1CmD, FSRBnob2CmD, and FCm petrofacies were recognized in this sequence of the lagoon-barrier sector.

This is the first sequence of the Amaral Formation, and it represents a depositional system of lagoon-barrier in a regressive context. Its upper surface records dinosaur footprints, which suggests proximity to the coastline at the end of this sequence. This sequence is also laterally recognized at the Cumieira Section of the lagoon sector.

High-Energy Transitional Sequence

The second sequence in the lagoon-barrier sector is recognized at the Praia do Salgado and Consolação sections. It is composed of amalgamated layers with external lenticular geometries and wide lateral continuity. They consist of packstones, wackstones, mudstones, and sandstones. It presents grooved and herringbone cross stratifications. Fragmented Anthozoa and Bivalvia representatives occur, predominantly supported by a medium to fine, carbonate sandy matrix, as well as carbonized fossil wood fragments and truncated wave stratification. FT, FM, FCm, FTC, FTCm, FSRUnobMH, FSRBnob1CmD, FSRBnob2CmD, FTSRBnob1CmD, FTSRBnob2CmD, and FSRBnob2Bob2CmD petrofacies were recognized in this sequence of the lagoon-barrier sector.

This second sequence is interpreted as characteristic of a transitional environment of moderate to high energy, subject to normal waves and occasional storms and to fluvial influence, that transported carbonized wood fragments to the lagoon. This sequence is also recognized laterally in the Cumieira Section in the lagoon sector.

Low-Energy Transitional Sequence

This sequence is represented in all the described sections of the lagoon-barrier sector. It is composed of layers displaying external lenticular geometry and wide lateral continuity including grainstones, packstones, wackstones, mudstones, and sandstones with plane-parallel, rippled, and crossed truncated stratification. Anthozoa, Bivalvia, Gastropoda, and Echinoidea fragments are supported by a silty-clay to very fine carbonate matrix, locally rich in oncolites up to 4.0 cm and fossil wood with cross-bedding and herring-bone cross-bedding. Bioturbations are found (Thalassinoides) with more incidence at the top of some layers, and accumulated corals occur in isolated levels. FM, FT, FCm, FSRUnobCmD, FSRBnob2MD, FSRBnob1Bob2MD, FSRBnob1Bob2CmD, FSRBnob1Bob2CD, FSRBnob2Bob2CmD, FSRBnob1CmD, FSRBnob2CmD, FSRBob1CmD, ISF 4 (TF 1.4), FSRBnob1Bob2CmH, FSRBnob1Bob1CmH, FTSRBnob2Bob1CmH, FTSRBnob1Bob2CmH, FTSRBnob1CmD, FTSRBnob1CmH, FTSRBnob2CmD, and FTSRBnob2Bob2CmD petrofacies were recognized in this sequence of the lagoon-barrier sector.

This sequence is characterized by the significant abundance of oncolites and is interpreted as representative of a low-energy context, under continuous action of normal waves. The input of fragments of fossil wood from the continent, located close to the west, persists; they were distributed among the sediments of the lagoon by the action of currents. The abundance of echinoderm spines suggests the occurrence of a transgressive event subsequent to the low-energy environment related to the previous sequence. Within this sector, it is recognized that there is a wide lateral variation over a relatively short distance. In a general context, the deposits described for the Praia do Salgado Section, in the northernmost portion of this sector, present facies with a calcarenitic matrix, without significative input of terrestrial constituents and with chaotic distribution of fossil elements. They represent a more protected region within the lagoon-barrier sector. On the other hand, the deposits described for the Praia da Consolação Section, in the southernmost portion of this sector, present facies with higher input of terrestrial constituents, higher cementation rates, and slight orientation of the fossil elements. They represent a more open region in the lagoon-barrier sector. This sequence is also recognized laterally in the Cumieira Section in the lagoon sector.

Lagoon-Barrier Sequence

This sequence can only be observed at the Praia da Consolação Section. It is composed of a thick set of layers displaying external lenticular geometries, with amalgamated lenses and wide lateral continuity. They consist of packstones, wackstones, and mudstones, and present plane-parallel and crossed stratifications truncated by waves. At some levels, desiccation cracks are seen at the top, and on one of these levels, dinosaur footprints are recognized. The occurrence of Anthozoa, Bivalvia, and Gastropoda representatives were also recognized, predominantly supported by a silty-clay, carbonated matrix. Rare fragments of fossil wood occur. FM, FCm, FTSRBnob2Bob2CmD, FTSRBnob2CmD, FSRBnob2Bob1MD, FTSRBnob1CH, and FTSRBnob2CH petrofacies were recognized in this sequence of the lagoon-barrier sector.

This sequence is interpreted as characteristic of a low-energy depositional lagoon environment, subject to the action of normal waves, and located near to the coast as suggested by the dinosaur footprints. This sequence is laterally recognized in all sectors apart from the Praia do Salgado Section.

Lagoon Alluvial-Deltaic Sequence (Transgressive Context)

The fifth sequence can only be recognized at the Praia do Salgado Section. It consists predominantly of mudstones intercalated in sandstones and calcarenites, containing Anthozoa and Bivalvia representatives. FT, FM, ISF 4 (TF 1.4), and ISF 2 (TF 1.2) petrofacies were recognized in this sequence of the lagoon-barrier sector.

This sequence is interpreted as characteristic of low-energy lagoon depositional environment, subject to normal waves and river systems, which would transport fragments of fossil wood to the lagoon. It is also recognized laterally at the Cumieira and Abiul Sections of the lagoon sector.

4.3.3. Marine-Distal Sector

The marine-distal sector is located in the southernmost portion of the large marine gulf; it corresponds to a failed and tilted basin bottom, with rift shoulders and highs produced by salt diapirism mainly during the beginning of Late Jurassic. These internal highs favor the formation of isolated platforms where coral bioconstructions and calcareous banks were installed, due to the intense action of waves and to the proximity of the open ocean. In this region, the Amaral Formation was analyzed based on the study of the Pedralvo (Arruda dos Vinhos) Section and on ten outcrops for lateral and vertical control of the facies (Figure 2f).

It consists of one sequence, defined considering paleoecological, taphonomic, sedimentological, and stratigraphic criteria, which reflects a specific depositional environment within the genetic context of the unit.

Lagoon-Barrier Sequence (in a Marine-Distal System)

This depositional sequence described for the Amaral Formation was recognized at the Pedralvo Section. It is composed of layers with external lenticular geometry and wide lateral continuity, easily observed in the thicker ones. They are constituted, from the base to the top, by mudstone with ripple marks and terrigenous constituents; wackstone containing Bivalvia representatives, ripple marks, and abundant terrestrial constituents; mudstone, containing rare echinoderm spines and bivalves; wackstone, with bivalves up to 7.0 cm, some of them with joined valves; grainstone Anthozoa, Bivalvia, and Gastropoda representatives; grainstone with rare echinoderm spines and abundant complete bivalves up to 4.0 cm; and grainstone with abundant echinoderm spines. FM, FTSRUnobMD, FSRUnobMD, FTSRUnobCmD, and FTSRBnob1Bob1CmD petrofacies were recognized in this sequence of the marine-distal sector.

The deposits described in this sequence are interpreted as generated in a context of fast sea level rise, which is corroborated by the occurrence of fragments of echinoderm spines. This sequence is laterally recognized in all sectors except for the Praia do Salgado Section.

4.4. Paleoenvironmental and Paleogeographic Reconstructions

The depositional systems that originated the sediments that comprise the Amaral Formation, in space and time, resulted from a quite-complex sedimentary dynamic with tectonic control and variation of the global sea level. The different depositional systems recognized at the Kimmeridgian–Tithonian transition in the Lusitanian Basin are conditioned by specific paleoenvironmental/paleogeographic contexts in each sector of the same basin. These constraints lead to a matrix of associations of characteristic facies (Table S3).

The stratigraphic correlation of the Amaral Formation sections, considering the transgressive–regressive cycles, allows for the understanding of the coquina record for this interval in a context of distinct depositional environments. These environments are framed in a large gulf paleogeography showing a very heterogeneous configuration (Figure 8).

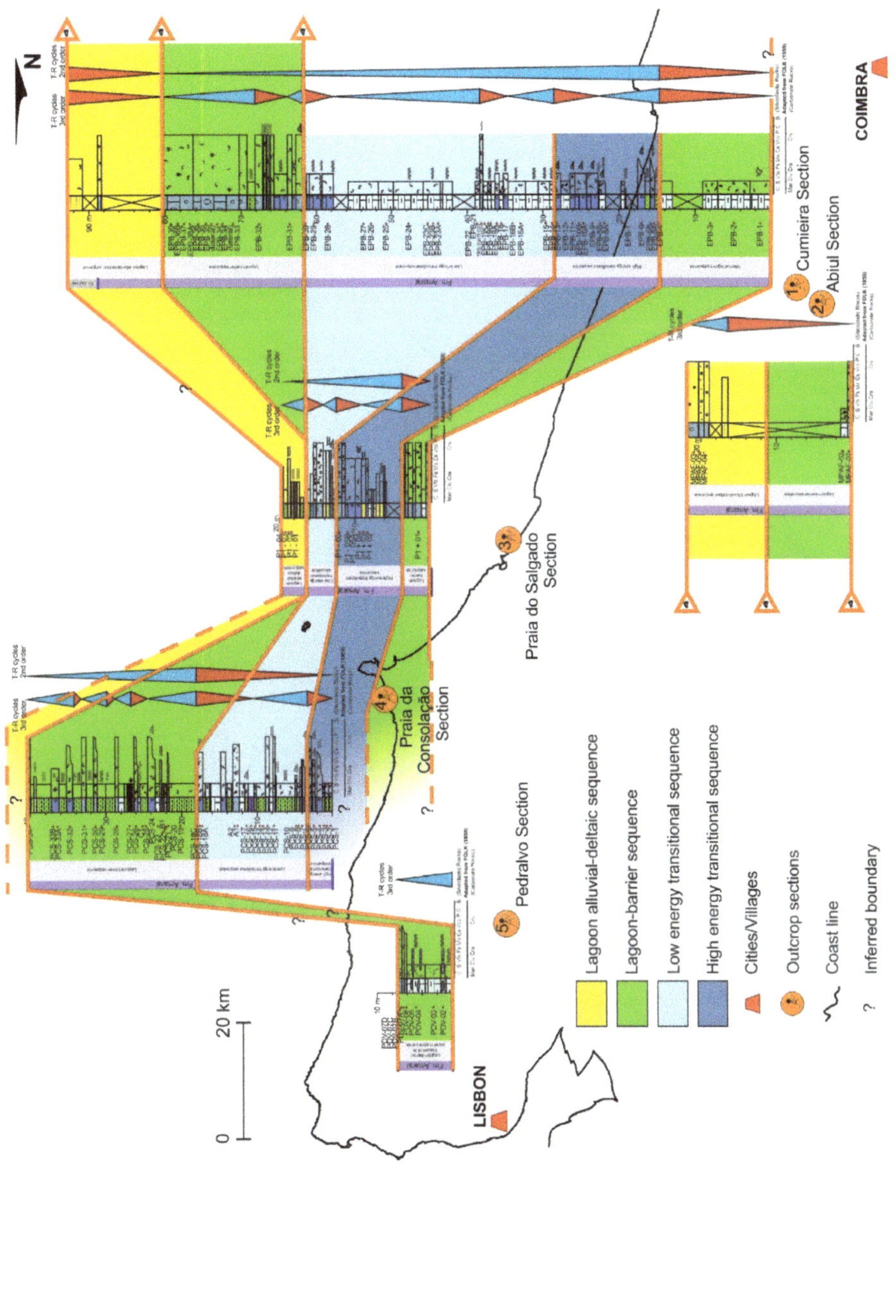

Figure 8. Regional correlation framework of the profiles described for the Amaral Formation with the correlation of the depositional sequences based on the T–R cycles and the fossiliferous diversity.

In each sector, the dominant energy conditions modeled different types of paleoenvironments (Figure 9). The lagoon, which corresponds to a low-energy environment protected by small and dispersed reef banks in the innermost part of the gulf, is represented by marginal and oncolitic facies with low input of terrestrial constituents (siliciclastic and carbonaceous).

The lagoon-barrier integrates two sub-environments. The first one is of low to medium energy, protected by a large number of aggregated and larger reef banks, located in the bay within the great gulf (north of the Berlengas structural high). It is represented by marginal oncolitic facies to reef facies (with in situ corals) with some contribution of terrigenous constituents. The second one corresponds to a high-energy environment, where a large number of coral banks was subject to major action by energetic agents (normal waves, storms, normal fluvial discharges, and runoffs). It corresponds to the proximal zone of the platform in the outer region of the gulf, and is represented by margins, reefs, washover, and alluvial-deltaic fans.

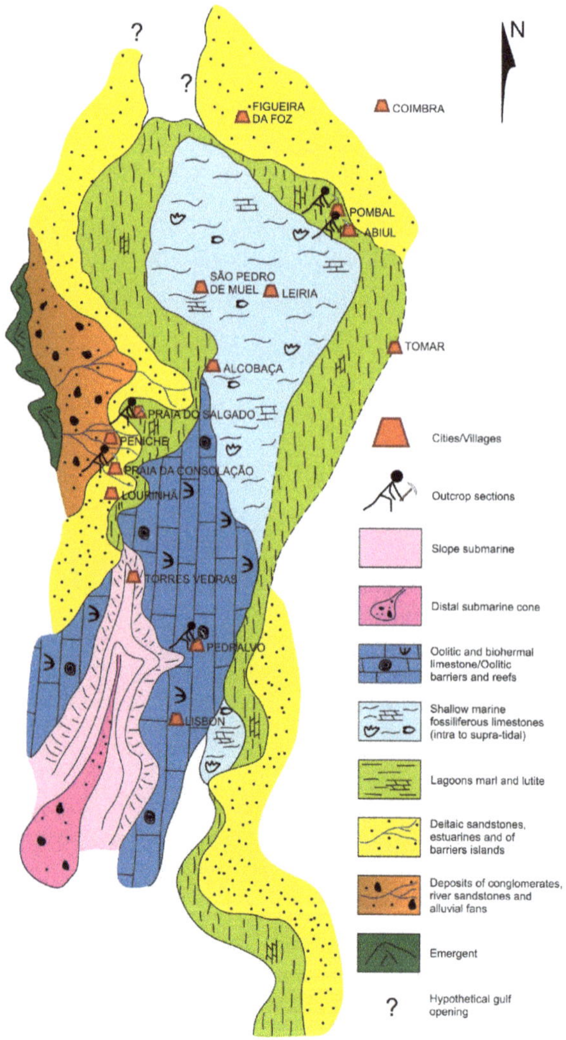

Figure 9. Paleoenvironmental sketch of the upper Kimmeridgian of the Lusitanian Basin (modified after Pena dos Reis et al. [38]; modified after Garcia et al. [39]).

The marine-distal sector corresponds to an environment with wide energy variation (from very low to high). It is represented by isolated reef and marshal facies, with significant terrestrial contribution, as a direct result of the proximity of the source areas, whose distribution is limited to high structures along a platform with a steep slope. This environment corresponds to the distal zone of the platform in the outer region of the gulf.

The distribution of the coquina petrofacies, conceived according to the energy relations between the active sedimentary agents (fluvial energy and wave energy) for each sector, allowed the construction of a 3D paleoenvironmental model for the coquina of the Amaral Formation. This was strongly controlled by the taphonomic mechanisms, namely bioestratinomics, and reflects an evolutionary trend from a lagoon-barrier depositional system, influenced by fluvial inputs and sea level rise, to a lagoon alluvial-delta system, through high- and low-energy transitional episodes (Figure 10).

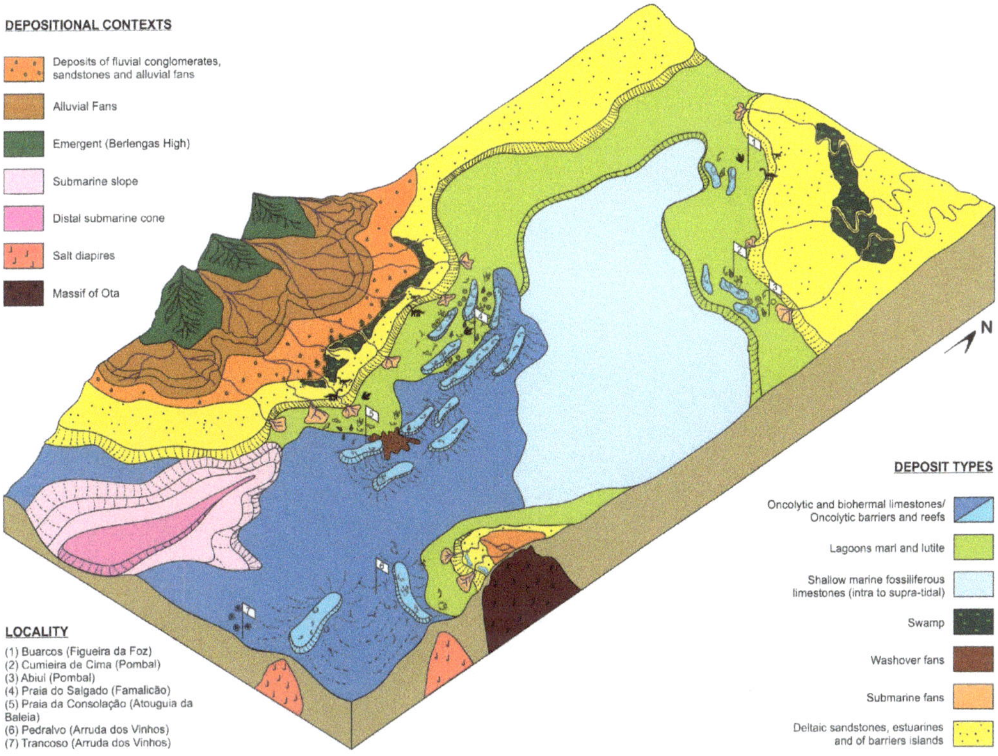

Figure 10. Paleoenvironmental/paleogeographic model showing the main types of sedimentary deposits and fossil records recognized in the studied areas.

The 3D model for the coquina of the Amaral Formation provides the basis for designing theoretical models of facies distribution. This model is capable of predicting the evolution of coquina facies within a mixed depositional system in the context of a heterogeneous platform (Figure 11). Such heterogeneity, due to the tectonic complexity of the Lusitanian Basin and its intense differential saliferous diapirism, results, regarding continuity, in two types of platforms: continuous (e.g., Pombal platform at SW, and Berlengas platform at E–SE), and isolated (e.g., Ota platform).

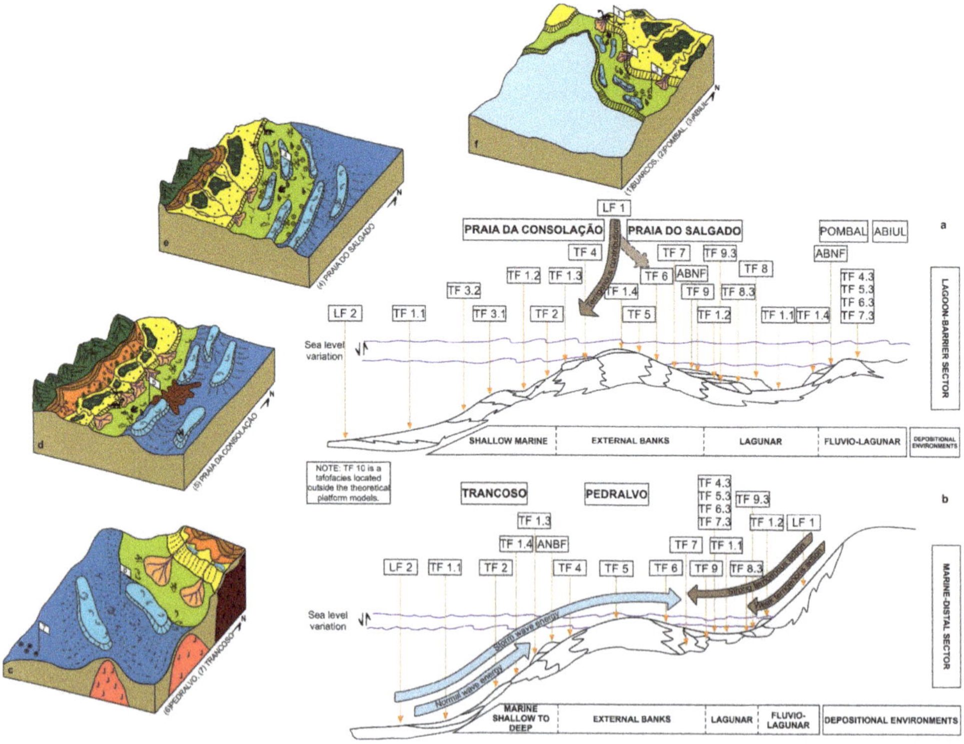

Figure 11. Theoretical models of distribution of taphofacies for the coquina deposits of the Amaral Formation in the context of a heterogeneous platform, namely, of two types: (**a**) continuous platform (e.g., Pombal platform for SW and Berlengas platform for E–SE); (**b**) isolated platform (e.g., Ota platform). Shown are 3D paleoenvironmental models of the depositional systems for the sectors studied based on the distribution of the taphofacies in the different sectors of the platform, namely, (**c**) depositional model of the marine-distal sector; (**d**,**e**) depositional models of the lagoon-barrier sector; (**f**) depositional model of the lagoon sector.

5. Discussion

The sedimentary infill of the Lusitanian Basin is related with the formation of a shallow/epicontinental sea at the beginning of the Jurassic [35], conditioned by intense tectonic activity and saline diapirism. The large lateral and temporal facies variations result from the structural complexity of the basin [20], which hinders the stratigraphic positioning of the recognized units; moreover, it makes it difficult to establish the lateral correlation between genetically distinct facies, which is particularly evident with regard to the units defined for the Upper Jurassic of the Basin, namely, for the Amaral Formation.

Choffat [40], studying the outcrops of the Torres Vedras and Montejunto region (i.e., the lagoon-barrier sector in this work), defined the Lusitanian stage as being constituted, from the base to the top, by three intervals: "Couches à Pholadomya protei", "Couches de Alcobaça", and "Couches à Lima alternicosta". Afterwards, he recognized that the marine sediments of the "Couches à Pholadomya protei" were analogous to those of the Oxfordian [41]; he redefines the Lusitanian stage as comprising only the "Couches de Alcobaça" and the "Couches à Lima alternicosta", to be of marine nature with "sequanian fauna", brackish, or lacustrine [41]. Later, Choffat [42] considers that the Lusitanian is composed of the "Calcaires de Montejunto" and "Marnes de Abadia", and defines, for the first time, the uppermost deposits of the Lusitanian, i.e., the upper part of the "Assise de Abadia", as "Corallien d'Amaral" [43,44]. The author describes the "Corallien d'Amaral" as a unit composed of carbonate deposits of small depth (intraclastic limestones, dolomitic limestones, oolitic limestones, and limestone conglomerates with reef limestone intercalations), represented by biofacies rich in fragments of echinoids, bivalves, stromatoporoids, spicules of spongiary, and microfauna (foraminifera, algae, and ostracods), and related to intertidal and coastal environments with clear marine influences [24].

In this work, the Amaral Formation is considered as composed of coquina, sandstones, and mudstones formed in the context of a mixed clastic-carbonate platform, spread within a wider region of the Lusitanian Basin, from Arruda dos Vinhos (south) to Pombal (north). The coquina deposits are made up of concentrations of elements that reflect different mechanisms of taphonomic alteration (fragmented, recrystallized, micritized, among others), namely, bivalves, gastropods, echinoderms (of echinoid spines and central body), corals, stromatoporoids, spongies, and microfossils (foraminifera, algae, and ostracods).

Leinfelder et al. [28] also recognized the faunal diversity of the Amaral Formation reefs in the Upper Jurassic record of the Lusitanian Basin, as well as the paleoenvironmental implications of their distribution in the context of a mixed carbonated platform. The authors consider that sequence stratigraphy seems to be the only method that can allow the design of depositional models for this interval in the basin [29–34]. However, this involves considering the different source areas and the structure of the basin. If these parameters are not considered, it may lead to paleoenvironmental interpretations that tend to define different lithostratigraphic units for deposits that, in fact, correspond to lateral variations of the same unit. This is the case, for example, of the Ota Limestone, considered so far as an independent unit [28], though a coeval of the Amaral Formation.

Finally, Fürsich et al. [35] consider the sedimentary record historically (and here also) assigned to the Amaral Formation as representing the Alcobaça Formation. Later, Fursich et al. [45] interpreted the Alcobaça Formation as related to a complex mosaic of low- to high-energy, carbonate- or siliciclastic-dominated shallow shelf settings; coastal embayments and lagoons; and coastal plains with rivers, lakes, and playas. Such interpretation regarding the depositional settings meets, in general, the different environments proposed here, but only for the central part of the Lusitanian Basin (between the Praia da Consolação and Praia do Salgado sections, i.e., the lagoon-barrier sector in this work).

As previously referred, the Amaral Formation presents a great diversity in terms of genetically distinct facies and fossil record. Thus, the application of the methodology to classify taphofacies and petrofacies developed for the Morro do Chaves Formation to the study of the Amaral Formation was adjusted in order to allow the integration of the interaction between the taphonomic mechanisms, the tectonostructural context of the Lusitanian Basin, the diversity of the recognized fossil record, and the depositional environments related to the coquina deposits.

As such, considering the paleoenvironmental distribution of the petrofacies model established by Garcia et al. [13] for the Morro do Chaves Formation, i.e., alluvial-fluvial/lacustrine-sabkha, and considering the particular characteristics of the Amaral Formation, it is possible to relate this unit with lagoon, lagoon-barrier, and marine-distal environments.

From the taphonomic point of view, the coquina deposits of the Amaral Formation present higher variety of allochemical components (bioclastic and nonbioclastic), including different types of bioclasts and oncolites/oolites. In this sense, the representativeness of each of these components in the studied outcrops of the Lusitanian Basin allowed us to understand the relationships and intensity of the energetic agents responsible for the taphonomic changes recorded among the analyzed fossil assemblages. Thus, it was possible to refine the spatial relationships of taphofacies/petrofacies distribution and the corresponding associations in the theoretical 2D models for the distribution of taphofacies for the coquina deposits of the Amaral Formation, and in the 3D models of the depositional systems for each studied sector.

The biostratigraphic correlation between the Amaral Formation and other units established for the Upper Jurassic of the Lusitanian Basin is still complex and controversial. However, through this investigation, it is possible to recognize different assemblages of preserved elements of the Amaral Formation throughout the Lusitanian Basin, which refers to the different depositional environments referred to above. In this context, the spatial distribution of taphofacies and the corresponding petrofacies represent an important contribution for the establishment of theoretical models for the distribution of taphofacies and 3D models of depositional systems. On the other hand, this work can help the improvement of the predictive capacity in exploration procedures of coquina deposits and the understanding of similar deposits of coquina in the Brazilian pre-salt.

6. Conclusions

The studies carried out in the coquina of the Amaral Formation (Lusitanian Basin, Portugal) allowed us to recognize the relevance of using knowledge in the field of evolutionary taphonomy to understand the genesis of the coquina deposits. In this context, different taphonomic processes/mechanisms were identified that led to the recognition of a variety of taphofacies/petrofacies.

The interpretation of the bioestratinomic and fossil-diagenetic processes inherent to the taphonomic alteration allowed for the elaboration of a facies classification and the spatial prediction of coquina facies in any analogous reservoir, as well as the design of theoretical facies distribution models in the context of mixed platforms in different morpho-structural contexts within the same basin.

According to the compositional, textural, and taphonomic aspects, ten taphofacies and four lithofacies were identified for the Amaral Formation, which allowed the definition of 84 petrofacies based on the types and intensity of the identified sedimentary and taphonomic processes and on the diversity of their constituents. Among these, five facies resulted from intense fossil-diagenetic processes, which led to the almost total replacement of the original constituents. Such data, combined with the lateral and vertical variations in each individual outcrop and in all the outcrops, supported the design of paleoenvironmental models of the coquina deposits of the Amaral Formation that illustrate the spatial distribution of the depositional sequences identified by sector (lagoon, lagoon-barrier, and marine-distal).

The analysis of taphofacies and petrofacies of the Amaral Formation allowed the development of theoretical models for their evolution, depending on the paleogeographic and structural contexts, the intensity of the energetic agents, and the fossil diversity. These reflect the rise and fall of the base level, identified through the interpretations of the transgressive–regressive cycles in the studied sections, in the context of a marine environment with the configuration of a gulf. The petrofacies distribution also enables the identification of different marine depositional environments, affected by the action of normal and storm waves and by normal and flood fluvial discharges, the latter responsible for the loading of terrestrial sediments into the carbonate system. In addition, due to the tectonic complexity of the Lusitanian Basin and its intense differential saliferous diapirism, that distribution also reflects the heterogeneity of the platform, which assumes two distinct types: continuous (e.g., Pombal platform at SW, and Berlengas platform at E–SE), and isolated (e.g., Ota platform).

The resulting geological models, related to the Morro do Chaves (Sergipe–Alagoas Basin, NE Brazil) and the Amaral (Lusitanian Basin, Portugal) formations supported the construction of theoretical models of coquina deposits for an ideal bank and platform that allow the adaptation for subsurface analogs, as in the case of contiguous coquina reservoirs in the pre-salt oil province of Brazil. However, the use of such theoretical models in other coquina deposits requires an adequate adjustment to the structural, stratigraphic, and petrographic–petrophysical aspects of the analogue; only then can they have the necessary robustness and reliability for simulation procedures with a low degree of uncertainty.

The in-depth analysis of the records, in the light of the geological model designed for the Amaral Formation, should lead to its expansion and detailing, making it possible to locate, in the Lusitanian Basin, facies of coquina similar to those already recognized in the Sergipe–Alagoas Basin and in the Brazilian pre-salt. As such, FSRBnob1CmD, FTSRBnob1CmD, FSRBnob1Bob1CmD, FTSRBnob1Bob1CmD, FSRBob1CmD, FTSRBob1CmD, FSRBnob1CmH, FTSRBnob1CmH, FSRBnob1Bob1CmH, FTSRBnob1Bob1CmH, FSRBob1CmH, FTSRBob1CmH, FSRBnob2CmD, FTSRBnob2CmD, FSRBnob2Bob1CmD, FTSRBnob2Bob1CmD, FSRBnob2Bob2CmD, FTSRBnob2Bob2CmD, FSRBob2CmD, and FTSRBob2CmD petrofacies, which present the best quality as reservoir facies, are predictably located in the offshore of the basin.

The methodology underlying the classification of coquina proposed for the Morro do Chaves Formation is valid for the Amaral Formation, with the necessary adaptations resulting from the particular characteristics of the petrofacies recognized in the latter, which exhibit higher diversity in the fossil record and are conditioned by regional tectonics and the most frequent occurrence of transgressive events. Therefore, the particularities of the facies described for the Amaral Formation enabled us to improve the classification of facies for other coquina deposits.

Supplementary Materials: The following are available online at https://www.mdpi.com/2077-1312/9/12/1319/s1, Table S1: Alphanumeric classification of facies types according to their petrographic and taphonomic characteristics identifying types and intensities of processes and products associated with coquina. Table S2: Lithofacies and taphofacies classification of the coquina from Amaral Formation. Table S3: Petrofacies recognized for each sector of the Lusitanian Basin, respective dominant energy conditions, diversity of the fossil record, and relevance of the terrigenous components.

Author Contributions: Conceptualization, G.G.G., A.J.V.G. and M.H.P.H.; methodology, G.G.G.; validation, G.G.G., A.J.V.G. and R.M.M.; investigation, G.G.G., A.J.V.G., M.H.P.H., R.M.M. and R.P.d.R.; writing—original draft preparation, G.G.G.; writing—review and editing, G.G.G., A.J.V.G., M.H.P.H. and R.M.M.; visualization, G.G.G., A.J.V.G., M.H.P.H., R.M.M. and R.P.d.R.; supervision, A.J.V.G., M.H.P.H. and R.P.d.R. All authors have read and agreed to the published version of the manuscript.

Funding: This research was funded by Portuguese funds by Fundação para a Ciência e a Tecnologia, I.P. (Portugal), grant number UIDB/00073/2020 and UIDP/00073/2020, projects of I & D unit Geosciences Center (CGEO).

Acknowledgments: The authors are grateful to Salomé Custódio and Yasir Shahzad (University of Coimbra, Portugal), for the revision of the manuscript.

Conflicts of Interest: The authors declare no conflict of interest. The funders had no role in the design of the study; in the collection, analyses, or interpretation of data; in the writing of the manuscript, or in the decision to publish the results.

References

1. Castro, J.C. Evolução dos conhecimentos sobre as coquinas-reservatório da Formação Lagoa Feia no trend Badejo-Linguado-Pampo, Bacia de Campos. *Geociências (UNESP)* **2006**, *25*, 175–186.
2. Garcia, A.J.V.; Pereira, F.C.; Araújo, W.B.; Rocha, L.M.; Ribeiro, D.D.M.; Leal, F.P.M.; Correia, G.L. Caracterização multiescalar de reservatórios carbonáticos análogos em afloramentos da Bacia de Sergipe-Alagoas, NE do Brasil. In *Boletim Técnico da Petrobras*; Neto, O.C.P., Ed.; Petrobras: Rio de Janeiro, Brasil, 2014; Volume 22, pp. 51–82.
3. Fernández-López, S. Criterios elementales de reelaboración tafonómica em ammonites de La Cordillera Ibérica. *Acta Geol. Hisp.* **1984**, *19*, 105–116.
4. Fernández-López, S. Ammonoid taphonomy, palaeoenvironments and sequence stratigraphy at the Bajocian/Bathonian boundary on the Bas Auran area (Subalpine Basin, south-eastern France). *Lethaia* **2007**, *40*, 377–391. [CrossRef]
5. Garcia, A.J.V.; Ribeiro, D.M.; Figueiredo, S.S.; Dantas, M.S.; Oliveira, I.; Leite, K.A.S.; Garcia, G.G. 3D Modeling of carbonate reservoir analogue outcrops using Camures methodology, Sergipe-Alagoas Basin (SEAL), NE, Brazil. In Proceedings of the AAPG Annual Convention & Exhibition, Denver, CO, USA, 31 May–3 June 2015.
6. Bellian, J.A.; Kerans, C.; Jennette, D.C. Digital outcrop models: Applications of terrestrial scanning lidar technology in stratigraphic modeling. *J. Sediment. Res.* **2005**, *75*, 166–176. [CrossRef]

7. Fabuel-Perez, I.; Hodgetts, D.; Redfern, J. Integration of digital outcrop models (DOMs) and high resolution sedimentology—Workflow and implications for geological modelling: Oukaimeden Sandstone Formation, High Atlas (Morocco). *Pet. Geosci.* **2010**, *16*, 133–154. [CrossRef]
8. Agada, S.; Chen, F.; Geiger, S.; Toigulova, G.; Agar, S.; Benson, G.; Shekhar, R.; Hehmeyer, O.; Amour, F.; Mutti, M.; et al. Deciphering the fundamental controls of flow in carbonates using numerical well-testing, production optimisation, and 3D high-resolution outcrop analogues for fractured carbonate reservoirs. In Proceedings of the EAGE Annual Conference & Exhibition Incorporating SPE Europe, London, UK, 10–13 June 2013.
9. Rodriguez, M.; Abad, F.; Rodriguez, L.; Gaibor, J.; Moran, M.; Verdezoto, A. Application of modern reservoir characterization in mature fields to unravel hidden reserves. In Proceedings of the SPE Latin American and Caribbean Petroleum Engineering Conference, Quito, Ecuador, 18–20 November 2015. [CrossRef]
10. Fernández-López, S. *Temas de Tafonomía*; Departamento de Paleontología, Faculdade de Ciencias Geológicas, Universidad Complutense de Madrid: Madrid, España, 2000; Unpublished work.
11. Folk, R.L. Spectral subdivision of limestones types. In *Classification of Carbonate Rocks—A Symposium*; Ham, W.E., Ed.; Geologists Memoir—American Association of Petroleum: Tulsa, OK, USA, 1962; Volume 1, pp. 62–84.
12. Choquette, P.W.; Pray, L.C. Geologic nomenclature and classification of porosity in sedimentary carbonates. *AAPG Bull.* **1970**, *54*, 207–250.
13. Garcia, G.G.; Henriques, M.H.; Garcia, A.J.V.; Dantas, M.V.S. Petrofacies and taphofacies analyses of coquinas as a tool for the establishment of a stratigraphic evolution model of the Morro do Chaves Formation (Sergipe-Alagoas Basin, NE Brazil). *Facies* **2020**, *67*, 4. [CrossRef]
14. Ribeiro, A.; Antunes, M.T.; Ferreira, M.P.; Rocha, R.B.; Soares, A.F.; Zbyszewski, G.; Moitinho de Almeida, F.; de Carcalho, D.; Monteiro, J.H. Introduction à la géologie générale du Portugal. *Servi. Geológ. Portug.* **1979**, *72*, 1–114.
15. Wilson, R.C.L. Mesozoic development of the Lusitanian Basin, Portugal. *Rev. Soc. Geol. Esp.* **1988**, *1*, 393–407.
16. Kullberg, J.C.; Rocha, R.B.; Soares, A.F.; Rey, J.; Terrinha, P.; Azerêdo, A.C.; Callapez, P.; Duarte, L.V.; Kullberg, M.C.; Martins, L.; et al. A Bacia Lusitaniana: Estratigrafia, Paleogeografia e Tectónica. In *Geologia de Portugal—Geologia Meso-Cenozóica de Portugal*; Dias, R., Araújo, A., Terrinha, P., Kullberg, J.C., Eds.; Livraria Escolar Editora: Forte da Casa, Portugal, 2013; Volume 2, pp. 317–368.
17. Rocha, R.B.; Marques, B.L.; Kullberg, J.C.; Caetano, P.C.; Lopes, C.; Soares, A.F.; Duarte, L.V.; Marques, J.F.; Gomes, C.R. *The 1st and 2nd Rifting Phases of the Lusitanian Basin: Stratigraphy, Sequence Analysis and Sedimentar Evolution*; Final Report; C. E. C. Proj. MILUPOBAS: Lisboa, Portugal, 1996.
18. Pena dos Reis, R.B.; Trincão, P.R.P.; Cunha, P.M.R.P.; Dinis, J.M.L. *Resumo do Estado Actual de Conhecimentos Sobre o Jurássico Superior da Bacia Lusitânica*; Relatório Técnico-Científico BAG: Coimbra, Portugal, 1993.
19. Garcia, A.J.V.; Pimentel, N.; Pena dos Reis, R. *Projeto Atlantis: "Modelo Geológico Evolutivo Para os Riftes Marinhos do Jurrásico da Bacia Lusitânica (Portugal)*; Relatório 04 PETROBRAS-UFS/FAPESE-UC/IPN-UL/FFCUL: Aracaju, Brasil, 2010; Unpublished work.
20. Pena dos Reis, R.B.; Pimentel, N.L.; Garcia, A.J.V. A Bacia Lusitânica (Portugal): Análise estratigráfica e evolução geodinâmica. In *Boletim de Geociências da Petrobras: Bacia Lusitânica*; Milani, E.J., Zalán, P.V., Spadini, A.R., Machado, D.L., Jr., Silva, E.F.F., Terra, G.J.S., Bueno, G.V., Eds.; Petrobras: Rio de Janeiro, Brasil, 2010; Volume 19, pp. 23–52.
21. Pena dos Reis, R.B.; Cunha, P.M.R.; Dinis, J.L.; Trincão, P. Geologic evolution of Lusitanian Basin during Late Jurrasic (Portugal). In *Advances in Jurassic Research*; Hall, R.L., Smith, P.L., Eds.; GeoResearch Forum—Trans Tech Pub: Zurich, Switzerland, 2000; Volume 6, pp. 345–356.
22. Pena dos Reis, R.B.; Pimentel, N.; Bueno, G. *III Curso de Campo na Bacia Lusitânica (Portugal)*, 3rd ed.; Impressão e Soluções, Laboratório de imagem, LDA: Coimbra, Portugal, 2008; p. 136. ISBN 978-989-20-0423-5.
23. Ramalho, M.M. Contribution à l'étude micropaléontologique et stratigraphique du Jurassique supérieur et du Crétace inférieur des environs de Lisbonne (Portugal). *Serv. Geol. Port.* **1971**, *19*, 1–212.
24. Choffat, P. Limite entre le Jurassique et le Crétacique. Notice Préliminaire. In *Extrait du Bulletin de la Société Belge de Géologie de Paléontologie et D'Hydrologie*; Hayez, Imprimeur de L'Académie Royale de Belgique: Bruxelles, Belgium, 1901; Volume XV, pp. 111–140.
25. Seifert, H. Beiträge zur geologie der Serra da Arrábida in Portugal. *Geol. Jahrb.* **1963**, *81*, 277–344.
26. Wilson, R.C.L. A reconnaissance study of Upper Jurassic sediments of the Lusitanian Basin. *Ciênc. Terra* **1979**, *5*, 53–85.
27. Teixeira, C.; Gonçalves, F. *Introdução à Geologia de Portugal*; Instituto Nacional de Investigação Científica: Lisboa, Portugal, 1980.
28. Leinfelder, R.R.; Nose, M.; Schmid, D.; Werner, W. Reefs and carbonate platforms in a mixed carbonate-siliciclastic setting. Examples from the Upper Jurassic (Kimmeridgian to Tithonian) of west-central Portugal. In Proceedings of the 23rd IAS Meeting of Sedimentology, Coimbra, Portugal, 15–17 September 2004; Duarte, L.V., Henriques, M.H., Eds.; Sedimentology and Society; Field Trip Guidebook—Carboniferous and Jurassic Carbonate Platforms of Iberia. Volume 1, pp. 95–123.
29. Leinfelder, R.R. *Upper Jurassic Reefs and Controlling Factos—A preliminar Report*; Profile: London, UK, January 1993; Volume 5, pp. 1–45.
30. Leinfelder, R.R. Distribution of Jurassic reef types: A mirror of structural and environmental changes during breakup of Pangea. In *Pangea: Global Environments and Resources*; Embry, A.F., Beauchamp, B., Glass, D.J., Eds.; Canadian Society of Petroleum Geologists Memoir: Calgary, AB, Canada, 1994; Volume 17, pp. 677–700.
31. Leinfelder, R.R. Jurassic Reef Ecosystems. In *The History and Sedimentology of Ancient Reef Systems*; Stanley, G.D., Jr., Ed.; Plenum Press: New York, NY, USA, 2001; pp. 251–309.

32. Leinfelder, R.R.; Krautter, M.; Nose, M.; Ramalho, M.M.; Werner, W. Siliceous sponge fácies from Upper Jurassic of Portugal. *Neues Jahrb. Geol. Paläontol. Abh.* **1993**, *189*, 199–254.
33. Leinfelder, R.R.; Nose, M.; Schmid, D.U.; Werner, W. The importance of microbial crusts in Upper Jurassic reef information. *Facies* **1993**, *29*, 195–230. [CrossRef]
34. Leinfelder, R.R.; Werner, W.; Nose, M.; Schmid, D.U.; Krautter, M.; Laternser, R.; Takacs, M.; Hartmann, D. Paleoecology, growth parameters and Dynamics of coral, sponge and microbolite reefs from the Late Jurassic. *Göttinger Arb. Geol. Paläont.* **1996**, *2*, 227–248.
35. Fürsich, F.T.; Werner, W.; Schneider, S. Autochthonous to parautochthonous bivalve concentrations within transgressive marginal marine strata of the Upper Jurassic of Portugal. *Palaeobiodiversity Palaeoenvironments* **2009**, *89*, 161–190. [CrossRef]
36. Fernández-López, S. Taphonomic concepts for a theoretical biochronology. *Rev. Esp. Paleontol.* **1991**, *6*, 37–49.
37. Fernández-López, S.R.; Meléndez, G. Fossilization of ammonites and sedimentary events in deep environments of carbonate platform (highest middle to lowest Upper Oxfordian, Iberian Range, Spain). *Riv. Ital. Paleontol. Stratigr.* **2004**, *110*, 219–229.
38. Pena dos Reis, R.; Trincão, P.; Cunha, P.P.; Dinis, J.L. *Relatório Final de Execução do Projecto "Estratigrafia Sequencial e Biostratigrafia do Jurássico Superior da Bacia Lusitânica"*; GPEP: Coimbra, Portugal, 1995.
39. Garcia, G.G.; Campos, M.F.; Henriques, M.H.; Pena dos Reis, R.; Garcia, A.J.V. As coquinas da formação amaral da Bacia Lusitânica (Kimmeridgiano; Portugal): Uma abordagem multidisciplinary. In *Livro de Resumos do Paleo Fall Meeting*, 1st ed.; Fialho, P., Silva, R., Eds.; Universidade de Évora: Évora, Portugal, 2019; Volume 1, p. 9. ISBN 978-972-778-124-9.
40. Choffat, P. Recueil de monographies stratigraphiques sur le Système Crétacique du Portugal. Première étude. Contrées de Cintra, Bella set de Lisbonne. *Mem. Sec. Trab. Geol. Port.* **1885**, *1*, 1–68.
41. Choffat, P. Description de la faune jurassique de Portugal. Mollusques lamellibranches, 2.ª ordre, Asiphonidae, 2ème livraison. *Mem. Dir. Trab. Geol. Port.* **1885**, *2*, 1–36.
42. Choffat, P. Recherches sur les Terrains Secondaires au Sud de Sado. *Com. Trab. Geol. Port.* **1887**, *1*, 222–312.
43. Choffat, P. Contributions à la connaissance géologique des sources minéro-thermales des aires mésozoïques du Portugal. *Impre Nacio.* **1893**, *1*, 1–136.
44. Choffat, P. Description de la faune jurassique du Portugal. Classe des Céphalopodes, 1ère série: Ammonites du Lusitanien de la contrée de Torres Vedras. *Mem. Dir. Trab. Geol. Port.* **1893**, *1*, 1–82.
45. Fürsich, F.T.; Schneider, S.; Werner, W.; Lopez-Mir, B.; Pierce, C.S. Life at the continental–marine interface: Palaeoenvironments and biota of the Alcobaça Formation (Late Jurassic, Central Portugal), with a formal definition of the unit appended. *Palaeobiodiversity Palaeoenvironments* **2021**, 1–65. [CrossRef]

Journal of
Marine Science and Engineering

MDPI

Article

Depositional Sedimentary Facies, Stratigraphic Control, Paleoecological Constraints, and Paleogeographic Reconstruction of Late Permian Chhidru Formation (Western Salt Range, Pakistan)

Syed Kamran Ali [1], Hammad Tariq Janjuhah [2,*], Syed Muzyan Shahzad [3], George Kontakiotis [4,*], Muhammad Hussain Saleem [1], Usman Khan [5], Stergios D. Zarkogiannis [6], Panayota Makri [4] and Assimina Antonarakou [4]

[1] Institute of Geology, University of Azad Jammu and Kashmir, Muzaffarabad 13100, Azad Kashmir, Pakistan; Kamran.ali@ajku.edu.pk (S.K.A.); hussainqadri6692@gmail.com (M.H.S.)
[2] Department of Geology, Shaheed Benazir Bhutto University, Dir (U), Sheringal 18050, Khyber Pakhtunkhwa, Pakistan
[3] School of Geoscience and Info-Physics, Central South University, Changsha 410083, China; syed.muzayyan799@gmail.com
[4] Department of Historical Geology-Paleontology, Faculty of Geology and Geoenvironment, School of Earth Sciences, National and Kapodistrian University of Athens, Panepistimiopolis, Zografou, 15784 Athens, Greece; pmakri@geol.uoa.gr (P.M.); aantonar@geol.uoa.gr (A.A.)
[5] Department of Earth Sciences, Université Lille 1, Cité Scientifique, CEDEX, 59655 Villeneuve d'Ascq, France; usman.khan.etu@univ-lille.fr
[6] Department of Earth Sciences, University of Oxford, Oxford OX1 3AN, UK; stergios.zarkogiannis@earth.ox.ac.uk
* Correspondence: hammad@sbbu.edu.pk (H.T.J.); gkontak@geol.uoa.gr (G.K.)

Citation: Ali, S.K.; Janjuhah, H.T.; Shahzad, S.M.; Kontakiotis, G.; Saleem, M.H.; Khan, U.; Zarkogiannis, S.D.; Makri, P.; Antonarakou, A. Depositional Sedimentary Facies, Stratigraphic Control, Paleoecological Constraints, and Paleogeographic Reconstruction of Late Permian Chhidru Formation (Western Salt Range, Pakistan). *J. Mar. Sci. Eng.* **2021**, *9*, 1372. https://doi.org/10.3390/jmse9121372

Academic Editor: János Kovács

Received: 20 November 2021
Accepted: 30 November 2021
Published: 3 December 2021

Abstract: The Upper Indus Basin, in Pakistan's western Salt Range, is home to the Zaluch Gorge. The sedimentary rocks found in this Gorge, belonging to the Chhidru Formation, were studied in terms of sedimentology and stratigraphy, and provide new insights into the basin paleogeographic evolution from the Precambrian to the Jurassic period. Facies analysis in the Chhidru Formation deposits allowed the recognition of three lithofacies (the limestone facies—CF1, the limestone with clay interbeds facies—CF2, and the sandy limestone facies—CF3) with five microfacies types (mudstone biomicrite—MF-1, wackestone-biomicrite—MF-2, wackestone-biosparite—MF-3, pack-stone-biomicrite—MF-4, and packstone-biosparite—MF-5), as well as their depositional characteristics. The identified carbonate and siliciclastic formations display various facies in a shallow marine environment, with different lithologies, sedimentary features, and energy conditions. It is thought that the depositional characteristics of these microfacies are closer to those of the middle to outer shelf. Because of the progressively coarsening outcrop sequence, this formation seems to be at the very top of the high stand system tract (HST). A modified dynamic depositional model of the Chhidru Formation is further built using outcrop data, facies information, and stratigraphy. According to this concept, the formation was deposited in the middle to inner shelf area of the shallow marine environment, during the Late-Permian period. The Permo-Triassic Boundary (PTB), which is the end of the type-1 series, is marked by this formation's top.

Keywords: Permian-Triassic boundary; sequence stratigraphy; high stand systems tract (HST); shallow marine deposits; microfacies types; stratigraphic correlations; subtidal-intertidal depositional environments; mixed siliciclastic-carbonate successions; sedimentary basin dynamics; paleoecology

1. Introduction

The Salt Range region of Pakistan has been the focus of geological study because of its complicated geology, which ranges in age from Precambrian to Recent [1], although the precise sedimentological sequence and its paleoenvironmental significance remain largely unclear. The Salt Range of Pakistan, situated in the southern Potwar Basin, is the southernmost portion of the Sub-Himalayan Mountains, which stretch east-west for more than 180 km between the Jhelum and Indus rivers (Figure 1). An unconformably thick sedimentary layer covers low-grade metamorphic and igneous rocks in the Salt Range from the Precambrian to modern eras (Figure 1). A long phase of nondeposition has persisted over the majority of the Indian subcontinent since the Late Proterozoic, when Pakistan and India's Permian sedimentation began [2].

The upper part of the Chhidru Group was given a Permian age [3–5]. The Permo-Triassic layers in the western Chhidru Nala were also measured by Wignall and Hallam [6]. The evidence supporting a significant split between the Permian and Triassic sequences, on the other hand, eluded him. As a result of their research, Kummel [7] identified the Permian-Triassic boundary as a "paraconformity of undetermined magnitude" and characterized the distribution of lithology and fauna of layers above and below it. For further information on the stratigraphic succession of this formation, we refer to the works of Theobald [8], Ghazi et al. [9], Schneebeli-Hermann et al. [10], and Waterhouse [11]. They classified it as the Permian period of the Upper Chhidru Group. No clear boundary existed between the Permian and Triassic period following these Salt Range studies.

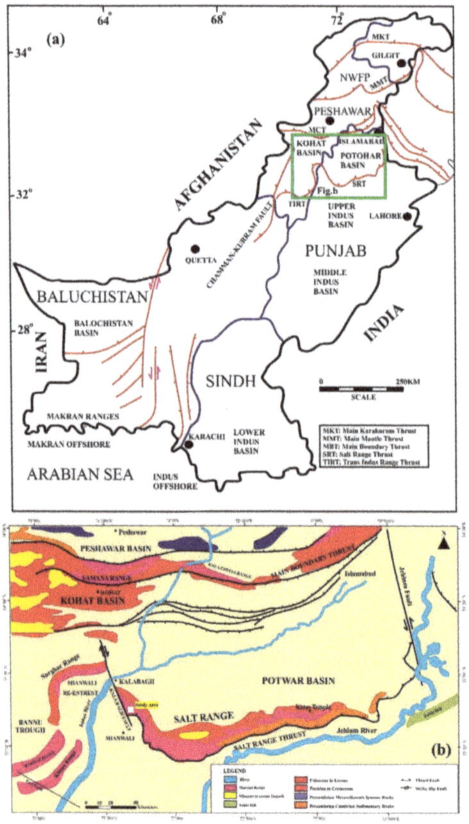

Figure 1. Pakistan's tectonic map. (**a**) showing Pakistan's overall geology and surrounding area; (**b**) the geological map of the Salt Range is shown as a green box in Figure 1a (Modified after Wadood et al. [12]).

Shen et al. [13] projected a proper age for the Chhidru's Formation, although the age is still debated. The most recent palynological and stratigraphic investigations detail the microfossil composition of the Middle Permian Sardhai Formation and the Permian to Middle Triassic assemblages [3,14,15]. Hermann et al. [3] used microfloral records from the Nammal and Chhidru (Salt Range), and Chitta-Landu and Narmia, to identify three palynological associations and five different palynostratigraphic zones from the top Permian to the middle Triassic (Surghar Range). To study palynological assemblages, scientists had to look at sedimentary gaps at the Permian-Triassic border and carbonate-dolomite lithology in the Lower Triassic. Mertmann [16] investigated the evolution of the Salt Range's marine Permian carbonate platform and described the various system tracts. Although general information about the stratigraphic and sedimentological aspects of the various formations in the Salt Range exists, the specific critical approach in terms of sedimentology and stratigraphy for the Chhidru Formation is lacking, which is crucial in understanding the Permo-Triassic (P-T) boundary in the Western Salt Range.

The present study has conducted a critical examination of facies analysis, microfacies types, and other aspects of the depositional environment, as well as the distribution of fauna in the Chhidru Formation; this is necessary to understand and identify the cyclical order in Pakistan's Western Salt Range stratigraphic framework, based on sedimentological and stratigraphic research on the Salt Range. The dynamic depositional model that was proposed will provide additional insights into the paleogeographic evolution of the basin and better correlation with coeval analogous tectono-stratigraphic successions along the south-eastern margin of the Paleo-Tethys Ocean.

2. Geological Setting

Pakistan is located on the Eurasian, Indian and Arabian Plates (Figure 1). A tertiary period convergence of the Eurasian and Indian plates is represented by this structure. Punjab is Pakistan's most populous province, followed by Sindh and Khyber Pakhtunkhwa (KPK) in terms of population [17]. Punjab's terrain is divided between the Plateau (Potwar), a mountain range straddling the Jhelum and Indus rivers, and the Punjab Plains, a lowland area. The hill system is also known as the Salt Range [18]. The stratigraphy of the Salt Range area has been identified in the Kohat-Potwar Plateau, as well as in the wells drilled in Punjab Plains. To create the Potwar sub-basin, geological force, such as the thrust belt, which runs north-south for around 150 km, is responsible for the development of the basin. Tectonically, the Potwar sub-basin is bounded by the Jhelum fault in the east and the Kalabagh fault in west [19,20]. The Potwar sub-basin is separated from the Main Boundary Thrust (MBT) on the north and the MBT on the south by the Salt Range thrust. (Figure 1). The Indian and Eurasian plates collision caused a 55-km shortening of the horizontal crust between Soan Syncline and NPDZ, and a 20-km shortening along the Salt Range [21–24]. From Precambrian to more recent geological strata, the Upper Indus Basin geology and sedimentary environment are diverse [25–27]. Eastward from the Salt Range, the thickness of pre-Cambrian and Cambrian rock formations rises, whereas westward from the Salt Range, it decreases [9,28,29]. A salt, marl, and gypsum-rich lithology first appeared in the Precambrian Salt Range Formation. East of the Salt Range is where it is at its peak. The Cambrian rocks include the Khewra Sandstone, the Kussak Formation, the Jutana Formation, and the Baghanwala Formation [30,31]. The eastern Salt Range has a well-developed Cambrian stratigraphic sequence. Large plate tectonic changes marked the Late Paleozoic–Early Mesozoic era. The Paleo-Tethys Ocean closed along the southern boundary of Eurasia in the Late Triassic, completing the Variscan assemblage into a single supercontinent, Pangea, in the Middle Cambrian period. During Gondwana's last stages, the Cimmerian Orogen, a prominent mountain range, was formed when the eastern boundary of Gondwana was broken by a strip of Gondwanan terranes called the Cimmerian terranes, which encompasses Iran, Afghanistan, the Karakoram, and Qiangtang [32]. According to the Late Permian–Early Triassic paleogeographic reconstruction, the Salt Range was located around 30–40° S along the Tethyan border and indicates an overall warm-temperate paleoclimate [33,34]. Limestones, dolostones, sandstones, and shales make up the Salt Range's Triassic strata. To study sediments below and across the Permian–Triassic Boundary (PTB), the Salt Range's exposures have made it possible.

The Permian deposit began several hundred years after the Cambrian deposit had finished. The Permian succession is divided into two stratigraphic successions, the Nammal Gorge and the Chhidru Nalah near Chhidru town, each of which has a different depositional environment (Figure 1) [35–37]. The Nilawahan Group is mainly a Gondwanan terrestrial succession, whereas the Zaluch Group is a shallow marine Tethyan succession [14,31,38,39]. The Tobra Formation's base has a disconformity, which scientists discovered. Several rock types lie above the Tobra Formation, including the Dandot Formation, Warchha Sandstone, and the Nilawahan Group's Sardhai Formation [40,41].

A well-exposed region of shallow to intertidal marine deposits may be found in the Zaluch Group of the Salt Range, which includes the Amb, Wargal, and Chhidru formations [7,16,42–46]. A discontinuity exists between the Amb and Sardhai formations at their base, while the upper Chhidru-overlying Triassic Mianwali Formation's contact has yet to be determined [7,9,36,47,48]. Permian rock formations may be found in the western Salt Range [49] (Figure 2).

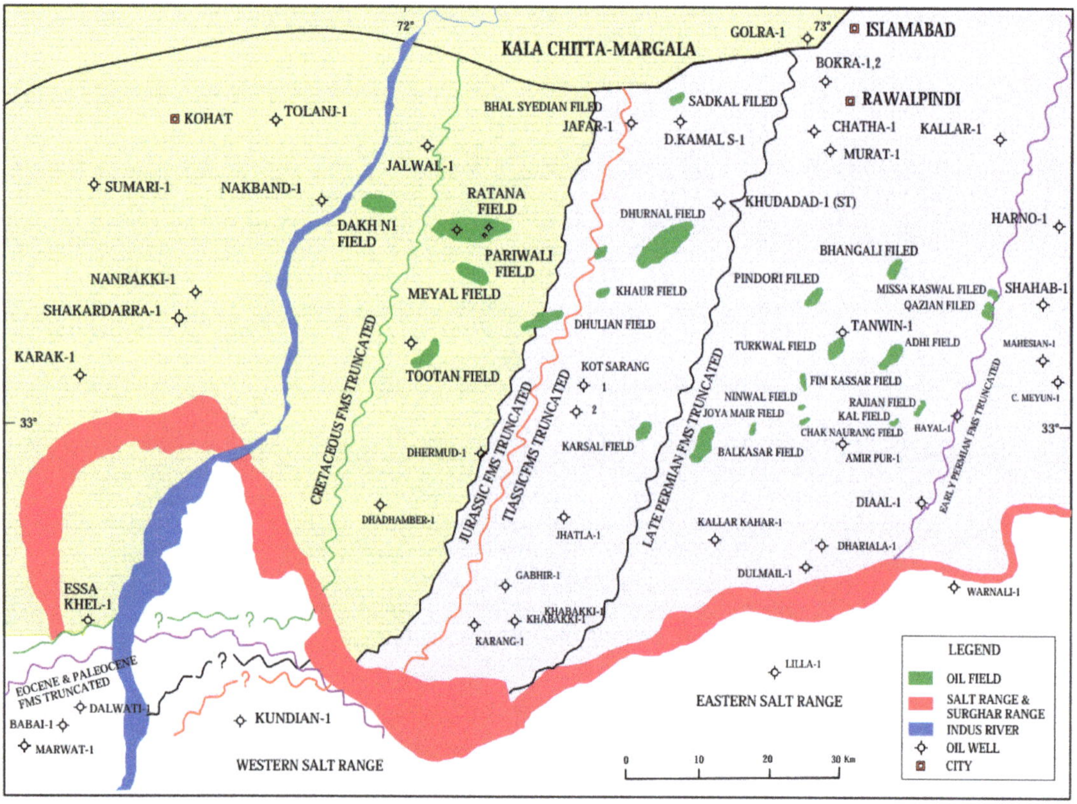

Figure 2. Lithology distribution map of the Salt Range area (Modified after Malkani and Mahmood [49]).

The rocks of the Mesozoic period can be characterized due to paraconformity. At the end of the Permian period, Hermann et al. [50] referred to the body as a para-unconformity showing subaerial exposure. The Permian rocks found in the Salt Range are rich in fauna and are related to Triassic rocks. There is a substantial faunal break at the boundary between the Permian and Triassic periods. Each of the Mesozoic formations may be found in the Upper Indus Basin and are part of the Triassic era [35]. The Mesozoic rock series may be found in abundance in the western Salt Range [51,52]. Samanasuk Limestone and the Shinawri Formation are part of the Jurassic period's Datta and Shinawri formations. The Cretaceous period is represented by the Chichali and Lumshiwal formations in Rajasthan (Figure 2). Unconformities in rock units may be seen across Africa in the Hangu Formation, which includes laterite and bauxite. The Cenozoic geologic sequence includes the Hangu Formation, Lockhart Formation, Patala Formation, Nammal Formation, Sakesar Limestone, and Chorgali Formation, respectively.

Surface and subsurface maps of the Potwar region and the area to the west of it are shown in Figure 2. The bright blue color of the Potwar region's subsoil indicates the lack of Jurassic and Cretaceous layers. The bright green color indicates the existence of Jurassic and Cretaceous formations under the surface of the western Potwar-Kohat region and the southern Hazara highlands. According to Alhubail and Ghahfarokhi [53] Pakistan's Upper Indus Basin Upper Jurassic to Early Cretaceous rocks are rich in geological and economic resources.

Typical Jurassic and Early Cretaceous formations, such as the Datta, Samana Suk, Chichali, and Lumshiwal formations, may be found as far east as the Abbottabad area's Thandiani Formation [54,55]. In the southern Hazara highlands, as well as across the Indus River, these rock units may be found in abundance. Since no Jurassic or Cretaceous ages were found in the Potwar region, a particular condition existed across the Main Boundary Thrust in the neighboring Potwar area (Figure 2).

2.1. *Stratigraphy of Zaluch Gorge-Chhidru Formation*

The study area is located in the Zaluch Gorge, in the westernmost section of the Salt Range. Sedimentary rocks from the Precambrian Salt Range Formation to the Jurassic Datta Formation make up the Zaluch Area (Table 1). The Nilawahan Group's Tobra Formation represents the beginning of the Permian period. Above the Tobra Formation is the Warchha Sandstone. Zaluch Gorge is devoid of the Dandot Formation [4,11]. The Warchha Sandstone gradually gives way to the Sardhai Formation, as it rises to the north [56]. Ascending toward the top from the Nilawahan Group is the Zaluch Group, which contains the Amb Formation, Wargal Limestone, and the Chhidru Formation. Fossil-rich Permian strata in the Salt Range are connected to rocks from the Triassic period; marine layers along the Permo-Triassic boundary are consistent, the fauna did not change much after contact with the marine layers along the Permo-Triassic boundary [43,57]. The Mianwali Formation is the first in the Triassic series, followed by the Tredian and Kingriali formations. The Kingriali Formation, which is younger than the Jurassic Datta Formation, rests erratically just above it.

Chhidru Formation

Originally, the Chhidru Formation was referred to as the Chhidru beds after early investigations by Waagen [58]. According to Dunbar [59], in the early nineteenth century, Chhidru Formation was characterized as a mix of siliciclastic and carbonate phases. There has been much debate about the age of the Chhidru Formation, and its relationship to Triassic rocks, since Teichert [60] first discovered it. Mei and Henderson [61] suggested the Wuchiapingian/Changhsingian boundary as the dividing line between the Wargal and Chhidru formations. Since the Chhidru Formation was deposited in a shallow subtidal to the intertidal environment, its thickness varies across the Salt Range, according to Mertmann [16]. The stratigraphic thickness, as determined by earlier researchers, varies widely. According to Mertmann [16], Jan and Stephenson [62], Sajjad Ahmed et al. [63] and Hussain et al. [64], Zaluch Nala, Zaluch Gorge and Chhidru Nala all recorded 58.4 m, 63.63 m, and 81.81 m of the Chhidru Formation, respectively. According to Zahid et al. [25], Hermann et al. [50] and Saboor et al. [65], the Chhidru Formation is a siliciclastic mixed carbonate series that is readily accessible in the research region under consideration.

Shallow sandstone and sandy limestone comprise most of the Chhidru Formation. The base of the Chhidru Formation is exposed, displaying shale units. Above the shale unit, a calcareous sandstone unit is evident. An oscillation ripple mark may be seen on the topmost layer of this formation, which might be a white sandstone bed [66,67]. This limestone-clay-sand interbeds formation may be found in Zaluch Gorge (Figure 3). The Chhidru Formation's lower contact with Wargal Limestone is conformable, while the upper contact is not visible, because of a paraconformity at the Permian-Triassic (P-T) boundary with upper contact of the Mianwali Formation (Figure 3). The Late Permian period is responsible for the development of Chhidru. [3,11,31].

Table 1. Stratigraphic successions of the Zaluch Gorge.

Formation	Lithology	Age
Datta Formation	Sandstone, Siltstone, Glass sand, Fire clay	Jurassic
Kingriali Formation	Dolomite, Limestone, Marl, Shale	Triassic
Tredian Formation	Sandstone, Shale	Middle Triassic
Mianwali Formation	Sandstone, Shale, Dolomite (Heterogeneous lithology)	Lower Triassic
	Unconformity (P-T Boundary)	
Chhidru Formation	Limestone, Sandstone	Late Permian
Wargal Limestone	Limestone, Dolomite	Late Permian
Amb Formation	Shale, Limestone	Middle to Late Permian
	Unconformity	
Sardhai Formation	Shale	Early Permian
Warchha Sandstone	Sandstone	Early Permian
Tobra Formation	Conglomerates having tillitic, fresh water and complex facies	Early Permian
	Unconformity	
Salt Range Formation	Salt, Marl, Gypsum	Precambrian

There is a rich fossil fauna in Pakistan's Salt Range and Surghar Range, which date back to the Permian-Triassic period, when carbonate-siliciclastic successions were combined with sandstones and clays. These sedimentary archives allow researchers to combine palynology with ammonoids, conodont, and chemostratigraphic age controls to better understand the origins of life on Earth [68,69]. Cyclosorus, a Permian ammonoid, was discovered in the Salt Range by Furnish and Glenister [70]. Wignall and Hallam [6] investigated the Salt Range and Trans Indus Range's topmost Permian and lower Triassic conodonts. The acritarchs and tasmanitids of the Chhidru Formation were explained by Ahmad et al. [67]. According to Qureshi et al. [57], the Salt Range and the Surghar Range are rich in palynology, which he investigated. Between 1981 and 1985, Pakistani and Japanese scientists studied the Permo-Triassic succession in the Salt Range. As opposed to the Chhidru Formation in Khan and Afzal's [71] research, the Zaluch Nalah and Nammal portions had a higher density of brachiopods. The Zaluch Gorge section of Pakistan's P-T boundary has received the most attention. Mertmann [16] studied the development of the Permian marine carbonate platform's salt range after discovering several system tracts in the Chhidru Formation.

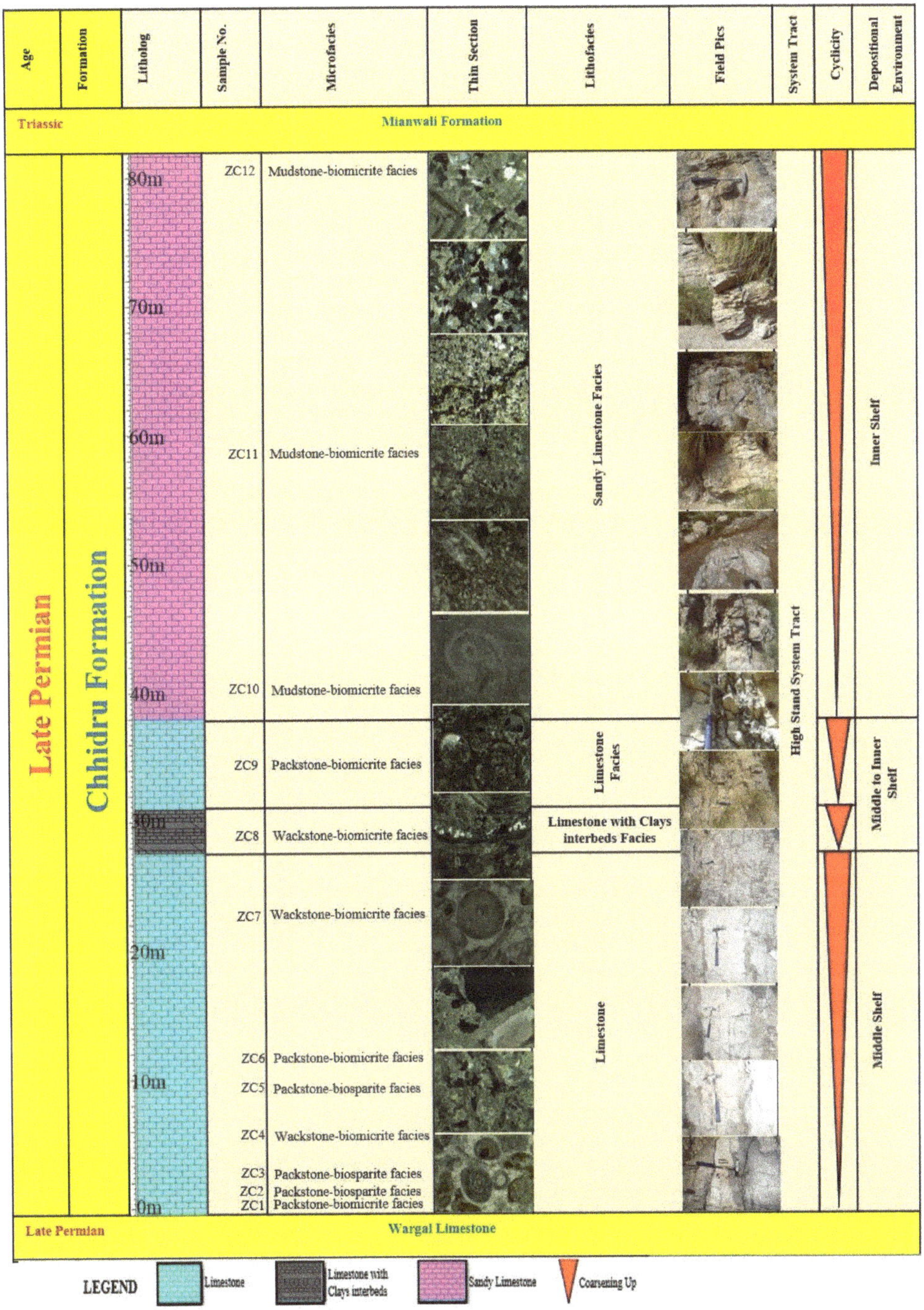

Figure 3. The Chhidru Formation's sedimentological log shows the outcrop description, facies distribution, sample site, cyclicity, and deposition interpretation.

3. Material and Methods

Fieldwork was carried out in the Salt Range area's Zaluch Gorge Chhidru Formation, which has a mix of siliciclastic and carbonates throughout its geological history. The existing outcrop was recorded, measured, and sampled using standard procedures [72]. An 82.15-m-thick sequence was researched and documented in the western Salt Range's Zaluch Gorge on a standard log sheet. Detailed analyses were performed to identify distinct lithofacies intervals based on sedimentological properties such as lithology, depositional texture as defined by Dunham [73], and carbonate components.

The section was then classified into three facies (Figure 3). The lithofacies in the region were identified and described using an outcrop study, important stratigraphic features, and microscopic hemicycles observed in the field. Thirty-four rock samples, each 2.8 cm long and 3.2 cm in diameter, were gathered from the same intervals investigated. These samples were selected to illustrate carbonate microfacies, and their textures ranged from mudstone to packstone. At the AJ&K University Laboratory, 12 thin sections were produced for chosen materials and then analyzed using reflected and transmitted light microscopy. The thin sections were described using an Olympus BX 51 camera with a DP-27 connected. Half of the thin sections were treated with an acid solution containing red alizarin S and potassium ferricyanide stains on a section of the slide. The microfacies definition and textural character analysis of the carbonate rocks, which included both biogenic and inorganic dominant components, were performed using Dunham [73] and Embry and Klovan [74] carbonate classification schemes, which were modified to include the Standard Microfacies Types (SMF) in the facies zones (FZ) of the rimmed carbonate platform model. Depositional environments were reconstructed based on the derived sedimentological features and through comparison with additional standard facies reconstructions [75–77].

To evaluate the depositional environment, both macroscopically and through lithofacies analyses, several petrographic observations and parameters were recorded and measured. Finally, a dynamic depositional model was developed, based on the dynamics of the paleo-deposition environment, emphasizing the paleogeographic relevance of the studied deposits. Figure 3 depicts a synthesis of the acquired findings.

4. Results

4.1. Lithofacies of Chhidru Formation

The investigated part is located entirely inside District Mianwali's Zaluch Gorge (Late Permian). Wargal Limestone has a conformable lower contact, whereas the Triassic Mianwali Formation, which lies above it, is paraconformable (Figure 4a–c). In the field, the following three lithofacies were found:

- Limestone facie (CF1)
- Limestone with clays interbeds facies (CF2)
- Sandy limestone facie (CF3)

4.1.1. Limestone Facies CF1

This facie (CF1) comprises ceramist grey to light brownish, thin-grained limestone (Figure 4a,d). The thickness of this facie is approximately 35.49 m in the studied region. This facies comprises more than 90% of limestone. The main grain size is medium grained. These facies are enriched in fossiliferous strata. Eight rock samples (ZC1, ZC2, ZC3, ZC4, ZC5, ZC6, ZC7, and ZC9) were obtained for the petrographic investigation (Figure 3). The petrographic study shows substantial grainstone depositional texture with moderate to poor preservation of allochems.

4.1.2. Limestone Interbedded with Clays Facie CF2

According to quantitative analysis, this facies (CF2) is composed of 80% limestone and 20% mudstone and has a greyish to light brownish color (Figure 4a). It also contains fossiliferous limestone and finely interbedded lamina. The grain is between fine and medium in texture. The petrographic analysis of CF2 (sample ZC8) reveals a mudstone depositional texture with limited preservation of the allomicritic matrix (Figure 4e). The diagenetic fabric of the mudstone fabric shows iron leaching at particular points of petrographic investigation. Fractures filled with stylolite and calcite have also been observed.

4.1.3. Sandy Limestone Facies CF3

The sandy limestone facies (CF3) are distinguished by their light grey to creamy white color (Figures 3 and 4a). In the studied region, the total thickness of this facie is about 44.36 m. The sandy limestone layers found are less fossiliferous than CF1 and CF2. The most often used grain size is medium grain. According to petrographic studies, this facie has a packstone depositional texture. Three rock samples (ZC10, ZC11, and ZC12) were collected for petrographic examination. The petrographic investigation also showed a random distribution of quartz, feldspar, and muscovite (Figure 5a). Calcite, stylolites, bioclasts (micritized), and recrystallized fossil pieces define the diagenetic fabric of this facies.

Figure 4. Images of outcrops depicting significant characteristics seen in the Zaluch Gorge, Western Salt Range. (**a**) lithology distribution and three distinct facies identified; (**b**) highlighting the lower contact of the Chhidru Formation with the Wargal Limestone; (**c**) highlighting the upper contact of the Chhidru Formation with the Mianwali Formation; (**d**) Facies-1 (CF1 limestone facies); (**e**) photomicrograph of Facies-2 (CF2 limestone with clay interbeds).

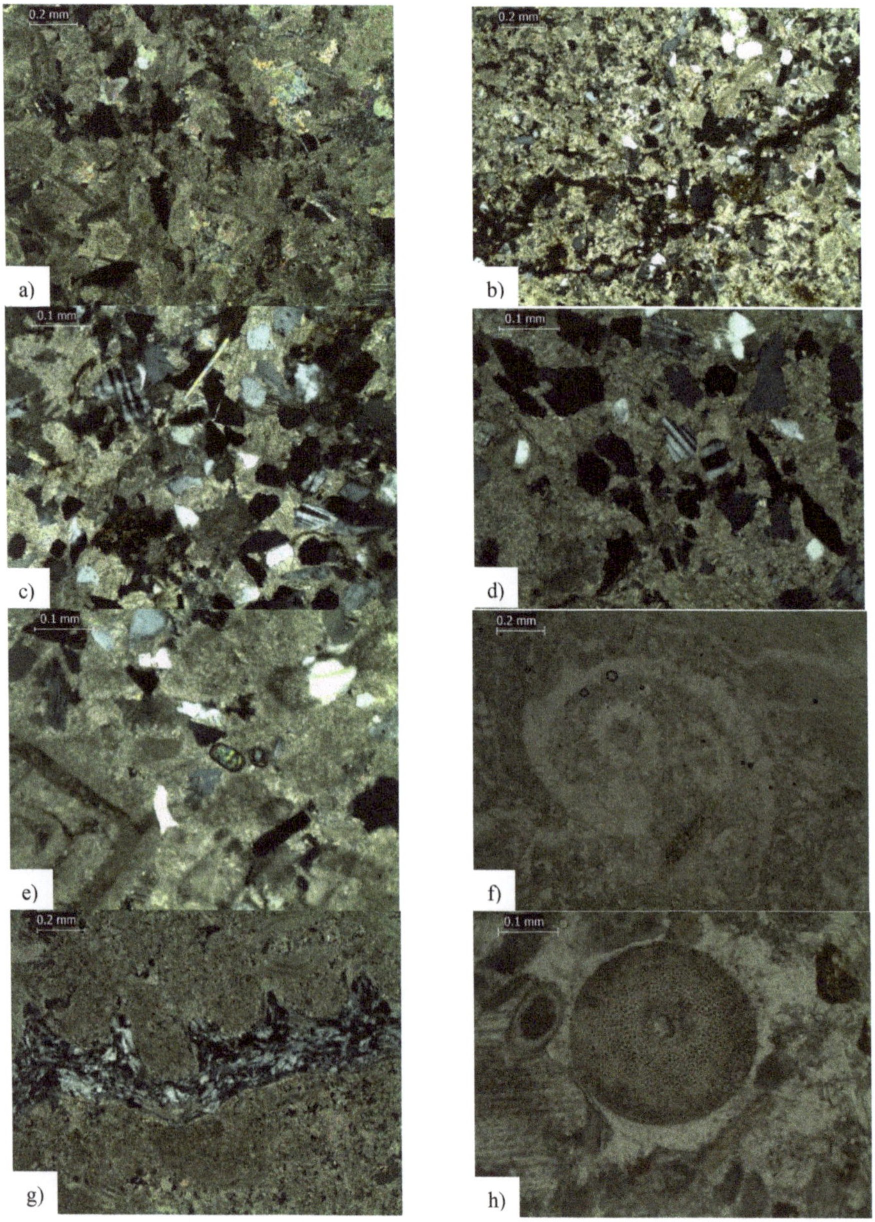

Figure 5. Photomicrographs representing different characteristics (**a**) Facies-3 (CF3 Sandy Limestone); (**b**) Sample-Zc10—Mudstone-biomicrite unit from the Chhidru Formation representing brachiopods and stylolites veins highlighted by hematite (cross Nicol 4×); (**c**) Sample-ZC11—Mudstone-biomicrite unit representing albite, microcline, muscovite, and nodosaria (cross Nicol 4×); (**d**) Sample-ZC-12—Mudstone-biomicrite unit showing albite; (**e**) Sample-ZC-12—Mudstone-biomicrite unit representing zircon which is colorless in plain polarized light (PPL 10×); (**f**) Sample-ZC-7—Wackstone-biomicrite unit representing ammonites (PPL 4×); (**g**) Sample-ZC-8—Wackstone-biomicrite unit showing bioclasts replaced by silica (Cross Nicol 4×); (**h**) Sample-ZC-4—Wackstone-biosparite unit representing radiolaria (PPL 10×).

4.2. Microfacies of Chhidru Formation

The microfacies found in the Chhidru Formation were identified and classified based on the classification scheme of Wilson (2008). Dominant microfacies are those made of multiple types of bioclasts, quartz, and feldspar components, showing textures from mudstone to packstone. A total of five microfacies have been identified; they are listed below.

Mudstone-biomicrite facies
Wackstone-biomicrite facies
Wackstone-biosparite facies
Packstone-biomicrite facies
Packstone-biosparite facies

4.2.1. Mudstone-Biomicrite Facies

The mudstone-biomicrite facies includes samples ZC10, ZC11, and ZC12 (Table 2). The petrographic examination reveals that these samples are composed of micrite, quartz, feldspar, and fairly preserved brachiopod and bryozoan bioclasts, among other minerals (Figure 5b,c).

A little quantity of hematite, around 1–2%, may be found in stylolite. Mineral zircon is present in very small amounts in ZC12 (Figure 5d). The most frequent kind of cement discovered in these microfacies is sparry calcite cement (Figure 5c,d). Dunham classifies samples ZC10, ZC11, and ZC12 as mudstone due to the presence of muddier carbonate (Figure 5b–e). The grains are subangular to subrounded and weakly to moderately sorted, and this microfacies represent marine fossil fragments containing these grains.

This microfacies is characterized by marine fossils fragments, subangular to sub rounded grains, moderate to poor sorting, and abundant clastic and ferruginous matrix. This is a low-energy deposition environment underneath the fair-weather wave base because of the presence of matrix. A close proximity to the inner shelf is suggested by the occurrence of tiny marine fossils and the predominant sandstone texture (such as quartz, feldspar, etc.) [78,79]. The same kind of microfacies was also regarded as inner shelf deposits by Zlobin et al. [80], Özgen-Erdem and Koç-Tasgin [81], and Moissette et al. [82].

4.2.2. Wackstone-Biomicrite Facies (MF-2)

The Wackstone-biomicrite facies includes samples ZC7 and ZC8 (Table 2). These samples include 30% bioclasts, 50% micrite, and 20% spar, as determined by petrographic examination (Figure 5f,g). Bioclasts in fine-grained cement is classified by Folk as such (micrite). ZC7 and ZC8 are classified by Dunham as Wackstone deposits with poor fusulinids and gastropod preservation (Figure 5f,g).

According to Sims and Belanger [83], Arefifard and Payne [84] and Huang et al. [85], Fusulinids (foraminifera) is most often seen in open sea. Bryozoans, on the other hand, can only exist in naturally salted water [86–89]. Marine subtidal habitat is indicated by the presence of a micrite matrix and a wide range of fauna. Because agitated water prevents lime mud from settling, a low energy depositional condition results in a large inflow or presence of mud. Low energy deposition, the presence of shallow marine creatures, poor preservation of fusulinids, and a large inflow of lime are believed to have contributed to the deposition of this microfacies below the middle shelf setting's fair-weather wave base condition.

4.2.3. Wackstone-Biosparite Facies (MF-3)

Sample ZC4 is classified as a Wackstone-biomicrite microfacies (Table 2). The petrographic analysis of ZC4 reveals that this rock sample includes 49% bioclasts, 16% micrite, and 35% spar. It depicts fragments of brachiopod shells (Figure 5h). The cement spar contains bioclasts that are coarsely crystalline. More than 10% of ZC4's carbonate grains are supported by mud, according to Dunham [73] categorization (Figure 5h).

The limestone of a shallow marine origin is especially rich in brachiopod fragments. Most of the organisms were benthonic and sessile [90–92]. However, even though brachiopods are all marine organisms, they have a wide salinity range, ranging from brackish to slightly hypersaline [93–95]. The presence of sparry cement indicates a high level of energy. These microfacies point to a deposition in the proximal middle shelf area below the storm wave base.

Table 2. Modal percentage composition of lithofacies of Chhidru Formation, Zaluch Gorge (Macro-scale Estimation).

Sample No.	Quartz %	Feldspar %	Calcite %	Bioclasts %	Others %	Classification	
						Dunham [73]	Folk [96]
ZC_{12}	29	5	63	2	Muscovite 1%, Zircon (Trace)	Mudstone	Biomicrite
ZC_{11}	63	4	28	1	Hematite 1%, Muscovite 3%	Mudstone	Biomicrite
ZC_{10}	41	4	51	2	Hematite 2%, Muscovite (Trace)	Mudstone	Biomicrite
	Bioclasts		Micrite	Spar			
ZC_9	31		43	26	——	Packstone	Biomicrite
ZC_8	19		77	10	——	Wackstone	Biomicrite
ZC_7	19		61	20	——	Wackstone	Biomicrite
ZC_6	73		18	9	——	Packstone	Biomicrite
ZC_5	14		36	50	——	Packstone	Biosparite
ZC_4	49		16	35	Hematite (Trace)	Wackstone	Biosparite
ZC_3	55		9	36	——	Packstone	Biosparite
ZC_2	43		14	37	Hematite 1%	Packstone	Biosparite
ZC_1	51		19	11	Hematite 5%	Packstone	Biomicrite

4.2.4. Packstone-Biomicrite Facies (MF-4)

The Packstone-biomicrite facies is shown by samples ZC1, ZC6, and ZC9 (Figure 6a–c; Table 2). Petrographic analysis of the C1, ZC6, and ZC9 samples shows that bioclasts make up more than half of the material, with micrite and spar coming in second and third in order. ZC1 has calcite veins as well (Figure 6a–c). Brachiopods, gastropods, and bryozoans are all found here. Since bioclasts are found in fine-grained cement (micrite), this is referred to as bio-micrite according to Folk's categorization. Because of mud-supported grains, ZC1, ZC6, and ZC9 are classified as Packstone by Dunham [73] (Figure 6a–c).

As stated by Tucker [97] and Cumming et al. [98], marine bryozoans are important carbonate sediment providers. They have helped create reefs and other limestones in the past. Even though brachiopods dwell in fresh water far from the offshore, Scholle and Ulmer-Scholle [94] and Karlsson et al. [99] describe them as marine invertebrates. A large inflow of continental sediments during deposition is thought to be the origin of the formation of this microfacies on a distal middle shelf.

4.2.5. Packstone-Biosparite Facies (MF-5)

Samples no. ZC2, ZC3, and ZC5 are categorized as Packstone-biosparite facies (Figure 6d–f; Table 2). The petrographic analysis of ZC2, ZC3, and ZC5 reveals that bioclasts are the most abundant, followed by micrite and spar. Bryozoans and brachiopods are the most commonly observed marine organisms. This microfacies indicates a proximal middle shelf deposition because of the poor preservation of marine fragments and the presence of a continental inflow during the deposition.

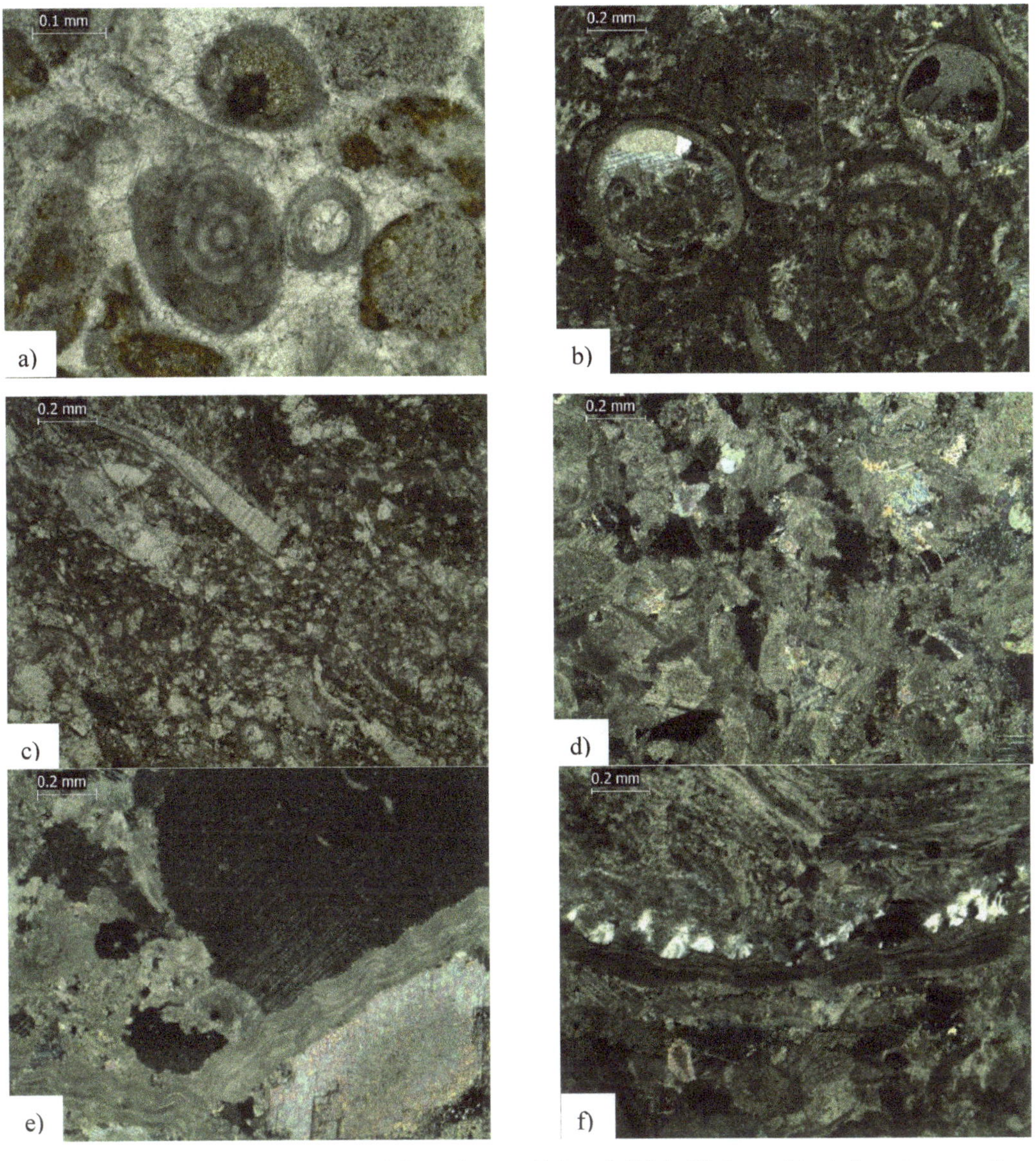

Figure 6. Photomicrographs representing different features (**a**) Sample-ZC-1—Wackstone-biomicrite unit representing ammonite (PPL 10×); (**b**) Sample ZC-6—Packstone-biomicrite unit representing ostracod and gastropod (Cross Nicol 4×); (**c**) Sample-ZC-9—Packstone-biomicrite unit representing brachiopod (PPL 4×); (**d**) Sample-ZC-2—Packstone-biosparite unit representing micrite and spar (PPL 4×); (**e**) Sample-ZC-3—Packstone-biosparite unit indicating brachiopod packed between spar (Cross Nicol 4×); (**f**) Sample-ZC-5—Packstone-biosparite representing bioclasts replaced by silica.

5. Discussion

5.1. Dynamic Depositional Model

For the dynamic depositional modeling, the sequence stratigraphy is a critical input tool for characterizing the sediments that fill the region's time and space. Rock relationships in a chronostratigraphic framework of repeated layers are studied using sequence stratigraphy. Unconformities or their corresponding conformities define the boundaries of these repeating layers [72,100,101]. Different system tracts (transgressive and regressive) represent a depositional sequence, which leads to the interpretation of understanding the deposition between changes in the sea level, according to Janjuhah et al. [102], Lin et al. [103], and Janjuhah and Alansari [104]. This is the sole method, according to Embry [100], to subdivide rock recordings into system tracts, which may be the only option in this specific instance.

The transgressive–regressive (T–R) sequence was established by Embry and Klovan [74] and is still in use. The unconformable boundary on the basin border is the subaerial unconformity, and the corresponding conformity farther out to sea is the maximum regressive surface (MRS). Through the use of this approach, we may avoid the problems associated with both the depositional and genetic stratigraphic sequences. The T–R sequence is divided into transgressive and regressive system tracts using maximum flooding surfaces. Regressive systems tracts, a basic method of separating the rock record into systems tracts, may be the only alternative in a given scenario when other methods are ineffective [100]. The T–R sequence model of Embry [105] was used in this investigation (Figure 7).

Figure 7. Field outcrop picture of a Zaluch Gorge outcrop displaying the following various cycles: (**a**) the stratigraphic cycle of the Sadri and Amb formations; (**b**) the stratigraphic cycles of Wargal Limestone; and (**c**) the stratigraphic cycle of the Chhidru Formation.

For this reason, the sequence boundary (SB) is located in Zaluch Gorge, near the base of the Sardhai Formation, where Warchha Sandstone and Sardhai Shale meet [16,47]. Sardhai formation has a fine ascending succession, as shown by the grain size. As a result, the water level rises, causing the Sardhai formation clays to be deposited [38]. The transgressive surface may be found at the very top of the Sardhai formation (TS) (Figure 7a). Underneath the Sardhai formation, the lower system tract (LST) is present. There is also a marine origin to the Amb Formation that covers the Sardhai formation (Figure 7a). It is clear from the lithology and grain size that sea level increased and ultimately reached at its highest point. Once the regression has begun, the sea level has dropped. The maximum flooding surface (MSF) is present at this point. It is located between the transgression surface (TS) and the MSF of the Amb Formation (Figure 7a). Following deposition (Amb Formation), regression takes place, and the facie begins to recede toward the basin's edge. Rocks, such as the Wargal Limestone and Chhidru formations, are deposited first (Figure 7b). When looking at the Chhidru Formation, the regressive system tract is composed of marine and non-marine fabric ranging from limestone to sandy limestones (Figures 3 and 7c) [66,67]. This group of facies was thought to have been deposited in an environment between the inner and the middle shelf.

The unconformity at the summit of the Chhidru Formation indicates a type-1 sequence boundary. The high stand system tract (HST) contains several geological formations. Figure 8 depicts the depositional hierarchy, stratigraphic cycles, and system tracts in this setting. Previous research has placed the age of the Chhidru Formation at various times. According to the biostratigraphy of brachiopods, Khan and Afzal [71] dated the Chhidru Formation as Guadeloupian. His objective was to use the brachiopod biostratigraphy as a means of correlating the fossil records of different species. His research also found that the Salt Range lacked Upper Permian strata. Using data from Furnish and Glenister [70], Nakazawa and Dickins [106], Balic and Malvic [107], Nugroho and Putra [108], and Moro et al. [109] utilized biostratigraphic dating of foraminifera fusulinids and ammonites to interpret the Chhidru Formation as having been deposited between 254 and 252.5 Ma. They proposed using the Upper Permian age as a baseline for cyclicity in their correlation studies. The relative sea-level curve was compared using the chronology of the eustatic sea-level curve developed by Chen et al. [110], Haq and Schutter [111], Ross and Ross [112]. The relative sea-level curve for Chhidru's Formation was derived using observations of various facies and microfacies, as well as their hypothesized depositional settings. It has been suggested that the Chhidru Formation was deposited 1.5 Ma ago, during two distinct periods. Six fourth-order cycles with a 0.1–0.4 Ma time duration are found inside these third-order cycles [113]. The Chhidru Formation's depositional environment was determined by combining data from lithofacies, microfacies, sequence stratigraphy, and preserved fauna. There is evidence from lithofacies (such as those found in the Chhidru Formation, which includes clastic as well as carbonate fragments) as well as microfacies (such as those found in the middle and inner shelf) that the Chhidru Formation was deposited in the middle to the inner shelf during periods of tectonic stability, with the distal source of the fine clastic sediments still providing the sediments for the shelf-diverse agents, representing two (3rd Order) cycles and six parasequence cycles (fourth order) of the T–R cycle. Samples ZC1, ZC3, ZC6, ZC7, and ZC9 are thought to have lived in clear water in a shallow marine environment, and fossils such as brachiopods and ammonites support this theory [114,115]. It is safe to assume that samples ZC2, ZC10, and ZC11, which all include varied fauna, were collected in a subtidal environment. However, while being mostly marine, brachiopods have a wide salinity range, being found in both brackish and somewhat hypersaline habitats. There is a middle-inner shelf depositional environment indicated by the fossils' poor preservation, which is due to a significant continental inflow during the deposit of the Chhidru Formation. Glacio-eustacy and tectonic subsidence influence the deposition in third-order cycles [116]. Global third order cycle deposits may have been formed by an increase in glacial ice volume in the late Paleozoic [117]. After sea inundation, local subsidence may have caused the Chhidru Formation's parasequences to appear cyclically.

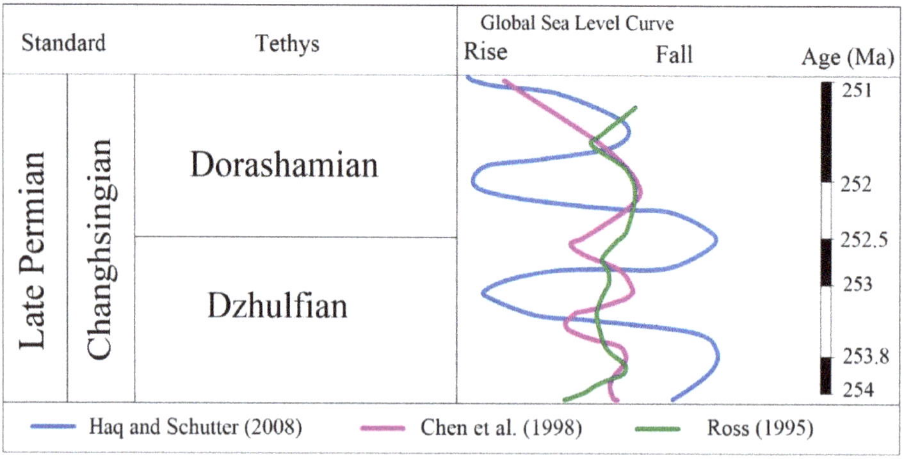

Figure 8. Global sea-level curves comparison of various researchers Chen et al. [110], Haq and Schutter [111], and Ross and Ross [112].

This system tract emerges during the part of the global sea-level rise, when rising rates outpace sedimentation rates. Comparison of the Chhidru Formation's sea level to global sea level in the Upper Permian time, these two medium-sized (25–75 m) decreases in the Upper Permian are associated with the sea level decline at the base and top of the Chhidru Formation, according to Haq and Schutter [111] (Figures 3 and 8). This means that the other two small cycles of falls are not present worldwide but are due to local tectonic activity (Figure 3). The Chen et al. [110] curve shows a fall of 253.8 Ma on the world chart, and this research is highly correlated, while a 252.5 Ma fall cannot be correlated as the Chen et al. [110] curve has a maximum flooding surface at that time. When compared to the sea-level curve of Ross and Ross [112], it seems that the worldwide impacts on the Neo-Tethys depositional facies on the Chhidru Formation, as well as local tectonic activity, played a role (Figure 8) [118]. We infer that the depositional environment of the Chhidru Formation is influenced by both global and local tectonics.

5.2. Permo-Triassic Boundary

The mixed siliciclastic and carbonate succession of the Permo-Triassic era may be found in Pakistan's salt. The sandstone bed known as the white sandstone bed is the uppermost lithological unit of the Chhidru Formation. The sandstone is fine to medium grained, thinly bedded with shale interbeds, and the upper contact with the Mianwali Formation's Kathwai member is dolomite. The P-T Boundary is marked by Ceratite beds from the Mianwali Formation (Table 1). The grain size rises from bottom to top (Chhidru Formation–Mianwali Formation), indicating a coarsening of material in increasing order (Figure 9). The limestone in the Chhidru Formation's lower section has been replaced by sandy limestone, which was deposited far from the coast on the middle to inner shelf during the Late Permian period. Grain size rises from bottom to top, indicating an increasingly coarsened layer. Later, the lithology changed dramatically, and the dolomitic unit of the Mianwali Formation started to rise sharply. During the Siberian Traps eruptions, seawater sulphate levels in the ocean dropped dramatically, which may have caused the early Triassic dolomite to accumulate [119]. Invertebrates, particularly brachiopods and ammonoids, were able to recover during the Spathian period due to continued cooling, which decreased marine anoxia and raised sulphate levels. During the P-T mass extinction event, there were fewer floral eaters and more piscivores in the environment [120]. It is argued that a sharp decline in the quantity of sediment-binding vegetation in the Early Triassic of South Africa, Australia, India, and Spain was responsible for the basinal shift from low-energy meandering rivers to high-energy braided rivers [121]. A variety of Permian flora was reintroduced in the Spathian Stage, resulting in the presence of algal bindstones in the South China sea. This demonstrates that the eastern Tethys Ocean's dominantly mixed siliciclastic-carbonate margins were dominated by sediment-binding organisms that played a key role in the formation of carbonates. Reduviasporonites, including the species *Reduviasporonites chalastus* and *Reduviasporonites catenulatus*, are common in Permian-Triassic palynological assemblages [122–124]. Reduviasporonites have been found in 'relatively high frequency' at the PTB, as reported by Yin et al. [125]. Reduviasporonites have been found in Pakistan's Lower Triassic by Schneebeli-Hermann and Bucher [4].

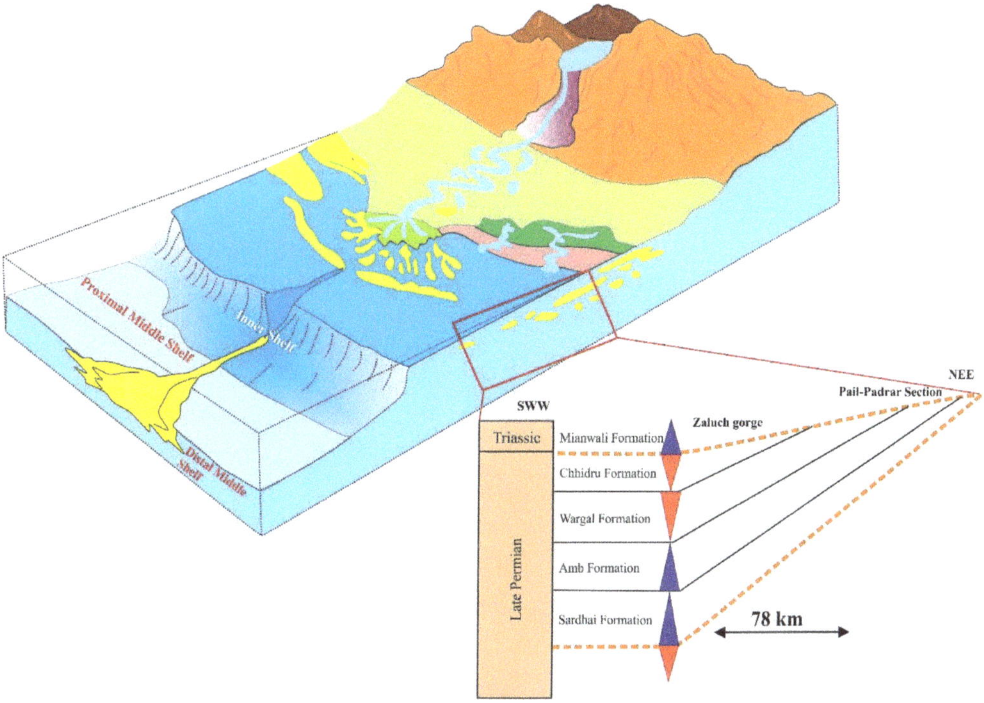

Figure 9. Zaluch Gorge's depositional model in addition to sequence stratigraphic cycles, surface, and system track data.

On the basis of the abrupt facies changes that occur between the Chhidru and Mianwali formations, as well as the record of the biostratigraphy, it seems that a P-T border exists between these two formations.

6. Conclusions

The Nilawahan and Zaluch formations date from the Permian, and they were formed in the Salt Range. During that period, the tendency of sediment deposition was east to west. As a result, the Permian succession's thickness rises from east to west over the Salt Range. These two formations exhibit progradation, while the Sardhai Formation and the Amb Formation in the Zaluch Gorge show retrogradation and may be found inside the low stand system tract or the transgressive system tract. The maximum flooding surface is located near the summit of the Amb Formation. Once the water level starts dropping, the regressive process may begin. There are two major deposits in the high stand system tract during this time period, which are the Wargal Limestone and Chhidru formations (HST); the border between Permian-Triassic and Triassic is seen at the apex of the Chhidru Formation. According to the outcrop data from the Chhidru Formation, three lithofacies have been observed, namely, the limestone facies (CF1), the limestone with clay interbeds facies (CF2), and the sandy limestone facies (CF3), while the petrographic study has classified the Chhidru Formation into five distinct microfacies, namely the mudstone biomicrite facies, the wackestone-biomicrite facies, and the wackestone-biosparite. Based on lithofacies, microfacies, stratigraphic control, and surviving fauna including brachiopods, ammonites, and gastropods, the Chhidru Formation was formed in a middle to inner shelf environment. The high stand system tract's sequence stratigraphic cycles indicate that this succession is situated in its uppermost portion. According to the sequence stratigraphic study, the Chhidru Formation exhibits a worldwide coursing-upward cycle. Using a laser level, we were able to map out where the transgressive surface and maximum flooding surface are in relation to each other.

Author Contributions: Conceptualization, S.K.A., H.T.J. and M.H.S.; methodology, M.H.S., S.K.A. and H.T.J.; software, M.H.S. and S.K.A.; validation, M.H.S., S.K.A. and H.T.J.; formal analysis, M.H.S. and S.K.A.; investigation, M.H.S., S.K.A., H.T.J., G.K., S.D.Z., P.M. and A.A.; resources, M.H.S.; data curation, M.H.S., S.K.A., H.T.J. and S.M.S.; writing—original draft preparation, M.H.S.; writing—review and editing, M.H.S., S.K.A., H.T.J., G.K., S.D.Z., P.M., U.K. and A.A.; visualization, S.K.A., H.T.J. and G.K.; supervision, S.K.A. and H.T.J.; project administration, S.K.A., H.T.J., G.K. and A.A.; funding acquisition, G.K. and A.A. All authors have read and agreed to the published version of the manuscript.

Funding: This research received no external funding.

Institutional Review Board Statement: Not applicable.

Informed Consent Statement: Not applicable.

Acknowledgments: We thank the director of the Institute of Geology, University of Azad Jammu and Kashmir, Muzaffarabad for his help and friendly attitude during our research work, which is always a source of inspiration for us.

Conflicts of Interest: The authors declare no conflict of interest.

References

1. Awan, R.S.; Khan, A.; Liu, C.; Yang, S.; Zang, Q.; Wu, Y.; Li, G.; Iltaf, K.H.; Tahir, M.; Ali, S. Subsurface geological model of sedimentary and metasedimentary wedge from Mansehra to Battal based on gravity data, Hazara area, Pakistan. *Energy Geosci.* **2021**, *2*, 229–237. [CrossRef]
2. Krishna, J. High-Resolution Intrabasinal to Inter-regional Geodynamic Chronicle During the Span of the Intra-Permian–Intra-Paleogene Mega-Sequence in and Around India on the GTM. In *The Indian Mesozoic Chronicle*; Springer: Berlin/Heidelberg, Germany, 2017; pp. 505–694.
3. Hermann, E.; Hochuli, P.A.; Bucher, H.; Roohi, G. Uppermost Permian to Middle Triassic palynology of the Salt Range and Surghar Range, Pakistan. *Rev. Palaeobot. Palynol.* **2012**, *169*, 61–95. [CrossRef]
4. Schneebeli-Hermann, E.; Bucher, H. Palynostratigraphy at the Permian-Triassic boundary of the Amb section, Salt Range, Pakistan. *Palynology* **2015**, *39*, 1–18. [CrossRef]
5. Khan, M.A.; Hussain, S.R.J.; Ali, B.; Ali, S.; Khan, M.J.; Mehmood, S. Microfacies analysis and depositional setting of the late Permian Wargal limestone exposed at Nammal Gorge, Western Salt Range, Pakistan. *Bahria Univ. Res. J. Earth Sci.* **2017**, *2*, 46–51.
6. Wignall, P.; Hallam, A. Griesbachian (Earliest Triassic) palaeoenvironmental changes in the Salt Range, Pakistan and southeast China and their bearing on the Permo-Triassic mass extinction. *Palaeogeogr. Palaeoclimatol. Palaeoecol.* **1993**, *102*, 215–237. [CrossRef]
7. Kummel, B. Stratigraphy and paleontology of the Permian-Triassic boundary beds, Salt Range and trans-Indus ranges, West Pakistan. In *Stratigraphic Boundary Problems: Permian and Triassic of West Pakistan*; Casa Camino Real: Las Cruces, NM, USA, 1970; Volume 4, pp. 1–110.
8. Theobald, W. Notes on geology of Salt Range. *Jl. ASB* **1854**, *23*, 651.
9. Ghazi, S.; Mountney, N.P.; Butt, A.A.; Sharif, S. Stratigraphic and palaeoenvironmental framework of the Early Permian sequence in the Salt Range, Pakistan. *J. Earth Syst. Sci.* **2012**, *121*, 1239–1255. [CrossRef]
10. Schneebeli-Hermann, E.; Kürschner, W.M.; Hochuli, P.A.; Bucher, H.; Ware, D.; Goudemand, N.; Roohi, G. Palynofacies analysis of the Permian–Triassic transition in the Amb section (Salt Range, Pakistan): Implications for the anoxia on the South Tethyan Margin. *J. Asian Earth Sci.* **2012**, *60*, 225–234. [CrossRef]

11. Waterhouse, J. Lopingian (Late Permian) stratigraphy of the salt range, Pakistan and Himalayan region. *Geol. J.* **2010**, *45*, 264–284. [CrossRef]
12. Wadood, B.; Aziz, M.; Ali, J.; Khan, N.; Wadood, J.; Khan, A.; Shafiq, M.; Ullah, M. Depositional, diagenetic, and sequence stratigraphic constrains on reservoir characterization: A case study of middle Jurassic Samana Suk Formation, western Salt Range, Pakistan. *J. Sediment. Environ.* **2021**, *6*, 131–147. [CrossRef]
13. Shen, S.Z.; Cao, C.-Q.; Henderson, C.M.; Wang, X.-D.; Shi, G.R.; Wang, Y.; Wang, W. End-Permian mass extinction pattern in the northern peri-Gondwanan region. *Palaeoworld* **2006**, *15*, 3–30. [CrossRef]
14. Jan, I.U. Progress in the Gondwanan Carboniferous–Permian palynology and correlation of the Nilawahan Group of the Salt Range, Pakistan: A brief review. *J. Earth Syst. Sci.* **2014**, *123*, 21–32. [CrossRef]
15. Mishra, S.; Aggarwal, N.; Jha, N. Palaeoenvironmental change across the Permian-Triassic boundary inferred from palynomorph assemblages (Godavari Graben, south India). *Palaeobiodivers. Palaeoenviron.* **2018**, *98*, 177–204. [CrossRef]
16. Mertmann, D. Evolution of the marine Permian carbonate platform in the Salt Range (Pakistan). *Palaeogeogr. Palaeoclimatol. Palaeoecol.* **2003**, *191*, 373–384. [CrossRef]
17. Anser, M.K.; Yousaf, Z.; Khan, M.A.; Nassani, A.A.; Abro, M.M.Q.; Vo, X.H.; Zaman, K. Social and administrative issues related to the COVID-19 pandemic in Pakistan: Better late than never. *Environ. Sci. Pollut. Res.* **2020**, *27*, 34567–34573. [CrossRef]
18. Jalil, A.; Khan, K. Preliminary appraisal of physio-chemical and bacteriological water contaminations in Rawalpindi/Islamabad catchment of Soan river, Potwar plateau (Punjab), Pakistan. *Punjab Univ. J. Zool* **2012**, *27*, 39–44.
19. Miraj, M.A.F.; Idrees, M.Z. Subsurface Structural Interpretation of Missa Keswal Area, Eastern Potwar, Pakistan. *Iraqi Geol. J.* **2021**, *54*, 146–156. [CrossRef]
20. Mehmood, W.; Aadil, N.; Jadoon, Y. 3-D Structural Modeling of Meyal Field, Potwar Sub-basin, Pakistan using Seismic and Well Data. *Nucleus* **2016**, *53*, 26–32.
21. Ghani, H.; Sobel, E.R.; Zeilinger, G.; Glodny, J.; Zapata, S.; Irum, I. Palaeozoic and Pliocene tectonic evolution of the Salt Range constrained by low-temperature thermochronology. *Terra Nova* **2021**, *33*, 293–305. [CrossRef]
22. Farooqui, M.A.; Umar, M.; Sabir, M.A.; Pervez, R.; Jalees, T. Geochemical attributes of late Neoproterozoic Salt Range Formation, Pakistan: Constraints on provenance, paleoclimate, depositional and tectonic settings. *Geosci. J.* **2019**, *23*, 201–218. [CrossRef]
23. Cotton, J.T.; Koyi, H.A. Modeling of thrust fronts above ductile and frictional detachments: Application to structures in the Salt Range and Potwar Plateau, Pakistan. *Geol. Soc. Am. Bull.* **2000**, *112*, 351–363. [CrossRef]
24. Yuan, D.Y.; Ge, W.P.; Chen, Z.W.; Li, C.Y.; Wang, Z.C.; Zhang, H.P.; Zhang, P.Z.; Zheng, D.W.; Zheng, W.J.; Craddock, W.H. The growth of northeastern Tibet and its relevance to large-scale continental geodynamics: A review of recent studies. *Tectonics* **2013**, *32*, 1358–1370. [CrossRef]
25. Zahid, M.; Khan, A.; ur Rashid, M.; Saboor, A.; Ahmad, S. Structural interpretation of Joya Mair oil field, south Potwar, Upper Indus Basin, Pakistan, using 2D seismic data and petrophysical analysis. *J. Himal. Earth Sci.* **2014**, *47*, 73.
26. Henderson, A.L.; Najman, Y.; Parrish, R.; Mark, D.F.; Foster, G.L. Constraints to the timing of India–Eurasia collision; a re-evaluation of evidence from the Indus Basin sedimentary rocks of the Indus–Tsangpo Suture Zone, Ladakh, India. *Earth Sci. Rev.* **2011**, *106*, 265–292. [CrossRef]
27. Khan, U.; Janjuhah, H.T.; Kontakiotis, G.; Rehman, A.; Zarkogiannis, S.D. Natural Processes and Anthropogenic Activity in the Indus River Sedimentary Environment in Pakistan: A Critical Review. *J. Mar. Sci. Eng.* **2021**, *9*, 1109. [CrossRef]
28. Abir, I.A.; Khan, S.D.; Ghulam, A.; Tariq, S.; Shah, M.T. Active tectonics of western Potwar Plateau–Salt Range, northern Pakistan from InSAR observations and seismic imaging. *Remote Sens. Environ.* **2015**, *168*, 265–275. [CrossRef]
29. Ullah, M.F.; Mahmood, K.; Akram, M.S. Coal mining trends and future prospects: A case study of Eastern Salt Range, Punjab, Pakistan. *J. Himal. Earth Sci.* **2018**, *51*, 87–93.
30. Sameeni, S.J. The Salt Range: Pakistan's unique field museum of geology and paleontology=[Le Salt Range: Un musée de géologie et de paléontologie à ciel ouvert au Pakistan]. In *PaleoParks: The Protection and Conservation of Fossil Sites Worldwide*; Carnets de Géologie/Notebooks on Geology, Brest, Book Chapter 6 (CG2009_BOOK_03/06); Université de Bretagne Occidentale Département des Sciences de la Terre: Brest, France, 2009; pp. 65–73.
31. Malkani, M.S.; Mahmood, Z. Revised stratigraphy of Pakistan. *Geol. Surv. Pak. Rec.* **2016**, *127*, 1–87.
32. Siehl, A. Structural setting and evolution of the Afghan orogenic segment—A review. *Geol. Soc. Lond. Spec. Publ.* **2017**, *427*, 57–88. [CrossRef]
33. Ueno, K. The Permian antitropical fusulinoidean genus Monodiexodina: Distribution, taxonomy, paleobiogeography and paleoecology. *J. Asian Earth Sci.* **2006**, *26*, 380–404. [CrossRef]
34. Shi, G.; Archbold, N. A quantitative analysis on the distribution of Baigendzhinian-Early Kungurian (Early Permian) brachiopod faunas in the western Pacific region. *J. Southeast Asian Earth Sci.* **1995**, *11*, 189–205. [CrossRef]
35. Khan, I.; Zhong, N.; Luo, Q.; Ai, J.; Yao, L.; Luo, P. Maceral composition and origin of organic matter input in Neoproterozoic–Lower Cambrian organic-rich shales of Salt Range Formation, upper Indus Basin, Pakistan. *Int. J. Coal Geol.* **2020**, *217*, 103319. [CrossRef]
36. Jan, I.U.; Iqbal, S.; Davies, S.J.; Zalasiewicz, J.A.; Stephenson, M.H.; Wagreich, M.; Haneef, M.; Hanif, M.; Ahmad, S. A periglacial palaeoenvionment in the upper carboniferous–lower permian tobra formation of the salt range, Pakistan. *Acta Geol. Sin. Engl. Ed.* **2017**, *91*, 1063–1078. [CrossRef]

37. Roohi, G.; Raza, S.M.; Schneebeli-Hermann, E.; Bucher, H.; Yaseen, A.; Imran, M. Permo-Triassic climate change and faunal turnover in the Salt and Surgharranges, Northern Pakistan. *J. Himal. Earth Sci.* **2017**, *50*, 1–12.
38. Shah, A.; Haneef, M.; Hanif, M.; Jan, I.U. Lithofacies and palaeoenvironments of the Carboniferous-Permian Nilawahan Group, Salt Range Pakistan. *J. Himal. Earth Sci.* **2012**, *45*, 136.
39. Mazumder, S.; Tep, B.; Pangtey, K.; Das, K.; Mitra, D. Probable existence of a Gondwana transcontinental rift system in western India: Implications in hydrocarbon exploration in Kutch and Saurashtra offshore: A GIS-based approach. *J. Earth Syst. Sci.* **2017**, *126*, 81. [CrossRef]
40. Shehzad, S.; Alam, I.; Mehmood, S.; Masood, F. Diagenetic History and Microfacies Analysis of Upper Permian Wargal Limestone in the Central Salt Range, Pakistan. *Pak. J. Sci. Ind. Res. Ser. A Phys. Sci.* **2018**, *61*, 163–172. [CrossRef]
41. Abbasi, I.A.; Haneef, M.; Khan, M.A. Early Permian Siliciclastic System of the North-Gondwanaland: A Comparison between Nilawahan Group of North Pakistan and Haushi Group of Oman. *Pak. J. Hydrocarb. Res.* **2011**, *21*, 19–33.
42. Shah, I. Stratigraphy of Pakistan: Memoirs of the Geological Survey of Pakistan. *Sci. Res.* **1977**, *12*, 138–140.
43. Ahmad, I.; Ahmad, S.; Ali, F. Structural analysis of the Kharthop and Kalabagh Hills area, Mianwali District, Punjab, Pakistan. *J. Himal. Earth Sci.* **2016**, *49*, 63–74.
44. Gee, E.; Gee, D. Overview of the geology and structure of the Salt Range, with observations on related areas of northern Pakistan. *Geol. Soc. Am. Spec. Pap.* **1989**, *232*, 95–112.
45. Jan, I.U.; Stephenson, M.H.; Khan, F.R. Palynostratigraphic correlation of the Sardhai Formation (Permian) of Pakistan. *Rev. Palaeobot. Palynol.* **2009**, *158*, 72–82. [CrossRef]
46. Abdulghani, A.; Ghazi, S.; Riaz, M.; Zafar, T. Sedimentary fabrics and diagenetic features of the Late Triassic Kingriali Formation, Khisor-Marwat ranges, Pakistan. *Indian J. Geo Mar. Sci.* **2020**, *49*, 954–964.
47. Khan, S.; Ahmad, S.; Hanif, M.; Jan, I.U.; Swati, M.A.F.; Khan, S.; Saboor, A. Lithofacies, paleoenvironments and sequence stratigraphic modelling of the Wargal Limestone: Implication for reservoir characterization in the Salt Range, northwest, Pakistan. *J. Himal. Earth Sci.* **2014**, *47*, 41.
48. Craig, J.; Hakhoo, N.; Bhat, G.; Hafiz, M.; Khan, M.; Misra, R.; Pandita, S.; Raina, B.; Thurow, J.; Thusu, B. Petroleum systems and hydrocarbon potential of the North-West Himalaya of India and Pakistan. *Earth Sci. Rev.* **2018**, *187*, 109–185. [CrossRef]
49. Malkani, M.S.; Mahmood, Z. Stratigraphy of Pakistan. *Geol. Surv. Pak. Mem.* **2017**, *24*, 1–134.
50. Hermann, E.; Hochuli, P.A.; Méhay, S.; Bucher, H.; Brühwiler, T.; Ware, D.; Hautmann, M.; Roohi, G.; Yaseen, A. Organic matter and palaeoenvironmental signals during the Early Triassic biotic recovery: The Salt Range and Surghar Range records. *Sediment. Geol.* **2011**, *234*, 19–41. [CrossRef]
51. Hussain, W.; Abbas, S.Q.; Hussain, S. 95 Structure Investigation, Economics and Stratigraphy of the Paleozoic, Mesozoic and Cenozoic Sequence in the Vicinity Eastern and Western side of the Salt Range, Punjab Pakistan. *J. Inf. Commun. Technol. Robot. Appl.* **2018**, *6*, 95–116.
52. Fahad, M.; Khan, M.A.; Hussain, J.; Ahmed, A.; Yar, M. Microfacies analysis, depositional settings and reservoir investigation of Early Eocene Chorgali Formation exposed at Eastern Salt Range, Upper Indus Basin, Pakistan. *Carbonates Evaporites* **2021**, *36*, 41. [CrossRef]
53. Alhubail, M.M.; Ghahfarokhi, R.B. Introduction to Unconventional Hydrocarbon Resources. *Unconv. Hydrocarb. Resour. Tech. Reserv. Eng. Anal.* **2020**, *1*, 1.
54. Hylland, M.D.; Yeats, R.S. Stratigraphic and structural framework of Himalayan foothills. In *Himalaya and Tibet: Mountain Roots to Mountain Tops*; Geological Society of America: Boulder, CO, USA, 1999; Volume 328, p. 257.
55. Kazmi, A.H.; Abbasi, I.A. *Stratigraphy & Historical Geology of Pakistan*; Department & National Centre of Excellence in Geology: Peshawar, Pakistan, 2008.
56. Ghazi, S.; Mountney, N.P. Petrography and provenance of the Early Permian Fluvial Warchha Sandstone, Salt Range, Pakistan. *Sediment. Geol.* **2011**, *233*, 88–110. [CrossRef]
57. Qureshi, M.N.; Talha, N.; Ahmad, M.; Zafar, M.; Ashfaq, S. Morpho-palynological investigations of natural resources: A case study of Surghar mountain district Mianwali Punjab, Pakistan. *Microsc. Res. Tech.* **2019**, *82*, 1047–1056. [CrossRef]
58. Waagen, W. Salt-range fossils; fossils from the Ceratite formation; Pisces–Ammonoidea: Ibid. *Mem. Palaeont. Indica Ser.* **1895**, *13*, 2.
59. Dunbar, C.O. *Stratigraphic Significance of the Fusulinids of the Lower Products [sic] Limestone of the Salt Range*; Geological Society of America: Boulder, CO, USA, 1933.
60. Teichert, C. Stratigraphic nomenclature and correlation of the Permian "Productus limestone". *Salt Range West Pak. Ibid. Recs* **1966**, *15*, 19.
61. Mei, S.; Henderson, C.M. Comments on some Permian conodont faunas reported from Southeast Asia and adjacent areas and their global correlation. *J. Asian Earth Sci.* **2002**, *20*, 599–608. [CrossRef]
62. Jan, I.U.; Stephenson, M.H. Palynology and correlation of the Upper Pennsylvanian Tobra Formation from Zaluch Nala, Salt Range, Pakistan. *Palynology* **2011**, *35*, 212–225. [CrossRef]
63. Ahmed, S.; Ahmad, I.; Khan, M.I. Structure and stratigraphy of the Paleozoic and Mesozoic sequence in the vicinity of Zaluch Nala, Western Salt Range, Punjab Pakistan. *Pak. J. Hydrocarb. Res.* **2005**, *15*, 1–8.
64. Hussain, H.S.; Jan, I.U.; Hanif, M.; Sadiq, I. Microfacies, palynofacies and depositional environment of the Upper Permian Wargal limestone, western Salt Range, Pakistan. *J. Himal. Earth Sci.* **2015**, *48*, 41–60.

65. Saboor, A.; Ul Haq, M.; Nawaz, A.; Swati, M.A.F.; Gohar, J.; Khan, S.; Ghaffari, A. Insight to the geological character of the Early Triassic strata in the Western Salt Range, Pakistan: A comparative study. *Geol. J.* **2021**, *56*, 4667–4684. [CrossRef]
66. Zia-ul-Rehman, M.; Masood, K.R.; Arshad, F.; Rehman, S.U.; Ullah, M.F. Biostratigraphy and Its Suggested Correlation with Lithostratigraphy, in Chhidru Formation, Salt Range, Pakistan. *Int. J. Agric. Appl. Sci.* **2015**, *7*, 12–19.
67. Ahmad, S.; Waqas, M.; Khan, S.S.; Swati, F.; Azhar, M.; Jan, I.U.; Ali, F.; Yaqoob, M.; Sadiq, A.; Khan, S. Facies analysis and dynamic depositional modelling of the Upper Permian Chhidru Formation, Salt Range, Upper Indus Basin, Pakistan. *J. Himal. Earth Sci.* **2015**, *48*, 54–70.
68. Schneebeli-Hermann, E. Regime shifts in an Early Triassic subtropical ecosystem. *Front. Earth Sci.* **2020**, *8*, 608. [CrossRef]
69. Chen, J.; Xu, Y.-G. Establishing the link between Permian volcanism and biodiversity changes: Insights from geochemical proxies. *Gondwana Res.* **2019**, *75*, 68–96. [CrossRef]
70. Furnish, W.; Glenister, B.F. Permian ammonoid Cyclolobus from the Salt Range, West Pakistan. In *Stratigraphic Boundary Problems: Permian and Triassic of West Pakistan*; Special Publication; Paleontological Institute, University of Kansas: Lawrence, KS, USA, 1970; Volume 4, pp. 153–175.
71. Khan, F.R.; Afzal, J. Comparative study of Brachiopods of Chhidru Formation from Zaluch and Nammal sections, Western Salt Range, Pakistan. *Pak. J. Hydrocarb. Res.* **2005**, *15*, 53–63.
72. Janjuhah, H.T.; Alansari, A.; Santha, P.R. Interrelationship Between Facies Association, Diagenetic Alteration and Reservoir Properties Evolution in the Middle Miocene Carbonate Build Up, Central Luconia, Offshore Sarawak, Malaysia. *Arab. J. Sci. Eng.* **2018**, *44*, 1–16. [CrossRef]
73. Dunham, R.J. Classification of carbonate rocks according to depositional textures. *Am. Assoc. Pet. Geol.* **1962**, 108–121.
74. Embry, A.F.; Klovan, J.E. A late Devonian reef tract on northeastern Banks Island, NWT. *Bull. Can. Pet. Geol.* **1971**, *19*, 730–781.
75. Flugel, E. *Microfacies of Carbonate Rocks: Analysis, Interpretation and Application*; Springer: Berlin/Heidelberg, Germany; New York, NY, USA, 2004; p. 979.
76. Flügel, E. Practical use of microfacies: Reservoir rocks and host rocks. In *Microfacies of Carbonate Rocks*; Springer: Berlin/Heidelberg, Germany, 2010; pp. 877–894.
77. Kontakiotis, G.; Moforis, L.; Karakitsios, V.; Antonarakou, A. Sedimentary Facies Analysis, Reservoir Characteristics and Paleogeography Significance of the Early Jurassic to Eocene Carbonates in Epirus (Ionian Zone, Western Greece). *J. Mar. Sci. Eng.* **2020**, *8*, 706. [CrossRef]
78. Hjálmarsdóttir, H.R.; Hammer, Ø.; Nagy, J.; Grundvåg, S.-A. Foraminiferal stratigraphy and palaeoenvironment of a storm-influenced marine shelf: Upper Aptian–lower Albian, Svalbard, Arctic Norway. *Cretac. Res.* **2021**, *130*, 105033. [CrossRef]
79. Wallet, E.; Slater, B.J.; Willman, S.; Peel, J.S. Small carbonaceous fossils (SCF s) from North Greenland: New light on metazoan diversity in early Cambrian shelf environments. *Pap. Palaeontol.* **2021**, *7*, 1403–1433. [CrossRef]
80. Zlobin, V.; Rosen, O.; Abbyasov, A. Two metasedimentary basins in the Early Precambrian granulites of the Anabar Shield (polar Siberia): Normative mineral compositions calculated by the MINLITH program and basin facies interpretations. In *Precambrian Sedimentary Environments: A Modern Approach to Ancient Depositional Systems*; Blackwell Publishing Ltd.: Algiers, Algeria, 2002.
81. Özgen-Erdem, N.; Koç-Tasgin, C. Microfacies and Depositional Environment of the Ilerdian Carbonates in the North-Western Tosya (SE Kastamonu) Region, Northern Turkey. *J. Geol. Soc. India* **2019**, *93*, 704–712. [CrossRef]
82. Moissette, P.; Cornée, J.-J.; Antonarakou, A.; Kontakiotis, G.; Drinia, H.; Koskeridou, E.; Tsourou, T.; Agiadi, K.; Karakitsios, V. Palaeoenvironmental changes at the Tortonian/Messinian boundary: A deep-sea sedimentary record of the eastern Mediterranean Sea. *Palaeogeogr. Palaeoclimatol. Palaeoecol.* **2018**, *505*, 217–233. [CrossRef]
83. Sims, E.R.; Belanger, C.L. Quantifying Late Pennsylvanian Multivariate Morphological Change in the Fusulinid Genus Triticites from the Central and Southwestern United States. *J. Foraminifer. Res.* **2021**, *51*, 165–181. [CrossRef]
84. Arefifard, S.; Payne, J.L. End-Guadalupian extinction of larger fusulinids in central Iran and implications for the global biotic crisis. *Palaeogeogr. Palaeoclimatol. Palaeoecol.* **2020**, *550*, 109743. [CrossRef]
85. Huang, X.; Aretz, M.; Zhang, X.; Du, Y.; Qie, W.; Wen, Q.; Wang, C.; Luan, T. Pennsylvanian—Early Permian palaeokarst development on the Yangtze Platform, South China, and implications for the regional sea-level history. *Geol. J.* **2018**, *53*, 1241–1262. [CrossRef]
86. Schwaha, T.; Bernhard, J.M.; Edgcomb, V.P.; Todaro, M.A. Aethozooides uraniae, a new deep-sea genus and species of solitary bryozoan from the Mediterranean Sea, with a revision of the Aethozoidae. *Mar. Biodivers.* **2019**, *49*, 1843–1856. [CrossRef]
87. Janjuhah, H.T.; Gamez Vintaned, J.A.; Salim, A.M.A.; Faye, I.; Shah, M.M.; Ghosh, D.P. Microfacies and depositional environments of miocene isolated carbonate platforms from Central Luconia, Offshore Sarawak, Malaysia. *Acta Geol. Sin. Engl. Ed.* **2017**, *91*, 1778–1796. [CrossRef]
88. Miranda, A.A.; Almeida, A.C.; Vieira, L.M. Non-native marine bryozoans (Bryozoa: Gymnolaemata) in Brazilian waters: Assessment, dispersal and impacts. *Mar. Pollut. Bull.* **2018**, *130*, 184–191. [CrossRef]
89. Moissette, P. Changes in bryozoan assemblages and bathymetric variations. Examples from the Messinian of northwest Algeria. *Palaeogeogr. Palaeoclimatol. Palaeoecol.* **2000**, *155*, 305–326. [CrossRef]
90. Tucker, M.E.; Wright, V.P. *Carbonate Sedimentology*; John Wiley & Sons: Hoboken, NJ, USA, 2009.
91. Bujtor, L.; Nagy, J. Fauna, palaeoecology and ecotypes of the Early Cretaceous sediment hosted hydrothermal vent environment of Zengővárkony (Mecsek Mountains, Hungary). *Palaeogeogr. Palaeoclimatol. Palaeoecol.* **2021**, *564*, 110179. [CrossRef]

92. Bergamin, L.; Ruggiero, E.T.; Pierfranceschi, G.; Andres, B.; Constantino, R.; Crovato, C.; D'Ambrosi, A.; Marassich, A.; Romano, E. Benthic foraminifera and brachiopods from a marine cave in Spain: Environmental significance. *Mediterr. Mar. Sci.* **2020**, *21*, 506–518. [CrossRef]
93. Fürsich, F.T.; Schneider, S.; Werner, W.; Lopez-Mir, B.; Pierce, C.S. Life at the continental–marine interface: Palaeoenvironments and biota of the Alcobaça Formation (Late Jurassic, Central Portugal), with a formal definition of the unit appended. *Palaeobiodivers. Palaeoenviron.* **2021**, 1–65. [CrossRef]
94. Scholle, P.A.; Ulmer-Scholle, D.S. *A Color Guide to the Petrography of Carbonate Rocks: Grains, Textures, Porosity, Diagenesis, AAPG Memoir 77*; AAPG: Tulsa, OK, USA, 2003; Volume 77.
95. Stock, C.W.; Sandberg, C.A. Latest Devonian (Famennian, expansa Zone) conodonts and sponge-microbe symbionts in Pinyon Peak Limestone, Star Range, southwestern Utah, lead to reevaluation of global Dasberg Event. *Palaeogeogr. Palaeoclimatol. Palaeoecol.* **2019**, *534*, 109271. [CrossRef]
96. Folk, R.L. Practical petrographic classification of limestones. *AAPG Bull.* **1959**, *43*, 1–38.
97. Tucker, M.E. *Sedimentary Petrology: An Introduction to the Origin of Sedimentary Rocks*; John Wiley & Sons: Hoboken, NJ, USA, 2009.
98. Cumming, R.L.; Gordon, D.P.; Gowlett-Holmes, K. Bryozoans in the marine benthos. In *Australian Bryozoa Volume 1: Biology, Ecology and Natural History*; Clayton, Missouri, USA, 2018; p. 145.
99. Karlsson, T.M.; Vethaak, A.D.; Almroth, B.C.; Ariese, F.; van Velzen, M.; Hassellöv, M.; Leslie, H.A. Screening for microplastics in sediment, water, marine invertebrates and fish: Method development and microplastic accumulation. *Mar. Pollut. Bull.* **2017**, *122*, 403–408. [CrossRef]
100. Embry, A. Practical sequence stratigraphy. *Can. Soc. Pet. Geol.* **2009**, *81*, 79.
101. Catuneanu, O.; Galloway, W.E.; Kendall, C.G.S.C.; Miall, A.D.; Posamentier, H.W.; Strasser, A.; Tucker, M.E. Sequence stratigraphy: Methodology and nomenclature. *Newsl. Stratigr.* **2011**, *44*, 173–245. [CrossRef]
102. Janjuhah, H.T.; Salim, A.; Mohammad, A.; Ali, M.Y.; Ghosh, D.P.; Hassan, A.; Hakif, M. Development of Carbonate Buildups and Reservoir Architecture of Miocene Carbonate Platforms, Central Luconia, Offshore Sarawak, Malaysia. In Proceedings of the SPE/IATMI Asia Pacific Oil & Gas Conference and Exhibition, Jakarta, Indonesia, 17–19 October 2017.
103. Lin, W.; Bhattacharya, J.P.; Stockford, A. High-resolution sequence stratigraphy and implications for cretaceous glacioeustasy of the Late Cretaceous Gallup System, New Mexico, USA. *J. Sediment. Res.* **2019**, *89*, 552–575. [CrossRef]
104. Janjuhah, H.T.; Alansari, A. Offshore Carbonate Facies Characterization and Reservoir Quality of Miocene Rocks in the Southern Margin of South China Sea. *Acta Geol. Sin. Engl. Ed.* **2020**, *94*, 1547–1561. [CrossRef]
105. Embry, A.F. Transgressive-regressive (TR) sequence stratigraphy. In Proceedings of the Gulf Coast SEPM Conference Proceedings, Houston, TX, USA, 8–11 December 2002; pp. 151–172.
106. Nakazawa, K.; Dickins, J.M. *The Tethys: Her Paleogeography and Paleobiogeography from Paleozoic to Mesozoic*; Tokai University Press: Tokyo, Japan, 1985.
107. Balic, D.; Malvic, T. Pliocene-Quaternary stratigraphy and sedimentation at the Neretva River Mouth, on Croatian Adriatic Coast. *Geol. Q.* **2013**, *57*, 233–241. [CrossRef]
108. Nugroho, S.H.; Putra, P.S. Determining textural and geochemical element characteristics of seafloor sediment using multivariate analysis along the Simeulue sub-basin, Indonesia. *Rud.-Geol.-Naft. Zb.* **2020**, *35*, 79–92. [CrossRef]
109. Moro, A.; Velić, I.; Mikuž, V.; Horvat, A. Microfacies characteristics of carbonate cobble from Campanian of Slovenj Gradec (Slovenia): Implications for determining the Fleuryana adriatica De Castro, Drobne and Gušić paleoniche and extending the biostratigraphic range in the Tethyan realm. *Rud.-Geol.-Naft. Zb.* **2018**, *33*, 1–13. [CrossRef]
110. Chen, Z.; Jin, Y.; Shi, G. Permian transgression-regression sequences and sea-level changes of South China. *Proc. R. Soc. Vic.* **1998**, *110*, 345–367.
111. Haq, B.U.; Schutter, S.R. A chronology of Paleozoic sea-level changes. *Science* **2008**, *322*, 64–68. [CrossRef]
112. Ross, C.A.; Ross, J.R. Permian sequence stratigraphy. In *The Permian of Northern Pangea*; Springer: Berlin/Heidelberg, Germany, 1995; pp. 98–123.
113. Strohmenger, C.; Voigt, E.; Zimdars, J. Sequence stratigraphy and cyclic development of Basal Zechstein carbonate-evaporite deposits with emphasis on Zechstein 2 off-platform carbonates (Upper Permian, Northeast Germany). *Sediment. Geol.* **1996**, *102*, 33–54. [CrossRef]
114. Bujtor, L. A Valanginian crustacean microcoprolite ichnofauna from the shallow-marine hydrothermal vent site of Zengővárkony (Mecsek Mts., Hungary). *Facies* **2012**, *58*, 249–260. [CrossRef]
115. Chen, Z.-Q.; Benton, M.J. The timing and pattern of biotic recovery following the end-Permian mass extinction. *Nat. Geosci.* **2012**, *5*, 375–383. [CrossRef]
116. Wynn, T.C.; Read, J.F. Sequence response of a distal-to-proximal foreland ramp to glacio-eustacy and tectonics: Mississippian, Appalachian Basin, West Virginia-Virginia, USA. *Am. Assoc. Pet. Geol.* **2003**, *83*, 11–34.
117. Fischer, A.G. The two Phanerozoic supercycles. In *Catastrophes and Earth History*; Princeton University Press: Princeton, NJ, USA, 2014; pp. 129–150.
118. Ahmad, S., Jr.; Ali, A.; Haneef, M.; Waqas, M.; Ali, J.; Khan, S.S.; Sadiq, A.; Yaqoob, M. Facies analysis, depositional model and sequence stratigraphy of the Upper Permian Chiddru Formation, Salt Ranges, Upper Indus Basin, Pakistan. *J. Himal. Earth Sci.* **2012**, *45*, 6.

119. Song, H.; Tong, J.; Algeo, T.J.; Song, H.; Qiu, H.; Zhu, Y.; Tian, L.; Bates, S.; Lyons, T.W.; Luo, G. Early Triassic seawater sulfate drawdown. *Geochim. Et Cosmochim. Acta* **2014**, *128*, 95–113. [CrossRef]
120. Sahney, S.; Benton, M.J. Recovery from the most profound mass extinction of all time. *Proc. R. Soc. B Biol. Sci.* **2008**, *275*, 759–765. [CrossRef] [PubMed]
121. Ward, P.D.; Montgomery, D.R.; Smith, R. Altered river morphology in South Africa related to the Permian-Triassic extinction. *Science* **2000**, *289*, 1740–1743. [CrossRef] [PubMed]
122. Foster, C.; Stephenson, M.; Marshall, C.; Logan, G.; Greenwood, P. A revision of Reduviasporonites Wilson 1962: Description, illustration, comparison and biological affinities. *Palynology* **2002**, *26*, 35–58. [CrossRef]
123. Liu, F.; Peng, H.; Bomfleur, B.; Kerp, H.; Zhu, H.; Shen, S. Palynology and vegetation dynamics across the Permian–Triassic boundary in southern Tibet. *Earth Sci. Rev.* **2020**, *209*, 103278. [CrossRef]
124. Kar, R.; Ghosh, A.K. First record of Reduviasporonites from the Permian–Triassic transition (Gondwana Supergroup) of India. *Alcheringa Australas. J. Palaeontol.* **2018**, *42*, 373–382. [CrossRef]
125. Yin, H.; Feng, Q.; Lai, X.; Baud, A.; Tong, J. The protracted Permo-Triassic crisis and multi-episode extinction around the Permian–Triassic boundary. *Glob. Planet. Chang.* **2007**, *55*, 1–20. [CrossRef]

Journal of
*Marine Science
and Engineering*

MDPI

Article

Fish Teeth Sr Isotope Stratigraphy and Nd Isotope Variations: New Insights on REY Enrichments in Deep-Sea Sediments in the Pacific

Fenlian Wang [1,2], Gaowen He [1,2,*], Xiguang Deng [1,2], Yong Yang [1,2] and Jiangbo Ren [1,2]

1 Southern Marine Science and Engineering Guangdong Laboratory (Guangzhou), Guangzhou 511458, China; wangfenlian@mail.cgs.gov.cn (F.W.); dxiguang@mail.cgs.gov.cn (X.D.); yyong@mail.cgs.gov.cn (Y.Y.); renjiangbo@mail.cgs.gov.cn (J.R.)

2 Key Laboratory of Marine Mineral Resources, Ministry of Natural Resources, Guangzhou Marine Geological Survey, China Geological Survey, Guangzhou 510075, China

* Correspondence: hgaowen@mail.cgs.gov.cn

Abstract: Rare earth elements and yttrium (REY) are widely recognized as strategic materials for advanced technological applications. Deep-sea sediments from the eastern South Pacific and central North Pacific were first reported as potential resources containing significant amounts of REY that are comparable to, or greater than, those of land-based deposits. Despite nearly a decade of research, quantitative abundances and spatial distributions of these deposits remain insufficient. Age controls are generally absent due to the lack of biostratigraphic constraints. Thus, the factors controlling the formation of REY-rich sediments are still controversial. In this study, the REY contents of surface sediments (<2 m depth) in 14 piston cores from the Central and Western Pacific were investigated. The results show that deep-sea sediments with high REY contents (>1000 μg/g) were mainly concentrated around seamounts (e.g., the Marshall Islands). The REY contents of surface sediments generally decreased with distance from the seamounts. Biostratigraphic and fish teeth debris (apatite) Sr isotopic stratigraphy of one piston core (P10) from the Central Pacific indicates that deep-sea sediments with high REY contents were aged from early Oligocene to early Miocene. Since the opening of the Drake Passage during the early Oligocene, the northward-flowing Antarctic Bottom Water (AABW) would have led to an upwelling of nutrients around seamounts with topographic barriers, and at the same time, AABW would delay the rate of sediment burial to try for enough time for REY entering and enriching in the apatite (fish teeth debris). Understanding the spatial distribution of fertile regions for REY-rich sediments provides guidance for searching for other REY resources in the Pacific and in other oceans.

Keywords: deep-sea sediment; rare earth element; enrichment mechanism; Sr isotope stratigraphy; Pacific Ocean

Citation: Wang, F.; He, G.; Deng, X.; Yang, Y.; Ren, J. Fish Teeth Sr Isotope Stratigraphy and Nd Isotope Variations: New Insights on REY Enrichments in Deep-Sea Sediments in the Pacific. *J. Mar. Sci. Eng.* **2021**, *9*, 1379. https://doi.org/10.3390/jmse9121379

Academic Editors: George Kontakiotis, Assimina Antonarakou and Dmitry A. Ruban

Received: 1 November 2021
Accepted: 21 November 2021
Published: 4 December 2021

1. Introduction

Rare earth elements and yttrium (REY) are widely recognized as strategic materials for advanced technological applications [1,2]. Global demands for REY are increasing rapidly [3]. At present, deposits associated with intrusive carbonate complexes and ion-adsorbed deposits are the world's most important source of REY [4,5]. Considering the supply risk, many countries have begun to explore REY resources beyond traditional terrestrial mines. Deep-sea sediments from the eastern South Pacific and central North Pacific were first reported as potential resources containing significant amounts of REY, comparable to, or greater than, those of land-based deposits [6]. Subsequently, REY-rich deep-sea sediments have also been found in the Indian Ocean [7]. The mineralogical, as well as major and trace elemental compositions, of deep-sea sediments have been investigated in the Pacific and India Ocean, with the goal of analyzing the host phase of REY elements and their provenance. It has been suggested that hydrothermal activity at mid-ocean ridges

(MORs) plays a role in the formation of REY-rich muds; in addition, phillipsite content and sedimentation rate may also affect the enrichment of REY in sediments [6].

REY-rich deep-sea sediments in the Pacific Ocean are mainly distributed in the eastern South Pacific and central North Pacific [6]. However, despite nearly a decade of research (since 2011), quantitative data on the abundances and spatial distributions of these deposits remain insufficient. In effect, the specific spatial distributions of REY-rich sediments in these two regions are enigmatic. Furthermore, precise age controls are generally absent due to the lack of biostratigraphic constraints. These factors restrict the study of REY enrichment mechanisms in deep-sea sediments.

Carbonate fossils are mostly absent in REY-rich pelagic clays because of their location below the carbonate compensation depth (CCD; ~4500 m). Moreover, the resolution of the magnetostratigraphic record is insufficient to accurately define the reversals [8]. Therefore, current high-resolution dating of pelagic clays is difficult when using conventional methods [9]. The $^{87}Sr/^{86}Sr$ record can be used to date and correlate marine sediments [10–15]. Furthermore, fish teeth debris, characterized by extremely high concentrations of REEs, occur in nearly all sediment types, including red clays [16,17]. The use of fish teeth Sr isotope stratigraphy for dating pelagic clays has been investigated [9,18]. The comparison of $^{87}Sr/^{86}Sr$ isotopic ratios to the marine Sr isotope curve [19] should, theoretically, constrain ages at a resolution of $\pm$0.5 myr for the time period of ~16–38 Ma, and $\pm$1.0 myr from ~16 Ma to the present [9].

The fractionation of Sm/Nd in continental crust relative to the mantle allows the $^{143}Nd/^{144}Nd$ ratio to be used to constrain Nd sources and trace the pathways of rare earth elements (REEs) in the ocean [20]. The residence time of Nd is shorter than the oceanic mixing cycle [21]; consequently, the oceans are poorly mixed with respect to Nd. Different oceans exhibit different Nd isotopic signatures [22], which allows isotopically distinct water masses to be traced [23,24]. Fish teeth debris can provide a record of the Nd isotopic signature of seawater, even in sediments that lack calcareous fossils [22].

In this study, the REY contents of surface sediments (<2 m depth) in 14 piston cores from the Central and Western Pacific were analyzed with the goal of identifying the spatial distribution of REY-rich sediments. In addition, in situ Sr and Nd isotopes of apatite (fish teeth detritus) at different depths in core P10—which contains both high REY content and low REY content sediments for comparison from the Central Pacific areas—were analyzed to try to constrain the age of the REY-rich sediments and to obtain the record of Nd isotopic signatures of seawater, which may have implications for constraining the source and pathways of the REY. These results may help in understanding the factors controlling REY enrichment in deep-sea sediments, which may provide guidance for searching for REY resources elsewhere in the Pacific or in other oceans.

2. Study Area and Methods

2.1. Study Area and Samples

The study area is located in the Central and Western Pacific (Figure 1), which are potential areas for REY-rich sediments [6]. The study area included the Pigafetta Basin, Eastern Mariana Basin, and parts of the Central Pacific Basin. The sediments were mainly pelagic clays, with minor radiolarian and calcareous oozes [25] at water depths of 4824–6251 m (Figure 2).

Sample information is provided in Table 1. Samples were taken for REY analysis at a sampling interval of 15 cm continuously from the top to the bottom of the cores. The average REY contents of the surface sediments (<2 m depth) of 14 piston cores were calculated (Figure 2). In addition, a 7.2 m core (P10) from the Central Pacific was selected for further in situ Sr and Nd isotopic analyses of apatite (fish teeth detritus) to constrain the age and the source of the high REY content sediment. The upper (0–210 cm) and lower (600–720 cm) sections of P10 mainly contained pelagic clay, whereas the middle layer (210–600 cm) was composed of zeolite clay with a higher REY content (REY > 1000 μg) than the other layers.

2.2. Major and Trace Elements

Analyses of major and trace elements were performed at the MNR Key Laboratory of Marine Mineral Resources, Guangzhou Marine Geological Survey. Major elements were analyzed using X-ray fluorescence (XRF; Axios), with a detection limit of 0.01–0.1% and a relative standard deviation (RSD) of <2%.

Trace elements were analyzed using inductively coupled plasma mass spectrometry (ICP-MS). Each sediment sample (0.1 g) was first placed in a beaker, to which 4 mL of 1:1 HCl, 10 mL of HF, and 1.5 mL of $HClO_3$ were added. The mixture was heated until it formed a dry paste. Then, 4 mL of 1:1 HCl was added to make a 25 mL solution; 1 mL of which was pipetted and diluted with 2% HNO_3 to a certain volume for analysis. Marine standard sediments GBW07313, GBW07315, and GBW07316 were used for quality control. The instrumental detection limit was 0.01–0.1 μg/mL and the RSD was <2%.

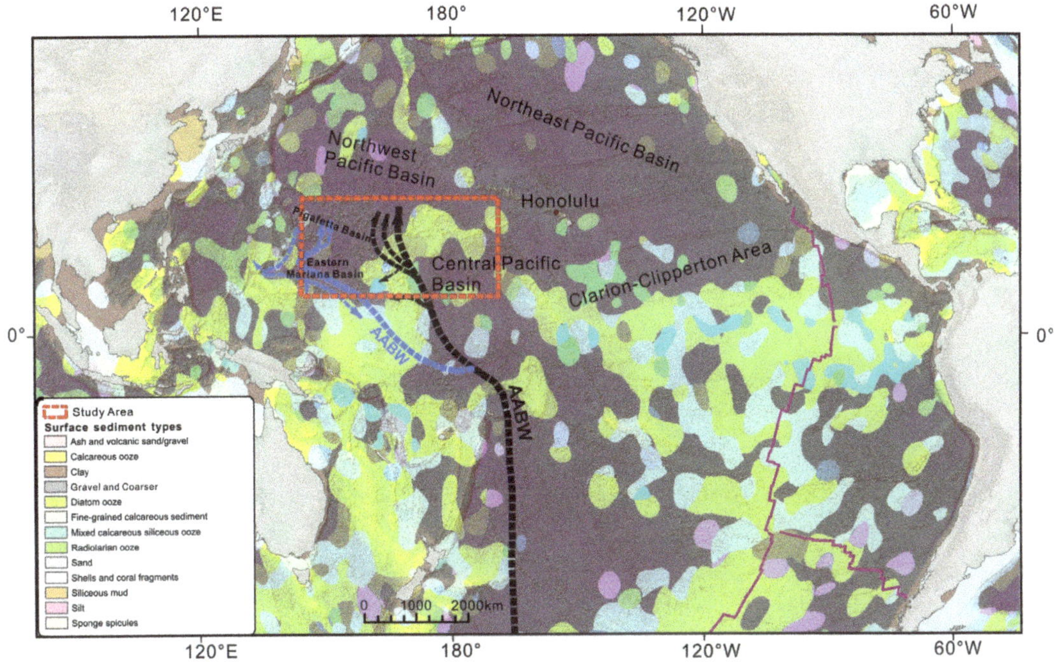

Figure 1. Surface sediment types and Antarctic Bottom Water (AABW) flow path. (Sediment types after [25], AABW path after [26]).

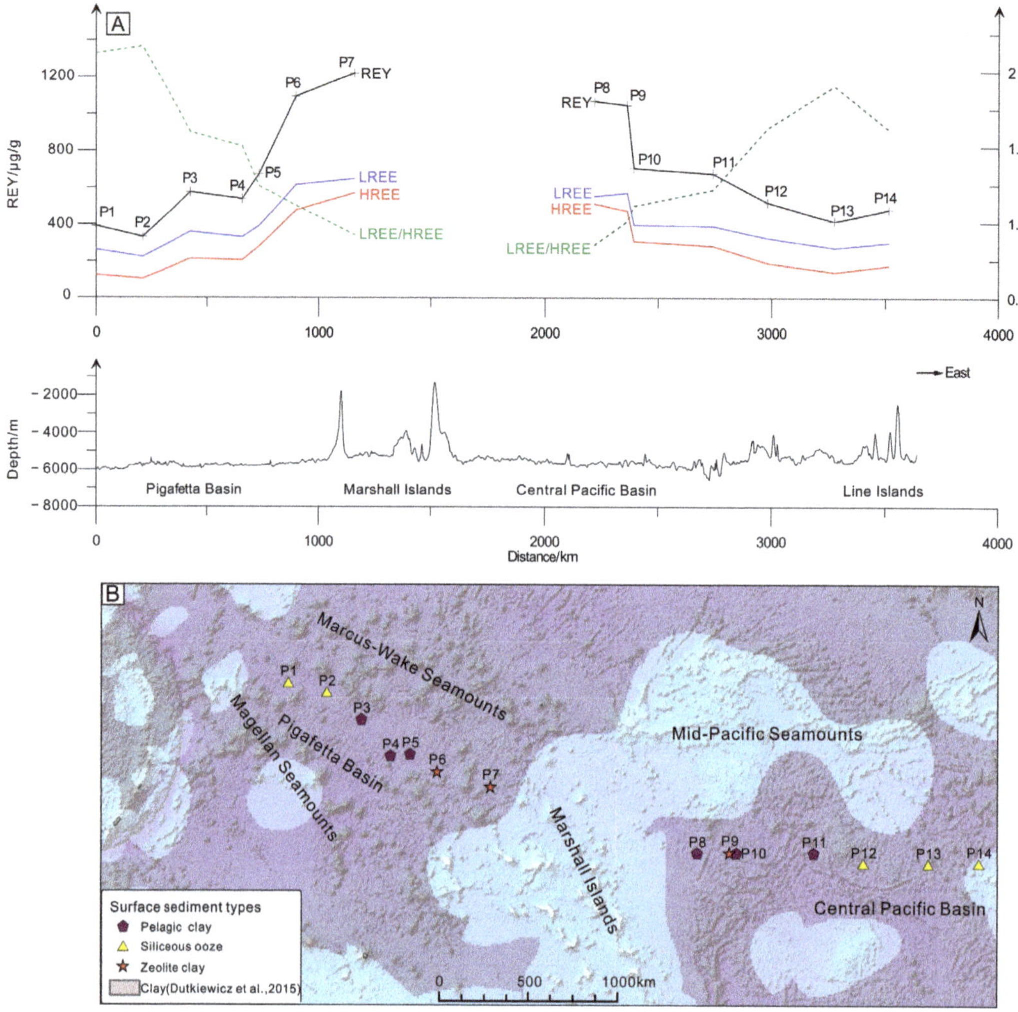

Figure 2. Characteristics of REY and sediment types in surface sediments (<2 m) in the Central and Western Pacific. (**A**) Characteristics of REY; (**B**) dominant surface sediment types (<2 m). P1–P14 correspond to piston cores.

Table 1. Water depth and length of the 14 piston cores, with type and REY content of their surface sediments (<2 m) in the study area.

Core	Sediment Type	REY (µg/g)	LREE (µg/g)	HREE (µg/g)	Depth (m)	Length (cm)
P1	Siliceous ooze	390	264	126	−5877	834
P2	Siliceous ooze	332	226	106	−5677	720
P3	Pelagic red clay	576	361	215	−5747	605
P4	Pelagic red clay	539	332	207	−5773	1445
P5	Pelagic red clay	675	393	282	−5696	732
P6	Zeolite clay	1095	617	478	−5368	715
P7	Zeolite clay	1221	649	572	−5163	802
P8	Pelagic red clay	707	398	309	−5495	765
P9	Pelagic red clay	1070	556	514	−5667	880
P10	Zeolite clay	1048	573	475	−5613	720
P11	Pelagic red clay	677	392	285	−5635	860
P12	Siliceous ooze	521	328	193	−5293	830
P13	Siliceous ooze	418	275	143	−5177	634
P14	Siliceous ooze	482	303	179	−5620	826

2.3. In Situ Sr and Nd Isotopes

The fish teeth debris were selected after sieving the wet sediment samples through a 63 μm mesh and then cleaned in an ultrasonic bath to remove any adhering clay particles and organic materials. After drying, the clean fish teeth were selected using a microscope and pasted on the annular target for in situ Sr and Nd isotopic analyses. In order to eliminate the influences of clay or other substances adhering to or infilling the fish teeth as much as possible, only triangular fish teeth or clear fish debris were selected for analysis (Figure 3). In addition, only fish teeth with high Sr and Nd contents (Sr > 400 μg/g; Nd >1 000 μg/g) were analyzed. Multiple fish teeth were analyzed in order to characterize the reproducibility in the same interval.

In situ Sr isotopic analyses were performed using an NWR 193 laser ablation system attached to a Thermo Fisher Scientific Neptune Plus multi-collector (MC) ICP-MS at the Beijing Createch Testing Technology Co., Ltd. A spot size of 38 μm was used, with a 6–8 Hz repetition rate and an energy density of 10 J/cm^2, depending on the Sr concentration of the sample. Sr isotopic data were acquired using static multi-collection in low-resolution mode with nine Faraday collectors. Prior to laser analysis, the Neptune MC-ICP-MS was tuned using a standard to obtain the maximum sensitivity. A typical data acquisition cycle consisted of a 40 s measurement of the Ar gas blank with the laser switched off, followed by 60 s of measurement with laser ablation. The Durango1 apatite reference material was analyzed every 10 samples for external calibration.

Data reduction was conducted offline, and potential isobaric interferences were accounted for in the following order: Kr, Yb^{2+}, Er^{2+}, and Rb. Initially, the interferences of ^{84}Kr and ^{86}Kr on ^{84}Sr and ^{86}Sr, respectively, were removed using the 40 s Kr gas baseline measurement. The presence of $^{167}Er^{2+}$, $^{171}Yb^{2+}$, and $^{173}Yb^{2+}$ at masses of 83.5, 85.5, and 86.5, respectively, were then monitored. Subsequently, the natural ratio of $^{85}Rb/^{87}Rb$ (2.5926) was used to correct for the isobaric interference of ^{87}Rb on ^{87}Sr using the exponential law, assuming that Rb has the same mass discrimination behavior as Sr. In addition, interferences from Ca argides or dimers and Ca–P–O were not considered for high Sr contents (>400 μg/g). The $^{87}Sr/^{86}Sr$ ratios were then calculated and normalized from the interference-corrected $^{86}Sr/^{88}Sr$ ratios using the exponential law. The data reduction procedure was performed using an Excel Visual Basic for Applications macro program [27,28].

All of the in situ Nd isotopic analyses were performed using a Neptune Plus MC-ICP-MS (Thermo Scientific, Waltham, MA, USA), coupled with a RESOlution M-50 193 nm laser ablation system (Resonetics, Nashua, NH, USA), hosted at the State Key Laboratory of Isotope Geochemistry, Guangzhou Institute of Geochemistry, Chinese Academy of Sciences. A detailed description of these instruments can be found in [29]. A small N_2 (2 mL L^{-1}) flow and X skimmer cone at the interface were used to improve the instrumental sensitivity. All isotope signals were detected using Faraday cups under static mode. Laser parameters were set as follows: a beam diameter of 82–112 μm; a repetition rate of 6 Hz; and an energy density of ~4 J cm^{-2}. Helium was chosen as the carrier gas (800 mL min^{-1}). Each analysis consisted of 250 cycles, with an integration time of 0.262 s per cycle. The first 29 s were used to detect the gas blank with the laser beam off, followed by 30 s of laser ablation for sample signal collection with the laser beam. During the measurements, gas blanks of ^{143}Nd were less than 0.2 mv. The interferences of ^{144}Sm on ^{144}Nd were derived from the ^{147}Sm intensities, with a natural $^{143}Sm/^{147}Sm$ ratio of 0.20504 [30]. The mass bias factor of Sm was calculated from the measured isotopic ratio of $^{147}Sm/^{149}Sm$ and its accepted value (1.08507; [30]). The mass bias of $^{143}Nd/^{144}Nd$ was normalized to $^{146}Nd/^{144}Nd$ = 0.7129 using an exponential law. A detailed description of the data reduction procedure can be found in [29]. In total, 40 analyses of McClure apatite and 25 analyses of Durango apatite yielded weighted means of $^{143}Nd/^{144}Nd = 0.512280 \pm 0.000055$ (2 SD) and $^{143}Nd/^{144}Nd = 0.512470 \pm 0.000060$ (2 SD), respectively, which are consistent (within error) with the values reported by [31].

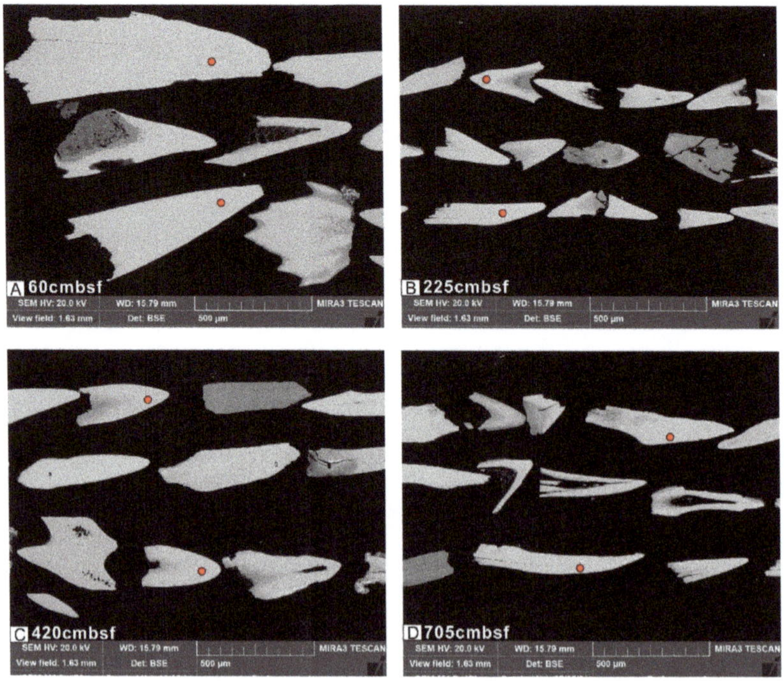

Figure 3. Representative BSE images of fish teeth debris grains from the core P10. Red circles show the location of LA-ICP-MS Sr analyses. (**A–D**) samples from different depths of the P10.

3. Results

3.1. REY Characteristics of Surface Sediments

The REY contents of the surface sediments from the Central and Western Pacific are listed in Table 1. The data indicate that the average REY contents of the surface sediments at different sites in the Pigafetta Basin (northwest of the Marshall Islands) varied from 332 to 1221 μg/g (Figure 2). Heavy REEs (HREEs), including Y, varied from 106 to 572 μg/g, while light REEs (LREEs) varied from 226 to 649 μg/g.

The average REY contents of the surface sediments at different sites in the Central Pacific basin (east of the Marshall Islands) varied from 418 to 1070 μg/g (Figure 2). HREE (including Y) contents varied from 143 to 514 μg/g, while LREE contents varied from 275 to 573 μg/g.

3.2. Vertical Distributions of REY

Vertical REY contents of sediments from the Central and Western Pacific varied from core to core. All of the analyzed piston cores contained multiple REY-rich sediment layers at different depths and with different thicknesses (Figure 4). For example, Core P7 contained three layers with REY contents of >1000 μg/g and two layers with REY contents ranging from 700 to 1000 μg/g. However, core P10 contained one layer with REY content of >1000 μg/g, two layers with REY contents ranging from 700 to 1000 μg/g, and two layers with REY contents ranging from 400 to 700 μg/g.

In North American shale composite (NASC)-normalized REE patterns (Figure 5), all of the sediments of P10 exhibited significant negative Ce anomalies and positive Y anomalies.

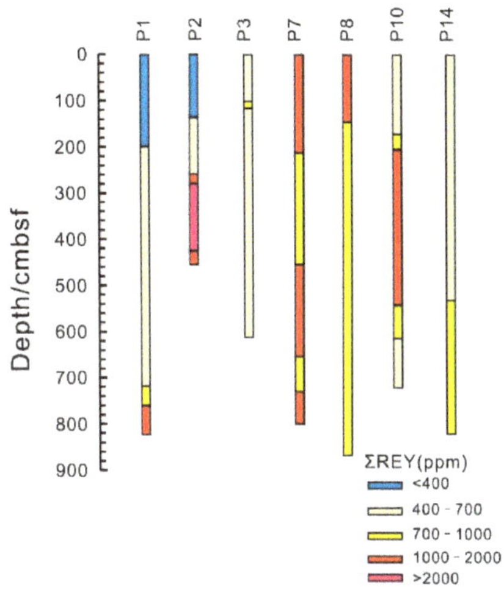

Figure 4. REY contents of typical piston cores in the study area.

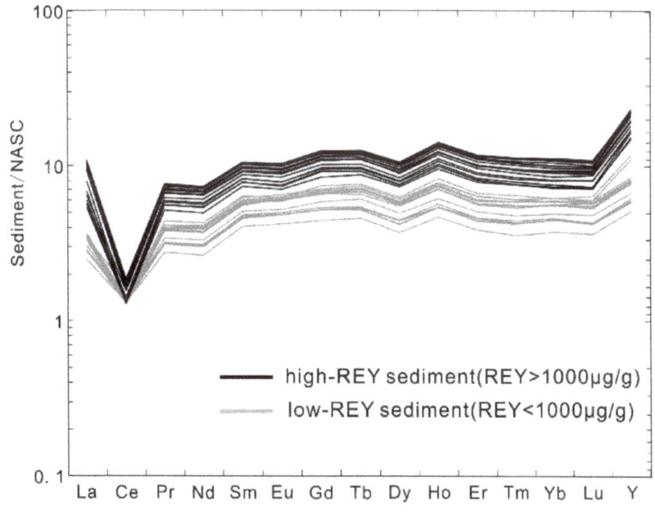

Figure 5. North American shale composite (NASC)-normalized REE patterns of deep-sea sediments in core P10. NASC after [32].

3.3. *Fish Teeth Sr Isotopic Compositions*

In situ fish teeth Sr values, at different depths in core P10, are listed in Table 2 and Figure 6. The initial $^{87}Sr/^{86}Sr$ ratios varied, ranging from 0.707960 to 0.709288, and multiple fish teeth from the same interval exhibited different isotopic characteristics.

The initial $^{87}Sr/^{86}Sr$ ratios ranged from 0.708961 to 0.709288 in the upper 210 cm of the core. Some fish teeth exhibited lower $^{87}Sr/^{86}Sr$ values (such as 0.708961) than the $^{87}Sr/^{86}Sr$ value of modern seawater (0.70924, [19]), indicating that they were old fish teeth that had been re-deposited [36]. The apatite $^{87}Sr/^{86}Sr$ values of the upper 210 cm in core P10 were generally similar to the modern seawater value, except for the unreasonable data described above.

Fish teeth $^{87}Sr/^{86}Sr$ ratios in the 210–600 cm layer and the lower 600–720 cm layer exhibited similar characteristics, ranging from 0.708005 to 0.708426 and from 0.707960 to 0.708481, respectively, indicating that these two layers were continuously deposited.

Theoretically, the fish teeth $^{87}Sr/^{86}Sr$ values should exhibit a decreasing trend in the vertical direction. However, the values fluctuated throughout the core length (Figure 6b). For example, some fish teeth $^{87}Sr/^{86}Sr$ values of lower layers (600–720 cm) were greater than those of the 210–600 cm layer (Table 2); moreover, different fish teeth particles at the same interval showed different $^{87}Sr/^{86}Sr$ values. These fluctuating $^{87}Sr/^{86}Sr$ values may have been raised by the clay materials adhering to the fish teeth [37] or reflect the overturning of the old and new sediments. Nevertheless, the fish teeth $^{87}Sr/^{86}Sr$ values in the upper sediment with lower REY contents, and in the middle layer sediment with higher REY contents (Figure 6a), showed significant difference, indicating that fish teeth $^{87}Sr/^{86}Sr$ values were of significance for the determination of sediment chronology.

Table 2. Strontium isotopic data for P10 fish teeth debris.

Depth (cm)	$^{87}Sr/^{86}Sr$ (corr)	2σ	Age (Ma)
30	0.709249	0.00008	0 [19]
30	0.709143	0.00009	0.5 [19]
45	0.709257	0.00010	0 [19]
60	0.709158	0.00011	0 [19]
60	0.709204	0.00007	0 [19]
75	0.709267	0.00013	0 [19]
90	0.709187	0.00015	0 [19]
120	0.709165	0.00012	0 [19]
135	0.708961	0.00008	5.2 [19]
150	0.709005	0.00008	4.5 [19]
180	0.709278	0.00012	0 [19]
195	0.709288	0.00010	0 [19]
210	0.709183	0.00012	1.8 [19]
225	0.708124	0.00007	26.77 [33]
225	0.708269	0.00016	23.71 [33]
240	0.708149	0.00009	25.31 [33]
285	0.708426	0.00011	20.08 [33]
285	0.708253	0.00011	23.36 [34]
300	0.708244	0.00009	23.48 [34]
300	0.708275	0.00011	21.9 [34]
315	0.708088	0.00011	26.29 [33]
315	0.708189	0.00010	23.96 [34]
330	0.708235	0.00015	23.23 [34]
330	0.708184	0.00011	23.96 [34]
345	0.708140	0.00010	25.31 [33]
360	0.708271	0.00014	23.71 [35]
360	0.708330	0.00010	22.61 [35]
375	0.708227	0.00011	24.11 [35]
390	0.708126	0.00008	25.7 [33]
390	0.708062	0.00012	27.67 [33]
405	0.708218	0.00010	24.33 [35]
405	0.708150	0.00007	25.31 [33]
420	0.708016	0.00007	28.68 [33]
420	0.708033	0.00010	28.18 [33]
435	0.708005	0.00009	29.22 [33]
450	0.708331	0.00013	22.61 [35]
465	0.708113	0.00010	26.67 [33]
465	0.708370	0.00011	21.06 [35]
480	0.708092	0.00014	27.46 [33]
480	0.708099	0.00009	26.97 [33]
495	0.708187	0.00009	24.87 [33]
495	0.708064	0.00010	27.67 [33]
510	0.708275	0.00010	23.71 [35]

Table 2. *Cont.*

Depth (cm)	$^{87}Sr/^{86}Sr$ (corr)	2σ	Age (Ma)
525	0.708196	0.00012	24.87 [33]
525	0.708385	0.00011	21.27 [35]
540	0.708041	0.00010	27.98 [33]
570	0.708178	0.00017	25.02 [33]
570	0.708239	0.00014	24.22 [35]
585	0.708106	0.00008	26.14 [33]
585	0.708066	0.00008	27.67 [33]
600	0.708096	0.00007	27.07 [33]
600	0.708056	0.00010	27.58 [33]
630	0.708254	0.00009	23.95 [35]
645	0.708008	0.00009	29.22 [33]
660	0.708184	0.00014	26.29 [33]
675	0.707960	0.00007	30.13 [33]
695	0.708309	0.00012	23.03 [35]
705	0.708350	0.00008	21.98 [35]
705	0.708477	0.00012	20.42 [35]
720	0.708481	0.00010	20.82 [35]

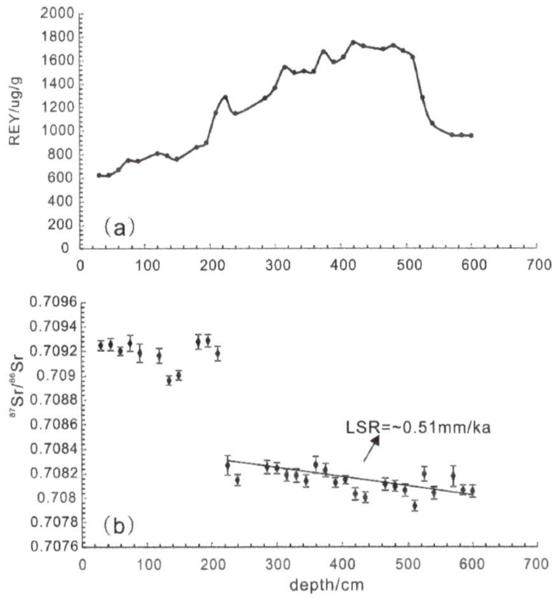

Figure 6. The REY content of sediment (**a**) and in situ fish teeth $^{87}Sr/^{86}Sr$ values (**b**) in core P10.

3.4. Fish Teeth Nd Isotopic Compositions

In situ fish teeth Nd data from core P10 are listed in Table 3. The $^{143}Nd/^{144}Nd$ varied from 0.512110 to 0.512533, corresponding to an ε_{Nd} range from −10.29 to −2.04 for the whole core. However, there were clear vertical variations in the Nd isotopes of apatite (Figure 7). Apatite ε_{Nd} values varied from −6.42 to −2.04, concentrated between −5 and −3, with an average value of −4.48 in the upper 210 cm. In the 210–600 cm layer, apatite ε_{Nd} values varied widely from −10.29 to −2.87 (i.e., significantly lower than in the upper layer). Apatite ε_{Nd} values in the middle layer were divided into two ranges: from −10.29 to −7 and −6 to −2. At 600–705 cm, the values ranged from −7.69 to −4.27, with an average of −5.97. Apatite ε_{Nd} is related to the REY content of the sediment. The REY contents of the 0–210 cm and 600–705 cm layers were lower than that of the 210–600 cm layer.

Table 3. Nd isotopic data for P10 fish teeth debris.

Depth (cm)	$^{143}Nd/^{144}Nd$	2σ	ε_{Nd}
15	0.512413	0.000039	−4.40
30	0.512374	0.000049	−5.14
45	0.512415	0.000032	−4.35
45	0.512371	0.000054	−5.21
45	0.512418	0.000076	−4.29
60	0.512309	0.000028	−6.42
75	0.512531	0.000037	−2.09
75	0.512497	0.000045	−2.75
90	0.512533	0.000035	−2.04
90	0.512320	0.000041	−6.20
105	0.512411	0.000037	−4.43
105	0.512372	0.000044	−5.19
135	0.512343	0.000077	−5.76
135	0.512387	0.000032	−4.90
150	0.512335	0.000051	−5.91
150	0.512416	0.000027	−4.34
165	0.512449	0.000038	−3.69
180	0.512436	0.000038	−3.95
180	0.512375	0.000029	−5.13
195	0.512397	0.000022	−4.71
195	0.512408	0.000024	−4.49
210	0.512183	0.000066	−8.88
210	0.512421	0.000031	−4.22
225	0.512124	0.000034	−10.03
225	0.512166	0.000032	−9.21
225	0.512187	0.000049	−8.79
240	0.512136	0.000032	−9.80
240	0.512127	0.000169	−9.97
255	0.512407	0.000024	−4.52
255	0.512155	0.000031	−9.43
270	0.512256	0.000039	−7.45
285	0.512110	0.000038	−10.29
285	0.512250	0.000042	−7.58
300	0.512212	0.000029	−8.31
300	0.512250	0.000042	−7.56
315	0.512256	0.000043	−7.46
315	0.512235	0.000034	−7.86
330	0.512241	0.000035	−7.75
330	0.512180	0.000039	−8.93
345	0.512406	0.000047	−4.53
360	0.512249	0.000028	−7.59
360	0.512269	0.000042	−7.19
375	0.512284	0.000042	−6.90
375	0.512200	0.000060	−8.55
390	0.512369	0.000039	−5.24
405	0.512238	0.000038	−7.80
405	0.512112	0.000054	−10.26
420	0.512264	0.000060	−7.29
435	0.512208	0.000048	−8.39
435	0.512349	0.000034	−5.64
465	0.512236	0.000036	−7.84
480	0.512137	0.000045	−9.78
495	0.512427	0.000060	−4.11
510	0.512491	0.000045	−2.87
525	0.512369	0.000020	−5.25
540	0.512250	0.000053	−7.56
555	0.512187	0.000030	−8.79
570	0.512247	0.000065	−7.64
570	0.512293	0.000028	−6.74
585	0.512373	0.000028	−5.18
600	0.512419	0.000036	−4.27
600	0.512358	0.000019	−5.47
630	0.512348	0.000023	−5.65
630	0.512315	0.000023	−6.30
660	0.512352	0.000023	−5.59
690	0.512352	0.000023	−5.59
690	0.512244	0.000049	−7.69
705	0.512329	0.000027	−6.03

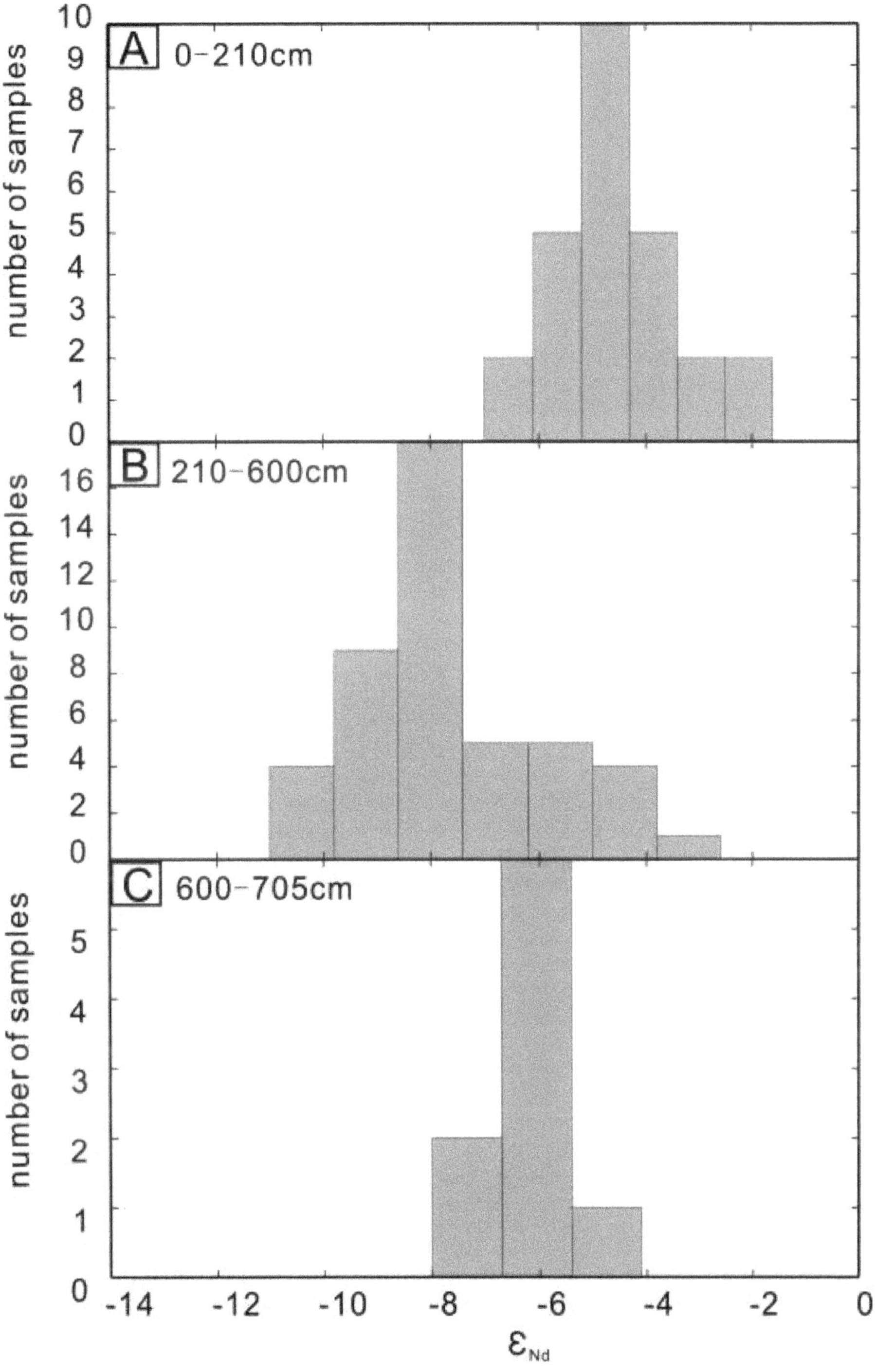

Figure 7. In situ fish teeth ε_{Nd} values in core P10. (**A–C**), the 0–210 cm layer, the 210–600 cm layer and 600–705 cm layer of P10.

4. Discussion

4.1. Distribution of REY-Rich Sediments in the Central and Western Pacific

As mentioned above, deep-sea sediments contain high concentrations of REY at numerous sites throughout the eastern South and central North Pacific [6]. However, the specific distributions of REY-rich sediments in the study area are still unknown.

In this study, REY-rich sediments in the surface layer (<2 m) of the Central and Western Pacific Ocean exhibited a regular distribution (Figure 2). Sediments with high REY contents (>1000 μg/g) were mainly concentrated around seamounts (e.g., the Marshall Seamounts and Mid-Pacific Seamounts). The REY contents of the surface sediments decreased with increasing distance from the seamounts. In the northwestern Pacific (northwestern Marshall Islands), the REY contents of the surface sediments gradually decreased from southeast to northwest. Surface sediments in the southeast had the highest REY contents (1000–1500 μg/g); whereas, those of the northwestern surface sediments were generally low (<400 μg/g). In the Central Pacific, the REY contents of the surface sediments also exhibited regular patterns, decreasing gradually with increasing distance from seamounts.

The REY content in the deeper core sediments were showed in Figure 4. Similar to those of the surface sediments, core sediments with high REY contents (e.g., >700 μg/g) also usually occurred around the Marshall Islands within a ~10 m depth range (i.e., cores P7, P8, and P10).

4.2. Dating REY-Rich Sediments

Core P10 was collected adjacent to Deep Sea Drilling Project (DSDP, the first of three international scientific ocean drilling programs that have operated over more than 40 years) site 170 in the Central Pacific. The DSDP initial reports (Volume 17, Site 170) indicate that the uppermost 16 m of zeolitic brown clay were late Oligocene–Quaternary in age; whereas, the 16–36 m layer was mainly Eocene–Oligocene cherts. Therefore, the deep-sea sediment in this area is not older than the Oligocene.

In this study, for each $^{87}Sr/^{86}Sr$ value, an age was assigned. Therefore, we were able to produce an age–depth curve for the clay (Table 2, Figure 8). Fish teeth Sr isotopic analyses from the upper 210 cm of the P10 yielded $^{87}Sr/^{86}Sr$ ratios of 0.7109143–0.709288 (with outliers of 0.709005 and 0.708961), which were identical (within error) to that of modern seawater (0.7092, [19]), except for the unreasonable data (mentioned in Section 3.3). In addition, common radiolarian fossils from the Quaternary period (e.g., *Druppatractus testudo*, *Euchitonia trianglulum*, and *Stylodictya validispina*) were observed in the upper 30 cm (internal report, unpublished data). According to the biostratigraphic and fish teeth Sr isotopic analyses, the age of the upper 210 cm of the core is Quaternary. We calculated a linear sedimentation rate (LSR) of ~1.0 mm/ky.

Fish teeth $^{87}Sr/^{86}Sr$ values in the middle (225–600 cm) layer, which had high REY contents (>1000 μg/g), ranged from 0.708005 to 0.708426, indicate a clear hiatus in deposition between the upper (210 cm) and middle layers. The fish teeth $^{87}Sr/^{86}Sr$ values of the lower (600–720 cm) layer were continuous with those of the middle layer (0.707960–0.708481), indicating that these two layers were continuously deposited. Fish teeth Sr ages indicate that the REY-rich sediments were Oligocene–early Miocene (30.13–20.08 Ma) in age. *Stichocorys delmontensis radiolarian*, which occurred in the early to late Miocene (internal report, unpublished data), were observed at 600 cm depth. Therefore, the radiolarian biostratigraphy is in agreement with the ages obtained from the fish teeth. We calculated an LSR of ~0.51 mm/ky, which was much slower than the upper layer.

From 84 Ma, the Drake Passage began to open [38]. Until the Oligocene and early Miocene (~29 Ma) [39,40], the opening of the Drake Channel led to the formation of deep circumpolar currents [41]. The evolution of the Miocene paleo-ocean was characterized by the opening and closing of channels, changes in ocean circulation, and the development of glaciers. Affected by underflows, extensive sedimentary discontinuities exist in Miocene sediments throughout the ocean basins [36,42]. In core P10, the Miocene deep-sea hiatuses are coincident with the global hiatuses. Therefore, the ages of the fish teeth obtained using in situ Sr isotopes are reliable and reasonable.

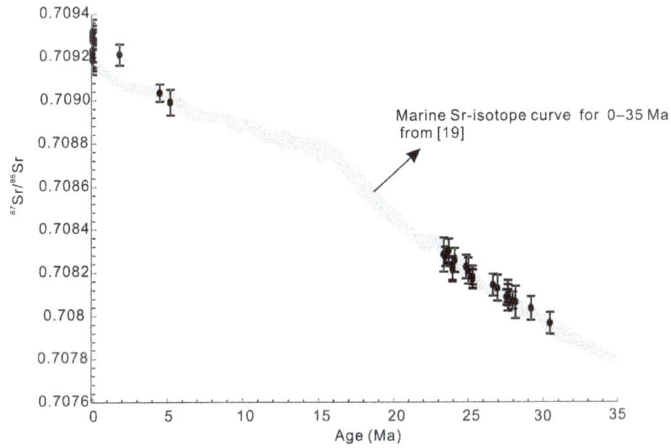

Figure 8. Fish teeth ages obtained from core P10.

4.3. Antarctic Bottom Water (AABW) and REY Enrichment

Clear vertical variations were observed in the Nd isotopes of fish teeth in core P10 (Figure 7), indicating that the sediments were affected by different water masses at different depths. The seawater ε_{Nd} value in the Northern and Central Pacific ranges from −5 to −3 at depths of 1000–5000 m [43]. In the upper 210 cm and lower 120 cm of core P10, the fish teeth ε_{Nd} values were consistent with the Pacific deep water ε_{Nd}, indicating that the REEs in these sediments were mainly derived from seawater.

However, the Nd isotopes of fish teeth in the middle layer (210–600 cm) were less radiogenic, exhibiting significantly reduced ε_{Nd} values with amplitudes exceeding 2 units, which is much lower than that of the seawater ε_{Nd}. This indicates that the REY in these sediments was affected by other factors. In general, ε_{Nd} values are mainly affected by continental and mantle materials. Overall, volcanic material from the seabed is more radiogenic and characterized by positive values; whereas, terrestrial materials are less radiogenic and characterized by negative values. For example, mid-ocean ridge basalts in the Pacific have an average ε_{Nd} value of +10; whereas, terrestrial silicates from the northern Central Pacific have an average ε_{Nd} value of −10.2 [44]. Therefore, significant inputs of terrestrial material will lower the ε_{Nd} value. However, Figure 9 shows that the Al_2O_3 and SiO_2 contents of the sediments in the 210–600 cm layer were slightly lower than those in the upper- and lowermost layers, indicating that the low apatite ε_{Nd} was not caused by an input of terrestrial materials. The changes in Nd isotopic compositions were therefore mainly caused by the mixing of ocean water masses with different Nd isotopic compositions under the condition that external inputs remained stable [45].

As aforementioned, during the early Oligocene, the opening of Drake Passage resulted in an established deep circumpolar current and an increase in AABW activity. The AABW forms two branches after inflowing into the Central Pacific Basin through the Samoa channel, of which, one branch flows to the Northwestern Pacific Basin through the Marshall Islands (Figure 1, [26,46]). The sediments in the study area are located in the area where the AABW flows. The AABW is a low-temperature, high salinity, high density, and oxygen-rich water mass. It controls deep ocean circulation and provides a strong oxidizing environment for crust formation at seamounts [47]. The environmental redox indicators of the Ce anomaly have been verified [48,49]. The NASC-normalized REE pattern (Figure 5) exhibited a significant negative Ce anomaly in the sediments of core P10, indicating that they were formed under strongly oxidizing conditions. In contrast, the fish teeth ε_{Nd}

values of the 210–600 cm layer of core P10 ranged from −10.29 to −2.87, with some values similar to the ε_{Nd} value of the AABW (≈−9, [24]), implying that the REY-rich sediments were affected by the AABW. However, where the AABW is strong, sediment is eroded or transported, which is not conducive for deposition. The fish teeth ε_{Nd} values of REY-rich sediment had a relatively broad range, indicating that they were partly affected by the AABW. As the thermal gradient between the polar region and the equator increased, the activity of the AABW strengthened during the Miocene. Correspondingly, sediments were eroded and transported away by the strong AABW, explaining the depositional hiatus observed in core P10 during this period.

In the modern ocean, enhanced primary productivity around seamounts is a result of upwelling generated by seamount–current interactions [50,51]. The intensified northward AABW would have led to the upwelling of nutrients in regions with topographic barriers that were steep and large enough to allow upwelling [52,53]. This supply of nutrients in oligotrophic pelagic regions may have been sufficient to increase local fish proliferation [53]. The phenomenon, whereby REY-rich sediments with high contents of fish teeth were observed mainly around the seamounts in this study, also supports the influence of the AABW on the enrichment of REY in the sediment.

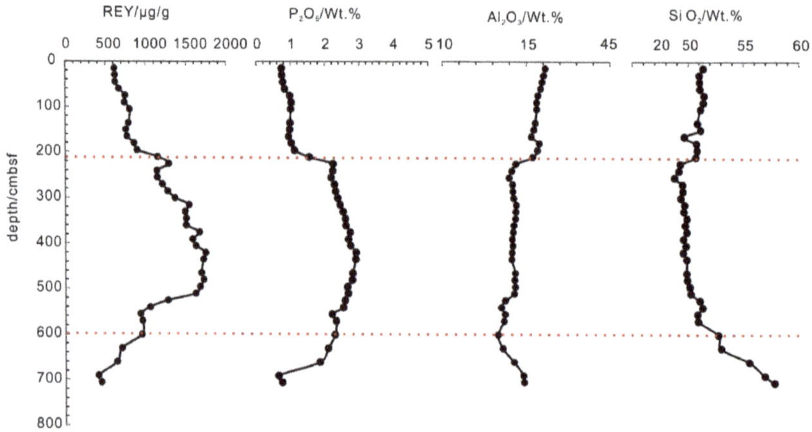

Figure 9. Variations in REY, P_2O_5, SiO_2, and Al_2O_3 contents in core P10.

5. Conclusions

This study indicates that the REY-rich sediments (>1000 μg/g) in the Central and Western Pacific have a clear spatial distribution and are mainly concentrated around seamounts. The chronological data from a typical piston core indicate that REY-rich sediments were mainly deposited during the early Oligocene to early Miocene. Therefore, since the opening of the Drake Passage in the early Oligocene, the AABW began to affect the Pacific Ocean. The upwelling of AABW induced fish proliferation around seamounts, simultaneously slowing sediment deposition and trying for enough time for REY entering and enriching in the apatite (fish teeth debris). During the early Oligocene to Miocene, when the AABW intensity was moderate, REY was enriched in the apatite (fish detritus) in the pelagic and zeolite clays. The specific spatial distribution of REY-rich sediments, described in this study, can provide guidance for future searches for REY-rich sediments in the Pacific, as well as in other oceans.

Author Contributions: Conceptualization and methodology, F.W. and G.H.; sample collection, F.W., X.D., Y.Y. and J.R.; formal analysis, F.W., X.D. and J.R.; figures, F.W. and Y.Y.; writing—original draft preparation, F.W. and G.H. All authors have read and agreed to the published version of the manuscript.

Funding: This research was funded by the National Natural Science Foundation of China (Grant No. 41702096), the Key Special Project for Introduced Talents Team of Southern Marine Science and Engineering Guangdong Laboratory (Guangzhou), grant number GML2019ZD0106, the National Natural Science Foundation of China (Grant No. 41803026,42002085), the PI Project of Southern Marine Science and Engineering Guangdong Laboratory (Guangzhou) (Project No: GML2020GD0802), and the China Geological Survey (DD20190629, DD20191009).

Institutional Review Board Statement: Not applicable.

Informed Consent Statement: Not applicable.

Data Availability Statement: The data presented in this study are available on request from the corresponding author. The data are not publicly available due to privacy.

Acknowledgments: All of the drill core samples in this study were collected by the "Hai Yang Di Zhi Liu Hao".

Conflicts of Interest: The authors declare no conflict of interest.

References

1. Chakhmouradian, A.R.; Wall, F. Rare Earth Elements: Minerals, Mines, Magnets (and More). *Elements* **2012**, *8*, 333–340. [CrossRef]
2. Hatch, G.P. Dynamics in the Global Market for Rare Earths. *Elements* **2012**, *8*, 341–346. [CrossRef]
3. Humphries, M. *Rare Earth Elements: The Global Supply Chain*; Congressional Research Service: Washington, DC, USA, 2010.
4. Liu, Y.; Chen, Z.; Yang, Z.; Sun, X.; Zhu, Z.; Zhang, Q. Mineralogical and geochemical studies of brecciated ores in the Dalucao REE deposit, Sichuan Province, southwestern China. *Ore Geol. Rev.* **2015**, *70*, 613–636. [CrossRef]
5. Liu, Y.; Chakhmouradian, A.R.; Hou, Z.; Song, W.; Kynicky, J. Development of REE mineralization in the giant Maoniuping deposit (Sichuan, China): Insights from mineralogy, fluid inclusions, and trace-element geochemistry. *Miner. Depos.* **2019**, *54*, 701–718. [CrossRef]
6. Kato, Y.; Fujinaga, K.; Nakamura, K.; Takaya, Y.; Kitamura, K.; Ohta, J.; Toda, R.; Nakashima, T.; Iwamori, H. Deep-sea mud in the Pacific Ocean as a potential resource for rare-earth elements. *Nat. Geosci.* **2011**, *4*, 535–539. [CrossRef]
7. Yasukawa, K.; Liu, H.J.; Fujinaga, K.; Machida, S.; Haraguchi, S.; Ishii, T.; Nakamura, K.; Kato, Y. Geochemistry and mineralogy of REY-rich mud in the eastern Indian Ocean. *J. Asian Earth Sci.* **2014**, *93*, 25–36. [CrossRef]
8. Gleason, J.D.; Moore, T.C., Jr.; Johnson, T.M.; Rea, D.K.; Owen, R.M.; Blum, J.D.; Pares, J.; Hovan, S.A. Age calibration of piston core EW9709-07 (equatorial central Pacific) using fish teeth Sr isotope stratigraphy. *Palaeogeogr. Palaeoclimatol. Palaeoecol.* **2004**, *212*, 355–366. [CrossRef]
9. Gleason, J.D.; Moore, T.C., Jr.; Rea, D.K.; Johnson, T.M.; Owen, R.M.; Blum, J.D.; Hovan, S.A.; Jones, C.E. Ichthyolith strontium isotope stratigraphy of a Neogene red clay sequence:calibrating eolian dust accumulation rates in the central North Pacific. *Earth Planet. Sci. Lett.* **2002**, *202*, 625–636. [CrossRef]
10. Hess, J.; Stott, L.D.; Bender, M.L.; Kennett, J.P.; Schilling, J.G. The Oligocene marine microfossil record: Age assessments using strontium isotopes. *Paleoceanography* **1989**, *4*, 655–680. [CrossRef]
11. McArthur, J.M.; Burnett, J.; Hancock, J.M. Strontium isotopes at K/T boundary. *Nature* **1992**, *355*, 28. [CrossRef]
12. McArthur, I.M.; Chen, M.; Gale, A.S.; Kenedy, W.J. Strontium isotope stratigraphy in the Late Cretaceous: Numerical calibration of the Sr isotope curve and intercontinental correlations for the Campanian. *Paleoceanogr. Paleoclimatol.* **1993**, *8*, 859–873. [CrossRef]
13. Dingle, R.V.; McArthur, J.M.; Vroon, P. Oligocene and Pliocene interglacial events in the Antarctic Peninsula dated using strontium isotope stratigraphy. *J. Geol. Soc. Lond.* **1997**, *154*, 257–264. [CrossRef]
14. Dingle, R.V.; Lavelle, M. Late Cretaceous–Cenozoic climatic variations of the northern Antarctic Peninsula: New geochemical evidence and review. *Palaeogeogr. Palaeoclimatol. Palaeoecol.* **1998**, *141*, 215–232. [CrossRef]
15. Crame, A.J.; McArthur, J.M.; Pirrie, D.; Riding, J.B. Strontium isotope correlation of the basal Maastrichtian stage in Antarctica to the European and U.S. standard biostratigraphic schemes. *J. Geol. Soc. Lond.* **1999**, *156*, 957–964. [CrossRef]
16. Arrhenius, G.; Bramlette, M.N.; Piciotto, E. Localization of radioactive and stable heavy nuclides in ocean sediments. *Nature* **1957**, *180*, 85–86. [CrossRef]
17. Wright, J.; Schrader, H.; Holser, W.T. Paleoredox variations in ancient oceans recorded by rare earth elements in fossil apatite. *Geochim. Cosmochim. Acta* **1987**, *51*, 631–644. [CrossRef]
18. Snoeckx, H.; Rea, D.K.; Jones, C.E.; Ingram, B.L. Eolian and Silica Deposition in the Central North Pacific: Results from Leg 145 sites 885/886. In *Proceedings of the Ocean Drilling Program. Scientific Results*; Rea, D.K., Basov, I.A., Scholl, D.W., Allan, J.F., Eds.; Ocean Drilling Progarm, Texam A&M University: College Station, TX, USA, 1995; Volume 145, pp. 219–230. Available online: http://www-odp.tamu.edu/publications/145_SR/VOLUME/CHAPTERS/sr145_14.pdf (accessed on 19 November 2021).
19. McArthur, J.M.; Howarth, R.J.; Bailey, T.R. Strontium isotope stratigraphy: LOWESS Version 3: Best fit to the marine Sr-isotope curve for 0–509 Ma and accompanying look-up table for deriving numerical ages. *J. Geol.* **2001**, *109*, 155–170. [CrossRef]
20. Palmer, M.R.; Elderfield, H. Variations in the Nd isotopic composition of foraminifera from Atlantic Ocean sediments. *Earth Planet. Sci. Lett.* **1985**, *73*, 299–305. [CrossRef]

21. Goldberg, E.D.; Koide, M.; Schmitt, R.A.; Smith, J. Rare earth distributions in the marine environment. *J. Geophys. Res.* **1963**, *68*, 4209–4217. [CrossRef]
22. Staudigel, H.; Doyle, P.; Zindler, A. Sr and Nd isotope systematics in fish teeth. *Earth Planet. Sci. Lett.* **1985**, *76*, 45–56. [CrossRef]
23. Piepgras, D.J.; Wasserburg, G.J. Nd isotopic variations in seawater. *Earth Planet. Sci. Lett.* **1980**, *50*, 128–138. [CrossRef]
24. Piepgras, D.J.; Wasserburg, G.J. Isotopic composition of neodymium in waters from the Drake Passage. *Science* **1982**, *217*, 207–214. [CrossRef] [PubMed]
25. Dutkiewicz, A.; Müller, R.D.; O'Callaghan, S.; Jónasson, H. Census of seafloor sediments in the world's ocean. *Geology* **2015**, *43*, 795–798. [CrossRef]
26. Kawabe, M.; Fujio, S.; Yanagimoto, D. Deep-water circulation at low latitudes in the western North Pacific. *Deep Sea Res. Part I Oceanogr. Res. Pap.* **2003**, *50*, 631–656. [CrossRef]
27. Yang, Y.H.; Wu, F.Y.; Xie, L.W.; Yang, J.H.; Zhang, Y.B. In-situ Sr isotopic measurement of natural geological samples by LA-MC-ICP-MS. *Acta Petrol. Sin.* **2009**, *25*, 3431–3441. (In Chinese)
28. Hou, K.J.; Qin, Y.; Li, Y.H.; Fan, C.F. In situ Sr-Nd Isotopic Measurement of Apatite Using Laser Ablation Multi-collector Inductively Coupled Plasma-Mass Spectrometry. *Rock Miner. Anal.* **2013**, *32*, 547–554. (In Chinese)
29. Zhang, L.; Ren, Z.Y.; Xia, X.P.; Li, J.; Zhang, Z.F. IsotopeMaker: A Matlab program for isotopic data reduction. *Int. J. Mass Spectrom.* **2015**, *392*, 118–124. (In Chinese) [CrossRef]
30. Wasserburg, G.J.; Jacobsen, S.B.; DePaolo, D.J.; McCulloch, M.T.; Wen, T. Precise determination of SmNd ratios, Sm and Nd isotopic abundances in standard solutions. *Geochim. Cosmochim. Acta* **1981**, *45*, 2311–2323. [CrossRef]
31. Yang, Y.H.; Wu, F.Y.; Yang, J.H.; Chew, D.M.; Xie, L.W.; Chu, Z.Y.; Zhang, Y.B.; Huang, C. Sr and Nd isotopic compositions of apatite reference materials used in U–Th–Pb geochronology. *Chem. Geol.* **2014**, *385*, 35–55. (In Chinese) [CrossRef]
32. Haskin, M.A.; Haskin, L.A. Rare earths in European shales: A redetermination. *Science* **1966**, *154*, 507–509. [CrossRef]
33. Mead, G.A.; Hodell, D.A. Controls on the $^{87}Sr/^{86}Sr$ composition of seawater from the middle Eocene to Oligocene: Hole 689B, Maud rise, Antarctica. *Paleoceanography* **1995**, *10*, 327–346. [CrossRef]
34. Hodell, D.A. Variations in the strontium isotopic ratio of seawater during the Miocene: Stratigraphic and geochemical implications. *Paleoceonography* **1994**, *9*, 405–426. [CrossRef]
35. Hodell, D.A.; Mueller, P.A.; Garrido, J.R. Variations in the strontium isotopic composition of seawater during the Neogene. *Geology* **1991**, *19*, 24–27. [CrossRef]
36. Dubois, N.; Mitchell, N.C. Large-scale sediment redistribution on the equatorial Pacific seafloor. *Deep Sea Res. Part I Oceanogr. Res. Pap.* **2012**, *69*, 51–61. [CrossRef]
37. Veizer, J.; Compston, W. $^{87}Sr/^{86}Sr$ composition of seawater during the Phanerozoic. *Geochim. Cosmochim. Acta* **1947**, *38*, 1461–1484. [CrossRef]
38. Vérard, C.; Flores, K.; Stampfli, G. Geodynamic reconstructions of the South America–Antarctica plate system. *J. Geodyn.* **2012**, *53*, 43–60. [CrossRef]
39. Barker, P.F.; Burrell, J. The opening of Drake Passage. *Mar. Geol.* **1977**, *25*, 15–34. [CrossRef]
40. Zachos, J.; Pagani, M.; Sloan, L.; Thomas, E.; Billups, K. Trends, Rhythms, and Aberrations in Global Climate 65 Ma to Present. *Science* **2001**, *292*, 686–693. [CrossRef] [PubMed]
41. Keller, G.; Barron, J. Paleoceaanographic implications of Miocene deep-sea hiatuses. *Geol. Soc. Am. Bull.* **1983**, *94*, 590–613. [CrossRef]
42. Barron, J.; Keller, G. Widespread Miocene deep-sea hiatuses: Coincidence with periods of global cooling. *Geology* **1982**, *10*, 577–581. [CrossRef]
43. Albarède, F.; Goldstein, S.L. World map of Nd isotopes in sea-floor ferromanganese deposits. *Geology* **1992**, *20*, 761–763. [CrossRef]
44. Jones, C.E.; Halliday, A.N.; Rea, D.K.; Owen, R.M. Neodymium isotopic variations in North Pacific modern silicate sediment and the insignificance of detrital REE contributions to seawater. *Earth Planet. Sci. Lett.* **1994**, *127*, 55–66. [CrossRef]
45. Zhao, K.D.; Jiang, S.Y.; Zheng, X.Y.; Chen, T.Y.; Ling, F.F. Nd isotope evolution of ocean waters and implications for paleo-ocean circulation. *Earth Sci. Front.* **2009**, *16*, 160–171. (In Chinese)
46. Kawabe, M.; Yanagimoto, D.; Kitagawa, S. Variations of deep western boundary currents in the Melanesian Basin in the western North Pacific. *Deep Sea Res. Part I Oceanogr. Res. Pap.* **2006**, *53*, 942–959. [CrossRef]
47. Han, J.; Wu, G.H.; Ye, Y.; Wu, D.D. Geochemical record of bottom water in ferromanganese crusts. *Miner. Depos.* **2006**, *25*, 620–628. (In Chinese)
48. De Carlo, E.H. Paleoceanographic implication of rare earth element variability in a marine Fe-Mn crust from the Hawaiian Archipelago. *Mar. Geol.* **1991**, *98*, 449–467. [CrossRef]
49. Banakar, V.K.; Pattan, J.N.; Mudholkar, A.V. Palaeoceano graphic conditions during the formation of a ferromanganese curst from the Afanasiy-Nikitin seamount, North Central Indian Ocean: Geochemical evidence. *Mar. Geol.* **1997**, *136*, 299–315. [CrossRef]
50. Rogers, A.D. The biology of seamounts. *Adv. Mar. Biol.* **1994**, *30*, 305–350.
51. Rogers, A.D. The biology of seamounts: 25 Years on. *Adv. Mar. Biol.* **2018**, *79*, 137–224.
52. Hein, J.R.; Yeh, H.W.; Gunn, S.H.; Sliter, W.V.; Benninger, L.M.; Wang, C.H. Two major Cenozoic episodes of phosphatization recorded in equatorial Pacific seamount deposits. *Paleoceanography* **2003**, *8*, 293–311. [CrossRef]
53. Ohta, J.; Yasukawa, K.; Nozaki, T.; Takaya, Y.; Mimura, K.; Fujinaga, K.; Nakamura, K.; Usui, Y.; Kimura, J.I.; Chang, Q.; et al. Fish proliferation and rare-earth deposition by topographically induced upwelling at the late Eocene cooling event. *Sci. Rep.* **2020**, *10*, 1–11. [CrossRef]

Journal of
Marine Science and Engineering

MDPI

Article

Integrated Porosity Classification and Quantification Scheme for Enhanced Carbonate Reservoir Quality: Implications from the Miocene Malaysian Carbonates

Hammad Tariq Janjuhah [1,2,*], George Kontakiotis [3], Abdul Wahid [4], Dost Muhammad Khan [4], Stergios D. Zarkogiannis [5] and Assimina Antonarakou [3]

1 Department of Geology, Shaheed Benazir Bhutto University, Sheringal 18050, KPK, Pakistan
2 Centre for Seismic Imaging, Department of Geosciences, University Technology PETRONAS, Sri Iskandar 32610, Malaysia
3 Department of Historical Geology-Paleontology, Faculty of Geology and Geoenvironment, School of Earth Sciences, National and Kapodistrian University of Athens, Panepistimiopolis, Zografou, 15784 Athens, Greece; gkontak@geol.uoa.gr (G.K.); aantonar@geol.uoa.gr (A.A.)
4 Department of Statistics, Abdul Wali Khan University Mardan, Mardan 23200, KPK, Pakistan; ab.wahid1996@gmail.com (A.W.); dostmuhammad@awkum.edu.pk (D.M.K.)
5 Department of Earth Sciences, University of Oxford, Oxford OX1 3AN, UK; stergios.zarkogiannis@earth.ox.ac.uk
* Correspondence: hammad@sbbu.edu.pk or hammadtariq013@gmail.com

Citation: Janjuhah, H.T.; Kontakiotis, G.; Wahid, A.; Khan, D.M.; Zarkogiannis, S.D.; Antonarakou, A. Integrated Porosity Classification and Quantification Scheme for Enhanced Carbonate Reservoir Quality: Implications from the Miocene Malaysian Carbonates. *J. Mar. Sci. Eng.* **2021**, *9*, 1410. https://doi.org/10.3390/jmse9121410

Academic Editor: Markes E. Johnson

Received: 19 November 2021
Accepted: 7 December 2021
Published: 10 December 2021

Abstract: The pore system in carbonates is complicated because of the associated biological and chemical activity. Secondary porosity, on the other hand, is the result of chemical reactions that occur during diagenetic processes. A thorough understanding of the carbonate pore system is essential to hydrocarbon prospecting. Porosity classification schemes are currently limited to accurately forecast the petrophysical parameters of different reservoirs with various origins and depositional environments. Although rock classification offers a way to describe lithofacies, it has no impact on the application of the poro-perm correlation. An outstanding example of pore complexity (both in terms of type and origin) may be found in the Central Luconia carbonate system (Malaysia), which has been altered by diagenetic processes. Using transmitted light microscopy, 32 high-resolution pictures were collected of each thin segment for quantitative examination. An FESEM picture and a petrographic study of thin sections were used to quantify the grains, matrix, cement, and macroporosity (pore types). Microporosity was determined by subtracting macroporosity from total porosity using a point-counting technique. Moldic porosity (macroporosity) was shown to be the predominant type of porosity in thin sections, whereas microporosity seems to account for 40 to 50% of the overall porosity. Carbonates from the Miocene have been shown to possess a substantial quantity of microporosity, making hydrocarbon estimate and production much more difficult. It might lead to a higher level of uncertainty in the estimation of hydrocarbon reserves if ignored. Existing porosity classifications cannot be used to better understand the poro-perm correlation because of the wide range of geological characteristics. However, by considering pore types and pore structures, which may be separated into macro- and microporosity, the classification can be enhanced. Microporosity identification and classification investigations have become a key problem in limestone reservoirs across the globe.

Keywords: carbonate reservoirs; porosity classifications; micro-macro-porosity; petrophysical properties; diagenetic processes; petroleum potential; grain sorting; sedimentary facies analysis; stratigraphic correlations; depositional environment

1. Introduction

Carbonate rocks account for 60% of the world's oil and gas reserves [1]. Most global oil reserves are in carbonate reservoirs [2–4]. These vast carbonate reserves are found in the Middle East, Libya, Russia, and Kazakhstan. Carbonate reservoirs are found in large conventional oil fields such as Saudi Arabia's Ghawar [5]. Despite their reputation, carbonate reservoirs are no more difficult to predict, interpret, or define than siliciclastic reservoirs [6,7]. The diverse nature of carbonate reservoirs complicates their growth, making it difficult to determine their quality, particularly at inter-well scales [8]. It is more difficult to overcome poor recovery factors. When depositional geometry and diagenesis coexist, very heterogeneous reservoirs are formed [9–11]. Carbonate succession stratigraphy, diagenesis, geochemical characteristics, and rock fabric variations influence development costs and ultimate recovery. Despite recent advances in stratigraphic and pore-scale carbonate reservoir characterization [12], numerous challenges remain in understanding and managing the uncertainties caused by heterogeneity. The use of modern petrophysics and computer modeling techniques simplifies solution development. Carbonate composition, diagenesis, microporosity prediction, microporosity effect on petrophysical properties, and depositional origins are yet unknown [13,14]. As a result, detecting carbonates is difficult since their characteristics vary within a reservoir.

These reservoirs, which have a permeability that is 1000 times greater than the surrounding rock matrix, might have a substantial impact on oil, gas, and water production [15,16]. To properly characterize carbonate deposits, quantitative physical parameters such as porosity and permeability must be considered. Because of their extreme variability, carbonate rocks are extremely difficult to identify, as shown by attempts to quantify petrophysical properties on different scales [17–19]. As a result of the enormous diversity of pore types, pore shape, and interconnectivity present in carbonates, little is known about their petrophysical properties. When working with carbonate reservoirs, poor relationships between porosity and other physical parameters such as permeability and sonic velocity are common [10].

The Central Luconia, offshore Sarawak, Malaysia research was used as a case study to address the limitations of several classifications. Malaysia's central Luconia carbonate reserves have great hydrocarbon potential. [20,21]. These carbonates contain 65 trillion cubic feet of gas and a modest oil deposit, making them commercially important [17]. Because it is a significant hydrocarbon province, the province of Luconia offshore Sarawak is a critical geological unit for understanding Malaysia's distribution of hydrocarbon resources. According to Janjuhah, Alansari, Ghosh and Bashir [21] commercial levels of an unrelated gas have been discovered in 56 carbonate platforms. These reservoirs are often associated with grain-rich facies types.

The main objectives of this research are to develop a qualitative understanding of carbonates, to highlight the challenges of current porosity classifications used in industry to identify carbonate reservoir properties, and to further emphasize the importance of categorizing the porosity classification, using a Miocene carbonate reservoir rock as an example.

2. Limitations of Existing Porosity Classifications and their Texture

Folk [22] was the first to attempt to analyze the differences in limestone texture and propose a classification scheme. Folk [22] divided limestone into three types (allochems, microcrystalline ooze, and sparry calcite cement) (Figure 1), as well as eight textural types (Figure 2). Furthermore, Folk [22] coined the concept "allochem,", which refers to chemical precipitation. In geology, the allochems' synonyms, "common chemical precipitation," do not have the same meaning as in pure chemistry. Rock complexes with a higher degree of order, which have been subjected to significant movement, may be found in geology. Based on their importance in carbonate rock, Folk divided allochems into four classes (intraclasts, oolites, fossils, and pellets). The allochems that make up the majority of limestone are reflected in the rock structure and non-skeletal grains. A clay-sized matrix is represented by the microcrystalline ooze. The sparry calcite cement, on the other hand, fills the pore spaces and leaves empty pores after the microcrystalline ooze has been washed away. Microcrystalline ooze and sparry calcite cement are fundamental features of rocks, Folk claims, based on their proportions.

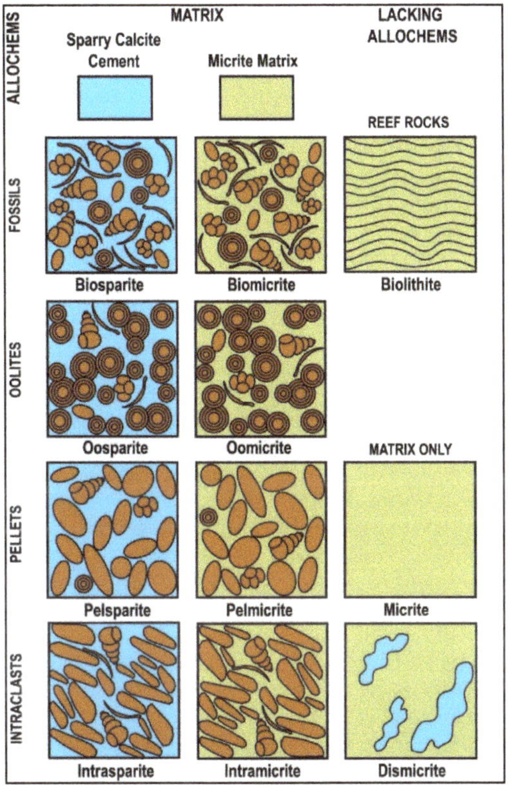

Figure 1. Representing the Folk classification of limestone, based on alluvial micrite matrix and sparry calcite cement.

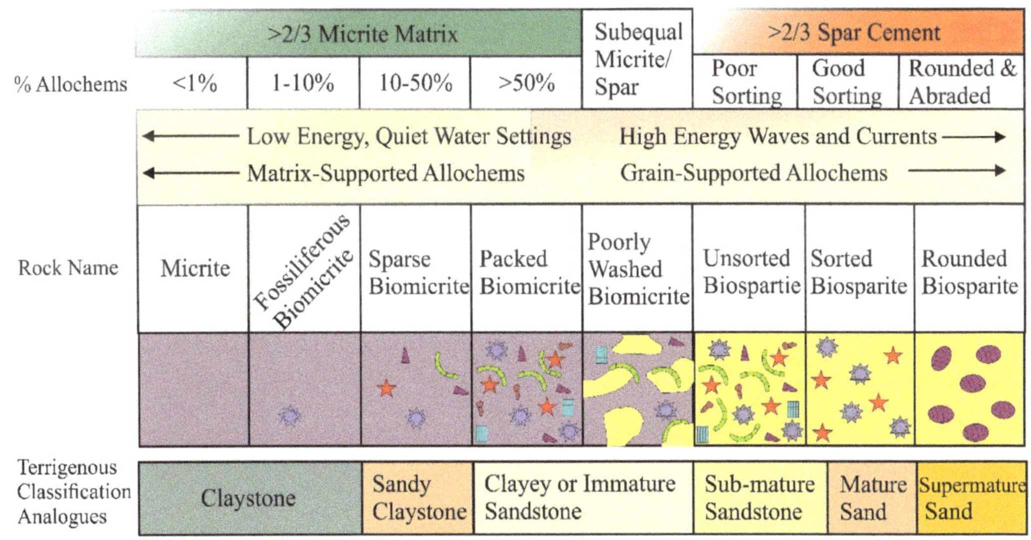

Figure 2. Folk classification of limestones based on textural parameters and implied depositional conditions.

Folk [22] also indicated that the textual structure of limestone had matured (Figure 2). Based on the ratio of grain- and mud-supported grains, this textural maturity is classified into eight groups. He employed a descriptive textural maturity classification, however, he left out certain genetic implications and repercussions. A thin section investigation is most suited to Folk [22] classification. Folk [22] suggested that micritic formation could be an exception to the norm. Lime mud may develop in a high-energy zone, if not eliminated by algae and waves. Sparite can only grow in a protected environment, such as a lagoon full of accumulating fossils. As a result of their chemical interaction, lime mud was not formed by precipitation.

However, since it is a well-known carbonate, the Dunham [23] rock classification, which was modified by Embry and Klovan [24] to enhance carbonate reef textures, is frequently used in the petroleum sector (Figures 3 and 4). This approach works well in explaining rocks using a hand lens or binocular microscope. If the limestone grains are close together and there is no mud present in the sediment, is the sediment is referred to as a grainstone. A packstone is a carbonate that is grain-supported and contains a small quantity of mud. The sediment is referred to as a wackestone if it is mud-supported yet contains more than 10% grains, while it is referred to as a mudstone if it contains less than 10% of grains. Folk [22] refers to carbonate mud-rich rocks as micrites, although Dunham [23] refers to them as mudstones or wackestone. "Grainstone" and "packstone" are terms used by both Folk [22] and Dunham [23] to describe rocks with a small matrix. Dunham's [23] classification was expanded by Embry and Klovan [24], who included coarse-grained carbonates. A wackestone with grains of a size greater than 2 mm has been renamed as floatstone, while a coarse grainstone is referred to as rudstone in the relevant scheme. However, applying these methodologies to ancient limestones is difficult because of the limits placed on the ability to judge an organism's function by diagenesis and sample size.

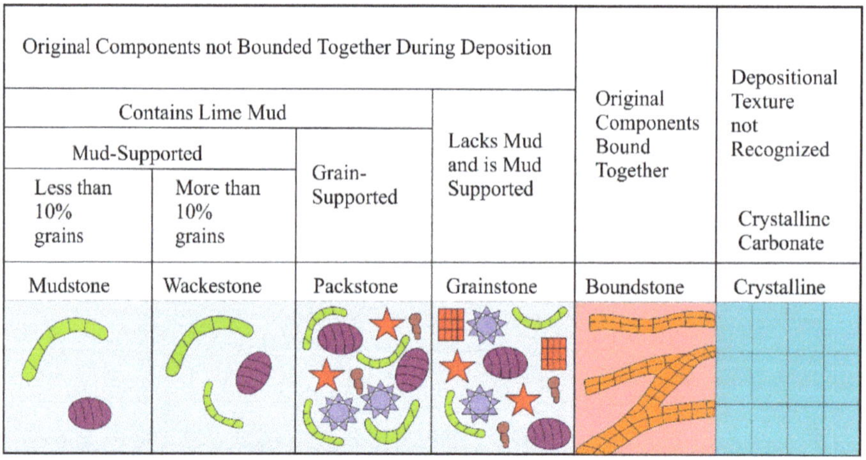

Figure 3. Representing the Dunham classification of limestone based on depositional texture.

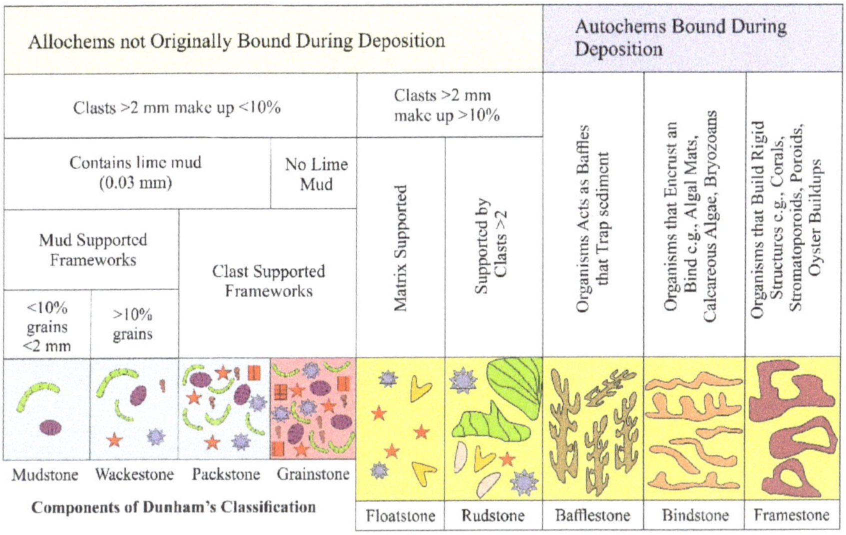

Figure 4. Representing the extended classification of limestone based on depositional texture.

Porosity Classifications

Carbonate rock has a broad range of pore sizes, and the network of connections within the rock is often complicated. There is usually a lack of correlation between porosity and other physical parameters such as permeability and velocity. Understanding how carbonate reservoirs form porosity is critical [25–28]. The structure of the pores has a significant influence on permeability and elastic properties [10,29,30]. Various rocks with the same porosity but varying permeability and acoustic velocity may be encountered [31]. The pore structure of carbonate rocks has been shown to affect the petrophysical properties in many studies [32–39]. Most studies support a dual-porosity model, since geologists and petrophysicists are unable to agree on how to define pore types. [31,40–44].

Archie [45] was the first to investigate the link between rock fabric and petrophysical parameters, highlighting the relevance of pore structure in defining pore types. Table 1 shows the matrix texture and pore type visibility used in the Archie [45] classification (Table 2).

Choquette and Pray [26] developed a descriptive classification scheme for carbonate pore types that is widely accepted and used in both the commercial and academic sectors. The majority of the system's genetic classification is made up of primary and secondary pore systems (Figure 5; Table 3). Intergranular pore spaces, which are frequent in terrigenous sand, are the major source of porosity [26,46]. Only one aspect of the carbonate pore system is shared with terrigenous sand. Primary pores occur within the skeletal chambers, and are well-protected by enormous skeletal framework textures, due to the diversity of carbonate grains and sediment texture. The preservation of primary pore systems depends on a variety of factors when a sediment is turned into limestone [26]. The majority of pores in carbonate rocks, on the other hand, are secondary pore systems in nature [26,47,48].

When carbonate sediment is deposited, it creates primary porosity [26,49]. In 1970, researchers Choquette and Pray [26] coined a term to characterize porosity's origins; its size, shape, and abundance. Cementation and dissolution are two ways of tracing the genesis. Carbonate sands and gravels lose their porosity due to the precipitation of cement and compaction [26,46,50,51]. Secondary porosity develops after the sediment has been deposited [26,28,49]. In this stage of diageneses, a variety of chemical and mechanical activities take place, including the conversion of carbonate sediment to limestone or the modification of limestone pores [26,28,52,53]. In addition to the fabric-selective secondary pores, there are also non-fabric-selective secondary pores [26].

Secondary pores that are fabric-selective are linked to the sediment or rock's initial textural components [26]. Moldic porosity is a prevalent form of pores in a fabric-selective dissolution [54]. Due to the dissolution of skeletal or nonskeletal grains, the grain mould changes into pores, resulting in this structure [26,28,46,49]. It is also typical to find inter-crystalline porosity between crystals that have developed in situ by recrystallizing or, more often, through dolomitization. Limestone's non-fabric secondary pores are unconnected to the limestone's textural components [26,28,49]. These pore systems crosscut the limestone's texture, such as fractures, vugs, and so on.

Table 1. Micrite texture Archie classification of carbonate rock.

Matrix Texture	Hand Sample Appearance	Microscopic Appearance
Type-I Compact Crystalline	Edges and surfaces that are both sharp and smooth.	The splitting of clusters of crystals in thin flanks produces "feather edges" in a matrix made up of crystals weakly interacting but with no obvious pore spaces between them.
Type-II Chalky	The small crystals are less closely interlocked and hence reflect light in diverse directions, or, because the material is composed of very fine granules, a dull, chalky, crystalline appearance is missing.	Less interlocking crystals interlock at various angles. This may cause a chalky texture, although others may become crystalline.
Type-III Granular	Sugary appearance (Sucrose). Very fine: 0.05 mm, Fine: 0.1 mm, Medium: 0.2 mm, Coarse: 0.4 mm.	Crystals interlocking at varying angles, allowing for significant porosity between crystals. This includes oolitic textures.

Table 2. Visible pore size classification of carbonate rock.

Class	Description
Class A	Pores smaller than 0.01 mm in diameter, not visible with a 10× resolution microscope.
Class B	Greater than or equal to 0.01 mm of visible porosity
Class C	More than 0.1 mm of visible porosity, however, the mess is less than the size of the cuttings
Class D	Secondary crystal development on the cutting surfaces, or "weathered-appearing" faces displaying signs of fracture or solution channels, is an indication of visible porosity.

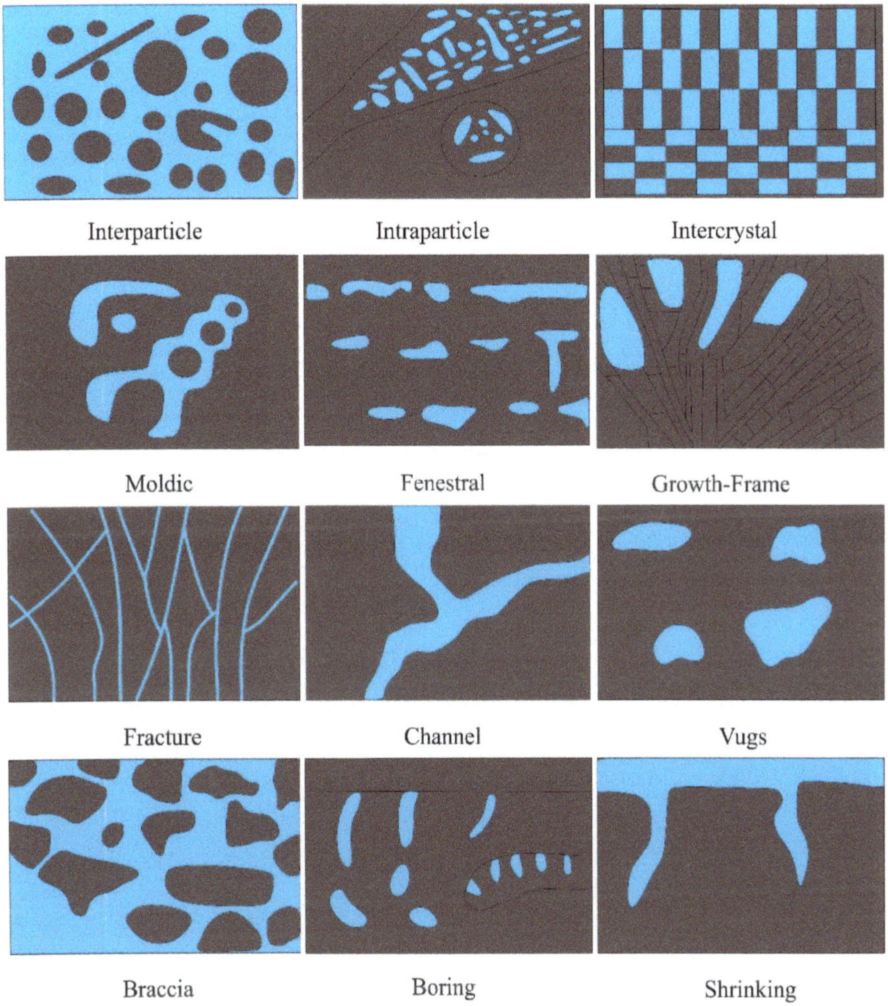

Figure 5. Choquette and Pray classification of limestone porosity.

The Lucia [38,55], classification uses both the Dunham [23] texture classification and the Choquette and Pray [26] classification to address the fundamental geological factors that reduce the uncertainty of petrophysical parameters of carbonate geological models. Lucia [38]'s classification is primarily composed of interparticle pores and vuggy pores. When applied to a geological model, the Archie [45] classification is still useful in determining petrophysical properties, but it does not provide comprehensive information regarding deposition and diagenetic processes. Dunham [23] and Choquette and Pray [26] classifications are also used often, according to Lucia [38]. Nonetheless, neither can be easily linked to quantitative reservoir properties that indicate the wellbore's local environment.

In an effort to close the gap, Lucia [38] presents a method for obtaining the basic mappable geological properties that allow for the petrophysical quantification of geological reservoir models. The Lucia [38] classification, controls permeability and saturation. Texture [23], grain size, pore type, and distribution are all factors that influence rock fabric. Grain-dominated, muddy grain-dominated, and mud-dominated are the three elements that make up the texture (Figure 6). The only two pore-size classes in the simplified Lucia [38] classification are intraparticle separated vugs and touching vugs.

Table 3. Description of porosity classification of limestone.

Interparticle	Between the grains, there is a space. The permeability and recovery of sand are generally excellent. Porosity and permeability are reduced due to the use of interparticle cement.
Intraparticle	Within particles, there is an area known as a pore space. The pore size and permeability of a fossil chamber, such as a foram's chamber, or a coral microstructure, determines its porosity and permeability. Water or oil in smaller pore networks may be irreducible.
Intercrystalline	Between the crystals of recrystallized limestone, there is porosity. The most prevalent kind of porosity in dolomites, which are often very porous and permeable reservoirs. If present in chalk mudstone, porosity may be significant due to the presence of a large amount of irreducible water, whereas permeability will be low until fractured.
Moldic	Porosity occurs when skeletal and non-skeletal grains leach. Effective porosity and permeability are determined by the texture, the degree to which grains leach, and the quantity of interparticle cement. Generally inefficient porosity and poor permeability when just isolated pore types are present, but great when inter-crystal or interparticle pore types are present.
Fracture	Porosity formed by the brittle deformation of rock. Porosity is usually low, but permeability will be higher.
Vugs	Porosity formed by an enlargement of fabric-selective pore such as moldic and intraskeletal pores.
Channel	Porosity formed by enlarged dissolution of early formed pore system.

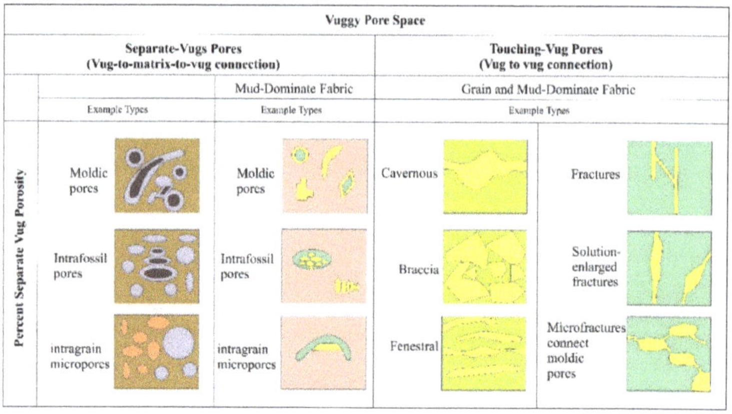

Figure 6. Geological and petrophysical classification of vuggy pore spaces based on vugs interconnection, after Lucia [38].

Lønøy [42] proposed a new classification of pores based on the nomenclature of Lucia [38] and Choquette and Pray [26] descriptive. There were 20 distinct types of pores for Lønøy [42]. It is feasible to attain good access to sedimentological and diagenetic properties of a sample using the Lønøy [42] classification. Unlike Lucia [38], Lønøy [42] emphasized that pore size, not grain size, is used to classify his samples. The Lønøy [42] classification is useful for working with a small data set, but it is not ideal for a huge dataset. To differentiate between macro and micropores, microporosity was previously included in the classification. Microporosity is present in several carbonate deposits across the world. Cantrell and Hagerty [41] and Janjuhah, Alansari and Vintaned [10] conducted extensive research on Middle East carbonates. With regard to microporosity definition, academics have differing viewpoints. Geologists, petrophysicists, and geophysicists have all reported microporosity in the literature. In this study, micropores were defined as pores with a diameter of less than 10 µm [31,56–59]. The addition of micropores to a reservoir has no effect on fluid migration and increases the risk of underestimating the reservoir's quality.

3. Application of Macro-Micro Classification from Miocene Carbonates

Material and Methods

Central Luconia's carbonate rocks from Cycle V are the focus of this research. To explain the core analysis, data was manually plotted onto a 1:40 scale core description sheet. This section focused on describing a variety of sedimentological properties, such as depositional textures, the nature of skeletal and non-skeletal grains, lithology, and diagenetic features such as leaching, stylolites, and visible porosity (Table 4). By examining the core, several lithofacies were identified.

Table 4. Flow of core description for this study.

	Grain Size	Lithology	Depositional Texture	Leaching	Stylolite Frequency	Visual Porosity Distribution
Core Description Sheet	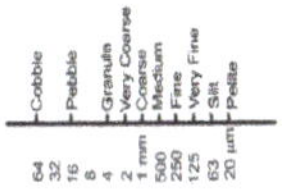	Lithology	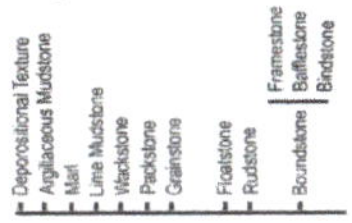	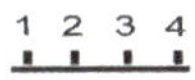	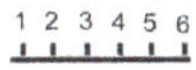	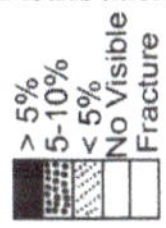
Resources	Wentworth [60]	HCL, Grain Density	Dunham [23] (Figure 4)	Visual Observation	Park, 1968	Visual Observation

The porosity, permeability, and grain density of the core were measured using 142 core plugs. Under transmitted light, 53 thin thin-sections were prepared and examined subjectively and statistically from these core plugs, and 700 counts per sample employed a point-counting technique (J. Microvision) to analyze thin sections' grains, matrix, cement, and visible porosity. Weger, Eberli, Baechle, Massaferro and Sun [35] developed the Digital Image Analysis (DIA) technique, which assesses microporosity, using four photographs to represent the whole thin section (Figure 7). However, due to the poor quality of the photographs, there is a significant amount of uncertainty. Each thin section in the present research is covered by 32 high-resolution petrographic images, which are combined into a photo panel in order to estimate the level of macroporosity, reducing the uncertainty associated with pore geometry variation (Figure 7). This means that any zone that falls below the threshold value will be colored red, indicating a zone with a high number of micropores. The entire red region is made up of very microporous sections, signifying the ideal areas. Macroporosity characterized pores with a diameter of greater than 10 µm. Total porosity was determined by injecting helium at 1800 to 2000 psi pressure into the sample using Vinci Technologies' poro-perm apparatus. Thin sections were utilized to quantify macroporosity, which was then deducted from the total porosity calculated from the core plug.

$$\text{Microporosity} = \text{Total porosity (central plug)} - \text{Macroporosity (thin section)}$$

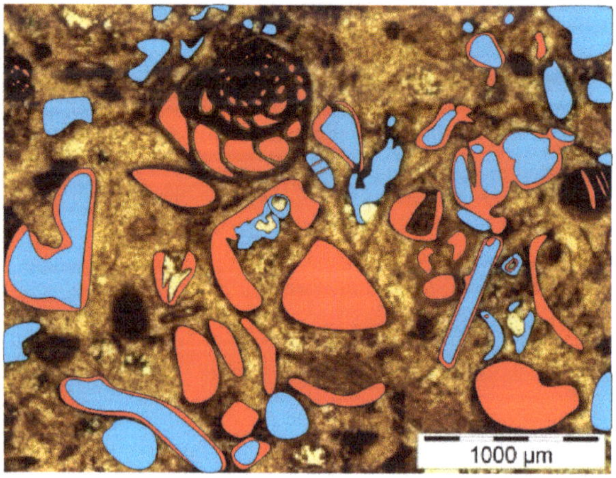

Figure 7. False color DIA image of mouldic, vuggy and intraparticle porosity developed in foraminifera and different bioclasts in Well A. Total porosity is 19%, the blue color represents macropores which are 8.5%, whereas red color represents microporosity which is 10.5%.

The well-observed pore types were found to be isolated in nature. Central Luconia carbonates were highly leached, causing partial or total disintegration of the basic rock structure and the formation of an expanded isolated molded porosity.

4. Results and Discussion

4.1. Facies Analysis

To describe the lithofacies, the standard terminology was used as provided by Janjuhah et al. [61]. They identified eight different lithofacies for Central Luconia carbonates, based on depositional texture and wireline-log response (Table 5) as follows: Coated Grain Packstone (F-1), Massive Coral Lime Grainstone (F-2), Oncolite Lime Grain-dominated Packstone (F-3), Skeletal Lime/dolo Packstone (F-4), Coral (p) Lime Mud dominated Packstone (F-5), Coral (B) Lime dominated Pack-Grainstone (F-6), Cross-Bedded Skeletal Lime Packstone (F-7) and Bioturbated Carbonate Mudstone (Chalk F-8). This work uses the facies system devised by Janjuhah, Salim, Mohammad, Ali, Ghosh, Hassan and Hakif [17].

According to facies scheme of Janjuhah, Alansari and Santha [9] and Janjuhah, Salim and Ghosh [61], five facies were identified in the studied well namely: Coated Grain Packstone (F-1), Massive Coral Lime Grainstone (F-2), Oncolite Lime Grain Dominated Packstone (F-3), Skeletal Lime Packstone (F-4) and Platy Coral Lime Mud Dominated Packstone (F-5) (Figure 8; Table 5).

Table 5. Lithofacies for Central Luconia carbonates based on depositional texture and wireline-log response (adopted from Janjuhah et al. [9]).

Lithofacies		Description
FA-1 Coated Grain Packstone		Texture: packstone (floatstone) Mineralogy: limestone Components: algae >50%, oncolite algae <40%, corals <30%, separate vugs, skeletal debris (angular—subangular), forams, echinoderms, gastropods, leaching Grain size/sorting: fine-medium gravel/moderately—poor

Table 5. *Cont.*

Lithofacies		Description
FA-2 Coral (m) Lime Pack-Grainstone	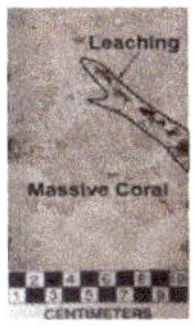	Texture: packstone–grainstone (rudstone) Mineralogy: limestone Components: corals (m) >50% (up to 8 cm in diameter), platy coral up to 20%, branching corals (15%), solitary coral <5%, algae, disconnected vugs, oncolite algae, skeletal grains (angular-subangular), gastropods, bivalves, echinoid spines Grain size/sorting: very coarse-granule/moderately poor
FA-3 Oncolite Lime Grain-dominated Packstone		Texture: packstone (rudstone) Mineralogy: limestone Components: oncolite algae >70% (diameter 2 to 6 cm), stylolite, corals >30%, separate vugs, algae, gastropods, bivalves, echinoid spines, skeletal grains (angular-subangular), leaching Grain size/sorting: medium-gravel/ moderately poor
FA-4 Skeletal Lime/dolo Packstone	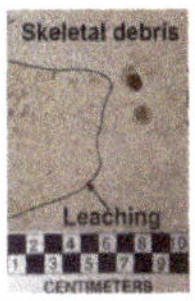	Texture: packstone (floatstone-rudstone) Mineralogy: limestone—dolomitic limestone Components: skeletal debris >60% (angular-subangular), bivalves, isolated gastropods, corals (m) <20%, Coral (p) <15%, leaching Grain size/sorting: fine-coarse grain/ moderately well sorted
FA-5 Coral (p) Lime Mud dominated Packstone		Texture: packstone (floatstone) Mineralogy: limestone Components: rich platy corals >70%, solitary coral up to 15%, algae, small fractures, disconnected small vugs, skeletal debris (angular-subangular), gastropod, forams, echinoid spines Grain size/sorting: fine-coarse/poor
FA-6 Coral (B) Lime dominated Pack-Grainstone		Texture: packstone-grainstone (floatstone) Mineralogy: limestone-dolomitic limestone Components: branching coral 50%, 20% red algae, 15% forams, 5% massive coral, 5% bivalve, 5% other skeletal debris (angular-subangular) Grain size/sorting: very coarse-granule/poor
FA-7 Cross Bedded Skeletal Lime Packstone		Texture: packstone (floatstone) Mineralogy: dolomitic limestone to limestone Structures: graded bedding Components: forams 65%, red algae 10%, coral fragments 10%, bivalve 5%, echinoderms 5%, other skeletal debris 5% Grain size/sorting: very coarse-pebble/poorly or moderately sorted

Table 5. *Cont.*

Lithofacies		Description
FA-8 Bioturbated Carbonate Mud stone (Chalk)		Texture: wackstone—packstone Mineralogy: dolomitic limestone Structures: bioturbation Components: burrowing & bioturbation 60%, forams 10%, coral debris 10%, red algae 5% Grain size/sorting: very fine-coarse/ moderately well sorted

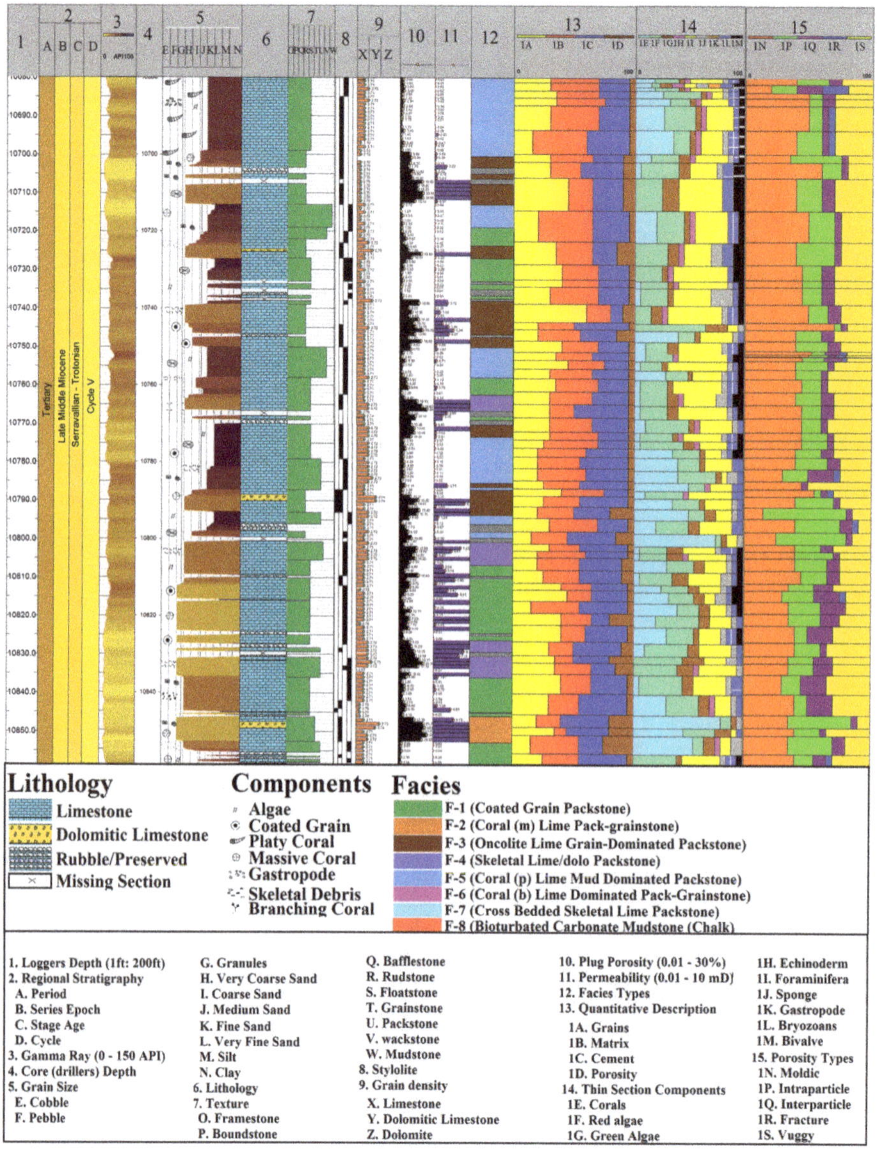

Figure 8. Sedimentological log is showing the qualitative and quantitative description of Well A in Central Luconia, offshore Sarawak, Malaysia.

4.2. Pore Types

The pore system in Central Luconia is dominated by isolated vugs [38,62]. The main rock fabric is thought to have been leached by skeleton allochems, resulting in partial or full dissolution. Carbonate mineralogies such as aragonite and high-magnesium calcite (corals and green algae, some echinoderms, bivalve, and sponges) were very unstable and dissolved quickly after burial, resulting in the selective breakdown of these organisms' skeletons [63]. Moldic pores were predominant in the thin section, according to first impressions (Figures 8 and 9). The porosity of mould was increased; however, it was not well-connected (Figure 9). Various pores can be observed, including moldic, intraparticle, interparticle, fractured, and vuggy porosity, according to the Choquette and Pray [26] classification (Figure 9). The observable pores are almost all secondary in origin.

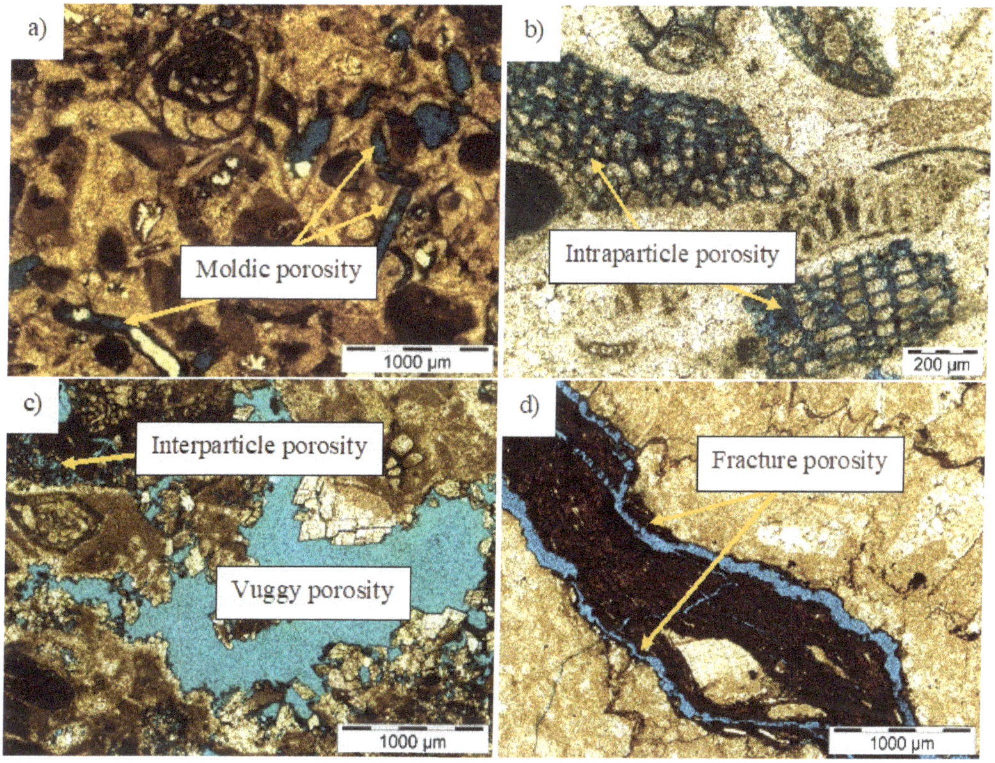

Figure 9. Photograph representing the different types of porosity (**a**) Intraparticle and mouldic porosity, (**b**) Fracture porosity, (**c**) Vuggy porosity and (**d**) Interparticle porosity and partial filling of dolomite, dolomite cement filling pore spaces and the rate of dolomitization increases with the time in Central Luconia, offshore Sarawak, Malaysia.

Facies 4 (F-4) dominates the five facies, accounting for 45% of the cored intervals, followed by 25% for Facies 1 (F-1), 15% for Facies 5, 10% for Facies 3, and 5% for Facies 2. (Figure 8). The thin sections are dominated by grain, which occupies 35% of the total area, followed by 30% matrix, 30% cement, and 5%porosity, as can be seen in Figure 8. Eight major grains are observed in Well A Central Luconia, offshore Sarawak, Malaysia (Figure 9). Coral, sponges, echinoderms, bivalve and bryozoans, and foraminifera make up most of the grains (Figure 8). Figure 3 shows that mouldic porosity is the most common, accounting for 50% of the total interval, followed by vuggy porosity (20%), intraparticle porosity (15%), interparticle porosity (10%), and fracture porosity (5%), respectively (Figure 8).

4.3. Quantification of Macro-Microporosity

To determine the presence of microporosity, the macroporosity of carbonate rocks was measured. When the diameter of moldic, intraparticle, interparticle, and vuggy pores exceeds 10 μm, they are identified as having macroporosity. The abundance of macroporosity was determined using two methods. Point counting and a digital image analysis with a threshold setting were used to collect data on macroporosity. DIA was utilized to produce a false-color image, based on the photomicrograph of each thin section to examine macroporosity distribution (Figure 7). The porosity of core plug measurements varies between 3% and 25% (Figure 8). The total porosity of a core-plug sample was subtracted from the macroporosity observed in the thin section analysis to calculate the microporosity values (point counting and DIA). In Central Luconia, most of the microporosity is located in limestone, and these pores are found between the cement of calcite microcrystals. Understanding the distribution and measurement of microporosity is critical when dealing with carbonates. Porosity in the core plug does not adequately account for fluid flow because the pore bodies have different pore throats. These pore throats have a detrimental influence on fluid flow [31,35,64,65]. According to this research, carbonate reservoir models overestimate matrix porosity, which has a major impact on reservoir evaluation. Moldic and vuggy porosity are frequently used in carbonate reservoir models, but matrix porosity is underestimated, which has a substantial impact on reservoir assessment. Depending on their size, these pores are classified as macropores or micropores. We studied the occurrence and distribution of microporosity in Central Luconia, since pores are divided into macropores and micropores (Figure 7). Cantrell and Hagerty [41], Anselmetti et al. [66], Ruzyla and Jezek [67], Yanguas and Dravis [68] and Milliken and Curtis [69], all agree that the quantitative observation of pore types on blue-impregnated thin sections is an effective method for improving the ability to detect the microporosity zone.

Following Lambert, Durlet, Loreau and Marnier [2], Kaczmarek, Fullmer and Hasiuk [59], Lucia [62], Pittman [64], Arribas et al. [70], Cox et al. [71], Flugel [72], Kaldi [73], Volery et al. [74], Morad et al. [75] and our observations, carbonate allochems are found to be frequently less porous than carbonate matrix material. According to Cantrell and Hagerty [41], the quantity of mud in limestone impacts the degree of microporosity and overall porosity. As mud content increases, the proportion of microporosity increases, resulting in a reduction in overall porosity. We observed that microcrystals are more abundant in foraminifera, coral, red algae, and sponge grains than in echinoderms, bivalves, and bryozoan grains, in our thorough investigation (Figure 10). Carbonate allochems are found to be less permeable than carbonate matrix material in the initial observation. Microcrystal concentrations are greater in foraminifera and red algae (Figure 10). Lucia and Loucks [43] also explicitly noted that forams, coral, and red algae contain more microcrystalline than echinoderms and bivalves. Dissolution pits and micro-fractures can include micron and submicron-sized pores, although they are less common. Micro- and submicron-scale pores in microporous limestone contribute less to overall porosity than inter-crystalline pores found in matrix and carbonate allochem [41].

Components	Microporosity in Component	FESEM Image of Component
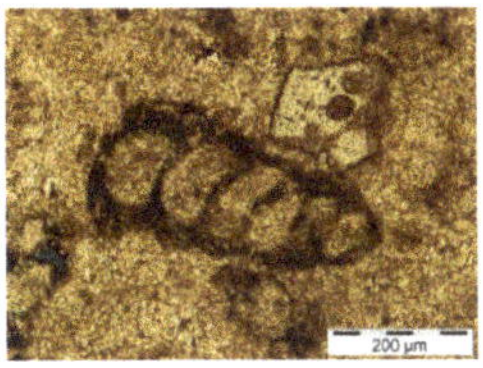	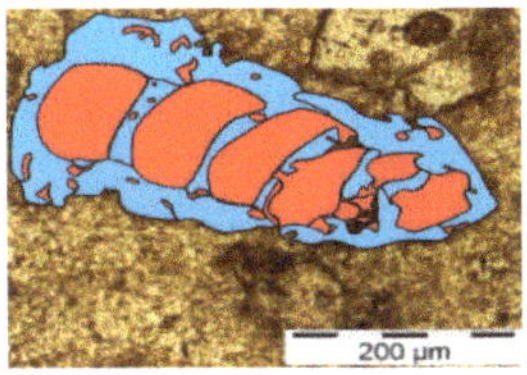	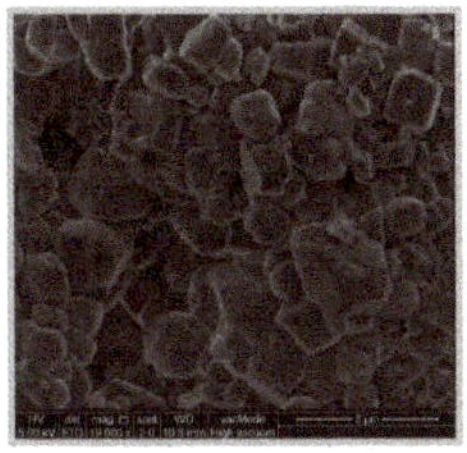
Foraminifera	Microporosity Percentage: 53%	

Figure 10. A systematic evaluation of microporosity in carbonate allochems (e.g., Foraminifera), Central Luconia, offshore, Sarawak, Malaysia.

4.4. Effect of Microporosity on Petrophysical Properties

Microporosity measurements show that the Miocene carbonates have an appropriate amount of microporosity, which may account for as much as 40–50% of the total porosity (Figure 11). Deposition and diagenesis are linked to the presence of this microporosity. Miocene carbonate build-ups in the studied Well-A underwent various episodes of diagenetic alterations, including micritization, cementation, compaction (physical and chemical), dissolution, and dolomitization, according to [9]. These diagenetic processes have specific alteration conditions that correspond to the changes that occur during the time. In their study, they claimed that micritization and calcite cementation impact reservoir quality and, in particular, carbonate rock permeability in the majority of cases. The higher the amount of microporosity, the more the area is micritized and cemented. Microporosity, on the other hand, reduces as dissolution increases [14,75]. Because transmitted light microscopy has a low resolution, this phenomenon was further confirmed using FESEM (Field Emission Scanning Electron Microscopy) images, which clearly revealed the micropores (Figure 12). In Well A, the degree of microporosity varies between 20 and 60%. The existence of macroporosity in different facies is related to the dominance of microporosity in Central Luconia, because each sample represents a distinct pore type, and these pore types are altered by various diagenetic processes.

The depositional texture is important in the development of microporosity. The microporosity of a rock sample is mostly controlled by the depositional textures as a whole. The proportion of microporous components increases as the depositional texture becomes muddier, increasing the abundance of microporosity in Central Luconia, offshore Sarawak, Malaysia. The floatstone and packstone depositional textures are dominated by grains with significant moldic porosity (large macropores), with small micropores present in the matrix (Figures 9–12). The positive contribution towards the quantification of microporosity is mostly limited to microporous grains, however, cement and matrix with micropores may also be present, but their contribution towards the quantification of microporosity generally are not considered in Central Luconia, as most of the core and thin sections revealed that the carbonate in Central Luconia is grain- rather than mud-dominated. The depositional texture of mudstone and wackestone is quite rare. The microporosity of these rocks (floatstone and packstone) is relatively high, ranging between 20 and 60% of total porosity (Figures 10–12). Macroporosity is less abundant in mud-dominated carbonate rocks because intraparticle and interparticle porosity is widespread and they are filled by the microporous matrix, resulting in reduced macroporosity but increased microporosity (Figures 11 and 12). The microporous matrix is abundant in mud-supported grains (Facies-1 and Facies-5) since macroporosity is often quite low. Microporous grains provide from 70 to 80% of the total porosity of these mud-supported grains (Facies-1, Facies-5). As pore spaces are filled with microporous grains and the microporous matrix, the percent of microporosity rises in total porosity.

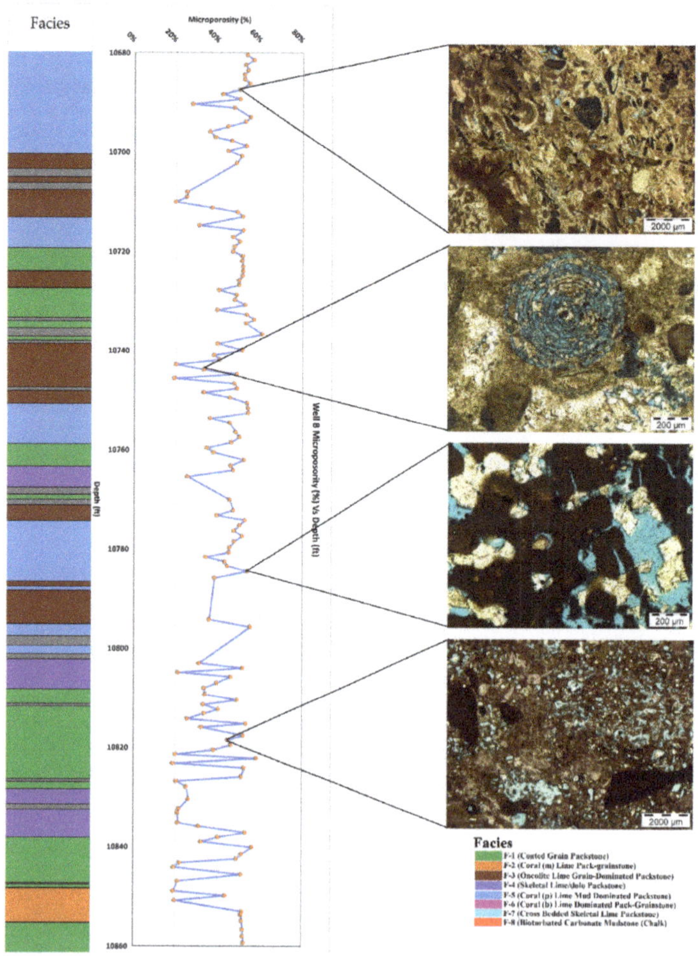

Figure 11. Macroporosity observation under transmitted light microscopy and distribution of microporosity at different depths of reservoir interval of Well A, Central Luconia, offshore, Sarawak, Malaysia.

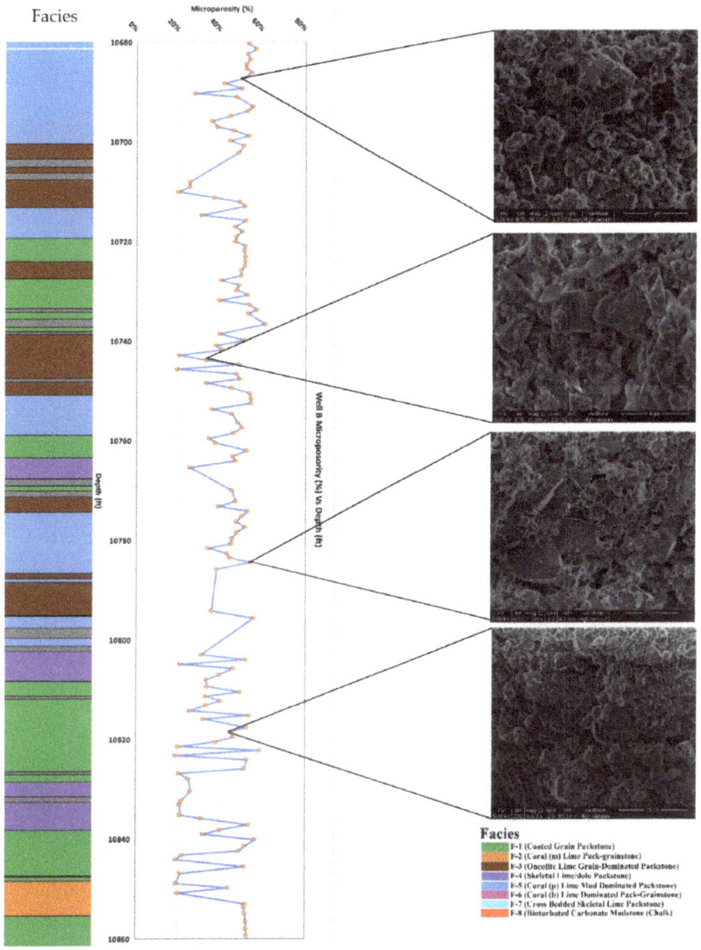

Figure 12. The observation of different microtextures with microporosity occurrence using FESEM at different depth of reservoir interval of Well A, Central Luconia, offshore, Sarawak, Malaysia.

In most cases, the relationship between porosity and permeability in carbonates is unclear. Understanding permeability behavior is critical for evaluating recoverable hydrocarbons in a reservoir because it controls fluid-flow properties [76]. Permeability is usually associated with porosity, however, when dealing with carbonate, the uncertainty is quite high due to the complicated pore structure [21,77]. The problems of porosity and permeability in hydrocarbon prediction in carbonate reservoirs have driven researchers to improve porosity and permeability correlation. Yu et al. [78], Okabe and Blunt [79], Chehrazi and Rezaee [80], Shah et al. [81] and Silin and Patzek [82] have proposed calculating permeability from porosity by classifying distinct pore types into macropores and micropores instead of focusing on total porosity. The difficulty of discerning the influence of pore types on permeability became an open challenge once the pore types were separated into macropores and micropores. The influence of microporosity on permeability was determined using data from Well A. The porosity of Well-A varies from 1% to 25% (Figure 13). By evaluating the influence of microporosity and deducing this from the total porosity indicated by the fact that the microporosity has a substantial impact on the

Miocene Luconian carbonates (Figure 13a). When compared to a cross-plot of total porosity versus permeability, the link between macroporosity and permeability has a better fit, with higher R2 (coefficient of correlation) values. The R^2 increased from 0.58 (total porosity vs. permeability) to 0.84 (macroporosity vs. permeability) (Figure 13). Baechle, Colpaert, Eberli and Weger [31], Lucia [38], Lønøy [42] and Archie [83] concluded that the scattering points in a cross plot of porosity and permeability account for most of the carbonate pore type classifications. These scattering points are the result of qualitative descriptions of pore size, pore types, and patchiness.

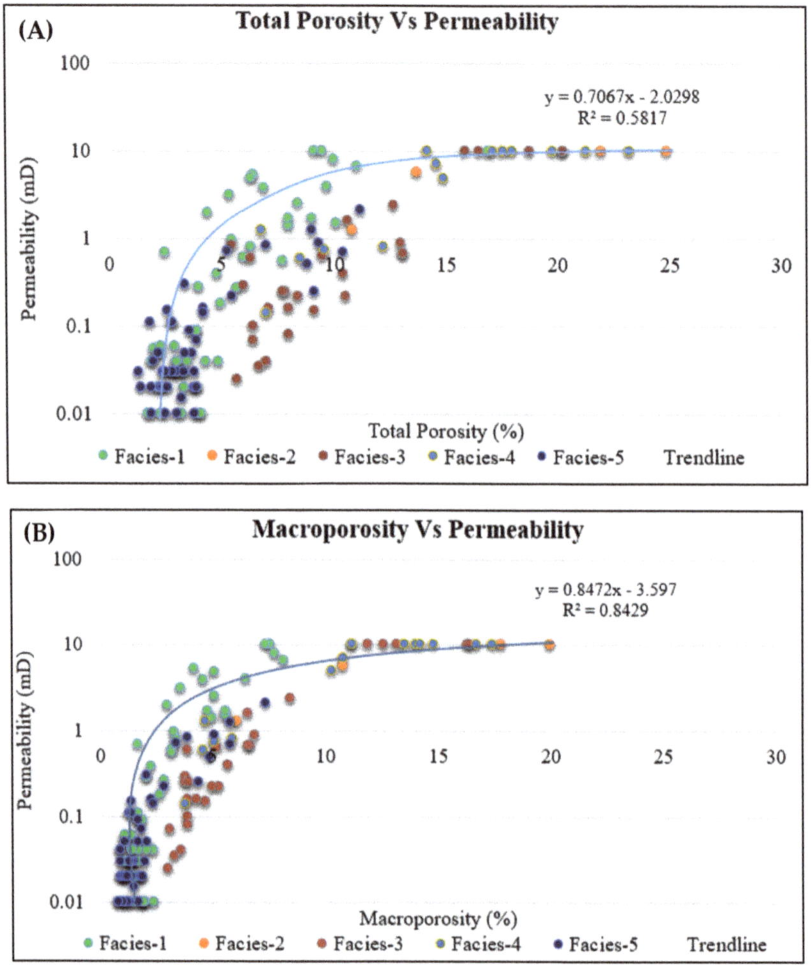

Figure 13. Total porosity, macroporosity and permeability cross plot for (**A**) total porosity versus permeability labelled with the different corresponding facies. The average coefficient of determination is R^2 = 0.58% (**B**) whereas the macroporosity versus permeability cross plot represents a better coefficient of determination which is R^2 = 0.84% in Well A Central Luconia, offshore, Sarawak, Malaysia.

It has also been suggested that the Miocene carbonates underwent a long period of diagenetic alterations [61]. As demonstrated in Figure 11, the reservoir's long diagenetic history might lead to repeated facies changes. The framework of porosity categorization may be improved by taking into consideration the pore types governed by diagenesis. The texture and lithotype of the rock might reveal the reservoir porosity development history. Microporosity has a clear influence on permeability, as seen in Figure 13, and hence cannot be neglected.

5. Conclusions

The carbonates of Central Luconia are mostly limestone, with some dolomitic limestone. The porosity–permeability measurement shows a good porosity but low permeability value at a given total porosity, which is interpreted by the presence of high isolated mouldic pores observed under the micropores. As reported, microporosity is important when analyzing carbonate deposits. Microporosity directly impacts reservoir quality. In Central Luconia, Offshore Sarawak, Malaysia, taking macroporosity into account may improve reservoir quality predictions. Although several case studies on limestone microtextures have been conducted, principally by Lambert, Durlet, Loreau and Marnier [2], Clerke et al. [84], Moshier [85], Domingo et al. [86] and Periere et al. [87], particle analysis and limestone petrophysical properties are required to better understand the carbonate pore system. Choquette and Pray [26] proposed a novel approach of classifying porosity that integrates pore structure and microporosity, which is essential.

Author Contributions: Conceptualization, H.T.J.; methodology, H.T.J.; software, H.T.J.; validation, H.T.J., G.K.; formal analysis, H.T.J.; investigation, H.T.J., G.K.; resources, PETRONAS, Shell, UTP; data curation, H.T.J.; writing—original draft preparation, H.T.J.; writing—review and editing, H.T.J., G.K, S.D.Z., D.M.K., A.A., A.W.; visualization, H.T.J., S.D.Z.; supervision, H.T.J.; project administration, H.T.J.; funding acquisition, G.K. All authors have read and agreed to the published version of the manuscript.

Funding: This research was funded by Yayasan University Technology PETRONAS, grant number YUTP 0153AA-A14.

Institutional Review Board Statement: Not Applicable.

Informed Consent Statement: Not Applicable.

Acknowledgments: The corresponding author would like to thank Petroliam Nasional Berhad (PETRONAS) and Sarawak Shell Berhad for kindly providing the data for this study. Deva Prasad Ghosh is acknowledged for his personal and financial support during the scientific work in the framework of the grant YUTP 0153AA-A14. HTJ would also like to thank Universiti Teknologi PETRONAS for the access to the research facilities needed for this project during his PhD study.

Conflicts of Interest: The authors declare no conflict of interest.

References

1. Janjuhah, H.T.; Alansari, A. Offshore carbonate facies characterization and reservoir quality of miocene rocks in the southern margin of south china sea. *Acta Geol. Sin.* **2020**, *94*, 1547–1561. [CrossRef]
2. Lambert, L.; Durlet, C.; Loreau, J.-P.; Marnier, G. Burial dissolution of micrite in Middle East carbonate reservoirs (Jurassic–Cretaceous): Keys for recognition and timing. *Mar. Pet. Geol.* **2006**, *23*, 79–92. [CrossRef]
3. Pal, S.; Mushtaq, M.; Banat, F.; Al Sumaiti, A.M. Review of surfactant-assisted chemical enhanced oil recovery for carbonate reservoirs: Challenges and future perspectives. *Pet. Sci.* **2018**, *15*, 77–102. [CrossRef]
4. Akbar, M.; Vissapragada, B.; Alghamdi, A.H.; Allen, D.; Herron, M.; Carnegie, A.; Dutta, D.; Olesen, J.-R.; Chourasiya, R.; Logan, D. A snapshot of carbonate reservoir evaluation. *Oilfield Rev.* **2000**, *12*, 20–21.
5. Han, M.; AlSofi, A.; Fuseni, A.; Zhou, X.; Hassan, S. Development of Chemical EOR Formulations for a High Temperature and High Salinity Carbonate Reservoir. In Proceedings of the IPTC 2013: International Petroleum Technology Conference, Beijing, China, 26–28 March 2013; European Association of Geoscientists & Engineers: Houten, The Netherlands, 2013; p. cp-350-00486.
6. Burchette, T.P. Carbonate rocks and petroleum reservoirs: A geological perspective from the industry. *Geol. Soc. Lond. Spec. Publ.* **2012**, *370*, 17–37. [CrossRef]
7. Perrin, C.; Marquez, X.; Flores, J.; Berthereau, G. Porosity depth saturation (PDS) model: Quantification of porosity preservation with burial in carbonate oil reservoirs, and application to infer oil charging time. *Mar. Pet. Geol.* **2020**, *120*, 104515. [CrossRef]
8. Yang, L.; Shenghe, W.; Jiagen, H.; Jianmin, L. Progress and prospects of reservoir development geology. *Pet. Explor. Dev.* **2017**, *44*, 603–614.

9. Janjuhah, H.T.; Alansari, A.; Santha, P.R. Interrelationship between facies association, diagenetic alteration and reservoir properties evolution in the Middle Miocene carbonate build up, Central Luconia, Offshore Sarawak, Malaysia. *Arab. J. Sci. Eng.* **2019**, *44*, 341–356. [CrossRef]
10. Janjuhah, H.T.; Alansari, A.; Vintaned, J.A.G. Quantification of microporosity and its effect on permeability and acoustic velocity in Miocene carbonates, Central Luconia, offshore Sarawak, Malaysia. *J. Pet. Sci. Eng.* **2019**, *175*, 108–119. [CrossRef]
11. Kontakiotis, G.; Karakitsios, V.; Cornée, J.-J.; Moissette, P.; Zarkogiannis, S.D.; Pasadakis, N.; Koskeridou, E.; Manoutsoglou, E.; Drinia, H.; Antonarakou, A. Preliminary results based on geochemical sedimentary constraints on the hydrocarbon potential and depositional environment of a Messinian sub-salt mixed siliciclastic-carbonate succession onshore Crete (Plouti section, eastern Mediterranean). *Mediterr. Geosci. Rev.* **2020**, *2*, 247–265. [CrossRef]
12. Janjuhah, H.T.; Sanjuan, J.; Alquadah, M.; Salah, M.K. Biostratigraphy, depositional and diagenetic processes in carbonate rocks form southern Lebanon: Impact on porosity and permeability. *Acta Geol. Sin.* **2021**, *5*, 1668–1683. [CrossRef]
13. Regnet, J.-B.; Fortin, J.; Nicolas, A.; Pellerin, M.; Guéguen, Y. Elastic properties of continental carbonates: From controlling factors to an applicable model for acoustic-velocity predictions. *Geophysics* **2019**, *84*, MR45–MR59. [CrossRef]
14. Janjuhah, H.T.; Salim, A.M.A.; Alansari, A.; Ghosh, D.P. Presence of microporosity in Miocene carbonate platform, Central Luconia, offshore Sarawak, Malaysia. *Arab. J. Geosci.* **2018**, *11*, 204. [CrossRef]
15. Peng, S. Gas relative permeability and its evolution during water imbibition in unconventional reservoir rocks: Direct laboratory measurement and a conceptual model. *SPE Reserv. Eval. Eng.* **2019**, *22*, 1346–1359. [CrossRef]
16. Bashir, Y.; Babasafari, A.A.; Biswas, A.; Hamidi, R.; Moussavi Alashloo, S.Y.; Tariq Janjuah, H.; Prasad Ghosh, D.; Weng Sum, C. Cohesive Approach for High-Resolution Seismic Using Inversion & Imaging in Malaysian Carbonate Field. In Proceedings of the International Petroleum Technology Conference, Beijing, China, 26–28 March 2019.
17. Janjuhah, H.T.; Salim, A.; Mohammad, A.; Ali, M.Y.; Ghosh, D.P.; Hassan, A.; Hakif, M. Development of Carbonate Buildups and Reservoir Architecture of Miocene Carbonate Platforms, Central Luconia, Offshore Sarawak, Malaysia. In Proceedings of the SPE/IATMI Asia Pacific Oil & Gas Conference and Exhibition, Jakarta, Indonesia, 17–19 October 2017.
18. Norbisrath, J.H.; Weger, R.J.; Eberli, G.P. Complex resistivity spectra and pore geometry for predictions of reservoir properties in carbonate rocks. *J. Pet. Sci. Eng.* **2017**, *151*, 455–467. [CrossRef]
19. Kontakiotis, G.; Moforis, L.; Karakitsios, V.; Antonarakou, A. Sedimentary facies analysis, reservoir characteristics and paleogeography significance of the early jurassic to eocene carbonates in epirus (Ionian Zone, Western Greece). *J. Mar. Sci. Eng.* **2020**, *8*, 706. [CrossRef]
20. Chacko, S.; Fainstein, R.; Li, C. Introduction to this special section: Southeast Asia. *Lead. Edge* **2020**, *39*, 541–542. [CrossRef]
21. Janjuhah, H.T.; Alansari, A.; Ghosh, D.P.; Bashir, Y. New approach towards the classification of microporosity in Miocene carbonate rocks, Central Luconia, offshore Sarawak, Malaysia. *J. Nat. Gas Geosci.* **2018**, *3*, 119–133. [CrossRef]
22. Folk, R.L. Practical petrographic classification of limestones. *AAPG Bull.* **1959**, *43*, 1–38.
23. Dunham, R.J. Classification of carbonate rocks according to depositional textures. *Am. Assoc. Pet. Geol.* **1962**, *1*, 108–121.
24. Embry, A.F.; Klovan, J.E. A late Devonian reef tract on northeastern Banks Island, NWT. *Bull. Can. Pet. Geol.* **1971**, *19*, 730–781.
25. Wardlaw, N. Pore geometry of carbonate rocks as revealed by pore casts and capillary pressure. *AAPG Bull.* **1976**, *60*, 245–257.
26. Choquette, P.W.; Pray, L.C. Geologic nomenclature and classification of porosity in sedimentary carbonates. *AAPG Bull.* **1970**, *54*, 207–250.
27. Moore, C.H. *Carbonate Diagenesis and Porosity*; Elsevier: Amsterdam, The Netherlands, 1989; Volume 46.
28. Maliva, R.G. Carbonate Facies Models and Diagenesis. In *Aquifer Characterization Techniques*; Springer: Berlin/Heidelberg, Germany, 2016; pp. 91–110.
29. Fabricius, I.L.; Røgen, B.; Gommesen, L. How depositional texture and diagenesis control petrophysical and elastic properties of samples from five North Sea chalk fields. *Pet. Geosci.* **2007**, *13*, 81–95. [CrossRef]
30. Fabricius, I.L.; Bächle, G.T.; Eberli, G.P. Elastic moduli of dry and water-saturated carbonates—Effect of depositional texture, porosity, and permeability. *Geophysics* **2010**, *75*, N65–N78. [CrossRef]
31. Baechle, G.T.; Colpaert, A.; Eberli, G.P.; Weger, R.J. Effects of microporosity on sonic velocity in carbonate rocks. *Lead. Edge* **2008**, *27*, 1012–1018. [CrossRef]
32. Anselmetti, F.S.; Eberli, G.P. Controls on Sonic Velocity in Carbonates. In *Experimental Techniques in Mineral and Rock Physics*; Springer: Berlin/Heidelberg, Germany, 1993; pp. 287–323.
33. Melim, L.A.; Anselmetti, F.S.; Eberli, G.P. The importance of pore type on permeability of Neogene carbonates, Great Bahama Bank. In *Subsurface Geology of a Prograding Carbonate Platform Margin, Great Bahama Bank: Results of the Bahamas Drilling Project*; SEPM: Tulsa, OK, USA, 2001; Volume 70, pp. 217–238.
34. Anselmetti, F.S.; Eberli, G.P. The velocity-deviation log: A tool to predict pore type and permeability trends in carbonate drill holes from sonic and porosity or density logs. *AAPG Bull.* **1999**, *83*, 450–466.
35. Weger, R.J.; Eberli, G.P.; Baechle, G.T.; Massaferro, J.L.; Sun, Y.-F. Quantification of pore structure and its effect on sonic velocity and permeability in carbonates. *AAPG Bull.* **2009**, *93*, 1297–1317. [CrossRef]
36. Kumar, M.; Han, D.-h. Pore Shape Effect on Elastic Properties of Carbonate Rocks. In *Proceedings of the 2005 SEG Annual Meeting*; Society of Exploration Geophysicists: Huston, TX, USA, 2005.
37. Fitch, P.J.; Lovell, M.A.; Davies, S.J.; Pritchard, T.; Harvey, P.K. An integrated and quantitative approach to petrophysical heterogeneity. *Mar. Pet. Geol.* **2015**, *63*, 82–96. [CrossRef]

38. Lucia, F.J. Rock-fabric/petrophysical classification of carbonate pore space for reservoir characterization. *Am. Assoc. Pet. Geol.* **1995**, *79*, 1275–1300.
39. Anselmetti, F.S.; Eberli, G.P. *Sonic Velocity in Carbonate—A Product of Original Composition and Postdepositional Porosity Evolution: Ann: AAPG-SEPMEMD-DPA-CSPG*; Conv. Abstracts; American Association of Petroleum Geologists: Tulsa, OK, USA, 1992.
40. Pittman, E.D. Relationship of porosity and permeability to various parameters derived from mercury injection-capillary pressure curves for sandstone (1). *AAPG Bull.* **1992**, *76*, 191–198.
41. Cantrell, D.L.; Hagerty, R.M. Microporosity in Arab formation carbonates, Saudi Arabia. *GeoArabia* **1999**, *4*, 129–154. [CrossRef]
42. Lønøy, A. Making sense of carbonate pore systems. *AAPG Bull.* **2006**, *90*, 1381–1405. [CrossRef]
43. Lucia, F.; Loucks, R. Microporosity in carbonate mud: Early development and petrophysics. *Gulf Coast Assoc. Geol. Socities* **2013**, *2*, 1275–1300.
44. Jobe, T.; Sarg, J. Microporosity Characterization of Mud-Rich Carbonate Rocks. In *Pore Scale Phenomena: Frontiers in Energy and Environment*; World Scientific: Singapore, 2015; pp. 67–89.
45. Archie, G.E. Classification of carbonate reservoir rocks and petrophysical considerations. *Aapg Bull.* **1952**, *36*, 278–298.
46. Tucker, M.E.; Wright, V.P. *Carbonate Sedimentology*; John Wiley & Sons: New York, NY, USA, 2009.
47. Wilson, J.C.; McBride, E.F. Compaction and porosity evolution of Pliocene sandstones, Ventura Basin, California. *AAPG Bull.* **1988**, *72*, 664–681.
48. Wilson, M.E. Global and regional influences on equatorial shallow-marine carbonates during the Cenozoic. *Palaeogeogr. Palaeoclimatol. Palaeoecol.* **2008**, *265*, 262–274. [CrossRef]
49. Scholle, P.A.; Ulmer-Scholle, D.S. *A Color Guide to the Petrography of Carbonate Rocks: Grains, Textures, Porosity, Diagenesis, AAPG Memoir 77*; AAPG: Tulsa, OK, USA, 2003; Volume 77.
50. Al-Dabbas, M.; Al-Jassim, J.; Al-Jumaily, S. Depositional environments and porosity distribution in regressive limestone reservoirs of the Mishrif Formation, Southern Iraq. *Arab. J. Geosci.* **2010**, *3*, 67–78. [CrossRef]
51. Purser, B.; Brown, A.; Aissaoui, D. Nature, Origin and Evolution of Porosity in Dolomites. In *Dolomites: A Volume in Honour of Dolomieu*; Purser, B., Tucker, M., Zenger, D., Eds.; Special Publication; International Association of Sedimentologists: Algiers, Algeria, 1994; pp. 283–308.
52. Xi, K.; Cao, Y.; Jahren, J.; Zhu, R.; Bjørlykke, K.; Haile, B.G.; Zheng, L.; Hellevang, H. Diagenesis and reservoir quality of the Lower Cretaceous Quantou Formation tight sandstones in the southern Songliao Basin, China. *Sediment. Geol.* **2015**, *330*, 90–107. [CrossRef]
53. Madden, R.H.; Wilson, M.E. Diagenesis of a SE Asian Cenozoic carbonate platform margin and its adjacent basinal deposits. *Sediment. Geol.* **2013**, *286*, 20–38. [CrossRef]
54. Wilson, J.L. *Carbonate Facies in Geologic History*; Springer Science & Business Media: Berlin/Heidelberg, Germany, 2012.
55. Lucia, F. Petrophysical parameters estimated from visual descriptions of carbonate rocks: A field classification of carbonate pore space. *J. Pet. Technol.* **1983**, *35*, 629–637. [CrossRef]
56. Rahman, M.H.; Pierson, B.J.; Yusoff, W.; Ismail, W. Classification of Microporosity in Carbonates: Examples from Miocene Carbonate Reservoirs of Central Luconia, Offshore Sarawak, Malaysia. In *Proceedings of the International Petroleum Technology Conference*; European Association of Geoscientists & Engineers: Houten, The Netherlands, 2011.
57. Rahman, M.H.; Pierson, B.J. Effects of Microporosity on Permeability and Sonic Velocity of Miocene Carbonates and an Approach to Relate Micrite Microtextures with Microporosity Occurrences in Miocene Carbonate Reservoirs of Offshore Sarawak, Malaysia. In *SEG Technical Program Expanded Abstracts 2011*; Society of Exploration Geophysicists: Tulsa, OK, USA, 2011; pp. 2059–2063.
58. Nole, M.; Daigle, H.; Milliken, K.L.; Prodanović, M. A method for estimating microporosity of fine-grained sediments and sedimentary rocks via scanning electron microscope image analysis. *Sedimentology* **2016**, *63*, 1507–1521. [CrossRef]
59. Kaczmarek, S.E.; Fullmer, S.M.; Hasiuk, F.J. A universal classification scheme for the microcrystals that host limestone microporosity. *J. Sediment. Res.* **2015**, *85*, 1197–1212. [CrossRef]
60. Wentworth, C.K. A scale of grade and class terms for clastic sediments. *J. Geol.* **1922**, *30*, 377–392. [CrossRef]
61. Janjuhah, H.T.; Salim, A.M.A.; Ghosh, D.P. Sedimentology and reservoir geometry of the Miocene Carbonate deposits in Central Luconia, Offshore, Sarawak, Malaysia. *J. Appl. Sci.* **2017**, *17*, 153–170. [CrossRef]
62. Lucia, F.J. *Carbonate Reservoir Characterization: An Integrated Approach*; Springer: Berlin/Heidelberg, Germany; New York, NY, USA, 2007.
63. Zhuravlev, A.Y.; Wood, R.A. Controls on carbonate skeletal mineralogy: Global CO_2 evolution and mass extinctions. *Geology* **2009**, *37*, 1123–1126. [CrossRef]
64. Pittman, E.D. Microporosity in carbonate rocks: Geological notes. *AAPG Bull.* **1971**, *55*, 1873–1878.
65. Ong, C.W.; Ho, P.; Leo, H.L. Effects of microporous stent graft on the descending aortic aneurysm: A patient-specific computational fluid dynamics study. *Artif. Organs* **2016**, *40*, E230–E240. [CrossRef]
66. Anselmetti, F.S.; Luthi, S.; Eberli, G.P. Quantitative characterization of carbonate pore systems by digital image analysis. *Am. Assoc. Pet. Geol. Bull.* **1998**, *82*, 1815–1836.
67. Ruzyla, K.; Jezek, D. Staining method for recognition of pore space in thin and polished sections: Research method paper. *J. Sediment. Res.* **1987**, *57*, 777–778. [CrossRef]
68. Yanguas, J.; Dravis, J.J. Blue fluorescent dye technique for recognition of microporosity in sedimentary rocks: Research method paper. *J. Sediment. Res.* **1985**, *55*, 600–602.

69. Milliken, K.L.; Curtis, M.E. Imaging pores in sedimentary rocks: Foundation of porosity prediction. *Mar. Pet. Geol.* **2016**, *73*, 590–608. [CrossRef]
70. Arribas, M.; Bustillo, A.; Tsige, M. Lacustrine chalky carbonates: Origin, physical properties and diagenesis (Palaeogene of the Madrid Basin, Spain). *Sediment. Geol.* **2004**, *166*, 335–351. [CrossRef]
71. Cox, P.; Wood, R.; Dickson, J.; Al Rougha, H.; Shebl, H.; Corbett, P. Dynamics of cementation in response to oil charge: Evidence from a Cretaceous carbonate field, UAE. *Sediment. Geol.* **2010**, *228*, 246–254. [CrossRef]
72. Flugel, E. *Microfacies of Carbonate Rocks: Analysis, Interpretation and Application*; Springer: Berlin/Heidelberg, Germany; New York, NY, USA, 2004; p. 979.
73. Kaldi, J. Diagenetic microporosity (chalky porosity), middle Devonian Kee Scarp reef complex, Norman wells, northwest territories, Canada. *Sediment. Geol.* **1989**, *63*, 241–252. [CrossRef]
74. Volery, C.; Davaud, E.; Foubert, A.; Caline, B. Lacustrine microporous micrites of the Madrid Basin (Late Miocene, Spain) as analogues for shallow-marine carbonates of the Mishrif reservoir Formation (Cenomanian to Early Turonian, Middle East). *Facies* **2010**, *56*, 385–397. [CrossRef]
75. Morad, D.; Paganoni, M.; Al Harthi, A.; Morad, S.; Ceriani, A.; Mansurbeg, H.; Al Suwaidi, A.; Al-Aasm, I.S.; Ehrenberg, S.N. Origin and evolution of microporosity in packstones and grainstones in a Lower Cretaceous carbonate reservoir, United Arab Emirates. *Geol. Soc. Lond. Spec. Publ.* **2016**, *435*, 47–66. [CrossRef]
76. Van Geet, M.; Swennen, R.; Durmishi, C.; Roure, F.; Muchez, P. Paragenesis of Cretaceous to Eocene carbonate reservoirs in the Ionian fold and thrust belt (Albania): Relation between tectonism and fluid flow. *Sedimentology* **2002**, *49*, 697–718. [CrossRef]
77. Ehrenberg, S. Porosity destruction in carbonate platforms. *J. Pet. Geol.* **2006**, *29*, 41–52. [CrossRef]
78. Yu, Y.; Visser, F.; Amro, M.M. Quantitative Effect of Microporosity on Permeability in Carbonate Reservoirs. In Proceedings of the International Petroleum Technology Conference, Kuala Lumpur, Malaysia, 10–12 December 2014.
79. Okabe, H.; Blunt, M. Predicting Permeability through 3D Pore-Space Images Reconstructed Using Multiple-Point Statistics. In Proceedings of the International Symposium of the Society of Core Analysts, Abu Dhabi, United Arab Emirates, 5–9 October 2004; pp. 5–9.
80. Chehrazi, A.; Rezaee, R. A systematic method for permeability prediction, a Petro-Facies approach. *J. Pet. Sci. Eng.* **2012**, *82–83*, 1–16. [CrossRef]
81. Shah, S.; Yang, J.; Crawshaw, J.P.; Gharbi, O.; Boek, E.S. Predicting Porosity and Permeability of Carbonate rocks from Core-scale to Pore-Scale using Medical CT, Confocal Laser Scanning Microscopy and Micro CT. In Proceedings of the SPE Annual Technical Conference and Exhibition, New Orleans, LS, USA, 20 September–2 October 2013.
82. Silin, D.B.; Patzek, T.W. Predicting Relative-Permeability Curves Directly from Rock Images. In Proceedings of the SPE Annual Technical Conference and Exhibition, New Orleans, LS, USA, 4–7 October 2009.
83. Archie, G.E. The electrical resistivity log as an aid in determining some reservoir characteristics. *Trans. AIME* **1942**, *146*, 54–62. [CrossRef]
84. Clerke, E.A.; Mueller, H.; Phillips, E.C.; Eyvazzadeh, R.Y.; Jones, D.H.; Ramamoorthy, R.; Srivastava, A. Application of Thomeer Hyperbolas to decode the pore systems, facies and reservoir properties of the Upper Jurassic Arab D Limestone, Ghawar field, Saudi Arabia: A "Rosetta Stone" approach. *GeoArabia* **2008**, *13*, 113–160. [CrossRef]
85. Moshier, S.O. Development of microporosity in a micritic limestone reservoir, Lower Cretaceous, Middle East. *Sediment. Geol.* **1989**, *63*, 217–240. [CrossRef]
86. Domingo, C.; García-Carmona, J.; Loste, E.; Fanovich, A.; Fraile, J.; Gómez-Morales, J. Control of calcium carbonate morphology by precipitation in compressed and supercritical carbon dioxide media. *J. Cryst. Growth* **2004**, *271*, 268–273. [CrossRef]
87. Periere, M.D.; Durlet, C.; Vennin, E.; Lambert, L.; Bourillot, R.; Caline, B.; Poli, E. Morphometry of micrite particles in cretaceous microporous limestones of the Middle East: Influence on reservoir properties. *Mar. Pet. Geol.* **2011**, *28*, 1727–1750. [CrossRef]

Journal of
Marine Science and Engineering

MDPI

Article

Geochemical Analysis of Cretaceous Shales from the Hazara Basin, Pakistan: Provenance Signatures and Paleo-Weathering Conditions

Abdul Ghaffar Fazal [1,*], Muhammad Umar [2], Faisal Shah [1,3], Muhammad Armaghan Faisal Miraj [4], Hammad Tariq Janjuhah [5,*], George Kontakiotis [6] and Abdul Khaliq Jan [7]

1 Department of Earth Sciences, Abbottabad University of Science and Technology, Havelian 22500, Pakistan; faisalshah@aust.edu.pk
2 Department of Earth Sciences, The University of Haripur, Haripur 22620, Pakistan; umarsani@uoh.edu.pk
3 National Centre of Excellence in Geology, University of Peshawar, Peshawar 25000, Pakistan
4 Institute of Geology, University of the Punjab, Lahore 54590, Pakistan; armghan.geo@pu.edu.pk
5 Department of Geology, Shaheed Benazir Bhutto University, Sheringal 18050, Pakistan
6 Department of Historical Geology-Paleontology, Faculty of Geology and Geoenvironment, School of Earth Sciences, National and Kapodistrian University of Athens, Panepistimiopolis, 15784 Athens, Greece; gkontak@geol.uoa.gr
7 Department of Chemistry, Shaheed Benazir Bhutto University, Sheringal 18050, Pakistan; abdukhaliq@gmail.com
* Correspondence: aghaffarfazal@gmail.com (A.G.F.); hammad@sbbu.edu.pk (H.T.J.)

Citation: Fazal, A.G.; Umar, M.; Shah, F.; Miraj, M.A.F.; Janjuhah, H.T.; Kontakiotis, G.; Jan, A.K. Geochemical Analysis of Cretaceous Shales from the Hazara Basin, Pakistan: Provenance Signatures and Paleo-Weathering Conditions. *J. Mar. Sci. Eng.* **2022**, *10*, 800. https://doi.org/10.3390/jmse10060800

Academic Editors: Antoni Calafat and Markes E. Johnson

Received: 27 April 2022
Accepted: 6 June 2022
Published: 10 June 2022
Corrected: 4 November 2022

Abstract: The geochemical investigation of shales from the Early to Middle Cretaceous Chichali Formation in the Hazara Basin was conducted to determine the origin, tectonic setting and evolution, paleo-weathering conditions, and paleo-oceanographic reconstruction. The research included a comprehensive field survey, sample collection, and analysis of a variety of main, trace, and rare-earth elements using an X-ray fluorescence spectrometer (XRF). Bivariate plots and ternary diagrams were used to determine the provenance, tectonic setting, and paleo-weathering conditions that existed during the development of the Chichali Formation in the Hazara Basin. The values of Ba/Sc, Ba/Co, Th/Sc, Cr/Th, Cr/Zr, Th/Co, Th/Cr, Sc/Th, bivariate plots of Al_2O_3 vs. TiO_2, TiO_2 vs. Zr, TiO_2 vs. Ni, Df1–Df2, Zr vs. Nb, and La/Sc vs. Th/Co, and ternary diagram of K_2O–Fe_2O_3–Al_2O_3 were used to illustrate the passive continental margin setting of Chichali Formation shales. The detailed chemical analysis also provides an understanding of the marine geochemical cycle, which reflects the origin of these sediments. The average K_2O/Al_2O_3 value is less than 0.4, indicating that the shale contains clay minerals. The Chichali Formation's Chemical Index of Alteration (mean = 71) and Index of Compositional Variation (mean = 1.12) values show a modest degree of chemical weathering in the source locations. From an environmental standpoint, the Chichali Formation is richer in toxic elements such as Ba, Zn, Ni, Cr, and Cu, which may be damaging to agricultural soils and drinking water when present in excess. These metals are incorporated into the formation during the weathering process.

Keywords: provenance; paleo-weathering; paleo-oceanographic reconstruction; tectonic evolution; marine geochemical cycle; passive continental margin; Chichali Formation; Pakistan

1. Introduction

The Hazara Basin is located in northern Pakistan's Lesser Himalaya and provides exceptional exposure to sedimentary strata ranging from the Late Proterozoic to the Holocene. It is a significant geological province due to its stratigraphic, sedimentological, and structural connections to the Himalayan orogeny [1]. The Chichali Formation is located in the Western Salt Range, Northern Kohat ranges, Trans Indus ranges, Kala-Chitta, Nizampur, and Southern Hazara. The Chichali Formation is mainly exposed in the Hazara area in

Riyala, Abbottabad Township, Hernoi, Galyat, and Balolia [2]. The formation is composed of yellowish-brown to dark grey to black shale with a few interbeds of marl and sandstone in the research region [3,4]. The Chichali Formation is thought to have been deposited in mid-to-upper-outer-shelf settings [1]. Its lower contact with the Samana Suk Formation is unconformable, whereas the upper contact with the Lumshiwal Formation is transitional [5]. The formation is highly fossiliferous and ranges in age from the Late Jurassic to the Early Cretaceous [6].

There is relatively limited published work on geochemical analysis of the Chichali Formation in Pakistan's Hazara Basin, and no major study on the geochemical investigation of these shales has been conducted so far. Shale is a common form of sedimentary rock that is exposed to the Earth's surface [7]. The concentration of elements in shale is influenced by a variety of processes, including weathering, transport, and sedimentation into the marine environment. The chemical composition of siliciclastic sediments has been extensively used to identify the features of the source rocks [8–10]. The relative variation of major components within shale determines the tectonic setting of sedimentary basins [11]. Major element analysis may be used to interpret both paleo-weathering and paleo-tectonic settings [12].

Water and soil are of prime concern to mankind as both are directly linked to human health. The incorporation of increased levels of trace elements in marine sedimentary deposits above their acceptable limits produces toxic effects on soils and marine fauna. Weathered agricultural soil may be toxic to crops, and when these crops are consumed by live organisms, they can have negative effects. It can be hazardous to aquatic life as well if it mixes with the surface streams. Surface runoff and erosion of sediments will ultimately have adverse effects if they reach the underground water by percolating through the soil deep into the ground, thus contaminating the groundwater and causing adverse effects on the environment. Overall, this research will explore the geochemical characterization of shales to obtain a better understanding of the study area's source rock potential, origin, tectonic context, and paleo-weathering environmental conditions.

2. Geological Setting and Stratigraphy

During the Cenozoic, the collision of India and Asia resulted in the formation of the Himalayas, a huge mountain chain [2,13]. The southern Hazara was formed as a result of the collision of the Indian and Eurasian plates. It is a part of the Himalayan fold and thrust belt [14,15]. The study area is in Hazara, which is situated within the Hazara Kashmir Syntaxis (Figure 1).

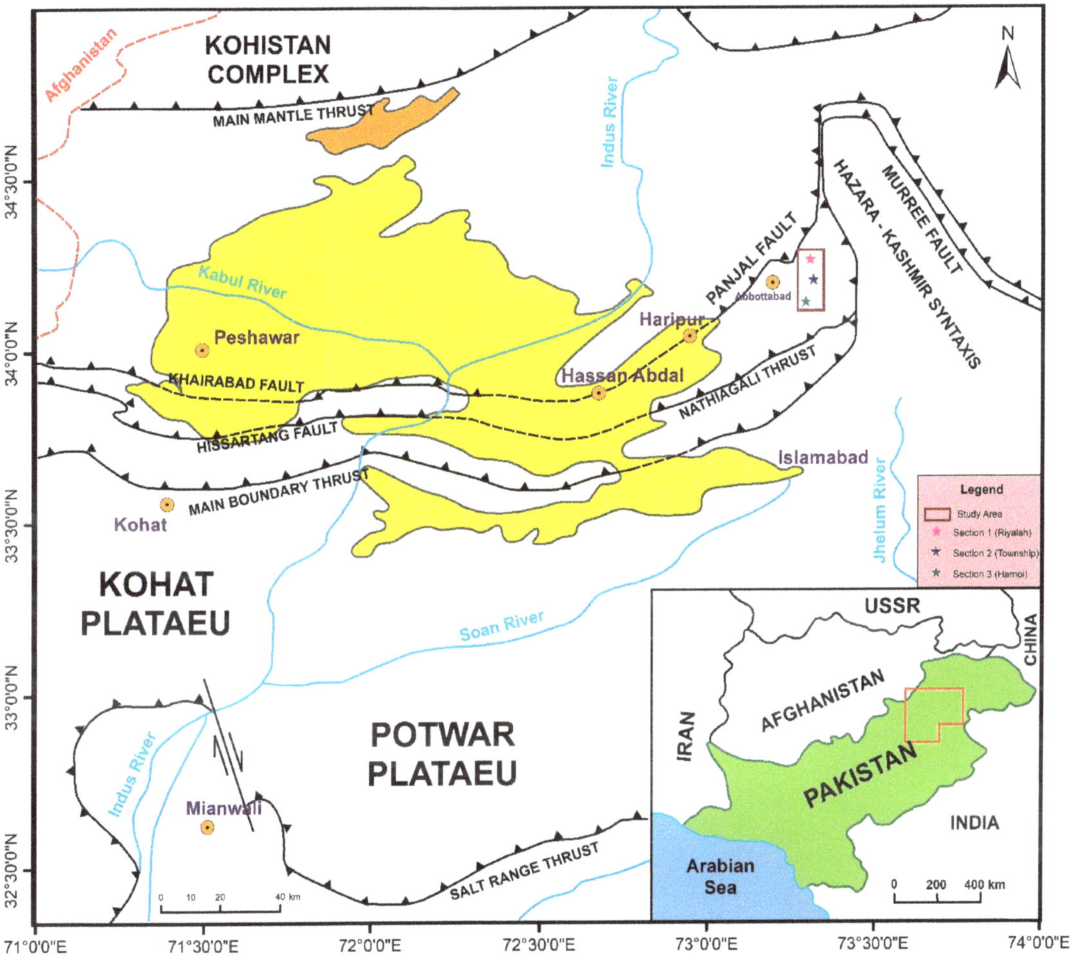

Figure 1. Tectonic map showing a segment of the Northwestern Himalayas' fold and thrust belt. The rectangle indicates the location of the research area.

The exposed rocks in the Hazara area vary in age from Precambrian to Holocene [1,16], except the middle and upper Paleozoic (Figure 2) [17,18].

The Chichali Formation is well exposed in the Hazara region, particularly at Riyala (Section 1), Township Abbottabad (Section 2), and Hernoi (Section 3). The formation's major composition is shale with intercalations of marl and sandstone. The thickness of the formation varies from 33 m in Section 1 to about 55 m in Sections 2 and 3. Shale is yellowish-brown, greenish to dark grey, and black. Due to the prevalence of glauconite, the weathered surface is yellowish-brown, but the fresh unweathered surface is greenish-grey to black. The base of Chichali, near Section 3, has coal seams, and some of those seams have been mined (Figure 3) [19,20].

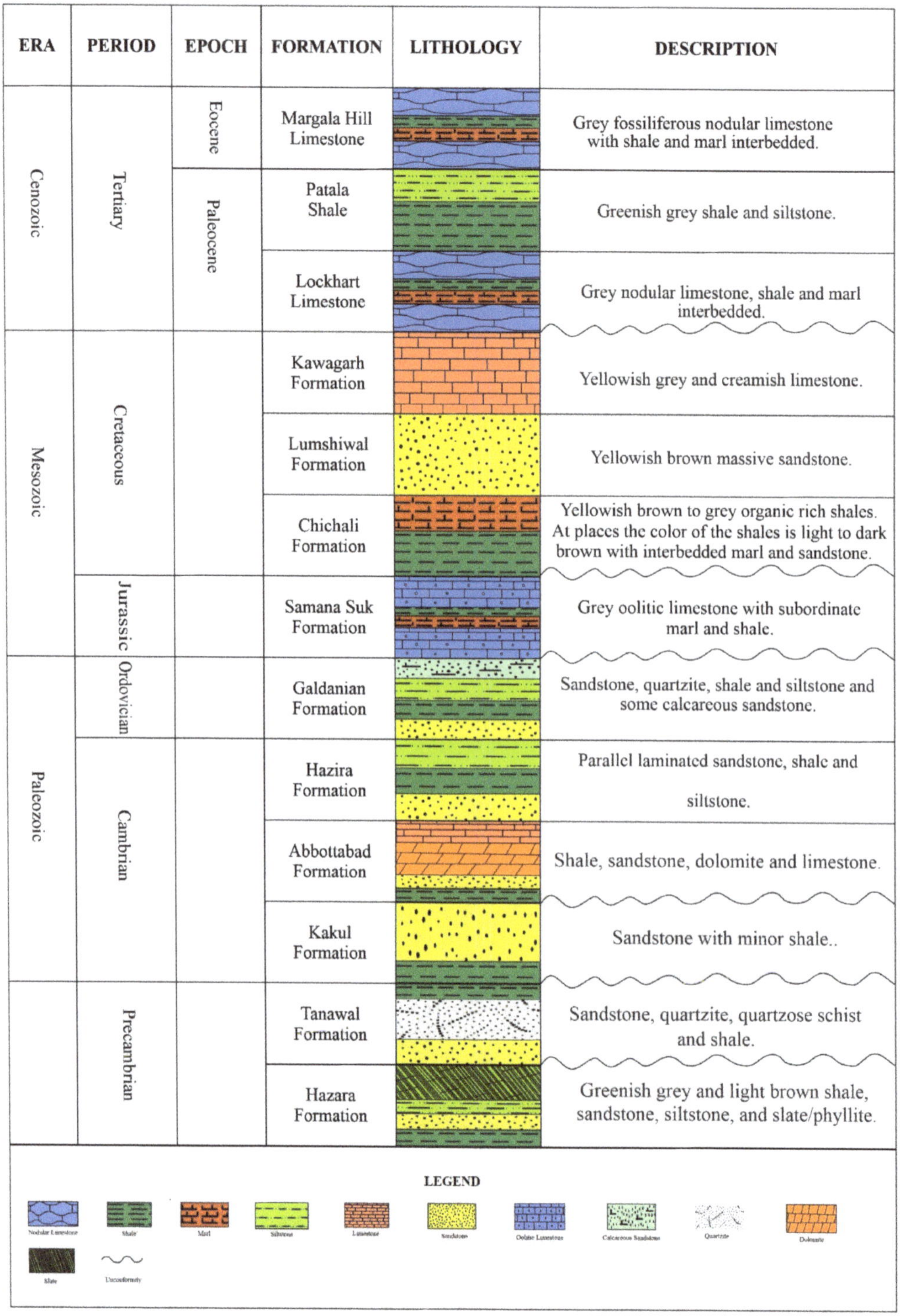

Figure 2. Representation of the generalized stratigraphic succession of the Hazara Basin.

Figure 3. Field photographs of the Chichali Formation: (**A**,**B**) Yellowish brown to grey shales (Section 1), (**C**–**E**) organic-rich shales (Section 2), (**F**–**H**) light to dark brown shales with interbedded marl and sandstone (Section 3).

3. Material and Methods

Fieldwork was conducted in detail, and three stratigraphic sections were selected for measurement and sampling. Three stratigraphic sections were measured, described, and sampled in the field (Riyala, Township Abbottabad, and Hernoi). In the field, a total of 40 shale samples were collected. The concentration of major and trace elements was determined for seventeen shale samples. An X-ray fluorescence spectrometer was used to determine the percentages of main and trace elements at Bestway Cements Limited in Hattar (PANalytical Cubix 2300 XRF), Almelo, The Netherlands.

Chemical Index of Alteration (CIA) values and the A–CN–K ternary plot reveal the paleo-weathering and tectonic history of sedimentary rocks [21]. This equation was used to calculate the Chemical Index of Alteration (CIA) for paleo-weathering analysis.

The Chemical Index of Alteration (CIA) was derived for paleo-weathering analysis using the following equation:

$$CIA = \{Al_2O_3/(Al_2O_3 + CaO + Na_2O + K_2O)\} \times 100 \quad (1)$$

The index of compositional variability (ICV) quantifies the maturity of the aluminosilicates transported to the depositional location, i.e., basin [22]. The following formula was used to calculate ICV [22]:

$$ICV = (Fe_2O_3 + K_2O + Na_2O + CaO + MgO + TiO_2)/Al_2O_3 \quad (2)$$

4. Results

4.1. Major and Trace Elements

The major and trace element studies of the Chichali Formation shales are summarized in Table 1. Among the principal elements analyzed, SiO_2 ranges from 17.89 to 60.14 wt%; Al_2O_3 ranges from 4.89 to 21.19 wt%; TiO_2 ranges from 0.5 to 1.48 wt%; Fe_2O_3 ranges from 1.26 to 9.27 wt%; K_2O ranges from 0.61 to 5.75 wt%; and Na_2O ranges from 0.01 to 0.17 wt%.

Table 1. Major and trace element data for whole-rock samples of the Chichali Formation. The concentration of major components is expressed as wt%, while trace element concentration is in ppm. CIA is an abbreviation for Chemical Index of Alteration, while ICV is an abbreviation for Index of Compositional Variability.

	Section 1			Section 2								Section 3					
Major Oxides/Trace Elements	**RC-1**	**RC-2**	**RC-3**	**TC-1**	**TC-2**	**TC-3**	**TC-4**	**TC-5**	**TC-6**	**TC-7**	**TC-8**	**HC-1**	**HC-2**	**HC-3**	**HC-4**	**HC-5**	**HC-6**
SiO_2	55.1	47.4	49.0	29.3	27.2	44.6	43.7	44.6	41.8	40.7	17.8	54.1	56.0	54.3	60.1	52.8	51.0
Al_2O_3	9.49	7.93	8.02	7.03	6.91	11.2	11.6	11.3	10.8	9.56	4.89	21.1	20.6	21.2	19.1	19.0	19.2
Fe_2O_3	3.51	3.74	3.61	3.74	3.17	4.75	3.24	4.77	4.82	4.58	1.36	5.9	6	5.7	5.34	4.57	9.27
CaO	2.04	0.99	1.77	1.24	2.06	1	1.83	1.9	1.66	1.52	2.88	1.02	1.77	1.47	1.73	1.48	1.79
MgO	5.85	7.27	7.1	1.4	1.26	1.52	1.82	1	1.46	1.54	1.34	1.4	1.39	1.42	1.09	1.14	1.22
SO_3	0.03	0.05	0.04	0.01	0.03	0.02	0.02	0.01	0.01	0.01	0	0.05	0.03	0.07	0.01	0.1	0.06
Na_2O	0.04	0.05	0.05	0.01	0.01	0.03	0.02	.012	0.12	0.01	0.01	0.13	0.13	0.1	0.17	0.13	0.11
K_2O	3.54	3.36	3.28	1.74	1.37	2.01	1.74	1.48	1.86	1.97	0.61	5.5	5.23	5.75	4.6	4.55	5.19
TiO_2	0.67	0.53	0.5	0.57	0.5	1	0.97	1.05	1.07	1.05	0.53	1.27	1.17	1.38	1.48	1.42	1.4
CIA	62.8	64.3	61.1	70.1	66.7	78.6	76.3	76.9	74.8	73.2	58.2	76.1	74.3	74.4	74.6	75.5	73.0
ICV	2.03	2.01	2.03	1.24	1.21	0.92	0.83	0.9	1.01	1.12	1.38	0.72	0.76	0.74	0.75	0.7	0.99
Sc	5.36	7	9.64	11.8	16	14.9	11	15.3	11.7	12.2	13.4	14.1	7.75	6.37	8.73	11.9	4.1
Cr	149	120	125	95.7	127	122	125	135	129	94	94	122	132	142	91	85	112
V	64	65	97	138	28	150	130	70	112	119	97	110	82	67	83	270	70
Co	6.41	4.21	3.71	5.99	5.03	6.05	4.46	5.54	3.9	4.17	5.01	11.4	7.96	6.16	9.01	5.29	8.39
Ni	9	9	10	15	19	11	18	22	20	19	21	28	14	13	17	12	21
Cu	65	65	67	57	42	45	24	44	18	36	29	46	33	32	56	79	66
Zn	112	169	89	86	41	78	47	42	57	46	51	81	89	121	114	154	139
Rb	190	119	193	198	169	248	184	199	131	175	183	151	194	159	213	211	227
Sr	130	211	148	232	261	103	200	142	350	212	222	144	234	124	135	136	139
Zr	211	347	321	331	259	100	300	200	190	241	210	388	485	476	438	454	327
Nb	74	64	54	45	47	85	39	55	70	57	48	68	50	61	73	44	54
Sb	0.6	0.38	0.25	0.23	0.65	0.15	0.46	0.32	0.39	0.36	0.41	0.59	0.37	0.29	0.39	0.3	0.37

Table 1. *Cont.*

	Section 1			Section 2								Section 3					
Major Oxides/Trace Elements	**RC-1**	**RC-2**	**RC-3**	**TC-1**	**TC-2**	**TC-3**	**TC-4**	**TC-5**	**TC-6**	**TC-7**	**TC-8**	**HC-1**	**HC-2**	**HC-3**	**HC-4**	**HC-5**	**HC-6**
Cs	2.02	2.99	2.59	3.63	2.42	2.46	2.5	2.47	2.8	2.72	2.45	3.59	2.16	2.26	2.19	3.62	2.43
Ba	299	153	194	184	292	243	208	189	176	165	198	205	187	272	631	210	290
La	21.1	29.3	19.8	17.6	23.5	18.5	15.6	33.5	27.8	31.9	34.1	17.8	26.9	17.4	21.3	39.9	30.3
Ce	306	157	77.9	116	120	135	152	139	118	144	137	76.6	54.9	117	154	63	117
Nd	125	51.9	37.2	56.4	52.6	63.6	73.4	59.5	55.6	61.3	51.8	30	21	51	78.6	30.7	42.9
Sm	19	7.98	5.06	8.56	7.51	9.38	10.5	8.09	9.7	7.98	8.32	6.23	4.3	7.97	11.2	4.3	8.45
Eu	3.17	1.38	1.41	1.58	1.34	1.61	2.03	1.46	1.7	1.33	0.98	1.2	0.9	1.46	2.2	0.86	1.61
Gd	9.7	5.13	5.31	4.17	4.29	5.75	4.49	5.44	5.24	4.9	4.2	4.49	2.73	4.63	7.97	2.67	5.04
Tb	1.65	0.92	0.96	0.81	0.79	0.99	0.73	0.89	0.94	0.85	0.81	1.04	0.62	1	1.25	0.58	1
Tm	0.89	0.6	0.61	0.41	0.45	0.48	0.61	0.53	0.58	0.57	0.51	0.64	0.42	0.64	0.75	0.35	0.64
Yb	6.57	4.16	3.99	3.38	3.5	4.21	3.78	3.6	3.2	4.1	3.9	4.48	2.73	4.21	5.47	2.4	4.35
Lu	1.03	0.64	0.61	0.54	0.55	0.65	0.56	0.57	0.54	0.61	0.59	0.68	0.44	0.65	0.83	0.39	0.66
Hf	3.46	5.29	3.33	5.29	3.33	3.74	2.9	4.1	2.76	3.12	3.4	3.62	4.7	2.61	3.6	7.64	5.72
Ta	5.03	4.03	3.35	3.08	3.15	5.9	2.17	3.77	4.23	2.78	3.12	4.93	3.09	4.27	4.58	2.32	3.6
Th	38.2	28.4	29	23	22.4	30	22.3	28.6	27.3	19.8	22.3	29.6	24.2	24.9	28.7	15.1	22.7
U	5.3	4.76	4.53	3.41	4.18	5.61	2.95	5.19	4.2	2.4	5.2	5.1	4.5	5	5.15	2.72	4.5

Sc ranges from 4.1–15.3 parts per million; V ranges from 64–270 ppm; Cr ranges from 85–135 ppm; Co ranges from 3.71–11.4 ppm; Ni ranges from 9–28 ppm; Cu ranges from 29–79 ppm; Zn ranges from 41–169 ppm; Rb ranges from 119–248 ppm; Sr ranges from 103–350 ppm; Zr ranges from 100–485 ppm; La ranges from 15.6–39.9 ppm; Th ranges 15.1–38.2 ppm; and U ranges from 2.4–5.3 ppm. CaO is present in small amounts in the sediments owing to the lack of carbonate cement. MgO and CaO are often found in association with dolomite, ferrous carbonates, and calcite [23]. Increased MgO and CaO concentrations indicate the presence of carbonate minerals, while a low concentration indicates the presence of clay minerals [24]. The reduced Na_2O demonstrates the source rock's recycling, moderate to severe weathering, and its removal during transportation [25]. The K_2O/Al_2O_3 ratio is used to determine the original composition of ancient mud rocks. The K_2O/Al_2O_3 ratios in the Hazara area's Chichali Formation shales are less than 0.4, indicating the presence of clay minerals.

The trace element ratio is frequently used to estimate the felsic and mafic composition of rocks. This research makes use of the Th/Sc, Th/Cr, Th/Co, and La/Sc ratios [26]. The results of the present study's elemental ratio analysis were compared to the conventional values for felsic and mafic rocks (Table 2). The findings indicate that all trace element ratios are in favor of a felsic composition (Table 2).

Table 2. The elemental ratios of shales from the Chichali Formation were compared to those of equivalent fractions obtained from felsic and mafic rocks in this research.

Elemental Ratio	Range of Shales from the Chichali Formation	Range of Sediments (Based on Cullers [27]).	
		Felsic Rocks	Mafic Rocks
Th/Sc	1.26–7.12	0.84–20.5	0.05–0.22
Th/Co	2.52–7.81	0.67–19.4	0.04–1.4
Th/Cr	0.17–0.31	0.13–2.7	0.018–0.046
La/Sc	1.24–7.39	2.5–16.3	0.43–0.86

4.2. Provenance and Source Rock Potential

The relationship between TiO_2 and Al_2O_3 was employed to determine the provenance [28]. The bivariate Al_2O_3 and TiO_2 graph shows a concentration of data in the granodiorite area and some in the granite and basalt region (Figure 4). The plot of TiO_2 vs. Ni is often used to classify rocks into acidic and basic igneous regions [29]. According to the TiO_2 vs. Ni plot in our model, the Chichali Formation is derived from intermediate igneous rocks (Figure 5).

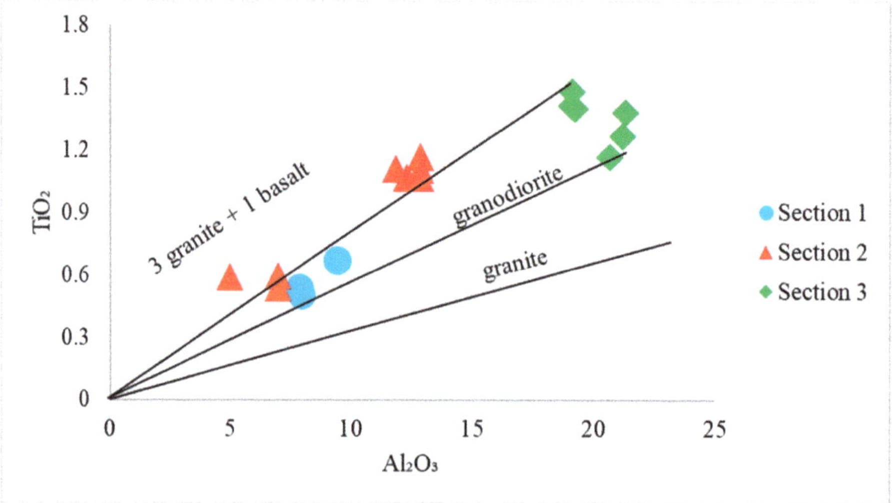

Figure 4. Plot of Al_2O_3 versus TiO_2 plot for Chichali Formation shales.

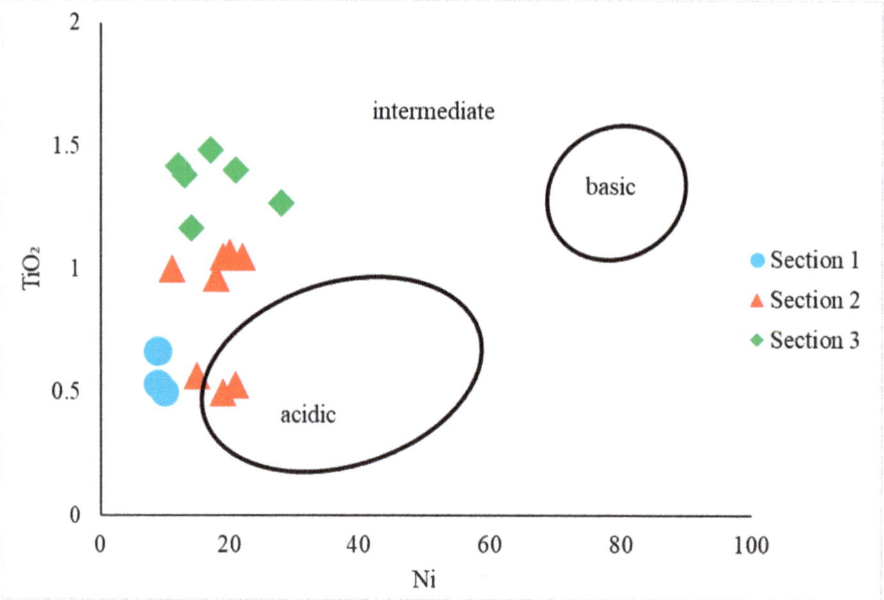

Figure 5. Diagram illustrating provenance, bivariate diagram of TiO_2 wt% vs. Ni (ppm).

The ternary diagram of K_2O–Fe_2O_3–Al_2O_3 [30] demonstrates that all shales of the Chichali Formation plotted around the Al_2O_3 apex, indicating Al_2O_3 enrichment. This finding implies that clay minerals controlled the element abundances in this shale (Figure 6) [29,31]. The plot of La/Sc vs. Th/Co may be used to identify the composition of the source rock [32]. The plotted data for La/Sc and Th/Co [32] indicate the presence of a silicic rock composition zone (Figure 7). The quantity of a trace element may be a reliable predictor of its origin [33]. The concentrations of Ni and Cr are used to indicate the presence of ultramafic rocks in the source location (Cr > 150 ppm and Ni > 100 ppm). The Chichali Formation samples show an average Cr value of 117.62 ppm and a Ni concentration of 16.35 ppm, indicating the lack of ultramafic rocks in the source region. The relationship between SiO_2/Al_2O_3 and K_2O/Na_2O [10,28] suggests that the majority of the Chichali shale data plotted in the field reflect Proterozoic–Phanerozoic shale composition, demonstrating that clay minerals regulate the main element composition (Figure 8) [10,28].

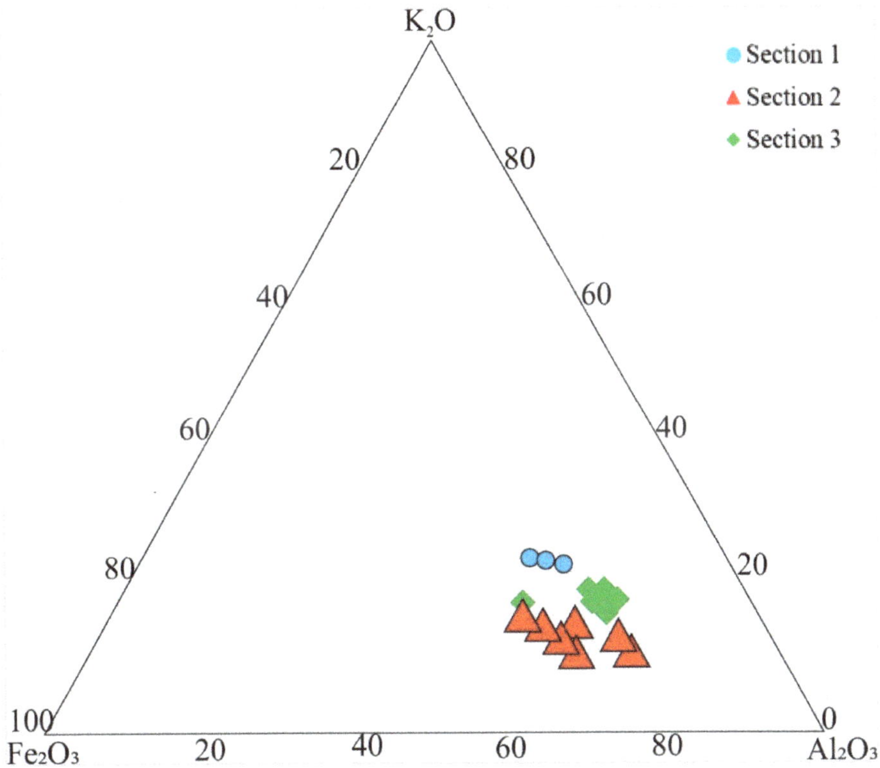

Figure 6. Ternary K_2O–Fe_2O_3–Al_2O_3 diagram for Chichali Formation shale.

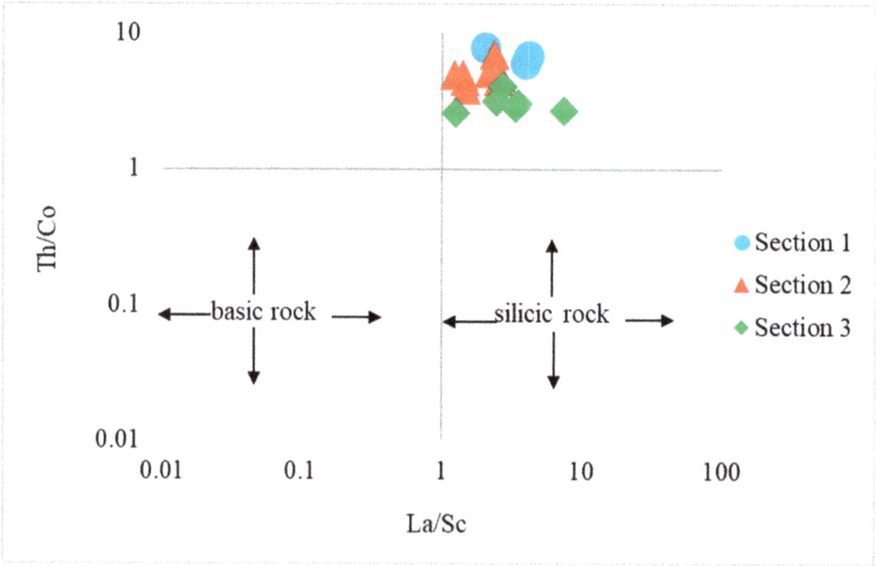

Figure 7. The plot of La/Sc vs. Th/Co for Chichali Formation shale.

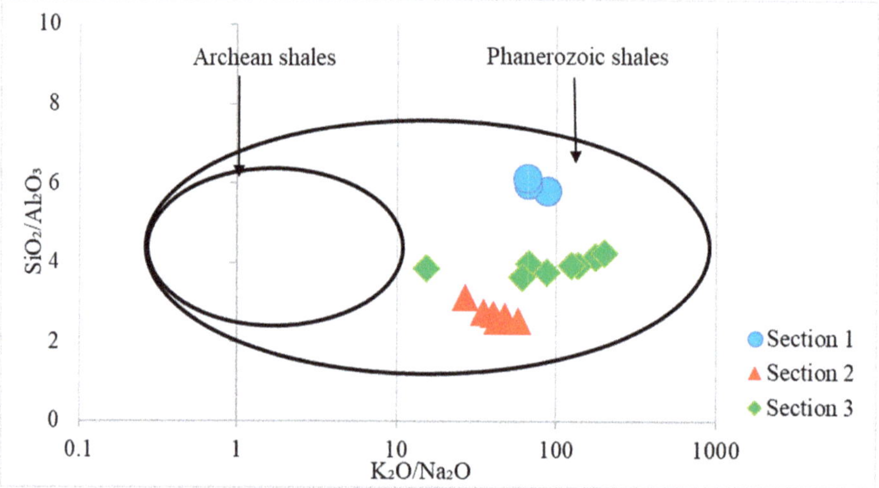

Figure 8. The plot of SiO_2/Al_2O_3 versus K_2O/Na_2O for shales of the Chichali Formation.

Numerous authors, such as Adriano [33], Eqani, et al. [34], Qasim, et al. [35], and Martini, Walter, Ku, Budai, McIntosh and Schoell [11], have presented a variety of diagrams for discriminating between major and trace elements. To establish the provenance, the discrimination diagram makes use of variably mobile major elements and static minor elements [36]. In the discriminating diagram, Sections 2 and 3 are mostly plotted in the field of intermediate igneous rock, while Section 1 is primarily plotted in the field of quartzose sedimentary provenance (Figure 9). Additionally, the ternary diagram of K_2O–Fe_2O_3–Al_2O_3 reveals important information concerning its origin [30].

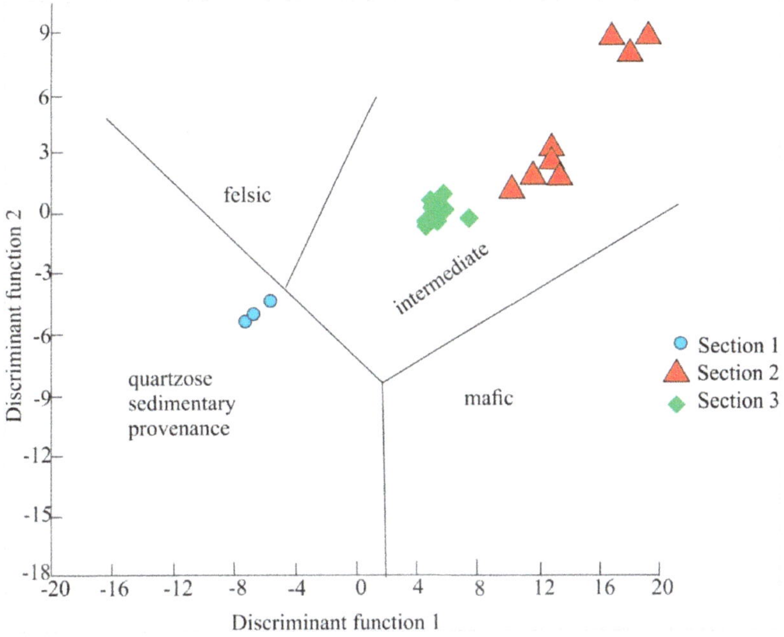

Figure 9. Provenance discrimination diagram for shales. Discriminant function 1 = (−1.773 × TiO_2%) + (0.607 × Al_2O_3%) + (0.76 × Fe_2O_3T%) + (−1.5 × MgO%) + (0.616 × CaO%) + (0.509× Na_2O%) + (−1.22 × K_2O%) + (−9.09). Discriminant function 2 = (0.445 × TiO_2%) + (0.07 ×Al_2O_3%) + (−0.25 × Fe_2O_3T%) + (−1.142 × MgO%) + (0.438 × CaO%) + (0.432 × Na_2O%) + (1.426 × K_2O%) + (−6.861).

4.3. Tectonic Setting

The binary diagram shows that all samples are plotted within the passive margin field, indicating that the Chichali Formation was deposited within a passive margin environment (Figure 10). The plot of SiO_2 vs. K_2O/Na_2O is used to determine whether the shale was deposited on passive continental edges, active continental margins, or oceanic island arc margins [37]. Additionally, by incorporating major and trace element data, the discrimination diagrams can be used to infer the tectonic settings of previous terrains [38–40], distinguishing oceanic island arc, active continental margin, passive continental margin, and continental island arc (Figure 11).

4.4. Paleo-Weathering in the Source Area

The findings indicate that the average CIA values for all three examined sections ranged from 58 to 78, with an average of 71, indicating that the weathering conditions for these shales were mild (Table 1). Al_2O_3 represents the immobile component in the CIA index, while CaO, Na_2O, and K_2O represent the mobile components. CaO, K_2O, and Na_2O are employed as mobile components due to their susceptibility to weathering and their propensity for leaching during weathering. The rise in CIA values indicates that CaO, K_2O, and Na_2O are being removed at a faster rate than the more stable Al_2O_3 (Figure 12).

For kaolinite and chlorite, the CIA value is close to 100; for shale, the average value is between 70 and 75 [41]; and for fresh granite, the average value is about 50 [42].

The ICV values of the shale studied in this research are modest, with an average of 1.12. ICV values greater than one indicate immature shale, which includes rock-forming minerals such as plagioclase, K-feldspar, amphiboles, and pyroxene. The present study's average ICV results indicate the existence of K-feldspar (0.8–1) [43]. A cross plot of the CIA and ICV values reveals an interesting picture of the chemistry of the key elements. According to Figure 13, the source rock is felsic to intermediate in composition. The classification of siliciclastic rocks is based on the plot between SiO_2/Al_2O_3 and Fe_2O_3/K2O [44]. Due to the samples' low SiO_2/Al_2O_3 and Fe_2O_3/K_2O contents, they all fall into the shale field when plotted in the SiO_2/Al_2O_3 and Fe_2O_3/K_2O binary plots (Figure 14).

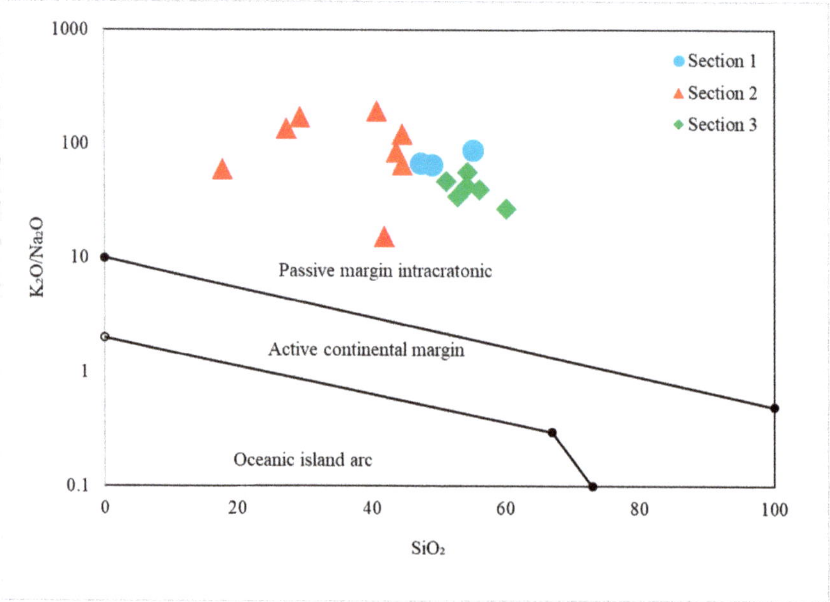

Figure 10. Plot of SiO_2 versus K_2O/Na_2O for shales of the Chichali Formation.

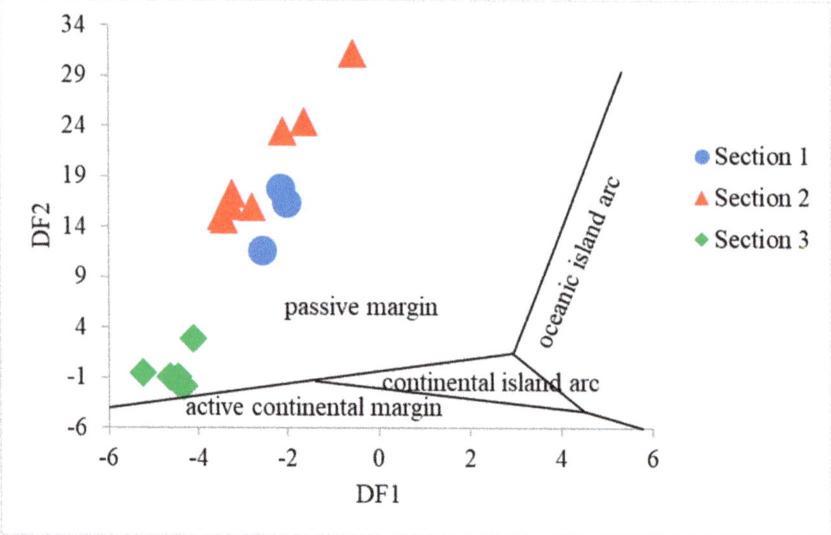

Figure 11. Discrimination function plot of the Chichali shales. Discrimination function 1 = 0.303 − $0.0447SiO_2$ $0.972TiO_2$ + $0.008Al_2O_3$ $0.267Fe_2O_3$ + 0.14MgO + 0.195CaO + $0.719Na_2O$ − $0.032K_2O$. Discrimination function 2 = 43.57 − $0.421SiO_2$ + $1.988TiO_2$ − $0.526Al_2O_3$ − $0.551Fe_2O_3$ + 0.881MgO − 0.907CaO $0.177Na_2O$ − $1.84K_2O$.

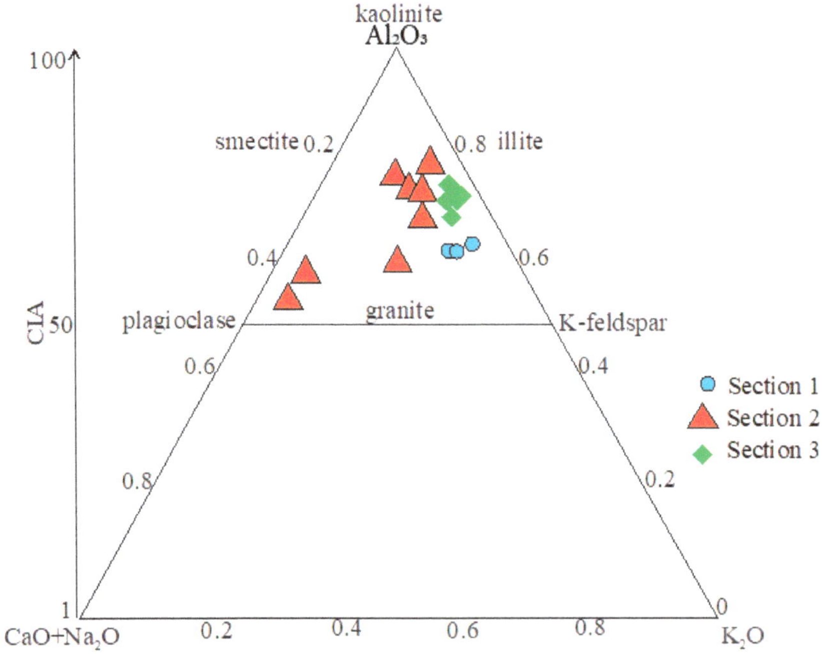

Figure 12. Al_2O_3–CaO+Na_2O–K_2O ternary diagram.

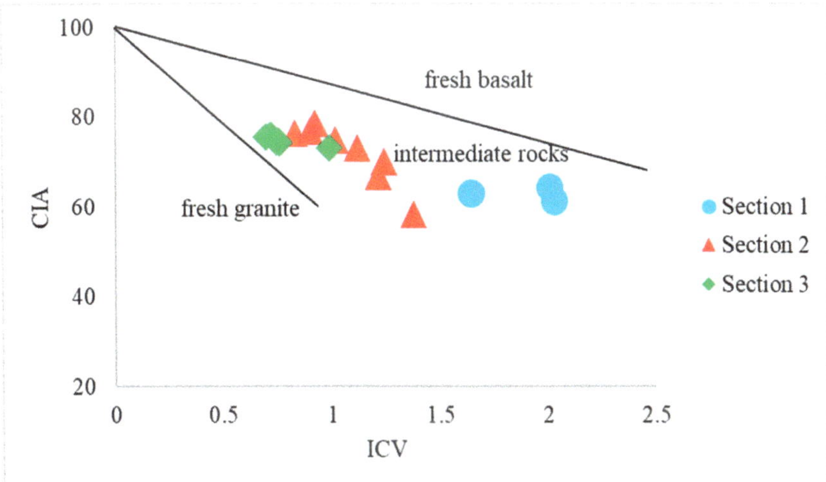

Figure 13. Weathering indicators, the Chemical Index of Alteration (CIA), and the Index of Chemical Variation (ICV) in Chichali Formation shales.

Th/Sc vs. Zr/Sc plots may be used to monitor sedimentary recycling and sorting. While first-order sediments exhibit a straightforward association between Th/Sc and Zr/Sc ratios, recycled sediments exhibit a significant rise in Zr/Sc relative to Th/Sc. The Th/Sc ratio is utilized to indicate chemical differentiation in the plot [40], while the Zr/Sc ratio indicates sediment recycling in the source area [45]. Zr is an abbreviation for zircon, a physically and chemically very stable mineral that may be used to determine the influence of recycling in the source location. All samples examined in the Th/Sc vs. Zr/Sc binary plot [45] exhibit a linearly rising compositional tendency toward a higher Zr/Sc ratio, and all samples plotted in the sediment recycling field (zircon added) exhibit a linearly increasing compositional trend toward a higher Zr/Sc ratio (Figure 15). These findings indicate that the rise in zircon content in the source region occurred before sediment transfer to the Hazara basin. Thus, a binary plot of Th/Sc vs. Zr/Sc is used to figure out the effect of recycling.

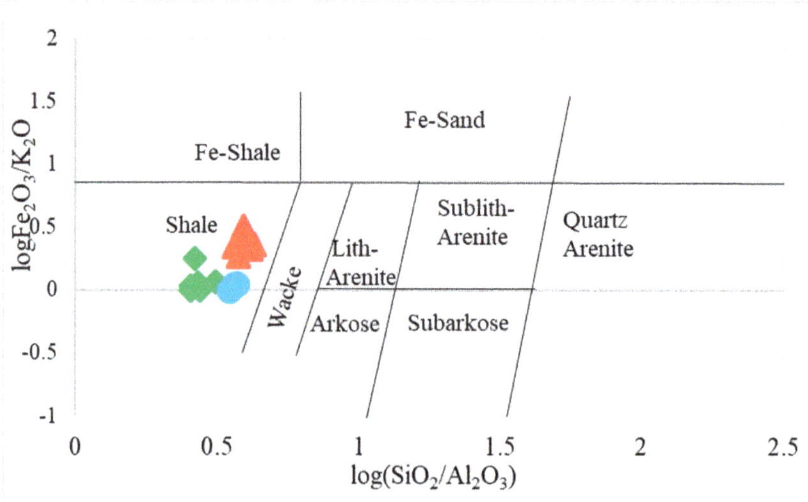

Figure 14. Classification of siliciclastic rocks.

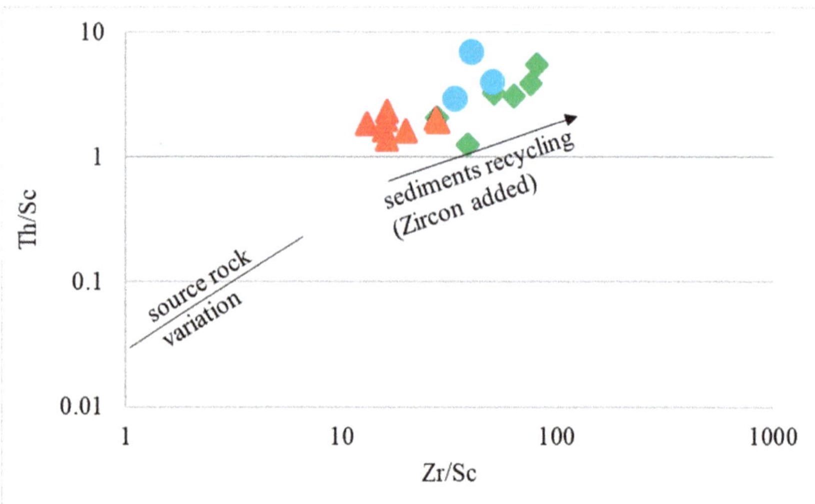

Figure 15. Plot of the Th/Sc ratio against the ratio of Zr to Sc for the Chichali Formation.

The findings indicate that the average Zr/Hf ratios in the examined samples are much higher (26.73–182.37, with an average of 81.24) than in UCC and PAAS (31.6 and 42, respectively) [46]. These increased ratios may be the outcome of recycling in the source location. The Al_2O_3/K_2O ratio exceeds 0.3, indicating the existence of K-feldspar [47].

Zr has a positive correlation value with Hf, r = 0.36. The Zr/Hf ratio of the Chichali Formation samples analyzed in the Hazara basin ranges from 26 to 182, with an average of 81. If the Zr/Hf ratio is greater than 40, zircon controls these elements. The favorable correlations between K_2O and Sc, K_2O and Co, Sc–Al_2O_3, and Co–Al_2O_3 indicate that these elements are concentrated in phyllosilicates after they have been broken down by the weather.

Al_2O_3 has a declining trend when plotted against SiO_2/Al_2O_3, Fe2O3/Al_2O_3, and K_2O/Al_2O_3 ratios for samples obtained from the Chichali Formation (Figure 16A–C). As a result of these plots, it can be deduced that as weathering proceeds, Al_2O_3 stays as a residue while Fe_2O_3, K_2O, and Na_2O weather away. As a result, it is indicative of the existence of clay minerals.

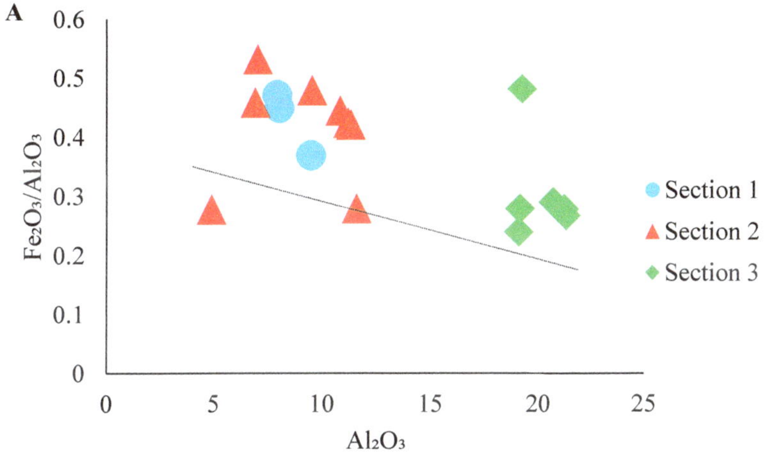

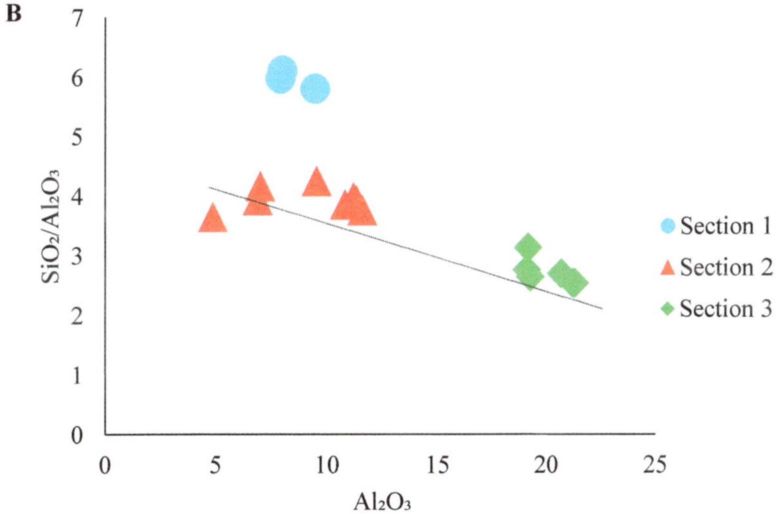

Figure 16. *Cont.*

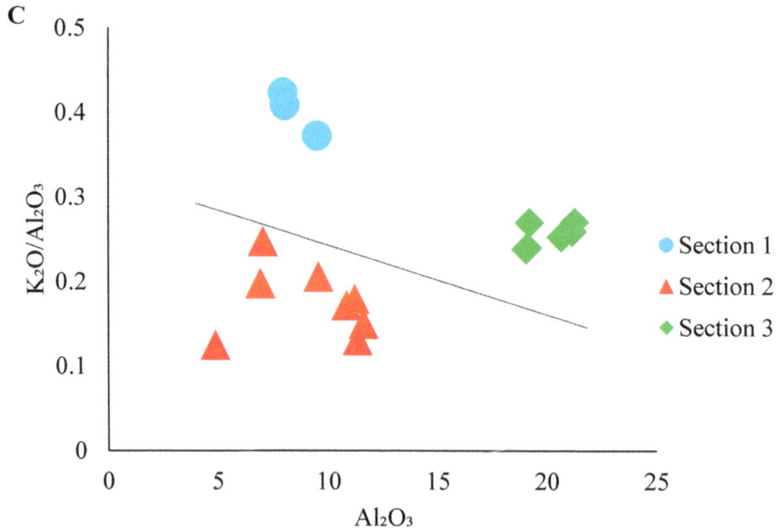

Figure 16. Plots of Al_2O_3 against (**A**): Fe_2O_3/Al_2O_3 ratio, (**B**): SiO_2/Al_2O_3 ratio, and (**C**): K_2O/Al_2O_3.

4.5. Geochemical Analysis of Chichali Formation

The quantities of various components in the samples investigated are summarized in Table 1. Soils are formed as a result of the weathering of rocks. In Figure 17, V, Cu, Ba, Zn, and U histograms are compared to the normal elemental concentrations of Cu, Ba, Zn, and U in soils (Figure 17) [48]. Cu levels greater than usual are seen in Sections 1 (65.67 ppm), 2 (36.88 ppm), and 3 (52 ppm) (Figure 17). Cu enrichment beyond the recommended level may result in stunting, discoloration of the root system, decreased growth, and chlorosis [49].

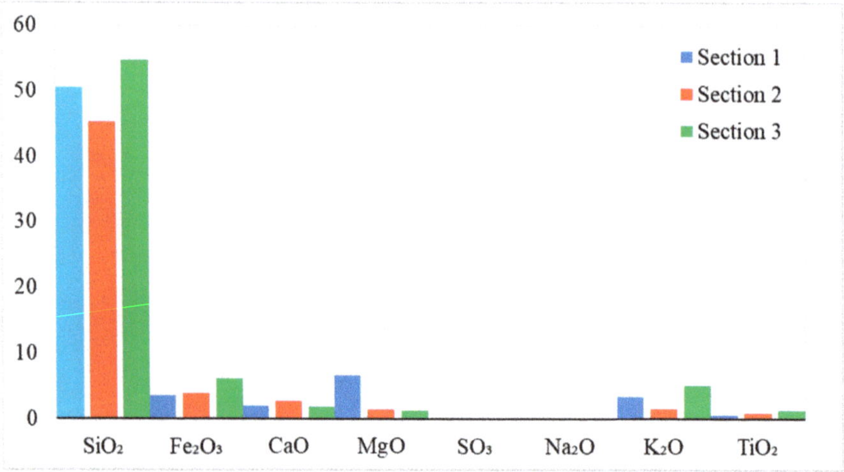

Figure 17. Histogram comparing the averages of the main oxides.

At 90 ppm, Zinc is recommended to be in typical soils around the globe [50]. Zn readings are greater than usual in Sections 1 (123.33 ppm) and 3 (116.33 ppm) but are within allowed limits in Section 2 (56 ppm) (Figure 17). These elevated levels cause toxicity in plants and crops. Ni levels are lower than usual in all three parts investigated, namely,

Section 1 (9.33 ppm), Section 2 (18.125 ppm), and Section 3 (17.5 ppm). The examined sections show that Cr levels are much higher than usual in all of them, including Section 1 (131 ppm), Section 2 (115 ppm), and Section 3 (114 ppm).

Comparison of the average values of main oxides in the analyzed portions is displayed in the histogram (Figure 18). Section-wise comparison of SiO_2 indicates that SiO_2 is practically similar in Sections 1 and 3 and less in Section 2. Fe_2O_3 concentration is greater in Section 3 compared to Sections 1 and 2. The peaks of CaO are extremely lower, suggesting the lack of carbonates in these sediments. MgO content in Section 1 is greater than Sections 2 and 3. The presence of mafic minerals suggests a high concentration of MgO, yet the concentration of MgO here is considerably smaller compared to the mafic minerals. Sections 2 and 3 contain about the same quantity of MgO. Na_2O concentration may also be observed in the histogram (Figure 18), which reveals extremely low concentration of Na_2O in all the sections. This may be rationalized by the weathering events which dissolved Na_2O to generate clay minerals in the examined portions.

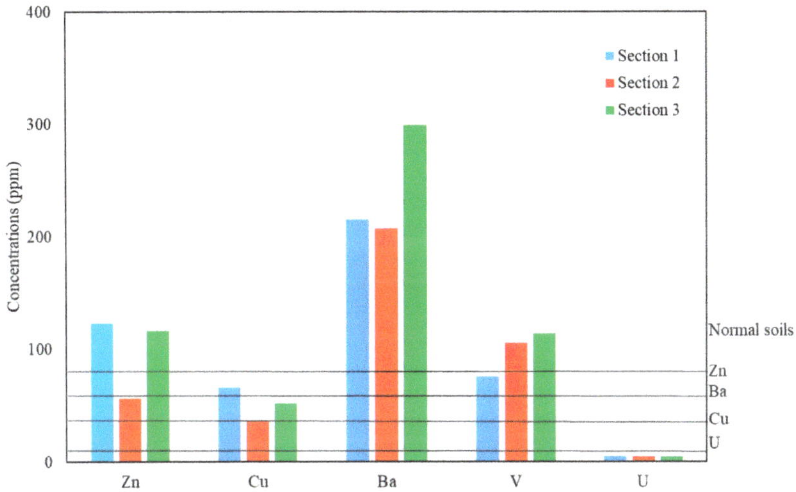

Figure 18. Concentrations of elements in soils generated from shales.

4.6. Statistical Analysis

Other intriguing aspects of the shale are shown by the ratios of key oxides and correlation coefficient data for major and trace elements. The Pearson coefficients of correlation are listed in Supplementary Table S1. Microsoft Excel was used to compute the correlation coefficients for the major and selected trace elements in the studied shale. Correlation coefficient data aided in the comprehension of the geochemical characteristics of the Chichali Formation.

Correlations between Al_2O_3 and TiO_2 (r = 0.90) and K_2O (r = 0.88) are statistically significant. Al_2O_3 has a negative Pearson correlation value of −0.30 and −0.39 with CaO and MgO, respectively. Al_2O_3 has a highly positive and significant association with Na_2O and Fe_2O_3 at 0.83 and 0.76, respectively. All of these correlations indicate that sediments arrange themselves during transportation. Al_2O_3 has a substantial positive correlation with K_2O (r = 0.88) and SiO_2 (r = 0.75), suggesting the existence of kaolinite and K-feldspar.

TiO_2 has a significant positive correlation (r = 0.90) with Al_2O3 and a negative association (r= −0.25) and MgO (r = −0.56) with CaO and MgO. TiO_2 correlates positively with both Na_2O and K_2O (r = 0.75 and 0.65, respectively). This indicates that TiO_2 is a key chemical element of clays, rather than mafic minerals. SiO_2 has a negative correlation with a number of key elements, suggesting that quartz has been dissolved. The strong relationship between TiO_2 and Al_2O_3 and the negative correlation between TiO_2 and other significant elements (CaO and MgO) indicate that TiO_2 is a key chemical ingredient of clays (kaolinite) rather than mafic minerals. Al_2O_3 correlates positively with K_2O (r = 0.88) and strongly positively with SiO_2 (r = 0.75). These linear lines indicate the traces of kaolinite. CaO has a negative correlation with SiO_2 (r = −0.44), indicating that calcite is somewhat dominant over other minerals.

Na_2O has a positive correlation with SiO_2 (r = 0.71), Al_2O_3 (r = 0.83), Fe_2O_3 (r = 0.62), TiO_2 (r = 0.75), and K_2O (r = 0.80), but correlates negatively with CaO and MgO (r = −0.23 and −0.19). These relationships imply the presence of smectite in the shale. Certain trace elements, such as Zr, Co, Rb, Ba, and Hf, have a positive correlation with Al_2O_3 (r = 0.70, 0.71, 0.16, 0.34, and 0.24, respectively), suggesting that they are likely fixed in K-feldspars and clays. On the other hand, some trace elements, such as Cr, V, Ni, Cu, Rb, Nb, Cs, and U, correlate positively with Al_2O_3 (r = 0.04, 0.16, 0.17, 0.08, 0.16, 0.13, 0.01, and 0.08, respectively) and negatively with Sc, Sr, Sb, La, Ce, Nd, Sm, Eu, Gd, Tb, Yb, and Th (r = −0.35, −0.41, −0.02) [31].

Correlations between Ni, Cr, and Co may be utilized to determine the contribution of mafic minerals to the formation of rock. Between Cr and Ni, the correlation value is negative (r = −0.15), whereas Co is negatively connected with Cr (r = −0.01). This association indicates that acidic minerals are more prevalent than basic and ultrabasic minerals.

The concentration of large-ion lithophile elements (LILEs) varies significantly among examined samples. The alkali element Rb varies between 131–248 ppm with a mean value of 185 ppm, 119–193 ppm with a mean value of 167 ppm, and 151–227 ppm with a mean value of 192 ppm for Sections 1–3. Sr concentrations range from 130–211 ppm with an average of 163 ppm in Section 1, 124–234 ppm with an average of 152 ppm in Section 3, and 103–350 ppm with an average of 215 ppm in Section 2. The concentration of Ba in samples obtained from Section 3 ranges between 187 and 631 parts per million, with an average of 299 ppm. Its concentration ranges between 153–299 ppm with an average of 215 ppm in Section 1, whereas it fluctuates between 165–292 ppm with an average of 206 ppm in Section 2. The association between K_2O and Rb is negligible, i.e., r = 0.03. The positive value indicates that the abundance of these elements is primarily controlled by clay minerals containing potassium (i.e., illite, muscovite, and biotite) [45,51]. The association between Al_2O_3 and Rb is positive, indicating that phyllosilicates regulate the distribution of these elements. Similarly, Al_2O_3 and Ba exhibit a positive correlation, indicating that phyllosilicates control the distribution of these elements as well.

5. Discussion

The findings and their interpretation in our models indicate that the Chichali Formation sediments were derived from intermediate igneous rock, namely, granodiorite. This may be validated geologically since the deposition of shelf glauconitic sand and shale of the Chichali Formation was observed during the Late Jurassic [52]. This is related to the Tethys transgression in the Late Jurassic as a result of the Indian Plate's northwestern margin expanding along rift faults. Flooding occurred during the Late Jurassic due to the rifting of the Indian Plate's passive continental edge. Clasts from the exposed portion of the Indian Shield were deposited in the Chichali and Sembar Formations on the shelf as a result of the thermal rising of the Indian Shield's eastern and southern margins [2,20].

The findings of U–Pb dating are consistent with the Tethyan and lower Himalayan values. The Indian Plate was believed to be the source of the Early Cretaceous Chichali Formation due to its similarities to the Tethyan and Lesser Himalayas. In this period, the present Higher Himalayas were not excavated because they were covered by Tethyan Himalayan sediments, which ruled out the possibility of the Chichali Formation being derived from them [53]. As a result of the detrital zircon provenance, the Indian Plate was the source of the Chichali Formation. The Indian Plate split apart from Gondwana between the Late Jurassic and Early Cretaceous period and began its journey north via the Somali and Mozambique basins [54,55]. The Indian Plate was also experiencing domal uplift during this period when it passed through the Ninety-East Kerguelen [56]. As a result of domal uplift-induced erosion, the Indian Plate basins contain a significant amount of detrital sediment [57]. The evidence from the detrital record of the Chichali Formation supports erosion from Cambrian and Ordovician plutons that are common in the Tethyan and Lesser Himalayas [58]. Thus, the Aravalli Ranges might be considered a prospective source of sediment inside the Indian Plate during the Cretaceous period.

According to the interpretations made in this study, the source area experienced moderate weathering. This can be justified further by the fact that during the Late Jurassic, the study area was located at $-35°$ paleo-latitude, which was a part of the sub-tropical climate belt, and a slightly humid to dry climate prevailed, resulting in moderate weathering in the source area. The Th/U ratio in the majority of upper crustal rocks is between 3.5 and 4. A Th/U ratio greater than 4 implies moderate to severe weather conditions in the source location. In the majority of instances, U is lost during weathering, increasing the Th/U ratio. The ratio of uranium to thorium in the Chichali Formation is more than 4, which means that the weathering of clasts from the parent rock happened in a somewhat wet-to-dry environment.

According to several tectonic setting discriminating diagrams, the Chichali Formation was deposited on the tectonically stable passive continental margin of the northern Indian Plate. At the period of the deposition of the Chichali Formation, the Indian Plate was in a state of calm, since no collisions with the Kohistan Island Arc (KIA) or the Eurasian Plate had occurred.

Nickel concentrations in typical soils have been reported to be as high as 40 ppm and as low as 25 ppm by various researchers [33,59]. Chromium levels in typical soils have been observed to range from trace to as high as 5.23% [60]. Cr concentrations of up to 4000 ppm are predicted in soils derived from igneous rocks, i.e., ultramafic rocks, while Cr concentrations of up to roughly 11 ppm are expected in sedimentary rocks [61]. The permitted limits for Zn, Ba, Ni, Cu, and Cr in drinking water are 5 ppm, 0.7 ppm, 0.02 ppm, and 0.05 ppm, respectively, as determined by Karasakal [62] (Table 3). This indicates that the average concentrations of Zn, Ba, Ni, Cu, and Cr in the examined sections that were looked at are higher than the acceptable limits set by US-EPA (1978) and Pakistan-EPA (1978).

Table 3. Criteria of heavy metal concentrations in normal soils and drinking water. A = Standard values for normal soils; B = Maximum contaminant level (MCL); C = Maximum contaminant level (MCL); D = Maximum contaminant level (MCL).

Parameter	Mean Sec-1	Mean Sec-2	Mean Sec-3	Min–Max Sec-1	Min–Max Sec-2	Min–Max Sec-3	Standard (Soils)A	(MCL)B	(MCL)C	(MCL)D
Copper (Cu)	65.6	36.8	52	65–67	18–57	32–66	20–40	1.3	1	1
Chromium (Cr)	131.3	115.2	114	120–149	94–135	85–142	90	-	0.05	0.05
Nickel (Ni)	9.33	18.1	17.5	44,814	44,887	44,923	40	-	0.02	<0.02
Zinc (Zn)	123.3	56	116.3	89–169	41–86	81–154	44,880	5	5	5
Barium (Ba)	215.3	206.8	299	153–299	165–243	187–631	-	-	0.7	0.7

The soil would be enriched with Cu, Zn, and Cr as a result of weathering and erosion. As Bowen [63] anticipated, weathering of the examined shale results in higher concentrations of Ba, Cu, and Zn than those seen in typical soils [64]. These elevated levels of trace elements have a toxic effect on soils. These are either agricultural or non-agricultural soils. If the soil formed by weathering is agricultural soil, it may be toxic to crops, and when these crops are consumed by living organisms, they may cause injury. The next section discusses some of the hazardous impacts of enriching these elements in crops, plants, and soils. While copper is required in minimal amounts for both plants and animals, its excessive concentration may be hazardous. Copper concentrations over a certain level may result in stunted development and chlorosis [65]. Although chromium is neither needed nor useful for plants, it is necessary for trace amounts for mammals. It is poisonous and may cause cancer in animals when present at high concentrations [66].

If surface runoff rises and erodes the Chichali Formation sediments, they will have hazardous consequences if they reach the underground water, which percolates deep into the soil pores. If it combines with surface streams, it may be toxic to aquatic life.

Copper levels in the examined portion exceed WHO (2004) and US-EPA (1978) acceptable drinking water limits [67]. If the copper in the examined portion combines with surface water or percolates down into the subsurface water table, it may cause hazardous consequences in animals and humans, such as gastrointestinal tract stimulation, irritation of stomach nerve endings, and induction of the vomiting reflex [68]. Chromium concentrations within the allowable range are required for human nutrition to sustain proper glucose metabolism. When that level exceeds a particular threshold, poisoning occurs. The deleterious consequences of consuming a high concentration of Cr in drinking water might result in nephritis and glycosuria. Nickel is reasonably safe since the US-EPA (1978) does not propose any limitations for drinking water [69]. When water containing dissolved nickel comes into contact with the skin, it may cause dermatitis, and inhalation of the water can induce lung cancer. Taking into account all of the trace elements, it has been determined that if the Chichali Formation weathers and mixes with streams or reaches the water table, a high concentration of trace elements would pollute the water, rendering it unfit for residential use.

6. Conclusions

SiO_2 (17.89–60.14 wt%) is the primary major oxide in the Chichali Formation in the Hazara Basin, followed by Al_2O_3 (4.89–21.19 wt%), MgO (1–7.27 wt%), Fe_2O_3 (1.26–9.27 wt%), CaO (0.99–2.88 wt%), K_2O (0.61–5.75 wt%), Na_2O (0.01–0.17 wt%), and TiO_2 (0.5–1.48 wt%). We computed ratios such as Ba/Sc, Ba/Co, Th/Sc, Cr/Th, Cr/Zr, Th/Co, Th/Cr, and Sc/Th, bivariate plots such as Al_2O_3 (wt%) versus TiO_2 (ppm), TiO_2 (wt%) vs. Zr (ppm), and TiO_2 (wt% vs. Ni (ppm), and discriminant function (Df1–Df2). The bivariate plots of SiO_2 vs. K_2O/Na_2O and discriminant function (Df1–Df2) indicate that the Chichali Formation's Early Cretaceous shales have mostly passive margin settings.

The computed values of CIA, ICV, Rb, and Cs indicate that the source regime experienced mild chemical weathering. The presence of heavy metals (Ba, Zn, Ni, Cr, and Cu) in the soils formed from Chichali Formation shale implies toxicity to agriculture and drinking water if the soils are combined with agricultural soils and/or drinking water.

Supplementary Materials: The following supporting information can be downloaded at: https://www.mdpi.com/article/10.3390/jmse10060800/s1, Table S1: Pearson correlation of major and minor oxides.

Author Contributions: Conceptualization, A.G.F. and M.U.; methodology, A.G.F.; software, A.G.F. and H.T.J.; validation, A.G.F., M.U. and H.T.J.; formal analysis, A.G.F., M.U. and H.T.J.; investigation, A.G.F., M.U. and H.T.J.; resources, A.G.F. and M.U.; data curation, A.G.F. and M.U.; writing—original draft preparation, A.G.F. and M.U.; writing—review and editing, H.T.J., F.S., M.A.F.M., G.K. and A.K.J.; visualization, M.U., H.T.J., F.S. and M.A.F.M.; supervision, M.U.; project administration, M.U.; funding acquisition, G.K. and H.T.J. All authors have read and agreed to the published version of the manuscript.

Funding: This research received no external funding.

Institutional Review Board Statement: Not Applicable.

Informed Consent Statement: Not Applicable.

Data Availability Statement: Not Applicable.

Acknowledgments: We would like to express our gratitude to Bestway Cements Limited in Hattar, Pakistan, for permitting us to utilize their facilities. The authors declare that they have no potential conflicts of interest related to the article's authorship and publication. We would like to thank the editor and the anonymous reviewers for their thorough and helpful reviews, which made our study much more valuable.

Conflicts of Interest: The authors declare no conflict of interest.

References

1. Umar, M.; Sabir, M.A.; Farooq, M.; Khan, M.M.S.S.; Faridullah, F.; Jadoon, U.K.; Khan, A.S. Stratigraphic and sedimentological attributes in Hazara Basin Lesser Himalaya, North Pakistan: Their role in deciphering minerals potential. *Arab. J. Geosci.* **2015**, *8*, 1653–1667. [CrossRef]
2. Ali, S.K.; Janjuhah, H.T.; Shahzad, S.M.; Kontakiotis, G.; Saleem, M.H.; Khan, U.; Zarkogiannis, S.D.; Makri, P.; Antonarakou, A. Depositional Sedimentary Facies, Stratigraphic Control, Paleoecological Constraints, and Paleogeographic Reconstruction of Late Permian Chhidru Formation (Western Salt Range, Pakistan). *J. Mar. Sci. Eng.* **2021**, *9*, 1372. [CrossRef]
3. Abbasi, I.A.; Haneef, M.; Obaid, S.; Daud, F.; Qureshi, A.W. Mesozoic deltaic system along the western margin of the Indian plate: Lithofacies and depositional setting of Datta Formation, North Pakistan. *Arab. J. Geosci.* **2012**, *5*, 471–480. [CrossRef]
4. Janjuhah, H.T.; Ishfaque, M.; Mehmood, M.I.; Kontakiotis, G.; Shahzad, S.M.; Zarkogiannis, S.D. Integrated Underground Mining Hazard Assessment, Management, Environmental Monitoring, and Policy Control in Pakistan. *Sustainability* **2021**, *13*, 13505. [CrossRef]
5. Qasim, M.; Khan, M.A.; Haneef, M. Stratigraphic characterization of the Early Cambrian Abbottabad Formation in the Sherwan area, Hazara region, N. Pakistan: Implications for Early Paleozoic stratigraphic correlation in NW Himalayas, Pakistan. *J. Himal. Earth Sci.* **2014**, *47*, 25.
6. Ahsan, N. Facies modeling, depositional and diagenetic environments of Kawagarh Formation, Hazara Basin, Pakistan. Doctoral Dissertation, College of Earth and Environmental Sciences, Punjab University, Lahore, Pakistan, 2007.
7. Hasterok, D.; Gard, M.; Webb, J. On the radiogenic heat production of metamorphic, igneous, and sedimentary rocks. *Geosci. Front.* **2018**, *9*, 1777–1794. [CrossRef]
8. Armstrong-Altrin, J.S.; Lee, Y.I.; Verma, S.P.; Ramasamy, S. Geochemistry of Sandstones from the Upper Miocene Kudankulam Formation, Southern India: Implications for Provenance, Weathering, and Tectonic Setting. *J. Sediment. Res.* **2004**, *74*, 285–297. [CrossRef]
9. Armstrong-Altrin, J.S.; Lee, Y.I.; Kasper-Zubillaga, J.J.; Carranza-Edwards, A.; Garcia, D.; Eby, G.N.; Balaram, V.; Cruz-Ortiz, N.L. Geochemistry of beach sands along the western Gulf of Mexico, Mexico: Implication for provenance. *Geochemistry* **2012**, *72*, 345–362. [CrossRef]
10. Zaid, S.M. Integrated petrographic, mineralogical, and geochemical study of the Late Cretaceous–Early Tertiary Dakhla Shales, Quseir–Nile Valley Province, central Egypt: Implications for source area weathering, provenance, and tectonic setting. *Arab. J. Geosci.* **2015**, *8*, 9237–9259. [CrossRef]
11. Martini, A.M.; Walter, L.M.; Ku, T.C.W.; Budai, J.M.; McIntosh, J.C.; Schoell, M. Microbial production and modification of gases in sedimentary basins: A geochemical case study from a Devonian shale gas play, Michigan basin. *AAPG Bull.* **2003**, *87*, 1355–1375. [CrossRef]
12. Lin, N.H.; Guo, Y.; Wai, S.N.; Tamehe, L.S.; Wu, Z.; Naing, N.M.; Zhang, J. Sedimentology and geochemistry of Middle Eocene-Lower Oligocene sandstones from the western Salin Sub-Basin, the Central Myanmar Basin: Implications for provenance, source area weathering, paleo-oxidation and paleo-tectonic setting. *J. Southeast. Asian Earth Sci.* **2019**, *173*, 314–335. [CrossRef]
13. Ding, L.; Qasim, M.; Jadoon, I.A.; Khan, M.A.; Xu, Q.; Cai, F.; Wang, H.; Baral, U.; Yue, Y. The India–Asia collision in north Pakistan: Insight from the U–Pb detrital zircon provenance of Cenozoic foreland basin. *Earth Planet. Sci. Lett.* **2016**, *455*, 49–61. [CrossRef]
14. Jan, I.U.; Iqbal, S.; Davies, S.J.; Zalasiewicz, J.A.; Stephenson, M.H.; Wagreich, M.; Haneef, M.; Hanif, M.; Ahmad, S. A Periglacial Palaeoenvionment in the Upper Carboniferous-Lower Permian Tobra Formation of the Salt Range, Pakistan. *Acta Geol. Sin. Engl. Ed.* **2017**, *91*, 1063–1078. [CrossRef]
15. Khan, U.; Janjuhah, H.T.; Kontakiotis, G.; Rehman, A.; Zarkogiannis, S.D. Natural Processes and Anthropogenic Activity in the Indus River Sedimentary Environment in Pakistan: A Critical Review. *J. Mar. Sci. Eng.* **2021**, *9*, 1109. [CrossRef]
16. Ishfaque, M.; Dai, Q.; Haq, N.U.; Jadoon, K.; Shahzad, S.M.; Janjuhah, H.T. Use of Recurrent Neural Network with Long Short-Term Memory for Seepage Prediction at Tarbela Dam, KP, Pakistan. *Energies* **2022**, *15*, 3123. [CrossRef]
17. Malkani, M.S. Stratigraphy, mineral potential, geological history and paleobiogeography of Balochistan Province, Pakistan. *Sindh Univ. Res. J. -SURJ (Sci. Ser.)* **2015**, *43*. Available online: https://sujo-old.usindh.edu.pk/index.php/SURJ/article/view/1383 (accessed on 26 April 2022).
18. Garzanti, E.; Liang, W.; Andò, S.; Clift, P.D.; Resentini, A.; Vermeesch, P.; Vezzoli, G. Provenance of Thal Desert sand: Focused erosion in the western Himalayan syntaxis and foreland-basin deposition driven by latest Quaternary climate change. *EarthSci. Rev.* **2020**, *207*, 103220. [CrossRef]
19. Malkani, M.S.; Mahmood, Z. Revised stratigraphy of Pakistan. *Geol. Surv. Pak. Rec.* **2016**, *127*, 1–87.

20. Malkani, M.S.; Mahmood, Z. Stratigraphy of Pakistan. Geological Survey of Pakistan. *Memoir* **2017**, *24*, 1–134.
21. Nesbitt, H.W.; Young, G.M. Early Proterozoic climates and plate motions inferred from major element chemistry of lutites. *Nature* **1982**, *299*, 715–717. [CrossRef]
22. Cox, R.; Lowe, D.R.; Cullers, R. The influence of sediment recycling and basement composition on evolution of mudrock chemistry in the southwestern United States. *Geochim. Cosmochim. Acta* **1995**, *59*, 2919–2940. [CrossRef]
23. Besly, B.; Burley, S.D.; Turner, P. The late Carboniferous 'Barren Red Bed' play of the Silver Pit area, Southern North Sea. In *Geological Society, London, Petroleum Geology Conference Series*; Geological Society of London: London, UK, 1993; Volume 4, pp. 727–740. [CrossRef]
24. Wang, L.; Cho, D.-W.; Tsang, D.C.; Cao, X.; Hou, D.; Shen, Z.; Alessi, D.; Ok, Y.S.; Poon, C.S. Green remediation of As and Pb contaminated soil using cement-free clay-based stabilization/solidification. *Environ. Int.* **2019**, *126*, 336–345. [CrossRef] [PubMed]
25. Hossain, I.; Roy, K.K.; Biswas, P.K.; Alam, M.; Moniruzzaman, M.; Deeba, F. Geochemical characteristics of Holocene sediments from Chuadanga district, Bangladesh: Implications for weathering, climate, redox conditions, provenance and tectonic setting. *Chin. J. Geochem.* **2014**, *33*, 336–350. [CrossRef]
26. Cox, P.; Wood, R.; Dickson, J.; Al Rougha, H.; Shebl, H.; Corbett, P. Dynamics of cementation in response to oil charge: Evidence from a Cretaceous carbonate field, U.A.E. *Sediment. Geol.* **2010**, *228*, 246–254. [CrossRef]
27. Deru, X.; Xuexiang, G.; Pengchun, L.; Guanghao, C.; Bin, X.; Bachlinski, R.; Zhuanli, H.; Gonggu, F. Mesoproterozoic–Neoproterozoic transition: Geochemistry, provenance and tectonic setting of clastic sedimentary rocks on the SE margin of the Yangtze Block, South China. *J. Southeast. Asian Earth Sci.* **2007**, *29*, 637–650. [CrossRef]
28. Paikaray, S.; Banerjee, S.; Mukherji, S. Geochemistry of shales from the Paleoproterozoic to Neoproterozoic Vindhyan Supergroup: Implications on provenance, tectonics and paleoweathering. *J. Southeast. Asian Earth Sci.* **2008**, *32*, 34–48. [CrossRef]
29. Howarth, R.J. Sources for a history of the ternary diagram. *Br. J. Hist. Sci.* **1996**, *3*, 337–356. [CrossRef]
30. Nagarajan, R.; Madhavaraju, J.; Nagendra, R.; Armstrong-Altrin, J.S.; Moutte, J. Geochemistry of Neoproterozoic shales of the Rabanpalli Formation, Bhima Basin, Northern Karnataka, southern India: Implications for provenance and paleoredox conditions. *Rev. Mex. De Cienc. Geológicas* **2007**, *24*, 150–160.
31. Cullers, R.L.; Barrett, T.; Carlson, R.; Robinson, B. Rare-earth element and mineralogic changes in Holocene soil and stream sediment: A case study in the Wet Mountains, Colorado, U.S.A. *Chem. Geol.* **1987**, *63*, 275–297. [CrossRef]
32. Adriano, D. *Trace Elements in Terrestrial Ecosystems*; Springer: New York, NY, USA, 1986.
33. Cullers, R.L. The controls on the major and trace element variation of shales, siltstones, and sandstones of Pennsylvanian-Permian age from uplifted continental blocks in Colorado to platform sediment in Kansas, USA. *Geochim. Cosmochim. Acta* **1994**, *58*, 4955–4972. [CrossRef]
34. Eqani, S.A.M.A.S.; Kanwal, A.; Bhowmik, A.; Sohail, M.; Ullah, R.; Ali, S.M.; Alamdar, A.; Ali, N.; Fasola, M.; Shen, H. Spatial distribution of dust–bound trace elements in Pakistan and their implications for human exposure. *Environ. Pollut.* **2016**, *213*, 213–222. [CrossRef] [PubMed]
35. Qasim, M.; Ding, L.; Khan, M.A.; Jadoon, I.A.K.; Haneef, M.; Baral, U.; Cai, F.; Wang, H.; Yue, Y. Tectonic Implications of Detrital Zircon Ages From Lesser Himalayan Mesozoic-Cenozoic Strata, Pakistan. *Geochem. Geophys. Geosystems* **2018**, *19*, 1636–1659. [CrossRef]
36. Roser, B.; Korsch, R. Provenance signatures of sandstone-mudstone suites determined using discriminant function analysis of major-element data. *Chem. Geol.* **1988**, *67*, 119–139. [CrossRef]
37. Keskin, Ş. Geochemistry of Çamardı Formation sediments, central Anatolia (Turkey): Implication of source area weathering, provenance, and tectonic setting. *Geosci. J.* **2011**, *15*, 185–195. [CrossRef]
38. Agrawal, S.; Guevara, M.; Verma, S.P. Tectonic Discrimination of Basic and Ultrabasic Volcanic Rocks through Log-Transformed Ratios of Immobile Trace Elements. *Int. Geol. Rev.* **2008**, *50*, 1057–1079. [CrossRef]
39. Verma, S.P.; Pandarinath, K.; Verma, S.K.; Agrawal, S. Fifteen new discriminant-function-based multi-dimensional robust diagrams for acid rocks and their application to Precambrian rocks. *Lithos* **2013**, *168*, 113–123. [CrossRef]
40. Bhatia, M.R. Plate Tectonics and Geochemical Composition of Sandstones. *J. Geol.* **1983**, *91*, 611–627. [CrossRef]
41. Yan, D.; Chen, D.; Wang, Q.; Wang, J. Large-scale climatic fluctuations in the latest Ordovician on the Yangtze block, South China. *Geology* **2010**, *38*, 599–602. [CrossRef]
42. Visser, J.N.; Young, G.M. Major element geochemistry and paleoclimatology of the Permo-Carboniferous glacigene Dwyka Formation and postglacial mudrocks in southern Africa. *Palaeogeogr. Palaeoclim. Palaeoecol.* **1990**, *81*, 49–57. [CrossRef]
43. Singh, P.K.; Khan, M.S. Geochemistry of Palaeoproterozoic Rocks of Aravalli Supergroup: Implications for Weathering History and Depositional Sequence. *Int. J. Geosci.* **2017**, *8*, 1278–1299. [CrossRef]
44. Herron, M.M. Geochemical Classification of Terrigenous Sands and Shales from Core or Log Data. *J. Sediment. Res.* **1988**, *58*, 820–829. [CrossRef]
45. McLennan, S.; Taylor, S.; Eriksson, K. Geochemistry of Archean shales from the Pilbara Supergroup, Western Australia. *Geochim. Cosmochim. Acta* **1983**, *47*, 1211–1222. [CrossRef]
46. Das, B.K.; Haake, B.-G. Geochemistry of Rewalsar Lake sediment, Lesser Himalaya, India: Implications for source-area weathering, provenance and tectonic setting. *Geosci. J.* **2003**, *7*, 299–312. [CrossRef]
47. Tulyaganov, D.; Agathopoulos, S.; Kansal, I.; Valério, P.; Ribeiro, M.; Ferreira, J. Synthesis and properties of lithium disilicate glass-ceramics in the system SiO_2–Al_2O_3–K_2O–Li_2O. *Ceram. Int.* **2009**, *35*, 3013–3019. [CrossRef]

48. Guagliardi, I.; Cicchella, D.; De Rosa, R. A Geostatistical Approach to Assess Concentration and Spatial Distribution of Heavy Metals in Urban Soils. *Water Air Soil Pollut.* **2012**, *223*, 5983–5998. [CrossRef]
49. Shabbir, Z.; Sardar, A.; Shabbir, A.; Abbas, G.; Shamshad, S.; Khalid, S.; Natasha; Murtaza, G.; Dumat, C.; Shahid, M. Copper uptake, essentiality, toxicity, detoxification and risk assessment in soil-plant environment. *Chemosphere* **2020**, *259*, 127436. [CrossRef]
50. Pak, E. *National Standards for Drinking Water Quality*; Pakistan Environmental Protection Agency, Ministry of Environment, Government of Pakistan: Isiamabad, Pakistan, 2008.
51. Podkovyrov, V.N.; Grazhdankin, D.V.; Maslov, A.V. Lithogeochemistry of the Vendian fine-grained clastic rocks in the southern Vychegda trough. *Lithol. Miner. Resour.* **2011**, *46*, 427–446. [CrossRef]
52. Banerjee, S.; Bansal, U.; Thorat, A.V. A review on palaeogeographic implications and temporal variation in glaucony composition. *J. Palaeogeogr.* **2016**, *5*, 43–71. [CrossRef]
53. Jonell, T.N.; Owen, L.A.; Carter, A.; Schwenniger, J.-L.; Clift, P.D. Quantifying episodic erosion and transient storage on the western margin of the Tibetan Plateau, upper Indus River. *Quat. Res.* **2017**, *89*, 281–306. [CrossRef]
54. Krishna, J. Applicability of the Sequence Framework in IEAP and GTM, with Brief Comments on the Hydrocarbon Prospects in the Indian Basins. In *The Indian Mesozoic Chronicle*; Springer: Singapore, 2017; pp. 367–459. [CrossRef]
55. Gakkhar, R.A.; Bechte, A.; Gratzer, R. Source-rock potential and origin of hydrocarbons in the Cretaceous and Jurassic sediments of the Punjab Platform (Indus Basin, Pakistan). *Pak. J. Hydrocarb. Res.* **2011**, *21*, 1–17.
56. Ghose, N.C.; Chatterjee, N.; Windley, B.F. Subaqueous early eruptive phase of the late Aptian Rajmahal volcanism, India: Evidence from volcaniclastic rocks, bentonite, black shales, and oolite. *Geosci. Front.* **2017**, *8*, 809–822. [CrossRef]
57. Shen, T.; Wang, G.; Leloup, P.H.; van der Beek, P.; Bernet, M.; Cao, K.; Wang, A.; Liu, C.; Zhang, K. Controls on Cenozoic exhumation of the Tethyan Himalaya from fission-track thermochronology and detrital zircon U-Pb geochronology in the Gyirong basin area, southern Tibet. *Tectonics* **2016**, *35*, 1713–1734. [CrossRef]
58. Naeem, M.; Burg, J.-P.; Ahmad, N.; Chaudhry, M.N.; Khalid, P. U-Pb zircon systematics of the Mansehra Granitic Complex: Implications on the early Paleozoic orogenesis in NW Himalaya of Pakistan. *Geosci. J.* **2016**, *20*, 427–447. [CrossRef]
59. Fulekar, M.H.; Jadia, C.D. Phytoremediation: The application of vermicompost to remove zinc, cadmium, copper, nickel and lead by sunflower plant. *Environ. Eng. Manag. J.* **2008**, *7*, 547–558. [CrossRef]
60. Dadar, M.; Adel, M.; Saravi, H.N.; Fakhri, Y. Trace element concentration and its risk assessment in common kilka (*Clupeonella cultriventris* caspia Bordin, 1904) from southern basin of Caspian Sea. *Toxin Rev.* **2017**, *36*, 222–227. [CrossRef]
61. Sarapää, O.; Lauri, L.S.; Ahtola, T.; Al-Ani, T.; Grönholm, S.; Kärkkäinen, N.; Lintinen, P.; Torppa, A.; Turunen, P. *Discovery Potential of Hi-Tech Metals and Critical Minerals in Finland*; Geological Survey of Finland: Espoo, Finland, 2015.
62. Karasakal, A. Determination of Trace and Major Elements in Vegan Milk and Oils by ICP-OES after Microwave Digestion. *Biol. Trace Element Res.* **2020**, *197*, 683–693. [CrossRef] [PubMed]
63. Bowen, H.J.M. *Environmental Chemistry of the Elements*; Academic Press: Cambridge, MA, USA, 1979.
64. de Caritat, P.; Reimann, C.; Team, N.P.; Team, G.P. Comparing results from two continental geochemical surveys to world soil composition and deriving Predicted Empirical Global Soil (PEGS2) reference values. *Earth Planet. Sci. Lett.* **2012**, *319*, 269–276. [CrossRef]
65. Cassanego, M.B.B.; Goldoni, A.; Heldt, F.H.; Osório, D.M.; Windisch, P.G.; Droste, A. Germination and sporophytic development of Regnellidium diphyllum Lindm. (Marsileaceae) in the presence of copper. *Acta Bot. Bras.* **2013**, *27*, 26–30. [CrossRef]
66. Pandey, G.; Madhuri, S. Heavy metals causing toxicity in animals and fishes. *Res. J. Anim. Vet. Fish. Sci.* **2014**, *2*, 17–23.
67. Sathawara, N.G.; Parikh, D.J.; Agarwal, Y.K. Essential Heavy Metals in Environmental Samples from Western India. *Bull. Environ. Contam. Toxicol.* **2004**, *73*, 756–761. [CrossRef]
68. Massa, F.; Monory, K. Endocannabinoids and the gastrointestinal tract. *J. Endocrinol. Investig.* **2006**, *29*, 47–57.
69. Christison, T.T.; Rohrer, J.S. Direct determination of free cyanide in drinking water by ion chromatography with pulsed amperometric detection. *J. Chromatogr. A* **2007**, *1155*, 31–39. [CrossRef] [PubMed]

Journal of
Marine Science and Engineering

MDPI

Article

The Tanais Bay of the Eastern Paratethys Sea at the Sarmatian–Maeotian Transition (Late Miocene): Widespread Desiccations and Local Uplifts in the Light of Historical Information

Dmitry A. Ruban

K.G. Razumovsky Moscow State University of Technologies and Management (The First Cossack University), Zemlyanoy Val Street 73, Moscow 109004, Russia; ruban-d@mail.ru

Abstract: The Late Miocene evolution of the Eastern Paratethys Sea was marked by significant palaeogeographical transformations. The knowledge of them should be improved with the information from the peripheral parts of this semi-enclosed marine basin. The study area corresponds to the Rostov Dome where the northern shore of the Eastern Paratethys is widely documented. The information from the previously published work, going back to the beginning of the 20th century, is collected. Its analysis allows us to document the spatial distribution of Middle Sarmatian–Late Maeotian (Tortonian–Messinian) deposits. The results shed light into the palaeogeographical changes in the Tanais Bay of the Eastern Paratethys Sea, which included the short-term hiatus at the Middle/Late Sarmatian boundary, the Early Maeotian regression, and the gradual Late Maeotian ingression when the bay re-established, but with a different configuration. These changes and the overall spatial distribution of the studied deposits cannot be explained by only the fluctuations in the level of the Eastern Paratethys and the desiccation episodes established in its central part. Most probably, the local tectonic uplifts were an important driver of the Late Miocene evolution of the Tanais Bay.

Keywords: Event Stratigraphy; Interior Palaeosea; Late Cenozoic; Local Tectonics; Rostov Dome

Citation: Ruban, D.A. The Tanais Bay of the Eastern Paratethys Sea at the Sarmatian–Maeotian Transition (Late Miocene): Widespread Desiccations and Local Uplifts in the Light of Historical Information. *J. Mar. Sci. Eng.* **2022**, *10*, 915. https://doi.org/10.3390/jmse10070915

Academic Editor: Dimitris Sakellariou

Received: 9 June 2022
Accepted: 29 June 2022
Published: 1 July 2022

1. Introduction

The northern periphery of the former Neo-Tethys Ocean experienced significant reorganizations, together with the growth of the Alpine tectonic belt, in the second half of the Cenozoic [1–3]. One outcome of these processes was the formation of a huge elongated water mass, namely the Paratethys, which stretched from West Europe to Central Asia [4–10]. The Paratethys was partly (and, temporarily, fully) isolated from the World Ocean, and it was also fragmented into several basins occupied by seas, some of which looked like megalakes [6,11]. Its eastern portion is known as the Eastern Paratethys Sea (palaeosea) [11–13]. In the Late Miocene, this was the well-shaped interior sea located between the Russian Platform in the north, Anatolia in the south, the Carpathian domain in the west, and the Turan Platform in the east (Figure 1). It consisted of several sedimentary basins [11].

To date, there has been significant progress in the understanding of the Late Miocene evolution of the Eastern Paratethys. Particularly, much has become known about its disconnections [13], strong unbalances of water budget [6], astronomical cycles [14,15], salinity changes [16], and interaction of eustasy and tectonic activity [17]. However, a significant part of this information has been accumulated in the central and southern parts of the palaeosea, i.e., in the Euxine Basin. The stratigraphical records of the peripheral, nearshore domains can provide another portion of knowledge, which can extend the understanding of this palaeosea's evolution. One very suitable locality for such investigations is situated on the northern periphery of the Ciscaucasian Basin, where a wide, shallow-water bay existed in the Late Miocene. This is known as the Tanais Bay (palaeobay). Tanais was the Ancient Greek name of the Don River and the ancient settlement near the mouth of this river [18,19]. At the end of the 19th century, the name Tanais Bay was introduced by Sokolov [20] to characterize the palaeogeographical feature where the specific Late Miocene deposits accumulated (the presence of this wide bay is reflected by the modern reconstructions [11]). During more than a century of subsequent research, the information about these deposits grew, although some questions remain unanswered, and even the configuration of the palaeobay is unclear. The current understanding of the stratigraphy of the Late Miocene deposits of the Tanais Bay is presented by Ruban [21–23]. It should be noted that lithologies and the related depositional environments in this palaeobay are already well-established, but serious biases exist in the understanding of the depositional architecture. These biases prevent accurate palaeogeographical reconstructions.

Figure 1. The Eastern Paratethys Sea in the Late Miocene (created by the author referring the palaeogeographical contours to [11]).

The aim of the present work is to reveal the palaeogeographical changes in the Tanais Bay at the Sarmatian–Maeotian (Tortonian–Messinian) transition through examination of the spatial distribution of the related sedimentary bodies. It is documented in regard to what has been learned about both reference sections and numerous small, similar-looking outcrops during about a century of research (attention is not paid to the correlation of the reference sections already completed by Ruban [22]). The outcomes of such an analysis permit comparison of the records from the palaeosea's periphery (Tanais Bay) and its internal part in order to form judgments about such notable events as basin desiccations [6]. In other words, the present paper clarifies the local evolution of the Tanais Bay for subsequently making more general conclusions. A specific and unusual aspect of this analysis is its focus on various information which has been accumulated during about a century and needs to be summarized. Many of the previous works are already too old and difficult to access. Moreover, they focus chiefly on small outcrops (the majority of these works result from geological mapping, not from stratigraphical investigations); as a result, these works were often missed. However, this old information is essential because the study area experienced significant anthropogenic modification linked to the urban sprawl in the second half of the 20th century (now, Rostov-on-Don has a population exceeding one million people, and there are other big settlements in its vicinity), and landslides also destroyed many outcrops. Thus, the earlier accumulated information cannot be recompensed with the new investigations. The present study is the first attempt to treat all these lines of historical evidence together, to systematize them, and to justify according to the latest stratigraphical developments.

2. Geological Setting

Geographically, the study area is situated in the southwestern part of the Rostov Region of the Russian Federation, where the large Don River enters the Taganrog Bay of the Azov Sea. It belongs to the southern edge of the Russian Platform (a Precambrian craton with rather thick sedimentary cover) and, particularly, the Rostov Dome, which is a Cenozoic anticline structure characterized by Belov et al. [24], Ivanitskaya and Pogrebnov [25], Kostyuchenko et al. [26], and Ruban [27]. The Precambrian crystalline basement is unconformably covered by Cretaceous mixed siliciclastic–carbonate deposits with a total thickness of ~400 m. They are overlaid by ~750 m of Paleogene–Middle Miocene sandstones and shales. Late Miocene deposits are represented by skeletal limestones, clays, and sands with a total thickness of ~100 m, and they are overlaid by Pliocene–Quaternary sands and clays exceeding 100 m in thickness. Pre-Pliocene deposits are chiefly marine, while the younger strata are continental.

The high right bank of the wide river's valley and the steep, landslide-affected slopes of the bay represent a lengthy "ribbon" of outcrops of the Late Miocene deposits [22]. There are also numerous small outcrops scattered in the study area. The Late Miocene deposits are represented by several lithostratigraphical units, namely the Taganrogskaya, Rostovskaya, Yanovskaya, Donskaya, Merzhanovskaya, and Aleksandrovskaya formations [21]. Their vertical and lateral relationships, composition, and depositional environments are summarized in Figure 2.

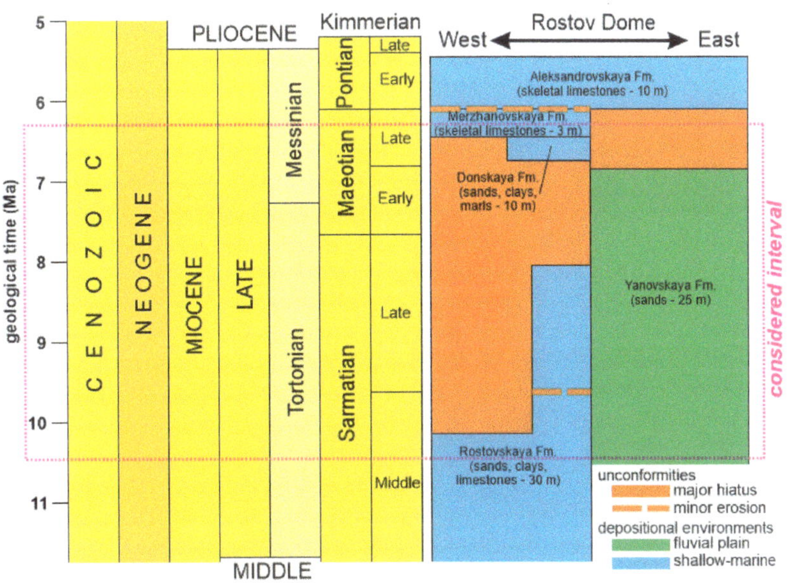

Figure 2. Stratigraphical framework of the Sarmatian–Maeotian transition; eteropic lateral relationships of units are shown (created by the author; some previous sources were referred for information: geological time scale after [28], regional units after [6,13], color codes follow [29], lithostratigraphy modified from [21,22]).

The age of the above-mentioned formations was established precisely with bivalves by Ruban [22,23]. Importantly, this age is justified to the regional stages, i.e., the geological time units developed for the Eastern Paratethys by generations of Soviet and Russian geologists. The relation of these regional stages to the standard stages of the Neogene formalized by the International Commission on Stratigraphy [28,29] has remained uncertain for decades. However, the present investigations, the outcomes of which are shown by Palcu et al. [6] and Popov et al. [13], allow us to correlate the regional and standard stages with precision (Figure 2).

Palaeogeographically, the study area belonged to the northern margin of the Eastern Paratethys (Figure 1). In the Late Miocene, the western and central parts of the Rostov Dome were occupied by the shallow-water Tanais Bay where skeletal limestones, sands, and clays deposited, whereas the eastern part was dominated by the fluvial plain and the delta of the Palaeo-Donets River, where relatively thick fluvial sands accumulated [22,23]. It should be noted that the local hydrology has experienced significant changes since the Miocene. The modern Don River flows from the east to the west to reach the Azov Sea. It has a right tributary, namely the Seversky Donets River, with its mouth located to the east of the study area. In the Miocene, these were two different river systems, namely the Palaeo-Don and the Palaeo-Donets [11]. They both ran from the north to the south to reach directly the Eastern Paratethys Sea. The Palaeo-Donets was situated relatively close to the Tanais Bay.

3. Materials and Methods

The present work employs two kinds of materials. First, these are observations of the spatial distribution of the Middle Sarmatian–Maeotian deposits made by the author in the course of field investigations in the Rostov Dome. Although the related descriptions and the correlations of fourteen reference sections were already published [21–23], this information can be re-considered together with the other data for more accurate palaeogeographical reconstructions. Second, the previous investigations (since the beginning of the 20th century) have accumulated vast knowledge about the spatial distribution of the Middle Sarmatian–Maeotian deposits, including those plots, which are densely-urbanized now and where the outcrops were either destroyed or became fully inaccessible due to infrastructure development. The outcomes of these previous investigations were published by Belov et al. [24], Ivanitskaya and Pogrebnov [25], Bogachev [30–32], Kolesnikov [33], Miroshnikov [34], Nalivkin and Sokolov [35], Paffengolts [36], and Vlasov [37–39]. They provide information about a few dozen individual outcrops and groups of outcrops. Additionally, these three palaeogeographical schemes, presented by Vlasov [37–39], are very important.

The collected information has been summarized so as to depict the spatial distribution of the Rostovskaya, Yanovskaya, Donskaya, and Merzhanovskaya formations, which represent the Sarmatian–Maeotian transition (Figure 2). In the case of the Rostovskaya Formation, its Middle Sarmatian and Late Sarmatian parts are considered separately due to the already known striking differences in their distribution [22]. For each unit, the areas of its presence and absence are documented on the basis of the available knowledge (see above). It should be noted that some old sources do not name outcrops, but often indicate the plots where the particular deposits are present, i.e., they provide the already generalized information (to avoid dealing with individual sections and outcrops was a very typical style of reporting lithostratigraphical data until the 1990s). Ignoring this information would be wrong, particularly, because the precision of these previous investigations was usually high (Figure 2). Synthesizing the available information permits us to establish contours of the spatial distribution of each above-mentioned formation in the Rostov Dome. This state-of-the-art synthesis also solves the other problem linked to the "scattered" representation of the formations in multiple outcrops, which are difficult (if possible) to correlate. The contours are shown on the same scheme of the Rostov Dome, which allows an understanding of the lateral and vertical relations of the sedimentary bodies. This scheme is essential for qualitatively deciphering the complex depositional architecture of the Late Miocene deposits of the study area.

4. Results

The performed synthesis of the available knowledge (see sources above) implies that the Rostovskaya, Yanovskaya, Donskaya, and Merzhanovskaya formations differ significantly by their spatial distribution (Figure 3). The entire geological time interval of their deposition lasted only ~4 Ma (Figure 2), which implies significant palaeogeographical changes and, thus, general dynamism of the Tanais Bay. Such a finding is expected, taking into account its peripheral position and very small depth [11,38,39].

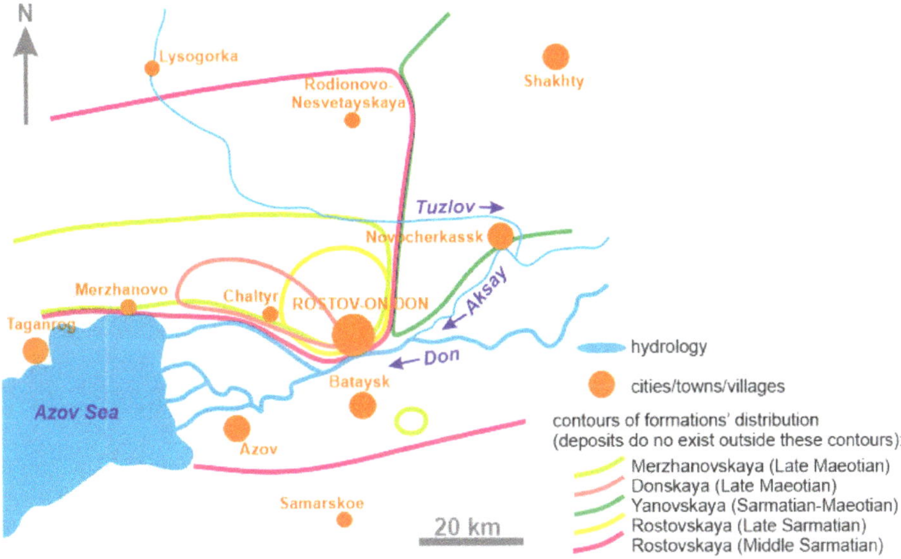

Figure 3. Synthesis of the knowledge of the present-day distribution of the Middle Sarmatian–Maeotian lithostratigraphical units within the Rostov Dome (created by the author; see text for additional explanations and data sources).

The Middle Sarmatian part of the Rostovskaya Formation is the most widely distributed. It occurs in the central, western, and southern parts of the study area (Figure 3). Most probably, these deposits also existed between Rostov-on-Don and Samarskoe, but they were totally eroded later, together with the deep incision of the wide valley of the modern Don River in the Pliocene–Quaternary (this incision exceeded 100 m). This formation is absent in the east of the study area, where the Yanovskaya Formation occurs. (Although one can hypothesize that the Yanovskaya Formation extended westwards, as a "tongue" between Rostov-on-Don and Samarskoe, there is not any evidence to support this idea.) The boundary between the areas of distribution of the Rostovskaya and Yanovskaya formations is a strait. The Late Sarmatian part of the Rostovskaya Formation and the Donskaya Formation are distinguished by very restricted distribution in the central part of the study area. The Merzhanovskaya Formation embraces a larger area, but its northern limit corresponds to the valley of the Tuzlov River, i.e., it is much less distributed than the Middle Sarmatian deposits (Figure 3). Notably, this formation crops out very locally to the southeast of Bataysk, which fact indicates its possible presence between Rostov-on-Don and Bataysk, where it was eroded in the Pliocene–Quaternary together with the Don River valley's development. In the southern part of the Rostov Dome, the Sarmatian deposits are unconformably overlaid by the Pontian deposits, i.e., the Maeotian deposits do not exist there.

In addition to the spatial distribution of the noted formations, the compilation of the information from the old sources indicates two hiatuses. The older one took place at the Middle–Late Sarmatian transition, where unconformity associates with intraformational conglomerate beds. The other hiatus started near the end of the Sarmatian and lasted until the mid-Maeotian. Importantly, these hiatuses are registered in the western and central parts of the study area, but not in its eastern part, where alluvial sands of the Yanovskaya Formation accumulated almost continuously until the Late Maeotian.

5. Discussion

5.1. Tracing Palaeogeographical Changes

The synthesized information allows reconstructing the local palaeogeography for four time slices, namely the Middle Sarmatian, the Late Sarmatian, the Early Maeotian, and the Late Maeotian. The lithologies and the related facies/palaeoenvironments are well-known [22,38,39] (Figure 2). The present study focuses on the configuration of the Tanais Bay.

As suggested by the wide distribution of the Middle Sarmatian part of the Rostovskaya Formation and its spatial contacts with the Yanovskaya Formation (Figure 3), the Tanais Bay was a wide feature on the northern periphery of the Eastern Paratethys. It occupied a flat space between the denudated land in the north and the fluvial plain of the Palaeo-Donets River in the east (Figure 4a). This corresponds to the reconstruction by Popov et al. [11]. The denudated land is interpreted because of long-term non-deposition in the northern part of the Rostov Dome and the presence of clastic material in the Sarmatian limestones derived from the adjacent land [39]. Apparently, the fluvial plain was a part of the propagated delta of the Palaeo-Donets River; it is also possible that the deltas of the Palaeo-Donets, the Palaeo-Don, and smaller rivers formed a single fluvial plain which extended far eastwards [11,39]. Notably, the Tanais Bay occupied generally the same (or an even smaller) territory of the Rostov Dome in the Middle Sarmatian relative to the Early Sarmatian (for instance, [37,39]), despite a rather strong transgression in the Eastern Paratethys [12]. This implies that the denudated land north of the Tanais Bay either remained high since the beginning of the Sarmatian or experienced gradual uplift (the mechanism of the latter cannot even be hypothesized). The second option seems to be more probable, taking into account the later scenario (see below).

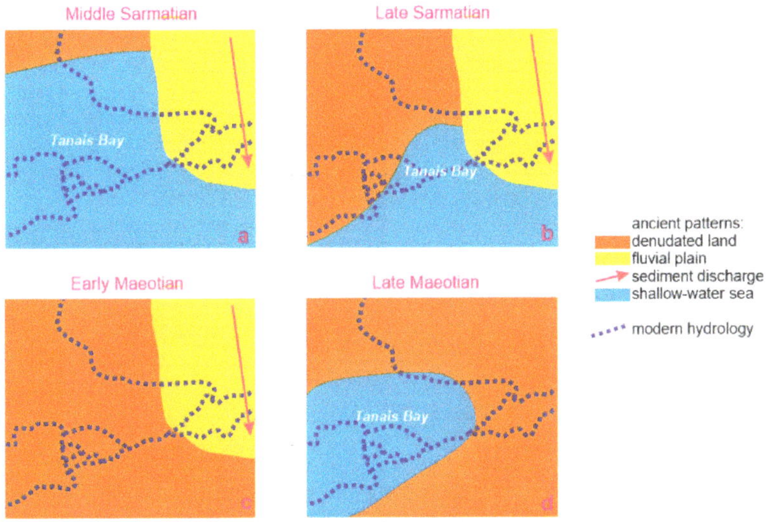

Figure 4. Principal palaeogeographical domains of the Rostov Done in the Middle Sarmatian (**a**), Late Sarmatian (**b**), Early Maeotian (**c**), and Late Maeotian (**d**) (created by the author; modern hydrology is shown to make a link to the present geography outlined on Figure 3).

The Middle–Late Sarmatian transition is marked by the unconformity, which implies regression and temporary bay disappearance. Marine sedimentation restarted very soon. The Late Sarmatian deposits are known locally in the central part of the study area (Figure 3). Apparently, they were also present in its southern part, but were later eroded. To argue this idea, it is necessary to stress that the preceding hiatus was too short to expect significant palaeogeographical re-organization, and, thus, the sea could ingress from the only south. It remains questionable whether the Tanais Bay reached the size comparable to that which it had in the Middle Sarmatian. Apparently, this hypothesis should be rejected, because Kolesnikov [33], who had the opportunity to deal with what is now a densely-urbanized area, explained that the Late Sarmatian conglomerates concentrate in Rostov-on-Don and its vicinities, and their upper surface also demonstrates signs of erosion. Most probably, these beds mark the position of the shoreline, which did not move far northwards. The observations made by Vlasov [38] in the entire northern periphery of the Eastern Paratethys also imply the lesser spatial extent of the Late Sarmatian marine basin. Therefore, the bay was restricted in the Late Sarmatian, but it continued to co-exist with the fluvial plain in the east (Figure 4b).

The sea regressed, and the hiatus corresponded to the entire Early Maeotian. The only Yanovskaya Formation is found in the eastern part of the Rostov Dome (Figure 3). The denudated land prevailed (Figure 4c). The Tanais Bay did not exist in this time slice. One should also note that the level of the Eastern Paratethys was higher in the Early Maeotian, when there was not marine deposition, than in the Late Maeotian, when marine deposition was active [12]. Only tectonic uplift of the dome could prevent bay development in the Early Maeotian. Hypothetically, this uplift started in the Middle Sarmatian, which explains the absence of the wider Middle Sarmatian transgression and the limited distribution of the Late Sarmatian deposits (see above).

The most principal palaeogeographical changes took place in the Late Maeotian (Figure 4d). The sea ingressed from the southwest, and the deposits of the Donskaya Formation accumulated on a restricted plot in the central part of the Rostov Dome (Figure 3). Evidently, their extent was higher, but a portion of them was eroded together with the development of the Don River valley in the Pliocene–Quaternary, and the other portion may exist beneath the Azov Sea. The ingression continued, and the shallow-marine skeletal limestones of the Merzhanovskaya Formation are also known in the western and northern parts of the Rostov Dome (Figure 3). The fluvial plain disappeared (the deposition of the Yanovskaya Formation stopped in the Early Maeotian). Hypothetically, the Palaeo-Donets River migrated eastwards. However, the most striking change was the appearance of some land in the south of the study area, which means the Tanais Bay changed its configuration (Figure 4). There, the Sarmatian deposits are overlain by the Pontian deposits, with unconformity in the southern part of the dome; neither this unconformity, nor the noted absence of the Late Maeotian terrestrial strata, can be explained by any post-Maeotian erosion. There are localities where the Maeotian and Pontian deposits lay conformably and form continuous succession (see discussion in Ruban [23]), and, thus, if there even was an episode of the post-Maeotian erosion, it was short and weak. The only possibility is that the southern and eastern parts of the Rostov Dome experienced tectonic uplifts during the Late Maeotian. The southern and eastern landmass existed until the Early Pontian, when the study area was drowned together with a widespread transgression [12]. Although this is out of scope of the present study, it can be hypothesized that the vertical tectonic motions continued on the northern periphery of the Eastern Paratethys. They could be responsible for the later palaeohydrological reorganizations, after which the Palaeo-Donets River became a tributary of the Don River, and the latter turned its flow direction to the west.

5.2. Basin-Scale Versus Local Events

The three principal palaeogeographical changes in the Tanais Bay at the Sarmatian–Maeotinan transition were (1) the erosion at the Middle/Late Sarmatian boundary, (2) the Early Maeotian sea retreat outside the study area, and (3) the mid-Maeotian ingression and bay reconfiguration. One would expect that the position of this bay on the northern, shallow-water, and tectonically passive periphery of the Eastern Paratethys means that such changes reflect the evolutionary history of the entire sea. This is especially so because the level of the latter experienced strong fluctuations [12]. However, the above-given interpretations indicate both consistencies and inconsistencies of the local and basin-scale records. For instance, the outstanding sea-level fall at the Middle/Late Sarmatian boundary has a direct analogue in the Tanais Bay, whereas the Early Maeotian transgression did not leave any local signature.

On the basis of the information from the central part of the Eastern Paratethys, Palcu et al. [6] established four desiccation episodes (Figure 5). The first and second episodes took place near the Middle/Late Sarmatian boundary. Evidently, these correspond to the sea retreat from the Rostov Dome, where the short-term hiatus took place (Figure 5). In the mid-Late Sarmatian, a rather minor desiccation episode cannot be registered in the study area due to the very thin sedimentary succession of the upper part of the Rostovskaya Formation [6]. The biggest was the episode at the Sarmatian/Maeotian boundary [6] (Figure 5). Apparently, it associated with the start of the long-term erosion in the Rostov Dome. All four episodes characterize the Late Sarmatian evolution of the entire sea. The Tanais Bay had experienced regression since the Middle Sarmatian, and it cannot be excluded that the latter can partly be explained by the general tendency to desiccations in the Eastern Paratethys. However, the interpretations presented above indicate local tectonic activity of an as yet unclear nature (for instance, along major basement faults or epeirogenic) that interplayed with the sea-level changes. For instance, the Eastern Paratethys experienced significant transgression in the Early Maeotian, when terrestrial conditions were typical to the Rostov Dome (Figure 5). The only possible explanation is the local tectonic uplift, which also triggered the later reconfiguration of the Tanais Bay. To trace local tectonic events along the entire northern margin of the Eastern Paratethys and to explain their nature are important tasks for future investigations.

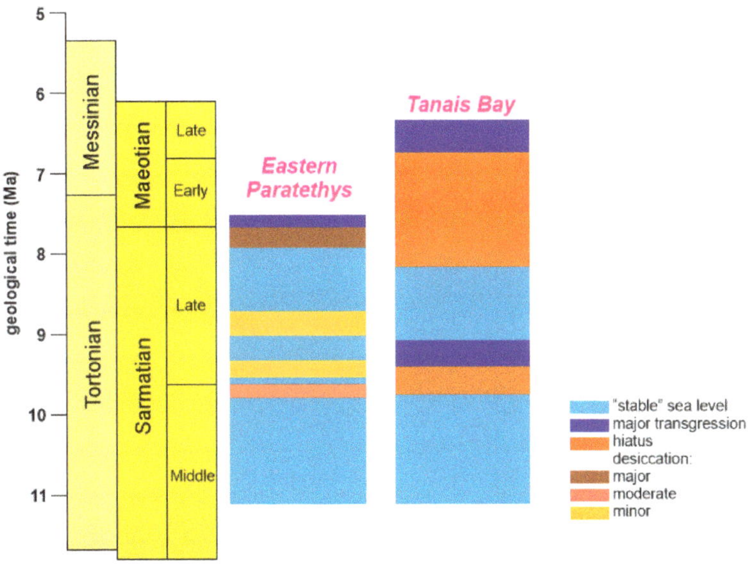

Figure 5. Correlation of the desiccation episodes in the Eastern Paratethys Sea (created by the author; the interpretations for the Eastern Paratethys are based on [6], and they are restricted to the time frame considered in the cited work; see Figure 2 for the stratigraphical framework).

The considered geological time interval corresponds to the late Tortonian (Figure 2), when the Mediterranean (connected, partly connected, or disconnected with the Eastern Paratethys) also experienced serious palaeogeographical transformations. What is really notable is that the local evolution of its peripheries was not only altered by hydrological changes, but demonstrated individual trajectories, often dictated by the local tectonic activity. This seems to be typical to the Mediterranean. For instance, this was established in the Las Minas Basin in Spain [40] and northeastern Tunisia [41]. It appears that the general changes in the water volume in enclosed and semi-enclosed marine basins and mega-lakes should be treated as characteristic to only entire basins, whereas signatures of these changes on basin margins could be altered tectonically. An important question is the mechanism of tectonic uplifts in the Rostov Dome at the Sarmatian–Maeotian transition. Large-scale mechanisms linked to regional evolution [42–44] or local vertical motions controlled by faults [26] could be related to such uplifts. However, the present state of the knowledge does not allow definite judgments, and further investigations in the Rostov Dome and the adjacent areas are necessary.

6. Conclusions

The re-examination of the lithostratigraphical knowledge of the Middle Sarmatian–Late Maeotian deposits of the Rostov Dome, undertaken in the light of historical evidence, implies that the Tanais Bay of the Eastern Paratethys Sea experienced significant changes at the Sarmatian–Maeotian transition. The long-term regressive trend culminated in the Early Maeotian when the sea retreated from the study area. Before this, there was a short-term hiatus at the Middle/Late Sarmatian boundary. A transgression started in the Late Maeotian, when the bay had a different configuration. These changes can only partly be linked to the sea-level changes and the desiccation episodes in the Eastern Paratethys. The Tanais Bay was shaped significantly by local forces, including tectonic uplifts.

The present study stresses the methodological importance of analyzing geological archives where old publications, maps, and reports can be stored. Although this old information can be too generalized and somewhat outdated, it can provide unique lines of evidence which cannot be re-collected by new field investigations. This is especially the case of those areas which have experienced dense urbanization and deep anthropogenic modification. It should also be noted that the "classical", simply-designed lithostratigraphical interpretations still matter in modern palaeogeography, and they seem to be equally important to more advanced microscope- and geochemistry-based analytical techniques.

Funding: This research received no external funding.

Institutional Review Board Statement: Not applicable.

Informed Consent Statement: Not applicable.

Data Availability Statement: Not applicable.

Acknowledgments: The author gratefully thanks the reviewers for their helpful recommendations and O.V. Dudnikova (Russia) for her outstanding help with finding the rare literature sources.

Conflicts of Interest: The author declares no conflict of interest.

References

1. Golonka, J. Plate tectonic evolution of the southern margin of Eurasia in the Mesozoic and Cenozoic. *Tectonophysics* **2004**, *381*, 235–270. [CrossRef]
2. Hässig, M.; Moritz, R.; Ulianov, A.; Popkhadze, N.; Galoyan, G.; Enukidze, O. Jurassic to Cenozoic Magmatic and Geodynamic Evolution of the Eastern Pontides and Caucasus Belts, and Their Relationship With the Eastern Black Sea Basin Opening. *Tectonics* **2020**, *39*, e2020TC006336. [CrossRef]
3. Van Hinsbergen, D.J.J.; Torsvik, T.H.; Schmid, S.M.; Matenco, L.C.; Maffione, M.; Vissers, R.L.M.; Gürer, D.; Spakman, W. Orogenic architecture of the Mediterranean region and kinematic reconstruction of its tectonic evolution since the Triassic. *Gondwana Res.* **2020**, *81*, 79–229. [CrossRef]
4. Il'ina, L.B. On connections between basins of the Eastern Paratethys and adjacent seas in the middle and late Miocene. *Stratigr. Geol. Correl.* **2000**, *8*, 300–305.
5. Magyar, I.; Geary, D.H.; Müller, P. Paleogeographic evolution of the Late Miocene Lake Pannon in Central Europe. *Palaeogeogr. Palaeoclimatol. Palaeoecol.* **1999**, *147*, 151–167. [CrossRef]
6. Palcu, D.V.; Patina, I.S.; Sandric, I.; Lazarev, S.; Vasiliev, I.; Stoica, M.; Krijgsman, W. Late Miocene megalake regressions in Eurasia. *Sci. Rep.* **2021**, *11*, 11471. [CrossRef]
7. Piller, W.E.; Harzhauser, M.; Mandic, O. Miocene Central Paratethys stratigraphy—Current status and future directions. *Stratigraphy* **2007**, *4*, 151–168.
8. Popov, S.V.; Rögl, F.; Rozanov, A.Y.; Steininger, F.F.; Shcherba, I.G.; Kovac, M. Lithological-paleogeographic maps of Paratethys. *CFS Cour. Forsch. Senckenberg* **2004**, *250*, 1–46.
9. Rogl, F. Mediterranean and Paratethys. Facts and hypotheses of an Oligocene to Miocene Paleogeography (short overview). *Geol. Carpathica* **1999**, *50*, 339–349.
10. Steininger, F.F.; Rögl, F. Paleogeography and palinspastic reconstruction of the Neogene of the Mediterranean and Paratethys. *Geol. Soc. Spec. Publ.* **1984**, *17*, 659–668. [CrossRef]
11. Popov, S.V.; Shcherba, I.G.; Ilyina, L.B.; Nevesskaya, L.A.; Paramonova, N.P.; Khondkarian, S.O.; Magyar, I. Late Miocene to Pliocene palaeogeography of the Paratethys and its relation to the Mediterranean. *Palaeogeogr. Palaeoclimatol. Palaeoecol.* **2006**, *238*, 91–106. [CrossRef]
12. Popov, S.V.; Antipov, M.P.; Zastrozhnov, A.S.; Kurina, E.E.; Pinchuk, T.N. Sea-level fluctuations on the northern shelf of the Eastern Paratethys in the Oligocene-Neogene. *Stratigr. Geol. Correl.* **2010**, *18*, 200–224. [CrossRef]
13. Popov, S.V.; Rostovtseva, Y.V.; Pinchuk, T.N.; Patina, I.S.; Goncharova, I.A. Oligocene to Neogene paleogeography and depositional environments of the Euxinian part of Paratethys in Crimean—Caucasian junction. *Mar. Pet. Geol.* **2019**, *103*, 163–175. [CrossRef]
14. Dzeboev, B.A.; Odintsova, A.A.; Rybkina, A.I.; Dzeranov, B.V. Assessment of the Influence of Astronomical Cyclicity on Sedimentation Processes in the Eastern Paratethys Based on Paleomagnetic Measurements Using Discrete Mathematical Analysis. *Appl. Sci.* **2022**, *12*, 580. [CrossRef]
15. Rybkina, A.I.; Kern, A.K.; Rostovtseva, Y.V. New evidence of the age of the lower Maeotian substage of the Eastern Paratethys based on astronomical cycles. *Sediment. Geol.* **2015**, *330*, 122–131. [CrossRef]
16. Merenkova, S.I.; Seregina, I.F.; Gabdullin, R.R.; Rostovtseva, Y.V.; Bol'shov, M.A. Reconstruction of the Paleosalinity and Paleobathymetry of the Yenikale Strait in the Eastern Paratethys in Sarmatian: Evidence from the Geochemical Data. *Mosc. Univ. Geol. Bull.* **2020**, *75*, 342–352. [CrossRef]
17. Ilgar, A. Miocene sea-level changes in northernmost Anatolia: Sedimentary record of eustasy and tectonism at the peri-Pontide fringe of Eastern Paratethys. *Sediment. Geol.* **2015**, *316*, 62–79. [CrossRef]
18. Arsen'eva, T.M. On the research of the antique city of Tanais near Don. *Eurasia Antiqua* **2005**, *11*, 78–85.

19. Khokhlova, O.S.; Dyuzhova, K.V.; Golyeva, A.A.; Trifonova, T.A.; Bunin, D.S.; Ilyashenko, S.M.; Khokhlov, A.A.; Shipkova, G.V. Paleoecology of the ancient city of Tanais (3RD century BC–5TH century AD) on the north-eastern coast of the sea of Azov (Russia). *Quat. Int.* **2019**, *516*, 98–110. [CrossRef]
20. Sokolov, N.A. On Neogene deposits of Lower Don and about the northern limit of distribution of the Pontian deposits in European Russia. *Izv. Geol. Kom.* **1891**, *10*, 29–51. (In Russian)
21. Ruban, D.A. Lithostratigraphy of the Upper Miocene deposits of the Rostov Dome. *Nautchnaja Mysl' Kavkaza. Prilozhenie* **2002**, *14*, 133–136. (In Russian)
22. Ruban, D.A. The Upper Miocene of the Rostov Dome (Eastern Paratethys): Implication of the chronostratigraphy and bivalvia-based biostratigraphy. *Geol. Anal. Balk. Poluostrva* **2005**, *66*, 9–15. [CrossRef]
23. Ruban, D.A. Stratigraphic evidence of a Late Maeotian (Late Miocene) punctuated transgression in the Tanais Palaeobay (northern part of the Eastern Paratethys, South-West Russia). *Geologos* **2010**, *16*, 169–181. [CrossRef]
24. Belov, F.A.; Egorov, A.I.; Pogrebnov, N.I. (Eds.) *Geology of the USSR, 46*; Nedra: Moscow, Russia, 1970; p. 667. (In Russian)
25. Ivanitskaya, V.B.; Pogrebnov, N.I. *Geological Structure of the Lower Don and Lower Volga*; RGU: Rostov-na-Donu, Russia, 1962; p. 64. (In Russian)
26. Kostyuchenko, S.L.; Morozov, A.F.; Stephenson, R.A.; Solodilov, L.N.; Vedrentsev, A.G.; Popolitov, K.E.; Aleshina, A.F.; Vishnevskaya, V.S.; Yegorova, T.P. The evolution of the southern margin of the East European Craton based on seismic and potential field data. *Tectonophysics* **2004**, *381*, 101–118. [CrossRef]
27. Ruban, D.A. *Dynamics of Palaeotectonic Position of the Rostov Block*; DGTU-Print: Rostov-na-Donu, Russia, 2018; p. 73. (In Russian)
28. International Commission on Stratigraphy (ICS). International Chronostratigraphic Chart. 2022. Available online: https://stratigraphy.org/chart (accessed on 4 May 2022).
29. Gradstein, F.M.; Ogg, J.G.; Schmitz, M.; Ogg, G. (Eds.) *Geologic Time Scale 2020*; Elsevier: Amsterdam, The Netherlands, 2020; p. 1390.
30. Bogachev, V.V. Preliminary report from the geological investigations in 1907 and 1908 years. *Izv. Geol. Kom.* **1910**, *29*, 765–837. (In Russian)
31. Bogachev, V.V. Miocene deposits of the town of Novocherkassk. *Ezhenedelnik Geol. Mineral. Ross.* **1911**, *13*, 61–71. (In Russian)
32. Bogachev, V.V. *Geological Description of the Taganrog District*; A. Ter-Abramyan: Rostov-on-Don, Russia, 1916; p. 32. (In Russian)
33. Kolesnikov, V.P. Facies of the Sarmatian of the Tanais bay. *Izv. Akad. Nauk SSSR. VII Seriya. Otd. Mat. Estestv. Nauk* **1934**, *2–3*, 217–234. (In Russian)
34. Miroshnikov, P.V. *From the Taganrog Bay to the Southern Donbass*; Rostovskiy Universitet: Rostov-na-Donu, Russia, 1958; p. 51. (In Russian)
35. Nalivkin, D.V.; Sokolov, B.S. *Neogene System, 2*; Nedra: Moscow, Russia, 1986; p. 420. (In Russian)
36. Paffengolts, K.N. *Geological Description of the Caucasus*; AN Armyanskoy SSR: Erevan, Armenia, 1959; p. 507. (In Russian)
37. Vlasov, D.F. Facies of the Lower Sarmatian deposits of the Rostov Region. *Uchenye Zap. Rostov. -Na-Donu Gos. Univ.* **1955**, *33*, 69–84. (In Russian)
38. Vlasov, D.F. Facies of the Pontian deposits of the Tanais Bay. *Uchenye Zap. Rostov. -Na-Donu Gos. Univ.* **1958**, *53*, 155–165. (In Russian)
39. Vlasov, D.F. Facies of the Middle Sarmatian deposits of the Tanais Bay. *Uchenye Zap. Rostov. -Na-Donu Gos. Univ.* **1959**, *44*, 33–41. (In Russian)
40. Pineda, V.; Gibert, L.; Soria, J.M.; Carrazana, A.; Ibáñez-Insa, J.; Sánchez-Román, M. Interevaporitic deposits of Las Minas Gypsum Unit: A record of Late Tortonian marine incursions and dolomite precipitation in Las Minas Basin (eastern Betic Cordillera, SE Spain). *Palaeogeogr. Palaeoclimatol. Palaeoecol.* **2021**, *564*, 110171. [CrossRef]
41. El Euch-El Koundi, N.; Barhoun, N. Discovery of late Tortonian incised valleys in the Saouaf Formation (northeastern Tunisia)—Evidence of high-frequency sea-level variations in the central Mediterranean. *Sediment. Geol.* **2020**, *398*, 105602. [CrossRef]
42. Kopp, M.L. Gravitational collapse of anteclises and its probable impact on the neotectonics of platforms and passive continental margins (by the example of the East European platform). *Russ. Geol. Geophys.* **2020**, *61*, 1156–1172. [CrossRef]
43. Makarova, N.V.; Sukhanova, T.V. Actual problems of studies of recent platform structures: A case study of the East European Craton and adjacent parts of the Scythian Plate. *Mosc. Univ. Geol. Bull.* **2017**, *72*, 245–254. [CrossRef]
44. Trifonov, V.G.; Sokolov, S.Y.; Sokolov, S.A.; Hessami, K. Mesozoic–Cenozoic Structure of the Black Sea–Caucasus–Caspian Region and Its Relationships with the Upper Mantle Structure. *Geotectonics* **2020**, *54*, 331–355. [CrossRef]

Journal of
Marine Science and Engineering

MDPI

Article

Bryozoan Diversity Dynamics at the Devonian–Carboniferous Transition: Evidence from Transcaucasia

Zoya A. Tolokonnikova [1] and Dmitry A. Ruban [2,*]

[1] Kuban State University, Stavropolskaja Street 149, 350040 Krasnodar, Russia; zzalatoi@yandex.ru
[2] K.G. Razumovsky Moscow State University of Technologies and Management (the First Cossack University), Zemlyanoy Val Street 73, 109004 Moscow, Russia
* Correspondence: ruban-d@mail.ru

Abstract: The Devonian–Carboniferous transition was marked by a series of perturbations in the geological and biological evolution. The palaeontological data from Transcaucasia allowed the bryozoan diversity dynamics on the northern Gondwanan margin (southern periphery of the Palaeo-Tethys Ocean) to be documented at this transition. Taxonomic ranges of 43 species, 26 genera, 19 families, and 4 orders were analysed to reveal changes in the total diversity, the number of appearances, the number of disappearances, and the turnover rates per substages. It is established that the bryozoan diversity was rather high in the beginning and the end of the Famennian, as well as in the Late Tournaisian. It declined significantly in the Middle–Late Famennian and the Early Tournaisian due to the combination of the high number of disappearances and the low number of appearances. The turnovers remained strong, and they peaked in the mid-Famennian. These regionally documented diversity changes match the patterns recorded globally and in Southern Siberia. Hypothetically, the Middle–Late Famennian and Early Tournaisian crises established in Transcaucasia were related to the global events (anoxia and mass extinctions), a series of which weakened the bryozoans' resistivity to negative external influences.

Keywords: biotic crisis; Gondwana; marine macroinvertebrates; middle Palaeozoic; Palaeo-Tethys Ocean; taxonomic diversity; turnover

Citation: Tolokonnikova, Z.A.; Ruban, D.A. Bryozoan Diversity Dynamics at the Devonian–Carboniferous Transition: Evidence from Transcaucasia. *J. Mar. Sci. Eng.* **2022**, *10*, 959. https://doi.org/10.3390/jmse10070959

Academic Editor: Anabela Oliveira

Received: 23 June 2022
Accepted: 11 July 2022
Published: 13 July 2022

1. Introduction

The Devonian–Carboniferous transition (~370–350 Ma) was one of the most critical intervals in the Palaeozoic geological and biological evolution. It was marked by anoxia [1–3], glacial advance [4,5], sea-level drop [6], and mass extinction [7–9]. However, many aspects of these events and their spatio-temporal appearance are yet to be fully understood, and new investigations bring "surprises" [10,11]. In order to document better the biotic events at the Devonian–Carboniferous transition, attention should be paid to the patterns of the diversity dynamics of many fossil groups and in many regions. Of special interest are organisms, preliminary investigations of which revealed uncertain reactions to the mid-Palaeozoic global crises, and regions, where diversity changes of these organisms were not studied earlier. In regard to these needs, investigation of Famennian–Tournaisian bryozoans from Transcaucasia seems to be urgent.

The previous analyses of diversity dynamics of bryozoans revealed their resistivity to the mid-Palaeozoic mass extinctions [8,12,13], which itself is important to consider when these biotic crises are addressed (notably, bryozoans were stressed significantly by the later catastrophes [14]). However, it remains unclear whether this resistivity is a global phenomenon. The Devonian–Carboniferous bryozoans have been studied in Transcaucasia for more than three decades [15–17]. A significant amount of the accumulated information would serve ideally for measuring various patterns of the bryozoan diversity dynamics. This regional record represents the northern Gondwanan margin where the knowledge of middle Palaeozoic bryozoans remains limited [18–20]. "The Paleobiology Database" [21] demonstrates the outstanding scarcity of the information about Devonian and Carboniferous bryozoans from the African-Arabian margin of mid-Palaeozoic Gondwana, which means the present knowledge of these organisms is seriously biased (the data from Transcaucasia are also absent in this source).

The objective of the present paper is to reconstruct the bryozoan diversity dynamics in Transcaucasia at the Devonian–Carboniferous transition. This is necessary to extend the general knowledge of bryozoan evolution at this transition with the information from the northern Gondwanan margin.

2. Materials and Methods

2.1. Study Area

The study area is located in the southern part of Transcaucasia (Figure 1). The latter has three meanings, namely administrative (Georgia, Armenia, and Azerbaijan taken together), geographical (Lesser Caucasus mountains and adjacent Rioni and Kura depressions), and geological (Lesser Caucasus as "exotic" terrane). The study area corresponds to the noted terrane (see the works [22,23] for more tectonic considerations).

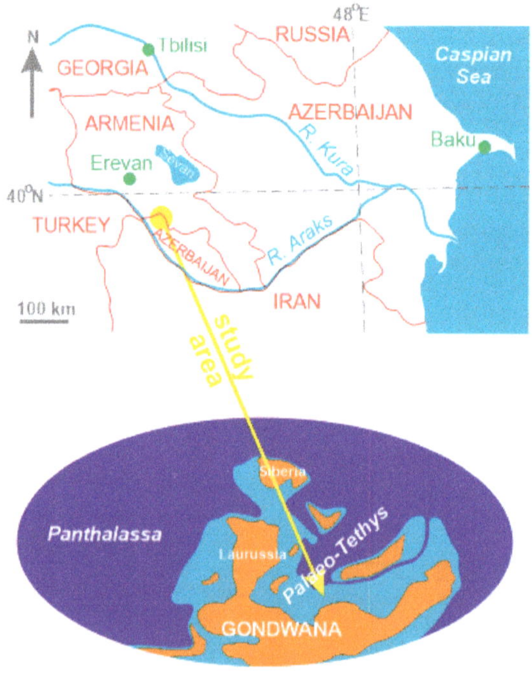

Figure 1. Present and Famennian location of the study area (created by the authors; the palaeogeographical reconstruction follows the contours shown by [24]).

The local stratigraphical framework was characterized by Grechishnikova and Levitskii [25], Mamedov [26,27], and Rzhonsnitskaya and Mamedov [20]. The transitional Devonian–Carboniferous deposits are represented in a dozen of closely related sections and smaller outcrops, which form together the full succession of the Famennian–Tournaisian strata. These are intercalating coarse and fine siliciclastics and limestones, with different proportions in different intervals (Figure 2). The total thickness reaches 900 m. Although formations have been introduced, the deposits are usually subdivided by biozones and substages (also in individual sections—for instance, see [25]), and thus, formations are not used in the present study. These strata bear abundant fossils, including marine macroinvertebrates. Brachiopods and conodonts are especially useful to establish the age with precision and to undertake accurate zonal correlations [22,25–28]. These deposits accumulated in shelf environments, with significant flux of siliciclastic matter and under slightly changed sea level [25].

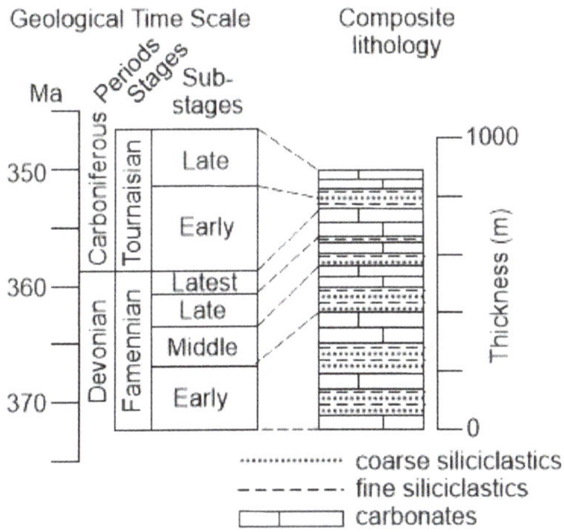

Figure 2. Stratigraphical framework of the Devonian–Carboniferous transition in the study area (created by the authors; the geological time scale follows [29], and the regional stratigraphy is compiled from [25–27]).

The palaeogeographical position of the study area can be derived from present interpretations of the pre-Jurassic history of the Lesser Caucasus terrane. Taking into account its relative position to the Greater Caucasus, which is thought to be a Galatian terrane [30,31], and the trajectories of the Middle Eastern blocks depicted on modern plate tectonic reconstructions [24,32–37], it appears that the Lesser Caucasus was a part of the northern (African-Arabian) Gondwanan margin until the Cimmerian terranes were detached from it in the late Palaeozoic together with the opening of the Neo-Tethys to reach the Laurasian margin in the early Mesozoic. If so, marine sedimentation in the study area can be attributed to the southern periphery of the Palaeo-Tethys Ocean, where there were also epeiric seas ingressed in Gondwana. The plate tectonic reconstructions noted above imply that the study area lied in the tropical latitudes and, most probably, in the Southern Hemisphere (Figure 1).

2.2. Palaeontological Information

The material for the present analysis was prepared in three steps. First, the sources devoted to the Famennian–Tournaisian bryozoans of the study area were collected. These included the works by Goryunova [15,38,39], Gorjunova and Lavrentjeva [40], Grechishnikova and Levitskii [25], Lavrentjeva [16,41], and Tolokonnikova [17,42]. These works summarize the outcomes of the previous research and make the considered palaeontological record suitable to the quantitative analysis. Importantly, the palaeontological data were collected in the "Soviet traditions", which means sampling was massive and aimed initially at representing the record of the entire area, not particular localities. Such an approach makes these data very suitable for diversity analysis.

Second, the stratigraphic ranges of all bryozoan taxa were compiled from the sources indicated above to compose a database for further quantitative analysis. For this purpose, the regional stratigraphical framework [22,25–28] was brought in correspondence to the modern version of the geological time scale [29,43] and, particularly, the stratigraphy of the Devonian–Carboniferous transition [1,44–49]. The presence of bryozoan taxa in the four substages of the Famennian and the two substages of the Tournaisian was established.

Indeed, the previously published data are heterogeneous, and some generalizations were required to avoid inconsistencies on the below-substage level. One can question whether the employed palaeontological data represent each substage adequately. The related uncertainties are generally unavoidable in palaeobiological studies because in only rare cases fossils of the same kind can be found in all layers of the studied stratigraphical interval. Nonetheless, it should be said that palaeontological collecting in the study area was massive, and thus, the bryozoan representation of each stage seems to be adequate. If bryozoans were not found at some levels, this means they did not exist, and thus, the diversity of the related substage was relatively small because there were environments unfavourable for bryozoans. At least, the sampling of the examined Famennian–Tournaisian succession during the previous studies, from which we compile the information, was massive and rather homogeneous.

Third, the preliminary information was reviewed and systematized to update the suprageneric taxonomy, to avoid synonyms, and to justify the stratigraphic ranges of species. This way, the final version of the database is prepared (Supplementary Materials). It registers the stratigraphic ranges of 43 bryozoan species, 26 genera, 19 families, and 4 orders in the study area.

2.3. *Diversity Dynamics Analysis*

The present analysis pays attention to two aspects of the bryozoan diversity dynamics at the Devonian–Carboniferous transition, namely principal patterns and turnover rates. The principal patterns include total diversity (total number of taxa), number of appearances, and number of disappearances, as done in previous investigations (for instance, [12]). The three principal patterns are measured for each substage (six time slices correspond to four Famennian and two Tournaisian substages) and four taxonomic levels of bryozoans, namely species, genera, families, and orders. Importantly, the number of appearances cannot be measured for the first time slice due to the absence of information from the preceding slice, and the number of disappearances cannot be measured for the last time slice due to the absence of information from the following slice.

Notably, appearances and disappearances should be distinguished from originations and extinctions because the former can be only regional (for instance, a given taxon could originate earlier in other regions) and/or temporary (for instance, a given taxon could disappear due to anoxia and then re-appear). Attention to the numbers of appearances and disappearances is reasonable when local palaeontological records are considered.

The turnover rates are measured by similarity indices using the approach tested previously [50]. The R index is the well-known Jaccard's similarity [51]. It is calculated for each two successive pairs of assemblages of bryozoan species, genera, families, and orders. The RST index is more complex because it reflects similarity of assemblages of higher-ranked taxa taking into account their diversity, i.e., the number of the related lower-ranked taxa. Two indices, namely the Czekanowski's quantified coefficient (C) [52] and the Gower' index (G) [53] can be employed. As they reflect essentially the same pattern, but on the basis of different mathematical procedures, they are mutually important, and thus, RST (C) and RST (G) are calculated for each of two successive pairs of assemblages of bryozoan orders taking into account their species diversity. All three indices (R, RST (C), and RST (G)) change from 0 to 1, where 0 indicates total turnover, and 1 indicates absence of any turnover. R reflects simple turnover rate, and RST (C) and RST (G) reflect turnovers in the taxonomic structure of diversity. Additionally, it appears reasonable to measure RST (C) and RST (G) for not only successive, but all possible pairs of bryozoan assemblages, which would permit us to understand whether a given turnover had relatively long-term effects. It cannot be excluded that the oldest and youngest assemblages demonstrate high similarity despite of strong turnovers in-between, which means some orders "re-gained" their importance for the species diversity after a certain pause.

3. Results

3.1. Principal Patterns

The diversity of bryozoans in the study area was measured for six geological time slices (Table 1), and this information allows for judgments of the principal patterns of its dynamics. The total diversity of species, genera, families, and orders changed rather similarly across the Devonian–Carboniferous transition (Figure 3). After the relatively high diversity in the beginning of the Famennian, a pronounced decline occurred in the midst of this stage. Diversification near its end was followed by small diversity drop (not registered for families) in the Early Tournaisian, after which a strong diversification happened (the latter did not involve orders).

Table 1. Taxonomic diversity of the studied bryozoans.

Parameter	Number of Taxa					
	Fm1	Fm2	Fm3	Fm4	To1	To2
Species (43 taxa)						
Total diversity	9	5	5	9	6	12
Appearances	-	5	5	8	6	11
Disappearances	9	5	4	9	5	-
Genera (26 taxa)						
Total diversity	7	4	2	7	5	11
Appearances	-	2	2	6	4	10
Disappearances	5	4	1	6	4	-
Families (19 taxa)						
Total diversity	6	4	2	5	5	9
Appearances	-	1	2	4	4	7
Disappearances	3	4	1	4	3	-
Orders (4 taxa)						
Total diversity	3	3	1	4	3	3
Appearances	-	0	0	3	0	0
Disappearances	0	2	0	1	0	-

Substage abbreviations: Fm1—Early Famennian, Fm2—Middle Famennian, Fm3—Late Famennian, Fm4—Latest Famennian, To1—Early Tournaisian, To2—Late Tournaisian.

The numbers of appearances rose synchronously in the Latest Famennian (Figure 3). They dropped slightly in the Early Tournaisian and rose significantly in the second part of this stage (this did not happen with orders). The numbers of disappearances were high in the Early and Latest Famennian, but low in the Late Famennian (Figure 3). The comparison of the principal patterns implies that the mid-Famennian total diversity decline occurred due to the high number of disappearances in the Early Famennian and the lower number of appearances in the Middle and Late Famennian. The Early Tournaisian total diversity drop was driven more by the high number of disappearances in the Latest Famennian.

3.2. Turnover Rates

The R, RST (C), and RST (G) indices are calculated for five successive pairs of assemblages corresponding to the substages, as well as to the non-successive assemblages (Table 2). The turnovers established with R were especially strong at the Middle–Late Famennian transition (Figure 4). The assemblages changed totally (except for orders). Another strengthening of turnovers occurred at the Latest Famennian–Early Tournaisian transition, but it was less pronounced than previously for genera, families, and orders. One should also note the relatively high rate of turnovers throughout the entire considered time interval.

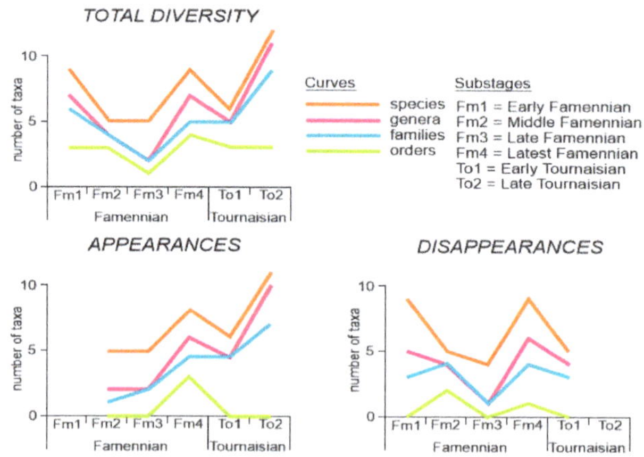

Figure 3. Principal patterns of the taxonomic diversity dynamics of the studied bryozoans (created by the authors, see Table 1 for numbers).

Table 2. R and RST indices for the studied bryozoan assemblages.

R					
Taxa	**Substage Transitions**				
	Fm1–Fm2	**Fm2–Fm3**	**Fm3–Fm4**	**Fm4–To1**	**To1–To2**
Species	0.00	0.00	0.08	0.00	0.06
Genera	0.22	0.00	0.13	0.09	0.07
Families	0.43	0.00	0.17	0.11	0.17
Orders	1.00	0.33	0.25	0.75	1.00
RST (C)					
Substages	**Fm2**	**Fm3**	**Fm4**	**To1**	**To2**
Fm1	0.43	0.29	0.44	0.53	0.86
Fm2	-	0.20	0.71	0.55	0.47
Fm3	-	-	0.43	0.73	0.47
Fm4	-	-	-	0.67	0.67
To1	-	-	-	-	0.67
RST (G)					
Substages	**Fm2**	**Fm3**	**Fm4**	**To1**	**To2**
Fm1	0.56	0.17	0.00	0.61	0.83
Fm2	-	0.07	0.00	0.58	0.50
Fm3	-	-	0.00	0.27	0.28
Fm4	-	-	-	0.00	0.00
To1	-	-	-	-	0.67

Substage abbreviations: Fm1—Early Famennian, Fm2—Middle Famennian, Fm3—Late Famennian, Fm4—Latest Famennian, To1—Early Tournaisian, To2—Late Tournaisian.

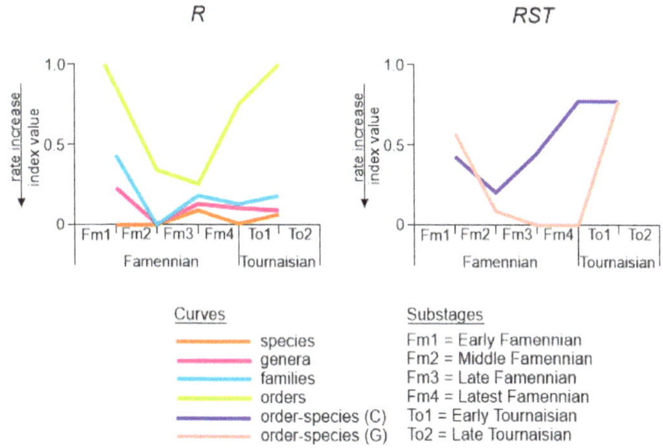

Figure 4. Rates of changes among the assemblages of the studied bryozoans (created by the authors, see Table 2 for numbers).

The turnovers in the order–species structure of bryozoan assemblages are revealed with less certainty (Figure 4). Both RST (C) and RST (G) indicate significant turnover at the Middle–Late Famennian transition, after which the intensity of turnovers either decreased gradually (RST (C)) or remained high (RST (G)) (Figure 4). Both indices indicate moderate-to-low turnovers within the Tournaisian. This pattern resembles what was registered by R (Figure 4). Consideration of non-successive pairs of assemblages indicates high similarity of the Early Famennian and Late Tournaisian assemblages measured by RST (C) and RST (G) (Table 2). This means limited effects of the Middle Famennian–Early Tournaisian turnovers on the order–species structure of the bryozoan diversity at the Devonian–Carboniferous transition.

4. Discussion

4.1. Global and Regional Records

Globally, the total species diversity of bryozoans changed when measured per stages: Horowitz and Pachut [54] noted 153 species in the Famennian and 405 species in the Tournaisian. "The Paleobiology Database" [21] indicates 160 genera in the Famennian and 154 genera in the Tournaisian, which means the almost comparable total generic diversity. The data from Transcaucasia (Supplementary Materials) implies the presence of 26 species and 15 genera in the Famennian and 17 species and 15 genera in the Tournaisian, which indicates on the impoverished species diversity and the unchanged generic diversity. The global and regional generic diversity changes at the Devonian–Carboniferous transition seem to be the same. If the estimates of the global species diversity by Horowitz and Pachut [54] remain correct, striking difference between the global and regional diversity changes should be explained anyhow (for instance, by the stronger regional biotic crises). However, these estimates are, most probably, outdated to certain degree because of accumulation of more palaeontological information in the beginning of the 21st century. If so, we should focus on the only significant similarity of the global and regional generic diversity changes (see above).

A detailed analysis of the Late Devonian bryozoan diversity dynamics was carried out in Southern Siberia [12]. The outcomes from this analysis indicate the same principal patterns as registered in Transcaucasia (Figure 3). Particularly, the total number of species, genera, and families was relatively high in both regions in the beginning and the end of the Famennian, and the mid-Famennian diversity decline was sharp. Such a similarity is somewhat surprising in regard to the big distance between Southern Siberia and Transcaucasia in the Late Devonian (Figure 1). The latter was located on the southern periphery of the Palaeo-Tethys, whereas the former was very far from the northern periphery of this ocean. If so, the similarity of the Famennian diversity dynamics of bryozoans between these regions reveals some general, global mechanisms discussed below.

4.2. Influence of Biotic Crises

The results of the present analysis allow for the recognition of two crises in the development of bryozoans in Transcaucasia at the Devonian–Carboniferous transition. The first encompasses the Middle–Late Famennian, and the second took place in the Early Tournaisian. These regional crises can be brought in correspondence to the global events at this transition. Their nomenclature was summarized by Becker et al. [1]. In addition to the "purely" Famennian events, it is necessary to consider the Fasnian/Famennian mass extinction and the related anoxia known as the Lower and Upper Kellwasser events [55–59]. A comparison of the two regional bryozoan crises and the global events reveals their complex temporal relationship (Figure 5). The bryozoan diversity was rather high in the Early Famennian. The subsequent global events in the Early Famennian were followed by the regional bryozoan crisis, and the latter could have been facilitated by the other global events in the Late Famennian. Similarly, the Hangenberg event and the related mass extinction [2,7–9,58] were followed by the bryozoan crisis in the beginning of the Tournaisian.

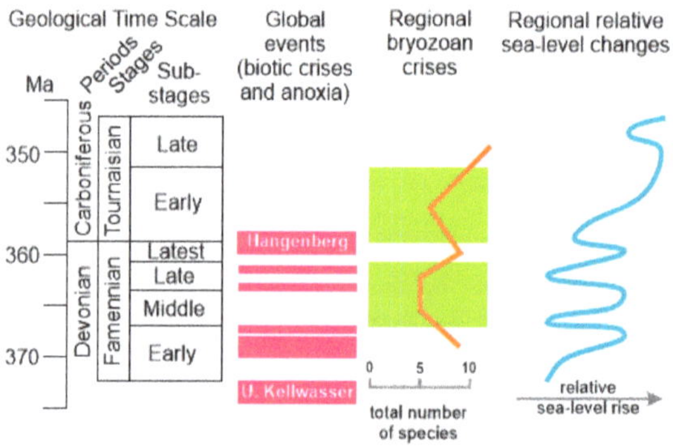

Figure 5. Correlation of the global and regional events (created by the authors; the geological time scale follows [29], the global events are shown according to [1], the regional bryozoan events are interpreted in this study (see text for more explanations), and the regional relative sea-level changes follow the reconstruction [25] and taking into account the information from [27]). All shown patterns are justified to substages.

Generally, it appears that there is some causal relationship between the global events and regional crises, and the above-mentioned similarity of the records of Transcaucasia and South Siberia also proves the idea of some general mechanisms. However, in this case, it is notable that the Frasnian/Famennian mass extinction did not stress bryozoans so strongly [8,12,13]. It is possible to hypothesize that the only series of several, closely happened events would stress bryozoan assemblages so as to make them vulnerable to subsequent negative influences (Figure 5). This hypothesis deserves further verification with data from different regional records.

It should be noted that the bryozoan diversity dynamics can be calculated only per entire substages. If so, one would wonder whether the established patterns may be influenced by the different duration of these time units. The present version of the geological time scale [29,43] implies that the Early Famennian was longer than the other Famennian substages (Figure 2). However, the latter were more or less comparable in duration, although the diversity differed significantly (Table 1). Moreover, the inequity of the Tournaisian substages cannot explain their differences in diversity (the shorter substage was characterized by the higher number of taxa). If so, the factor of the substage duration does not seem to be significant.

It is also necessary to pay attention to the regional palaeoenvironmental factors, namely relative sea-level changes and oxygen depletion. Sea level fluctuated in the Famennian, and trend to deepening established in the Tournaisian (Figure 5). The comparison of these fluctuations with the total diversity changes (Figure 3) indicates the absence of any clear correspondence. Moreover, it is very notable that the total generic diversity is equal in the Famennian and the Tournaisian (when calculated per stages—see above), although the sea was generally deeper in the latter. However, one may wonder whether the decline of species diversity from 26 species in the Famennian and 17 species in the Tournaisian can be linked to the noted deepening. As for anoxia, geochemical investigations are yet to be conducted in the study area. However, grey-to-black shales are noted among the Famennian deposits [25,27]. Their presence permits us to hypothesize episodes of oxygen depletion, which generally correspond to the multiple Famennian anoxic events [1]. Indeed, geochemical investigations coupled with high precision dating of potential oxygen depletion beds are necessary in the study area for their accurate correlations. However, if even the global anoxia did not leave signatures in the study area, bryozoans could be stressed by the global events. They evolved in the open sea connected to the world's water masses (Figure 1), and thus, the global-scale collapses of fossil communities linked to anoxia and mass extinctions were able to stress organisms even where the palaeoenvironmental triggers of the stress did not "work".

4.3. *Morphological Notes on Cryptostomata*

Representatives of the Order Cryptostomata played important role in the bryozoan assemblages in Transcaucasia at the Devonian–Carboniferous transition; several genera and species appeared in the study area in the Famennian and the Tournaisian (Supplementary Materials). Preparation of the material for the purposes of the present study has permitted to note some interesting morphological peculiarities of the originated taxa. Mentioning these peculiarities together with diversity dynamics is reasonable to demonstrate that the changes considered in the present study associated with some peculiarities of the development of bryozoans.

New cryptostomate bryozoans were characterized by tiny little, two layer-symmetrical or branching colonies 1–2 mm in size, which are spiral in plan. Evidently, such morphological adaptations as bundle of axial zooids (*Pseudoascopora*) and mesotheca (*Geranopora*, *Mysticella*, and *Taeniodictya*) facilitated strength and stability of the colonies growing up. The colonies allowed cryptostomate bryozoans to inhabit small places in densely populated palaeoecosystems. In order to decrease competition with bryozoans of other orders and other benthic organisms, cryptostomate bryozoans implemented the strategy of increase in the efficiency of feeding via highly regular location of apertures by diagonally crossed rows in all originated genera and development of monticules in *Taeniodictya*.

Apparently, bryozoans possessed a specific form of camera (chamber with an arched roof) of autozooids and actinotostyles for prohibiting their colonies from predators. A knee-curved camera of autozooids with lower and/or upper hemisepta developed at the edge of the inflection protected soft body of polypide from any invasion. Many originated genera (except for *Taeniodictya*) demonstrate development of actinotostyles with different size and frequency in the exozone of colonies. These often bounded each autozooid in significant amount for creation of a kind of barrier ("stockade"). All new taxa have a developed network of microstyles linking structural elements of colonies for better feeding and gas exchange. For scaring away predators, regulating water flows, and swamping with sand, colonies beard crests and ribs (*Geranopora*, *Mysticella*, and *Taeniodictya*) separating rows of apertures in local lows.

5. Conclusions

The present study of the bryozoan diversity dynamics in Transcaucasia at the Devonian–Carboniferous transition allows making three conclusions. First, the regional bryozoan assemblages experienced significant changes in the total diversity, the number of appearances, and the number of disappearances during the Famennian–Tournaisian, and the turnover rate remained chiefly high. Second, there were two crises in the regional development of bryozoans, namely the Middle–Late Famennian and Early Tournaisian crises. Third, a series of global events could decrease the resistivity of bryozoans to palaeoenvironmental stress, and thus, the established regional crises can be related hypothetically to widespread anoxic events and mass extinctions.

This paper stresses the importance of regional palaeontological records for deciphering mechanisms of critical transitions in the Earth's history. Further investigations of bryozoan diversity dynamics are necessary in different regions (not only remote from one another, but also located in different palaeoenvironmental conditions) in order to understand the influence on the global perturbations at the Devonian–Carboniferous transition on this fossil group. More investigations in the other parts of northern Gondwanan margin can improve the knowledge of mid-Palaeozoic bryozoans from this palaeogeographical domain.

Supplementary Materials: The following supporting information can be downloaded at: https://www.mdpi.com/article/10.3390/jmse10070959/s1, Supplement: Taxonomic ranges of the Famennian–Tournaisian bryozoans from the study area; see text for sources, numbers: 1—presence, 0—absence.

Author Contributions: Conceptualization, Z.A.T. and D.A.R.; methodology, D.A.R.; data curation, Z.A.T. and D.A.R.; formal analysis, Z.A.T. and D.A.R.; writing, Z.A.T. and D.A.R. All authors have read and agreed to the published version of the manuscript.

Funding: The work by Z.A.T. is supported by RSF (project 22-27-00030).

Institutional Review Board Statement: Not applicable.

Informed Consent Statement: Not applicable.

Data Availability Statement: Not applicable.

Conflicts of Interest: The authors declare no conflict of interest. The funder had no role in the design of the study; in the collection, analyses, or interpretation of data; in the writing of the manuscript; or in the decision to publish the results.

References

1. Becker, R.T.; Königshof, P.; Brett, C.E. Devonian climate, sea level and evolutionary events: An introduction. *Geol. Soc. Lond. Spec. Publ.* **2016**, *423*, 1–10. [CrossRef]
2. Caplan, M.L.; Bustin, R.M. Devonian-Carboniferous Hangenberg mass extinction event, widespread organic-rich mudrock and anoxia: Causes and consequences. *Palaeogeogr. Palaeoclimatol. Palaeoecol.* **1999**, *148*, 187–207. [CrossRef]
3. Zhang, F.; Dahl, T.W.; Lenton, T.M.; Luo, G.; She, S.; Algeo, T.J.; Planavsky, N.; Liu, J.; Cui, Y.; Qie, W.; et al. Extensive marine anoxia associated with the Late Devonian Hangenberg Crisis. *Earth Planet. Sci. Lett.* **2020**, *533*, 115976. [CrossRef]
4. Scotese, C.R.; Song, H.; Mills, B.J.W.; van der Meer, D.G. Phanerozoic paleotemperatures: The earth's changing climate during the last 540 million years. *Earth-Sci. Rev.* **2021**, *215*, 103503. [CrossRef]
5. Streel, M.; Caputo, M.V.; Loboziak, S.; Melo, J.H.G. Late Frasnian-Famennian climates based on palynomorph analyses and the question of the Late Devonian glaciations. *Earth-Sci. Rev.* **2000**, *52*, 121–173. [CrossRef]
6. Haq, B.U.; Schutter, S.R. A chronology of Paleozoic sea-level changes. *Science* **2008**, *322*, 64–68. [CrossRef] [PubMed]
7. Assemat, A.; Girard, C.; Joachimski, M.M.; Adnet, S. Vertebrate diversity reveals perturbations in faunal communities prior to the Hangenberg event in the Montagne Noire (France). *Bull. Geosci.* **2022**, *97*, 109–122. [CrossRef]
8. Kaiser, S.I.; Aretz, M.; Becker, R.T. The global Hangenberg Crisis (Devonian–Carboniferous transition): Review of a first-order mass extinction. *Geol. Soc. Lond. Spec. Publ.* **2016**, *423*, 387–437. [CrossRef]
9. Racki, G. A volcanic scenario for the Frasnian–Famennian major biotic crisis and other Late Devonian global changes: More answers than questions? *Glob. Planet. Chang.* **2020**, *189*, 103174. [CrossRef]
10. Li, S.; Wignall, P.B.; Poulton, S.W.; Hedhli, M.; Grasby, S.E. Carbonate shutdown, phosphogenesis and the variable style of marine anoxia in the late Famennian (Late Devonian) in western Laurentia. *Palaeogeogr. Palaeoclimatol. Palaeoecol.* **2022**, *589*, 110835. [CrossRef]
11. Young, A.; Flament, N.; Williams, S.E.; Merdith, A.; Cao, X.; Müller, R.D. Long-term Phanerozoic sea level change from solid Earth processes. *Earth Planet. Sci. Lett.* **2022**, *584*, 117451. [CrossRef]
12. Gutak, J.M.; Tolokonnikova, Z.A.; Ruban, D.A. Bryozoan diversity in Southern Siberia at the Devonian-Carboniferous transition: New data confirm a resistivity to two mass extinctions. *Palaeogeogr. Palaeoclimatol. Palaeoecol.* **2008**, *264*, 93–99. [CrossRef]

13. Tolokonnikova, Z.A.; Ernst, A. Richness of Famennian-Tournaisian (Late Devonian-Eraly Carboniferous) bryozoans in shallow areas of Palaeotethys and Palaeoasian oceans. *Palaeobiodiversity Palaeoenvironments* **2021**, *101*, 885–906. [CrossRef]
14. Powers, C.M.; Bottjer, D.J. The effects of mid-Phanerozoic environmental stress on bryozoan diversity, paleoecology, and paleogeography. *Glob. Planet. Chang.* **2009**, *65*, 146–154. [CrossRef]
15. Gorjunova, R.V. New cryptostomide bryozoans from the Upper Devonian of Transcaucasia and some aspects of the evolution of the order Cryptostomida. *Paleontol. J.* **2007**, *41*, 600–613. [CrossRef]
16. Lavrentjeva, V.D. Upper Devonian Bryozoans of Transcaucasia. *Izvestiya vuzov. Geologiya i Razvedka* **1985**, *8*, 12–18. (In Russian)
17. Tolokonnikova, Z.A. New data on the Tournaisian bryozoans (Lower Carboniferous) from Azerbaijan and Armenia. *Paleontol. J.* **2016**, *50*, 388–395. [CrossRef]
18. Ernst, A.; Kora, M.; El-Desouky, H.; Herbig, H.-G.; Wyse Jackson, P.N. Stenolaemate bryozoans from the Carboniferous of Egypt. *J. Afr. Earth Sci.* **2020**, *165*, 103811. [CrossRef]
19. Sakagami, S.; Sciunnach, D.; Garzanti, E. Late Paleozoic and Triassic bryozoans from the Tethys Himalaya (N India, Nepal and S Tibet). *Facies* **2006**, *52*, 279–298. [CrossRef]
20. Webb, G.E. Quantitative analysis and paleoecology of earliest Mississippian microbial reefs, Gudman Formation, Queensland, Australia: Not just post-disaster phenomena. *J. Sediment. Res.* **2005**, *75*, 877–896. [CrossRef]
21. The Paleobiology Database. Available online: https://paleobiodb.org (accessed on 23 May 2022).
22. Aristov, V.A. Conodonts of the Devonian–Lower Carboniferous of Eurasia: Assemblages, zonation, correlation of polyfacial deposits. *Trudy GIN RAN* **1994**, *484*, 1–192. (In Russian)
23. Rustamov, M.I. Plate tectonics of Kur depression within Meso-Tethys geodynamic evolution. *ANAS Trans. Earth Sci.* **2018**, *2*, 57–71. (In Russian) [CrossRef]
24. Kocsis, Á.T.; Scotese, C.R. Mapping paleocoastlines and continental flooding during the Phanerozoic. *Earth-Sci. Rev.* **2021**, *213*, 103463. [CrossRef]
25. Grechishnikova, I.A.; Levitskii, E.S. The Famennian–Lower Carboniferous reference section Geran-Kalasi (Nakhichevan Autonomous Region, Azerbaijan). *Stratigr. Geol. Correl.* **2011**, *19*, 21–43. [CrossRef]
26. Mamedov, A.B. Zonal subdivision of the Tournaisian Stage of the Lower Carboniferous of the Nakhichevan ASSR. *Trudy Akademii nauk Azerbaidzhanskoy SSR* **1980**, *12*, 53–57. (In Russian)
27. Mamedov, A.B. The Famennian stage of the southern Transcaucasia. *Stratigr. Geol. Correl.* **2002**, *10*, 57–68.
28. Rzhonsnitskaya, M.A.; Mamedov, A.B. Devonian stage boundaries in the southern Transcaucasus. *CFS Cour. Forsch. Senckenberg* **2000**, *225*, 329–333.
29. International Commission on Stratigraphy (ICS). International Chronostratigraphic Chart. 2022. Available online: https://stratigraphy.org/chart (accessed on 24 May 2022).
30. Ruban, D.A. The Greater Caucasus—A Galatian or Hanseatic Terrane? Comment on "The formation of Pangea" by G.M. Stampfli, C. Hochard, C. Vérard, C. Wilhem and J. von Raumer [Tectonophysics 593 (2013) 1–19]. *Tectonophysics* **2013**, *608*, 1442–1444. [CrossRef]
31. Ruban, D.A.; Al-Husseini, M.I.; Iwasaki, Y. Review of Middle East Paleozoic plate tectonics. *GeoArabia* **2007**, *12*, 35–56. [CrossRef]
32. Cao, W.; Zahirovic, S.; Flament, N.; Williams, S.; Golonka, J.; Müller, R.D. Improving global paleogeography since the late Paleozoic using paleobiology. *Biogeosciences* **2017**, *14*, 5425–5439. [CrossRef]
33. Domeier, M.; Torsvik, T.H. Plate tectonics in the late Paleozoic. *Geosci. Front.* **2014**, *5*, 303–350. [CrossRef]
34. Matthews, K.J.; Maloney, K.T.; Zahirovic, S.; Willims, S.E.; Seton, M.; Müller, R.D. Global plate boundary evolution and kinematics since the late Paleozoic. *Glob. Planet. Chang.* **2016**, *146*, 226–250. [CrossRef]
35. Metcalfe, I. Multiple Tethyan ocean basins and orogenic belts in Asia. *Gondwana Res.* **2021**, *100*, 87–130. [CrossRef]
36. Van Hinsbergen, D.J.J.; Torsvik, T.H.; Schmid, S.M.; Matenco, L.C.; Maffione, M.; Vissers, R.L.M.; Gürer, D.; Spakman, W. Orogenic architecture of the Mediterranean region and kinematic reconstruction of its tectonic evolution since the Triassic. *Gondwana Res.* **2020**, *81*, 79–229. [CrossRef]
37. Williams, S.; Wright, N.M.; Cannon, J.; Flament, N.; Müller, R.D. Reconstructing seafloor age distributions in lost ocean basins. *Geosci. Front.* **2021**, *12*, 769–780. [CrossRef]
38. Goryunova, R.V. *Morphology, System and Phylogeny of Bryozoans (Order Rhabdomesida)*; Nauka: Moscow, Russia, 1985; p. 151. (In Russian)
39. Gorjunova, R.V. Family Coelotubuliporidae fam. nov. and morphological parallelism in the evolution of bryozoans. *Paleontol. J.* **2011**, *45*, 510–624. [CrossRef]
40. Gorjunova, R.V.; Lavrentjeva, V.D. New bryozoans from the Devonian-Carboniferous boundary beds of Transcaucasia. *Paleontol. J.* **2007**, *41*, 146–155. [CrossRef]
41. Lavrentjeva, V.D. New bryozoan species of the family Atactotoechidae from the Devonian of Transcaucasia. *Paleontol. J.* **2001**, *35*, 152–156.
42. Tolokonnikova, Z.A. New Famennian bryozoans (Upper Devonian) from Azerbaijan and Armenia. *Paleontol. J.* **2016**, *50*, 245–254. [CrossRef]
43. Gradstein, F.M.; Ogg, J.G.; Schmitz, M.; Ogg, G. (Eds.) *Geologic Time Scale 2020*; Elsevier: Amsterdam, The Netherlands, 2020; p. 1390.

44. Aretz, M.; Corradini, C.; Denayer, J. The Devonian-Carboniferous Boundary around the globe: A complement. *Palaeobiodiversity Palaeoenvironments* **2021**, *101*, 633–662. [CrossRef]
45. Devuyst, F.-X.; Hance, L.; Hou, H.; Wu, X.; Tian, S.; Coen, M.; Sevastopulo, G. A proposed Global Stratotype Section and Point for the base of the Viséan Stage (Carboniferous): The Pengchong section, Guangxi, Sputh China. *Episodes* **2003**, *26*, 105–115. [CrossRef] [PubMed]
46. Harrigan, C.O.; Schmitz, M.D.; Over, D.J.; Trayler, R.B.; Davydov, V.I. Recalibrating the Devonian time scale: A new method for integrating radioisotopic and astrochronologic ages in a Bayesian framework. *Geol. Soc. Am. Bull.* **2021**, *134*, 1931–1948. [CrossRef]
47. Klapper, G.; Feist, R.; Becker, R.T.; House, M.R. Definition of the Frasnian/Famennian Stage boundary. *Episodes* **1993**, *16*, 433–441. [CrossRef]
48. Paproth, E.; Feist, R.; Flajs, G. Decision on the Devonian-Carboniferous boundary stratotype. *Episodes* **1991**, *14*, 331–336. [CrossRef]
49. Streel, M.; Brice, D.; Degardin, J.-M.; Derycke, C.; Dressen, R.; Groessens, E.; Hance, L.; Legarnd-Blain, M.; Lethiers, F.; Loboziak, S.; et al. Proposal for a Struntian Substage and a subdivision of the Famennian Stage into four Substages. *Subcomission Devonian Stratigr. Newsl.* **1998**, *15*, 47–49.
50. Ruban, D.A.; Radulović, B.V.; Radulović, V.J. Diversity dynamics of Early Cretaceous brachiopods in the tectonic units of Serbia: Regional versus global patterns. *Proc. Geol. Assoc.* **2016**, *127*, 691–698. [CrossRef]
51. Jaccard, P. Etude comparative de la distribution florale dans une portion des Alpes et du Jura. *Bull. Société Vaud. Sci. Nat.* **1901**, *37*, 547–579.
52. Sepkoski, J.J., Jr. Quantified Coefficients of Association of Measurement of Similarity. *Math. Geol.* **1974**, *6*, 135–152. [CrossRef]
53. Gower, J.C. A general coefficient of similarity and some of its properties. *Biometrics* **1971**, *27*, 857–871. [CrossRef]
54. Horowitz, A.S.; Pachut, J.F. The fossil record of bryozoan species diversity. In *Proceedings of the 11th International Bryozoology Association Conference*; Smithsonian Tropical Research Institute: Calzada de Amador, Panama, 2000; pp. 245–248.
55. Bond, D.; Wignall, P.B.; Racki, G. Extent and duration of marine anoxia during the Frasnian-Famennian (Late Devonian) mass extinction in Poland, Germany, Austria and France. *Geol. Mag.* **2004**, *141*, 173–193. [CrossRef]
56. Buggisch, W. The global Frasnian-Famennian »Kellwasser Event«. *Geol. Rundsch.* **1991**, *80*, 49–72. [CrossRef]
57. Cohen, P.A.; Junium, C.K.; King Phillips, E.; Uveges, B.T. Carbon cycle dynamics and ecology revealed by the carbon isotopic composition of single organic microfossils during the Late Devonian Biotic Crisis. *Geobiology* **2022**, *20*, 346–362. [CrossRef] [PubMed]
58. Racki, G.; Wignall, P.B. Devonian global changes—Recent advances and challenges in different domains. *Glob. Planet. Chang.* **2020**, *191*, 103200. [CrossRef]
59. Sandberg, C.A.; Morrow, J.R.; Ziegler, W. Late Devonian sea-level changes, catastrophic events, and mass extinctions. *Spec. Pap. Geol. Soc. Am.* **2002**, *356*, 473–487.

Journal of
Marine Science and Engineering

MDPI

Article

Contourite and Turbidite Features in the Middle Caspian Sea and Their Connection to Geohazards Derived from High-Resolution Seismic Data

Vsevolod Yutsis [1,*], Oleg Levchenko [2] and Victoria Putans [2]

[1] División de Geociencias Aplicadas, Instituto Potosino de Investigación Científica y Tecnológica (IPICYT), Camino a la Presa San José 2055, Lomas 4ª Sección, San Luis Potosí 78216, S.L.P., Mexico
[2] Shirshov Institute of Oceanology, Russian Academy of Sciences, 36, Nakhimovsky Prospect, 117997 Moscow, Russia; olevses@gmail.com (O.L.); vitapu@mail.ru (V.P.)
* Correspondence: vsevolod.yutsis@ipicyt.edu.mx; Tel.: +52-444-834-2000

Abstract: High fluvial input combined with specific topographic and oceanographic settings in the Caspian Sea create favorable conditions for contourite deposition. For the first time in its middle portion, contourite deposits have been observed in high-resolution seismic profiles. Various types of contourite drifts and mixed depositional systems have been revealed on the lower slope and in the adjacent basin, some of which are accompanied by sediment wave fields. The deposition of contourites or turbidites and their lateral distribution is controlled by sea-floor topography and oceanographic processes, as well as the modern activity of gravity flows downslope on the western Caucasian slope and in the channel system on the Mangyshlak Sill. The contourite drifts and sediment wave fields form several contourite depositional systems, which seem to merge in the Caspian contourite depositional complex. This occurs near the foot of slopes of the Derbent Basin and is related to the counterclockwise circum-Caspian current in the Middle Caspian Sea. The fact that the Caspian Sea is the largest lake in the world makes this region a significant area for research into the "lake contourites" issue. The Caspian Sea is an important oil-producing area, and sedimentary processes related to the contourite and turbidite can be a source of potential geohazards in the construction and exploitation of underwater engineering structures

Keywords: contourites; drift; sediment waves; bottom currents; turbidite; geohazards; high-resolution seismic data

Citation: Yutsis, V.; Levchenko, O.; Putans, V. Contourite and Turbidite Features in the Middle Caspian Sea and Their Connection to Geohazards Derived from High-Resolution Seismic Data. *J. Mar. Sci. Eng.* **2022**, *10*, 990. https://doi.org/10.3390/jmse10070990

Academic Editors: Assimina Antonarakou, George Kontakiotis and Dmitry A. Ruban

Received: 3 June 2022
Accepted: 12 July 2022
Published: 20 July 2022

1. Introduction

Contourites are deposits of bottom thermohaline-driven geostrophic contour currents [1–3]. Numerous studies in recent years have revealed their presence in a wide range of settings, such as the open ocean, along continental margins, in seas and even in lakes. Despite their apparent abundance, many aspects with respect to the formation, internal structure, and sedimentology of contourites still remain poorly understood in general. The most progress in understanding contourite deposits has been made by studies in deep-water environments, whereas shallow water contourites, including lacustrine contourites, are still largely unknown.

The Caspian Sea (CS) is the largest lake of the world (Figure 1), although it is composed of salt water. The maximum water depth in the Middle Caspian is approximately 700 m. The specific topographic and oceanographic settings of the sea, as well as its high fluvial input, are favorable conditions for contourite deposition; however, bottom circulation and its impact on sedimentation and the sea floor topography in the Caspian Sea still requires further investigation, and appropriate special sedimentological studies for contourite identification in the Caspian have not been carried out. The Caspian Sea is one of the oldest gas- and oil-producing areas, with some sedimentary processes related to the contourites and turbidites perhaps being sources of potential geohazards in the construction and exploitation of underwater engineering structures. The downslope gravity and turbidity flows causing seafloor erosion and intensive sediment input are treacherous processes for submarine pipelines that could cause pipe rupture, while sediment waves should serve as a warning.

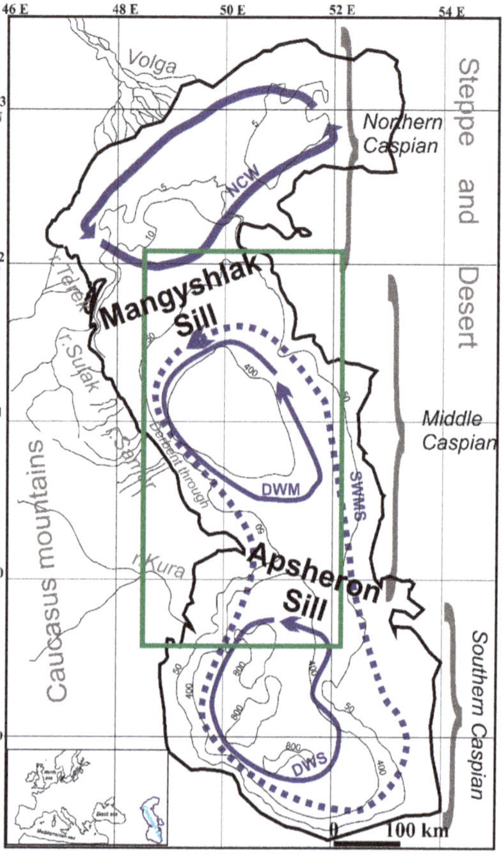

Figure 1. Bathymetric map (m) of the Caspian Sea with regional divisions (northern, middle, and southern) and major geographical sites (mountains, rivers, and steppe). Major bottom currents: NCW (north CS water mass which takes all water levels), SWMS (surface-water mass of the middle and south Caspian sub-basins), DWM (deep water mass of the middle Caspian sub-basin), DWS (deep-water mass of south Caspian sub-basin); river system of the Caucasus Mountains in the west and the Volga River delta in the north are shown. Rectangle shows the area of seismic surveys performed between 2004 and 2012.

Lacustres contourite drift studies could be used to better understand all current-controlled sedimentation processes. Current-controlled sedimentary features have been discovered in Lake Superior (USA), some of the larger West-African Rift lakes, Lake Lago Cardiel (Argentina) and others. For the first time, lacustrine contourite drifts with dimensions comparable to those of their deep-water counterparts were discovered in Lake Baikal (Russia) over an intra-basin high; however, they do not occur in the deep basin or along its margins [4]. Current-controlled sedimentary features in the Caspian Sea are generally located in the lower part of the continental slope and in a deep basin at its foot, which is in keeping with such deep-water features. The morphology and dimensions of these Caspian Sea features are also comparable to those of many of their oceanic deep-water counterparts. This general similarity may be explained by the fact that some features that are particular to oceans remain in the evolution of the closed Caspian Sea. Nevertheless, giant contourite drifts and fields of sedimentary waves, such as those in the Argentine Basin, have not been found in the Caspian Sea.

The initial recognition of contourite deposits in the marine setting is generally based on reflection seismic data, which is now considered a standard method in most contourite studies. This study is based on a new dataset of mostly high-resolution (HR) and ultra-high-resolution (UHR) seismic reflection profiles collected respectively with the sparker and the parametric sub-bottom profiler SES by the Shirshov Institute of Oceanology in the Caspian Sea between 2004 and 2012 (Figure 2). The results obtained significantly clarified our understanding of the structure of the upper Pliocene-Quaternary sedimentary strata. It was these new data that made it possible to identify, for the first time, accumulative and erosive features in the Middle Caspian, similar to contourite ones [5–8]. This paper focuses on the description of the most prominent erosive and accumulative forms of bottom relief and the Late Cenozoic sedimentary cover of the Derbent Basin (DB) in the Middle Caspian Sea, as well as the processes responsible for the interaction of three different processes. Apsheron Sill (AS) with the along-slope bottom contour currents, the formation of the Late Cenozoic sedimentary cover is affected by the downslope turbidite and gravity flows.

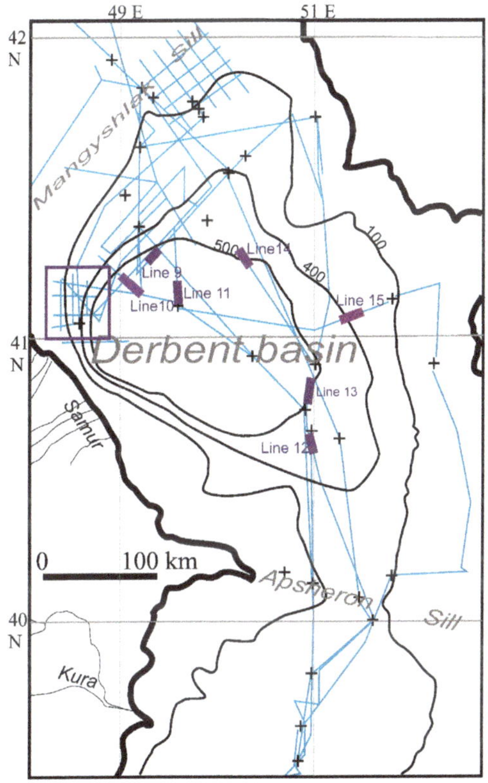

Figure 2. Scheme of seismic reflection profiles (high (sparker) and ultra-high (SES) resolution), collected by Shirshov Institute of Oceanology from 2004 to 2012. The straight lines show the positions of the seismic profiles illustrated in Figures 3–15. The rectangle shows the position of the seismic profiles shown in Figure 4. Crosses for oceanology stations.

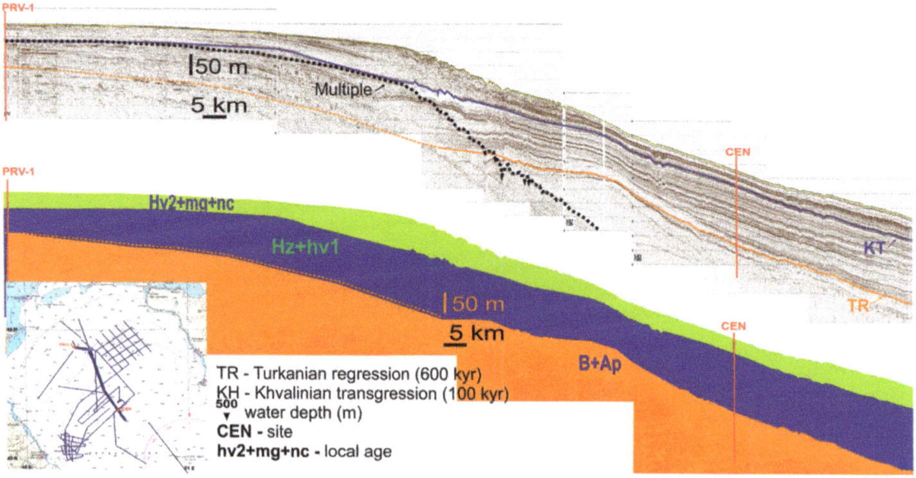

Figure 3. Middle Caspian sparker data and stratigraphy scheme. Regional unconformities: TR—Turkanian regression (600 kyr), KH—Khvalinian transgression (100 kyr). Local age: B + Ap—Bakinian and Apsheron transgressions, Hz+Hv1—Khazar and Khvalinian (first step) transgressions, hv2 + mg + nc—Khvalinian (second step) transgression, Mangyshlak regression (7–10 kyr) and new-Caspian transgression. CEN—Site Centralnaya. PRV-1—Site PRV-1.

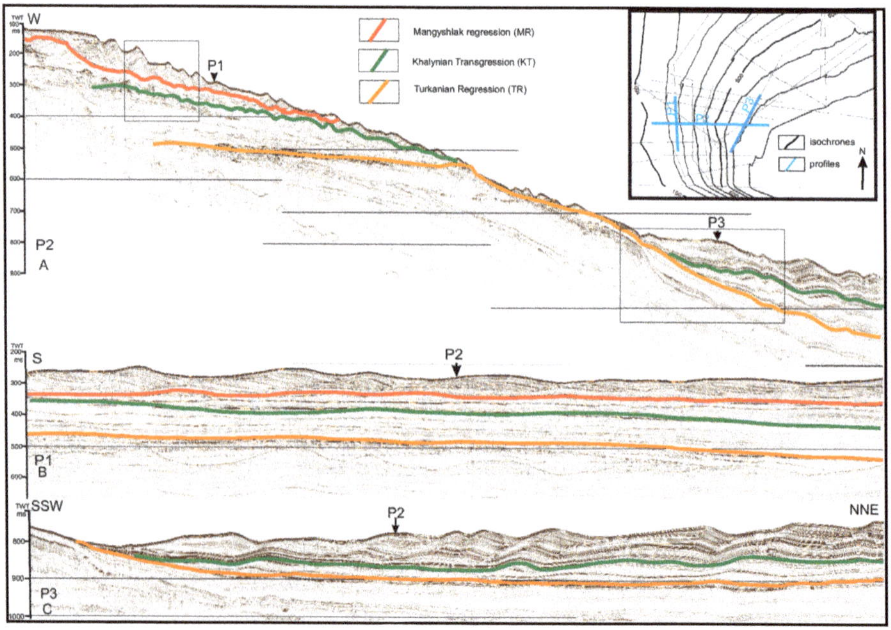

Figure 4. High resolution seismic profiles (sparker) on the western slope of the Derbent Basin off the Caucasus Mountains. (**A**) P2—Seismic profile P2, W-E across the slope; (**B**) P1—Seismic profile P1, S-N along the slope; (**C**) P3—Seismic profile P3, SW-NE along foot of the slope. Location in inset. The rectangle at the top left of Figure 4A shows a fragment of this seismic profile, illustrated in detail in Figure 5. The rectangle at the bottom right of Figure 4A shows a fragment of this seismic profile, illustrated in detail in Figure 6.

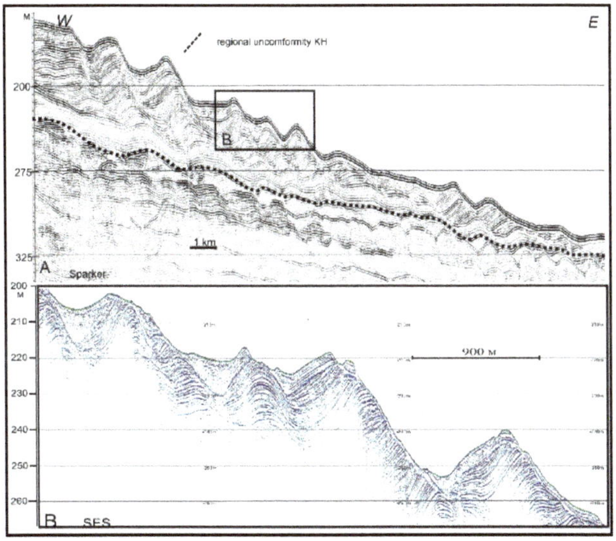

Figure 5. Sediment waves of turbidity + contourite genesis on the western slope of the Derbent Basin. (**A**) Fragment of the profile (sparker) from Figure 4A. KT—regional unconformity Khvalinian transgression. (**B**) Enlarged fragment of ultra-high-resolution seismic reflection profile (SES) showing in detail uppermost sedimentary layer with evidence of the young to present-day erosion activity.

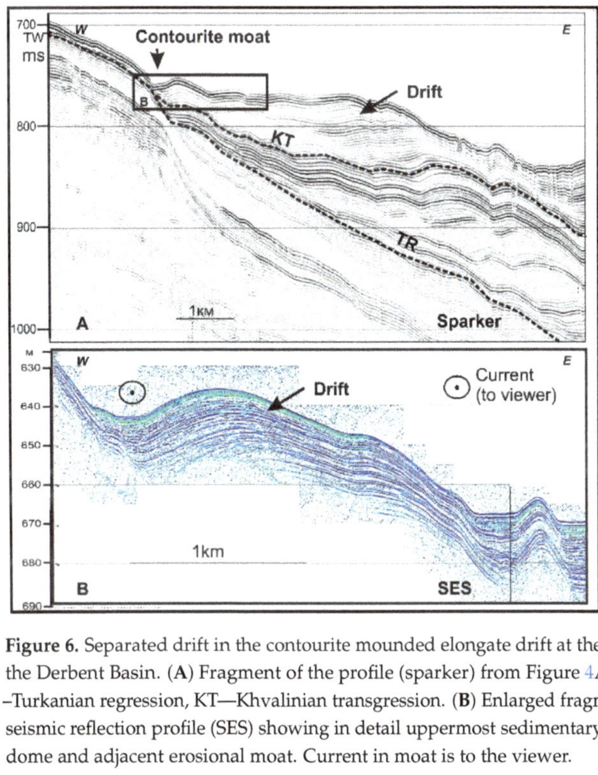

Figure 6. Separated drift in the contourite mounded elongate drift at the foot of the western slope of the Derbent Basin. (**A**) Fragment of the profile (sparker) from Figure 4A. Main unconformities: TR –Turkanian regression, KT—Khvalinian transgression. (**B**) Enlarged fragment of ultra-high-resolution seismic reflection profile (SES) showing in detail uppermost sedimentary layer with a migration drift dome and adjacent erosional moat. Current in moat is to the viewer.

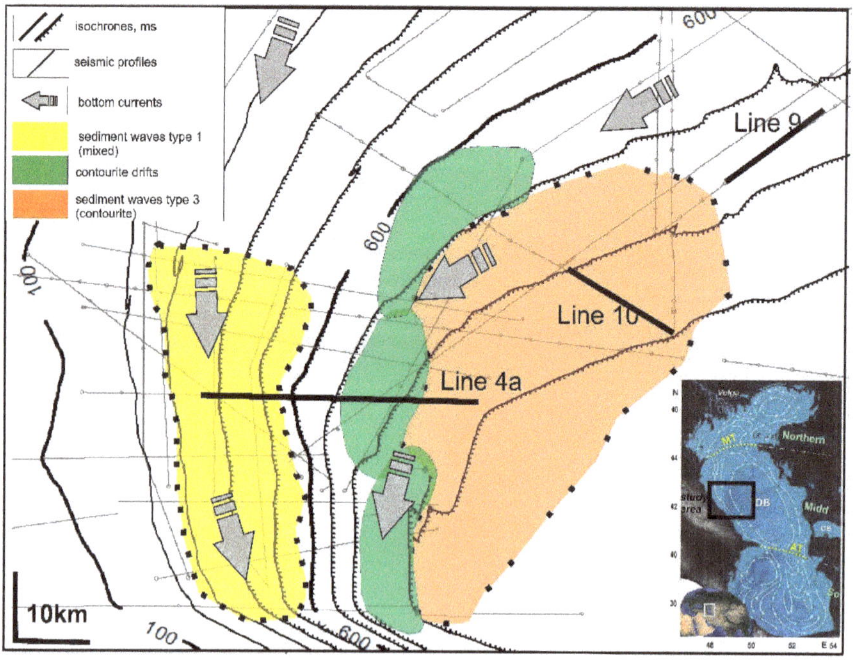

Figure 7. Contourite deposits near the western slope of the Middle Caspian Sea. Location in insert. Arrows for near bottom water currents. Lines 4a, 9, and 10 show the positions of the seismic profiles shown in the corresponding figures.

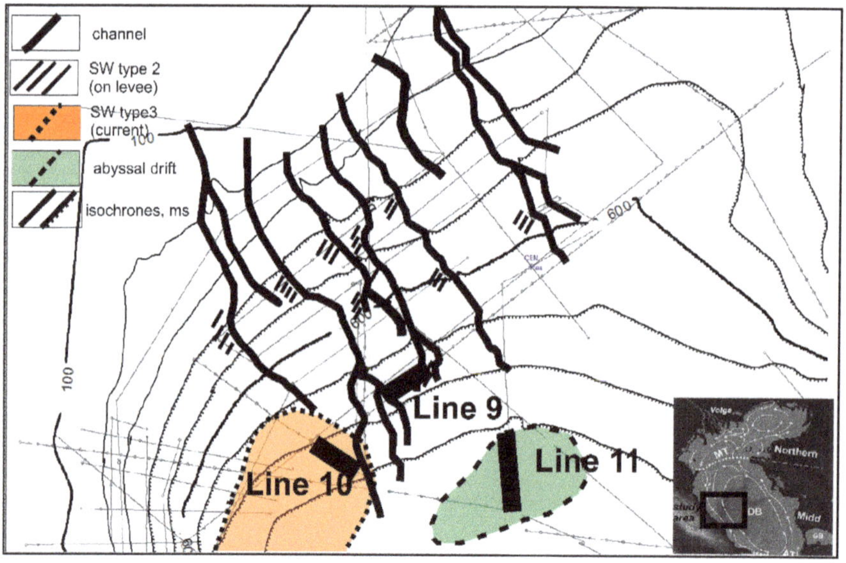

Figure 8. Scheme of modern channel system over the Mangyshlak Sill and contourites off the northern slope of the Middle Caspian Sea: sediment wave types 2 (on levee) and 3 (contourite) and abyssal drift. Lines 9, 10 and 11 show the positions of the seismic profiles shown in the corresponding figures.

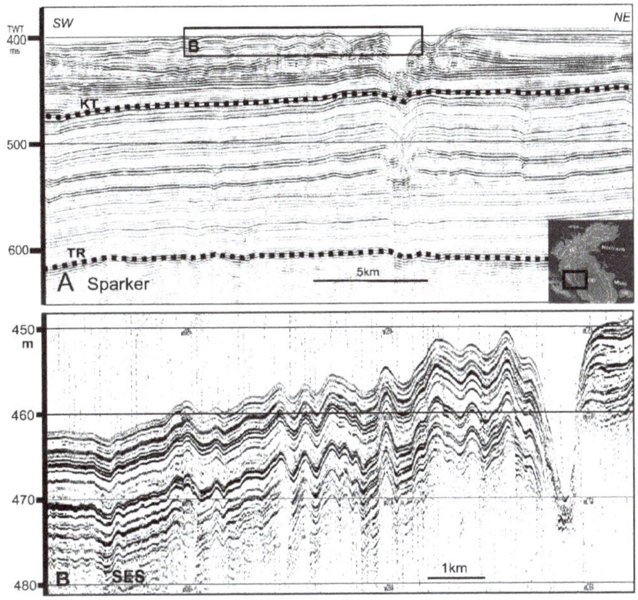

Figure 9. Sediment waves type 2 (on channel levees) on the northern slope of the Derbent Basin near the Mangyshlak Sill. (**A**) Fragment of the profile (sparker) SW-NE trending along foot of the slope. Main unconformities: TR—Turkanian regression, KT—Khvalinian transgression. (**B**) Enlarged fragment of ultra–high-resolution seismic reflection profile (SES) showing in detail uppermost sedimentary layer. Location of the profiles is shown in Figures 2 and 8.

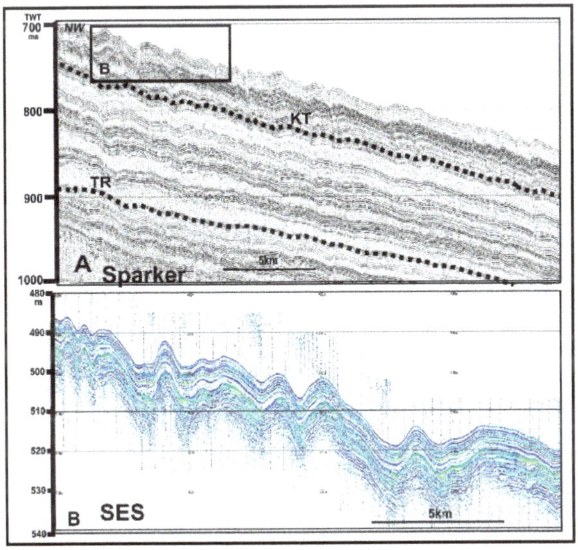

Figure 10. Sediment waves type 3 (contourite) on the northern slope of the Derbent Basin near the Mangyshlak Sill. (**A**) Fragment of the profile (sparker) NW-SE trending across foot of the slope. Main unconformities: TR—Turkanian regression, KT—Khvalinian transgression. (**B**) Enlarged fragment of ultra-high-resolution seismic reflection profile (SES) showing in detail uppermost sedimentary layer. Location of the profiles is shown in Figures 2 and 8.

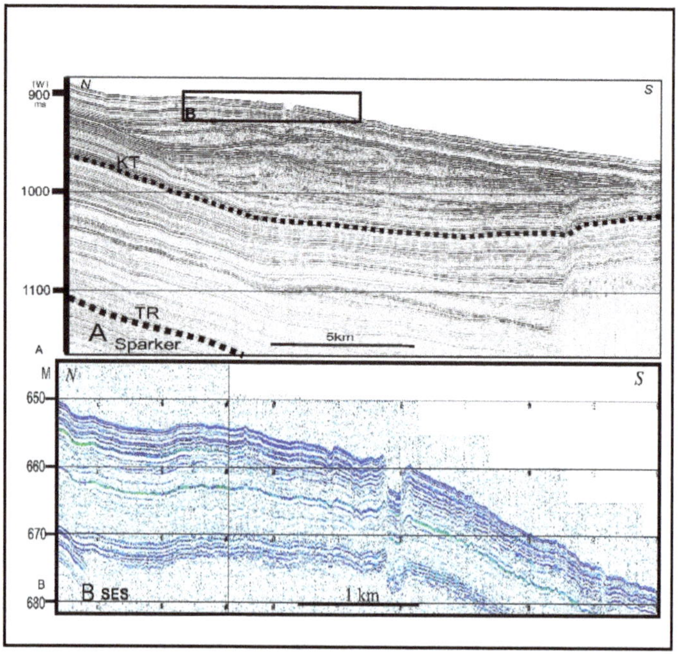

Figure 11. Contourite detached drift in the northern Derbent Basin off the Mangyshlak Sill. (**A**) Fragment of the profile (sparker) NW-SE across foot of the northern slope. Main unconformities: TR—Turkanian regression, KT—Khvalinian transgression. (**B**) Enlarged fragment of ultra-high-resolution seismic reflection profile (SES) showing in detail uppermost sedimentary layer with present-day channel. Location in Figures 2 and 8.

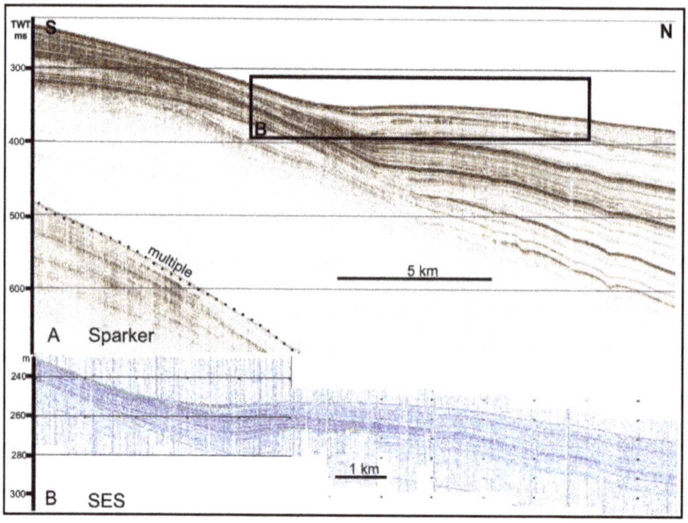

Figure 12. Contourite plastered drift on the southern slope of the Derbent Basin near the Apsheron Sill. Fragment of the profile (sparker) roughly S-N across the slope. (**A**) Fragment of the profile (sparker). (**B**) Enlarged fragment of ultra-high-resolution seismic reflection profile (SES) showing in detail uppermost sedimentary layer. Location in Figure 2.

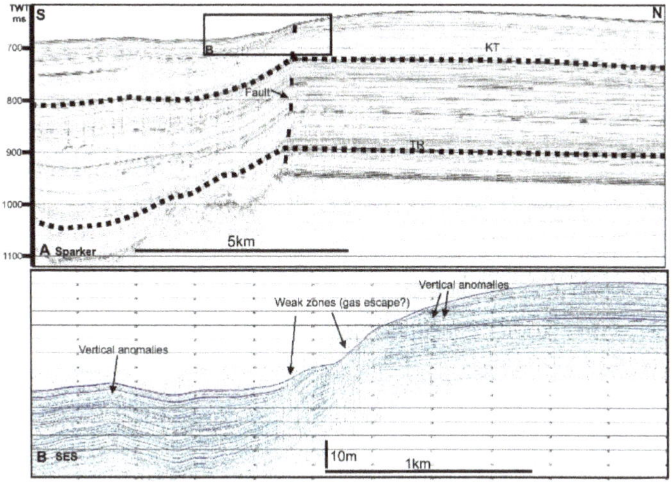

Figure 13. Contourite fault-controlled drift on the southern slope of the Derbent Basin near the Apsheron Sill. (**A**) Fragment of the profile (sparker) roughly S-N along foot of the slope. Main unconformities: TR—Turkanian regression, KT—Khvalinian transgression. (**B**) Enlarged fragment of ultra-high-resolution seismic reflection profile (SES) showing in detail uppermost sedimentary layer with weak acoustical transparent gas escape zones. Location in Figure 2.

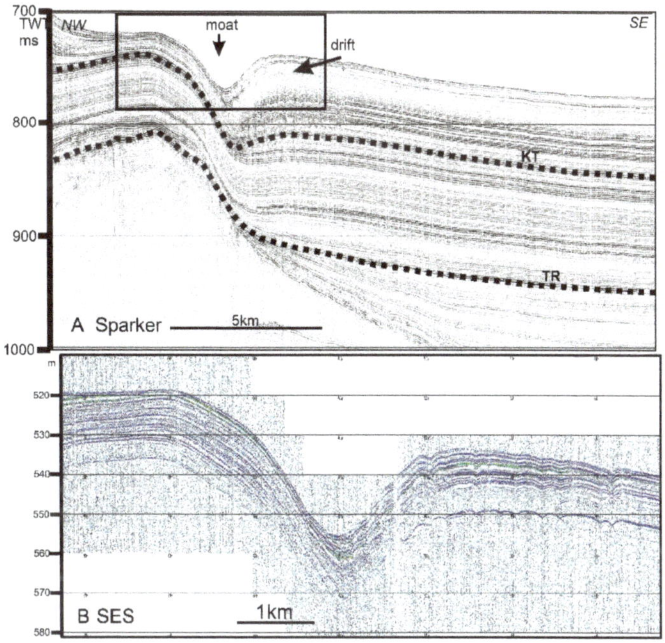

Figure 14. Contourite separated drift and deep erosional moat in the Derbent Basin off its eastern slope. (**A**) Fragment of the profile (sparker) NW-SE along foot of the slope which shows buried ancient creep deformation. Main unconformities: TR—Turkanian regression, KT—Khvalinian transgression. (**B**) Enlarged fragment of ultra-high-resolution seismic reflection profile (SES) showing in detail uppermost sedimentary layer with lens of the drift. Location in Figure 2.

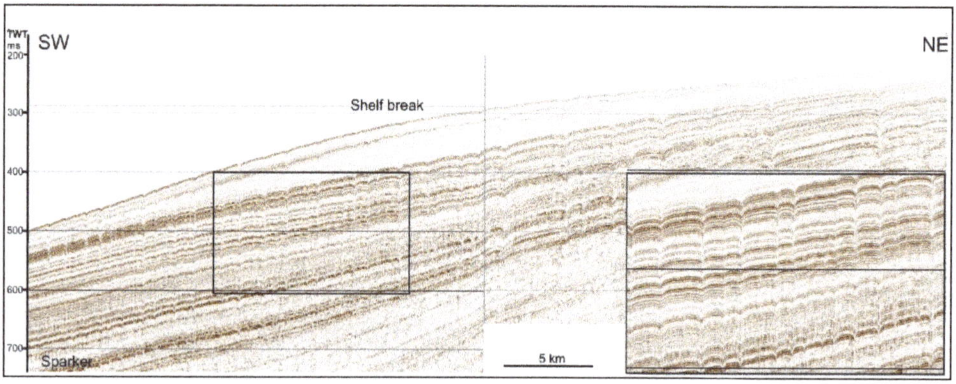

Figure 15. Exogenic–gravitational folds or creep deformation on the eastern slope of the Derbent Basin. Fragment of the profile (sparker) SW-NE trending across foot of the slope. Location of the piles is shown in Figure 2.

High regional seismicity initiates numerous landslides and density flows on the slopes of the Derbent Basin. The exogenic–gravitational folds in the Quaternary clayey and calcareous–clayey sediments were previously defined along the eastern slope of the Derbent Basin, which were produced by the plastic flow of poorly consolidated sediments under the influence of gravitational forces [9]. Our HR and UHR surveys revealed similar slump box-shaped rootless folds on the northern slope of the Derbent Basin [10]. These new seismic reflection data made it possible to reveal the structure of the upper sedimentary layer of the western slope of the Derbent Basin in greater detail. Areas with a chaotic pattern of the acoustic record are marked here, which are interpreted as a set of separate slump and landslide blocks [11]. They appear to have moved in the recent geologic past (Late Quaternary) down a relatively steep slope. These landslide deformations and gravitational features were previously considered [10,11]; therefore, in this paper, we focus on the description of turbidite features and, mostly, contourite ones.

This paper focuses on the bottom and sub-bottom features in the central part of the Caspian Sea (the Middle Caspian) and the processes of near bottom water circulation which formed these features. Our main objectives are (1) to test the contourite paradigm on special sea-lake conditions, where the morphology is similar to an oceanic one, but all oceanography processes occur in far smaller time-space scale, and (2) to explore potential geohazards of which contouritic features can be signs.

2. Regional Settings

The Caspian Sea (CS) is located between Europe and Asia and is the largest lake in the world; however, its physical environment and its floor have mixed oceanic-continental characteristics [12]. The CS was created tectonically when the Paratethys began splitting into smaller water bodies some 5.6 million years ago [13]. Now the sea is confined between the Caucasus Mountains in the south and west, the vast steppe of Central Asia in the east and an extensive alluvial plain in the north (Figure 1). The water surface is almost 28 m below the World Ocean level [14]. The sea extends for about 1030 km from north to south and for 196–433 km in the east-west direction (total area 386,400 km^2 in 2017 [15]).

2.1. Geological Setting

The Caspian Sea is divided into three parts (Figure 1): northern, middle, and southern [16]. The northern part is the shallowest with a maximum depth of 25 m and smooth bottom topography. The Middle Caspian Sea represents a large depression separated from the northern and southern parts by the Mangyshlak Sill (a bow-shaped depositional body) and the Apsheron Sill (a large structural high), respectively. A major part of the Middle

Caspian area is occupied by the Derbent Basin, which is contoured by an isobath of 400 m and has a maximum depth of 788 m. The Southern Caspian Basin is the deepest part of the sea, with a maximum depth of 1025 m [17].

The seismotectonic provinces were mapped by Ivanova and Trifonov [18] using the criteria of Earth's crustal structure, specific features of neotectonic (Pliocene–Quaternary) evolution, the pattern and kinematics of active faults, as well as the distribution of earthquake focal zones and the dynamics of their activation during the last 160 years. Average seismic energy values were computed for individual seismotectonic provinces that differ in the Earth's crustal structure as regards the intensity of neotectonic movements and seismicity. Our study concerns an area of contrasting combination of the Caucasus Mountains with the Derbent Trough in the west of the Middle Caspian region (Figure 1). The structures originated simultaneously in the Pliocene–Quaternary, and their formation was accompanied by a reworking of the Earth's crust in the post-Paleozoic era [19]. The main earthquake focal zone, which features events of up to M = 6.3 with hypocenter depths of up to 110 km, extends along the southwestern slope of the Derbent Trough. Significant geological processes on a regional scale are neotectonic movements and deep-focus earthquakes. Seismicity triggers mud volcano activity and mass movements on the slopes of both deep basins in the Middle and Southern Caspian.

The western slope of the Derbent Basin is characterized by a slope angle from 0° to 4° and a "double break" at water depths of 100 m and 400 m (Figure 3). The upper break correlates with isobath 100 m and marks the outer edge of the modern shelf; the lower break goes with isobath 400 m, which outlines the Derbent Basin. The Mangyshlak Sill and the northern slope of the Derbent Basin are intersected by a channel system, which relates to the overall fan of the Volga, Terek, and Ural rivers.

The Caspian Sea's most unique feature is the unpredictable and fast level changes, which are impossible to correlate either with the World Ocean or with the history of glaciations [20,21]. Furthermore, seismicity variations for the Caspian region do not point to any unique correlation with the sea level oscillations. At the same time, tectonic processes, which are partly reflected in seismicity, make a significant contribution to the contemporary CS level oscillations [18]. This manifests itself in sea level rises during large earthquakes and, more convincingly in a correlation between values of the total seismic energy released in individual seismotectonic provinces in the region and sea level changes. The tectonic impacts produce an integral effect of interaction between the deformation processes, causing changes in the sea basin volume (depression deepening or transverse shortening, and local anticline growths), which may be supplemented by an intermittent water supply from the deep-seated layers of the sedimentary cover [22]. This supply could be more significant in the deep and rapidly subsiding South Caspian depression, where it is enhanced by faults revived by earthquakes [18]. Since 2016, the CS level seems to have stabilized to an equilibrium at −28 m [14]. As it is an endorheic lake, CS level variations are mostly controlled by water discharge from rivers, direct precipitation and evaporation over the CS [14,23].

The Caspian deep-water environment benefited from the water-level fall until 2016 and the consequent accelerated circulation that causes deep-water ventilation and nutrient exchange; however, most changes are felt at the surface level [14].

2.2. Sedimentation Setting

The maximum thickness of sediments in the Derbent Trough is 14 km, while the thickness of Pliocene–Quaternary sediments is >5 km. It should be noted that the most intense subsidence only began at the end of the Pliocene and is continuing at the present time, remaining uncompensated by sediment accumulation [19,24].

Sedimentation processes in the Caspian Sea are controlled by bottom topography and sources of sedimentary material (river runoff first), as the Caspian Sea is asymmetric in terms of sediment input. Its western part is dominated by a strong alluvial input: a slope adjoins the high chain of the Great Caucasus Mountains with many mountain rivers, most notably, The Terek, Sulak, Samur and Kura. Runoff sediments move across the shelf and down the western slope; nevertheless, the highest surface concentration suspensions (41 mg/L) are measured on the northern Caspian and in the delta of the Volga River [16]. The overall sediment discharge of more than 130 rivers exceeds 110 tons per year [25], while the eastern part of the sea is adjacent to the desert and is characterized by the absence of river input and a prevalence of aeolian processes [26].

2.3. Hydrological Settings

The differences in water masses, which are mainly created by the climate, lead to relatively distinctive physical and chemical specifications. To a lesser extent, geological forces may cause a mixing of water masses, e.g., by the dispensing of materials and energy through the water column. Four flows of water masses circulation seem to be identified in the Caspian Sea (Figure 1): (1) the north CS water mass, which takes all water levels (NCW), (2) the surface-water mass of the middle and south Caspian sub-basins (SWMS), (3) the deep-water mass of the middle Caspian sub-basin (DWM) and (4) the deep-water mass of south Caspian sub-basin (DWS) [27]. The Apsheron Sill separates the two deep-water masses of middle and south sub-basins and prevents free mixing between them. The middle Caspian deep-water mass has a lower temperature and salinity and higher dissolved oxygen content, compared to the southern one [14].

Vertical exchange in the middle CS is partially caused by wind forcing that forms a cyclonic gyre in the center and upwelling along the east and west coasts of the middle Caspian sub-basin, which leads to contrasting temperatures. Moreover, the intense evaporation at the eastern coast may cause local saline and dense warm water to penetrate deeper waters [28]. Extensive water exchange among the water masses occurs during the cold season. The winter convection mixes the water column vertically, and during mild winters, it penetrates down to around 200 m in the middle Caspian masses. During severe winters, it reaches down to enable the vertical mixing of oxygen and other biochemical elements [29,30]. The vertical mixing is forced by wind, evaporation and winter convection; this, however, is not affected by current water-level fluctuations. The Arctic-type mixing occurs annually in the CS, and it affects the whole Caspian water mass [17]. The strength of this type of mixing is closely related to the severity of winter and the Caspian water-level status. During water-level lowstand, mixing would be stronger and deeper; however, in the highstand, it would hardly reach deep-water masses.

The rate of the CS deep-water circulation has changed significantly during water-level fluctuations [17,31]. The Middle Caspian Sea is characterized by a counterclockwise circulation of surface currents (Figure 1). Intense current sweeps fine-grain sediments from the northern area down to the Derbent Basin, where abundant mass-wasting deposits form and sometimes cause gravity flow activity. Although deep-water currents are poorly studied there, it is proposed [25,32] that the bottom circulation has the same pattern as the surface circulation; this was proven by recent hydrological and mineralogical studies [26,33], where an active bottom current along the western slope of the Derbent Basin was revealed. Near the eastern slope, surface currents form vortices and water masses sink along with pycnocline at the deep-sea level and the velocity of bottom currents increases up to 1 m/s [33]. Thus, a contour current is flowing along the western slope roughly from NNW to SSE, while along the eastern slope, it is flowing from SSE to NNW. A strong upwelling of cold deep water is observed near the eastern slope [33]. The experimental material shows the seasonal variability of the flow of the cyclonic cycle along the western coast of the Middle Caspian. The observations showed that the cyclonic circulation in the Middle Caspian is, essentially, a seasonal contour current flowing around the slope of the Derbent Depression. In winter (December–February), the current in the circulation is relatively fast (up to 70 cm/s), while in summer (June–August), it practically disappears everywhere, except for its western branch, which carries water into the Southern Caspian [34].

The formation of the bottom current in the Caspian Sea is associated with the Volga River's giant outflow of water into the shallow northern part of the Caspian Sea, and its paths are determined by the relief of the Middle Caspian Sea bottom. Moving southward, this water rolls into the deep water Derbent Basin, forming a flow along its western slope. This is confirmed by the observed channels in the lower part of the slope and adjacent basin. Furthermore, when reaching the Apsheron Sill, the bottom water flow turns eastward along the southern slope of the Derbent Basin, then turns northward along the eastern slope of the Derbent Basin.

In the open ocean, bottom contour currents are associated with thermohaline circulation created by the difference in water density resulting from the inhomogeneous distribution of temperature and salinity in the ocean. This circulation creates a global ocean conveyor belt in the world's oceans. In the closed Caspian Sea, the formation of the bottom current is associated with the giant Volga River outflow, and the geometry of this bottom current distribution is determined by the topography of the Middle Caspian Seabed.

2.4. Stratigraphy Setting

The stratigraphy of the upper sedimentary sections on the shallow-water shelf and the upper slope areas in the Middle Caspian Sea is determined mostly by regional transgressions and regressions [20]. On long time scales, Caspian Sea level oscillations are dramatic. During the last 8000 years, the sea level has fluctuated repeatedly with amplitudes up to at least 25 m, even dropping from a Last Glacial highstand at +50 m down to possibly −113 m in the early Holocene [35,36]. The causes of Caspian Sea level change are not yet fully understood. Caspian Sea levels depend largely on the balance between the influx of the Volga River water and evaporation over the sea surface [15,37]. However, the correlation of Caspian Sea level with global and regional circulation patterns, which cause precipitation in the Volga basin, is often surprisingly poor, or only significant for specific intervals with relatively stable sea levels [38]. In spite of great advances in understanding of our climate system, the predictive power of our Global Circulation Models, and accurate monitoring by satellite systems, such as Topex-Poseidon/Jason, opinions about future Caspian Sea level trends diverge. Some authors do not believe in climatic forcing at all, referring instead to tectonics, geochemical causes, or chaotic behavior [39].

Previous data interpretation showed three regional acoustic sequences [40]. Due to two deep drilling sites, it was possible to refer the acoustic sequences to lithological ones and to the regional stratigraphic scale and absolute age. Such correlation proves the TR (Turkanian regression) and KT (Khvalinian transgression) horizons to be one of the largest regional stratigraphic markers. These are as follows: TR, the most dramatic Caspian regression in the last 1 million years (occurred about 600–700 kyr ago); and KT, the most dramatic transgression of the whole Caspian history (occurred about 100 kyr ago) (Figure 3). There is also the M horizon, which is correlated to the Mangyshlak regression (10 kyr) (vertical resolution of Figure 3 is not enough to show the MR horizon). For details see Section 4.1 of this paper.

3. Materials and Methods

The study is based on high- and ultra-high-resolution seismic reflection profiles collected by the Shirshov Institute of Oceanology in the Caspian Sea between 2004 and 2012 (Figure 2). The techniques used provide both high resolution and penetration of acoustic signals below the seafloor.

3.1. High-Resolution Seismic Reflection Profiling (Sparker)

The high-resolution (HR) seismic reflection profiling was carried out using a multi-electrode (64–114 pieces) sparker (500–1000 J) towed at a depth of 0.5–1.0 m. The shot interval was set to 1 or 2 s, and the bandwidth of the transmitted seismic signal was 100–1000 Hz. The 25 m long single-channel streamer was towed at water depth of 0.5–1.0 m, with the collected data being visualized in real-time and recorded on a PC in the SEG-Y format with simultaneous processing (initial muting, filtration). More advanced processing of the seismic data was carried out after the field campaign in the RadExPro software package (more frequency filtering, muting, static correction, equalization of the amplitudes, deconvolution, etc.). Depending on sediments, penetration varied between tens and hundreds of meters with a resolution from a minimum of 2 m. The resulting SEG-Y files were imported into a Kingdom Suite™ project for interpretation.

3.2. Ultra-High-Resolution Parametric Profiling (SES UHR)

The ultra-high-resolution seismic reflection profiling (UHR) was carried out using the two-channel (high-frequency (HF), 100 kHz, and low-frequency (LF), 4–15 kHz) parametric acoustical system SES-2000 standard (Innomar Technology, Germany). The hardware includes an echosounder and sub-bottom profiler, and the HF (100 kHz) channel is used for bathymetric surveying.

Depending on the selected registration mode, the shooting rate can be up to 30 shots per second and depending on the sediments, penetration and resolution, are 50 m and 5 cm, respectively. The motion sensor MRU-H was used to resolve the sea swell. Data were collected and displayed in real time by original software SESWIN. For further processing, the special software ISE 2 was used.

4. Results

The high-resolution seismic reflection surveys revealed expressive erosional and depositional features on slopes and the rise of the Derbent Basin and in the middle part of the Caspian Sea [5,7,41]. These features had not been distinguished earlier as contourites because of the low resolution of air gun seismic profiling, and similar ones were considered here as slump or landslide features, tectonic folds, or creep deformations [9,32]. The high-resolution seismic reflection data presented in this paper were collected mainly in the western part of the Middle Caspian Sea and significantly less in the eastern part (Figure 2).

4.1. Western (Derbent) Slope

The sparker seismic profiles on the western slope of the Derbent Basin show some strong reflectors (Figure 4), which emphasize the complex history of transgressions and regressions in the Caspian Sea [40,42]. The strong sub-horizontal reflector TR (see Section 2.4 of this paper) in the base of the upper sedimentary wedge between edge of the shelf (water depth ~100 m) and sharp bend of the slope (~400–550 m) (Figure 4) correspond to the erosional unconformity formed during the great Turkanian regression of around 700–800 ka, when the sea level dropped by ~180 m. Another strong rough KT reflector (see Section 2.4 of this paper) expresses the dramatic Khvalinian transgression of around 20–30 ka, which was largest during the Quaternary of the Caspian Sea when the sea level rose by ~50 m. The upper non-smoothed strong MR reflector corresponds to the erosional unconformity formed during follow-up the Mangychlak regression about 7–10 ka, when the sea level dropped again by ~60 m [9,32]. Therefore, the sedimentary wedge of its upper part between edge of the shelf (water depth ~100 m) and sharp bend of the slope (~550 m) is composed of sediments from the Middle Pleistocene–Holocene.

The expressive seafloor and subsurface undulating features, which are observed in the high-resolution seismic reflection profiles (Figures 4 and 5), were interpreted as sediment waves [43]. We call them SW Type 1; they are 300–1600 m (on average 1 km) in wavelength across the continental slope and 5–40 m high (on average 20–25 m). The highest amplitudes are observed in the upper slope just behind the shelf edge, and they decrease in a gradual downslope seaward. The angles of the limbs range from some minutes to 3–4°, with the maximum angles being in the upper slope and decreasing downslope. These regularly spaced wave-like accumulative features have steep western coastward limbs and gentle eastern seaward ones. It seems that the top of the accumulative features moves sequentially toward the coastline, with superposed individual layers representing a migrating up slope wave configuration. The detailed bathymetry survey showed that the N–S trending sediment waves are not elongated features, but rather ellipsoid-shaped ones with an axis ratio of 1:3 (like brachyanticlines), which strike along the continental slope [40,42]. These sediment waves appear to form a vast field covering an area of ~7500 km^2. The ultra-high resolution SES profiles collected show erosion seabed forms (Figure 4B).

At the base of the western slope, at water depths from 450 to 750 m, there is a large accumulative sedimentary body (interpreted as separated drift, see Section 5.2 of this paper) ~70 m thick with uneven internal reflectors (Figure 6A). The upper thin-layered sequence with numerous parallel curved up reflectors conformal to the seabed has a thickness of ~15 m (Figure 6B). On the contrary, the deeper internal reflectors are curved down following the geometry of the main unconformity KT (Figure 6A). This sedimentary accumulative body that extends along the regional contours for about 80 km is subdivided into three depositional mounds (Figure 7), which have almost the same dimensions: 8–11 km in width, 25–30 km in length, and their average length/width ratio is 3:1. All of them are separated from the slope foot by the shallow moat around 1–2 km deep, with their depth increasing southward from 4 to 12 m. Seismic data demonstrate the absence of modern unconsolidated sediments in the moat (Figure 6B).

The profiles show a good range of downslope features. According to previous research [10,11], uplifts in the upper part of the western slope of the Derbent Basin do not represent individual slump bodies moving along the seafloor; nevertheless, the internal structure of the Neopleistocene sequence in this area suggests the presence of numerous intraformational slump duplexes. In the steepest part of the slope, clear indications of the disintegration of bottom deposits and collapsing and slumping processes were revealed. However, an extensive discussion of downslope features here does not fit into the main context; therefore, it would be more appropriate to continue the discussion in another publication.

4.2. Northern Slope (Mangyshlak Sill)

The upper part of the northern slope of the Derbent Basin is called the Mangyshlak Sill. Its main geomorphologic feature is a vast channel system built by the Volga River and some rivers from the Caucasus (Figure 1). The channels run across the slope (Figure 8). The undulated, wave-like acoustic facies that climb toward the channel axes are usually observed on the channel's western flanks (Figure 9A). The dimensions of these wavy features (height and wavelength) decrease laterally from the channel axes (Figure 9B); we have labeled them SW—Type 2.

Relatively deep-water sediment waves are observed in the corner between the northern and western slopes of the Derbent Basin (Figure 10). They are characterized by the absence of migration and very slight and unpredictable downslope decreasing in height (from 10 to 4 m) and wavelength (from 1.7 to 0.5 km). The field of these migrating sediment waves starts at the western slope and ends below the northern one, so it occupies several tens of square kilometers (Figure 7). We have labeled them SW—Type 3.

A large lenticular accumulative sedimentary body (detached drift) is observed in the Derbent Basin near the base of the northern slope at water depths from 670 to 720 m (Figure 11). It is over 30 km long and 15 km wide (length/width ratio 2:1), rising 20 m above the seafloor. The maximum thickness of the body down to the reflector KT is about 100 m. Its visual inner seismic reflection pattern is very complicated with different types of reflections (except conformal horizontal/parallel ones). The thin-layered sequence of parallel smooth reflectors directly below this unconformity KT differs dramatically from the irregular architecture of the overlying deposits. In the bottom relief, the drift is complicated by a channel-looking feature (Figure 11); however, this feature can be interpreted as a crater because canyons are normally V-shaped and have new sedimentation at the bottom, but the materials at the bottom of this structure seem to collapse down. Gerivani et al. [44] interpret this structure as a crater formed by a mechanism such as that suggested and described by [45,46] for "hill-hole pairs". In this crater, it seems that the mixture of high-pressure gas, sediments and water were first emitted from probably weak zones in both sides of the crater, and then the upper beds collapsed down.

4.3. Southern Slope (Apsheron Sill)

Two prominent large accumulative sedimentary bodies are observed on the southern slope the Derbent Basin in the eastern part of the Apsheron Sill. The upward convex lenticular gently sloped depositional body (plastered drift) extends for about 20 km at depths from 250 to 300 m and is related to an erosional moat-like feature with a depth of 10 m and width of around 2 km (Figure 12). The pinching-out upslope of the internal reflectors and local unconformities within thin-layered strata evidence lateral migration of the sedimentary feature.

The second body (interpreted as fault-controlled drift, see Section 5.2 of this paper) observed at a depth range of 470–520 m is interrupted by a fault that significantly changes its overall geometry. The northern part of the fault seismic record is characterized by continuous, parallel, roughly horizontal reflectors, while in the southern part, reflectors demonstrate distinct undulation (Figure 13). The fault influence is a syn-fault sedimentary migration of a drift.

4.4. Eastern Slope

On the eastern slope of the Derbent Basin, there are no such clear contourite features as are observed on its western and northern slopes and discussed above. At the base of the eastern slope (water depths 550–600 m), seismic profiles cross a depositional body (Figure 14) and extend across the slope for around 150 km. Their orientation is controlled by a specific slope morphology that is likely affected by a deep fault. The thickness of the body (down to the KT reflector) is about 50 m, while the upper part of the seismic section is characterized by a distinct undulation of reflectors appearing in a hummocky surface of the body. Reflector undulation is also slightly expressed below the KT. The mounded depositional feature is separated from the slope by a moat-like asymmetrical feature with a depth of 30 m and a width up to 2 km. In a big stretch, we can assume that this is a separated drift (see Section 5.2 of this paper) with a corresponding erosion moat; however, the orientation perpendicular to the contour along-slope flow is unusual for contour structures. Below the top of this depositional body, the lower stratified sequence is crumpled into small symmetrical wavy features (Figure 14).

Another atypical characteristic is box-shaped rootless fold features on the eastern slope of the Derbent Basin (Figure 15). Previously, such features were interpreted here as exogenic–gravitational folds or creep deformation produced by the plastic flow of poorly consolidated sediments under the influence of gravitational forces [9].

5. Discussion

The processes controlling recent sedimentation in closed water basins are similar in some cases, while differing significantly in others. For example, high-resolution seismic studies in Lake Saint-Jean (Québec, Canada) revealed the presence of sedimentary structures controlled by currents, which owe their origin to the movement of air masses [47,48]. In particular, they refer to the formation of sedimentary drifts and intense erosion within the shelf, which are caused by wind-related hydrodynamics. The peculiarity of these processes, according to the authors, is that they manifest themselves in the central part of the lake at depths below the wave base. The authors note that the change in the lake hydrodynamics is associated with the post-glacial processes that took place 8.5 cal. ka BP. An analysis of modern sedimentation in the tropical Lake Ossa and its relation to climatic changes, morphology and hydrology of this large water reservoir in West Africa is the subject of the article by Giresse et al. [49]. Of interest for us in this work is the fact that the hydrological features of the lake are related to the Sanaga River. Another example of studying the relationship between hydrodynamic processes and sedimentation is the Lake Superior sediment study. The description of samples and their seismostratigraphic identification served as the basis for the conclusion that the distribution of surface sediments was related to the glacial and post-glacial evolution of the lake [50,51].

Sedimentation processes in the Middle Caspian Sea are quite complicated. On the one hand, there are the turbidity currents in the western part, where the high fluvial input and deposition rate of unconsolidated sediments provide slope instability, while seismic activity triggers gravity flows and structures caused by local earthquakes [52]. We consider the results of the hyperpycnal flows as well. On the other hand, there are strong bottom currents forming a giant counterclockwise gyre that embraces the central part of the sea [33,53]. The activity of these processes has resulted in a wide range of erosional and depositional features, as illustrated above.

All these accumulative features fall into two groups: sediment waves (types 1, 2, 3) and large elongated sediment mounds (drifts). Erosional moats and channels are related to some of these sedimentary accumulative features. In general, the collected high-resolution seismic reflection data suggest the existence of contourite accumulative and erosive forms around the Derbent Basin in the Middle Caspian Sea. Below, we consider the possibility of combining them within the framework of regional physiography and oceanographic processes.

Possible accumulative and erosive sedimentary contourite features were first identified on the northern and especially western slopes of the Derbent Basin [5,41]. Its southern and eastern slopes are much less studied, but sedimentary features similar to the contourite ones appear to be observed there as well [6]. Such features over all slopes are illustrated below sequentially, in order to then consider their possible association with a single contourite sedimentary complex, according to the contourite paradigm [2].

5.1. Genesis of Wavy Features

In the study area, three major types of sediment waves (SW) were distinguished: SW Type 1, mixed waves [40,42] that come from an environment which favored both turbidite and contourite processes; SW Type 2, formed by turbidity flows, including those on channel meanders; and SW Type 3, sediment waves of contourite origin formed by bottom currents.

SW Type 1. The sparker seismic profiles on the western slope of the Derbent Basin show strong undulating reflectors (Figures 4 and 5), which recall the complex history of transgressions and regressions in the Caspian Sea. The most peculiar features are spectacular seafloor and subsurface undulations (Figures 4 and 5). They are interpreted as sediment waves but no syn- or post-sedimentary gravity deformation within the Quaternary sediments or creep was possible [40,43]. Along the western slope, the numerous rivers from the adjacent Caucasus represent the main regional sediment source. The sediment wave field is only 30–40 km from the coast, and the supply of terrigenous sediments is abundant here. Such waves seem to comply with these conditions of migrating sediment waves and are believed to have formed from sedimentary material transported and deposited by cyclic high density turbidity currents [54–56]. In particular, the thickness of the sediment wave sequences decreases naturally downstream, that is, downslope because the volume of splayed sediment is reduced away from its source. However, oceanographic data indicate that active bottom currents strongly influenced the seafloor, controlling the morphology of the sediment waves (see erosion on Figure 5B). The erosion exists just nearby an active branch of the bottom current [33]; the vertical profile of water is shown on Figure 16A1 (ellipsoids mark current maximum zones). The diagram in Figure 16C1 demonstrates the relation between wavelength and water depth. The form of graph of this function is sinusoidal. Its maximum extremes are for greater wavelength, and on seismic data (Figure 16B1) they correlate to areas of modern erosion. Previous work interprets this erosion places caused by bottom current activity [40]. Lithological data show coarse sediments on both sides of the waves, which together with the oblique orientation of wave crests and two-directional migration (upslope, i.e., upstream the turbidity flows, and along slope, upstream the bottom currents) appear to suggest the mixed origin of these sediment waves.

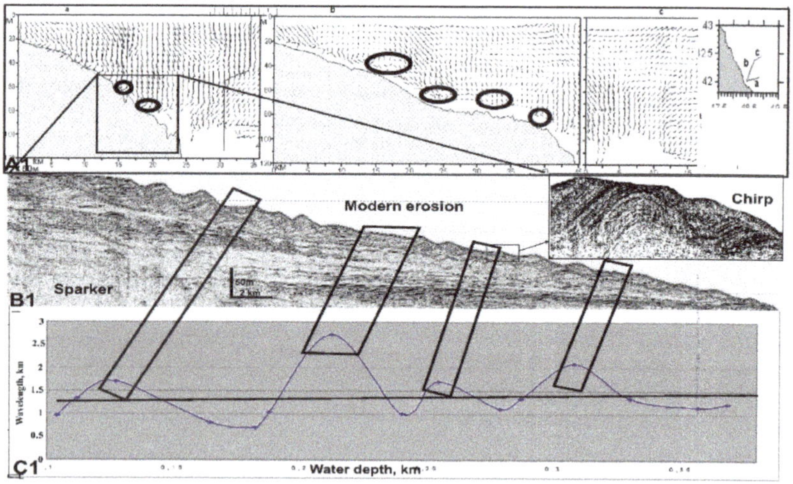

Figure 16. Modern erosion and resuspension on upper part of the western slope (sediment waves field, Figure 5A). Wavelength—Water depth relation, sinusoid, and linear trend (**C1**) correlated with hydrophysical profile (**A1**) and sparker section (**B1**) (seismic from Figure 5A). Circles for current branches. Details of modern erosion are shown on Chirp data (sweep seismoacoustic signal, inset).

SW Type 2. Several fields of wavy features exist on the Mangyshlak Sill (Figures 7 and 17). These features are interpreted as channel levee sediment waves migrating toward the channel. The sediment waves on levees are believed to have formed by classic outspill on channel meanders [57].

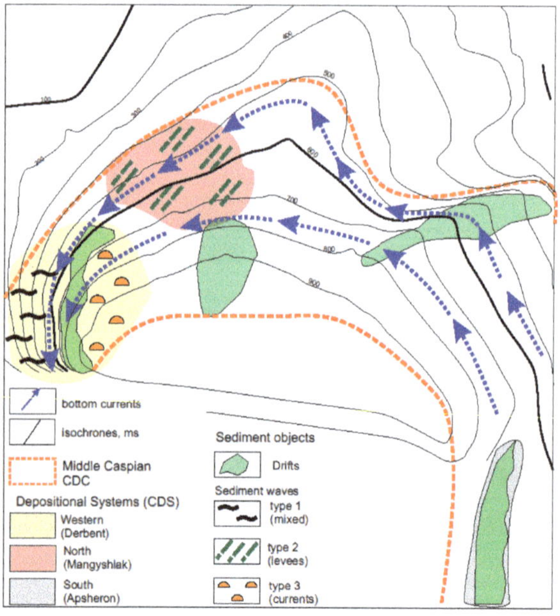

Figure 17. Contourite deposits (sediment waves and contourite drifts), contourite depositional complex (CDC) and contourite depositional systems (CDS) in the Middle Caspian Sea. Arrows show large circumcaspian current (Modified from [6]).

An atypical characteristic is the box-shaped rootless fold features on the eastern slope of the Derbent Basin (Figure 15). They were interpreted as exogenic–gravitational folds or creep deformation produced by the plastic flow of poorly consolidated sediments under the influence of gravitational forces [9].

SW Type 3. The vast field of undulations is located between the western and northern slopes, where the western slope turns eastward to the Mangyshlak Sill (Figure 8). Unlike sediment waves on the western slope (Figures 4 and 5), there appears to be no pronounced migration of crest and no symmetry in crest steep here (Figure 9). Such patterns together with the absence of alluvial sediments supply suppose that the waves here seem to be "contouritic" genesis. The "corner" looks like a natural "trap" for the bottom current. Its flow is activated by material coming from the channel system and is hampered by a sharp change of topography from the fairly gentle slope of the Mangyshlak Sill to the steep Derbent slope. As a result, the current throws off the "excess weight" of sediments and forms the sediment waves. After this, the current follows the topography, turns to the south, and erodes the western slope.

5.2. Contourite Sedimentary Drifts

Classic contourite sedimentary drifts are found in the Derbent Basin near the base of each of its slopes. At the base of the western slope, at water depths from 450 to 750 m, there are three large accumulative sedimentary bodies ~70 m thick with inconsistent internal reflectors (Figures 6 and 17). The overall geometry and visual seismic reflection pattern (down-current elongation, extensive regional unconformities, broadly lenticular, upward-convex seismic units, extensive sub-parallel moderate to low amplitude reflectors typically with gradual changes between seismic facies [58]), allow us to attribute these three depositional mounds to elongated mounded drift. It is believed that such drifts are formed as a result of the migration and aggradation of sediments [59,60]. The sedimentation rate of around 25 cm/kyr was calculated for these bodies [33]. According to seismic data, they are typical separated drifts [3]. As three of these individual separated drifts partly overlap each other, they build up a giant mounded elongate drift that extends along the western slope of the Derbent Basin.

The large lenticular accumulative sedimentary body is observed at water depths from 670 to 720 m nearby the Mangyshlak Sill (Figure 11). The body looks like a large, detached drift. Its complicated seismic reflection pattern appears to evidence the existence of several internal small detached and plastered drifts. This large contourite drift is located opposite the mouths of manifold channels that intersect the slope (Figure 10). Typical buried channel levels are found inside this body (Figure 11A), while the spectacular channel-looking feature is at its top. In general, this indicates that the formation of this body was strongly affected by turbidity currents from the adjacent channel system. Lenticular seismic units with distinct channels represent deposits of gravity flows. We suggest a contourite origin of the under- and overlying deposits with slightly undulating, parallel reflectors. However, in the bottom relief, the drift is complicated by a channel-looking structure, implying the modern activity of gravity flows in this area and the interplay between down-slope and along-slope processes. The studied depositional body is interpreted as a mixed turbidite-contourite system.

Two contourite drifts seem to have formed in the southern slope of the Derbent Basin off the Apsheron Sill. The southern one is an upward convex lenticular gently sloped depositional body located in upper slope at a water depth of ~350 m (Figure 12). It is characterized by the pinching-out upslope of the internal reflectors and local unconformities within the thin-layered strata. We interpret the unconformities as a lateral migration of the sedimentary features, which could be plastered drift. The northern contourite feature is divided by the obvious fault into two segments which together form typical fault-controlled drift (Figure 13). The acoustical transparent anomalies indicative of gas escape are widely distributed in the Caspian Sea [44,61]. Similar weak acoustic anomalies are observed in this area (Figure 13B).

The high-resolution seismic reflection profiles intersect a coupled erosive moat and depositional body at the base of the eastern slope of the Derbent Basin (Figure 14). The across-slope orientation of the depositional body is interpreted to have been caused by the interplay between regional bottom topography and specific circulation pattern. Seismic and bathymetric data suggest a deep fault modifying the slope morphology. The rapid change in slope trend makes the bottom current turn sharply to the east. This results in a significant increase in current intensity. Bottom circulation here is also complicated due to a thermo-cline upwelling gyre [33]. The Caspian Sea is filled with salty sea water. This is due to the connection in the recent geological past (Pleistocene) of the Caspian Sea with the Black Sea and the Mediterranean Sea through the Manich and Bosporus Straits, respectively, whereas other large, enclosed lakes such as the Great Lakes and Baikal are freshwater. It is the thermohaline structure of the waters that determines their density with which the formation of bottom currents is connected. Lack of data concerning this erosion-depositional system allows only preliminary interpretation, where the depositional body is initially interpreted as a separated drift, while a local depression detaching the drift likely represents an erosive moat.

5.3. *Middle Caspian Contourite Depositional Complex*

The association of various individual depositional drifts and related moats form part of a contourite depositional system (CDS) [3,62]. In turn, distinct but connected CDS within the same water mass are considered a contourite depositional complex (CDC). According to this idea, the whole investigated area in the Middle Caspian Sea with various contourite drifts and fields of sediment waves seems to contain several CDS.

One CDS is probably located in the western part of the Middle Caspian Sea, covering the western continental slope and rise and extending deeper in the adjacent basin (Figure 17). The most prominent feature in this CDS is elongated mounded and separated drift near the base of the continental slope; other prominent features here are sediment waves on the slope and in the basin. Another CDS located in the northern part near the Mangyshlak Sill and the adjacent basin is related to a widespread channel system with which expressive contourite drift and sediment waves are associated. In general, both these contourite sedimentary systems (western and northern) are apparently parts of a more extensive contourite sedimentary complex spanning the respective parts of the Middle Caspian Sea. The contourite drifts revealed on the southern slope and adjacent basin near the Apsheron Sill suggest that CDS exists here too. However, we currently do not have enough high-resolution seismic reflection data to be certain. Regarding the eastern slope, available high-resolution seismic reflection data show no CDS at this time.

The synthesis of contouritic deposits in the Middle Caspian Sea evidence that their position coincides with the general route of the circumcaspian current (Figure 17). In this case, a large single contourite depositional complex related to this circumcaspian current would exist along the whole continental rise of central part of the Middle Caspian Sea at a depth range from 250 to 720 m. This CDC should integrate both contourite depositional systems revealed in its western and north-western parts, the proposed CDS in the southern part and confidently expected CDS in the eastern part (Figure 17).

According to analysis and stratigraphic interpretation of seismic profile formation of the Middle Caspian, the contourite depositional complex started at least 250 kyr ago (since the Khvalinian transgression). To have a better idea of the existence of this circular contourite depositional complex, purposeful seismic reflection surveys should be carried out in the Middle Caspian Sea.

Contourite drifts can develop accumulations with important seal potential for trapping hydrocarbons; however, the subject of academic studies based on traditional shallow penetration high-resolution seismic 2D data and sea-floor piston coring are not contourites associated with oil-bearing reservoirs [63]. The modern contourites and turbidites of the Caspian Sea contain large amount of organic plant matter, which can subsequently become a source of hydrocarbons under the appropriate temperature-pressure regime. Our high-resolution seismic reflection profiles do not penetrate deep enough to answer the question: do these modern sedimentary systems act as possible analogues for any of the subsurface hydrocarbon discoveries in the region? It is also necessary to have an idea of the paleoenvironment and paleowater circulation in the Caspian Sea during the accumulation of ancient sedimentary oil and gas-bearing sediments. Therefore, while any conclusions about such an analogy are premature, perhaps a dedicated study in the future will answer this question.

5.4. Geohazards

The Caspian Sea is a large hydrocarbon basin of great value, with its most remarkable geohazards, along with shallow gas accumulation within sediments and regional seismicity, being the bottom water circulation and its sediment transport. That is, both downslope gravity mass transport and transport by turbidity and contour currents. Every natural process is responsible for a geological seafloor feature, which can be seen on seismic sections.

The hydrodynamic regime of the Middle Caspian Sea is a very complicated one; velocities of near-bottom currents are high (15–100 cm/s) and have a different direction here [33]. The most dangerous appear to be submarine pipelines, which stretch along all sediment forms and pathways of various natural processes. Thus, it is very important to know the slope stability and sediment transport pattern.

The downslope gravity and turbidity flow on the western slope, which causes erosion of the seafloor and intensive sediment input, are treacherous for submarine pipelines and could cause ruptures. The SES seismic profiling data indicate that the processes of such turbidite sedimentation in the studied area of the western slope of the Derbent Basin are probably accompanied by active underwater slumping. This fact should be considered in the designation of engineering structures within the petroliferous Yalama-Samur structure (Figure 18). This issue is especially important because of the substantial thickness (more than 200 m) deposits west of the studied area [10,11].

Another danger, related to the canyon-channel system in Mangyshlak Sill, is turbidity sediment waves on their levees, which are indicators of active hydrodynamics. One more such danger relates to contourite sediment waves in the Derbent Basin near all its slopes. Thus, any pipeline construction will face issues, making it necessary to obtain more data regarding the current picture and the bottom relief. It should be noted that the gravitational folding and faulting on the northern slope of the Derbent Basin were most intense precisely in the Holocene [11].

The investigation of high-pressure fluids including shallow gas and gas hydrate is important for the risk assessment of marine geohazards and for localizing conventional and unconventional hydrocarbon reservoirs [64]. Gas-related processes may cause geohazards or may be considered a natural source of energy in the near future. For example, pipelines often have to wind their way through pockmarks fields, adding to the cost of construction. Furthermore, for other structures, they may represent a significant hindrance and may severely limit the freedom of the areal use of construction [65].

As a result of the analysis of our seismic profiles, a map of geohazards in the Middle Caspian was compiled, and the most pertinent geo-hazard threats in each region were diagnosed [40], to which we made additions and reworked (Figure 18). They should be taken into consideration when designing, building, and operating submarine constructions for the oil and gas industry in order to prevent potential natural hazards and reduce their consequences.

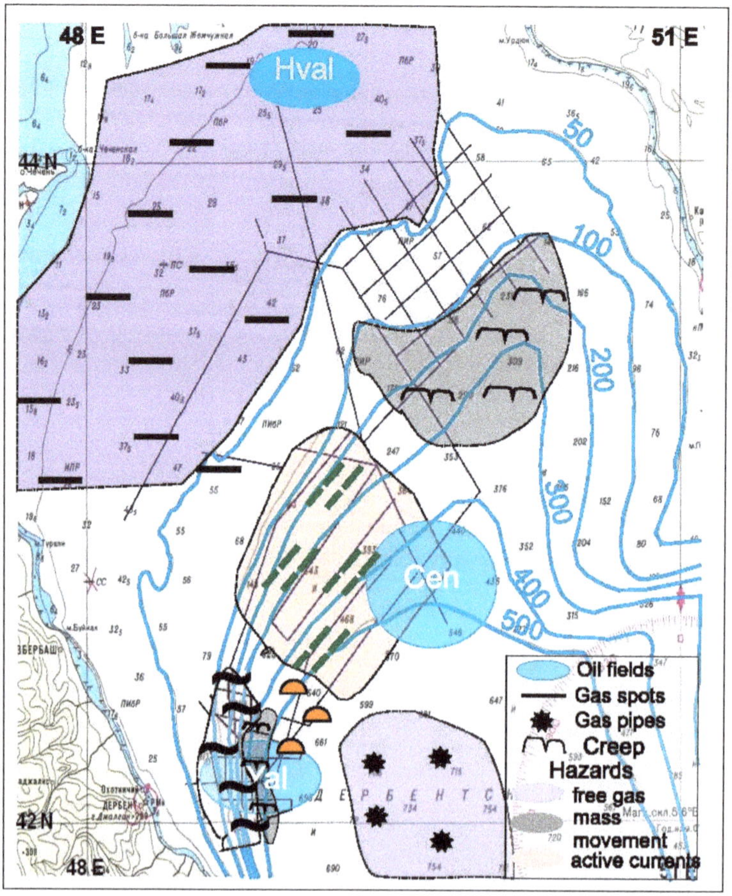

Figure 18. Map of sedimentary features and geohazards. Bathimetry in meters. Oil fields: Hval—Khvalynskoe, Cen—Centralnaya, Yal—Yalama-Samur. For sediment features legend see Figure 17.

6. Conclusions

The Caspian Sea is an important region for the investigation of lake contourites. This was confirmed by a wide range of contourite erosional and depositional features revealed in the Middle Caspian Sea for the first time. Contourite drifts, sediment waves and moats build a large contourite depositional complex covering a major part of the Middle Caspian Sea. This complex is formed by the strong counterclockwise gyre of bottom currents and was developed during the last 250 kyr (at least). In the north, the complex is affected by turbidity currents, while the interplay between along-slope and down-slope sedimentation processes resulted in the formation of mixed depositional features, such as sediment waves and sediment mounds.

The modern contourites and turbidites of the Caspian Sea contain a large amount of organic plant matter, which could subsequently become a source of hydrocarbons under the appropriate temperature–pressure regime. Our high-resolution seismic reflection profiles do not penetrate deep enough to answer the following question: do these modern sedimentary systems act as possible analogues for any of the subsurface hydrocarbon discoveries in the region? It is also necessary to have an idea of the paleoenviroment and paleowater circulation in the Caspian Sea during the accumulation of ancient sedimentary oil- and gas-bearing sediments. Therefore, while any conclusions about such an analogy are premature; perhaps a dedicated study in the future will answer this question.

These results should be taken into consideration when designing, building, and operating submarine constructions for the oil and gas industry to prevent potential natural hazards and reduce their consequences.

Author Contributions: Conceptualization, methodology, V.Y., O.L.; validation, V.Y., O.L., V.P.; investigation, V.Y., O.L., V.P.; writing—original draft preparation, V.Y., O.L., V.P.; writing—review and editing, V.Y., O.L., V.P.; discussion, interpretation of results and conclusions, all authors; visualization (preparation of figures), O.L., V.P.; supervision, V.Y., O.L.; project administration, V.Y., O.L. All authors have read and agreed to the published version of the manuscript.

Funding: Financial support for this research was partially provided by Instituto Potosino de Investigación Científica y Tecnológica, A.C. The research was funded by the state assignment of IO RAS: FMWE-2021-0005.

Institutional Review Board Statement: Not applicable.

Informed Consent Statement: Not applicable.

Data Availability Statement: The original data are property of Shirshov Institute of Oceanology, it could be available to third parts after consideration of official request.

Acknowledgments: The authors are grateful to Graham Tippett for the editing the English version. The authors are grateful to three anonymous reviewers and to the editor for valuable observations that have improved the manuscript.

Conflicts of Interest: The authors declare no conflict of interest.

References

1. Duarte, C.S.L.; Viana, A.R. Santos Drift System: Stratigraphic organization and implications for late Cenozoic palaeocirculation in the Santos Basin, SW Atlantic Ocean. In *Economic and Palaeoceanographic Significance of Contourite Deposits*; Viana, A.R., Rebesco, M., Eds.; Geological Society, Special Publications: London, UK, 2007; Volume 276, pp. 171–198.
2. Rebesco, M.; Camerlenghi, A. (Eds.) *Contourites*; Elsevier: Amsterdam, The Netherlands, 2008; p. 666.
3. Rebesco, M.; Hernández-Molina, F.J.; Van Rooij, D.; Wåhlin, A. Contourites and associated sediments controlled by deep-water circulation processes: State-of-the-art and future considerations. *Mar. Geol.* **2014**, *352*, 111–154. [CrossRef]
4. Ceramicola, S.; Rebesco, M.; De Batist, M.; Khlystov, O. Seismic evidence of small-scale lacustrine drifts in Lake Baikal (Russia). *Mar. Geophys. Res.* **2001**, *22*, 445–464. [CrossRef]
5. Levchenko, O.V.; Roslyakov, A.G.; Polyakov, A.S.; Zverev, A.S.; Merklin, L.R. New data about sedimentary waves on western continental slope of the Caspian Sea. *Dokl. Earth Sci.* **2008**, *420*, 537–542. [CrossRef]
6. Levchenko, O.; Putans, V.; Borisov, D. Contourites in the Middle Caspian Sea? In Proceedings of the 2nd Deep-Water Circulation Congress, Ghent, Belgium, 10–12 September 2014; pp. 65–66. Available online: https://www.vliz.be/imisdocs/publications/264048.pdf (accessed on 10 May 2022).
7. Levchenko, O.V.; Putans, V.A.; Borisov, D.G. Contourites in the Derbent Basin, Caspian Sea (Geophysical Data). *Dokl. Earth Sci.* **2018**, *482*, 1239–1243. [CrossRef]
8. Putans, V.A.; Levchenko, O.V.; Borisov, D.V. Circum Middle Caspian Contourite Depositional Complex. In *Abstracts of the VIII Simposio MIA15*; Universidad de Málaga: Malaga, Spain, 2015.
9. Maev, E.G. Exogenous Folding in Quaternary Deposits of the Continental Slope of the Caspian Sea. *Dokl. Earth Sci.* **1999**, *365*, 323–325.
10. Verzhbitsky, V.E.; Levchenko, O.V.; Lobkovsky, L.I. New Data on Quaternary Processes of Underwater Slumping on the Western Slope of the Derbent Basin (Caspian Sea). *Dokl. Earth Sci.* **2007**, *416*, 1085–1089. [CrossRef]
11. Verzhbitsky, V.E.; Lobkovsky, L.I.; Roslyakov, A.G.; Merklin, L.R.; Polyakov, A.S.; Levchenko, O.V.; Kovachev, S.A.; Zverev, A.S.; Garagash, I.A.; Mar, G.N. Slump Structures in Quaternary Slope Sediments of the Northern Derbent Basin (Caspian Sea). *Oceanology* **2009**, *49*, 396–404. [CrossRef]
12. Mir, N. *English Version of Explanatory Notes of International Tectonic Map of the Caspian Sea Region*; Scale 1:2, 500,000; Khain, V.Y., Bogdanov, N.A., Eds.; Russian Academy of Science: Moscow, Russia, 2005.
13. Hinds, D.J.; Aliyeva, E.; Allen, M.B.; Davies, C.E.; Kroonenberg, S.B.; Simmons, M.D.; Vincent, S.J. Sedimentation in a diascharge dominated fluvial-lacustrine system: The Neogene Productive Series of the South Caspian Basin, Azerbaijan. *Mar. Pet. Geol.* **2004**, *20*, 613–638. [CrossRef]

14. Leroy, S.A.G.; Lahijani, H.A.; Cretaux, J.-F.; Aladin, N.V.; Plotnikov, I.S. Past and current changes in the largest lake of the World: The Caspian Sea. In *Large Asian Lakes in a Changing World. Springer Water*; Mischke, S., Ed.; Springer: Cham, Switzerland, 2020; pp. 65–107. [CrossRef]
15. Arpe, K.; Bengtsson, L.; Golitsyn, G.S.; Mokhov, I.I.; Semenov, V.A.; Sporyshev, P.V. Connection between Caspian Sea level variability and ENSO. *Geophys. Res. Lett.* **2000**, *27*, 2693–2696. [CrossRef]
16. Lebedev, L.I. Structure of upper part of sedimentary cover according to geoacoustical profiling. In *Caspian Sea: Geology and Oil and Gas Resources*; Nauka: Moscow, Russia, 1987; pp. 105–114.
17. Kostyanoy, A.G.; Kosarev, A.N. *The Caspian Sea Environment*; Springer: Berlin, Germany, 2005.
18. Ivanova, T.P.; Trifonov, V.G. Seismotectonics and modern regime of Caspian Sea level fluctuations. *Geotectonics* **2002**, *2*, 27–42.
19. Yutsis, V.V.; Kalinin, V.V. Pliocene-Quaternary geodynamics of the Apsheron shelf of the Caspian Sea revealed by geological-geophysical data. In Proceedings of the Geodynamic Basis of Prognostication of Oil and Gas, Moscow, Russia, 6–8 September 1988.
20. Leonov, Y.G.; Antipov, M.P.; Bobylova, E.E.; Volozh, Y.A.; Lavrushin, Y.A.; Spiridonova, E.A. *Geological History of Quaternary Sedimentary Basins within Caspian Region during Last 700,000 Years: Sedimentation Its Water Regimen and Geodynamics Events [Comments to "Map of Quaternary (Neo Pleistocene) Sediments within Caspian Area with Elements of Paleogeography", Scale 1:2,500,000]*; Scientific World: Moscow, Russia, 2005; Volume 3. (In Russian)
21. Chen, J.L.; Pekker, T.; Wilson, C.R.; Tapley, B.D.; Kostianoy, A.G.; Cretaux, J.-F.; Safarov, E.S. Long-term Caspian Sea level change. *Geophys. Res. Lett.* **2017**, *44*, 6993–7001. [CrossRef]
22. Golubov, B.N.; Ismagilov, D.F. Pipe-looking structures under the Northern Caspian fluid regime. In Proceedings of the All-Russia Conference of Oil and Gas Genesis, Moscow, Russia, 15–18 April 2003; pp. 78–80.
23. Kosarev, A.N.; Kostyanoy, A.G.; Zonn, I.S. Kara-Bogas-Gol Bay: Physical and chemical evolution. *Aquat. Geochem.* **2009**, *15*, 223–236. [CrossRef]
24. Leonov, Y.G.; Antipov, M.P.; Volozh, Y.A. Geological Aspects of Caspian Sea Level Fluctuations. In *Global Changes of Environment*; Siberian RAS: Novosibirsk, Russia, 1988; pp. 30–57.
25. Kravchishina, M.D.; Novigatskii, A.N.; Politova, N.V.; Zernova, V.V.; Mosharov, S.A.; Dara, O.M.; Klyuvitkin, A.A. Studying the biogenic and abiogenic parts of suspended particulate matter in the Volga delta during spring flood of May 2008. *Water Resour.* **2013**, *40*, 143–156. [CrossRef]
26. Kozina, N.V.; Putas, V.A.; Zhdan, M.I. Elaboration of sediment transport pathways by integrated interpretation of geological and high-resolution seismoacoustic data (Caspian Sea). In *Proceedings of the Dialogue between Contourite and Oceanology Processes International Workshop*; Hull, UK, 28–26 January 2013, 2013; p. 34.
27. Lahijani, H.; Abbasian, H.; Naderi-Beni, A.; Leroy, S.A.G.; Haghani, S.; Habibi, P.; Hosseindustn, M.; Shahkeremi, S.; Yeganeh, S.; Zandinasab, Z.; et al. Sediment distribution pattern of South Caspian Sea: Possible hydroclimatic implications. *Can. J. Earth Sci.* **2018**, *56*, 637–653. [CrossRef]
28. Kosarev, A.N. *Hydrology of the Caspian and Aral Seas*; Moscow State University: Moscow, Russia, 1975; p. 372.
29. Terziev, F.S.; Kosarev, A.N.; Kerimov, A.A. *Hydrometeorology and Hydrochemistry of Seas*; Caspian Sea, Hydrometeorological Conditions, Gidrometeoizdat: St. Petersburg, Russia, 1992; Volume 6. (In Russian)
30. Ghaffari, P.; Lahijani, H.A.K.; Azizpour, J. Snapshot observation of the physical structure and stratufucation in deep-water of the South Caspian Sea (western part). *Ocean Sci.* **2010**, *6*, 877–885. [CrossRef]
31. Sapozhnikov, V.V.; Mordasova, N.V.; Metreveli, M.P. Transformations in the Caspian Sea ecosystem under the fall and rise of the sea level. *Oceanology* **2010**, *50*, 488–497. [CrossRef]
32. Glumov, I.F.; Malovitsky, Y.P.; Novikov, A.A.; Senin, B.V. *Regional Geology and Oil and Gas Content of the Caspian Sea*; Nedra: Moscow, Russia, 2004; p. 342. (In Russian)
33. Ambrosimov, A.K.; Ambrosimov, E.C.; Libina, N.V. Dynamic structure of currents near western slope of the Derbent Basin in the Caspian Sea. *Eng. Phys.* **2010**, *10*, 31–45.
34. Ambrosimov, A.K.; Klyuvitkin, A.A.; Lisitsyn, A.P. Season variations of currents over the western slope of the middle Caspian bed. *Water Resour.* **2018**, *45*, 685–694. [CrossRef]
35. Kroonenberg, S.B.; Badyukova, E.N.; Storms, J.E.A.; Ignatov, E.I.; Kasimov, N.S. A full sea-level cycle in sixty-five years: Barrier dynamics along Caspian shores. *Sediment. Geol.* **2000**, *134*, 257–274. [CrossRef]
36. Hoogendoorn, R.M.; Boels, J.F.; Kroonenberg, S.B.; Simmons, M.D.; Aliycva, E.; Babazadeh, A.D.; Huseynov, D. Development of the Kura delta, Azerbaijan; a record of Holocene Caspian Sea level changes. *Mar. Geol.* **2005**, *222–223*, 359–380. [CrossRef]
37. Klige, R.K.; Myagkov, M.S. Changes in the water regime of the Caspian Sea. *GeoJournal* **1992**, *27*, 299–307. [CrossRef]
38. Meshcherskaya, A. Analysis of global climatic characteristics in relation to Caspian Sea Level. In *TA-CIS/Caspian Environmental Programme, Report WLF*; Centre for Water Level Change: Almaty, Kazakhstan, 2001; p. 63. (In Russian)
39. Kroonenberg, S.B.; Abdurakhmanov, G.M.; Badyukova, E.N.; van der Borg, K.; Kalashnikov, A.; Kasimov, N.S.; Rychagov, G.I.; Svitoch, A.A.; Vonhof, H.B.; Wesselingh, F.P. Solar-forced 2600 BP and Little Ice Age Highstands of the Caspian Sea. *Methods Ecol. Res. South Russ. Ecol. Dev.* **2008**, *173–174*, 12–21. [CrossRef]
40. Putans, V.A.; Merklin, L.R.; Levchenko, O.V. Sediment waves and other forms as evidence of geohazards (Caspian Sea). *Int. J. Offshore Polar Eng.* **2010**, *20*, 1–4.
41. Putans, V.A. Sediment waves: Geohazard or geofeature? *Hydro Int.* **2013**, *10*, 25–29.

42. Levchenko, O.V.; Roslyakov, A.G. Cyclic sediment waves on western slope of the Caspian Sea as possible indicators of main transgressive/regressive events. *Quat. Int.* **2010**, *225*, 210–220. [CrossRef]
43. Levchenko, O.V.; Gainanov, V.G.; Merklin, L.R.; Polyakov, A.S.; Roslyakov, A.G. New data about seismic stratigraphy and sedimentological proscesses over western continental slope of the Middle Caspian Sea. *Dokl. Earth Sci.* **2006**, *355*, 671–673.
44. Gerivani, H.; Putans, V.A.; Merklin, L.R.; Modarres, M.H. Characteristics of features formed by gas hydrate and free gas in the continental slope and abyssal plain of the Middle Caspian Sea. *Mar. Georesourc. Geotechnol.* **2020**, *39*, 419–430. [CrossRef]
45. Andreassen, K.; Hubbard, A.; Winsborrow, M.; Patton, H.; Vadakkepuliyambatta, S.; Plaza-Faverola, A.; Gudlaugsson, E.; Serov, P.; Deryabin, A.; Mattingsdal, R. Massive blow-out craters formed by hydrate-controlled methane expulsion from the Arctic seafloor. *Science* **2017**, *356*, 948–953. [CrossRef]
46. Nixon, F.C.; Chand, S.; Thorsnes, T.; Bjarnadóttir, L.R. A modified gas hydrate-geomorphological model for a new discovery of enigmatic craters and seabed mounds in the Central Barents Sea, Norway. *Geo-Mar. Lett.* **2019**, *39*, 191–203. [CrossRef]
47. Nutz, A.; Schuster, M.; Ghienne, J.-F.; Roquin, C.; Hay, M.B.; Retif, F.; Certain, R.; Robin, N.; Raynal, O.; Cousineau, P.A.; et al. Current-controlled Sedimentary Features into Lake Saint-Jean (Québec, Canada): A Record of Wind-driven Processes? In Proceedings of the AGU Fall Meeting, San Francisco, CA, USA, 15–19 December 2014.
48. Nutz, A.; Ghienne, J.-F.; Schuster, M.; Certain, R.; Robin, N.; Roquin, C.; Raynal, O.; Bouchette, F.; Duringer, P.; Cousineau, P.A. Seismic-stratigraphic record of a deglaciation sequence: From the marine Laflamme Gulf to Lake Saint-Jean (late Quaternary, Québec, Canada). *BOREAS* **2014**, *43*, 309–329. [CrossRef]
49. Giressea, P.; Maley, J.; Kossoni, A. Sedimentary environmental changes and millennial climatic variability in a tropical shallow lake (Lake Ossa, Cameroon) during the Holocene. *Palaeogeogr. Palaeoclimatol. Palaeoecol.* **2005**, *218*, 257–285. [CrossRef]
50. Thomas, R.L.; Dell, C.I. Sediments of Lake Superior. *J. Great Lakes Res.* **1978**, *4*, 264–275. [CrossRef]
51. Tonello, M.S.; Hebner, T.S.; Sterner, R.W. Geochemistry and mineralogy of southwestern Lake Superior sediments with an emphasis on phosphorus lability. *J. Soils Sediments* **2020**, *20*, 1060–1073. [CrossRef]
52. Putans, V.A.; Merklin, L.R.; Zelenin, E.A. Signs of modern tectonic events in Late-Quaternary sediments of middle Caspian. *Adv. Curr. Nat. Sci.* **2018**, *4*, 139–144. [CrossRef]
53. Ambrosimov, A.K.; Klyuvitkin, A.A.; Kozina, N.V.; Kravchishina, M.D.; Libina, N.V.; Filippov, A.S.; Artamonova, K.V.; Torgunova, N.I.; Baranov, V.I.; Pol'kin, V.V. Complex studies of the Caspian Sea system during the 41st cruise of the R/V Rift. *Oceanology* **2014**, *54*, 671–676. [CrossRef]
54. Lee, H.J.; Syvitski, J.P.M.; Parker, G.; Orange, D.; Locat, J.; Hutton, E.W.H.; Imran, J. Distinguishing sediments waves from slope failure deposits: Field examples, including the "Humboldt slide", and modeling results. *Mar. Geol.* **2002**, *192*, 79–104. [CrossRef]
55. Normark, W.R.; Piper, D.J.W.; Posamentier, H.; Pirmez, C.; Migeon, S. Variability in form and growth of sediment waves on turbidity channel levees. *Mar. Geol.* **2002**, *192*, 23–58. [CrossRef]
56. Cattaneo, A.; Correggiari, A.; Marsset, T.; Thomas, Y.; Marsset, B.; Trincardi, F. Seafloor undulation pattern on the Adriatic shelf and comparison to deep-water sediment waves. *Mar. Geol.* **2004**, *213*, 121–148. [CrossRef]
57. Wynn, R.B.; Piper, D.J.; Gee, W.; Martin, J.R. Generation and migration of coarse-grained sediment waves in turbidity current channels and channel lobe transition zones. *Mar. Geol.* **2002**, *192*, 59–78. [CrossRef]
58. Wynn, R.B.; Stow, D.A.V. Classification and characterization of deep-water sediment waves. *Mar. Geol.* **2002**, *192*, 7–22. [CrossRef]
59. Hernandez-Molina, F.J.; Stow, D.A.V. Continental slope contourites. In *Contourites: Developments in Sedimentology 60*; Rebesco, M., Camerlenghi, A., Eds.; Elsevier: Amsterdam, The Netherlands, 2008; pp. 379–408.
60. Rebesco, M. Contourites. In *Encyclopedia of Geology*; Selley, R.C., Cocks, L.R.M., Plimer, I.R., Eds.; Elsevier: Oxford, UK, 2005; pp. 513–527.
61. Ivanov, A.Y.; Golubov, B.N.; Zatyagalova, V.V. On oil and gas potential and subsoil fluid discharge in southern part of Caspian Sea by satellite geolocation data. *Earth Satell. Res.* **2007**, *2*, 62–81.
62. Hernandez-Molina, F.J.; Paterlini, M.; Violante, R.; Marshall, P.; de Isasi, M.; Samoza, L.; Rebesco, M. Contourite depositional system on the Argentine slope: An exceptional record of the influence of Antarctic water masses. *Geology* **2009**, *37*, 507–510. [CrossRef]
63. Viana, A.R.; Almeida, J.R.W.; Nunes, M.C.V.; Bulhoes, E.M. The economic importance of contourites. In *Economic and Palaeoceanographic Significance of Contourite Deposits*; Viana, A.R., Rebesco, M., Eds.; Geological Society, Special Publications: London, UK, 2007; Volume 276, pp. 1–23.
64. Roy, S.; Hovland, M.; Braathen, A. Evidence of fluid seepage in Grønfjorden, Spitsbergen: Implications from an integrated acoustic study of seafloor morphology, marine sediments and tectonics. *Mar. Geol.* **2016**, *380*, 67–78. [CrossRef]
65. Hovland, M.; Gardner, J.V.; Judd, A.G. The significance of pockmarks to understanding fluid flow processes and geohazards. *Geofluids* **2002**, *2*, 127–136. [CrossRef]

Journal of
Marine Science and Engineering

MDPI

Article

Simulation of the Nearshore Sediment Transport Pattern and Beach Morphodynamics in the Semi-Enclosed Bay of Myrtos, Cephalonia Island, Ionian Sea

Alexandros Petropoulos [1,*], Vasilios Kapsimalis [2], Niki Evelpidou [1], Anna Karkani [1] and Katerina Giannikopoulou [1]

[1] Faculty of Geology and Geoenvironment, National and Kapodistrian University of Athens, Panepistimiopolis, 15784 Athens, Greece; evelpidou@geol.uoa.gr (N.E.); ekarkani@geol.uoa.gr (A.K.); katagiann@geol.uoa.gr (K.G.)

[2] Institute of Oceanography, Hellenic Centre for Marine Research, P.O. Box 712, 19013 Anavyssos, Greece; kapsim@hcmr.gr

* Correspondence: alexpetrop@geol.uoa.gr

Citation: Petropoulos, A.; Kapsimalis, V.; Evelpidou, N.; Karkani, A.; Giannikopoulou, K. Simulation of the Nearshore Sediment Transport Pattern and Beach Morphodynamics in the Semi-Enclosed Bay of Myrtos, Cephalonia Island, Ionian Sea. *J. Mar. Sci. Eng.* **2022**, *10*, 1015. https://doi.org/10.3390/jmse10081015

Academic Editors: Harshinie Karunarathna, Markes E. Johnson, George Kontakiotis, Assimina Antonarakou and Dmitry A. Ruban

Received: 4 April 2022
Accepted: 20 July 2022
Published: 25 July 2022

Abstract: Myrtos Beach (Cephalonia Island, Ionian Sea, Greece) represents a pocket beach with strong touristic, economic and natural interest. In this research, the morphodynamic behavior of the coastal area (e.g., hydrodynamic and sedimentary state, morphology, orientation, etc.), the current wave conditions (extreme and dominant waves, wave exposure), and also external factors, such as human impact and the geotechnical condition of the wider area, are examined. Short- and medium-to-long-term analysis took place, such as mapping, sediment analysis, wave/wind analysis, numerical modeling, and satellite monitoring, in order to identify the dynamic forcing parameters related to geomorphology, sedimentology, and hydrology that prevail in the area. Additionally, the intense tectonics, the karstified limestones, and the steep slopes of the cliffs in combination with the frequent seismic events on the island set up a geotechnically unstable area, which often cause landslides on the beach of Myrtos; these supply the beach with a large amount of aggregates, constituting the main sediment supply. Wave exposure forcing conditions, longshore–rip current direction, and other hydrodynamic processes are stable with high values in the area, causing notable sediment transport within the bay boundaries. As a result, at Myrtos Bay there is a dynamic balance of the natural system, which is directly affected by human interventions. Taking also into consideration that Myrtos is one of the most famous beaches in Greece and one of the main attractions of Cephalonia Island with thousands of visitors every year, beach management must be focused on preserving the natural system of the coastal area.

Keywords: pocket beach; sand transport; longshore–rip current; hydrodynamic and sedimentary modeling; beach rotation; shoreline displacement

1. Introduction

A pocket beach is defined as a limited beach that is laterally bounded by two headlands [1–5]. The headlands significantly inhibit the sediment transport of nearby areas, restricting any hydrosedimentary processes between their boundaries and making them an autonomous and independent ecosystem, ideal for investigating beach morphodynamics [3,6–11]. Hydrodynamic conditions (longshore and rip currents, etc.) and beach characteristics (slope, grain size, nearshore bars, etc.) determine the morphological behavior of pocket beaches, which present a different circulation pattern than open water beaches [6,7,12–14]. An area with high wave exposure and increased wave radiation stress leads to coastal erosion in the exposed part and aggression in the sheltered part [7], making the determination of nearshore circulation (e.g., rip currents and longshore currents) and sediment transport trends important. Hence, pocket beaches are characterized as areas of limited sediment supply or as closed systems with restricted sediment exchange between them [3,15]. Under the worst-case scenarios, pocket beaches can lose significant volumes of sediment (temporarily or even permanently) if the sediment is transported outside the closure depth, which threatens their existence, posing significant risks landwards [16] (e.g., landslides and inundations). New sediment input can also occur through nearshore sediment sources [17], such as erosion of the backshore [18] (e.g., cliff and dunes) or input from local streams during stormy periods [19]. It is obvious that pocket beaches are particularly important areas for the ecosystem and the local tourism economy, yet, they are extremely sensitive to any change in existing conditions; therefore, their vulnerability to any dynamic change is of great significance at a global level [1,12,16,20,21]. Human activities are often the reason for the disruption of the coastal system, such as through civil engineering structures (e.g., river dredging, river damming and armoring, and anchored steel meshes) [22]. These factors reduce the available sediment that is naturally deposited in the coastal area, determining the inception of the erosion state.

The study of pocket beaches provides useful information on the prevailing pattern of nearshore sediment transport during different environmental conditions and its impact on beach evolution, but also regarding beach management [23]. Beach orientation, sediment classification, morphology, bottom composition, currents, and prevailing wind-generated waves are the main parameters for the morphodynamic variability that acts along and across the shore. The identification of the equilibrium in the beach planform could be characterized as static equilibrium, dynamic equilibrium, or unstable or natural reshaping [24]. Such parameters can help to determine the wave effect on sediment transport, to determine the duration of wave action, and to identify if there is a potential threat to the wider area.

In the present work, by identifying these processes, we aim to specify the coastal system formation of a pocket beach and to understand the nearshore hydrodynamics by analyzing the impact of the geomorphological parameters and the induced variability on circulation patterns. Numerical modeling with the MIKE 21 Flow Model (HD, SW, ST, and Shoreline Morphology modules) in addition to extensive subaerial and subaqueous field work and laboratory analysis may give promising predictions of beach evolution [25,26]. The ultimate goal of our work is to assess the dynamic balance of the natural system of the coastal area in a sustainable manner. Quantitative and qualitative results of sediment transport, and morphodynamic and hydrosedimentary results are presented alongside representative/extreme wave model simulations that were taken into account for the coast of Myrtos.

Myrtos Beach is a unique national monument of nature with high touristic and economic value, visited every year by thousands of tourists, and thus offering significant benefits to the local community of Cephalonia Island. Any negative evolution in the coastal system, such as coastal erosion, has a negative social and economic impact on coastal areas that revolve around tourism. Through the results of our study, the competent authorities will be able to implement better long-term management in the specific area of significant economic importance and take preventive actions. Nature-based strategic management activities are carried out to ensure long-term success in almost all sectors (environmental, touristic, economic, social, etc.). They can achieve effective long-term management by focusing on the preservation of natural and cultural resources while promoting the economic contribution to local communities.

2. Geological and Geomorphological Setting

Cephalonia Island is in the Ionian Sea (Greece). Myrtos Bay is located at the northwestern part of the island, between the northern section of Mt. Aenos and the southern part of Erissos peninsula (Figure 1). The stratigraphy of the area consists of carbonate sequences, marl formations, and turbidite limestones [27]. The tectonism in the area is intense and is characterized by faults in the NE–SW and NW–SE direction. The reverse fault of Agia Efimia, with a NW–SE direction, crosses the northern part of the bay, presenting a strike-slip component, and is characterized by uplift and erosion. The fault of Agia

Efimia, along with other active faults with a NE–SW and NW–SE direction, cut across Myrtos Bay and form a geotechnically unstable area [28,29]. Therefore, limestones on the slopes of Myrtos are strongly fragmented and locally karstified, and in some places, they turn into tectonic lattice due to the intense tectonism from the fault; they are also covered by colluvial deposits that derive from the weathering of steep slopes and consist of sand, gravel, and irregular limestone fragments [30]. Frequent seismic sequences on the island and weathering processes often cause landslides and rock falls on the steep slopes of the cliffs, causing damages on the beach of Myrtos and at various points on the road network [28]. The occurrence of landslides forms scree landforms, which also supply a large amount of aggregates to the beach.

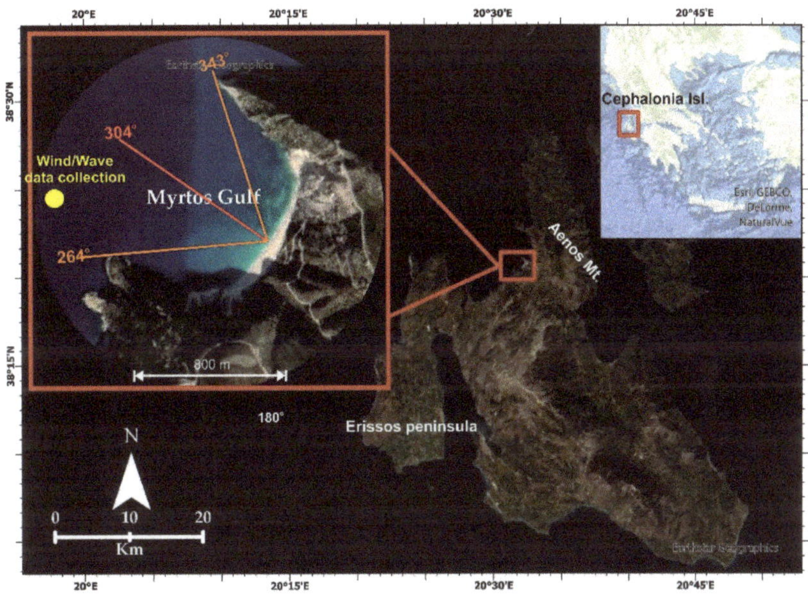

Figure 1. The island of Cephalonia located in the Ionian Sea. Myrtos Beach is located at the north-western part of the island. The direction of Myrtos Bay ranges between 265 and 343°, whereas the orientation of the coastline is 304°.

The hydrographic network of the area is not well developed due to the karstic carbonate formations and the intense tectonics of the wider area. Independent and isolated seasonal streams of limited length and sediment supply flow in the bay; these streams are characterized by significant momentum during periods of heavy rainfall (flash floods), and deep erosion due to the steep slopes [28].

Geologically, the Myrtos area is part of the Pre-Apulian zone (external Hellinides, Paxos unit) and the lithological formations of which it is comprised are Upper Cretaceous limestone, Paleocene limestone, Eocene limestone, Miocene deposits, and Plio-Quaternary deposits (from the earliest to the latest). The approximate values that characterize the beach, considering its variability, are as follows: a length of 850 m, and a width of 100 m in the center and 30 m on both sides, whereas at the backshore a steep cliff of 30 to 300 m delimits the beach (Figure S1).

The direction of Myrtos Bay ranges between 265° and 343°, whereas the orientation of the coastline is 304° facing the NW, with a total width of 800 m from the entrance of the bay to the coast; it is bounded by two headlands preventing any sediment exchange with the neighboring coasts.

Based on the geographical position, the orientation of the bay, and the wind/wave data collection, resulting in the numerical reconstruction located at lat: 193,980.992 m/lon: 4,250,162.324 m (coordinates from Greek Grid) for Myrtos Bay for the period 1995–2004 [31], the coastline is mainly exposed to incoming waves from the N, NW, and W direction, with a fetch length extending to 266 km for the northwest direction, limited by the Apulian peninsula of Italy. The main annual wind intensity is classified between 3.3 and 7.9 m/s of the order of 3 to 4 Beaufort, with a frequency of ~13% and a NW direction. Extreme wind velocity values are also noticed from the same direction (10.7 to 19.3 m/s, 6 to 8 Beaufort) with a frequency of less than 2% [31]. Regarding the annual wave conditions, the most common wave approach direction to the beach is from the NW with a ~30% frequency, with prevailing waves of 0.5 to 1 m (~12.5% frequency). Extreme wave height values (2 to 5.2 m) also have a NW approach direction with a frequency of less than ~1.5% [31] (Figure S2).

3. Materials and Methods

The current coastal morphodynamic analysis was determined with two distinct approaches related to the duration. In situ field observations and measurements took place in October 2018 and March 2019, and allowed the short-term analysis of Myrtos Beach characteristics (e.g., topographic sections, sediment samples, bathymetric data, etc.), such as seasonal variations that change in short times. On the other hand, the medium-to-long-term analysis of the beach was accomplished using data on the wind and wave conditions for the period 1995–2017 [31–34] in order to identify the prevailing wave conditions and develop simulation scenarios. These scenarios were applied to the sediment and hydrodynamic transport numerical MIKE 21 Flow Model, which was coupled with the HD, SW, ST, and Shoreline Morphology modules in order to simulate the prevailing conditions through a dense array of measurements. Satellite images were also used (Google images from 2003 to 2019) to estimate the shoreline displacements through time. Medium-to-long-term analysis gives a sense how the coastal area has evolved over time and provides information on how it is likely to evolve in the near future (Figure 2).

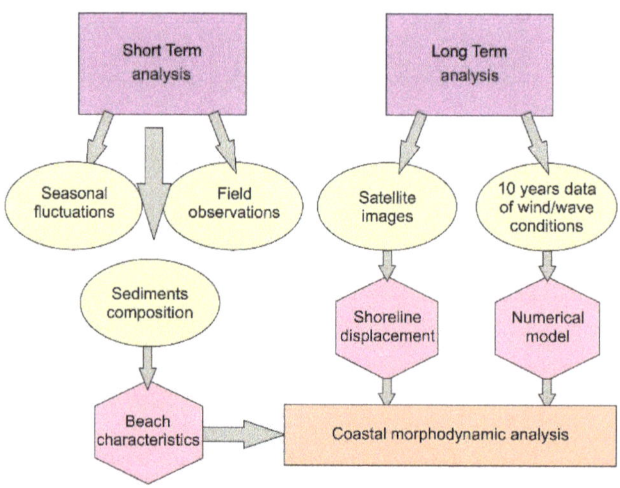

Figure 2. Flow chart that displays the methodology of the study area.

3.1. Short-Term Analysis and Measurements

In situ field surveys and measurements were carried out in two different periods, in October 2018 and March 2019, in order to estimate the representative characteristics of the beach in each season (summer/winter). Specifically, for each period the following were measured: (a) the evolution of seven sedimentary cross-shore profiles with 32 sediment samples (equally spaced in distance, ~80 m), (b) 55 topographic shore-sections at ~10 m distance and perpendicular to the coastline, which allowed a high resolution of the beach and c) recording of the coastline. A marine survey was also carried out, which included seabed morphological mapping of the bay, a bathymetric survey, and collection of 11 seabed sediment samples at a depth from 2 m to 15 m during each time period (Figure S3).

The elevation and geographical position were precisely determined byusing a Real-Time Kinematic (RTK) Differential Global Positioning System (DGPS). The detailed topo-bathymetric data were collected by the acquisition of side-scan sonar imagery (StarFish 450) and a single-beam echo sounder (Lowrance LCX-15MT). The substrate component mapping of the seabed was also performed with in situ field observations through a diving survey and underwater photography. The sediment samples for both periods were analyzed by dry granulometry and were sorted according to Folk and Ward's (1957) [35] nomenclature using the GRADISTAT (version 8.0) software by Simon J. Blott and Kenneth Pye (Royal Holloway University of London, UK) [36].

3.2. Medium-to-Long-Term Analysis

3.2.1. Digital Shoreline Analysis System (DSAS)

The comparison of the satellite images from 2003 to 2019 led to the quantification of the long-term shoreline displacements using the Digital Shoreline Analysis System (DSAS) application and the Net Shoreline Movement (NSM) tool in the ArcGIS (version 10.3) software by Esri (Redlands, California, U.S.) [37]. This tool offers the ability to calculate the distance between the oldest and the newest coastline and provides a value of the retreat or advance [38]. The satellite images were from the periods 06/2003, 12/2005, 01/2008, 04/2013, 07/2013, 03/2016, 04/2017, 10/2018, and 03/2019, and were selected because of their clearance at the land–water interface. They were georeferenced from WGS 84′ to WGS 84/UTM zone 34N. The National Cadastre was used as a reference base map, as it is a unified and constantly updated information system that records legal, technical, and other additional information on real estate and rights that are under the responsibility of the state.

3.2.2. Wind and Wave Data

The role of wind forcing is crucial for pocket beaches as, in most cases, it is the major forcing factor of wind-generated waves. The prevailing significant waves with high-energy flux, depending on their approaching direction, produce nearshore currents (longshore/rip current) that cause sediment transportation along the coast [39].

The wind and wave data for Myrtos Bay concern the period 1995–2017. The 10 years of data (1995–2004) are the result of the numerical reconstruction of wind and wave conditions [31] located at lat: 193,980.992 m/lon: 4,250,162.324 m (coordinates from Greek Grid). The wind regeneration data were derived from the atmospheric nonhydrostatic model SKIRON/ETA and the wave data from a combination of the above model with the WAM-Cycle 4 wave model [31]. The values of the quantities, since they derive from a numerical reconstruction, always involve a margin of error in relation to the actual wave and wind conditions. The wave data for 2005–2017 derive from the Copernicus Climate Change Service (ERA-Interim reanalysis dataset), providing a consistent European dataset for wave conditions that is produced by the ECMWF (European Centre for Medium-Range Weather Forecasts) [32–34].

3.2.3. Modeling

The numerical MIKE 21 Flow Model, coupled with the HD, SW, ST, and Shoreline Morphology modules, takes into account the space and time period of the prevailing and extreme conditions of the phenomena of interest by simulating the morphodynamics of an embayed beach with sediment transport and bed level changes due to currents or combined waves/currents. The MIKE 21 Flow Model has the efficiency to study the wave transformation over different temporal and spatial scales; it also allows repeating running tests with different wave exposure forcing conditions along different values of parameters [40] in order to better understand the nearshore circulation.

The numerical model estimates the coastal erosion processes and the natural variations in sand budgets, and allows for an assessment of marine spatial planning and the study of the impact and effectiveness of shore interference works. The reliable key calibration parameters, such as sediment grading, grain diameter, manning, and bed thickness, were calculated and used in the MIKE 21 Coupled Model FM simulations (Table 1) [41] in combination with bathymetry, sediment analysis, and substrate component data from the bay. The values of the main parameters used by the model were set based on the program manuals along with empirical validation through the repeated running of tests. The model was successfully applied and tested in a number of basic, idealized, realistic, and complicated situations from which the output results can be compared with analytical solutions or information from the literature [42–44].

Table 1. Main parameters of MIKE 21 Coupled Model FM used in simulations.

Parameters	Values
Sediment Grading ($\sqrt{(d\_84 \div d\_16)}$)	1.1–2
Mean Grain Size (Mz)	0.07–1.9 mm
Manning Number	10–32 $m^{1/3}/s$
Bed Thickness	0.05–2 m

4. Results

The results of this work concern two distinct approaches, related to the in situ measurements and the statistical and numerical model results, which are presented below. The results from the topographic, sedimentological, and morphodynamic seasonal analysis of the subaerial part of the morphological profile sections reveal the difference between the two seasonal periods. Additionally, the statistical results of the DSAS-NSM from 2003 to 2019 allowed to identify the displacement of the coastline. Finally, the results from the MIKE 21 Coupled Model FM simulations examined the hydrodynamic and sediment transport for each scenario/situation.

4.1. *Morphological and Nearshore Characteristics and Grain Size Analysis*

The coastline length of Myrtos Beach is 860 m, with a limited width at both ends having an average value of 30 m, whereas the central part of the coastal width reaches 80 m. Vegetation is observed, during both periods, only at the central part of the coast, at the backshore area, and near the cliffs (~60 m from the coastline). Human impact is limited at the base of the cliff through the construction of a local road and a parking area, and also at the central area of the backshore with a human structure.

The sedimentary materials in both periods show some similarities in terms of grain distribution. Figure 3a,b depicts that, for marine sediments, the finest material is concentrated in the SW area, developing a long tongue of fine sediments (mud, sand) surrounded by hard substrate with coarse-grained sediments (granule, pebbles), whereas in the coastal area there is a difference in the grain distribution, as in summer the backshore and foreshore area consists mainly of fine sediments (sandy sediments) in contrast to the winter period, where the sediment distributed in the area is characterized as coarse (gravel).

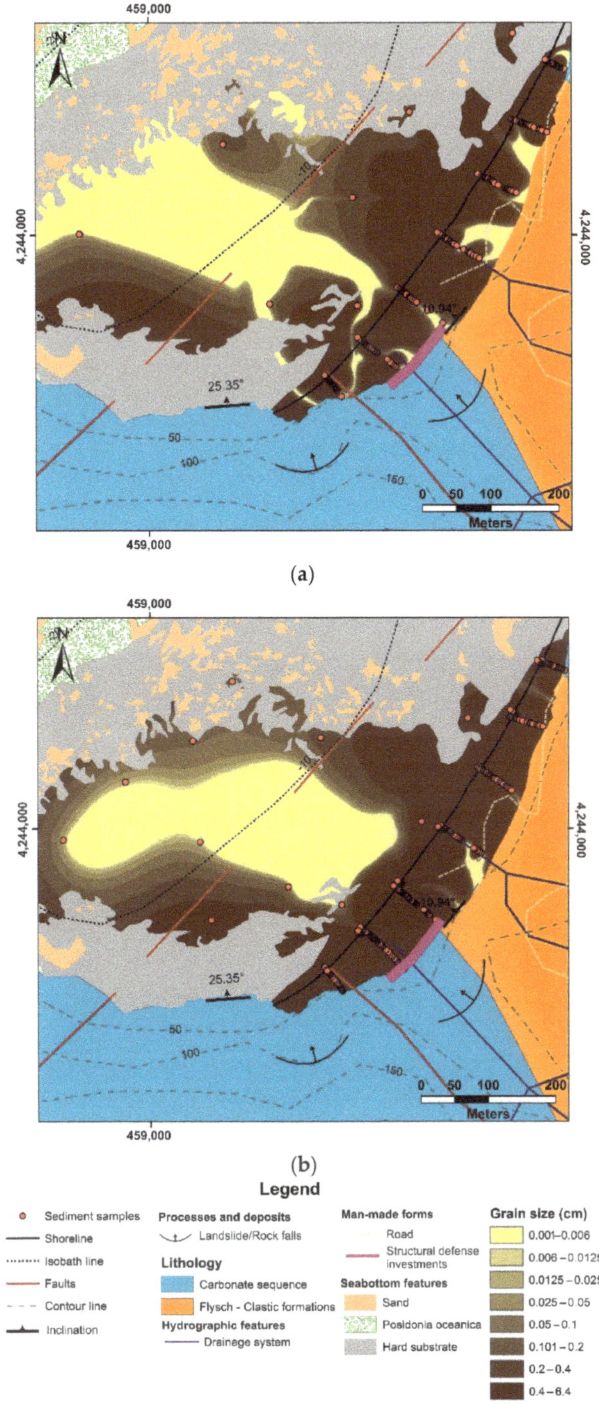

Figure 3. Grain size distribution in the summer–October 2018 (**a**) and winter–May 2019 (**b**) period.

Figure 4 depicts the temporal evolution of the beach profiles from October 2018 to May 2019, located from the backshore until the water breaker zone. Significant variability is present at the surf and swash zone where the seasonal fluctuation is clear, with sediment accumulation during the summer months and erosion during the winter months. The summer Beach Profile 3 displays the highest sediment supply compared to that of the winter period. The other beach profiles follow the same pattern (compared to the summer and winter profile) on a smaller scale. Additionally, morphological formations such as beach berms are formed along the beach in both periods, but in the winter months they are smaller and more in number in contrast to the summer months. Sediment composition in both periods is characterized as gravelly sand to coarse gravel, confirming that the sediment supplies is a combination derived from landslide materials off nearby cliffs and the coastal high-energy environment. The beach area is composed of different spatial and quantitative values, with the coarser sediments located landward and well sorted, whilst at the dynamic swash zone, different sizes of sediments are located and poorly sorted. Most fine sediments are absent due to high wave energy, whereas the coarser remains. The winter profiles are characterized by coarser sediments that are found mainly at the swash zone in contrast to the summer profiles.

Beach Profile 2 to Beach Profile 5 display a smoother profile with finer sediments than the rest of the beach profiles (BP 1, BP 6–7), especially during the summer period. Beach Profile 4 and 5 are the most exposed to the wave conditions, as topographic anomalies occur along their entire length.

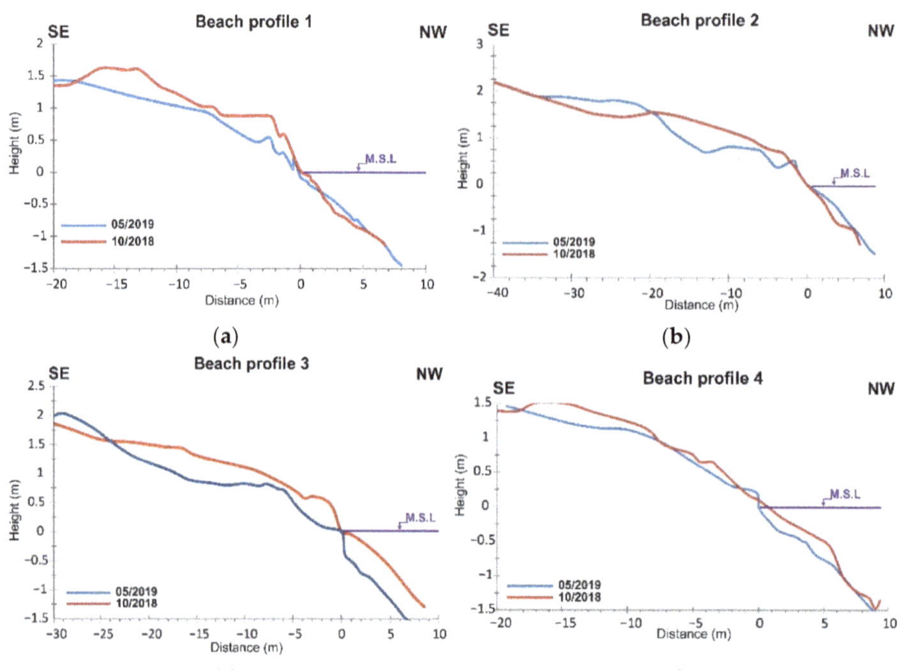

Figure 4. *Cont.*

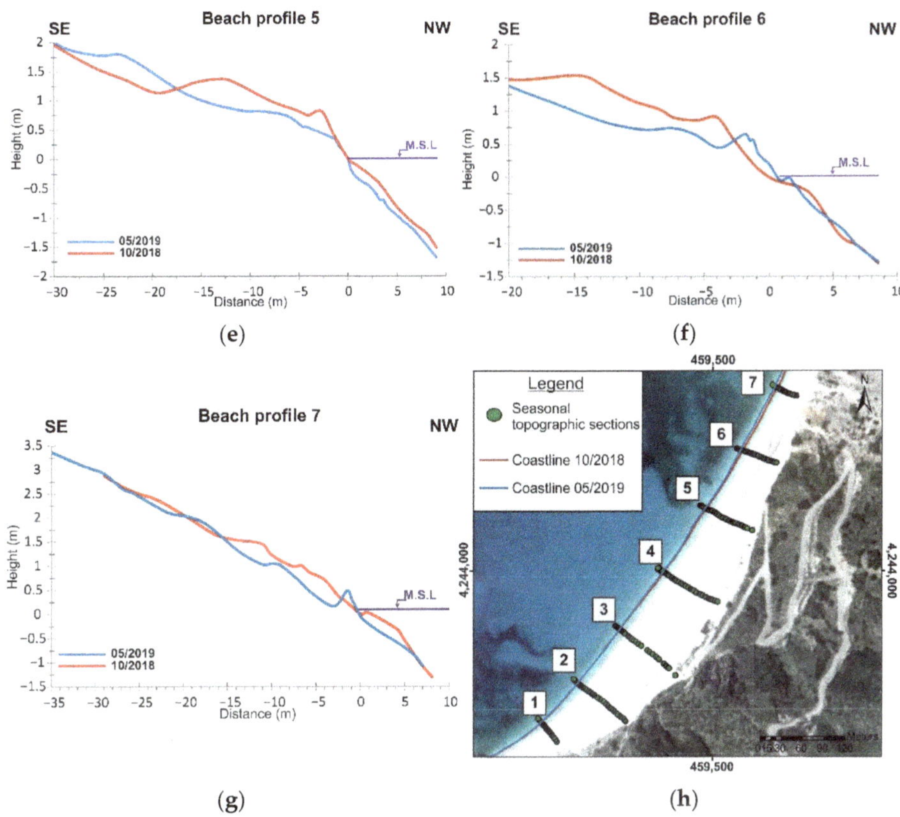

Figure 4. Seasonal evolution of beach profiles. (**a**–**g**) display the corresponding beach profile, focusing on the surf and swash zone. (**h**) specifies the geospatial location of the beach profiles sections.

Myrtos Bay presents a topographic anomaly in the slope at the bottom, as between the depths of 2 and 8 m the slope is steep (20 m from the coastline); then, until the depth of 20 m, the slope becomes relatively gentle, and up to a depth of 34 m it becomes steep again (Posidonia *Oceanica* meadow area). Afterwards, the seabed presents a variation of inclination from steeper to mild values. The closure depth is estimated at 10 m with a nearshore slope at 5°. The calculation of the closure depth is supported by the Hallermeier equation [45]. The maximum measured depth of the enclosed bay is 45 m and is located at the entrance of the bay. The bathymetric contours are parallel to the coastline from the entrance of the bay until the isobaths of 18 m, whereas from this area until the shoreline, the bay is divided into two parts. The first includes the SW area where the isobaths are parallel to the coastline, and the second, the NE area where the isobaths are affected by the hard substrate, creating a swallow environment (Figure 5a). The Myrtos pocket beach can be exposed to high wave energy due to the steeply sloping sea bottom near the coast.

Wave radiation stress is significant to detect the circulation pattern in the coastal area, where longshore and cross-shore currents act, especially at the surf zone. Wave radiation stress acts on mean flow, causing wave setup and wave-induced current [46]. Figure 6 shows the isobaths, the gradient spatial distribution, and the sea bottom composition, which specify the fluctuation of the radiation stress areas. At the central part of this area, the radiation stress is generally normal towards the coastline and reaches maximum when the bottom topography changes abruptly, as at the wave break zone, where the significant wave height changes most (hard substrate, isobaths of 8 m).

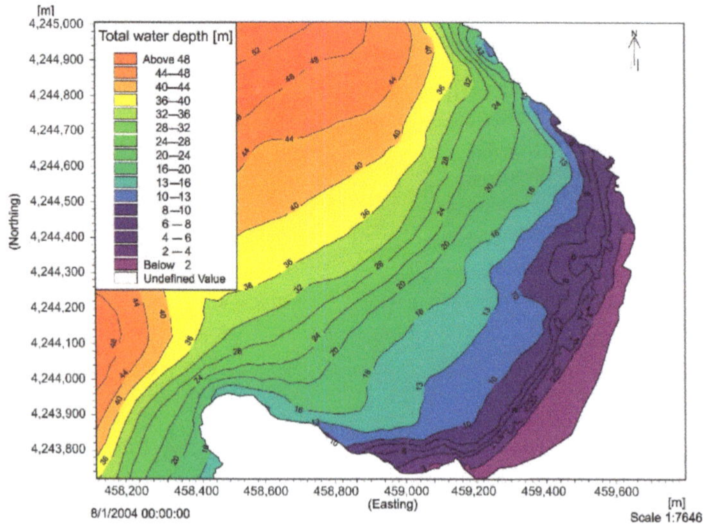

(**a**)

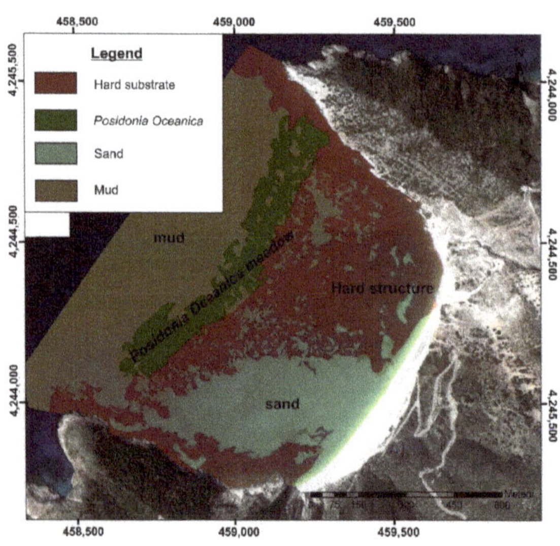

(**b**)

Figure 5. (**a**) Bathymetry map of Myrtos Bay, (**b**) substrate component mapping of the seabed of Myrtos Bay.

The total surface area of the bay is about 1 km^2 and the seafloor consists of noncohesive material (0.58 km^2), rocky outcrops (0.32 km^2), and a seagrass (Posidonia Oceanica) meadow (0.1 km^2). The sandy, gravelly, and hard substrate is located from the nearshore zone to a depth of 22 m. The meadow is developed at depths between 22 and 32 m, whereas the deepest part of the bay is covered by fine-grained sediment (mud) (Figure 5b).

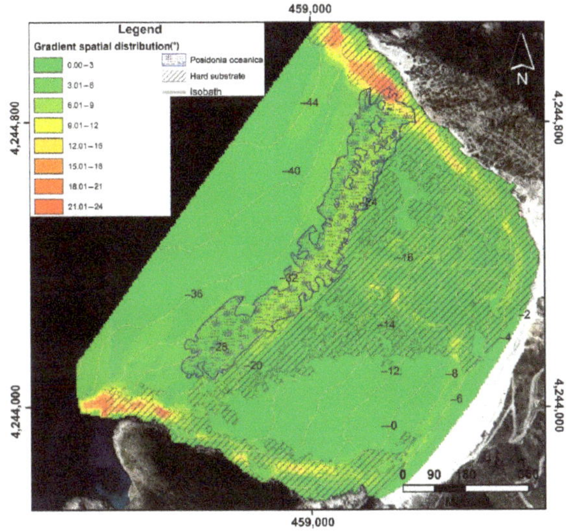

Figure 6. Gradient spatial distribution of the seabed of Myrtos Bay.

4.2. Digital Shoreline Analysis (DSAS)

The use of satellite images from 2003 to 2019, in combination with the DSAS-NSM (ArcMap) software tool, led to the quantification of the long-term shoreline displacements. The statistical results show that the earthquake event of 2014 [28,29,47–50] had a significant role in the evolution of the coast, as from 2003 to 2013 the average retreat of the shoreline reached 8 m for the central and southern part of the beach due to the limited sediment supply from the land, with an average regression rate of 1.3 m/year. In the northern part, sediment deposition results showed an average coastal advance of 4 m, as a result of the beach rotation mechanism. After the earthquake event and until 2019, the statistical results demonstrate that deposition phenomena prevailed in the area, with an average shoreline progradation of about 15 m (Figure 7). The shoreline progradation had an average rate of 1.4 m/year at both edges of the coast and 2.5 m in the central part of the beach.

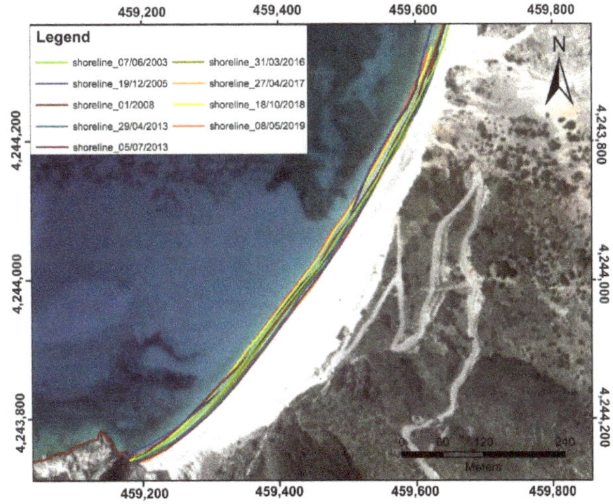

(a)

Figure 7. *Cont.*

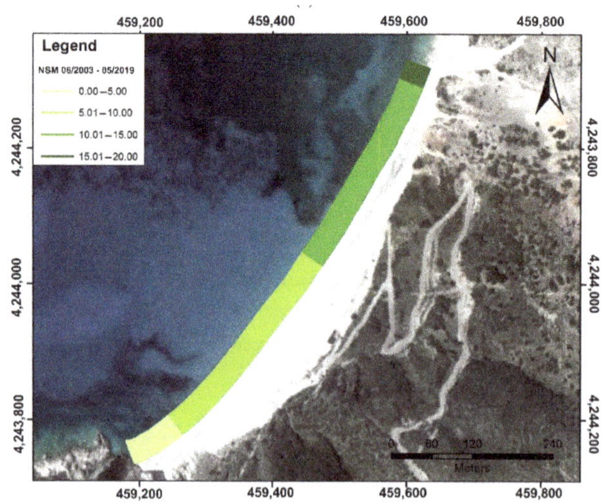

(b)

Figure 7. (**a**) Shoreline evolution. Nine different shoreline displacements for the period 2003 to 2019. (**b**) Net shoreline movement at Myrtos Beach. The figure displays the coastline fluctuation for the period 2003 to 2019.

The result for the seasonal displacement of the shoreline (10/2018–05/2019) reflects a stable situation with little variation between summer and winter, based mainly on the seasonal beach rotation mechanism.

4.3. Model Examination

The boundary conditions applied to the model were obtained from two directions, depending on the significant wave's direction and the orientation of the coastline, as the sediment transportation takes place in both directions. For each part of the coast, one scenario corresponds to the prevailing conditions, depending on their approaching direction, which contribute to the sediment transport from one side to the other, and the second scenario corresponds to the opposite direction. The prevailing conditions correspond to the representative significant waves for each scenario according to the maximum wave energy flux in each direction annually [41,51–55], estimating the potential and total sediment transport at the time of simulation. The direction of Myrtos Bay ranges between 265° and 343°, whereas the orientation of the coastline is 304°; a scenario for each direction of the coast was determined, estimating the pure and total sediment transport at the time of simulation (scenario 1: 264–304°, scenario 2: 305–343°) (Figure S4). The duration of each simulation scenario was determined from the total flow of wave energy at an annual frequency to the maximum energy flux (Table 2).

Table 2. Representative scenarios taken into account for Myrtos Bay.

Scenario	H Sign. (m)	T(s)	MWD (Deg N)	Duration (Days)
1	2.25	8.25	300	7
2	0.75	5.25	315	203

Additionally, notable is the sediment transport that takes place during extreme wave events. Taking into account the wind and wave data of the study area [31–34], the most extreme event was identified as occurring between the 265 and 343° direction [56] (Table 3).

Table 3. Extreme wave event taken into account for Myrtos Bay.

Scenario	Hmean (m)	Hmax (m)	Tmax (s)	MWD (Deg N)	Duration (Days)	Start Date—Time
Extreme Event	2.7	5.19	10.26	304	2.2	24/11/2001—00:00

Wave radiation stress is significant where the depth bottom decreases drastically, leading to an increase in wave energy dissipation through bottom friction and wave breaking, especially within the surf zone area, and a decrease in wave height.

The circulation patterns (longshore/cross-shore currents) for each wave radiation stress in the respective scenarios and each extreme wave event are the result of the contribution of wave breaking due to depth change, wave refraction, wave diffraction, and sea bottom component effects (Figures 5 and 6) [46]. The prevailing conditions corresponding to the annual representative waves from 265 to 304° were simulated for scenario 1, and they have a mean energy flux direction of 300°, significant wave height of 2.25 m, wave period of 8.25 s, and time of simulation of 7 days (Figure 8a). The coastal currents of the study area are characterized by an average velocity of 0.19 m/s and a maximum value of 1.09 m/s, located in the southern part of the coast and parallel to the coastline in a NE direction. It is worth noting the formation of two longshore currents parallel to the coastline, which are oriented from SW to NE and from NE to SW; they meet each other in the middle of the coastline, where they create a rip current with an average velocity of 0.11 m/s (maximum value 0.28 m/s), up to a depth of ~10 m, at a distance of 240 m from the coastline (Figure 8b). Sediment transportation and seabed variation values are directly affected by the way coastal currents act on the area. The first is oriented to the NE and the other to the SW, whereas they meet each other in the middle of the coastline, with a direction to the NW and to the deeper parts. The largest volumes of coastal sediment transportation occur along the coastline with values ranging between 1.1×10^{-6} and 5.7×10^{-4} m^3/sec/meter (average value 4×10^{-5} m^3/sec/meter). Affected by the coastal currents and the sediment transportation, seabed variation concerns the transport and deposition of sediments from the coast, and especially from the surf zone to the deeper parts (7 m), creating longshore bars at a distance of 60 m from the coastline, with a range of values between −2 m (surf/swash zone) and −3.4 m (breaker zone) (Figure 9).

Scenario 2 is characterized by a mean energy flux direction of 315°, significant wave height of 0.75 m, wave period of 5.25 s, and time of simulation of 203 days in order to simulate the prevailing conditions corresponding to the annual representative waves between 305 and 343°(Figure 10a). The energy flux that reaches the coast is significantly reduced compared to Scenario 1. Coastal currents have an average velocity of 0.06 m/s and a maximum value of 0.47 m/s in the southern part of the coast, perpendicular to it. A typical longshore current is formed at the northern part of the coast with an orientation to the SW, whereas at the southern part of the coast the currents act perpendicular to the coast (Figure 10b). The values of sediment transportation and seabed variations depend on coastal current values; sediment transportation takes place in two areas: (a) from north to south until the middle of the coast, with an average value of 1.3×10^{-5} m^3/sec/meter, and (b) perpendicular to the coast, in its southern part, with an average value of 1.9×10^{-5} m^3/sec/meter. The prevailing trend of seabed variation concerns the transport and deposition of sediments by displaying erosion at the northern part of the coast, with average values of −0.15 m, and deposition at the southern part with average values at 0.88 m. Maximum values of seabed variations are estimated as −1.96 m for the eroded area and 3.97 m for the deposition area (Figure 11).

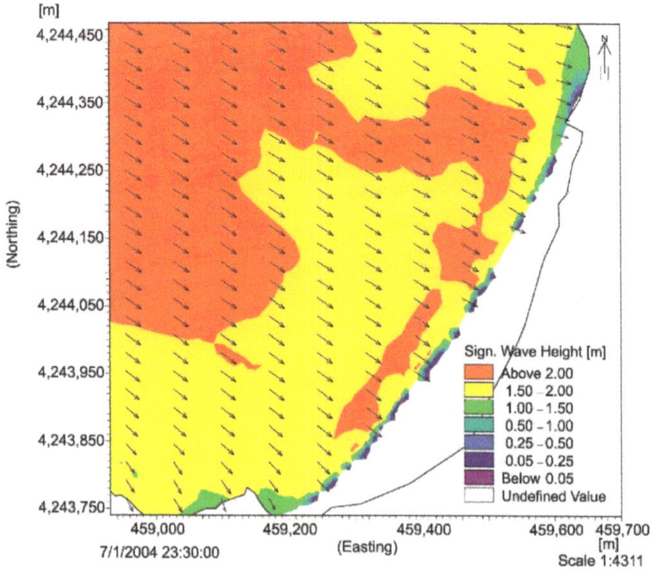

(**a**)

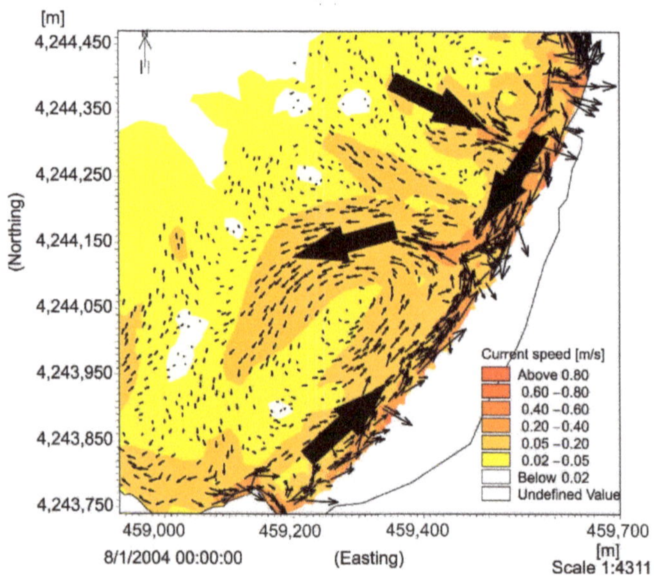

(**b**)

Figure 8. Model simulation for scenario 1. (**a**) Map of significant wave height distribution. (**b**) Map of coastal current velocity. The arrows show the pattern of behavior of the (**a**) Sign. wave and (**b**) current.

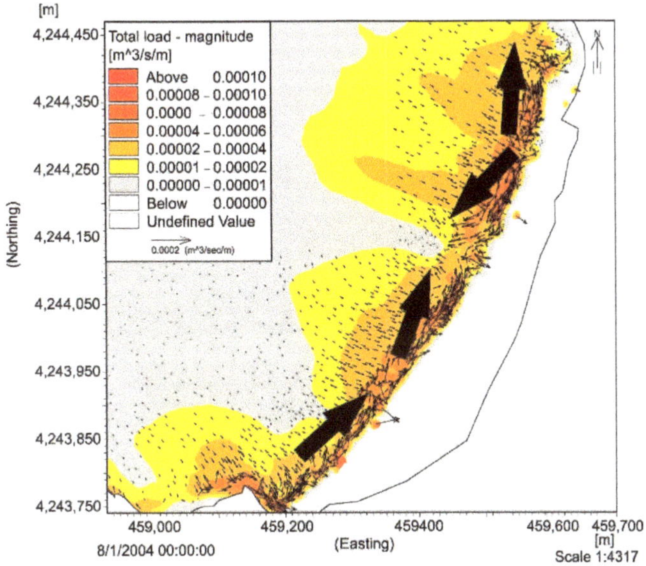

(**a**)

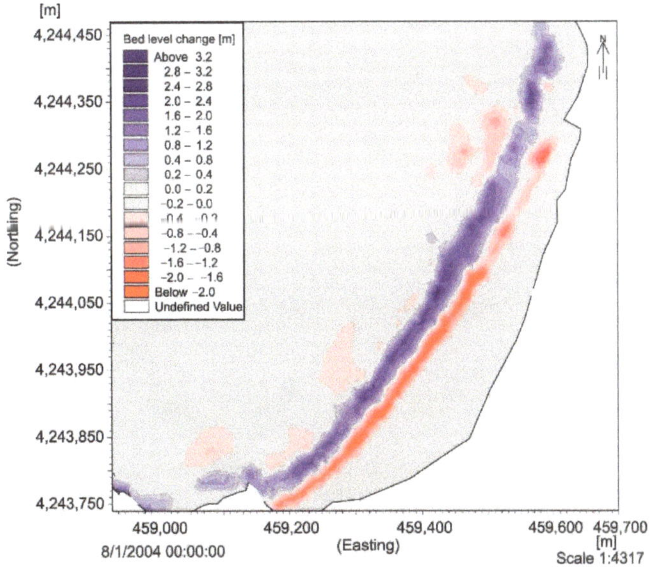

(**b**)

Figure 9. Model simulation for scenario 1. (**a**) Stereo transport map of sediments. (**b**) Map of seabed-level change. The arrows show the pattern of behavior of the (**a**) coastal sediment transportation.

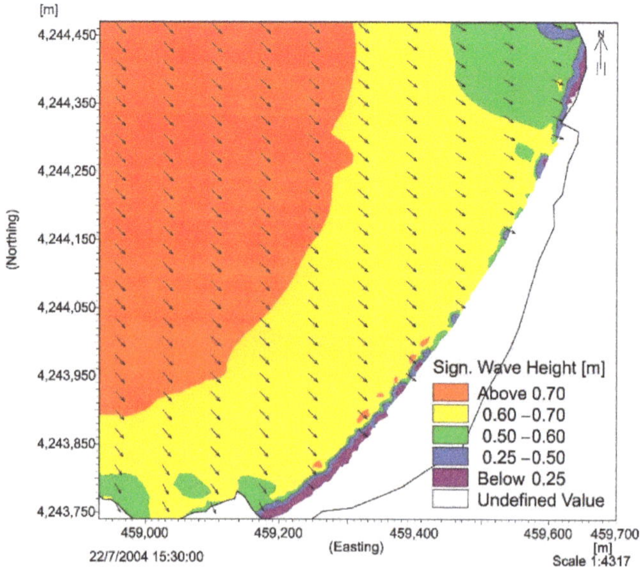

(a)

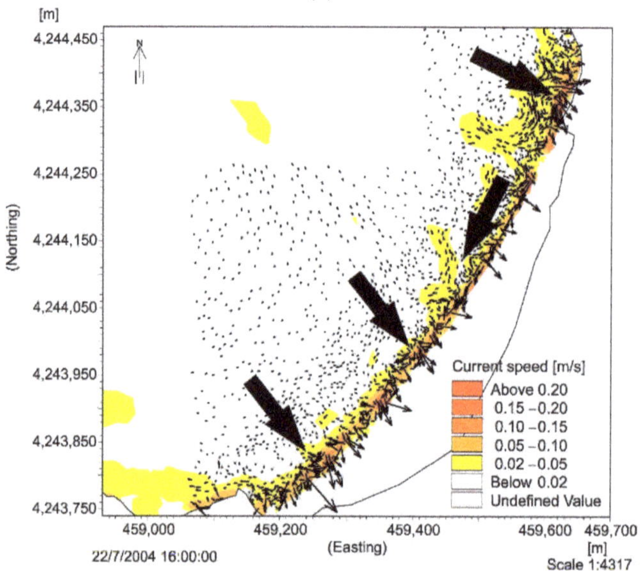

(b)

Figure 10. Model simulation for scenario 2. (**a**) Map of significant wave height distribution. (**b**) Map of coastal current velocity. The arrows show the pattern of behavior of the (**a**) Sign. wave and (**b**) current.

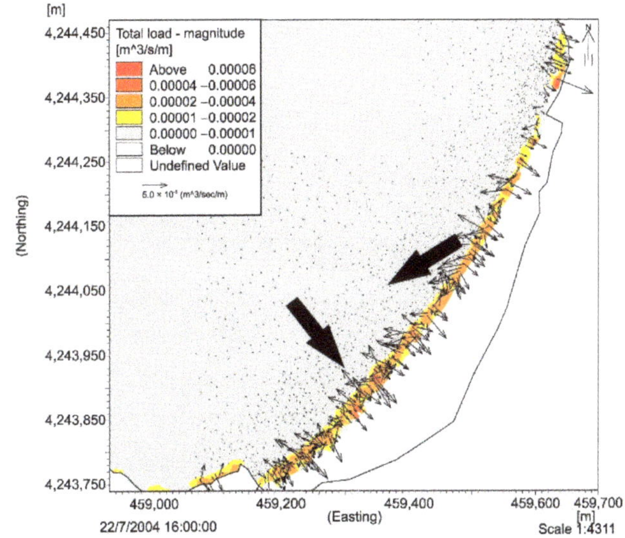

(a)

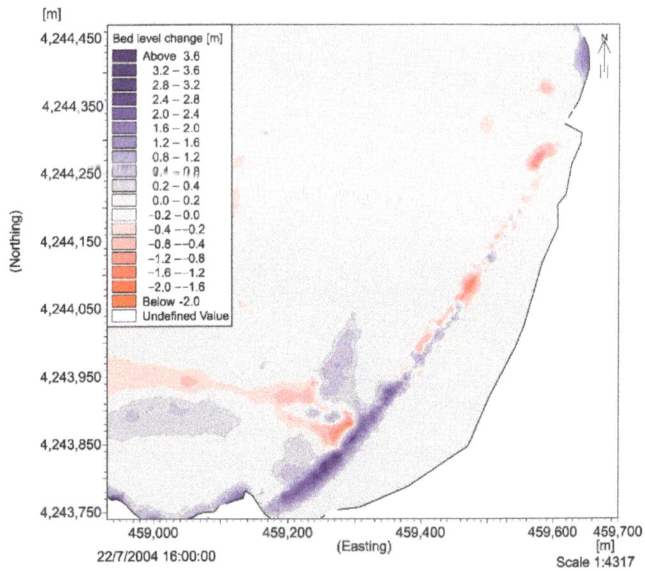

(b)

Figure 11. Model simulation for scenario 2. (**a**) Stereo transport map of sediments. (**b**) Map of seabed-level change. The arrows show the pattern of behavior of the (**a**) coastal sediment transportation.

In order to estimate the sediment volumes that were transported based on the two scenario simulations, the study area was separated into three sections. For each section, the sediment volumes transported on either side of each section for each scenario were quantified annually. The total sediment volume for Section 1 was estimated at 3047 m^3, directed to the southwest, and for Section 2, at 4744 m^3 in the same direction, causing erosion in the intermediate area. In Section 3, the transported sediment volumes were decreased (745 m^3), indicating a depositional environment in the southern part of the study area, between Sections 2 and 3 (Figure 12).

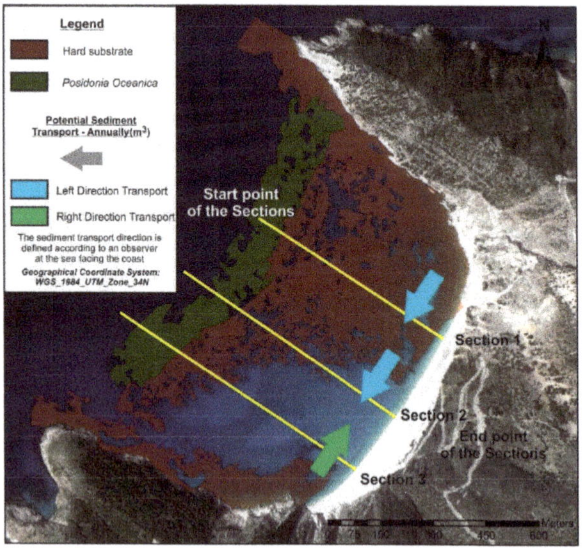

(a)

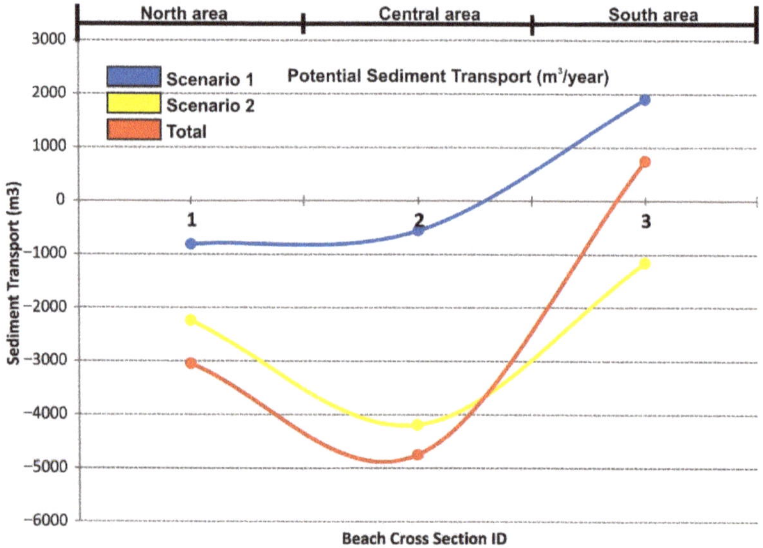

(b)

Figure 12. (**a**) A potential sediment direction that cross each section on an annual basis for scenarios 1 and 2, by dividing the study area in three sections. (**b**) An overall estimate (total) and volume of sediments that cross each section on an annual basis for scenarios 1 and 2, by dividing the study area in three sections.

Taking into account the data of wind and wave conditions for the study area [31–34] and the orientation of the coast, the most extreme event was on 24/11/2001 at 00:00; it was characterized by a mean wave height of 2.7 m, maximum wave height of 5.19 m, maximum wave period of 10.26 m, and mean wave direction of 304°, with a duration of 2.2 days. The highest values of wave height and wind speed were recorded on 24/11/2001 between 12:00 p.m. and 22:00 p.m. (wave height > 4 m, wind speed > 15 m/s) with a general wave direction of 307° and wind direction of 325° (Figure 13). For the specific time period, the average velocity of the coastal current was 0.22 m/s and the maximum value was 2.95 m/s, located at the southern part of the area with a NE direction. The actions of the coastal currents have the same impact at the coastal area as that which prevailed in scenario 1, with the formation of two longshore currents parallel to the coastline that are oriented from SW to NE and from NE to SW. The morphodynamics of the area favors the longshore current with a SW direction, and hence, it prevails over the one from NE direction; at the contact point a new current is created in the deeper parts of the bay with a SW direction. The average velocity of the longshore current with a SW direction is 0.4 m/s with a maximum of 2.5 m/s, whereas the average velocity for the current with a SW direction at the deepest parts of the bay is 0.45 m/s (maximum 0.83 m/s), up to the depth of 20 m (Figure 14a). Sediment transportation has a tendency to move from the surf zone to the deeper part perpendicular to the coast, especially in the area where the currents act. The largest volumes of coastal sediment transportation range from 1×10^{-5} to 3×10^{-3} m^3/sec/meter, with a prevailing orientation towards the deeper parts of the bay. Affected by the sediment transportation, seabed variation concerns the transport and deposition of sediments from the coast, and especially from the surf/wave zone to greater depths (10 m) at the central and south area, creating longshore bars at a distance of 100 m from the coastline, with a range of values between −2 m and 3.6 m (Figure 14b,c). It is worth noting that for the considered extreme event, small-scale hydrodynamic and sedimentary procedure occurred in the deepest parts of the bay and, specifically, in the SW sandy part, up to a depth of 33 m and at a distance of 750 m from the coastline.

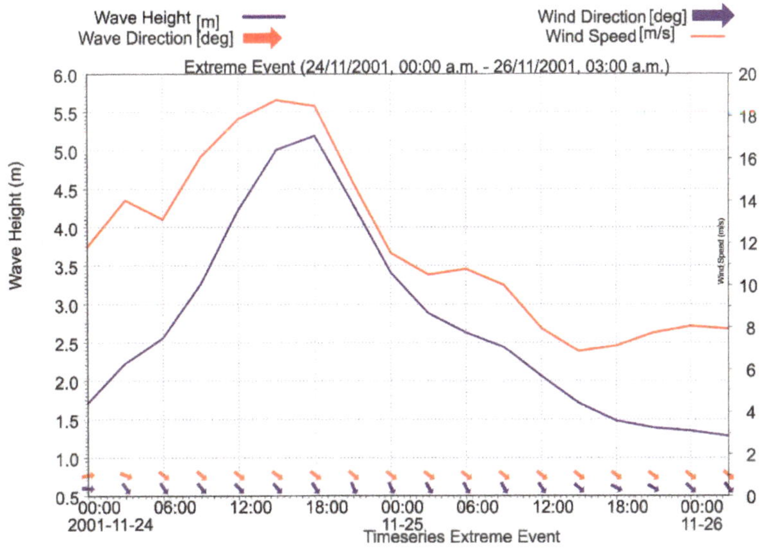

(a)

Figure 13. *Cont.*

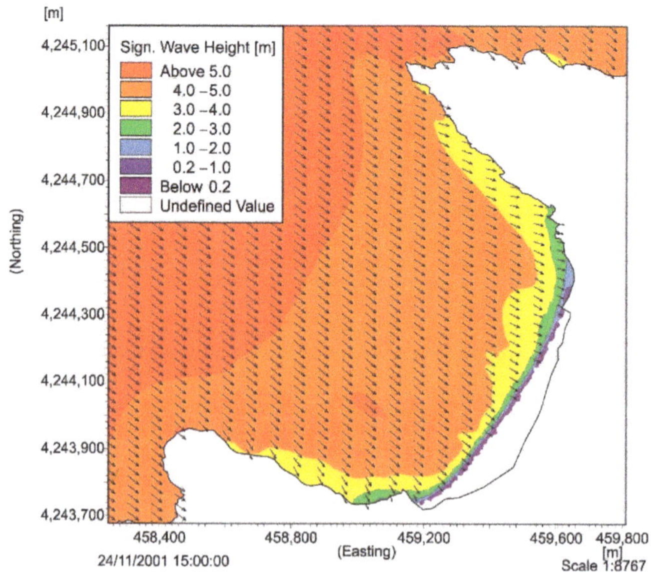

(**b**)

Figure 13. Model simulation for the maximum extreme event. (**a**) Diagram of the wind and wave conditions of the time period in which the extreme event occurred. (**b**) Map of significant wave height distribution. The arrows show the pattern of behavior of the (**b**) Sign. wave.

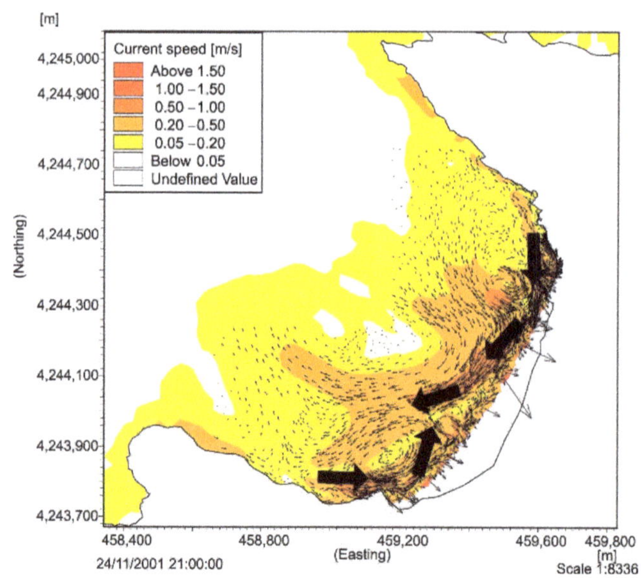

(**a**)

Figure 14. *Cont.*

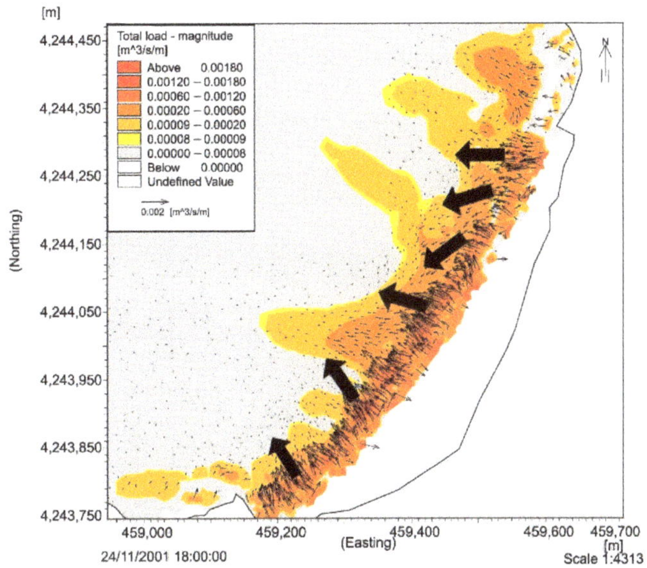

(b)

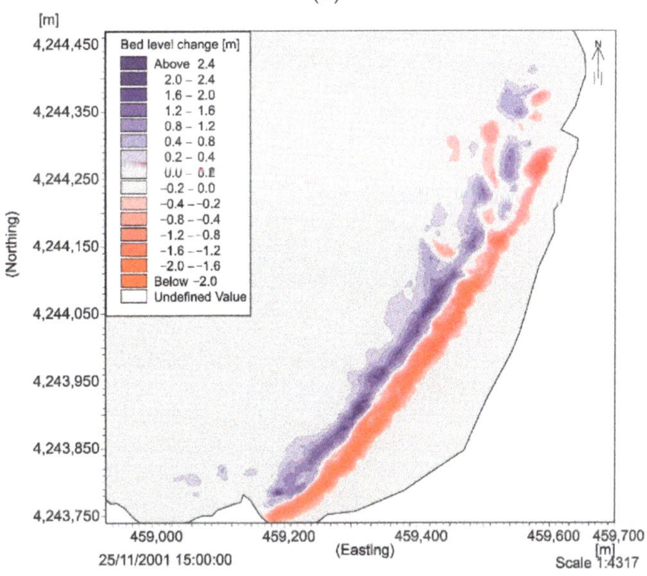

(c)

Figure 14. Model simulation for the maximum extreme event. (**a**) Map of coastal current velocity. (**b**) Stereo transport map of sediments. (**c**) Map of seabed-level change. The arrows show the pattern of behavior of the (**a**) current and the (**b**) coastal sediment transportation.

5. Discussion

The wider area of Myrtos is a part of a constantly evolving process, controlled by both the intense geodynamic neotectonic processes (onshore/offshore) and hydrometeorological phenomena. These processes form a geotechnically unstable area, affecting its morphology, such as bathymetry and topography, whereas landslides that are caused mainly at the northern part of the cliff constitute the main sediment supply to Myrtos Beach [28,29,47,50].

The topographic, sedimentological, and morphodynamic seasonal analysis of the coastal area during 10/2018 and 03/2019 and the morphological profile sections revealed the seasonal fluctuations between those two periods, with erosion patterns in the winter months by the more energetic waves, and sedimentation during the summer months [26,57–60]. In all sections and mainly in Beach Profile 3 (Figure 4), the sediment deposition during the summer period was recorded over the winter deposits, revealing a progression in sediment loss or gain along successive beach profiles. Additionally, beach berms were noticed along the cross sections at the beach face and were characterized by steep slopes and coarse sediments that followed the weather profile, becoming also a frequent practice to buffer coastal erosion. The gravelly sand to coarse gravel sediment composition of the beach is characterized by heterogeneous properties, and the sediments are located in high-energy environments [61], with the coarser sediments located landward and well sorted, and at the dynamic swash zone different sediment sizes are found which are poorly sorted [4,62,63].

The summer beach berms present a smoother topography, formed as a result of low energy swelling waves, whereas the winter beach berms are formed in large numbers with a limited extent due to storm events. The fluctuations between the topography and sedimentology indicates the importance of morphodynamic stages and, consequently, energy levels [64,65]. The different behavior of the beach profiles, mainly at the surf zone, is related to the cross-shore and longshore transportation perpendicular or at an angle to the shore, caused by the combined action of wind and waves and shore currents; the coarser material suggests its origin from landslides, which, in combination with the prevailing hydrodynamic conditions, play an important role in the sedimentation of the area [2,66]. The evaluation of the shoreline displacement at a seasonal scale (short-to-medium-term analysis) reveals that beach rotation (erosion at one end of the coast with deposition at the other end) is common and can result from seasonal fluctuations that affect waves [66–68]. The influence of landslides in the area can be identified through the evolution of the shoreline, as from 2003 to 2013 there was a tendency of retreat in most of the beach (Figure 15a); then, in 2014 two earthquakes occurred in the area, causing landslides and rock falls on Myrtos Beach [28,29,47,48,50]. This resulted in the supply of the beach with large volumes of material, causing shoreline progradation of about 15 m during the period 2013–2019 (Figure 15b). The advance rate varied between 1.4 m/year at both edges of the coast and 2.5 m/year in the central part of the beach. These rates are indicative for the period 2013–2019, as after each landslide the values change.

The numerical model indicates that for both annual representative scenarios and the extreme event, the prevailing morphodynamic and hydrodynamic conditions are identical. As the wind-induced waves are decisive for the hydrodynamic circulation of the pocket beach, wind direction and beach orientation, are two crucial parameters for the produced longshore currents.. The only difference is in the value range for each category (e.g., coastal current velocity, total load of sediment transported, seabed-level change) and the duration of the simulation, as depending on wave energy conditions, the hydrodynamic circulation can rapidly change. Scenario 2, although with low values, prevailed in the morphodynamic formation of the bay due to the long-term simulation (203 days), according to the annually maximum wind-induced wave energy flux from that direction (315°) [41,51–55] (Table 4). The hydrodynamic conditions formed by the specific scenarios create a typical circular surface current pattern developed in pocket beaches by two wave-induced longshore currents parallel to the coast in opposite directions (SW to NE for the southern part of the beach and NE to SW for the northern part) [12], making them the major contributor to sediment transport along the shoreface. Due to the morphology of the seabed and the dominance of the hard substrate in the northern area, the currents coming from the north prevail over the others, with the sediment transportation taking place mainly in the central part, but also in the deeper parts of the southern bay. This confirms that the geomorphological substrate controls the location and intensity of coastal currents. An anti-clockwise circulation pattern was also observed in the southern part of the area, enhancing the assumption of sedimentation at that certain area. In fact, the sediment transportation at the bay is determined by the combination of the current and wave direction. The currents that dominated at the coastal area are produced according to the wave-induced current. Therefore, the morphological changes of the beach (e.g., longshore bar, beach berm, erosion/deposition area) and the direction of those features demonstrate that the prevailing significant waves with high-energy flux, depending on their approaching direction, play a major role in sediment transportation along the coast [39,69]

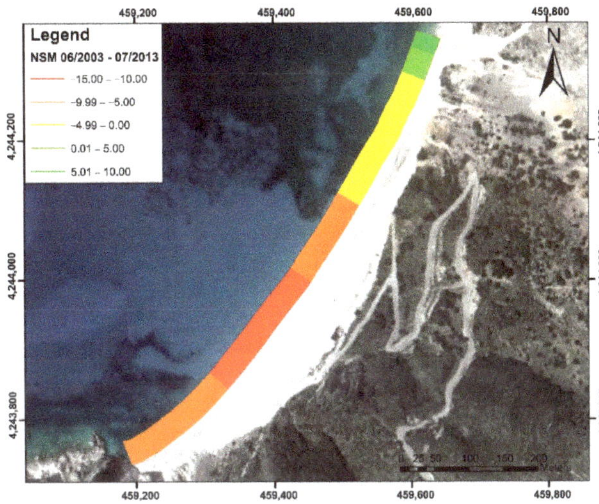

(a)

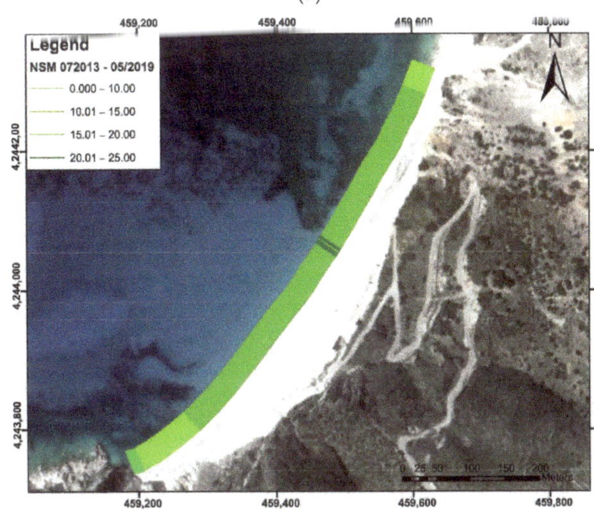

(b)

Figure 15. Net shoreline movement at Myrtos Beach. The figures display the size of coastline fluctuation for the periods (**a**) 2003–2013 and (**b**) 2013–2019. In 2014 two earthquakes occurred in the area, causing landslides and rock falls on Myrtos Beach [28,29,47,48,50], causing shoreline progradation.

Table 4. Potential sediment transport (m^3/year). The sediment transport direction is defined according to an observer on the sea, facing the coast. The negative values indicate a transport to the right and positive values indicate a transport to the left.

	Scenario 1	Scenario 2	Extreme Event
Section 1	−811.848	−2235.61	−638.776
Section 2	−558.566	−4184.97	−872.509
Section 3	1900.41	−1155.65	5680.17

According to Table 4 and and Figure 16a,b the current patterns from the simulations, the largest sediment volume is transported from the northern (Section 1) to the central and southern part of the coast (Sections 2 and 3); a large percentage of sediment is deposited near the coast down to a 10 m depth between Sections 2 and 3, whereas in extreme events it can reach deeper parts of the bay, down to a 34 m depth. In addition, the formation of longshore bars indicates sediment removal and transportation from the coast, causing sea bottom fluctuations mainly up to the break zone. In these areas, the breaking waves increase the accumulation of sandy sediments due to increased local bed shear stress and form longshore bars [70].

These observations suggest that beach morphodynamic, hydrodynamic, topographic, and sedimentological functions of a pocket beach (limited by headlands) are a part of the geological/geodynamic condition of the area (e.g., coast orientation, substrate mapping, coastal plain morphology, sediment source) and hydrodynamic factors (e.g., wind/wave condition, currents, sediment transportation, longshore bar). As the headlands act as a physical barrier to any renourishment from neighboring coasts, Myrtos Bay shows a certain ability to retain sediments, enhancing their movement between this area despite the high intensity wind-induced waves and currents, creating a local morphodynamic/hydrodynamic environment [1,3,10,12]. Essentially, the Myrtos pocket beach is considered to behave with its own limited sediment supply, recycling sediments between the nearshore, foreshore, and backshore. This work showed that the sediment fluxes carried mainly by longshore and rip currents until water depths of 10 m for the annual scenario and 33 m for the extreme wave event were a function of the direction and energy of wave-induced flux, of the substrate type, of the coast morphology, and of the sediment characteristics. The sedimentary material coming from the landward part is also notable; the sediment supply derives almost entirely from the landslide material from the neighboring cliffs and in relation to the prevailing hydrodynamic/topographic factors; they are deposited in the central and southern area of the bay. Landslides are characterized as episodic events that occur relatively often because the force of the earthquake, heavy rainfall (flash floods), human activity, and wave erosion disturb the natural stability of a slope (gravity moves), offering significant volumes of sediment to the Myrtos pocket beach during these events [18].

It is generally accepted that climate change has resulted in a further rise in global sea-levels [71–73] and, consequently, many coastal areas' susceptibility to floods has increased [74]. The coastal zone is among the most densely populated and fastest-growing areas on Earth, and these areas need to be viewed as interactive systems, including both human and physical components [75,76]. The sustainability of coastal environments depends on understanding these interactions. The pressures that intense human use brings are exacerbated by climate change, particularly the observed and anticipated sea-level rise, which increases coastal erosion and inundation, resulting in habitat loss and causing what has been called "coastal squeeze" as well as impacting local populations and tourism [77]. The overall rise in the sea-level is expected to directly affect coastal human activities and infrastructure, as well as cultural and geological heritage, agriculture, and biodiversity [78–80].

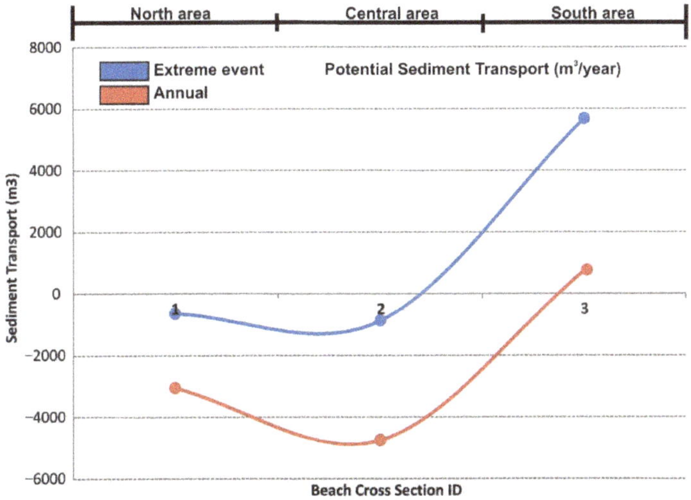

(a)

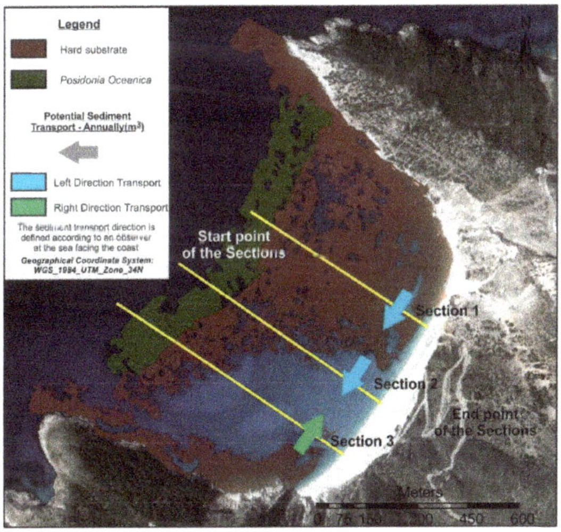

(b)

Figure 16. (**a**) An overall estimate (total) and volume of sediments that cross each section on an annual basis and in an extreme event, by dividing the study area in three sections. (**b**) A potential sediment direction that cross each section on an annual basis by dividing the study area in three sections.

In this sense, our findings between hydrodynamic and sediment circulation, geomorphology, and the geological context (embayed setting, landslides, rocky shoreface) may be crucial for better management by the local authorities, as any inappropriate method will degrade the quality of the environment and directly affect the environmental and economic sustainability of the wider area. Risk management policies focus on structural defense investments, such as reinforcing slope material, installation of structures such as piles and retaining walls, diversion of debris pathways, and rerouting of surface and underwater drainage. Such methods aim to protect the affected area from landslides, but they also significantly reduce the sedimentary material offered by landslides.

Any method of a "hard/soft technique" that is going to be used must be environmentally acceptable in order to maintain a dynamic balance of the natural system of the coastal area [81–83]. The investment in structural defense should be focused on soil management in relation to landslide prevention, combining this with the economic and tourist development of the wider area with regard to security and sustainability.

6. Conclusions

The coastal morphodynamics of the Myrtos pocket beach was investigated through in situ measurements and statistical and numerical models, with the aim to evaluate its short and medium-to-long-term dynamics. Our study revealed a dynamic condition with a high-energy environment and results showing an environment under deposition. Specifically, in a short-term analysis, the beach profiles allowed the characterization of a high-energy environment with coarse sediments and with seasonal fluctuations in the topography and the granulometry, responding to the different weather conditions prevailing in the area for each period. Through the satellite images, shoreline progradation was noted with an advance rate close to 2 m/year for the period 2013–2019 and after two landslides that occurred from the 2014 earthquakes. The extensive field work, laboratory analysis, and hydrographical and sediment transport model simulations presented in this work confirmed sediment transportation and deposition that is mainly at the coastal region from the northern to the central and southern part of the bay, developing the expected circular pattern that characterizes a typical pocket beach. In an extreme event, this transportation and sedimentation can take place in deeper areas. However, all the sediment dynamics take place between the two headlands that isolate the bay, whereas the sediment sources are exclusively material deriving from landslides.

Our findings may be used for practical approaches to achieve a better environmental and economic management of the coastal area seeking dynamic equilibrium by creating a static bay pocket beach that can remain stable without requiring external contribution. Any activity that is to be implemented, such as a defense structure to stabilize landslides, should not have any negative consequences for the development of the beach, as any change to beach sediment budget, sediment characteristics, morphology, and hydrodynamics is critical for long-term changes and the preservation of pocket beaches for future human use considering climate change.

Supplementary Materials: The following supporting information can be downloaded at: https://www.mdpi.com/article/10.3390/jmse10081015/s1, Figure S1: Lithological and geomorphological map of the study area. Geology based on geological map of Greece (1:50,000), Cephalonia Island (northern part), Institute of Geology and Mineral Exploration; Figure S2: (a) Wind and (b) wave rose diagram for the study area from1995 to 2004; Figure S3. Field survey that was carried out during 10/2018 and 05/2019; Figure S4: The direction of Myrtos Bay ranges between 265° and 343°, whereas the orientation of the coastline is 304°; the two representative scenarios that contribute to the sediment transport from one side to the other and in the opposite direction are scenario 1: 264–304° and scenario 2: 305–343°.

Author Contributions: Conceptualization, A.P., V.K. and N.E.; methodology, V.K., A.P.; software, A.P.; validation, V.K., A.P.; formal analysis, V.K., N.E., A.K. and A.P.; investigation, A.P. and K.G.; data curation, A.P. and K.G.; writing—original draft preparation, A.P.; writing—review and editing, V.K., A.K., N.E. and A.P.; visualization, A.K. and A.P.; supervision, V.K., A.K., N.E. and A.P. All authors have read and agreed to the published version of the manuscript.

Funding: Alexandros Petropoulos have been awarded with a scholarship for their PhD research from the "1st Call for PhD Scholarships by HFRI"–"Grant Codes 1812", from the General Secretariat for Research and Technology (G.S.R.T.) and the Hellenic Foundation for Research and Innovation (H.F.R.I.).

Acknowledgments: Special thanks to the Hellenic Centre for Marine Research (HCMR) for the supply of scientific equipment.

Conflicts of Interest: The authors declare no conflict of interest.

References

1. Short, A.D.; Masselink, G. Embayed and structurally controlled beaches. In *Handbook of Beach and Shoreface Morphodynamics*; Short, A.D., Ed.; John Wiley and Sons Ltd.: Chichester, UK, 1999; pp. 230–249.
2. Bowman, D.; Guillén, J.; López, L.; Pellegrino, V. Planview Geometry and morphological characteristics of pocket beaches on the Catalan coast (Spain). *Geomorphology* **2009**, *108*, 191–199. [CrossRef]
3. Dehouck, A.; Dupuis, H.; Sénéchal, N. Pocket beach hydrodynamics: The example of four macrotidal beaches, Brittany, France. *Mar. Geol.* **2009**, *266*, 1–17. [CrossRef]
4. Vousdoukas, M.I.; Velegrakis, A.F.; Dimou, K.; Zervakis, V.; Conley, D.C. Wave run-up observations in microtidal, sediment-starved pocket beaches of the Eastern Mediterranean. *J. Mar. Syst.* **2009**, *78*, S37–S47. [CrossRef]
5. Storlazzi, C.D.; Field, M.E. Sediment distribution and transport along a rocky, embayed coast: Monterey Peninsula and Carmel Bay, California. *Mar. Geol.* **2000**, *170*, 289–316. [CrossRef]
6. Short, A.D. *Handbook of Beach and Shoreface Morphodynamics*; John Wiley & Sons: New York, NY, USA, 1999; ISBN 0-471-96570-7.
7. Klein, A.H.F.; Ferreira, Ó.; Dias, J.M.A.; Tessler, M.G.; Silveira, L.F.; Benedet, L.; de Menezes, J.T.; de Abreu, J.G.N. Morphodynamics of structurally controlled headland-bay beaches in southeastern Brazil: A review. *Coast. Eng.* **2010**, *57*, 98–111. [CrossRef]
8. Hsu, J.R.C.; Yu, M.J.; Lee, F.C.; Benedet, L. Static bay beach concept for scientists and engineers: A review. *Coast. Eng.* **2010**, *57*, 76–91. [CrossRef]
9. Daly, C.J.; Bryan, K.R.; Gonzalez, M.R.; Klein, A.H.F.; Winter, C. Effect of selection and sequencing of representative wave conditions on process-based predictions of equilibrium embayed beach morphology. *Ocean Dyn.* **2014**, *64*, 863–877. [CrossRef]
10. Oliveira, S.; Moura, D.; Horta, J.; Nascimento, A.; Gomes, A.; Veiga-Pires, C. The morphosedimentary behaviour of a headland–beach system: Quantifying sediment transport using fluorescent tracers. *Mar. Geol.* **2017**, *388*, 62–73. [CrossRef]
11. Short, A.D.; Trembanis, A.C.; Turner, I.L. Beach oscillation, rotation and the Southern Oscillation, Narrabeen Beach, Australia. In Proceedings of the Coastal Engineering 2000—27th International Conference on Coastal Engineering, Sydney, Australia, 16–21 July 2000; pp. 2439–2452. [CrossRef]
12. Castelle, B.; Coco, G. The morphodynamics of rip channels on embayed beaches. *Cont. Shelf Res.* **2012**, *43*, 10–23. [CrossRef]
13. Goda, Y. Reanalysis of regular and random breaking wave statistics. *Coast. Eng.* **2010**, *52*, 71–106. [CrossRef]
14. Balouin, Y.; Belon, R. Evolution of Corsican pocket beaches. In Proceedings of the 13th International Coastal Symposium, Durban, South Africa, 13 April 2014; Green, A.N., Cooper, J.A.G., Eds.; pp. 96–101.
15. Brunel, C.; Sabatier, F. Potential influence of sea-level rise in controlling shoreline position on the French Mediterranean Coast. *Geomorphology* **2009**, *107*, 47–57. [CrossRef]
16. Horta, J.; Oliveira, S.; Moura, D.; Ferreira, Ó. Nearshore hydrodynamics at pocket beaches with contrasting wave exposure in southern Portugal. *Estuar. Coast. Shelf Sci.* **2018**, *204*, 40–55. [CrossRef]
17. Norcross, Z.M.; Fletcher, C.H.; Merrifield, M. Annual and interannual changes on a reef-fringed pocket beach: Kailua Bay, Hawaii. *Mar. Geol.* **2002**, *190*, 553–580. [CrossRef]
18. Sallenger, A.H.; Krabill, W.; Brock, J.; Swift, R.; Manizade, S.; Stockdon, H. Sea-cliff erosion as a function of beach changes and extreme wave runup during the 1997-1998 El Niño. *Mar. Geol.* **2002**, *187*, 279–297. [CrossRef]
19. Anthony, E.J.; Gardel, A.; Dolique, F.; Guiral, D. Short-term changes in the plan shape of a sandy beach in response to sheltering by a nearshore mud bank, Cayenne, French Guiana. *Earth Surf. Process. Landf.* **2002**, *27*, 857–866. [CrossRef]
20. Dalrymple, R.A.; MacMahan, J.H.; Reniers, A.J.H.M.; Nelko, V. Rip currents. *Annu. Rev. Fluid Mech.* **2011**, *43*, 551–581. [CrossRef]
21. Van Dongeren, A.; Lowe, R.; Pomeroy, A.; Trang, D.M.; Roelvink, D.; Symonds, G.; Ranasinghe, R. Numerical modeling of low-frequency wave dynamics over a fringing coral reef. *Coast. Eng.* **2013**, *73*, 178–190. [CrossRef]
22. Billi, P.; Fazzini, M. Global change and river flow in Italy. *GPC* **2017**, *155*, 234–246. [CrossRef]
23. Hsu, J.R.-C.; Benedet, L.; Klein, A.H.F.; Raabe, A.L.A.; Tsai, C.P.; Hsu, T.W. Appreciation of static bay beach concept for coastal management and protection. *J. Coast. Res.* **2008**, *24*, 198–215. [CrossRef]
24. Dean, R.G. Equilibrium beach profiles: Characteristics and applications. *J. Coast. Res.* **1991**, *7*, 53–84.
25. Karunarathna, H.; Horrillo-Caraballo, J.M.; Reeve, D.E. Prediction of cross-shore beach profile evolution using a diffusion type model. *Cont. Shelf Res.* **2012**, *48*, 157–166. [CrossRef]
26. Anfuso, G.; Pranzini, E.; Vitale, G. An integrated approach to coastal erosion problems in northern Tuscany (Italy): Littoral morphological evolution and cell distribution. *Geomorphology* **2011**, *129*, 204–214. [CrossRef]
27. Papanikolaou, D. *Geology of Greece*, 1st ed.; Pataki: Athens, Greece, 2015; ISBN 978-960-16-6343-2.
28. Lekkas, E.; Mavroulis, S. Fault zones ruptured during the early 2014 Cephalonia Island (Ionian Sea, Western Greece) earthquakes (January 26 and February 3, Mw 6.0) based on the associated co-seismic surface ruptures. *J. Seismol.* **2016**, *20*, 63–78. [CrossRef]
29. Lekkas, E.L.; Mavroulis, S.D. Earthquake environmental effects and ESI 2007 seismic intensities of the early 2014 Cephalonia (Ionian Sea, western Greece) earthquakes (January 26 and February 3, Mw 6.0). *Nat. Hazards* **2015**, *78*, 1517–1544. [CrossRef]
30. Saroglou, H. Rockfall susceptibility in Greece and study of coseismic events. In Proceedings of the 15th Symposium on Engineering Geology & Geotechnics, Belgrade, Serbia, 22–23 September 2016; pp. 19–35.
31. Soukisian, T.; Hatzinaki, M.; Korres, G.; Papadopoulos, A.; Kallos, G.; Anadranistakis, E. *Wind and Wave Atlas of the Hellenic Seas*, 1st ed.; Hellenic Centre for Marine Research: Athens, Greece, 2007; ISBN 987 960 86651 9-4.

32. Haiden, T.; Janousek, M.; Bauer, P.; Bidlot, J.-R.; Dahoui, M.; Ferranti, L.; Prates, F.; Richardson, D.; Vitart, F. *Evaluation of ECMWF Forecasts, including 2014-2015 Upgrades*; ECMWF: Reading, UK, 2015; 7p.
33. *Deltares Product User Guide for Sea Level and Ocean Wave Products-Time Series and Indicators C3S_422_Lot2_Deltares-European Services*; ECMWF: Reading, UK, 2019.
34. Sifnioti, D.; Soukissian, T.; Poulos, S.; Nastos, P.; Hatzaki, M. Evaluation of in-situ wind speed and wave height measurements against reanalysis data for the Greek Seas. *Mediterr. Mar. Sci.* **2017**, *18*, 486–503. [CrossRef]
35. Folk, R.L.; Ward, W.C. Brazos River bar [Texas]; a study in the significance of grain size parameters. *J. Sediment. Res.* **1957**, *27*, 3–26. [CrossRef]
36. Blott, S.J.; Pye, K. GRADISTAT: A grain size distribution and statistics package for the analysis of unconsolidated sediments. *Earth Surf. Processes Landf.* **2001**, *26*, 1237–1248. [CrossRef]
37. Thieler, E.R.; Himmelstoss, E.A.; Zichichi, J.L.; Ergul, A. *The Digital Shoreline Analysis System (DSAS) Version 4.0—An ArcGIS Extension for Calculating Shoreline Change*; U.S. Geological Survey: Reston, VA, USA, 2009.
38. Anfuso, G.; Bowman, D.; Danese, C.; Pranzini, E. Transect based analysis versus area based analysis to quantify shoreline displacement: Spatial resolution issues. *Environ. Monit. Assess.* **2016**, *188*, 568. [CrossRef]
39. Sedrati, M.; Anthony, E.J. A brief overview of plan-shape disequilibrium in embayed beaches: Tangier bay (Morocco). *J. Mediterr. Geogr.* **2007**, *108*, 125–130. [CrossRef]
40. Balas, L.; Inan, A. A Numerical Model of Wave Propagation on Mild Slopes. *J. Coast. Res.* **2002**, *36*, 16–21. [CrossRef]
41. Petropoulos, A.; Evelpidou, N.; Kapsimalis, V.; Anagnostou, C.; Karkani, A. Sediment transport patterns and beach morphodynamics in the semi-enclosed bay of Platis Gialos, Sifnos Island, Aegean Sea. *Z. Geomorphol.* 2021; *in press*. [CrossRef]
42. MIKE powered by DHI. *MIKE 21 Wave Modelling, MIKE 21 Spectral Waves FM*; DHI: Horsholm, Denmark, 2017.
43. Bulhoes, E.; Fernandez, G.B. Analysis of shallow water wave propagation and coastal response in embayed beaches. Case study in Cape Buzios, Rio de Janeiro, Brazil. *J. Coast. Res.* **2011**, *SI64*, 2022–2026.
44. Jose, F.; Kobashi, D.; Stone, G.W. Spectral wave transformation over an elongated sand shoal off south-central Louisiana, USA. *J. Coast. Res.* **2007**, *SI50*, 757–761.
45. Hallermeier, R.J. A profile zonation for seasonal sand beaches from wave climate. *Coast. Eng.* **1980**, *4*, 253–277. [CrossRef]
46. Zou, Q.; Xie, D. Tide-surge and wave interaction in the Gulf of Maine during an extratropical storm. *Ocean. Dyn.* **2016**, *66*, 1715–1732. [CrossRef]
47. Valkaniotis, S.; Ganas, A.; Papathanassiou, G.; Papanikolaou, M. Field observations of geological effects triggered by the January–February 2014 Cephalonia (Ionian Sea, Greece) earthquakes. *Tectonophysics* **2014**, *630*, 150–157. [CrossRef]
48. Papathanassiou, G.; Valkaniotis, S.; Ganas, A. Evaluation of the macroseismic intensities triggered by the January/February 2014 Cephalonia, (Greece) earthquakes based on ESI-07 scale and their comparison to 1867 historical event. *Quat. Int.* **2017**, *451*, 234–247. [CrossRef]
49. Saroglou, H.; Asteriou, P.; Tsiambaos, G.; Zekkos, D.; Clark, M.K.; Manousakis, J. Investigation of Two Co-Seismic Rockfalls During the 2015 Lefkada and 2014 Cephalonia Earthquakes in Greece. In Proceedings of the 3rd North American Symposium on Landslides, Roanoke, VA, USA, 4–8 June 2017; pp. 521–528.
50. Theodoulidis, N.; Karakostas, C.; Lekidis, V.; Makra, K.; Margaris, B.; Morfidis, K.; Papaioannou, C.; Rovithis, E.; Salonikios, T.; Savvaidis, A. The Cephalonia, Greece, January 26 (M6.1) and February 3, 2014 (M6.0) earthquakes: Near-fault ground motion and effects on soil and structures. *Bull. Earthq. Eng.* **2016**, *14*, 1–38. [CrossRef]
51. Coastal Engineering Research Center (CERC). *Shore Protection Manual*; US Army Corps of Engineers, Ed.; Scientific Research Publishing: Washington, DC, USA, 1984; Volume I-I.
52. Borah, D.K.; Balloffet, A. Beach Evolution Caused by Littoral Drift Barrier. *J. Waterw. Port Coast. Ocean. Eng.* **1985**, *111*, 645–660. [CrossRef]
53. *IEC TS 62600-101*; Marine Energy—Wave, Tidal and Other Water Current Converters—Part 101: Wave Energy Resource Assessment and Characterization. 1st ed. International Electrotechnical Commission (IEC): Geneva, Switzerland, 2015; ISBN 978-2-8322-2724-4.
54. Tucker, M.J.; Malcolm, J.; Pitt, E.G. *Waves in Ocean Engineering*, 1st ed.; Tucker, M.J., Malcolm, J., Pitt, E.G., Eds.; Elsevier Science & Technology: Oxford, UK, 2001; ISBN 9780080435664.
55. Kapsimalis, V.; Gad, F.K. Research, Study of the Coastal Zone and the Marine Area of the Gulf of Chania. Coastal Engineering Investigation of the Coastal Zones of Kolimbari-Tavronitis & Platania-Kato Stalou. In *Study of Erosion and Protection of the Coasts of the Gulf of Chania*; H.C.M.R.: Athens, Greek, 2016.
56. Gad, F.K.; Chatzinaki, M.; Vandarakis, D.; Kyriakidou, C.; Kapsimalis, V. Assessment of wave storm-induced flood vulnerability in rhodes island, greece. *Water* **2020**, *12*, 2978. [CrossRef]
57. Cooper, N.J.; Barber, P.C.; Bray, M.J.; Carter, D.J. Shoreline management plans: A national review and engineering perspective. In *Proceedings of the Institution of Civil Engineers—Water and Maritime Engineering*; Thomas Telford Ltd.: London, UK, 2002; Volume 154, pp. 221–228. [CrossRef]
58. Carter, R.W.G. *Coastal Environments: An Introduction to the Physical, Ecological and Cultural Systems of Coastlines*; Academic Press Inc.: Cambridge, MA, USA, 1988; ISBN 0121618552.
59. Taylor, J.A.; Murdock, A.P.; Pontee, N.I. A macroscale analysis of coastal steepening around the coast of England and Wales. *Geogr. J.* **2004**, *170*, 179–188. [CrossRef]

60. Dornbusch, U.; Williams, R.B.G.; Moses, C.A.; Robinson, D.A. Foreshore narrowing along the coast of southeast England, UK—A reevaluation. *J. Coast. Res.* **2008**, *24*, 14–24. [CrossRef]
61. Orford, J.; Anthony, E. 10.9 Coastal Gravel Systems. In *Treatise on Geomorphology*; Elsevier: Amsterdam, The Netherlands, 2013; pp. 245–266.
62. Simeone, S.; Palombo, L.; De Falco, G. Morphodynamics of a nontidal embayed beach: The case study of is arutas (Western Mediterranean). *J. Coast. Res.* **2013**, *29*, 63–71. [CrossRef]
63. Alexandrakis, G.; Ghionis, G.; Poulos, S. The Effect of Beach Rock Formation on the Morphological Evolution of a Beach. The Case Study of an Eastern Mediterranean Beach: Ammoudara, Greece. *J. Coast. Res.* **2013**, *69*, 47–59. [CrossRef]
64. Bascom, W.N. The relationship between sand size and beach-face slope. *Trans. Am. Geophys. Union* **1951**, *32*, 866. [CrossRef]
65. Wright, L.D.; Short, A.D. Morphodynamic variability of surf zones and beaches: A synthesis. *Mar. Geol.* **1984**, *56*, 93–118. [CrossRef]
66. Ranasinghe, R.; McLoughlin, R.; Short, A.; Symonds, G. The Southern Oscillation Index, wave climate, and beach rotation. *Mar. Geol.* **2004**, *204*, 273–287. [CrossRef]
67. Thomas, T.; Phillips, M.R.; Williams, A.T.; Jenkins, R.E. Medium time-scale behaviour of adjacent embayed beaches: Influence of low energy external forcing. *Appl. Geogr.* **2012**, *32*, 265–280. [CrossRef]
68. Short, A.D.; Trembanis, A.C. Decadal scale patterns in beach oscillation and rotation Narrabeen beach, Australia—Time series, PCA and wavelet analysis. *J. Coast. Res.* **2004**, *20*, 523–532. [CrossRef]
69. Kuang, C.; Han, X.; Zhang, J.; Zou, Q.; Dong, B. Morphodynamic Evolution of a Nourished Beach with Artificial Sandbars: Field Observations and Numerical Modeling. *J. Mar. Sci. Eng.* **2021**, *9*, 245. [CrossRef]
70. Davidson-Arnott, R. 10.6 Nearshore Bars. In *Treatise on Geomorphology*; Shroder, J., Sherman, D.J., Eds.; Academic Press: San Diego, CA, USA, 2013; Volume 10, pp. 130–148. ISBN 9780080885223.
71. Oppenheimer, M.; Glavovic, B.C.; Hinkel, J.; van de Wal, R.; Magnan, A.K.; Abd-Elgawad, A.; Cai, R.; Cifuentes-Jara, M.; DeConto, R.M.; Ghosh, T.; et al. *Sea Level Rise and Implications for Low-Lying Islands, Coasts and Communities*; Pörtner, H.-O., Roberts, D.C., Masson-Delmotte, V., Zhai, P., Tignor, M., Poloczanska, E., Mintenbeck, K., Alegría, A., Nicolai, M., Okem, A., et al., Eds.; Cambridge University Press: Cambridge, UK; New York, NY, USA; Geneva, Switzerland, 2019; pp. 321–446. [CrossRef]
72. Evelpidou, N.; Petropoulos, A.; Karkani, A.; Saitis, G. Evidence of Coastal Changes in the West Coast of Naxos Island, Cyclades, Greece. *J. Mar. Sci. Eng.* **2021**, *9*, 1427. [CrossRef]
73. Karkani, A.; Evelpidou, N.; Giaime, M.; Marriner, N.; Morhange, C.; Spada, G. Late Holocene sea-level evolution of Paros Island (Cyclades, Greece). *Quat. Int.* **2019**, *500*, 139–146. [CrossRef]
74. Defeo, O.; McLachlan, A.; Schoeman, D.S.; Schlacher, T.A.; Dugan, J.; Jones, A.; Lastra, M.; Scapini, F. Threats to sandy beach ecosystems: A review. *Estuar. Coast. Shelf Sci.* **2009**, *81*, 1–12. [CrossRef]
75. Smith, K. We are seven billion. *Nat. Clim. Change* **2011**, *1*, 331–335. [CrossRef]
76. Brown, S.; Nicholls, R.J.; Goodwin, P.; Haigh, I.D.; Lincke, D.; Vafeidis, A.T.; Hinkel, J.; Brown, S.; Nicholls, R.J.; Goodwin, P.; et al. Quantifying Land and People Exposed to Sea-Level Rise with No Mitigation and 1.5 °C and 2.0 °C Rise in Global Temperatures to Year 2300. *EaFut* **2018**, *6*, 583–600. [CrossRef]
77. Orford, J.D.; Pethick, J. Challenging assumptions of future coastal habitat development around the UK. *Earth Surf. Processes Landf.* **2006**, *31*, 1625–1642. [CrossRef]
78. WHO. *COP24 Special Report: Health and Climate Change*; World Health Organization: Geneva, Switzerland, 2018.
79. IPCC. *Global Warming of 1.5 °C The Context Of Strengthening The Global Response To The Threat Of Climate Change, Sustainable Development, and Efforts to Eradicate Poverty*; IPCC: Geneva, Switzerland, 2018.
80. Norwegian Red Cross (NRC). *Overlapping Vulnerabilities: The Impacts of Climate Change on Humanitarian Needs*; Norwegian Red Cross: Oslo, Norway, 2019.
81. Chapin, F.S.; Carpenter, S.R.; Kofinas, G.P.; Folke, C.; Abel, N.; Clark, W.C.; Olsson, P.; Smith, D.M.S.; Walker, B.; Young, O.R.; et al. Ecosystem stewardship: Sustainability strategies for a rapidly changing planet. *Trends Ecol. Evol.* **2010**, *25*, 241–249. [CrossRef]
82. Cantasano, N.; Pellicone, G.; Ietto, F. The Coastal Sustainability Standard method: A case study in Calabria (Southern Italy). *Ocean. Coast. Manag.* **2020**, *183*, 104962. [CrossRef]
83. Del-Rosal-Salido, J.; Folgueras, P.; Bermúdez, M.; Ortega-Sánchez, M.; Losada, M. Flood management challenges in transitional environments: Assessing the effects of sea-level rise on compound flooding in the 21st century. *Coast. Eng.* **2021**, *167*, 103872. [CrossRef]

Journal of
Marine Science and Engineering

MDPI

Article

Prediction of Overpressure Zones in Marine Sediments Using Rock-Physics and Other Approaches

Nikita Dubinya [1,2,*], Irina Bayuk [1,*], Alexei Hortov [3], Konstantin Myatchin [3], Anastasia Pirogova [1,4] and Pavel Shchuplov [5]

1 Laboratory of Fundamental Problems of Petroleum Geophysics and Geophysical Monitoring, Schmidt Institute of Physics of the Earth of the Russian Academy of Sciences, 123242 Moscow, Russia
2 Science and Technology Center of Geophysics and Mineral Resources, Moscow Institute of Physics and Technology (National Research University), 141701 Dolgoprudny, Russia
3 NOVYE PROEKTY Limited Liability Company (NOVYE PROEKTY LLC)., 119034 Moscow, Russia
4 Laboratory of Seismoacoustic Exploration Methods, Faculty of Geology, Lomonosov Moscow State University, 119234 Moscow, Russia
5 Seismic Data Analysis Center Limited Liability Company, Lomonosov Moscow State University, 119992 Moscow, Russia
* Correspondence: dubinia.nv@mipt.ru (N.D.); ibayuk@ifz.ru (I.B.)

Abstract: The paper discusses the problem of localizing zones of high pore pressure in sub-bottom sediments (first tens of meters under the seafloor). Prediction of the overpressure zones in the near-surface is required for the mitigation of risks at the early stages of the offshore hydrocarbon field exploration and development. The results of seismic data interpretation generally serve as the main source of information for this kind of problems, yet there are other methods to predict overpressure zones in the subsurface. The paper presents the results of the overpressure zone prediction using a set of methods including empirical ones, and the approach based on rock-physics modeling that features the soft-sand model of unconsolidated media effective properties. While the rock-physics modeling grants the most reliable result, it is also the most demanding method to the input data. Hence, it can be used to verify other methods of the overpressure zone prediction. We present the results of the overpressure zone prediction at the research site on the Black Sea shelf. The mitigation of the drilling risks via changing the drilling conditions is discussed in detail. As the drilling through the overpressure zones is often a necessity, the engineering solutions proposed in the paper can be applied elsewhere when facing similar problems typical for offshore exploration.

Keywords: overpressure zones; pore pressure prediction; offshore hydrocarbon fields; Black Sea shelf

Citation: Dubinya, N.; Bayuk, I.; Hortov, A.; Myatchin, K.; Pirogova, A.; Shchuplov, P. Prediction of Overpressure Zones in Marine Sediments Using Rock-Physics and Other Approaches. *J. Mar. Sci. Eng.* 2022, *10*, 1127. https://doi.org/10.3390/jmse10081127

Academic Editors: George Kontakiotis, Assimina Antonarakou and Dmitry A. Ruban

Received: 15 July 2022
Accepted: 14 August 2022
Published: 17 August 2022

1. Introduction

Pore pressure prediction is an important step in hydrocarbon reservoirs exploration as it is often associated with several geological risks [1]. These risks include, but are not limited to deepwater kicks [2], well blowouts [3], activation of faults [4], and mud volcanoes [5]. There are risks of geohazards caused by unpredicted overpressure zones during hydrate-bearing sediments exploitation [6], as pressure affects the conditions of the hydrate stability zone [7] and potential recovery [8]. Realization of such risks is not extraordinary: according to [9], roughly 10% of the drilling time in deepwater of the Gulf of Mexico was associated with incidents related to overpressure zones. Failure to reliably predict overpressure zones—zones with fluid pressure exceeding hydrostatic pressure—in subsea sediments can lead to a number of results considerably increasing costs of offshore hydrocarbon field exploration and lead to induced geologic hazards and disasters, highlighting the importance of the usage of various blowout preventer systems [10].

There is a general understanding of overpressure zones' origins. As formulated in [7], continuous sedimentation causes an increase in the overburden, resulting in consolidation of sediments over geological times. Overpressure can develop in such systems if the saturating fluid cannot be expelled from the pore space fast enough and, instead, supports some of the overburden. As permeability controls this rate of pore water expulsion, sediments with low permeability can develop overpressure in this way. Alternatively, overpressure can also develop in sediments with relatively high permeability if the sedimentation rate is fast, i.e., increase in the overburden is faster than rate of pore water expulsion—the latter case is closer to the focus of the current paper.

There are many studies presenting methods for pore pressure prediction using geophysical data of varying spatial and temporal scale following the pioneering paper [11], where acoustic travel time, velocity, and resistivity data were used to estimate pore pressure of fluid-saturated rocks. Well-known and extensively applied approaches to evaluate the pore pressure were proposed in papers [12,13], where electrical resistivity [12], compressional transit time [13], and overburden stress [14] were used as the basis for overpressure zone prediction. A number of specific correlations for particular fields have been proposed by various authors, their comprehensive overviews have been provided by [15,16]. In this brief overview, we would like to mention several recent trends in the pore pressure prediction based primarily on seismic data obtained at offshore sites, as this is the focus of the current study.

The Gulf of Mexico, as already mentioned above, is an important site from the perspective of pore pressure prediction as several studies proposing novel approaches deal with this particular area known for a number of issues [9]. While Eaton's [12,13] and Bowers' [14] are among the most widely used methods of pore pressure prediction on shore [16], offshore reservoirs, especially the upper layers of seafloor sediments, are unconsolidated fluid-saturated masses, frequently, in the state of under-compaction. As a result, correlations developed for different reservoirs can be inapplicable for pore pressure characterization at the upper layers of seafloor sediments. Alternative approaches have been developed based on analysis of compaction trends for unconsolidated sediments [17]: the relationship between pore pressure of fluid-saturated sediments, their porosity, and effective stress has been established based on experimental data. This relationship, however, demands the presence of direct measurements of the pore pressure at certain depths to provide a reliable evaluation of overpressure [18]. Although methods of direct estimation, e.g., borehole penetrometers, are being developed [19] and successfully applied at various sites [20,21], direct pore pressure measurements require long equilibration times [18]. Whenever direct estimations of the pore pressure at a particular depth are known, the porosity trend can be used to reliably predict pore pressure and overpressure zones even in unconsolidated sediments [22,23] with an emphasis on zones near the seafloor—up to several dozens of meters from the seafloor [18,24]. This method, however, is limited to unconsolidated zones, so greater depths require application of other methods, primarily based on well logging data [25].

While well logging data remain an important source of information for estimation of pore pressure and prediction of overpressure zones, early stages of offshore fields exploration are characterized by lack of these data. At the same time, the need to estimate pore pressure emerges at pre-drilling stages [26]. Logging-while-drilling (LWD) data can be used as well for estimation of pore pressure [27,28] before drilling the main exploratory wells.

While porosity, resistivity, and acoustic travel time are generally used to predict pore pressure from well logs, there are also other sources of information worth mentioning, especially, in the context of the following sections. The results obtained at offshore site located in the Eastern India [29] suggest the presence of interrelationships between pore pressure and inner structure of the sediments, including grain size distribution. These interrelations give a chance to believe that rock physics approaches can be helpful in pore pressure prediction. The majority of rock physics methods gives a possibility to relate the effective (measured) physical properties of rocks with the parameters describing the physical properties of components, their volume concentration, shape, and mutual distribution in the rock volume. Among the rock physics approaches, methods exist that give a possibility to incorporate the effect of pore pressure on the macroscopic elastic properties of granular media in an explicit form [30–33]. These methods can be applied for granular media where the pore pressure plays a dominant role in the effective elastic properties [34]. Thus, in the works [35,36], the authors proposed a heuristic combination of the Hertz–Mindlin method [30] and the Hashin–Shtrikman bounds [37] in order to consider two different types of granular rocks—with and without cement between the grains—the so-called stiff-sand and soft-sand models. The authors noted that instead of the Mindlin–Hertz method any method suitable for granular medium can be applied. However, in practice, the former is the most popular.

To sum up, there are a variety of methods for overpressure zone prediction based on geophysical data. A number of correlations have been proven to be reliable for different sites—and exploitation of a new site is complicated due to the lack of information on which correlation should be used to obtain reasonable estimations of pore pressure. Moreover, it is generally necessary to have direct or indirect measurements of the pore pressure at certain depths to find the proper correlation coefficients for the current site. Moreover, different methods of the pore pressure estimation can be the most suitable at different zones even at the same site: compaction trend-based methods provide reliable results for unconsolidated sediments near the seafloor, while widely used estimations based on resistivity and acoustic travel time are more suitable for the greater depths. On the other hand, the rock-physics modeling provides another perspective on pore pressure estimation—while it generally requires more data for reliable estimations, the corresponding workflow remains independent on the specific site. Comparison of usage the same data to predict pore pressure using these two approaches seems of both scientific and practical interest: for example, it remains unknown at the current stage of pore pressure estimation methods, which of them are more, and which of them are less conservative. This comparison is among the theoretical goals of the current study—it is carried out at a certain site with available data, making the application of both methods possible.

In the current study, we aim at estimating the pore pressure at a certain site, located in the eastern Black Sea, in a zone with expected drilling for offshore hydrocarbon reservoir development. Seismic data act as the main source of information for the pore pressure estimation—only a limited number of engineering wells have been drilled at the site, but drilling results provided both an understanding of the need to reliably predict pore pressure due to a risk of well blowout, and some insight on the potential properties of overpressure zones. The pore pressure prediction procedure was completed using various methods: as far as a certain indirect estimation of pore pressure at a particular depth was performed during drilling of engineering wells, the compaction trend [18,24] could be successfully applied to locate overpressure zones. Rock-physics modeling was carried out as well to check the agreement between estimations obtained from different methods. Simultaneous usage of different methods of pore pressure estimation provided an opportunity to get an extra verification of predictions and adjust the following strategy of exploration of the field.

2. Overview

2.1. Geological Setting

The study area is located in the northwestern part of the Black Sea within the Kerch-Taman shelf. In the west, it is limited by the meridian of the southern tip of the Crimea; in the northwest by the Crimean, and in the east—by the Caucasian coast of the Krasnodar Territory.

Tectonically, a significant part of the area under consideration is located within the Kerch-Taman trough, which includes the Taman Peninsula, the southeastern part of the Kerch Peninsula and the adjacent Kerch-Taman Shelf of the Black Sea. The southern part of the area covers the outer zone of the shelf and adjacent areas of the slope of the Black Sea basin. There are several ideas about the tectonic nature of the trough noted above. It is classified either as a transverse or interpericlinal trough. A number of geologists tend to consider it as a southwestern branch from the Indolo-Kuban trough [38] located to the north.

It is assumed that the southeastern and southern limits of the described trough are the folded formations of the Northwestern Caucasus and the Anapa ledge in the sea, composed of Mesozoic and Paleogene deposits. In the west, in the area of the city of Feodosiya, the Kerch-Taman trough is limited by shallow Mesozoic formations and, in general, the eastern closure of the Crimean mountain structure, although the features of its eastern limit cause discrepancies. If we assume that the Crimean mountain structure is a block uplifted on one side in the Cenozoic, the eastern limit of which can be a system of transverse strike-slip faults. In the south and southwest, the Kerch-Taman trough is limited by the Barrier anticline zone in the outer part of the shelf of the same name, which, in our interpretation, is a system of fold-thrust dislocations [39]. In the north, the trough does not have a clear tectonic boundary. Only in the northwest of the Taman Peninsula (the area of the Fontalovsky uplift) is a relatively shallow position of the Mesozoic strata noted, which can be considered as one of the fragments of the northern limit of the Kerch-Taman trough. According to seismic studies, the thickness of the sedimentary fulfillment of the trough exceeds 10 km. Most of the section of its central sections is composed of Maikop-Neogene deposits. The Paleogene-Mesozoic (pre-Maikop) part of the section is much less studied.

Marine seismic surveys have established a system of linearly extended anticlinal zones in the trough. Their southwestern strikes within the shelf are replaced by sublatitudinal ones on the Taman Peninsula. High-amplitude brachianticlines, composed mainly of Maikop and supra-Maikop deposits, take part in the structure of anticlinal zones. Until recently, clay diapirism played the main role in the structure and formation of these brachyanticlines [40,41]. The common depth point (CDP) materials of recent years, obtained in the described area, indicate a significantly greater significance of fault tectonics and, in particular, fold-thrust dislocations in the formation of the structural plan of the Mesozoic-Cenazoic deposits.

In 1997–2005, at the new technical level of marine seismic exploration, various promising areas of the Kerch-Taman shelf and adjacent sections of the slope were covered by 2D CDP studies using a uniform network of profiles in order to clarify the structure and nature of local structures previously identified in the supra-Maikop deposits, study the pre-Maikop sedimentary complexes, and resolve the issue of dominant types of hydrocarbon traps within the shelf [42]. 3D CDP surveys were carried out in a number of areas.

The deep exploration well (Rifovaya-302) on the Anapa shelf was drilled in 1985 in the crest of the Rifovoye uplift. The well penetrated the top of the Maikop deposits at a depth of 600 m and was stopped in the same deposits at a depth of 2000 m.

Another exploratory well, drilled at the northwestern end of the Shatsky swell at the Yuzhno-Doobskaya (Maria) structure, opened the top of the Mesozoic deposits to a depth of more than the Zapadno-Chernomorskaya area on the Black Sea shelf, the press service of Rosneft OJSC reports. The depth of the sea at the drilling point reaches 2109 m. The actual depth of the well was 5265 m. As a result of the work, a unique structure was discovered, in the roof of which a Cretaceous carbonate stratum with a thickness of more than 300 m was discovered.

The main object of the study area is a seismic complex, including deposits of the Oligocene-Lower Miocene (Maikop series). On seismic profiles, the presence of a fairly thick thickness of Maikop deposits is established. Here, the deep erosion of these deposits by the post-Maikop erosion processes is clearly recorded. On the map of the thickness of the Maikop deposits, a band of their maximum erosion is highlighted to the west and south of the study area of the ledge. Thus, the area of accumulation of Maikop deposits occupied a vast area, including the modern deep-water part of the Black Sea. Based on the available materials, it is difficult to judge tectonic movements in the Maikop time, but, apparently, they were not as intense as in the post-Maikop time, and especially in the Pliocene-Pleistocene.

Most of the hydrocarbon deposits on the adjacent Kerch and Taman peninsulas are associated precisely with the Oligocene-Miocene, mainly Maikop and, to a lesser extent, Chokrak-Karagan deposits.

The newest CDP materials, obtained by VO "Chernomorneftegaz", unambiguously resolved the issue of the fold-thrust nature of the structures in this part of the shelf [42].

It can be said with confidence that the asymmetric folds complicated by reverse thrusts in this part of the shelf can be associated mainly with domed reservoir deposits. Seismic data also indicate that such deposits in most cases should be tectonically screened and, in some places, stratigraphically screened. At the same time, deposits on the arches of many uplifts were subjected to erosion during periods of breaks in sedimentation and, therefore, the possibility of erosion of the main Middle Miocene oil and gas complex is not excluded. However, it was preserved in the subthrust parts of these uplifts. Therefore, we highly appreciate the prospects of Miocene deposits within the uplifts identified here, mainly in the subthrust parts of the structures. The asymmetric folds of this part of the shelf may also be associated with hydrocarbon deposits in the pre-Maikop (Paleocene-Eocene) deposits in their arched parts.

In 2005, in the Ukrainian zone of the Prykerchensky shelf, VO "Chernomorneftegaz" drilled the parametric well Subbotina 403, in which commercial hydrocarbon inflows were obtained from Maikop deposits [43,44]. According to well logging data, promising objects in the Tortonian, Eocene, and Paleocene deposits were identified in the well, but they were not tested for technical reasons.

The bottom hole is located at a depth of 4300 m in the Early Eocene deposits (according to N.V. Maslun). In 2008, in exploration well 1, positive results were obtained when testing objects in the deposits of the lower Maikop (layer M3). In 2009, well 2 was drilled with a design depth of 3150 m in the Subbotin area in order to search for productive horizons in the Torton, Maikop, and Eocene deposits. The bottom of the well is located at a depth of 3200 m in the Eocene deposits (after reaching the design depth, it was decided to deepen the well by 50 m).

The well uncovered deposits of the Neogene, Maikop series, Eocene, represented by limestones, marls, argillaceous sandstones, siltstones, clays, mudstones; from the interval 2030–2047, an oil inflow of 13.1 m^3/d was obtained, from the interval 2006–2047 m, an oil inflow of 19.1 m^3/day was obtained (productive horizon M-1).

In the Lower Eocene, according to electrometry data, a repetition of the section of the interval 3890–4038 m is clearly observed. The well probably passed a tectonic fault of the reverse-thrust type twice, at a depth of 3890 and 4162 m. This is also indicated by the geological complications that were noted at these depths during drilling, as well as by the finds of cores of redeposited crumpled Cretaceous and Danish foraminiferal forms. According to the new interpretation of seismic data, taking into account the drilling data of parametric well No. 403, along the reflecting horizon Ib, confined to the lower Maikop formation, the fold is an anticline of latitudinal strike [42].

Its dimensions within the limits of the extremely closed isohypse minus 3000 m are 10×2–4 km, the amplitude is 650 m. The fold in the north and south is complicated by tectonic faults of the latitudinal strike-thrust type. The geological structure of the Subbotin area is significantly more complicated along the deeper horizon IIa, confined to the roof of the Paleocene-Eocene formations. The crest of the fold is displaced to the south of well No. 403 and is complicated by tectonic disturbance. The fold is contoured by isohypse minus 3800 m.

2.2. *Data*

At the research site, 2D multichannel reflection very-high-resolution seismic data were available. The frequency range of the seismic data was from 100 to 600 Hz with the central frequency of 250 Hz. The range of offsets was from 50 to 250 m which facilitated the range of incident angles of reflected P-waves from 5 to 45° in the near-surface. The maximum propagation depth of the seismic survey was about 400 m, and the vertical resolution of the resulting seismic images was about 2–5 m.

The signal processing of the seismic data included a bandpass filtering, FK-filtering, an amplitude correction for the spherical divergence of the wavefront, suppression of multiples, a pre-stack Kirchhoff migration in time-domain, construction of the supergathers by averaging of the neighbor gathers, and stacking. The figure below shows an example of the processed seismic image (a seismic stack) along the 2D profile that crosses the project borehole. The position of the project borehole coincides with the projection of X engineering well to the 2D seismic profile—X projection (X proj in Figure 1), for which the pore pressure prediction was conducted.

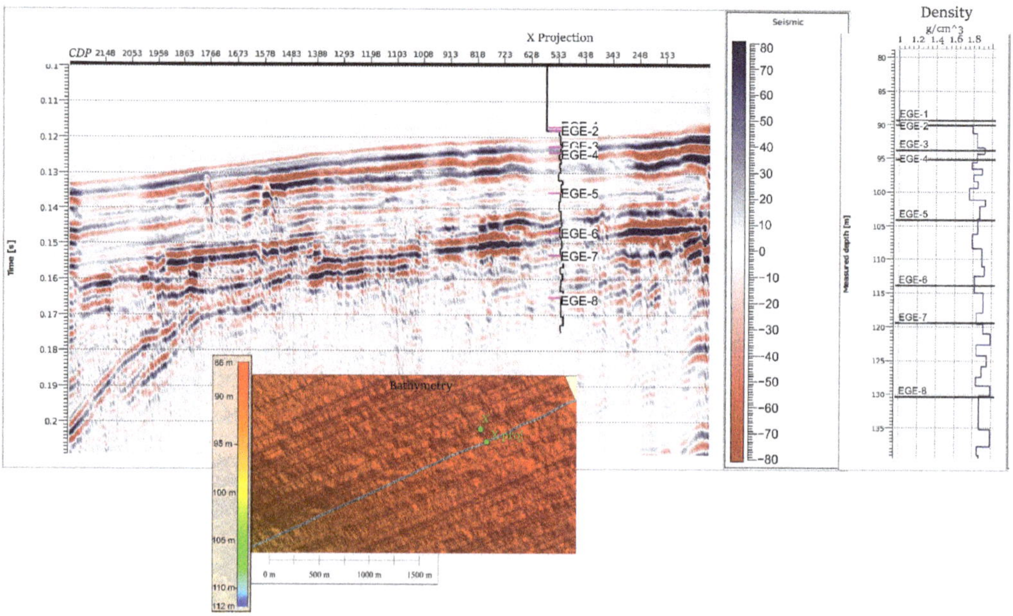

Figure 1. The 2D seismic profile along the project well (X-proj is the projection of the engineering well X). The 1D model of the bulk density measured in the laboratory for the well X is shown on the seismic profile and on the right.

At the site, several engineering wells, up to 50 m deep, were also available. Soil tests and laboratory measurements on the soil samples were conducted in each engineering well. The lab measurements were aimed at the estimation of physical and mechanical properties of the soils and their classification. Among the soil properties that were measured and estimated in the laboratory were the bulk density (cutting ring method), moisture (drying to permanent mass method), flow (standardized cone method, plastic limit method), and elastic moduli. The static Young modulus of the soils were measured under static loading by triaxial and compression tests. Note that the Poisson ratio v was estimated using the laboratory data on Liquidity index I_L [45]:

$$v = 0.05 + 0.45I_L \quad (1)$$

As a result, seven engineering and geological elements (EGE) were identified at the research site. Their description is given in Table 1. The tops of the EGE in the well X are shown in Figure 1.

Table 1. Description of engineering geologic elements identified at the research site. The depth and thickness ranges are given according to several engineering wells.

Engineering Geologic Element (EGE)	Depth (from the Seafloor), m	Thick-Ness, m	Description	Poisson's Ratio	Den-sity, g/cc	Young's Modulus, MPa
EGE 1	0.15–1.7	0.15–1.80	Silt	-	1.54	1
EGE 2	4.65–5.50	3.0–3.90	Fluid-saturated plastic clay with an admixture of organic matter	0.43	1.87	3
EGE 3	5.90–6.20	0.40–1.55	Fluid-saturated plastic clay loam with an admixture of organic matter	0.46	1.86	6
EGE 4	14.75–15.40	8.80–9.40	Calcareous fluid-saturated plastic clay with an admixture of organic matter	0.40	1.81	4
EGE 5	24.50–24.90	9.40–10.15	Calcareous high-plastic clay with an admixture of organic matter	0.28	1.85	7
EGE 6	41.25	16.55	Calcareous high-plastic clay with an admixture of organic matter	0.31	1.90	9
EGE 7	>50.5	>9.25	Calcareous high-plastic clay with an admixture of organic matter	0.30	1.91	-

2.3. Seismic Inversion

To estimate the elastic properties of the sediments in the near-surface at the research site, we applied a pre-stack seismic data inversion on the available 2D reflection seismic datasets. The conducted inversion was based on the Zoeppritz equations for the reflection coefficients of an incident P-wave.

The pre-stack amplitude inversion of reflection seismic gathers allows one to simultaneously estimate dynamic elastic moduli and density of the subsurface. Typically, we estimate the acoustic impedance, $Z_p = \rho V_p$, the bulk density ρ, and the body wave velocity ratio V_p/V_s (where V_p—compression wave velocity, V_s—shear wave velocity), by the pre-stack amplitude inversion. From these parameters, we can calculate the dynamic elastic moduli such as bulk and shear moduli, Young's modulus, and Poisson's ratio.

The main challenge for the inversion of the very-high-resolution seismic data is posed by the absence of the low-frequency information up to 80–120 Hz [46,47]. The absence of the low frequencies in the seismic data results in poor estimates of low-frequency trends of the elastic properties.

To fill this gap in low frequencies and stabilize the inversion results, we utilized the results of seismic P-wave velocity analysis and the laboratory measurements of the bulk density and the static elastic moduli of the soils available at the engineering wells at the research site.

We calculated the model of interval P-wave velocities from RMS P-wave seismic velocities using a Dix formula. The static Young modulus obtained in the laboratory was corrected for its dynamic analog. For the correction, we applied the coefficient 20 (the dynamic modulus is 20 times greater than its static analog), commonly used for soils. We assume that the Poisson ratio provided by the Liquidity index test is similar to its dynamic analog. The estimated dynamic Young modulus and Poisson ratio are used to predict the low-frequency S-wave velocity:

$$V_S = (E/2\rho(1+v))^{0.5}, \tag{2}$$

where E is dynamic Young's modulus, ρ is the bulk density and v is Poisson's ratio.

Based on these preliminary estimates, we built the low-frequency models of the P-wave, S-wave velocities, and density. Then, we conducted an acoustic post-stack inversion on the seismic stacked sections to clarify the models of the acoustic impedance, and finally, we inverted the seismic gathers to detail the models of the acoustic impedance, density, and V_p/V_s. The low-frequency models and the results of the seismic pre-stack inversion are shown in Figure 2. These results are further used for the pore pressure inversion.

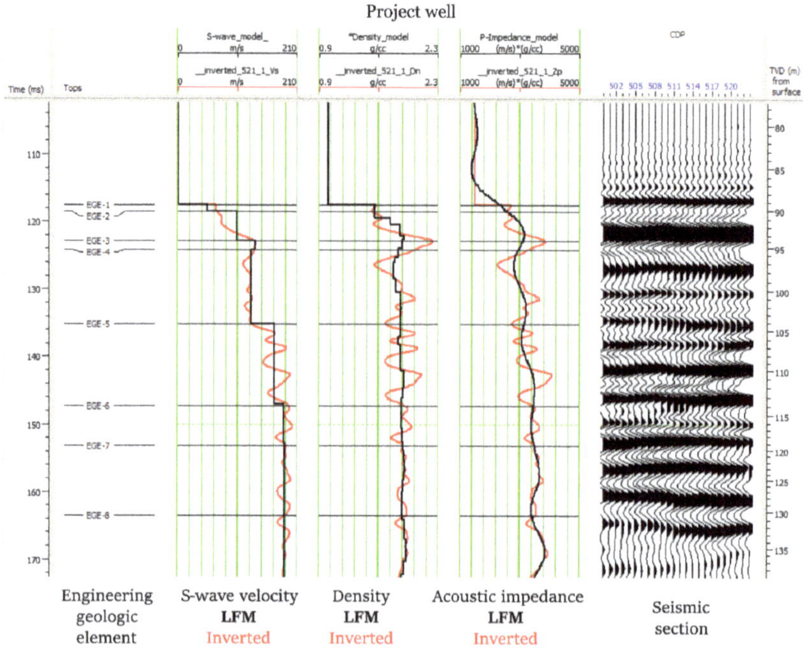

Figure 2. The low-frequency and the inverted models of S-wave velocity, density, and acoustic impedance along the project well. The seismic section in the vicinity of the well is shown on the right.

2.4. Model of Mechanical Properties

The seismic data obtained in the previous section allow us to derive the P-wave velocities from the density and acoustic impedance. From P- and S-velocities and density, the dynamic moduli were calculated including the bulk and shear moduli, Young's modulus and Poisson's ratio. Moreover, knowing the density of the solid grain material and fluid, the porosity can be estimated. The laboratory measurements carried out for the marine sediments provides a rather high density of the mineral grain material which varies from 2.70 to 2.74 g/cm^3. The density of fluid is equal to 1.024 g/cm^3 that is the sea water density. The porosity, bulk modulus, and Poisson's ratio along the project well are shown in Figure 3. Note that the Poisson ratio is rather high varying from 0.493 to 0.498 that is very close to the sea water value (0.5). These values are much greater than those obtained in the laboratory test on Liquidity index (see Table 1). For these values of Poisson's ratio, the shear modulus does not exceed 0.08 GPa that is comparable to the experimental error of its determination. This forced us to assume that this modulus may have higher values than provided by the seismic data.

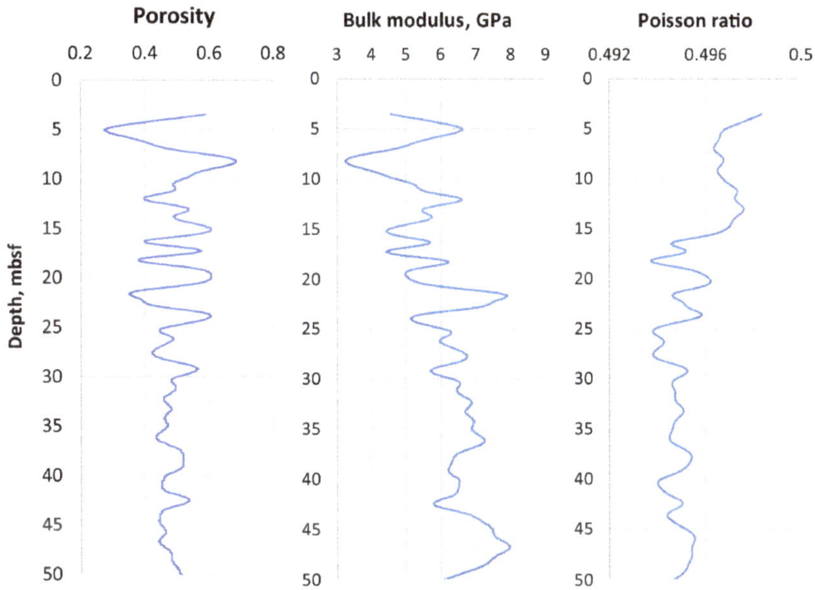

Figure 3. Porosity, bulk modulus, and Poisson's ratio along the project well calculated from the seismic data (P- and S-wave velocities and density) and density of solid grain material and sea water.

The elastic moduli and porosity of the sediments were further used for the rock-physics modeling and preconsolidation analysis.

As far as overpressure zone prediction is carried out in order to estimate various risks, problems with drilling have been analyzed as well. It is widely known [48] that seismic data without any additional information on the mechanical properties of sediments limits the opportunity to predict drilling risks. Although special methods of estimating risks solely from seismic data using data and system analysis methods have been developed [49], there are still considerable uncertainties in quantitative prediction of risks associated with overpressures. Fortunately, a number of samples have been extracted from the engineering drilling site, so there was an opportunity to carry out typical triaxial loading tests to evaluate static elastic moduli of the extracted samples.

The tests have been performed for 44 samples extracted from the depths between 1 and 50 m below seafloor. Cylindrical samples have been subjected to constant radial stress and gradually increasing vertical stress. Due to the rheological features of the studied seafloor sediments, the obtained "stress vs. strain" rheological curves were characterized by high nonlinearity. The obtained critical differential stresses (maximal obtained differences between axial and radial stress for each sample) were as low as 0.2 MPa, corresponding to low internal friction angles and cohesions of the samples—which is typical of unconsolidated seafloor sediments.

The typical procedure of analyzing drilling risks via geomechanical modelling starts from construction of the model of mechanical properties, which typically consists of: (1) transition from dynamic elastic moduli to their static analogues; (2) transition from static elastic moduli to strength properties. With the seismic data described above and available triaxial loading test results, these steps have been completed to establish the foundation of drilling risks analysis.

Figure 4 represents the transition between dynamic and static Young's moduli. The solid blue line stands for the dynamic Young modulus calculated from seismic data described in Section 2.3 along the trajectory of the studied well. Black circles represent the results of triaxial test data processing: the linear part of the "stress vs. strain" rheological curve was analyzed to obtain its slope, which was considered as Young's modulus for the sample at the depth of its extraction. Horizontal dotted lines are plotted to highlight 6 interfaces between layers characterized by similar mineral compositions of the extracted samples according to laboratory analysis of the extracted samples.

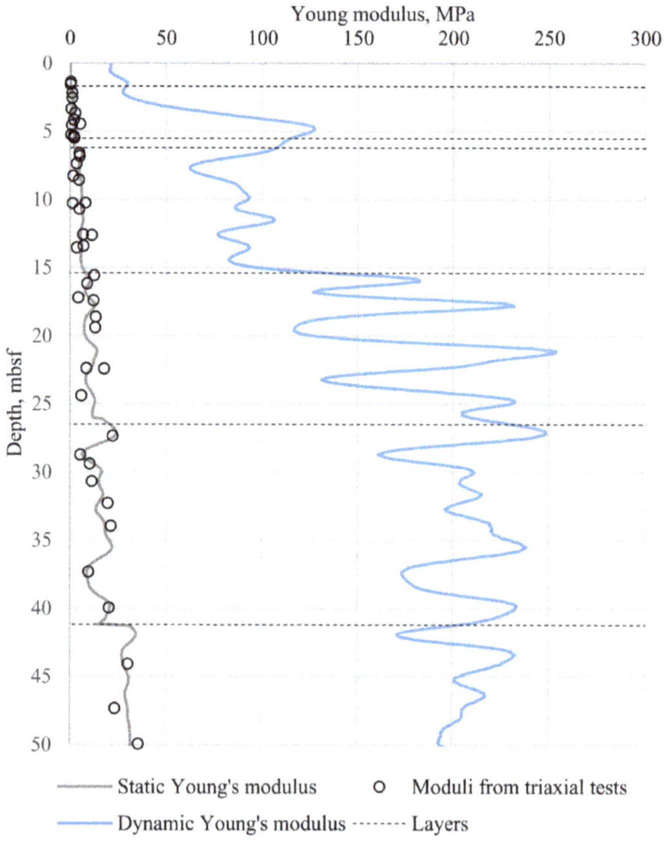

Figure 4. Young's modulus for the studied well.

With the dynamic Young modulus profile and a set of dots for static Young's modulus, one gets the opportunity to solve the linear regression problem: static Young's modulus is expected to be linearly related to dynamic elastic modulus; the coefficients in this dependency are found from solving the optimization problem to reduce the difference between the calculated static Young modulus and its analogues from triaxial tests. The specific feature of the current case is associated with the importance of the layers: it can be clearly seen that typical dynamic Young's moduli of the selected layers differ a lot between each other. As a result, usage of one linear correlation between static and dynamic Young's moduli would result into considerable errors. As a result, each layer was characterized by its own solution of the optimization problem, resulting into reasonable correlation coefficients.

Poisson's ratio for the considered sediments was rather high: it exceeded 0.4 for the studied samples from layers EGE2–EGE4. Meanwhile, for layers EGE5–EGE7, the Liquidity index test produced lower values (Table 1). However, the rock-physics modeling (see Section 3.2) suggested high values of Poisson ratio along the whole well (greater than 0.43). Therefore, the equality between static and dynamic Poisson's ratio was considered to construct the model of mechanical properties.

Completed triaxial loading tests had also provided an opportunity to deal with strength properties of the studied samples. The linear Mohr–Coulomb failure criterion was used to deal with the strength properties of the sediments:

$$\tau_n = C + \mu_f \sigma_n, \quad (3)$$

where τ_n and σ_n are the tangential and normal stress leading to the failure, C is the inherent shear strength (cohesion), and μ_f is the internal friction coefficient (tangent of internal friction angle φ_i).

With a small number of samples extracted from each layer, the following assumption was used to evaluate failure criterion parameters: as the friction coefficients of all samples were found to be very low, while cohesion differed significantly, the internal friction coefficient was reconstructed as a step function: each layer was characterized by its own value of μ, which was evaluated from the sets of stresses leading to the failures of the samples related to each layer. After averaging the internal friction coefficients for all layers, cohesion was evaluated from the Mohr circle constructed for the failure conditions with the known slope of the Mohr–Coulomb linear failure line. While this method of dealing with two parameters independently can lead to problems with other reservoir geomechanics problems—multi-stage triaxial loading tests are necessary for the evaluation of the internal friction angle—in this particular case, averaging of the internal friction coefficient is backed by the very low strength of the sediments and low stresses, providing that cohesion alters the risk evaluation results to a higher degree compared to the internal friction coefficient.

Cohesion C, estimated using the described methodology, can be plotted as a function of static Young's modulus obtained from triaxial tests—Figure 5. Different colors of the dots represent different layers mentioned above, yet it is clear that in this particular case, layers' differentiation does not play an important role for transition from dynamic moduli to static: a linear dependency between static Young's modulus and cohesion seems to be a reasonable correlation for strength property evaluation.

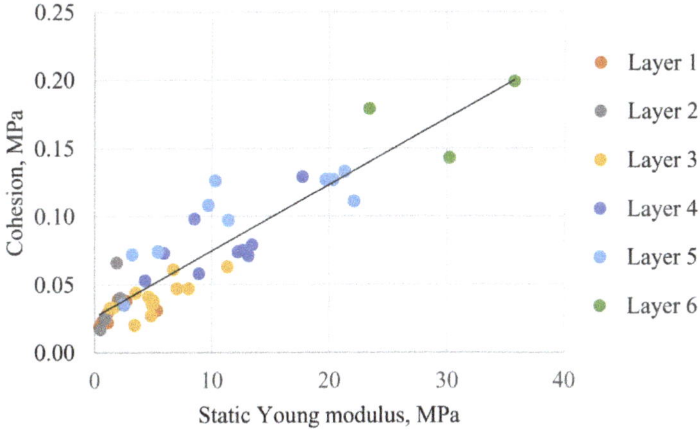

Figure 5. Reconstruction of cohesion using the static Young modulus.

Once again, the parameters applied to evaluate the cohesion are obtained using the solution of the linear regression problem.

Estimation of the static elastic moduli and strength properties sums up the construction of the model of mechanical properties which will be consequently used for risk evaluation. The next steps in the geomechanical analysis of risks are associated with the reconstruction of the pore pressure and stresses acting in the studied sediments. While the latter is a relatively easy task—at least for non-consolidated sediments with known density and elastic properties—evaluation of pore pressure deserves special attention and will be described below in detail.

3. Results

The following sections provide the main results of pore pressure prediction using two methods—pressure prediction from the void ratio [18,24] and rock-physics modeling [36]. The final part contains the comparison of the results obtained via these methods.

3.1. Pore Pressure Prediction from Void Ratio

As it was already highlighted above, the void ratio can be used based on the porosity of the sediments. Reconstruction of the porosity was described in Section 2.4: rock-physics methods can be successfully applied on data in the presence to reconstruct the porosity profile along the well (Figure 3).

In the current section, we focus on the pore pressure evaluation and prediction of overpressure zones based on the void ratio e which is in a simple relationship with the porosity φ:

$$e = \varphi/(1 - \varphi), \quad (4)$$

A specific method of interpreting the porosity profile from the perspective of consolidation was described in detail by [18,24]. Consolidation analysis considers a specific preconsolidation stress as a parameter corresponding to deformation processes taken place for the considered sediments. This term was proposed to describe the maximum vertical effective stress that the soil had been subjected to during the history of its elastic and elasto-plastic deformation [50]. While specific laboratory experiments based on reconsolidation of extracted samples were developed by various researchers [51,52], there still remains the problem of estimating the true preconsolidation stress due to significant sample disturbance that leads to the risk of underestimating the preconsolidation stress from these tests [53].

Nevertheless, equations used for the pressure prediction from the void ratio discussed by [18] can still be used for this purpose: preconsolidation tests provide necessary information for finding the parameters of pore pressure prediction equations. Various data can be used to check the validity of the pore pressure prediction, including direct measurements. Engineering drilling at the studied site was characterized by an intensive gas leak visible at sea surface when a specific depth interval was drilled. Given the known depth of drilling, sea depth at the drilling site, and the intensity of the leak, the evaluation of pore pressure at the layer was completed: pore pressure in the gas-saturated zone at the depth of 20 mbsf was proved to be 1.2 MPa.

This estimation was used to find the parameters of the exponential relationship between void ratio and effective vertical stress [18] that was assumed to be valid at the studied site:

$$e = e_0 - C_c \log \sigma_V, \quad (5)$$

where e_0 is the void ratio at a vertical effective stress of unity (1 MPa), C_c is the specific compression index describing deformation along the yield surface, and σ_V is the effective vertical stress. Proper rearrangement of this equation provides the following dependency for the pore pressure P_{por}:

$$P_{por}(z) = S_V(z) - 10^{-(e(z) - e_0)/\,C_c}, \quad (6)$$

where S_V is the total stress, z is the depth below the seafloor. If the sea depth is known alongside with the density profile, the total stress can be obtained as:

$$S_V(z) = \rho_w g h_w + \int_0^z \rho(z) g dz, \quad (7)$$

where ρ_w is the water density, h_w is the sea depth, and $\rho(z)$ is the estimated density profile. With all these parameters described in the previous sections, the total vertical stress reconstruction is a straightforward procedure. It is worth mentioning that effective horizontal stresses for these unconsolidated sediments tend to be close to the vertical effective stress, due to both high Poisson's ratio and considerable elasto-plastic deformation.

A block-scheme of the forward problem solution is given in Appendix A (Figure A1).

It is clear that two independent measurements are necessary for successful application of Equation (6) for pore pressure prediction. While one measurement (1.2 MPa at 20 mbsf) was obtained from direct observations, the second point was the expected hydrostatic pressure near the seafloor. As a result, two depths and two measurements were used in Equation (6) to get two parameters: C_c = 0.18, e_0 = 0.88, typical of the considered site at the Black Sea. The analogous values for the Gulf of Mexico were reported to be C_c = 0.54 and e_0 = 0.47 [18].

As a result, Equation (6) was used to provide a simple, preliminary evaluation of the pore pressure at the site with mentioned parameters. Nevertheless, it is clear that there are some drawbacks of using this estimation: firstly, specific consolidation tests were not carried out, so the preconsolidation stress was not known for sure; secondly, the exponential dependency between the void ratio and the effective vertical stress was not backed by experiments; finally, the assumption of hydrostatic pressure near the seafloor was drawn without a solid proof as well. These problems led to the need of using extra approaches to check the reconstructed pore pressure profiles. Rock-physics modeling can provide an approach, which is described in the following section.

3.2. Pore Pressure Prediction from Rock-Physics Modeling

To estimate the pore pressure along the project well from the elastic wave velocities and density provided by the seismic inversion, we applied the well-known soft-sand model of Mavko et al. [36] combined with the Gassmann fluid substitution [54]. We used this scheme since as was already noted in Introduction, the soft-sand model based on the Hertz–Mindlin approach makes it possible to incorporate the pore pressure effect on the elastic moduli in explicit form. In turn, this gives an opportunity to invert the pore pressure from the elastic wave velocities and density of marine sediments. Note that we also analyzed the stiff-sand model but came to the conclusion that this model gives greatly overestimated velocities. The soft-sand model assumes that the rock is composed of isotropic particles of a solid material and a granular medium surrounding the particles. The granular medium is a dense pack of the spherical grains of the solid material and the grains have the same size. The porosity of the granular medium (so-called "critical porosity") is larger than the rock porosity. The formulas of the soft-sand model for effective elastic moduli of dry rock are as follows:

$$\begin{aligned} K^*_{dry} &= \left[\frac{\widetilde{\varphi}}{M(K_{HM}, \mu_{HM})} + \frac{1-\widetilde{\varphi}}{M(K, \mu_{HM})}\right]^{-1} - \frac{4}{3}\mu_{HM}, \\ \mu^*_{dry} &= \left[\frac{\widetilde{\varphi}}{\mu_{HM}+\frac{\mu_{HM}}{6}R(K_{HM}, \mu_{HM})} + \frac{1-\widetilde{\varphi}}{\mu+\frac{\mu_{HM}}{6}R(K_{HM}, \mu_{HM})}\right]^{-1} - \frac{\mu_{HM}}{6}R(K_{HM}, \mu_{HM}), \end{aligned} \quad (8)$$

where

$$\widetilde{\varphi} \equiv \frac{\varphi}{\varphi_0}, \ M(K, \mu) \equiv K + \frac{4}{3}\mu, \ R(K, \mu) \equiv \frac{9K+8\mu}{K+2\mu} \quad (9)$$

In Equations (8) and (9), K^*_{dry} and μ^*_{dry} are the effective bulk and shear moduli of the dry rock; φ is the porosity, φ_0 is the critical porosity, K and μ are the bulk and shear moduli of the grain material, K_{HM} and μ_{HM} are the bulk and shear moduli of the dry granular medium calculated with the help of the Hertz–Mindlin (HM) method. According to the HM method, these moduli are calculated by the formulas:

$$K_{HM} = \left[\frac{D(P)}{18}\right]^{1/3}, \ \mu_{HM} = \frac{5-4\nu}{5(2-\nu)}\left[\frac{3}{2}D(P)\right]^{1/3}, \ D(P) \equiv P\left[\frac{\widetilde{C}(1-\varphi_0)\mu}{\pi(1-\nu)}\right]^2 \tag{10}$$

In Equation (10), P is the effective pressure equal to the difference of the vertical stress and pore pressure; μ and v are the shear modulus and Poisson's ratio of the grain material; φ_0 is the critical porosity; $\widetilde{C}$ is the average number of contacts per grain (so-called coordination number). Definitely, $\varphi_0 \geq \varphi$. If $\varphi_0 = \varphi$, then the soft-sand model is reduced to the Hertz–Mindlin model.

After the effective bulk and shear moduli for dry rock are calculated, the Gassmann fluid substitution is applied and the resulting effective moduli of marine sediments saturated with sea water are found:

$$\begin{aligned} K^*_{sat} &= \tfrac{AK_0}{1+A}, \ A \equiv \tfrac{K^*_{dry}}{K_0-K^*_{dry}} + \tfrac{K_{fl}}{\varphi\left(K_0-K_{fl}\right)}, \\ \mu^*_{sat} &= \mu^*_{dry} \end{aligned} \tag{11}$$

where K^*_{sat} and μ^*_{sat} are the effective bulk and shear moduli of saturated rock, K_0 and K_{fl} are the bulk moduli of the grain material and fluid. After the effective moduli of the saturated rock are found, the elastic wave velocities Vp and Vs are calculated by the formulas:

$$V_P = \sqrt{\frac{K^*_{sat} + 4/3\mu^*_{sat}}{\rho^*}}, \ V_S = \sqrt{\frac{\mu^*_{sat}}{\rho^*}} \tag{12}$$

where ρ^* is the rock density. The rock density is calculated by the exact formula

$$\rho^* = (1-\varphi)\rho + \varphi\rho_{fl} \tag{13}$$

where ρ is the density of grain material and ρ_{fl} is the density of fluid.

The coordination number $\widetilde{C}$ takes values from 5 to 9 [36]. However, when fitting the theoretical and experimental velocities, larger values can be used (for example, 15).

A workflow for calculating the effective elastic moduli (forward problem) according to the described rock-physics model is as follows. First, the input data on the elastic moduli of grains, critical porosity, coordination number, and pore pressure are used to calculate the HM moduli by Formula (10). Then, these moduli together with the elastic moduli of grains and rock porosity are used to estimate the dry moduli of sediments (Formulas (8) and (9) are used). Finally, the effective elastic moduli of fluid-saturated sediments are calculated with help of the Gassmann Formula (11). These moduli together with the rock density are used to calculate the elastic wave velocities Vp and Vs in the marine sediments. A block-scheme of the forward problem solution is given in Appendix A (Figure A2).

Formulas (8)–(10) contain the following unknown parameters: the coordination number, critical porosity, and pore pressure. Moreover, the elastic moduli of solid grains are rather uncertain since commonly in marine sediments, such as those in this study, this is a mixture of mud and clay with additions of carbonate material at some depth intervals [18]. As follows from the literature, the clay elastic moduli and density may vary in a considerably wide range [55]. A sensitivity study of the above formulas to the elastic moduli of grain material shows that the bulk modulus of the grain material changing from 20 to 80 GPa forces to vary the P-wave velocity only within 150 m/s, keeping Vs almost constant. The shear modulus changing from 7 to 35 GPa causes variations in Vp and Vs around 100 and 200 m/s, respectively. Note that the chosen interval for the moduli variation corresponds to the possible values of clay and carbonate minerals reported in the literature.

Since the grain density is rather high (averaging 2.71 g/cm^3), we also chose rather high values of the grain material for the rock-physics modeling: the bulk and shear moduli equal to 52 GPa and 32 GPa, respectively. These moduli are close to the Voight–Reuss–Hill average of the stiffness matrices of illite, chlorite, and kaolinite addressed in the work of Katahara [56]. The corresponding density of these clay minerals are 2.79, 2.69, and 2.59 g/cm^3 which, on average, gives the value close to the measured density of the grain material.

A variation in the coordination number $\widetilde{C}$ from 5 to 12 increases both Vp and Vs by 100 m/s for the critical porosity $\varphi_0 = 0.76$. The respective increase in the velocities is 50 m/s for Vp and 150 m/s for Vs at the critical porosity $\varphi_0 = 0.5$.

In order to estimate the pore pressure along the project well, all the model parameters should be chosen such that the discrepancy between the theoretical and experimental velocities does not exceed a specified level. We specified this level as 20% difference for P-wave velocities and 100% for S-wave velocities. The latter value is too large due to high uncertainty in velocities of S-waves. This is not a very large difference in the absolute values of S-wave velocities of marine sediments. Thus, if the experimental velocity is 200 m/s, the theoretical value can reach 600 m/s. In this case, the absolute value of difference in these velocities is only 400 m/s. Moreover, the Vs values for unconsolidated rocks are often reported as varying up to 600 m/s.

To solve the inverse problem on pore pressure estimation for each depth along the project well, we fixed different combinations of $\widetilde{C}$ and φ_0. Then, for every combination, we found a set of pore pressure values satisfying the above conditions. To do this, we applied the n-dimensional mesh method. This method assumes that the range of possible values of every unknown parameter is divided into intervals. The forward problem on the determination of P- and S-wave velocities was solved for each node. Then, "good" solutions, i.e., the solutions satisfying the accepted misfit between experimental and theoretical velocities were collected. Among the "good" solutions, the most probable solution was chosen. The most probable solution is the solution having the minimum value of "misfit measure".

In our case, $n = 1$ since we had only one unknown parameter—the pore pressure. The range of pore pressure variation is from the hydrostatic to overburden pressure S_V. The "misfit measure" was chosen to be the relative difference between the theoretical and experimental P-wave velocity. A workflow for the inverse problem solution on the pore pressure estimation is shown in Appendix A (Figure A3).

Note that different combinations of $\widetilde{C}$ and φ_0 show different solutions for the behavior of pore pressure along the project well. Among all combinations of these two parameters, we chose the combination providing the "reference" pore pressure 1.2 MPa at the depth 20 mbsf. These values are $\widetilde{C} = 12$ and $\varphi_0 = 0.76$. Note that without the knowledge of a "reference" value, the solution for pore pressure is rather uncertain and may exhibit different values between the hydrostatic pressure and overburden stress. This fact should be taken into account when inverting the pore pressure with the use of similar approaches.

A comparison of experimental and theoretical velocities and Poisson's ratio obtained as the most probable solution is shown in Figure 6. As seen, the S-wave velocities are greater than experimental ones and may attain 600 m/s in the depth range 22–52 mbsf. At the same time, the rock-physics model provides, generally, underestimated P-wave velocity (within 20%). In the upper part of the depth interval down to 8 mbsf, the experimental and theoretical velocities are very close to one another. The Poisson ratio provided by the rock-physics modeling varies from 0.43 to 0.49 that is lower than the experimental values but higher than those obtained in the Liquidity index test for the respective layers (called engineering geologic elements) (Table 1). As seen, the difference between the Poisson ratio obtained with different methods is large. The difference between the Poisson ratio calculated from the theoretical and experimental velocities (dynamic Poisson's ratio) is explained by the fact that the Poisson ratio is controlled by the squared ratio *Vp/Vs*: $\upsilon = 0.5\left[(V_P/V_S)^2 - 2\right] / \left[(V_P/V_S)^2 - 1\right]$. The difference between theoretical and experimental shear wave velocities my attain 400 m/s. However, the shear wave velocities have rather small values. This leads to the sufficiently large difference in the squared *Vp/Vs* ratio (4–10 times).

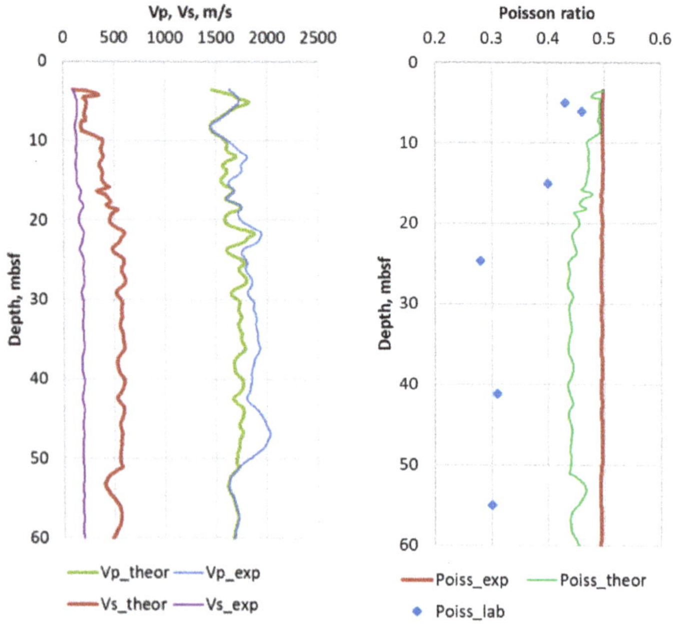

(a) (b)

Figure 6. Comparison of seismic data (exp), results of rock-physics modeling (theor), and laboratory data (lab): (**a**) velocities of P- and S-waves and (**b**) Poisson's ratio.

The difference in the experimental dynamic Poisson ratio and its laboratory analog provided by the Liquidity index test is due to the fact that these values are determined from different types of action on the rock. When considering the dynamic Poisson ratio, this is a short-time action of low amplitude (excitation of elastic waves). The Poisson ratio provided by the Liquidity index test is based on the measurements of the water content in the clayey sample. The water content is compared with the so-called liquid and plastic limits. The liquid limit is specified as a water content at which the plastic body starts to behave as a liquid. The opposite limit—the plastic limit is a water content that allows the clayey sample to exhibit plasticity (the rock becomes less brittle). Note that the Poisson ratio provided by the Liquidity index test was used for estimating the shear wave velocity in our study. The discrepancy in the Poisson ratio determined by these two different experimental methods suggests rather high uncertainty in shear wave velocity estimation for the studied unconsolidated rocks.

Among the solutions of the inverse problem, a solution exists that provides a good correspondence between both P- and S-wave velocities. However, this solution produces a pore pressure almost equal to the overburden stress within the whole depth interval under study. This situation seems to be unrealistic.

Note that according to the sensitivity study of the rock-physics model, S-wave velocities are more sensitive to the pore pressure compared to P-waves. This fact draws attention to the problem of increasing the reliability of the S-wave velocity determination.

3.3. Pore pressure Prediction Comparison

The previous two sections covered the methods used for pore pressure prediction based on consolidation analysis and rock-physics modeling. Comparison of the obtained results appears to be a logical way to summarize the drawbacks and advantages of these methods.

Figure 7 represents the comparison between the predicted pore pressure profiles obtained from the two methods. The dotted line represents the pore pressure predicted from consolidation analysis – from Equation (6), while the solid line represents the best solution of the inverse problem postulated in the previous section devoted to rock-physics modeling. It should be noted that Equation (6) was used solely in cases when it provided estimations of pore pressure exceeding the hydrostatic pressure, so the hydrostatic pressure can be visualized as the lower boundary of the dotted line with a corresponding slope.

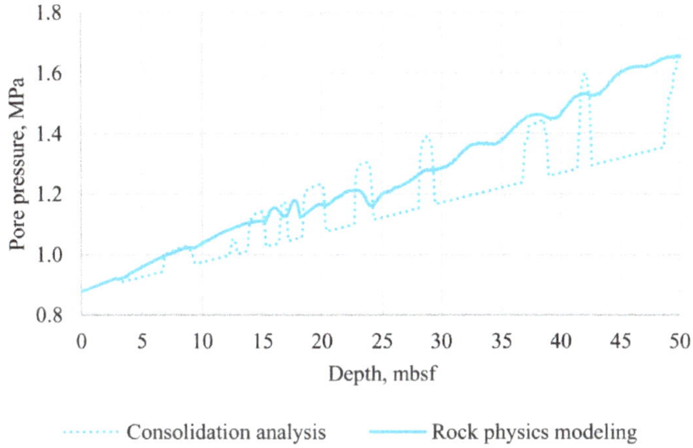

Figure 7. Pore pressure predictions using two approaches.

Several conclusions can be drawn from analyzing Figure 7. The first and the most general result is that both solutions are in qualitative agreement with each other. They clearly coincide at the depth of 20 mbsf, where the pore pressure should be equal to 1.2 MPa. Nevertheless, one specific difference is clear: the rock-physics modeling provides considerably smoother results. While the dotted line is close to the hydrostatic pressure trend almost in the whole interval with a number of secluded overpressure zones (7~10, 14~15, 16~17, 18~20, 28~30, 36~39, and 42~43 mbsf), the rock-physics modeling provides a more linear trend with a slope of approximately 0.8 MPa per 50 m, which is 16 kPa/m—recall that the hydrostatic pressure gradient is as low as 10 kPa/m, meaning that the whole zone should be considerably overpressured. At the same time, the local maxima of pore pressure from consolidation analysis exceed the estimations based on the rock-physics modeling results. It is also worth mentioning that local variations of the pore pressure tend to be present in both profiles, so the overpressured zones still exist even at the smoother profile obtained from rock-physics modeling.

Nevertheless, the deviation between pore pressure profiles evaluated using two different approaches remains clear and deserves additional discussion. As preconsolidation analysis is relatively straightforward compared to the rock-physics approach, its variations remain at the same level as the variations of porosity—this can be clearly seen from Equations (4)–(7). At the same time, rock-physics modeling suggests a solution of an inverse problem. So, while a minimization problem is solved for smoother data on seismic velocities, one could expect a smoother result compared to preconsolidation analysis. Given that the solution shown in Figure 7 as rock-physics modeling result is in fact only the best solution of the inverse problem, it is shown here as a profile obtained from rock-physics modeling. There are still lesser probable solutions that cannot be as smooth as shown in Figure 7; so, the apparent effect of this solution being close to the averaging of the consolidation analysis may be explained by this particular "best solution" choice. In general, the results shown in Figure 7 suggest that the two methods agree with each other, but the consolidation analysis tends to highlight the overpressured zones, with possible overestimation of the pore pressure. At the same time, the rock-physics modeling suggests that the whole depth interval is slightly overpressured. This means that analysis of risks using the rock-physics modeling should be completed from the perspective of integral analysis of the site.

A worrying result is associated with depth intervals where the rock-physics analysis highlights the local maxima of pore pressure, which are not correlated with variations of the profile obtained from consolidation analysis (intervals 31~34 and 44~47 mbsf). The origin of these zones remains unknown, so only direct observations of drilling will make it possible to assure whether these variations are related to real overpressure zones.

Nevertheless, the pore pressure profiles themselves cannot be considered as an engineering result directly associated with the drilling risks. Gas leaks and blowouts are most likely to occur in the overpressure zones, but their effect can differ a lot—and blowout prevention systems can be used to drill the well at this particular site: in fact, the largest pore pressure gradient in the most overpressure zone is roughly 20 MPa, and this zone is relatively small. Extra calculations with the help of geomechanics can be used to compare the obtained results from the perspective of drilling risks.

4. Discussion

The pore pressure profiles shown in Figure 7 can be used to deal with the drilling risks. A model of mechanical properties described in the corresponding paper consists of static and dynamic elastic moduli alongside with strength properties. It is widely known [48] that these data can be used to evaluate drilling risks: whenever effective stresses in the sediments are high enough to achieve the failure criterion, a number of problems start.

Geomechanical analysis of drilling risks at offshore sites with considerably low depths have some specific points compared to other problems. First of all, one needs both the total stress and pore pressure in order to calculate the effective stress. These data were obtained in the previous sections: the pore pressure was discussed above in detail, while the total effective stress was calculated with the use of Equation (7).

The failure criterion generally deals with all components of the stress tensor, but the considered sediments are characterized by a considerable portion of plastic flow in their total deformation, and very high Poisson's ratio, leading to the insignificant difference between the principal components of the effective stress tensor.

At the same time, drilling procedures alter the stress state of the sediments. In the current case, we considered a pessimistic scenario when drilling procedures alter only the vertical stress (σ_V turns into $\sigma_V + \Delta\sigma$, where $\Delta\sigma$ is an extra stress caused by the drilling operation, the weight of the drilling rig can be a simple source of this extra load), with the horizontal stress σ_H remaining untouched. The Mohr–Coulomb failure criterion (1) can be rearranged for this case of triaxial loading to the following form:

$$\frac{\sigma_V + \Delta\sigma - \sigma_H}{2}\sin 2\theta = C + \left(\frac{\sigma_V + \Delta\sigma + \sigma_H}{2} + \frac{\sigma_V + \Delta\sigma - \sigma_H}{2}\cos 2\theta\right)\mu, \quad (14)$$

where $2\theta = \pi/2 + \arctan(\mu)$. This is true for a stress state leading to failure and consequent plastic flow of the seafloor sediments due to the extra load $\Delta\sigma$. So, the critical value can be found for the extra load $\Delta\sigma$ as:

$$\Delta\sigma = (\sigma_H + Cctg\varphi)\frac{2\sin\varphi_i}{1-\sin\varphi_i} - (\sigma_V - \sigma_H), \quad (15)$$

where φ_i is the internal friction angle (its tangent is equal to the internal friction coefficient). The horizontal stress can be estimated as:

$$\sigma_H = v\sigma_V/(1 - v), \tag{16}$$

where v is the Poisson ratio. Note that effective stresses are used in this equation.

Equation (15) was used to evaluate the critical extra vertical stress $\Delta\sigma$ necessary to exceed the bearing load of the sediments using the two pore pressure profiles discussed in the previous section. The following figures (Figures 8 and 9) represent the final results associated with drilling risks. Note that due to Poisson's ratio used in Equation (16) being close to 0.5, the effective vertical and horizontal stresses tend to be very close to each other, so the corresponding profiles in Figures 8 and 9 almost completely overlap.

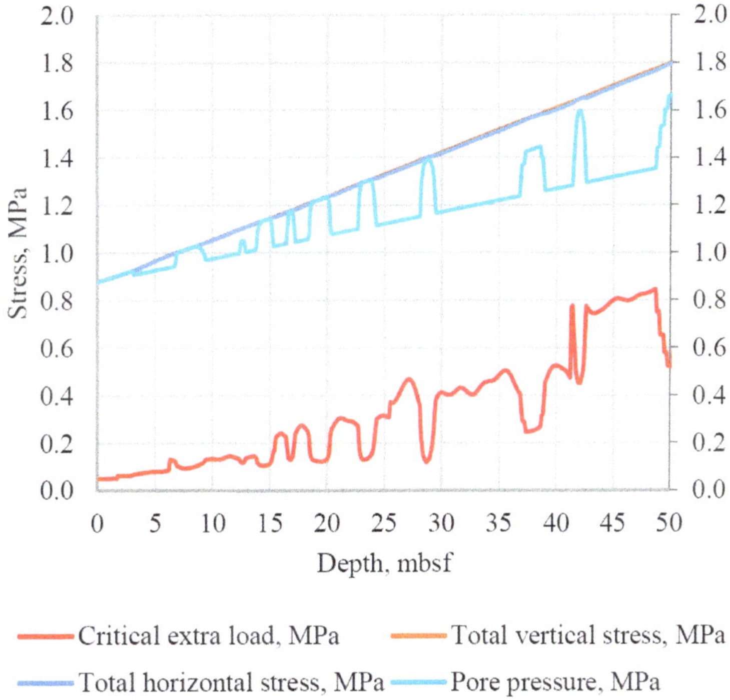

Figure 8. Critical load estimation for consolidation analysis.

Critical extra load is a useful parameter to analyze bearing strength of the sediments. It is in fact the maximum extra vertical stress (addition to the existing vertical stress shown in the upper part of the figure) that can be applied to the sediments without failure—or cataclastic flow for the unconsolidated sediments. Both the pore pressure and strength of the sediments affect it, so one can consider it as a combination of different factors leading to potential drilling risks. The red curves in Figures 8 and 9 should be used to optimize the weight of the drilling rig and its working conditions to maintain safe exploitation.

The smoothing of the pore pressure estimated from the rock-physics modeling discussed in the previous section finds itself at the critical load estimation as well. In general, the pore pressure estimation from the consolidation analysis is related to a more variable critical extra load for the sediments, and it is in most places higher than the critical load corresponding to the rock-physics modeling. It is worth mentioning that the changes in sediment strength parameters (primarily in cohesion) are highlighted in these figures as well, due to their presence in Equation (15). As it was already mentioned, the horizontal stress remains very close to vertical stress in the whole interval.

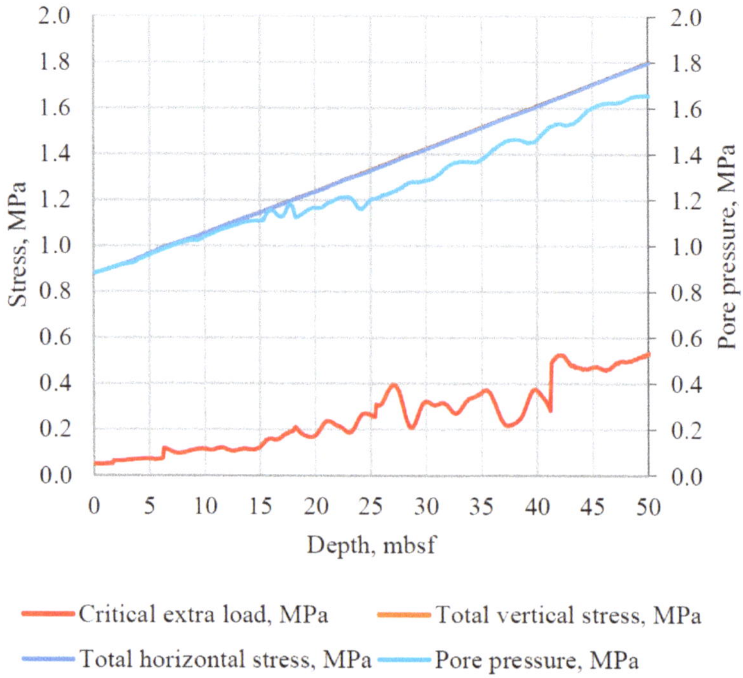

Figure 9. Critical load estimation for rock-physics modeling.

These results highlight the general tendency mentioned above: the rock-physics modeling seems to be a proper tool to predict overpressure and deal with its degree, while the consolidation analysis appears to be a tool aimed at overpressure zone localization, and the degree of evaluated overpressure remains questionable.

The critical extra load itself is an important result for drilling risks assessment. First, the magnitude of this extra stress is not extremely high, so that usage of a special blowout prevention system can mitigate these drilling risks. These values of stresses are relatively low, so the special redistribution of the stress state due to drilling procedures can be used to reduce drilling risks: in this particular case, a change in the drill bit diameter was altered to redistribute stresses in vicinity of the drilled well.

The drilling rig itself is an important source of extra load—and it does not have to be vertical. In fact, numerical modeling can be performed to analyze the stress state of the sediments surrounding the drilling rig legs taking into account the predicted overpressure zones, static elastic moduli, and parameters of plastic flow, which can be obtained from triaxial loading test data [57]. A detailed numerical modeling of the stress state of the sediments surrounding drilling rig legs and drill bit can provide the explicit conditions leading to a gas leak, drilling rig loss of stability, failure of weak layers, etc. This analysis suggests usage of detailed data on mechanical properties of the sediments, but can provide way more than engineering solutions based on correlations that can be unreliable for a specific site. While a detailed numerical analysis of these processes of drilled seafloor sediments is not within the scope of the current research work, the results discussed above can provide insight on the degree of reliability of numerical modeling results and discrepancy of the input data needed for it.

Detailed analysis of the sediments relatively close to the seafloor provides important insights on the lower sediments as well. Availability of seismic data inversion results for deeper layers can be used in a similar way to predict and localize overpressure zones using the discussed methods. Nevertheless, the models used for the pore pressure prediction discussed here have certain limits for application: they only provide reliable results in weak unconsolidated seafloor sediments. Other methods mentioned in the introductory section are generally used for unconsolidated rocks—search for the smooth transition from one type of prediction to another appears to be an intriguing direction of further research of this type of problems.

Another direction is associated with evolution of overpressure zones: there is an opportunity to carry out seismic surveys at the same site, but at different moments of time. Preliminary results obtained for this site suggest that some of the overpressure zones localized in the current study remain at their places, but some disappear and reappear. This rises a specific question: does drilling risk prediction remain reliable during all periods between seismic surveys and drilling procedures? Depending on the answer to this question, several ways can be suggested to drill as fast as possible after seismic surveys, or to carry out an external seismic survey at the site if preliminary seismic data prove that the potential risks can be handled.

Our study also draws attention to the problem of reliability of S-wave velocity determination for marine sediments. According to results of the rock-physics modeling, these velocities demonstrate high sensitivity to the pore pressure in this type of rocks. If the accuracy of the S-wave velocities is poor, reference values of the pore pressure are required to decrease uncertainty in the pre-pressure prediction based on the rock-physics modeling.

5. Conclusions

Pore pressure prediction remains an important practical problem, as overpressure zones can cause problems during exploitation and development of offshore hydrocarbon and gas hydrate reservoirs. While empirical methods have been developed a lot during the last decades, their application is usually limited to certain sites. To modify the existing correlations for overpressure zone localization for specific sites, one has to use external data, which can be obtained during drilling or special surveys. On the other hand, rock-physics modeling makes it possible to estimate pore pressure using a general workflow, independent of the particular site. There is still need to utilize a great amount of data—and requirements on the data quality are relatively strict. The comparison of pore pressure prediction based on these two groups of approaches remains of high practical and scientific interest.

In the current paper, we have applied these two groups of methods for a specific site at the Black Sea shelf. Pore pressure has been evaluated along the trajectory of a well to be drilled at the zone of potential hydrocarbon accumulation using two methods: preconsolidation analysis of seafloor sediments and soft-sand model with Gassmann fluid substitution. High-quality seismic data were used for this estimation alongside with indirect local measurement of pore pressure. Only the upper layers of seafloor sediments have been analyzed: the greatest depth of 50 mbsf was studied.

The results of pore pressure estimation using these two methods proved to be in qualitative agreement with each other: the same overpressure zones were proven to be in the corresponding intervals of depth. Nevertheless, rock-physics modeling provided a smoother result: it appears that preconsolidation analysis is more suitable for finding localized highly overpressure zones, while rock-physics modeling gives a better idea on total risks for the whole interval of consideration. Regardless, both methods provided the results of pore pressure evaluation to conclude that expected overpressure zones are not associated with high risks in the studied area; so, a typical blowout prevention system can deal with them.

Depending on the available data, one can still have freedom in choosing the proper empirical model of pore pressure prediction at any other site. The question of finding the optimal model remains open. Rock-physics modeling is capable of giving a hint to answer that question—but there are still different models even within the rock-physics workflow. While only a limited number of parameters within one of these models was altered during pore pressure prediction described in the current paper, there are still other parameters to be analyzed. We assume that a continuation of this study, including variation of a number of rock-physics model parameters during inverse problem solution and application of different empirical methods of pore pressure estimation, can decrease the overpressure-associated risks and geohazards.

Author Contributions: Conceptualization, A.H., A.P., I.B., K.M. and N.D.; methodology, A.H., A.P., I.B., K.M. and N.D.; software, A.P., I.B. and P.S.; validation, K.M., N.D. and A.H.; investigation A.H., A.P., I.B., K.M. and N.D.; writing A.H., A.P., I.B., K.M. and N.D.; visualization A.H., A.P., I.B., K.M., N.D. and P.S.; supervision A.H. and K.M. All authors have read and agreed to the published version of the manuscript.

Funding: Methodological aspects of the research were funded by Russian Science Foundation, grant number 19-77-10062.

Institutional Review Board Statement: Not applicable.

Informed Consent Statement: Not applicable.

Data Availability Statement: Data on simulated porosity, overburden stress, hydrostatic pressure and results of rock physics modeling including the pore pressure and other model parameters are available at https://cloud.mail.ru/public/hYeT/hy4ywu1oK (accessed on 14 July 2022).

Acknowledgments: The authors are grateful to editor and anonymous reviewers who provided the valuable comments that led to considerable increase in quality of the paper.

Conflicts of Interest: The authors declare no conflict of interest. The funders had no role in the design of the study; in the collection, analyses, or interpretation of data; in the writing of the manuscript; or in the decision to publish the results.

Appendix A

A flowchart for the preconsolidation analysis—determination of the pore pressure from porosity is shown in Figure A1.

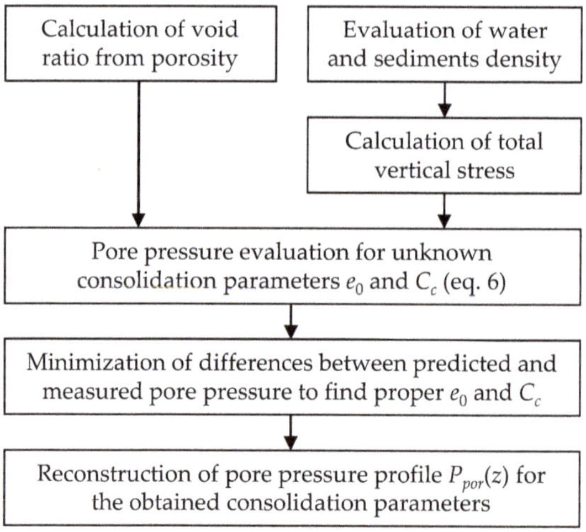

Figure A1. Flowchart for the determination of pore pressure through consolidation analysis.

A flowchart for the forward problem solution—determination of elastic wave velocities from the model parameters—is shown in Figure A2.

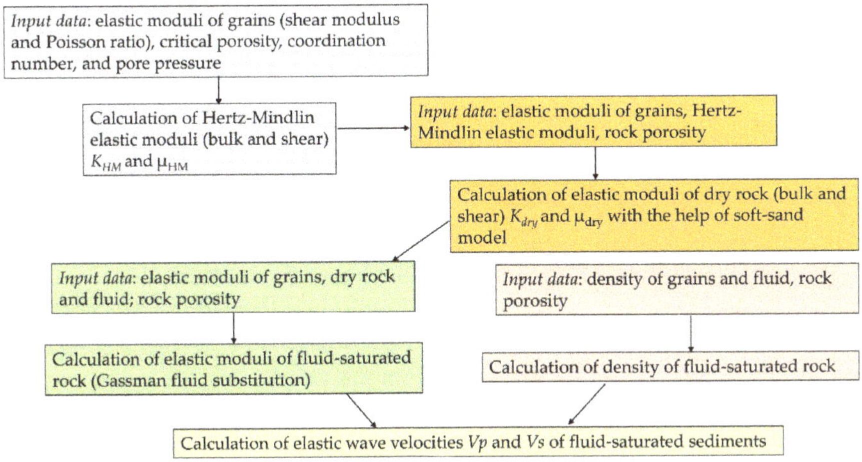

Figure A2. Flowchart for the determination of elastic wave velocities from the model parameters.

A flowchart for the inverse problem—determination of the pore pressure from seismic data is shown in Figure A3.

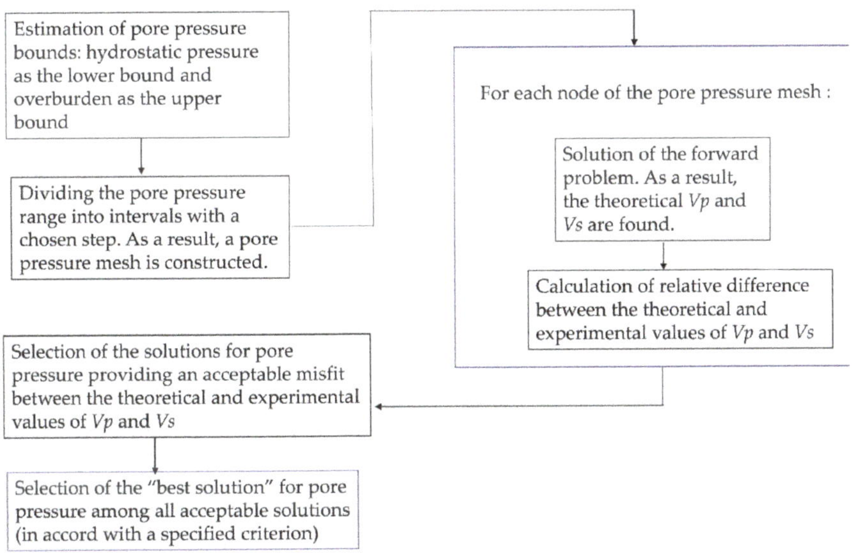

Figure A3. Flowchart for the determination of the pore pressure from seismic data.

References

1. Zhang, J. Effective stress, porosity, velocity and abnormal pore pressure prediction accounting for compaction disequilibrium and unloading. *Mar. Pet. Geol.* **2013**, *45*, 2–11. [CrossRef]
2. Holand, P.; Skalle, P. *Deepwater Kicksand and BOP Performance*; SINTEF Report for U.S.; Minerals Management Service: Washington, DC, USA, 2001; 123p.
3. Skalle, P.; Podio, A.L. Trends extracted from 1200 Gulf Coast blowouts during 1960–1996. *World Oil* **1998**, *219*, 67–72.
4. Tobin, H.J.; Saffer, D.M. Elevated fluid pressure and extreme mechanical weakness of a plate boundary thrust, Nankai Trough subduction zone. *Geology* **2009**, *37*, 679–682. [CrossRef]

5. Tingay, M.R.P.; Hillis, R.R.; Swarbrick, R.E.; Morley, C.K.; Damit, A.R. Origin of overpressure and pore-pressure prediction in the Baram province, Brunei. *AAPG Bull.* **2009**, *93*, 51–74. [CrossRef]
6. Yin, Z.; Linga, P. Methane hydrates: A future clean energy source. *Chin. J. Chem. Eng.* **2019**, *27*, 2026–2036. [CrossRef]
7. Bhatnagar, G.; Chapman, W.G.; Hirasaki, G.J.; Dickens, G.R.; Dugan, B. Effect of Overpressure on Gas Hydrate Distribution. In Proceedings of the 6th International Conference on Gas Hydrates (ICGH 2008), Vancouver, BC, Canada, 6–10 July 2008.
8. Ren, J.; Liu, X.; Niu, M.; Yin, Z. Effect of sodium montmorillonite clay on the kinetics of CH_4 hydrate-implication for energy recovery. *Chem. Eng. J.* **2022**, *437*, 135368. [CrossRef]
9. York, P.; Prithard, D.; Dodson, J.K.; Dodson, T.; Rosenberg, S.; Gala, D.; Utama, B. Eliminating Non-productive Time Associated Drilling Trouble Zone. In Proceedings of the Offshore Technology Conference, Houston, TX, USA, 4–7 May 2009. [CrossRef]
10. Holand, P.; Awan, H. *Reliability of Deepwater Subsea BOP Systems and Well Kicks*; ExproSoft AS Report no. 201252/02; Exprosoft: Trondheim, Norway, 2012; 164p.
11. Hottmann, C.E.; Johnson, R.K. Estimation of Formation Pressures from Log-Derived Shale Properties. *J. Pet. Technol.* **1965**, *17*, 717–722. [CrossRef]
12. Eaton, B.A. The Effect of Overburden Stress on Geopressures Prediction from Well Logs. *J. Pet. Technol.* **1972**, *24*, 929–934. [CrossRef]
13. Eaton, B.A. The Equation for Pore Pressure Prediction from Well Logs. In Proceedings of the Fall Meeting of the Society of Petroleum Engineers of AIME, Dallas, TX, USA, 28 September–1 October 1975.
14. Bowers, G.L. Pore Pressure Estimation from Velocity Data; Accounting for Overpressure Mechanisms Besides Undercompaction. *SPE Drill. Compl.* **1995**, *10*, 89–95. [CrossRef]
15. Zhang, J. Pore pressure prediction from well logs: Methods, modifications, and new approaches. *Earth-Sci. Rev.* **2011**, *108*, 50–63. [CrossRef]
16. Wang, Z.; Wang, R. Pore pressure prediction using geophysical methods in carbonate reservoirs: Current status, challenges and way ahead. *J. Nat. Gas Sci. Eng.* **2015**, *27*, 986–993. [CrossRef]
17. Mondol, N.H.; Bjorlykke, K.; Jahren, J.; Hoeg, K. Experimental mechanical compaction of clay mineral aggregates–changes in physical properties of mudstones during burial. *Mar. Pet. Geol.* **2007**, *24*, 289–311. [CrossRef]
18. Schneider, J.; Flemings, P.B.; Dugan, B.; Long, H.; Germaine, J.T. Overpressure and consolidation near the seafloor of Brazos-Trinity Basin IV, northwest deepwater Gulf of Mexico. *J. Geophys. Res.* **2009**, *114*, B05102. [CrossRef]
19. Long, H.; Flemings, P.B.; Germaine, J.T. Interpreting in situ pressure and hydraulic properties with borehole penetrometers in ocean drilling: DVTPP and Piezoprobe deployments at southern Hydrate Ridge, offshore Oregon. *J. Geophys. Res.* **2007**, *112*, B04101. [CrossRef]
20. Sultan, N.; Lafuerza, S. In situ equilibrium pore-water pressures derived from partial piezoprobe dissipation tests in marine sediments. *Can. Geotech. J.* **2013**, *50*, 1294–1305. [CrossRef]
21. Radwan, A.E.; Abudeif, A.M.; Attia, M.M.; Mohammed, M.A. Pore and fracture pressure modeling using direct and indirect methods in Badri Field, Gulf of Suez, Egypt. *J. Afr. Earth Sci.* **2019**, *156*, 133–143. [CrossRef]
22. Rostami, S.A.; Kinik, K.; Gumus, F.; Kirchhoff, M. Dynamic Calibration of the Empirical Pore Pressure Estimation Methods Using MPD Data. In Proceedings of the Offshore Technology Conference, Houston, TX, USA, 4–7 May 2015. [CrossRef]
23. Ganguli, S.S.; Sen, S. Investigation of present-day in-situ stresses and pore pressure in the south Cambay Basin, western India: Implications for drilling, reservoir development and fault reactivation. *Mar. Pet. Geol.* **2020**, *118*, 104422. [CrossRef]
24. Long, H.; Flemings, P.B.; Germaine, J.T.; Saffer, D.M. Consolidation and overpressure near the seafloor in the Ursa Basin, Deepwater Gulf of Mexico. *Earth Planet. Sci. Let.* **2011**, *305*, 11–20. [CrossRef]
25. Han, T.; Pervukhina, M.; Clennell, M.B.; Dewhurst, D.N. Model-based pore-pressure prediction in shales: An example from the Gulfof Mexico, North America. *Geophysics* **2017**, *82*, M37–M42. [CrossRef]
26. Ziegler, F.E.; Jones, J.F. Predrill pore-pressure prediction and pore pressure and fluid loss monitoring during drilling: A case study for a deepwater subsalt Gulf of Mexico well and discussion on fracture gradient, fluid losses, and wellbore breathing. *Interpretation* **2014**, *2*, SB45–SB55. [CrossRef]
27. Zhang, J.; Yin, S. Real-Time Pore Pressure Detection: Indicators and Improved Methods. *Geofluids* **2017**, *2017*, 3179617. [CrossRef]
28. Oloruntobi, O.; Adedigba, S.; Khan, F.; Chunduru, R.; Butt, S. Overpressure prediction using the hydro-rotary specific energy concept. *J. Nat. Gas Sci. Eng.* **2018**, *55*, 243–253. [CrossRef]
29. Tanikawa, W.; Hirose, T.; Hamada, Y.; Gupta, L.P.; Ahagon, N.; Masaki, Y.; Abe, N.; Wu, H.Y.; Sugihara, T.; Nomura, S.; et al. Porosity, permeability, and grain size of sediment cores from gas-hydrate-bearing sites and their implication for overpressure in shallow argillaceous formations: Results from the national gas hydrate program expedition 02, Krishna-Godavari Basin, India. *Mar. Pet. Geol.* **2019**, *108*, 332–347. [CrossRef]
30. Mindlin, R.D. Compliance of elastic bodies in contact. *J. Appl. Mech.* **1949**, *16*, 259–268. [CrossRef]
31. Brandt, H. A study of the speed of sound in porous granular media. *J. Appl. Mech.* **1955**, *22*, 479–486. [CrossRef]
32. Digby, P.J. The effective elastic moduli of porous granular rocks. *J. Appl. Mech.* **1981**, *48*, 803–808. [CrossRef]
33. Walton, K. The effective elastic moduli of a random packing of spheres. *J. Mech. Phys. Sol.* **1987**, *35*, 213–226. [CrossRef]
34. Jenkins, J.; Johnson, D.; La Ragione, L.; Makse, H. Fluctuations and the effective moduli of an isotropic, random aggregate of identical, frictionless spheres. *J. Mech. Phys. Sol.* **2005**, *53*, 197–225. [CrossRef]
35. Dvorkin, J.; Prasad, M.; Sakai, A.; Lavoie, D. Elasticity of marine sediments. *Geophys. Res. Lett.* **1999**, *26*, 1781–1784. [CrossRef]

36. Mavko, G.; Mukerdji, T.; Dvorkin, J. *Rock Physics Handbook*, 3rd ed.; Cambridge University Press: Cambridge, UK, 2020; pp. 220–308.
37. Hashin, Z.; Shtrikman, S. A variational approach to the elastic behavior of multiphase materials. *J. Mech. Phys. Sol.* **1963**, *11*, 127–140. [CrossRef]
38. Polukhtovich, V.M.; Popadok, I.V.; Samarski, A.D.; Khnykin, V.I. Features of the geological structure and prospects for oil and gas potential in the southwestern part of the Indolo-Kuban trough. *Geol. Oil Gas* **1981**, *1*, 43–47.
39. Gorshkov, A.S.; Meisner, L.B.; Soloviev, V.V.; Tugolesov, D.A.; Khakhalev, E.M. *Album of Structural Maps and Thickness Maps of the Cenozoic Deposits of the Black Sea Depression, m 1:1,500,000*; Tugolesov, D. A.M., GUGK pri CM USSR: Moscow, Russia, 1989.
40. Bogaets, A.T.; Bondarchuk, G.K.; Les'kiv, I.V.; Novociletskiy, R.M.; Pavlyuk, M.I.; Paliy, A.M.; Panchenko, D.E.; Samarski, A.D.; Khnykin, V.I.; Chir, N.M.; et al. *Geology of the Shelf of the Ukrainian SSR. Oil and Gas Potential*; Naukova Dumka: Kyiv, Ukraine, 1986; 152p.
41. Terekhov, A.A.; Shimkus, K.M. Young sediments and thrust structures in the Prikrymskaya and the Caucasian zones of the Black Sea depression. *Geotectonics* **1989**, *1*, 72–79.
42. Bagriy, I.D.; Voitsitsky, Z.I.; Maslun, N.V.; Naumenko, U.Z.; Aksem, S.D.; Grieg, M.Y. Integrated geological-structural-thermo-atmo-geochemical studies—A tool for forecasting and searching for hydrocarbons and methane hydrates on the continental slope of the Black Sea. In *Geology and Mineral Resources of the World Ocean*; National Academy of Sciences of Ukraine: Kyiv, Ukraine, 2014; Volume 4, pp. 24–47.
43. Gozhik, P.F.; Maslun, N.V.; Plotnikova, L.F.; Ivanik, M.M.; Yakushin, L.M.; Ishchenko, I.I. *Stratigraphy of the Meso-Cenozoic Clades of the Peninsular-Western Shelf of the Black Sea*; Institute of Geological Sciences of the National Academy of Sciences of Ukraine: Kyiv, Ukraine, 2006; 171p.
44. Maslun, N.V.; Ivanik, M.M.; Tsikhotska, N.N.; Klyushina, G.V. *Detailed Stratification of Maikop Clade in the Peninsula-Zahodny Shelf of the Black Sea*; Biostratigraphic Criteria for the Analysis and Correlations in the Clade of the Phanerozoic of Ukraine; Institute of Geological Sciences of the National Academy of Sciences of Ukraine: Kyiv, Ukraine, 2005; pp. 153–159.
45. Tsytovich, N.A. *Soil Mechanics*; Gosstroiizdat: Moscow, Russia, 1963; pp. 73–76. (In Russian)
46. Vardy, M.E. Deriving shallow-water sediment properties using post-stack acoustic impedance inversion. *Near Surf. Geophys.* **2014**, *13*, 143–154. [CrossRef]
47. Pirogova, A.S.; Tikhotskii, S.A.; Tokarev, M.Y.; Suchkova, A.V. Estimation of Elastic Stress-Related Properties of Bottom Sediments via the Inversion of Very- and Ultra-High-Resolution Seismic Data. *Izv. Atmos. Ocean. Phys.* **2019**, *55*, 1755–1765. [CrossRef]
48. Zoback, M.D. *Reservoir Geomechanics*, 1st ed.; Cambridge University Press: Cambridge, UK, 2007; 490p.
49. Dubinya, N.V.; Vershinin, A.V.; Pirogova, A.S.; Tikhotsky, S.A. Usage of Imitational Geological-Petrophysical Models to Reduce Drilling Risks for Offshore Reservoirs Exploration. In Proceedings of the SPE Russian Petroleum Technology Conference 2020, Moscow, Russia, 26–29 October 2020. [CrossRef]
50. Becker, D.E.; Crooks, J.H.A.; Been, K.; Jefferies, M.G. Work as a criterion for determining in situ and yield stresses in clays. *Can. Geotech. J.* **1987**, *24*, 549–564. [CrossRef]
51. Saffer, D.M.; Silver, E.A.; Fisher, A.T.; Tobin, H.; Moran, K. Inferred pore pressures at the Costa Rica subduction zone: Implications for dewatering processes. *Earth Planet. Sci. Lett.* **2000**, *177*, 193–207. [CrossRef]
52. Dugan, B.; Germaine, J. Near-seafloor overpressure in the deep-water Mississippi Canyon, northern Gulf of Mexico. *Geophys. Res. Lett.* **2008**, *35*, L02304. [CrossRef]
53. Jamiolkowski, M. New Developments in Field and Laboratory Testing of Soils. In Proceedings of the 11th International Conference on Soil Mechanics and Foundation Engineering, San Francisco, CA, USA, 12–16 August 1985.
54. Gassmann, F. *Über die Elastizität Poröser Medien*; Vierteljahrsschrift der Naturforschenden Gesellschaft in Zurich: Zurich, Switzerland, 1951; Volume 96, pp. 1–23.
55. Bayuk, I.; Ammerman, M.; Chesnokov, E. Elastic moduli of anisotropic clay. *Geophysics* **2007**, *72*, D107–D117. [CrossRef]
56. Katahara, K.W. Clay minerals elastic properties. *SEG Tech. Program Exp. Abstr.* **1996**, 1691–1694. [CrossRef]
57. Vasilev, I.; Dubinya, N.; Tikhotskiy, S.; Nachev, V.; Alekseev, D. Numerical model of jack-up rig's mechanical behavior under seismic loading. *Comput. Res. Model.* **2022**, *14, accepted*.

Journal of
Marine Science and Engineering

Article

Islands in the Caucasian Sea in Three Mesozoic Time Slices: Novel Dimension of Geoheritage and Geotourism

Dmitry A. Ruban [1,2]

[1] Department of Organization and Technologies of Service Activities, Higher School of Business, Southern Federal University, 23-ja Linija Street 43, 344019 Rostov-on-Don, Russia; ruban-d@mail.ru
[2] K.G. Razumovsky Moscow State University of Technologies and Management (the First Cossack University), Zemlyanoy Val Street 73, 109004 Moscow, Russia

Abstract: Framing geoheritage thematically is important to reveal its diversity. Field investigations in the western part of the Greater Caucasus orogen have allowed for the characterization of three localities representing palaeoislands of the Caucasian Sea, which evolved as a semi-enclosed, marginal palaeosea during the Mesozoic. The Gosh locality represents coarse siliciclastics formed on the cliffed shore of the early Induan (Early Triassic) island. The Lipovy locality exhibits conglomerates accumulated on the shore of the early Toarcian (Early Jurassic) island due to erosion of the uplifted crystalline rocks. The Shakhan locality boasts a representative section of cross-bedded sandstones deposited on the alluvial plain of the Hauterivian (Early Cretaceous) island. All these localities are interpreted as geoheritage points, which are parts of larger geosites. Taken together, these points constitute thematic geoheritage sites reflecting the existence of palaeoislands in all Mesozoic periods, which is essential for the understanding of the evolution of the Caucasian Sea during this era. These localities are perfectly accessible, but visiting them requires professional interpretation. The importance of the thematic geoheritage makes its adequate management urgent (particularly, maintenance of geoheritage points, the installation of interpretive panels, and promotion). A geoexcursion route is proposed to facilitate geotouristic exploitation of the characterized geoheritage sites.

Keywords: geosites; Induan; Hauterivian; palaeosea; siliciclastics; Toarcian; Western Caucasus

Citation: Ruban, D.A. Islands in the Caucasian Sea in Three Mesozoic Time Slices: Novel Dimension of Geoheritage and Geotourism. *J. Mar. Sci. Eng.* **2022**, *10*, 1300. https://doi.org/10.3390/jmse10091300

Academic Editor: János Kovács

Received: 12 August 2022
Accepted: 13 September 2022
Published: 15 September 2022

1. Introduction

The spectrum of unique geological phenomena is very wide, which determines a high diversity of geoheritage. On the one hand, this diversity itself strengthens the need for geoheritage conservation in general (indeed, each case also needs close attention). On the other hand, it challenges effective geoheritage management and use in tourism, which are difficult to organize properly in heterogeneous geological environments. If so, it is very reasonable to find approaches for the systematic, logical treatment of unique geological phenomena to receive true benefits from their diversity. One approach is putting geoheritage into thematic frames. Poiraud et al. [1] paid attention to five thematic spaces (glaciotectonic, palaeogeomorphological, hydrogeological, morphotectonical, and stratigraphical) of the Vercors massif in the French Alps for the better development of geoeducation and geotourism. Migoń et al. [2] thematically grouped the granite geomorphological heritage of the Waldviertel region in Lower Austria. Herrera-Franco et al. [3] analyzed thematic travel itineraries in the Santa Elena province of Ecuador.

Indeed, grouping geoheritage features along the geological time arrow [4] or the sequence of depositional environments [5] seems to be very promising for communicating local geological knowledge. Particularly, such solutions allow one to present this knowledge logically for better comprehension. The gradually developed idea of palaeogeographical geoheritage, which is understood as either a particular geoheritage type [6] or highly complex phenomena [7], helps to justify such thematic frames. Nonetheless, more examples from different geological domains are necessary to realize the full potential of this approach and the spectrum of options for its application. Particularly, rare but often spectacular legacies of palaeoislands [6,8–11] constitute a promising theme for geoheritage treatment and geotourism development.

The Caucasus is a large, geologically rich domain in the northern flank of the Cenozoic active tectonic belt stretching through Eurasia for thousands of kilometers [12–19]. Its largest, northern part, namely, the Greater Caucasus, can serve as a "natural laboratory" for studying Mesozoic sedimentary complexes, which include widely distributed Triassic, Jurassic, and Cretaceous deposits of different natures. These deposits formed chiefly in marine palaeoenvironments of the vast Caucasian Sea, the deepest part of which corresponded to the "axis" of the present-day mountain chain. The Mesozoic marine evolution of the Greater Caucasus is well-documented and characterized, particularly in the synthetic work by Yasamanov [20]. The palaeogeographical schemes in this work depict an elongated island (or chain of islands) at the southern border of the mentioned palaeosea, which existed more or less continuously through the Mesozoic; the idea of the Caucasian island(s) corresponds well with the reconstructions of the long-lived island arc along the southern periphery of the present-day Greater Caucasus [15,21–25]. However, too little is known about this island mass, chiefly on the basis of indirect evidence. Filling this gap in the geological knowledge is essential for the understanding of the configuration and openness of the Mesozoic Caucasian Sea, and some new information would be precious in regard to the regional geological oceanography. Moreover, geological features related to the Mesozoic islands would extend the thematic frames of geoheritage and geotourism in the Greater Caucasus. If several lines of evidence of palaeoislands are found at different geosites, the latter can be grouped for the representation of an important palaeogeographical theme, which seems to be potentially attractive to geotourists.

New field investigations in the western part of the Greater Caucasus have allowed for the documentation of geological objects, the palaeogeographical interpretation of which indicates the existence of island mass in three time slices representing all periods of the Mesozoic, namely, the Induan (Triassic), the Toarcian (Jurassic), and the Hauterivian (Cretaceous). This new knowledge is valuable not only geologically due to the scarcity of information about palaeoislands from the study area but also in regard to geoheritage management and geotourism development, which can become more definite when particular themes are established. The study area boasts rather active exploitation of its rich geoheritage resources [26], and, thus, the theme of Mesozoic palaeoislands can be demanded there. The objective of the present contribution is to characterize the newly established geoheritage related to the Mesozoic island(s) in the Caucasian Sea. A new, palaeogeography-related geoheritage and geotourism theme is proposed. Although Mountainous Adygeya has been investigated in regard to geoheritage and geotourism for many years [4,5,26], this study presents new lines of evidence and offers novel interpretations; thus, it extends the available knowledge and proposes a new vision for geoheritage management and geotourism development.

2. Geological Setting

The study area corresponds to Mountainous Adygeya, a distinctive area of the Western Caucasus (western part of the Greater Caucasus) in southwestern Russia (Figure 1a) where geoheritage resources and geotourism concentrate [26]. It stretches along the valleys of the Belaya River and its main tributaries such as the Syryf River and the Dakh River. The area is crossed by several subparallel ranges, which trend from the west to the east. The Pastbischny Range is located to the north of Kamennomostsky, the Skalisty Range is located between Kamennomostsky and Dakhovskaya, the Burelom Range is located between Dakhovskaya and Khamyshki, and the Inzhenerny Range is located between Khamyshki and Guzeripl. Where the Belaya River crosses these ranges, it forms narrow gorges and canyons. The heights of mountains are below 1500 m. Although this area is covered by dense forests and is not populated densely, there are several settlements (these also serve as important touristic centers), the biggest of which is Kamennomostsky town. The high-quality paved and unpaved roads cross the area and determine the perfect accessibility of its geoheritage (Figure 1a).

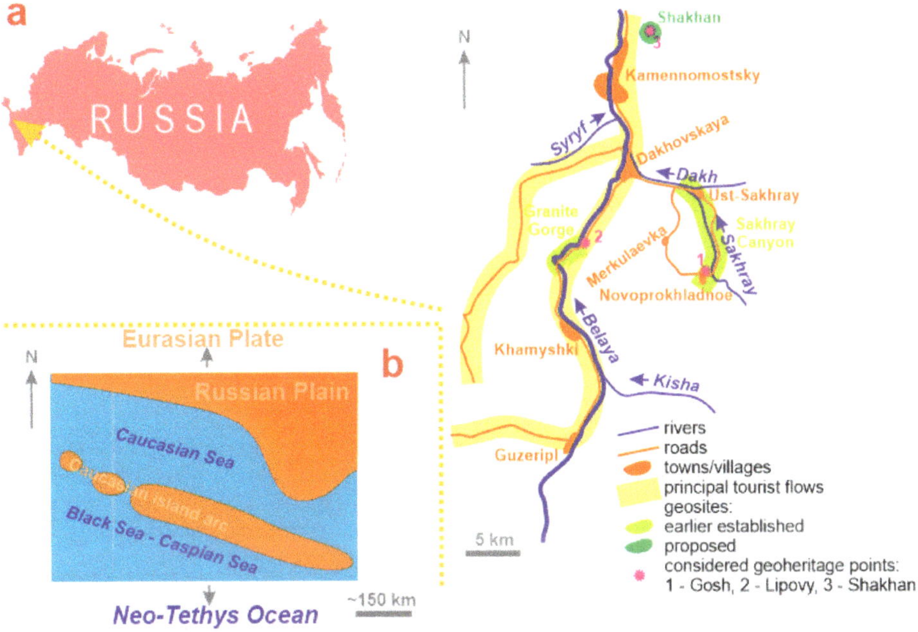

Figure 1. Geographical location of the study area, with focus on touristic infrastructure and geoheritage (**a**), and the general outline of the Mesozoic Caucasian Sea between the rather stable landmass in the north and the active island arc in the south (**b**).

Geologically, this area represents a Late Cenozoic orogen (fold-thrust belt) formed as a result of the collision of the Eurasian and Arabian plates with the interaction of smaller microplates [12,15,16,19]. According to the latest developments, the Greater Caucasus represents a deformation zone with a certain individuality related to the Eurasian plate [27]. Apparently, this individual domain corresponds to the Paleozoic Greater Caucasus terrain [28], and, if so, it is possible that the latter remains rather active in the regional plate tectonic framework.

The stratigraphical framework of the study area was summarized, particularly, by Plyusnina et al. [4]. The oldest are Precambrian metamorphic rocks (gneisses, amphibolites, and schists). Together with Early-Middle Paleozoic serpentinites and Late Paleozoic granitoids, these are represented in several small massifs (Dakh, Sakhray, and Rufabgo), which are uplifted basement blocks. The sedimentary complexes include the Early-Middle Permian molassic sequence (up to 1000–2000 m in thickness), Triassic carbonate and siliciclastic deposits (~1700 m in thickness), Early-Middle Jurassic siliciclastic, often dark shale-dominated deposits (up to 10,000 m in thickness), Late Jurassic carbonate and evaporite deposits (~3000 m in thickness), Early Cretaceous mixed carbonate and siliciclastic deposits (up to 500 m in thickness), and Quaternary deposits (alluvial, deluvial, colluvial, and locally glacial). The Jurassic complexes are spread most widely.

The Mesozoic deposits accumulated in the tropical, elongated, semi-enclosed Caucasian Sea with rich ecosystems, which corresponded to the back-arc basin between the rather stable platform domains in the north and the active island arc in the south (Figure 1b). Although its configuration changed due to global eustatic fluctuations and regional tectonic movements (both horizontal and vertical), this palaeosea remained generally the same since at least the Jurassic period [20]. The Triassic palaeogeography remains almost unknown, although it is very likely that the general situation (marginal palaeosea setting) was the same. The Cenozoic orogeny led to closure of the marine basin and the general uplift of the area. The related deformations triggered the intense folding and faulting of the Mesozoic sedimentary complexes.

3. Materials and Methods

The materials for the present study were collected in the course of field investigations in Mountainous Adygeya in July–August 2022. A total of three localities representing terrestrial palaeoenvironments were examined, namely, Gosh, Lipovy, and Shakhan (Figure 1a). Each of them was visited, described, and photographed. Attention was paid to both the geological peculiarities and geoheritage properties (see below). The local geological setting was mapped carefully to facilitate further interpretations, which are really challenging in deformed and poorly known domains (this is especially the case for the Gosh locality, with highly complex contacts of strata of different ages). In regard to geoheritage, field investigations required (among other things) the evaluation of accessibility and the approximate time and distances for reaching these localities.

The collected materials were first interpreted in terms of stratigraphy and palaeogeography. For these purposes, both original data and the earlier published developments were used (these are indicated below, in each particular case). The geological time scale developed by the International Commission on Stratigraphy [29] and the regional palaeogeographical reconstructions by Yasamanov [20] were helpful in these interpretations. The geological analysis is usually brief, simple, and somewhat preliminary to match the purposes of the present contribution aimed at thematic geoheritage.

New field observations coupled with the results of the previous assessment of unique geological phenomena of Mountainous Adygeya [26] allow for judgments of the three studied localities in terms of geoheritage, which would facilitate subsequent geotourism-related interpretations. Two main foundations of the present study should be noted. First, direct evidence of palaeoislands seems to be rare and thus unique; the related objects can always be attributed to the palaeogeographical type of geoheritage [6,7]. Second, although the basic elements of geoheritage are geosites, the latter are often large (up to several square kilometers in size) and complex (due to the coexistence of several geoheritage types), and, thus, particular unique features can be represented in their fragments, which are known as geoheritage points [30] that have something in common with geotopes [31]. Dealing with such fragments seems to be very suitable when single-theme geoheritage is considered and geosites/geoparks are "diffused" [30,32]. Of course, the correspondence of geoheritage points to geosites should be traced.

Various approaches have been proposed for the assessment of geoheritage and geosites [33–39]. In Mountainous Adygeya, a comprehensive assessment of geoheritage resources has been realized recently by Ruban et al. [26]. Three geoheritage points considered in the present study are only fragments of geosites, and, thus, it is not reasonable to undertake their quantitative analysis with the noted approaches aimed at geosites. Nonetheless, they require qualitative description, which can be based on the principles specified by Ruban [38], with certain additions and minor modifications. A template for descriptions is composed (Table 1), and it is used for each considered locality. Several elements and criteria of description are related to the correspondence of geoheritage localities to geosites. Spatial correspondence (the overlap between points and geosites) and contribution to geosite uniqueness (the "weight" of a given point relative to the value of the entire geosite) should be distinguished from cohesion. The latter signifies the isolation of a given point from the rest of the geosite.

Table 1. Template for characterizing geoheritage points.

Elements/Criteria	Essence	Significance
Name	Name of geoheritage point, which may or may not correspond to the name of the geosite	Basic
Geological outline	Brief geological characteristics (descriptions and interpretations) based on field investigations and literary evidence	
Thematic outline	Relevance of geological peculiarities to thematic geoheritage	
Spatial correspondence to geosite	Full correspondence, partial correspondence, or a tiny portion of a geosite	Essential
Novelty of geosite *	Earlier-established geosite, newly proposed geosite, tentatively proposed geosite needing further examination	
Contribution to geosite uniqueness	Whether this locality determines geosite uniqueness fully, partly, or minimally	
Complexity	Coexistence of several geoheritage types	Technical
Cohesion	Isolated position within the geosite or continuous links to the geosite	
Accessibility	Outer and inner accessibility (sensu [40]) and seasonality of access	
Vulnerability	No danger, potential danger, partial damage, or significant damage; danger/damage factor active or inactive	
Interpretation needs	Clear to non-experts or only experts; professional guidance required or not	
Aesthetics	Aesthetic-related features and their relative appeal to visitors' sense of beauty (see aesthetic properties in [41])	
Importance	Interesting to research audience, students or schoolchildren, and tourists	Utilitarian

Note: all items are related to a given geoheritage point, except for that marked by an asterisk and related to the entire geosite.

4. Results

4.1. Gosh

4.1.1. Description

The Gosh locality is named after the Gosh River, which is a left tributary of the Sakhray River belonging to the Belaya River watershed (Figure 1a). It is situated in the northeastern periphery of Novoprokhladnoe village, where Triassic deposits crop out in the valley of the Gosh River. The local geology is highly complex (Figure 2). The area is dominated by Early Jurassic shales dipping to the southwest. They overlay transgressively Early Triassic limestones dipping generally northward (in fact, the dipping direction changes from northwestern to northeastern due to the intense deformation of beds) (Figure 2). This situation is represented well in the valley of the Kamennaya River, which is a right tributary of the Gosh River. However, the detailed investigation of the valley of the Gosh River has led to the discovery of two unusual beds of coarse siliciclastics near the contact between the Triassic and the Jurassic. These beds differ strikingly from the Early Jurassic deposits and dip in the same direction as the Early Triassic limestones. Evidently, they underlay the latter, and their absence in the valley of the Kamennaya River can be explained by the non-exposure of the basal part of the Triassic sequence. The lower bed, which is ~10 m in thickness, is represented by conglomerate-breccia with poorly sorted coarse particles (pebbles and cobbles) of very different shapes (from angular to perfectly rounded) consisting of pre-Triassic igneous, metamorphic, and sedimentary rocks known locally. The upper bed, which is <3 m in thickness, consists of coarse and very coarse sandstone bearing flat pebbles and cobbles, as well as irregular particles of a larger size (megaclasts) (Figure 2).

Figure 2. Gosh locality: geological scheme and principal Triassic lithologies (hammer's length is 40 cm).

The Triassic stratigraphy of the Western Caucasus has been developed for many decades (ages of formations were established on the basis of massive palaeontological findings, including ammonoids, brachiopods, and foraminifers), and the main outcomes of these investigations were summarized by Chaitsky et al. [42], Gaetani et al. [43], and Rostovtsev et al. [44]. According to these developments, the conglometrate-breccia and megaclast-bearing beds can be attributed to the Bambak Formation (early Induan), and the overlaying limestones can be attributed to the Yatyrgvarta Formation (Induan–Olenekian). The accumulation of coarse siliciclastics took place on ancient shores. Probably, the lower bed formed at the toe of cliffs where angular slope debris particles mixed with cobbles and pebbles rounded by wave activity. The upper bed marks cliff retreat and beach extension. Hypothetically, megaclasts were emplaced by tsunamis or severe tropical storms—analogues are known from both present and ancient sedimentary archives [45–51]. The accumulation of coarse siliciclastics and the subsequent shift to carbonate deposition mark the transgression of the palaeosea.

Although the Bambak Formation overlays crystalline rocks, it also bears clasts consisting of Permian rocks [42] cropped out in the neighboring areas. The latest Permian deposits are Changhsingian in age, and marine deposition continued at the Permian/Triassic boundary [52]. If so, it is possible that the relative sea-level fall, which preceded the above-mentioned transgression that started already in the early Induan, was very short-term but pronounced. It cannot be brought in correspondence to any global eustatic fall [53], and, therefore, it should be explained by local uplifts immediately after the Permian/Triassic boundary. Notably, the Bambak Formation formed only locally in Mountainous Adygeya, and marine sandstones and probably even limestones accumulated contemporaneously. Hypothetically, the palaeosea did not retreat but only changed its configuration because of the noted uplifts. If so, and taking into account the palaeogeographical position of the study area far from the "stable" craton, the ancient cliffed shore interpreted at the Gosh locality can be attributed to any palaeoisland. The latter was quickly submerged together with the Induan transgression, as implied by the wide distribution of the generally monotonous Yatyrgvarta Formation.

4.1.2. Geoheritage

The characteristics given above imply that the Gosh locality allows for interesting insights into the earliest Triassic palaeogeography of the Western Caucasus, and it is related directly to the thematic geoheritage considered in the present study. This locality is a geoheritage point of a large, earlier-established geosite, namely, the Sakhray Canyon (Figure 1a), which is famous for waterfalls and its extensive Triassic section [26]. The small locality represents the edge of this geosite. The features reported from the valley of the Gosh River contribute to the uniqueness of the entire geosite, but they do not determine its value. The dominant geoheritage type in this point is palaeogeographical, but one should also note the existence of the stratigraphical type (a section of the Bambak Formation outside of where it is reported commonly) and the sedimentary type (megaclasts). The Gosh geoheritage point occupies an isolated position relative to the rest of the geosite, although one can reach the Sakhray Canyon from there either by going down the stream or preferring an unpaved road.

The outer accessibility of this geoheritage point is excellent because an unpaved but well-maintained road leads directly to the Gosh River in Novoprokhladnoe village (Figure 3). The inner accessibility is more or less perfect: the river valley is rather wide and free of woody debris, although one should be prepared for hiking directly in streams with masses of pebbles and cobbles. In other words, some experience is required for reaching the target outcrops. Indeed, visits there can be organized in only good weather (without rain) in summer. This locality is not vulnerable to any negative factors, although the occasional accumulation of woody debris may make visiting the outcrops challenging. Indeed, expert knowledge and/or professional guidance are necessary for visiting this geoheritage point with profit, i.e., the highly complex local geology, including the stratigraphical framework and the palaeogeographical hypotheses, should be communicated. The aesthetic properties of this geoheritage point, determined by its genuine wilderness, unusual view of megaclast-bearing sandstone, and small waterfalls, recompense interpretation needs.

Figure 3. An unpaved scenic road leading to Novoprokhladnoe from Dakhovskaya.

The scientific importance of the Gosh locality is undisputable because it allows for the undertaking of research in the Triassic palaeogeography (*sensu lato*) of the Greater Caucasus, which remains poorly known. Although this research seems to be only regionally important at a first glance, one should take into account the "key" position of the Greater Caucasus for revealing the palaeogeographical setting between the domains of South Europe and the Middle East. The previous investigation of an international research group [43] proves the significance of such studies. Therefore, this geoheritage point seems to be interesting to the international research audience. The educational importance is determined by the possibility to demonstrate examples of deposits of ancient shores to university students (at least three large Russian universities organize summer field practices for their students in geology and geography, and some others do this irregularly), as well as to offer them some tasks in structural geology in highly deformed sedimentary domains. The touristic importance is discussed separately (see below).

4.2. Lipovy

4.2.1. Description

The Lipovy locality is named after the Lipovy River, which is a right tributary of the Belaya River (Figure 1a). It is located near the northern entrance to the Granite canyon, where there is a sharp boundary between Mesozoic sedimentary complexes and the Precambrian–Paleozoic crystalline rocks, and both crop out in the valley of the Lipovy River. The local geology has been deciphered previously by Ruban [54], Nenakhov et al. [55], and Chepurnoy [56]. The area is dominated by the rocks of the Dakh Crystalline Massif. Although it is composed chiefly of Late Paleozoic granitoids, Precambrian metamorphics with sporadic bodies of Early–Middle Paleozoic serpentinites constitute its northern periphery (Figure 1a). Near the mouth of the Lipovy River, these rocks are overlain transgressively by horizontally lying conglomerates (Figure 4) with a thickness of ~2 m, bearing thin (5–15 cm) sandstone interbeds. The conglomerates consist of poorly sorted, chiefly rounded (but the angularity of some particles is also evident) pebbles, cobbles, and boulders representing igneous and metamorphic rocks of the Dakh Crystalline Massif [56].

Figure 4. Lipovy locality: geological scheme and principal Jurassic lithology (hammer's length is 40 cm). The nature of the contact between crystalline rocks and overlying Jurassic deposits shown on the scheme is not clear everywhere (it can be transgressive and tectonic).

The Jurassic stratigraphy of the Western Caucasus has been developed for more than a century, and the present stratigraphical framework (ages of formations were established with ammonites, brachiopods, foraminifers, and other fossils) is given in the synthetic work by Rostovtsev et al. [57]. The latter specifies the Bagovskaya Formation (early Toarcian) with basal conglomerate beds. The conglomerates from the valley of the Lipovy River were attributed to these basal beds [54]. They accumulated on the shore of a small palaeoisland, which corresponded to the Dakh Crystalline Massif uplifted during the Late Triassic–Early Jurassic phase of tectonic activity [6,54]. This palaeoisland submerged together with the development of the Toarcian transgression and the related quick growth of the Caucasian Sea. These interpretations generally match the reconstructions undertaken by Panov et al. [58], who described the Early Jurassic evolution of the Western Caucasus and adjacent areas as a time of multiple graben development and related palaeogeographical reorganizations, as a result of which palaeoislands appeared and disappeared.

4.2.2. Geoheritage

The Lipovy locality, which is the best studied among the other considered localities, offers important information for the understanding of the Early Jurassic palaeogeography of the Western Caucasus when the Caucasian Sea evolved (Figure 1b). It is directly related to the thematic geoheritage considered in the present study. This is a geoheritage point of a larger, linear geosite, namely, the Granite Gorge. The latter was characterized comprehensively by Karpunin et al. [59], Mikhailenko et al. [60], and Ruban et al. [26], and it is famous for lengthy exposures of granitoids, metamorphic formations, and a deep river incision. The Lipovy geoheritage point is very small in size and marks the very northern edge of this geosite. Although the former is very important, its geoheritage value is incomparably less than that of the entire geosite. The Early Jurassic conglomerate beds overlay metamorphic rocks in the same outcrop where their most representative section is available. Some interesting and very rare mineral assemblage and highly specific rodingite rock are also found there. As a result, this locality represents, at least, four geoheritage types (palaeogeographical, metamorphic, mineralogical, and igneous—see Mikhailenko et al. [60] for more details), which indicates its significant complexity. The cohesion of the Lipovy geoheritage point to the Granite Gorge geosite is moderate. On the one hand, the conglomerate outcrop is linked physically to the neighboring outcrops of metamorphic rocks available in the Lipovy River valley and close to it along the road. On the other hand, the group of outcrops near the northern entrance to the gorge demonstrates individual features, and this group can be differentiated from the main chain of outcrops representing granitoids.

The outer accessibility of the geoheritage point is excellent because it is located close to the principal, paved, and well-maintained road. The inner accessibility is limited by the necessity to travel several dozens of meters along the Lipovy River, the valley of which has steep slopes and a lot of large clasts on the bottom (Figure 5). Moreover, one needs to climb several meters to reach the upper part of the slope where the conglomerates are exposed. Principally, this challenge can be addressed via creating a temporary trail with stairs, which would require no more than 1–2 h of work. The geoheritage point can be visited during the entire year, except for winter and spring days with bad weather conditions or high water levels. The dense tree cover protects visitors from heavy rain. A true problem is that the accommodation space in the locality and, thus, its carrying capacity are too limited; no more than three persons can stay near the conglomerates and no more than fifteen persons can stay at the slope's toe. The Lipovy locality is vulnerable to two negative factors. First, the maintenance of the road can lead to the slope cutting in the near future, which will either damage or even destroy the outcrop. Second, the valley of the Lipovy River is prone to landslides and rockfalls. One of them occurred a few years ago and already damaged the conglomerate outcrop (a part of it was detached and slid down). The geoheritage point can be clear to both experts and non-experts with elementary geological knowledge because conglomerates and their contact with the underlying crystalline rocks are easy-to-realize. However, professional guidance is required to learn about the age and the palaeogeographical interpretations. The locality does not boast any significant aesthetic properties, and it is permanently shadowed by tree cover (Figure 5).

Figure 5. The "entrance" to the valley of the Lipovy River from the road; the dense vegetation and tree cover should be noted.

The Lipovy geoheritage point has already been investigated intensively [54–56]. Indeed, some additional material for subsequent, international-level research can be gathered there. The educational importance of the locality is significant because it is an ideal site to learn about basal conglomerates. However, it is diminished by the limited carrying capacity. The touristic importance is discussed separately (see below).

4.3. *Shakhan*

4.3.1. Description

The Shakhan locality is named after the Shakhan Mountain, which is situated to the north of Kamennomostsky town and borders the valley of the Belaya River from the east (Figure 1a). The area is dominated by the Late Jurassic and Early Cretaceous deposits (Figure 6). The former are represented by variegated, often red siliciclastics, whereas the latter consist of carbonate (limestones) and siliciclastic (shales and sandstones) rocks, which often form mixed packages. Near the top of the Shakhan Mountain, on its southwestern slope, an abandoned sandstone quarry is located. There, a thick succession (up to 30 m) of yellowish-white, cross-bedded sandstones is represented. These rocks consist of coarse and very coarse sand particles, and they also include rare rounded gravels. Some layers exhibit brown patterns due to iron enrichment. Coalified plant remains (particularly, stem fragments) are rather numerous.

Figure 6. Shakhan locality: geological schemes and principal Cretaceous lithology with well-visible layering and cross-bedding (tape's length is 2 m).

The Cretaceous stratigraphy of the Western Caucasus has been developed for a long time, although some questions remain unresolved. The available stratigraphical frameworks justified by biostratigraphical developments (with ammonites and foraminifers) were characterized, particularly, by Drushits and Mikhailova [61], Prosorovskaya [62], and Mikerina and Pinchuk [63]. They assign the cross-bedded sandstones from the Shakhan locality to the Gubs Formation. The age of the latter is defined provisionally as Hauterivian. The sandstones formed, most probably, on the alluvial plain. However, the latter was close to the seashore because the Gubs Formation includes typically marine facies and fossils in the neighboring areas [62].

Taking into account the palaeogeographical setting of the Western Caucasus depicted by Yasamanov [20], it is evident that the Caucasian Sea separated the study area from the "northern" land masses, i.e., its general configuration (Figure 1b) remained in the Hauterivian. If so, the alluvial plain where the cross-bedded sandstones accumulated can be attributed to any large palaeoisland along the southern periphery of the palaeosea [20]. The sandstones dip to the northeast, although cross-bedding demonstrates almost northward dipping (Figure 4). This means that the initial (before tectonic deformation) direction of cross beds dipping was northwestern, which is the palaeocurrent direction. The latter proves the affinity of the alluvial plain to the palaeoisland. Alluvial plains on large islands are not so uncommon, and, particularly, these are known from the geological records of the Japanese Islands [64]. Although the information about the Hauterivian deposits of the Western Caucasus remains limited [61–63], it appears that the cross-bedded sandstones occur sporadically, and the related alluvial plain was relatively small in size and, thus, matched the dimension of the hypothesized palaeoislands.

4.3.2. Geoheritage

Information from the Shakhan locality advances the understanding of the Early Cretaceous palaeogeography of the Western Caucasus, when the Caucasian Sea remained the dominant feature (Figure 1b). It is directly related to the thematic geoheritage considered in the present study. This geoheritage point constitutes about half of the potential Middle Khadzhokh geosite (Figure 1a). The latter seems to be a new addition to the geoheritage of Mountainous Adygeya, and it was not considered in its previous inventories [26]. The noted geosite includes artificial outcrops (road cuttings and a quarry) of Late Jurassic and Early Cretaceous deposits on the slopes of the Shakhan Mountain, their natural outcrops in the Middle Khadzhokh River, and some geomorphological features. Although the area is close to densely populated territories, it is covered by dense forests and roads, and trails are few. Further investigations are necessary to propose this geosite formally, which is beyond the scope of the present work, but the knowledge from the abandoned quarry and the brief, very preliminary observations in its vicinities are enough to propose this geosite tentatively and put the question of its further examination on the agenda. Apparently, the Shakhan geoheritage point is the most valuable element of the proposed geosite, and it contributes significantly to the uniqueness of the latter. Although the palaeogeographical geoheritage type dominates the Shakhan locality, it coexists with the sedimentary type (peculiar structures of sandstones) and the economical type (the abandoned quarry itself), which is a sign of certain complexity. The geoheritage point takes an isolated position within the geosite because the other outcrops can be found in other places; nonetheless, the quarry is indivisible from the dominant landform (the Shakhan mountain), which is evidence of a certain cohesion.

The outer accessibility of the geoheritage point is more or less perfect. It is linked by a lengthy, unpaved road to Kamennomostsky town. Apparently, only heavy rain or snowfall may prohibit the use of this road, which is suitable to cars of all types. The inner accessibility is perfect (Figure 7), and one can move easily in the quarry, except for those parts where serious caution is necessary (the sudden collapse of walls cannot be excluded). Revitalizing this quarry in the future may restrict access there. Presently, the vulnerability of this geoheritage point is minimal. Potential damage may occur when many walls are covered by slumped debris. Visiting this geoheritage point requires elementary geological knowledge to understand what sandstones and cross-bedding are. However, professional guidance is necessary for stratigraphical and palaeogeographical interpretations. The aesthetic properties of the locality are rather significant. These are determined by spectacular patterns related to cross-bedding and scenic landscapes (the contrast between the yellowish sandstones and green forest should be noted).

Figure 7. The lower part of the Shakhan geoheritage point, with an unpaved road leading to it from Kamennomostsky (the road is visible on the right).

The scientific importance of the Shakhan geoheritage point is significant. It represents an excellent reference section for studying the Gubs Formation and interpreting its depositional environment. Special attention should be paid to the accurate dating of sandstones, and palynological approaches may help. The educational importance is also high because the locality offers a textbook example of cross-bedding in sandstones. The touristic importance is discussed separately (see below).

5. Discussion

5.1. Geoheritage Management Implications

The descriptions of the geoheritage points representing palaeoislands in the Caucasian Sea can be summarized as follows. First of all, it is necessary to stress that the Gosh, Lipovy, and Shakhan localities represent palaeoislands of all three periods of the Mesozoic. These pieces of geoheritage imply that palaeoislands were rather typical palaeogeographical elements of the study area, if each particular locality sheds light on a tiny "thin" time slice. If so, the discussed thematic framing is not only reasonable but also very logical. Importantly, the reported palaeogeographical features are well connected to the other geoheritage of Mountainous Adygeya [26] (Figure 1a), although their individuality is also evident. The localities are more or less accessible and not too vulnerable to negative influences. The main challenge is interpretation needs, but this is a common issue when the palaeogeographical type of geoheritage is addressed [6,65,66]. Two of three geoheritage points are aesthetically attractive, and they are also useful for geological research and education. These characteristics imply that the identification and special management of the palaeoisland-related thematic geoheritage is very promising.

Various approaches are used for the purposes of effective geoheritage management. These were reviewed by Bratton et al. [67], Cayla [68], de Oliveira et al. [69], Erikstad [70], Fuertes-Gutiérrez et al. [71], Lech et al. [72], Lima et al. [73], Mucivuna et al. [74], Pál and Albert [75], Prosser et al. [76], Reynard and Brilha [77], and Spyrou et al. [78]. In regard to the state of three considered geoheritage points (see above) and the entire geoheritage of Mountainous Adygeya [26], five managerial actions can be recommended. First, geosites, which include these points, should be investigated with precision to delineate the spatial distribution of geoheritage within them and, thus, to specify geoheritage points. Additionally, the Middle Khadzhokh geosite should be proposed formally. Second, it is important to evaluate the stability of slopes at the Lipovy locality in order to judge the urgency of technical solutions for protecting it from possible destruction. Some slope debris cleaning and establishing a kind of trail to the conglomerate outcrop is required. Third, although geoheritage points do not need any official status, the installation of simple signs informing about their names is necessary. Moreover, it is reasonable to start an initiative aimed at thematic geoheritage conservation, which can be administrated at the municipal level, with consultancy support from professional geologists working locally. Fourth, it appears reasonable to install interpretive panels explaining the nature of each geoheritage point and communicating stratigraphical and palaeogeographical knowledge in a form suitable to both experts and non-experts in geology. These panels should be designed similarly to maintain the integrity of the thematic geoheritage. The general principles of geoheritage interpretation [79–81] should also be followed. Fifth, the thematic geoheritage should be promoted actively to attract the attention of the research groups working locally, several universities organizing summer field practices for their students in geology and geography, and tourism authorities. A special webpage can be dedicated to this geoheritage. Indeed, all these actions need moderate funding, which can be provided by administrative authorities, the tourism industry, or any external investors. The development of a plan for funding/investment attraction is another important task.

5.2. Geotourism Implications

Mountainous Adygeya is an important tourist destination of the Russian South, which attracts hundreds of thousands of visitors from the Southern Federal District and other territories of the country [82]. Geotourism develops there, although its resources are far from being fully exploited [26]. Indeed, the further expansion of these activities will help to diversify the services offered by the regional industry of tourism and recreation and, therefore, to attract more visitors and increase their satisfaction. The established thematic geoheritage related to the Mesozoic palaeoislands may be attractive to geotourists because of three reasons:

(1) the very idea of islands in ancient tropical seas in what is now the mountainous domain sounds intriguing;
(2) the thematic geoheritage makes geotourism intentions less vague and dispersed and allows for attention to be paid to a particular tourism project as a real opportunity;
(3) the considered localities are linked to larger geosites (Figure 1a), accessible, and located in natural domains with significant aesthetic properties.

The potential target audience includes university students and their lecturers (Mountainous Adygeya is used actively for field geosciences programs by several large Russian universities), "pure" geotourists (geology amateurs, for whom the study area is very famous), and other tourists (occasional geotourists, either organized or individual) needing to diversify their experience and/or enrich their knowledge (besides the others, these include ecotourists and groups of schoolchildren). Moreover, the thematic geoheritage can be used for organizing irregular pre- and post-conference trips for senior scholars.

The development of geotourist routes is essential in geotourism [83–88], and it is especially important when several localities represent the same unique phenomenon and need to be visited in logical order. Earlier, Migoń et al. [89] emphasized the idea of the development of thematic geotourist trails. The development of a geoexcursion route for the discussed thematic geoheritage seems to be an urgent task because such a route can be employed as a basis for local geotourism planning and related geoheritage management. Two main reasons should be taken into account. First, a route should connect the three geoheritage points in logical order, i.e., from the Triassic to the Cretaceous. Second, the geoexcursion should be planned in regard to the location of the main accommodation places, the accessibility of geoheritage points (see above), and the options of access (excursion bus, individual car, and hiking). Field investigations permit the proposal of the geoexcursion route (Figure 8) and the specification of its main parameters (Table 2).

Table 2. Parameters of the proposed geoexcursion.

Elements (see Figure 8 for More Information)	Approximate Time, h (Taking into Account the Quality of Roads)			Distance, km	Height Range, m
	Bus *	Car	Hiking, Examining, and Discussing		
K -> Attraction 1	1.5	1	-	30	<200
D -> Attraction 1	1	0.7	-	20	<200
Attraction 1	-	-	1	0.3	<25
Sub-route A	1.5	1	-	30	~100
Attraction 2	-	-	0.5	0.1	<10
Sub-route B	1.5	1	-	35	~200
Attraction 3	-	-	1.5	0.5	<25
Attraction 3 -> D	1	0.7	-	15	~200
Attraction 3 -> K	0.5	0.5	-	5	~200

Note: a relatively small bus suitable for driving unpaved roads and well-experienced, local drivers are preferred.

Figure 8. The proposed geoexcursion route connecting the considered geohertage points representing Mesozoic islands of the Caucasian Sea.

The excursion starts in Kamennomostsky or Dakhovskaya, depending on the geotourists' choice for accommodation. The local road to Novoprokhladnoe village is used to reach the Gosh geoheritage point representing the Early Triassic palaeoisland. Visiting the latter takes up to an hour. Then, geotourists should return to Dakhvskaya village and follow the main road to the northern entrance to the Granite Gorge, where the Lipovy geoheritage point representing the Early Jurassic palaeoisland is located. It is small and simple enough, and visiting it takes no more than half an hour. The excursion follows the principal road to the northern periphery of Kamennomostsky and then the unpaved road to the abandoned quarry, which is the Shakhan geoheritage point representing the Early Cretaceous palaeoisland. This locality is rather large, and visitors need up to 1.5 h to comprehend it. The excursion ends at the accommodation place. The entire geoexcursion requires ~8 h by bus or ~6.5 h by car. In fact, it can last a bit more or less depending on the driving experience and weather conditions. One hour can be reserved for dinner.

The proposed route can be used from May to October. Although the excursion is planned for the mountainous domain, the height range is minimal (Table 2). An apparently "weak" aspect of the proposal is the significant time required for travelling between the localities (sub-routes A and B in Table 2). However, the entire route crosses a very scenic domain with continuous panoramic views of the geoheritage landscapes of Mountainous Adygeya. Moreover, outcrops of Mesozoic (especially Jurassic) marine deposits are visible in many places along the proposed route (these can be visited briefly if the visitors have interest), and they give an idea of the dominance of the Caucasian Sea in the Mesozoic in the study area. Indeed, the proposed geoexcursion is not restricted to the three geoheritage points in regard to the information available to its participants.

"Visiting" Mesozoic islands of the Caucasian Sea may be an interesting experience to tourists and may also be very informative, enriching their geological knowledge. First of all, the proposed geoexcursion route can be tested with groups of university students who have their field practice in Mountainous Adygeya. When well established, it can be offered to local tourism firms. The latter, however, should receive professionally treated and popularly explained geological information about the geoheritage points. The installation of interpretative panels would also help significantly. The promotion of the route can be carried out by the administrative authorities responsible for tourism development and interested in revenues to local budgets from tourism growth and diversification.

6. Conclusions

The present study of the palaeogeographical features in Mountainous Adygeya allows for three general conclusions:

(1) evidence of the Induan (Early Triassic), Toarcian (Early Jurassic), and Hauterivian (Early Cretaceous) islands in the Mesozoic Caucasian Sea is gathered;
(2) the related localities are valuable geoheritage points which open novel thematic perspectives for the geoheritage of Mountainous Adydeya;
(3) these geoheritage points can be employed for the purposes of new geoexcursion route development, which seems to be important for facilitating the growth of geotourism in the study area.

These conclusions stress the importance of focusing on thematic, sometimes highly specific geoheritage for its effective, logical treatment and touristic exploitation. Indeed, more research is necessary to conceptualize thematic geoheritage and to offer templates for its analysis. Generally, thematic framing seems to be very promising for dealing with marine geoheritage.

Some questions about the features reported in this work remain unresolved:

(1) Do these three geoheritage points represent the same or different palaeoislands?
(2) How many palaeoislands were there in the Caucasian Sea and how did they change?
(3) What was the exact tectonic setting of these palaeoislands?
(4) Which practical (local infrastructural) solutions are required to exploit these geoheritage points effectively (also in regard to serious interpretation needs)?
(5) Is it possible to use these localities for geotourists with functional diversity (and if yes, how)?

Indeed, more research is necessary to answer these questions. Notably, performing further research will facilitate interest in the considered localities, which is a factor stimulating academic geotourism. In other words, the very presence of unresolved geological questions makes establishing the related thematic geoheritage important and urgent.

Funding: This research received no external funding.

Institutional Review Board Statement: Not applicable.

Informed Consent Statement: Not applicable.

Data Availability Statement: Not applicable.

Acknowledgments: The author thanks Natalia V. Ruban (Russia) for the field assistance.

Conflicts of Interest: The authors declare no conflict of interest.

References

1. Poiraud, A.; Chevalier, M.; Claeyssen, B.; Biron, P.-E.; Joly, B. From geoheritage inventory to territorial planning tool in the Vercors massif (French Alps): Contribution of statistical and expert cross approaches. *Appl. Geogr.* **2016**, *71*, 69–82. [CrossRef]
2. Migoń, P.; Różycka, M.; Michniewicz, A. Conservation and Geotourism Perspectives at Granite Geoheritage Sites of Waldviertel, Austria. *Geoheritage* **2018**, *10*, 11–21. [CrossRef]
3. Herrera-Franco, G.; Mora-Frank, C.; Kovács, T.; Berrezueta, E. Georoutes as a Basis for Territorial Development of the Pacific Coast of South America: A Case Study. *Geoheritage* **2022**, *14*, 78. [CrossRef]
4. Plyusnina, E.E.; Ruban, D.A.; Zayats, P.P. Thematic dimension of geological heritage: An evidence from the Western Caucasus. *J. Geogr. Inst. Jovan Cvijic SASA* **2015**, *65*, 59–76. [CrossRef]
5. Gnezdilova, V.V.; Ruban, D.A.; Bruno, D.E.; Perrotta, P.; Crowley, B.E.; Oheim, K.B.; Zayats, P.P. Geoheritage sites with palaeogeographical value: Some geotourism perspectives with examples from Mountainous Adygeja (Russia). *Geološki Anal. Balk. Poluostrva* **2015**, *76*, 93–104. [CrossRef]

6. Bruno, D.E.; Crowley, B.E.; Gutak, J.M.; Moroni, A.; Nazarenko, O.V.; Oheim, K.B.; Ruban, D.A.; Tiess, G.; Zorina, S.O. Paleogeography as geological heritage: Developing geosite classification. *Earth Sci. Rev.* **2014**, *138*, 300–312. [CrossRef]
7. Henriques, M.H.; Pena dos Reis, R.; Garcia, G.G.; João, P.; Marques, R.M.; Custódio, S. Developing paleogeographical heritage concepts and ideas through the Upper Jurassic record of the Salgado and Consolação geosites (Lusitanian Basin, Portugal). *Resour. Policy* **2022**, *76*, 102594. [CrossRef]
8. Baarli, B.G.; Webb, G.E.; Johnson, M.E.; Cook, A.G.; Walsh, D.R. Shoal-water dynamics and coastal biozones in a sheltered-island setting: Upper Devonian Pillara Limestone (Western Australia). *Lethaia* **2016**, *49*, 507–523. [CrossRef]
9. Johnson, M.E. Paleoislands in the stream: Paleogeography and expected circulation patterns. *Geobios* **2002**, *35*, 96–106. [CrossRef]
10. Ollive, V.; Petit, C.; Garcia, J.-P.; Reddé, M. Rhine flood deposits recorded in the Gallo-Roman site of Oedenburg (Haut-Rhin, France). *Quat. Int.* **2006**, *150*, 28–40. [CrossRef]
11. Schuster, M.; Duringer, P.; Ghienne, J.-F.; Vignaud, P.; Beauvilain, A.; MacKaye, H.T.; Brunet, M. Coastal conglomerates around the Hadjer El Khamis inselbergs (Western Chad, Central Africa): New evidence for Lake Mega-Chad episodes. *Earth Surf. Processes Landf.* **2003**, *28*, 1059–1069. [CrossRef]
12. Adamia, S.; Zakariadze, G.; Chkhotua, T.; Sadradze, N.; Tsereteli, N.; Chabukiani, A.; Gventsadze, A. Geology of the Caucasus: A review. *Turk. J. Earth Sci.* **2011**, *20*, 489–544. [CrossRef]
13. Khain, V.E. Mesozoic-Cenozoic accretionary complexes of the Greater Caucasus. *Dokl. Earth Sci.* **2007**, *413*, 376–379. [CrossRef]
14. Koronovskiy, N.V. he Caucasus: Unsolved problems of geology. *Vestn. Mosk. Univ. Seriya Geol.* **1990**, *4*, 3–11.
15. Mosar, J.; Mauvilly, J.; Koiava, K.; Gamkrelidae, I.; Enna, N.; Lavrishev, V.; Kalberguenova, V. Tectonics in the Greater Caucasus (Georgia—Russia): From an intracontinental rifted basin to a doubly verging fold-and-thrust belt. *Mar. Pet. Geol.* **2022**, *140*, 105630. [CrossRef]
16. Rolland, Y. Caucasus collisional history: Review of data from East Anatolia to West Iran. *Gondwana Res.* **2017**, *49*, 130–146. [CrossRef]
17. Sharkov, E.; Lebedev, V.; Chugaev, A.; Zabarinskaya, L.; Rodnikov, A.; Sergeeva, N.; Safonova, I. The Caucasian-Arabian segment of the Alpine-Himalayan collisional belt: Geology, volcanism and neotectonics. *Geosci. Front.* **2015**, *6*, 513–522. [CrossRef]
18. Sosson, M.; Stephenson, R.; Adamia, S. Tectonic Evolution of the Eastern Black Sea and Caucasus: An introduction. *Geol. Soc. Spec. Publ.* **2017**, *428*, 1–9. [CrossRef]
19. Van Hinsbergen, D.J.J.; Torsvik, T.H.; Schmid, S.M.; Matenco, L.C.; Maffione, M.; Vissers, R.L.M.; Gürer, D.; Spakman, W. Orogenic architecture of the Mediterranean region and kinematic reconstruction of its tectonic evolution since the Triassic. *Gondwana Res.* **2020**, *81*, 79–229.
20. Yasamanov, N.A. *Landscape-Climatic Conditions of the Jurassic, the Cretaceous and the Paleogene of the South of the USSR*; Nedra: Moskva, Soviet Union, 1978. (In Russian)
21. Adamia, S.; Alania, V.; Chabukiani, A.; Kutelia, Z.; Sadradze, N. Great Caucasus (Cavcasioni): A long-lived north-tethyan back-arc basin. *Turk. J. Earth Sci.* **2011**, *20*, 611–628. [CrossRef]
22. Kazmin, V.G.; Tikhonova, N.F. Evolution of Early Mesozoic back-arc basins in the Black Sea—Caucasus segment of a Tethyan active margin. *Geol. Soc. Spec. Publ.* **2006**, *260*, 179–200. [CrossRef]
23. Lordkipanidze, M.B.; Adamia, S.A.; Asanidze, B.Z. Evolution of the active margins of the ocean Tethys (by example of the Caucasus). In *Oceanology: Reports. 27 International Geological Congress, 3*; Lisitsin, A.P., Ed.; Nauka: Moscow, Russia, 1984; pp. 72–83. (In Russian)
24. McCann, T.; Chalot-Prat, F.; Saintot, A. The Early Mesozoic evolution of the Western Greater Caucasus (Russia): Triassic-Jurassic sedimentary and magmatic history. *Geol. Soc. Spec. Publ.* **2010**, *340*, 181–238. [CrossRef]
25. Saintot, A.; Brunet, M.-F.; Yakovlev, F.; Sebrier, M.; Stephenson, R.; Ershov, A.; Chalot-Prat, F.; McCann, T. The Mesozoic-Cenozoic tectonic evolution of the Greater Caucasus. *Geol. Soc. Mem.* **2006**, *32*, 277–289. [CrossRef]
26. Ruban, D.A.; Mikhailenko, A.V.; Yashalova, N.N. Valuable geoheritage resources: Potential versus exploitation. *Resour. Policy* **2022**, *77*, 102665. [CrossRef]
27. Hasterok, D.; Halpin, J.A.; Collins, A.S.; Hand, M.; Kreemer, C.; Gard, M.G.; Glorie, S. New Maps of Global Geological Provinces and Tectonic Plates. *Earth Sci. Rev.* **2022**, *231*, 104069. [CrossRef]
28. Ruban, D.A. The Greater Caucasus—A Galatian or Hanseatic Terrane? Comment on "The formation of Pangea" by G.M. Stampfli, C. Hochard, C. Vérard, C. Wilhemand J. von Raumer Tectonophysics 593 (2013) 1–19]. *Tectonophysics* **2013**, *608*, 1442–1444. [CrossRef]
29. Gradstein, F.M.; Ogg, J.G.; Schmitz, M.; Ogg, G. (Eds.) *Geologic Time Scale 2020*; Elsevier: Amsterdam, The Netherlands, 2020; p. 1390.
30. Ruban, D.A.; Zorina, S.O.; Rebezov, M.B. Dispersed geoheritage points of the Lagonaki Highland, SW Russia: Contribution to local geoheritage resource. *Geosciences* **2019**, *9*, 367. [CrossRef]
31. Panek, N. Geotope protection and Geoparks in Germany—A critical inventory. *Nat. Und Landsch.* **2018**, *50*, 182–191.
32. Varriale, R.; Genovese, L.; Aldighieri, B. "Diffused Geoparks": Territorial Integration as Solution for a Shared Sustainable Growth Based on Geotourism in Italy, Japan and Tunisia. *Heritage* **2022**, *5*, 2083–2105. [CrossRef]
33. Brilha, J. Inventory and quantitative assessment of geosites and geodiversity sites: A review. *Geoheritage* **2016**, *8*, 119–134. [CrossRef]
34. Bruschi, V.M.; Cendrero, A. Geosite evaluation; can we measure intangible values? *Alp. Mediterr. Quat.* **2005**, *18*, 293–306.

35. Carcavilla, L.; Díaz-Martínez, E.; Erikstad, L.; García-Cortés, Á. Valuation of geoheritage in Europe. *Bol. Parana. Geosci.* **2013**, *70*, 28–40.
36. Carrión-Mero, P.; Dueñas-Tovar, J.; Jaya-Montalvo, M.; Berrezueta, E.; Jiménez-Orellana, N. Geodiversity assessment to regional scale: Ecuador as a case study. *Environ. Sci. Policy* **2022**, *136*, 167–186. [CrossRef]
37. Mucivuna, V.C.; Motta Garcia, M.D.G.; Reynard, E. Comparing quantitative methods on the evaluation of scientific value in geosites: Analysis from the Itatiaia National Park, Brazil. *Geomorphology* **2022**, *396*, 107988. [CrossRef]
38. Ruban, D.A. On the duality of marine geoheritage: Evidence from the Abrau area of the Russian Black Sea coast. *J. Mar. Sci. Eng.* **2021**, *9*, 921. [CrossRef]
39. Štrba, L.; Rybar, P.; Balaz, B.; Molokac, M.; Hvizdak, L.; Krsak, B.; Lukac, M.; Muchova, L.; Tometzova, D.; Ferencikova, J. Geosite assessments: Comparison of methods and results. *Curr. Issues Tour.* **2015**, *18*, 496–510. [CrossRef]
40. Mikhailenko, A.V.; Ruban, D.A.; Ermolaev, V.A. Geoheritage meaning of artificial objects: Reporting two new examples from Russia. *Heritage* **2021**, *4*, 2721–2731. [CrossRef]
41. Kirillova, K.; Fu, X.; Lehto, X.; Cai, L. What makes a destination beautiful? Dimensions of tourist aesthetic judgment. *Tour. Manag.* **2014**, *42*, 282–293. [CrossRef]
42. Chaitsky, V.P.; Popkov, V.I.; Popkov, I.V.; Pinchuk, T.N. Triassic of the Northern Caucasus. *Geol. Geogr. I Glob. Energiya* **2020**, *77*, 11–21. (In Russian)
43. Gaetani, M.; Garzanti, E.; Poline, R.; Kiricko, Y.; Korsakhov, S.; Cirilli, S.; Nicora, A.; Rettori, R.; Larghi, C.; Bucefalo Palliani, R. Stratigraphic evidence for Cimmerian events in NW Caucasus (Russia). *Bull. Société Géologique Fr.* **2005**, *176*, 283–299. [CrossRef]
44. Rostovtsev, K.O.; Savel'eva, L.M.; Efimova, N.A.; Shvemberger, Y.N. *Decision of the 2nd Interdepartmental Regional Stratigraphical Maeeting on the Mesozoic of the Caucasus (Triassic)*; VSEGEI: Leningrad, Russia, 1979. (In Russian)
45. Callahan, G.; Johnson, M.E.; Guardado-France, R.; Ledesma-Vázquez, J. Upper Pleistocene and Holocene storm deposits eroded from the granodiorite coast on Isla San Diego (Baja California Sur, Mexico). *J. Mar. Sci. Eng.* **2021**, *9*, 555. [CrossRef]
46. Dewey, J.F.; Ryan, P.D. Storm, rogue wave, or tsunami origin for megaclast deposits in western Ireland and North Island, New Zealand? *Proc. Natl. Acad. Sci. USA* **2017**, *114*, E10639–E10647. [CrossRef] [PubMed]
47. Johnson, M.E.; Ledesma-Vázquez, J. Evaluation of boulder deposits linked to late Neogene hurricane events. *J. Mar. Sci. Eng.* **2021**, *9*, 1278. [CrossRef]
48. Lorang, M.S. A wave-competence approach to distinguish between boulder and megaclast deposits due to storm waves versus tsunamis. *Mar. Geol.* **2011**, *283*, 90–97. [CrossRef]
49. Noormets, R.; Felton, E.A.; Crook, K.A.W. Sedimentology of rocky shorelines: 2 Shoreline megaclasts on the north shore of Oahu, Hawaii—origins and history. *Sediment. Geol.* **2002**, *150*, 31–45. [CrossRef]
50. Oliveira, M.A.; Andrade, C.; Freitas, M.C.; Costa, P.; Taborda, R.; Janardo, C.; Neves, R. Transport of large boulders quarried from shore platforms of the Portuguese west coast. *J. Coast. Res.* **2011**, *64*, 1871–1875.
51. Terry, J.P.; Lau, A.Y.A.; Nguyen, K.A.; Liou, Y.-A.; Switzer, A.D. Clustered, Stacked and Imbricated Large Coastal Rock Clasts on Ludao Island, Southeast Taiwan, and Their Application to Palaeotyphoon Intensity Assessment. *Front. Earth Sci.* **2021**, *9*, 792369. [CrossRef]
52. Ruban, D.A. The Permian/Triassic mass extinction among brachiopods in the Northern Caucasus (northern Palaeo-Tethys): A tentative assessment. *Geobios* **2010**, *43*, 355–363. [CrossRef]
53. Haq, B.U. Triassic Eustatic Variations Reexamined. *GSA Today* **2018**, *28*, 4–9. [CrossRef]
54. Ruban, D.A. Basal Liassic of the Belaya River watershed (North-Western Caucasus). *Nautchnaja Mysl' Kavkaza. Mezhdistsiplinarnye I Spetsial'nye Issled.* **2007**, *2*, 95–97. (In Russian)
55. Nenakhov, V.M.; Zhabin, A.V.; Nikitin, A.V.; Bondarenko, S.V. Internal structure of the tectonic zone of the northern frame of the Dakh Crystalline Massif (Western Caucasus). *Vestn. Voronezhskogo Gos. Universiteta. Seriya Geol.* **2019**, *1*, 5–14. (In Russian)
56. Chepurnoy, E.A. Sedimentary rocks of the northern contact of the Dakh Crystalline Uplift (Adygeya). In *Praktika Geologov na Proizvodstve*; Granovskaya, N.V., Ed.; SFU: Rostov-on-Don, Russia, 2021; pp. 132–134. (In Russian)
57. Rostovtsev, K.O.; Agaev, V.B.; Azarian, N.R.; Babaev, R.G.; Besnosov, N.V.; Hassanov, N.A.; Zesashvili, V.I.; Lomize, M.G.; Paitschadze, T.A.; Panov, D.I.; et al. *Jurassic of the Caucasus*; Nauka: St. Petersburg, Russia, 1992. (In Russian)
58. Panov, D.i.; Stafeev, A.N.; Yutsis, V.V. Early Jurassic stage in the development of the Northern Caucasus and the Ciscaucasus. *Bull. Mosc. Soc. Naturalists. Geol. Ser.* **1996**, *6*, 3–14. (In Russian)
59. Karpunin, A.M.; Mamonov, S.V.; Mironenko, O.A.; Sokolov, A.R. *Geological Monuments of Nature of Russia*; Lorien: Sankt-Peterburg, Russia, 1998. (In Russian)
60. Mikhailenko, A.V.; Ruban, D.A.; Yashalova, N.N.; Rebezov, M.B. The Unique Granite Gorge in Mountainous Adygeya, Russia: Evidence of Big and Complex Geosite Disproportions. *Geosciences* **2019**, *9*, 372. [CrossRef]
61. Drushits, V.V.; Mikhailova, I.A. *Biostratigraphy of the Lower Cretaceous of the Northern Caucasus*; MGU: Moscow, Soviet Union, 1966. (In Russian)
62. Prosorovskaya, E.L. (Ed.) *Stratigraphical Dictionary of the USSR. Triassic, Jurassic, Cretaceous*; Nedra: Leningrad, Soviet Union, 1979. (In Russian)
63. Mikerina, T.B.; Pinchuk, T.N. Facies of Cretaceous deposits of the northern part of the Western Ciscaucasus. In *Melovaya Sistema Rossii i Blizhnego Zarubezh'ya: Problemy Stratigrafii i Paleogeografii*; Baraboshkin, E.Y., Guzhikov, A.Y., Eds.; MAOBTI: Magadan, Russia, 2020; pp. 171–174. (In Russian)

64. Wakita, K. Geology and tectonics of Japanese islands: A review—The key to understanding the geology of Asia. *J. Asian Earth Sci.* **2013**, *72*, 75–87. [CrossRef]
65. Seghedi, A.; Andrășanu, A.; Rădan, S. The Transylvanian dinosaur museum project: The contribution of Geoecomar to valorize and promote the paleontological heritage of Romania. *Geo. Eco. Mar.* **2017**, *23*, 145–164.
66. Wolniewicz, P. Bringing the History of the Earth to the Public by Using Storytelling and Fossils from Decorative Stones of the City of Poznań, Poland. *Geoheritage* **2019**, *11*, 1827–1837. [CrossRef]
67. Bratton, A.; Smith, B.; McKinley, J.; Lilley, K. Expanding the Geoconservation Toolbox: Integrated Hazard Management at Dynamic Geoheritage Sites. *Geoheritage* **2013**, *5*, 173–183. [CrossRef]
68. Cayla, N. An Overview of New Technologies Applied to the Management of Geoheritage. *Geoheritage* **2014**, *6*, 91–102. [CrossRef]
69. de Oliveira, G.H.; Servidoni, L.E.; Spalevic, V.; Gaspar Junior, L.A.; Mincato, R.L. Geological heritage management and preservation instruments: Case study of the municipality of Conceicao da Aparecida, Minas Gerais—Brazil. *Agric. For.* **2021**, *567*, 71–82.
70. Erikstad, L. Geoheritage and geodiversity management—the questions for tomorrow. *Proc. Geol. Assoc.* **2013**, *124*, 713–719. [CrossRef]
71. Fuertes-Gutiérrez, I.; García-Ortiz, E.; Fernández-Martínez, E. Anthropic Threats to Geological Heritage: Characterization and Management: A Case Study in the Dinosaur Tracksites of La Rioja (Spain). *Geoheritage* **2016**, *8*, 135–153. [CrossRef]
72. Lech, R.R.; Marcus, A.; Reinoso, J.R. Science and tourism together in the management for the preservation of the geological heritage. *Ser. Correl. Geol.* **2018**, *34*, 35–41.
73. Lima, E.A.; Machado, M.; Guerreiro, M.; Nunes, J.C.; Costa, M.P. Geological Heritage Management in Small Islands: The Example of the Azores UNESCO Global Geopark (Portugal). *Geoheritage* **2018**, *10*, 659–671. [CrossRef]
74. Mucivuna, V.C.; Garcia, M.D.G.M.; Reynard, E.; Rosa, P.A.D.S. Integrating geoheritage into the management of protected areas: A case study of the Itatiaia National Park, Brazil. *Int. J. Geoheritage Parks* **2022**, *10*, 252–272. [CrossRef]
75. Pál, M.; Albert, G. Examining the Spatial Variability of Geosite Assessment and Its Relevance in Geosite Management. *Geoheritage* **2021**, *13*, 8. [CrossRef]
76. Prosser, C.; Murphy, M.; Larwood, J. *Geological Conservation: A Guide to Good Practice*; English Nature: Peterborough, UK, 2006.
77. Reynard, E.; Brilha, J. (Eds.) *Geoheritage: Assessment, Protection, and Management*; Elsevier: Amsterdam, The Netherlands, 2018.
78. Spyrou, E.; Triantaphyllou, M.; Tsourou, T.; Vassilakis, E.; Asimakopoulos, C.; Konsolaki, A.; Markakis, D.; Marketou-Galari, D.; Skentos, A. Assessment of Geological Heritage Sites and Their Significance for Geotouristic Exploitation: The Case of Lefkas, Meganisi, Kefalonia and Ithaki Islands, Ionian Sea, Greece. *Geosciences* **2022**, *12*, 55. [CrossRef]
79. Began, M.; Visnic, T.; Djokic, M.; Vasiljevic, D.A. Interpretation Possibilities of Geoheritage in Southeastern Serbia—Gorge and Canyon Study. *Geoheritage* **2017**, *9*, 237–249. [CrossRef]
80. Bruno, B.C.; Wallace, A. Interpretive Panels for Geoheritage Sites: Guidelines for Design and Evaluation. *Geoheritage* **2019**, *11*, 1315–1323. [CrossRef]
81. Hose, T.A. 3G's for Modern Geotourism. *Geoheritage* **2012**, *4*, 7–24. [CrossRef]
82. Dzhimova, S.B.; Bayanduryan, G.L. Evaluation of tourism complex functioning in the Republic of Adygeya. *Ekon. I Predprin.* **2021**, *127*, 418–421. (In Russian)
83. Abratis, M.; Viereck, L.; Büchner, J.; Tietz, O. Route to the volcanoes in Germany—conceptual model for a geotourism project interconnecting geosites of Cenozoic volcanism. *Z. Der Dtsch. Ges. Fur Geowiss.* **2015**, *166*, 161–185. [CrossRef]
84. Bentivenga, M.; Palladino, G.; Prosser, G.; Guglielmi, P.; Geremia, F.; Laviano, A. A Geological Itinerary Through the Southern Apennine Thrust Belt (Basilicata—Southern Italy). *Geoheritage* **2017**, *9*, 1–17. [CrossRef]
85. Bucci, F.; Tavarnelli, E.; Novellino, R.; Palladino, G.; Guglielmi, P.; Laurita, S.; Prosser, G.; Bentivenga, M. The History of the Southern Apennines of Italy Preserved in the Geosites Along a Geological Itinerary in the High Agri Valley. *Geoheritage* **2019**, *11*, 1489–1508. [CrossRef]
86. Dóniz-Páez, J.; Alonso, C.Q. Urban geoturism rutes in Icod de Los Vinos (Tenerife, Canary Islands, Spain): A proposal. *Cuad. Geogr.* **2016**, *55*, 320–343.
87. Gałka, E. Geotourist potential of the loess relief of the projected Kamienna Valley Geopark. *Prz. Geol.* **2018**, *66*, 55–64.
88. Palladino, G.; Prosser, G.; Bentivenga, M. The Geological Itinerary of Sasso di Castalda: A Journey into the Geological History of the Southern Apennine Thrust-belt (Basilicata, Southern Italy). *Geoheritage* **2013**, *5*, 47–58. [CrossRef]
89. Migoń, P.; Duszyński, F.; Jancewicz, K.; Różycka, M. From Plateau to Plain—Using Space-for-Time Substitution in Geoheritage Interpretation, Elbsandsteingebirge, Germany. *Geoheritage* **2019**, *11*, 839–853. [CrossRef]

Journal of
Marine Science and Engineering

MDPI

Article

Fragmentation Characteristics of Seafloor Massive Sulfides: A Coupled Fluid-Particle Flow Simulation

Huan Dai [1], Hao Li [1] and Yan Li [1,2,*]

[1] College of Mechanical and Electrical Engineering, Central South University, Changsha 410083, China
[2] National Key Laboratory of Deep-Sea Mineral Researches Development and Utilization Technology, Changsha 410013, China
* Correspondence: yanlicsu@csu.edu.cn

Abstract: The research on the fragmentation mechanism of seabed minerals under high ambient pressure significantly contributes to the exploitation of seafloor massive sulfides (SMS). In this paper, the uniaxial compressive strength (UCS) test and triaxial compressive strength (TCS) test were carried out on two kinds of SMS samples to obtain the key mechanical properties of minerals, including cohesion, internal friction angle, compressive strength, and elastic modulus. Then, based on these mechanical parameters, the fluid-solid coupling cutting model of two SMS samples at high ambient pressure is established by using the coupling method of discrete elements and smooth particles. A mixed-bond model is selected, and the microscopic parameters are determined by a repeated calibration process. Meanwhile, the cutting force and debris information are monitored and collected in real time during the whole cutting process. The results show that under different confining pressure environments, the model shows the transformation of minerals from brittleness to ductility. The cutting force increases with the increasing ambient pressure. Due to the fluid pressure, the crushing mechanism tends to shear failure, which is more likely to produce mud and finer fragments.

Keywords: fluid-solid coupling; seafloor massive sulfides; cutting mechanism; discrete element method; smoothed particle

Citation: Dai, H.; Li, H.; Li, Y. Fragmentation Characteristics of Seafloor Massive Sulfides: A Coupled Fluid-Particle Flow Simulation. *J. Mar. Sci. Eng.* **2022**, *10*, 1306. https://doi.org/10.3390/jmse10091306

Academic Editors: George Kontakiotis, Assimina Antonarakou and Dmitry A. Ruban

Received: 22 August 2022
Accepted: 10 September 2022
Published: 15 September 2022

1. Introduction

Over the past decades, deep-sea mining has received renewed attention due to the continuous development of the global economy and a shortage of raw materials [1–3]. Currently, the most valuable seafloor solid minerals known are seafloor massive sulfides (SMS, or black smoker), polymetallic nodules, and cobalt-rich crusts [4,5]. Among them, SMS deposits form on and below the seabed through thousands of years of hydrothermal activities along ridges, island arcs, and in rifted back-arc basins behind active subduction zones. By August 2022, 721 hydrothermal spots related to SMS minerals had been discovered around the world, with the majority of them located in water depths ranging from 1000 to 3000 m [6]. And how to strip SMS from the deposit in a high-pressure environment is the primary problem before mining, and it is also one of the key technologies of seabed mining.

The material properties of seabed minerals are the basis of studying their crushing properties. At present, some scholars have conducted a series of tests on SMS minerals and obtained valuable test data. For example, Yamazaki [7] obtained data such as porosity, elastic modulus, compressive strength, and tensile strength by using samples from the Izena Cauldron at Okinawa Trough, Japan. Spagnoli et al. [8] summarized the mechanical properties of 12 groups of polymetallic sulfide samples. They analyzed the relationship between the geotechnical properties of the samples and the mineral types and compositions. The results showed that porosity, mineral phase composition, and internal texture are important factors affecting the mechanical properties of minerals. The above studies only consider the mechanical properties of minerals under an atmospheric environment and ignore their mechanical reactions under high pressure. Liu et al. [9,10] team conducted uniaxial and triaxial tests on deep-sea polymetallic sulfide samples from China's Indian Ocean Exploration Contract Area and obtained the tensile strength, compressive strength, cohesion, internal friction angle, and other data of the samples. However, these samples are collected from different voyages, so they are not suitable for direct study of mineral crushing characteristics. Thus, a deeper understanding of its material properties is necessary for the study of SMS crushing.

In the past decades, some international consortia and research institutions have carried out in situ cutting experiments for seabed deposits and proved that it is feasible to extract seabed ore by traditional methods [11–13]. Numerous rock-cutting experiments have also been carried out to obtain the cutting mechanism and the variation regularities with various influencing factors. Kaitkay et al. [14] conducted experiments on the influence of hydrostatic pressure on rock cutting on Carthage marble. They found that with the increase in ambient pressure, the cutting process changes from a brittle to a ductile-brittle failure mode, and the cutting force increases. Grima et al. [15] conducted the cutting experiment on limestone in a high-pressure vessel. The results showed that the brittle cutting process changes into an apparent ductile mode due to the effect of dilatancy strengthening. Meanwhile, the force required to cut the material will increase due to the combined effect of high cutting speed and high pressure, and smaller fragments and narrower grooves will be obtained. Various studies have shown that the cutting process can be significantly influenced by the presence of (pore) fluid pressure [15,16]. However, these tests are all based on the cutting process of non-SMS materials under confining pressure.

Currently, more and more researchers use numerical methods to study the rock cutting process [17,18]. Among them, the discrete element method (DEM) is a powerful method widely used to simulate the rock cutting process because it can deal with rock deformation, fracturing, and fragmentation at the same time. Traditionally, rock-cutting models for land-based applications assume that the rock is cut under atmospheric and dry conditions. This assumption is not valid for deep-sea applications, especially because hydrostatic pressure may be of the same order of magnitude as the unconfined compressive strength of the rock. Thus, it is particularly difficult in the simulation of underwater rock cutting because it must deal with the whole range from intact rock to debris, also combined with fluid pressure. However, several approaches have been developed by extending DEM with fluid pressure effects. The first group is to simulate the flow of fluid in pores along the contact/connection bond of elements [19–21]. This is a discontinuous method, which is often used in the simulation of the hydraulic fracturing process and is not suitable for the large-deformation rock cutting process. The second group of hydro-mechanical coupling methods is the method of applying an adaptive confining pressure boundary condition [22–24]. However, these methods ignore the pore pressure inside the material, so they are only suitable for dense materials, obviously not for porous materials such as SMS. LV et al. [25] used the discrete element method based on the Voronoi particle binding cluster model to investigate the influence of cutting speed and hydrostatic pressure on crack propagation characteristics, cutting resistance, and chip distribution. This method considers the limitation of the surface confining pressure on the rock crushing process but does not consider the infiltration of water and the diffusion of internal confining pressure during the cutting process. Therefore, there is a large gap between the above methods and the deep-sea mineral crushing process. Helmons et al. [26,27] used the method of combining discrete element and smooth particle flow to establish the fluid structure coupling model of rock cutting and simulated the rock cutting process with different ambient pressure. The combined method takes into account the diffusion of pore pressure in the cutting process and is capable of predicting the average cutting force well. This method provides a new idea for studying the influence of pore pressure and can be easily extended to 3D.

SMS minerals are attached to the seabed with a depth of several thousand meters and must be broken before collection in submarine mining. Due to the peculiarity of the formation mechanism, the interior of SMS is full of pores with a porosity of 20–50% [7,8]. The pores of rock existing in the water depth of several thousand meters are full of seawater, which will be squeezed in the high-speed crushing process, thus affecting the cutting force and the generation of debris. The difference between the study of rock fragmentation in deep-sea mining and land mining is mainly reflected in two aspects. On the one hand, seawater will produce a huge hydrostatic pressure on deep-sea rocks; on the other hand, the pore pressure in the rock will also affect the cutting process. Therefore, it is necessary to study the crushing process of SMS minerals under high ambient pressure so as to provide a basis for the research and design of seabed mining equipment. In this study, the mineralogical study and geotechnical tests of SMS minerals were carried out, and some important mechanical parameters required for numerical analysis were obtained. Although mineralogical investigation is important for an in-depth understanding of SMS, it is not the focus of this study. Then, a two-dimensional simulation model of SMS under hydrostatic pressure is established by using the method of combining discrete element and smooth particle flow. The cutting process of SMS under high hydrostatic pressure is simulated, and the crack evolution, stress distribution, and load characteristics in the crushing process are analyzed. It provides a certain theoretical basis and technical support for deep-sea mining.

2. Mechanical Property Test

2.1. Sample

While studying the fragmentation properties of SMS, we need to have a more comprehensive understanding of their material properties. Therefore, we applied for samples from the China Ocean Sample Repository for material and mechanical property tests and received two pieces of SMS ores. Both samples were collected from the southwest Indian Ocean ridge with a grab bucket, and the sample segments were 30iv-swir-s012tvg01 and 34iv-swir-s035tvg08, respectively.

We used a vhx-5000 ultra-high magnification lens zoom 3D microscope to observe the polished sections of two SMS samples. The 3D microstructure and morphology of the SMS sample are shown in Figure 1. It can be seen that the polished surface of the sample shows many pits of different sizes, which strongly confirms the porous characteristics of SMS. The porosity of the black and white samples is 20.94% and 25.86%, respectively. Generally speaking, porosity is inversely proportional to the stability of minerals. The micro-cracks produced by the pores are easy to expand during the compression process. Then the rock samples become unstable and damaged, affecting the bearing capacity and reducing the strength.

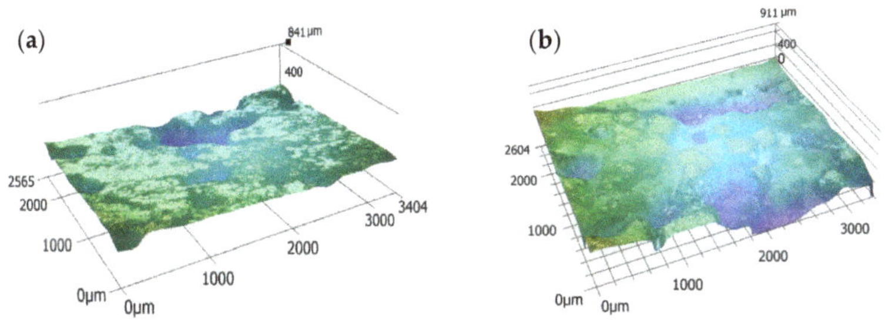

Figure 1. Micrograph of the polished surface in three dimensions:(**a**) black sample, (**b**)white sample.

In order to explore the relationship between mechanical properties and mineralogical properties of minerals, the quantitative composition of minerals was determined by XRD. As shown in Table 1, the basic mineral phases in the black sample are pyrite, marcasite, and chalcopyrite. In contrast, the iron in the white sample basically exists in the form of pyrite. Based on the elemental test of SMS minerals in the same mining area [10] and the research results of other scholars [8], we have reason to speculate that other components (including amorphous) are mainly silica.

Table 1. Main mineralogical compositions of the SMS samples.

Sample Type	Pyrite (FeS_2)	Marcasite (FeS)	Chalcopyrite ($CuFeS_2$)	Others (Including Amorphous)
Black	25.1%	31.9%	5.4%	37.6%
White	46.3%	—	4.2	49.5%

2.2. Mechanics Test

Due to the limitation of the size, the sample in the geotechnical mechanics test is processed into a non-standard size. The sample in the triaxial compression strength (TCS) test is processed into 25 × 50 mm, and the sample size in the Brazilian splitting test is 25 × 25 mm, as shown in Figure 2. It shows that there are a large number of pores and defects in the sulfide sample so the sample is sealed in a heat-shrinkable tube in the triaxial compression test to prevent hydraulic oil from entering the sample during the test and affecting the determination of mechanical properties. It should be noted that before the triaxial compression test, it is necessary to repair the obvious holes on the sample surface with lime or cement to prevent the high confining pressure oil from penetrating the heat-shrinkable tube and causing the test failure.

Figure 2. Processed test samples.

The triaxial compressive strength test determines the triaxial compressive strength of rock specimens under different lateral pressures. Equal lateral pressure conditions ($\sigma_2 = \sigma_3$) used in the triaxial compression test are determined according to the engineering needs and rock characteristics. The load acting on the unit area of the specimen is called the compressive strength, which is defined as the ratio of the critical load to the cross-sectional area perpendicular to the load direction. This relationship can be expressed as follows:

$$R = \frac{P}{A'} \quad (1)$$

with the TCS of SMS R, MPa; the failure load P, N; and the cross-section area perpendicular to the loading direction A', mm^2.

In the triaxial compression test, 11 specimens were prepared for both samples, and lateral pressure was set at 2 MPa, 5 MPa, 8 MPa, and 10 MPa, respectively. Because the porosity of the white sample is higher and the internal structure of the mineral is looser, the white sample fails more times in the experiment.

Because the Brazilian splitting test is simple and the results are close to those measured using the direct tensile method, it is used to measure the tensile strength of SMS specimens. The maximum tensile stress acting in the center of the specimen is calculated according to Equation (2).

$$\sigma_t = \frac{2P}{\pi dt} \quad (2)$$

where σ_t is the maximum tensile stress in the center of the specimen, MPa; P is peak tensile force at failure, N; and d and t are diameter and thickness of the pressure-bearing disc, mm.

Figure 3 shows the failure modes of the two samples after the TCS test and the Brazilian splitting test, indicating that SMS conforms to the failure characteristics of typical brittle materials. Table 2 illustrates the final mechanical property parameters of the two samples, including the UCS, tensile strength, elastic modulus, internal friction angle, and cohesion.

Figure 3. The failure samples in: (**a**) TCS test; (**b**) Brazilian splitting test.

Table 2. Mechanical parameters of SMS sample.

Sample Type	Tensile Strength/(MPa)	Compressive Strength /(MPa)	Internal Friction Angle φ/(°)	Cohesion c/MPa	Elastic Modulus/(GPa)
Black	3.84	33.73	32.19	9.31	9.85
White	2.25	20.57	29.67	5.98	1.98

3. Fluid-Structure Coupling Method

In underwater rock cutting simulation, how to treat the separate phases, solid and fluid, and the coupling between these two has always been one of the key difficulties. Therefore, we established a numerical model of rock drainage effects through DEM-SP coupling. It should be noted that the two methods of DEM and SP have similar algorithms, but the interaction and scope of interaction in the algorithms are different. Among them, the coupling model of pore fluid pressure between particles and rock is transformed into a pore pressure diffusion equation, which is solved by the method based on the smooth particle (SP) technique. The coupling between the two phases is based on the force generated by volume deformation and pressure gradient.

3.1. Rock Model—DEM

The discrete element method can treat the rock mass as a discontinuous discrete medium, in which there may be large displacement, rotation, sliding, and even block separation, so it can truly simulate the discontinuous and large deformation characteristics of the rock mass. Itasca-PFC is widely used in geotechnical engineering, geological engineering, mechanical engineering, and other fields. In PFC, rock material is represented as rigid spherical (3D) or circular (2D) discrete elements (particles) [28]. The interaction between any two particles is realized by the contact constitutive model between particles. The constitutive model is composed of the stiffness model, the slip model, and the bond model, which control the deformation, separation, and movement of particles. The translation and rotation of particles are controlled by the standard equations for rigid body mechanics:

$$m_i \ddot{\mathbf{u}}_i = \mathbf{F}_i \tag{3}$$

$$I_i \dot{\omega}_i = \mathbf{T}_i \tag{4}$$

where $\mathbf{F}_i$ and $\mathbf{T}_i$ are the sums of force and moment of particle i, respectively, m and I are the mass and moment of inertia, $\mathbf{u}$ is the position vector of particle centroid in a fixed coordinate system, ω is the angular velocity. The vectors $\mathbf{F}_i$ and $\mathbf{T}_i$ are calculated by Formulas (5) and (6).

$$\mathbf{F}_i = \mathbf{F}_i^{ext} + \sum_{j=1}^{n_i^c} \mathbf{F}_{ij}^{cont} + \mathbf{F}_i^{damp} + \mathbf{F}_i^{\nabla p} \tag{5}$$

$$\mathbf{T}_i = \mathbf{T}_i^{ext} + \sum_{j=1}^{n_i^c} S_{ij}^c \times \mathbf{F}_{ij}^{cont} + \mathbf{T}_i^{damp} \tag{6}$$

where, $\mathbf{F}^{ext}$ and $\mathbf{T}^{ext}$ are external loads, $\mathbf{F}_{ij}^{cont}$ is the interaction between particle i and adjacent particle, $j = 1, \ldots, \mathrm{n}_i^c$, with n_i^c is the number of neighboring particles in contact with the particles i, Numerical damping load $\mathbf{F}_i^{damp}$ and $\mathbf{T}_i^{damp}$, $\mathbf{S}_{ij}^c$ are the vector connecting the centroid of particle i and the contact point of particle j. The contact force $\mathbf{F}_{ij}^{cont}$ can be decomposed as normal component and tangential component. $\mathbf{F}_i^{\nabla p}$ is the coupling force generated by the fluid pressure gradient on the particle, which will be described in Section 3.2. the numerical damping is defined by Formulas (7) and (8).

$$\mathbf{F}_i^{damp} = -\alpha \left| \mathbf{F}_i^{ext} + \mathbf{F}_i^{cont} \right| \frac{\dot{\mathbf{u}}_i}{|\dot{\mathbf{u}}_i|} \tag{7}$$

$$\mathbf{T}_i^{damp} = -\alpha \left| \mathbf{T}_i^{ext} + \mathbf{T}_i^{cont} \right| \frac{\dot{\omega}_i}{|\dot{\omega}_i|} \tag{8}$$

with α, the numerical damping coefficient.

3.2. Fluid Model—Smooth Particle

Owing to the low permeability of rock-like materials, the contribution of hydrodynamic effects to mechanical rock properties mainly depends on the pressure gradient [29]. In the case of deep-sea excavation with a high cutting speed, the influence of internal fluid speed on the mechanical behavior of rock can be ignored. Furthermore, the thermal effects are negligible relative to the fluid pressure effect. Then, the final pressure diffusion equation is described as follows.

$$\frac{Dp}{Dt} - M\nabla \cdot \left(\frac{\kappa}{\mu} \nabla p\right) = -\alpha M \frac{D\varepsilon_V}{Dt} \tag{9}$$

with pore pressure p, Biot modulus M, intrinsic permeability κ, dynamic viscosity of fluid μ, effective stress coefficient α, volumetric strain ε_V.

Due to the discontinuity of DEM, the continuity Equation (9) cannot be solved directly. Therefore, the pressure diffusion equation is coupled with DEM by applying the interpolation technology based on the smooth particle (SP) method, which weights and accumulates the attributes of each particle around by kernel function. Here we also select Wendland C2, a kernel function with good consistency, a small compact domain, and smoothness, for correction. The calculation method of field quantity A is shown in Formula (10).

$$A(\mathbf{u}_i) \frac{\sum_j A_j m_j W(\mathbf{r}_i - \mathbf{r}_j, h)}{\sum_j m_j W(\mathbf{r}_i - \mathbf{r}_j, h)} \tag{10}$$

where m is particle mass, W is kernel function, h is smoothing length, with index i for the particle under consideration and index j for the neighboring particles (including particle i).

However, the higher derivative of Formula (9) cannot be calculated directly. Therefore, the diffusion term is discretized by using a method similar to that of Cleary and Monaghan [30] for simulating heat conduction. The diffusion terms can be expressed as

$$\nabla \cdot (\frac{\kappa}{\mu} \nabla p) = \sum_j \frac{m_j(\kappa_i + \kappa_j)}{\rho_i(\mu_i + \mu_j)}(p_i - p_j)\frac{\mathbf{n}_{ij} \cdot \nabla W(\mathbf{r}_i - \mathbf{r}_j, h)}{|\mathbf{r}_i - \mathbf{r}_j|} \tag{11}$$

According to the pore pressure distribution, the local pressure gradient of the fluid is calculated. Then, the pressure gradient on the particles is added as an interaction force to the external force acting on the particles.

$$\mathbf{F}_i = -\nabla p \pi r_i^2 \tag{12}$$

As illustrated in Figure 4, the coupling force generated by the fluid pressure gradient on the particle builds a bridge between the fluid phase and the discrete phase. The pressure gradient on a particle is added as an interaction force to the sum of forces acting on the particles as Equation (12). Then DEM is advanced half a timestep and the volumetric strain rate is calculated for the fluid based on the intermediate velocities of the DEM particles. It is then used to promote pore pressure diffusion once again. In the whole simulation, the interpolation points of SP coincide with the centroid of DEM particles and move with the movement of DEM particles. In addition, bidirectional coupling is applied at each time step.

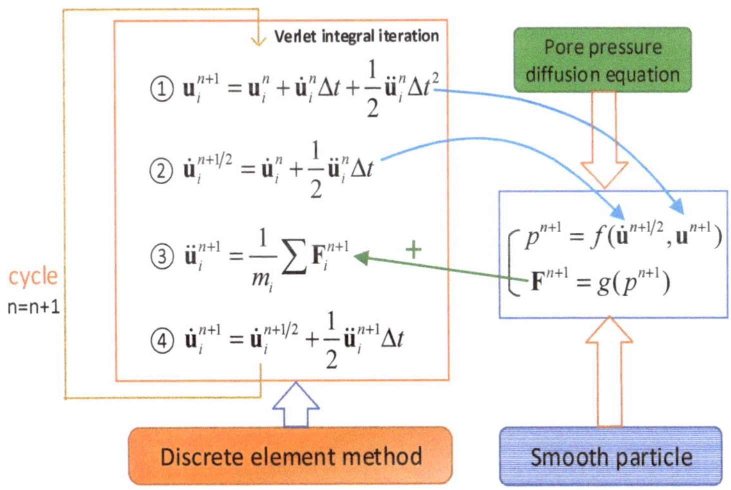

Figure 4. Flowchart of the coupling in DEM-SP.

3.3. Boundary Identification

In the process of underwater cutting, the water around the mineral boundary flows freely, while the internal liquid is in a restricted flow state due to the restriction between particles. Therefore, the boundary particles in high-pressure fluid should be identified first in numerical simulation. In addition, due to the random fragmentation of rock cuttings and the fact that the internal particles may also become the external boundary, the boundary recognition should be real-time and dynamic. Thus, a method that can deal with disordered particle structure and uneven particle size and mass distribution is needed in the process of boundary recognition. These requirements can be achieved by using the position divergence formula proposed by Muhammad et al. [31].

$$\nabla \cdot \mathbf{r} = \frac{\partial \mathbf{r}_x}{\partial x} + \frac{\partial \mathbf{r}_y}{\partial y} = 2 \tag{13}$$

Equation (13) is also applicable to Lucy's (1977) standard smooth particle method to prevent adjusting position divergence to particle defects at the boundary. As shown in Figure 5, it is assumed that Equation (13) is valid within the blue search radius r. However, if the search range (such as the magenta search circle) is at the boundary, $\nabla \cdot \mathbf{r}$ will be less than 2 due to fewer particles. In this work, particles with $\nabla \cdot \mathbf{r} < 1.5$ are considered to be boundary particles. In fact, when the search radius is given in the model, the average number N of adjacent particles of each particle within the search radius is constant. Specifically, particles with fewer than $(1.5/2) \times$ N particles in the search domain are judged as boundary particles. Eventually, all boundary particles are identified, and their pressures are fixed to the target value. Generally speaking, a larger search radius leads to more accurate particle judgment, but the computational overhead also increases correspondingly. Therefore, it is necessary to balance the accuracy and computational expense of the research.

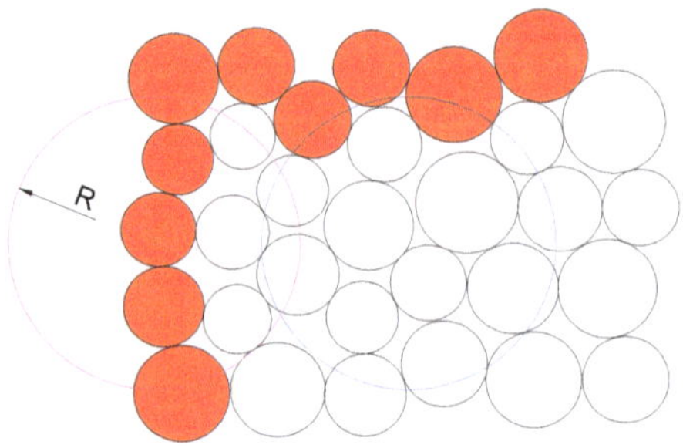

Figure 5. Boundary recognition.

3.4. Implementation Method

PFC provides the Python environment and the interface API between these two. Figure 6 shows the DEM-SP model implementation architecture. Python first calls the specific API of PFC software to obtain the velocity vector and position vector of discrete particles in PFC at n + 1/2 steps. After Python preprocessing, it is submitted to the CUDA computing core on the GPU for parallel computing. By substituting the data into the fluid pressure diffusion equation, the fluid pressure of the particle located and the coupling force generated are calculated. Then, the data is transferred back to PFC through a specific API, and the external force is applied to the particles to realize the coupling of the discrete phase and fluid phase. After that, PFC updates step n + 1, and iteration was achieved according to the above process.

The DEM part of the simulation is written in the fish language provided by PFC. In contrast, the fluid part is programmed with the CUDA module provided by the Numba library and accelerated through real-time compilation. Before calling the Python program every time, the kernel function written by CUDA was compiled, and then GPU was called to accelerate the above program calculation. The Python program is called by PFC software in the form of a plug-in. Once the Python file is called and run, Python will read the parameters provided by PFC and perform some steps, such as parameter initialization, variable initialization, GPU initialization, etc. Then the CUDA kernel function, data exchange, and coordination functions of the GPU are defined, and finally, the iterative function is inserted into the PFC operation step through a callback. Each iteration of PFC will automatically call the CPU data exchange and coordination functions defined by Python and GPU calculation functions for hydrodynamics calculation so as to realize coupling calculation. Since two-dimensional cutting is more convenient for analyzing the cutting mechanism and stress from the meso point of view, the algorithms and simulations are set up in 2D.

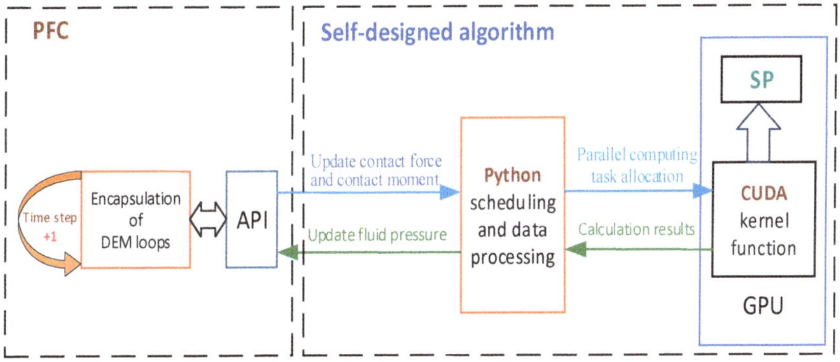

Figure 6. DEM-SP model implementation architecture.

4. Material Calibration and Model Establishment

4.1. Material Calibration

The difference in mineral formation environments and conditions result in significant differences in mechanical parameters measured by different types of samples in different regions. As described in Section 1, we obtained the basic mechanical parameters of two SMS samples. However, the macro mechanical parameters of ore samples cannot be directly used in simulation. They should be calibrated repeatedly by biaxial compression, uniaxial compression, and other simulation tests to determine the appropriate meso parameters so that the numerical model has the same macro performance as the real experiment.

For SMS, its macroscopic characteristics are not only able to bear the action of tension and compression but can also resist bending and torsion. Meanwhile, based on the consideration of nonlinear forces in ore, we chose the parallel bond model and linear contact model as the contact models between particles and between particles and tools. In the calibration, particles obey a uniform size distribution with a radius of 0.5–0.75 mm. Practice has proved that it is easy to simply meet the requirements of compressive strength or tensile strength. However, no matter how to adjust the combination of normal and tangential bond strength, it cannot meet the ratio of tensile strength to compressive strength. This is because the actual rock is usually composed of particles with significantly different stiffness. Different particles can coordinate the deformation in the deformation process while using uniform, same-bonding parameters and stiffness parameters cannot reflect this phenomenon. Therefore, a mixed bond parameters model was applied in parameter calibration. Soft particles account for 40% of the total, and their tangential and normal strengths are 5% of the hard particles. The micromechanical parameters of SMS ore calibrated by multiple numerical simulation tests are shown in Table 3, and the corresponding macro parameter error is less than 10%.

4.2. Model Establishment

Deep sea mining is a high-speed cutting process with a large cutting depth. As shown in Figure 7, according to the method described in Section 3, a two-dimensional model of rock breaking with a single cutter is established by using PFC2D software. A linear cutting simulation was carried out on a block with a size of 300 mm × 120 mm and a total number of 24,300 particles, and the particles at the right end and bottom of the ore body model are fixed. In this model, the cutter is modeled as a rigid body by using the wall element, and the cutter breaks the ore at a constant speed v. The cutter is placed at an inclination of 45° with a tip angle of 60° and a cutting depth of 15 mm. In the 2D simulation, the cutting process is carried out with a pressure of 0 MPa, 10 MPa, 20 MPa, and 30 MPa, respectively, according to different mineral occurrence depths.

Table 3. Solid parameters of SMS sample.

Meso Parameters	Value	
	Black	White
$\rho/kg/m^3$	3050	3150
Kn	1	0.8
Ks	1	0.8
Tn/Pa	13.4×10^6	7.8×10^6
Ts/Pa	9.8×10^6	6.8×10^6
Emod/Pa	3.5×10^9	1.2×10^9
Pb_emod/Pa	34×10^9	5×10^9
Fluid parameters (SP)		
Permeability coefficient	1×10^{-15}	1×10^{-15}
Biot modulus M/GPa	2	2
Dynamic viscosity/Pa s	0.001	0.001
Macro parameters		
Tensile strength/MPa	4.01	2.32
Compressive strength/MPa	38.6	20.1
Poisson's ratio	0.14	0.17
Modulus of elasticity/GPa	9.45	2.13

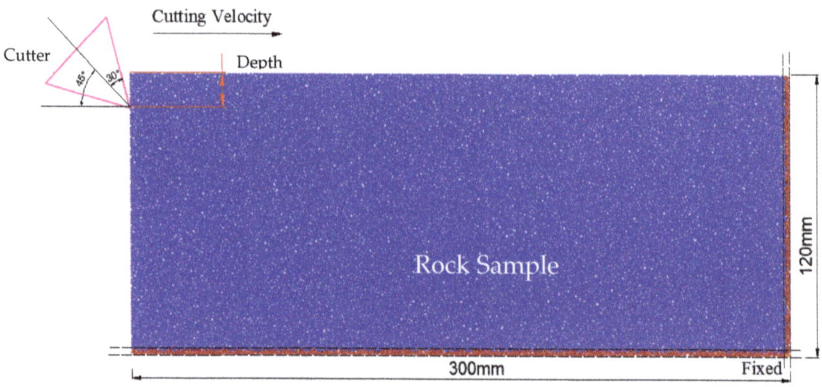

Figure 7. Schematic of the rock cutting model.

We build the model according to the framework shown in Figure 6 and set the liquid phase parameters after coupling with PFC2D through Python and a CUDA self-made algorithm, as shown in Table 3. The dynamic viscosity of water is 0.001 pa·s, and the permeability of SMS rock is set at 1×10^{-15}. The real-time pore liquid pressure in the numerical calculation process is transmitted to PFC2D through API to promote iterative calculation. During the simulation, the model status was saved every 0.005 s, and the contact force of particles, the information of particles, cracks, and chips were also recorded.

5. Intact Coupling Cutting Process

To investigate the cutting mechanism of deep-sea minerals, several intact simulations of the coupled cutting process were performed. According to the method described in Section 3.3, the radius of the boundary recognition domain is set to 2.8 mm. Consequently, particles with fewer than 13 particles within the research domain are considered to be fully surrounded by hydrostatic pressure, and the coupling force of these particles is zero. It should be noted that in order to avoid large deviations in the number of particles in each interpolation domain and reduce the possibility of particles being misjudged, the ratio of maximum radius to the minimum radius of particles should not be too large. Figure 8 shows the results of boundary recognition, in which the red circles represent the boundary particles recognized by the algorithm. Although several misjudged particles were found in the model, the overall effect is good.

Figure 8. Boundary particles recognized in cutting process.

The strong impact of the cutter forces the rock particles to disconnect and move away, and the separate particles represent the rock fragments that have been cut away. In the cutting process, the pore fluid pressure is applied to the particle center by PFC2D in the form of an external force, which is constantly updated by calculating the fluid pressure diffusion equation. According to the theory introduced in Section 3.2, the external force is formed by the pressure diffusion caused by the increase or decrease of the pore volume. Consequently, the external force exerted on particles is mainly concentrated at the front and bottom of the contact surface, while it is zero on the particles at other positions (Figure 9). Due to the relative motion between the parallel bonded particles, the contact force or moment along the relative direction will be generated. The bond will break, and microcracks form in the case where the contact force exceeds the bond strength. The particles separate and move under the combined action of the strong impact of the cutter and the fluid pressure. As the cutter moves forward, the pore pressure is also updated iteratively.

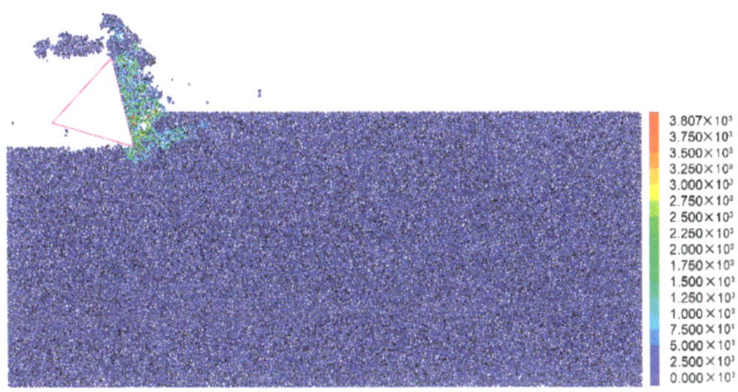

Figure 9. Fluid force applied to particles.

More interestingly, with the use of the DEM-SP coupling method, it is possible to investigate effects that are difficult to measure, such as the pore pressure distribution. Figure 10 shows the pore fluid pressure diffusion process in local rock samples around the tip. When the tip of the cutter wedges into the rock, local pore compression occurs due to particle extrusion, resulting in an increase in local fluid pressure, as shown in Figure 10. With the gradual advancement of the cutting tool, the bond between particles is destroyed, and cracks appear, which leads to local pore expansion and further leads to the reduction of pore pressure. The low pore pressure area in front of the tool tip indicates that the rapid deformations applied to the rock might result in cavitation of the pore fluid.

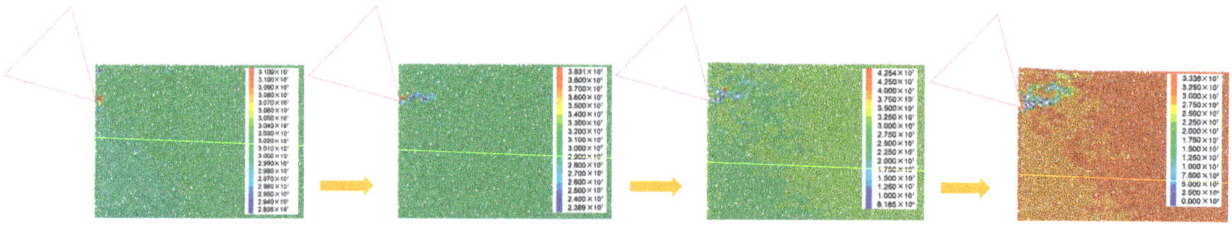

Figure 10. Pore pressure diffusion process (ambient pressure of 30 MPa).

Figure 11 shows a variation history of the x and y components of tool force in cutting the black sample, where positive Fx represents the force in the opposite direction of tool advance, and positive Fy represents the tool being pulled into the rock. Since the model is two-dimensional, the force refers to the force on the tool per unit thickness. The rock-cutting process is a typical cycle of elastic deformation, crack initiation, propagation, and failure. Accordingly, the cutting force oscillates as saw-teeth in the cutting process. The upward stage of the curve represents that the tool resists the adhesion between particles. The downward stage represents that the pick penetrates the rock to form microcracks, which are combined into macro cracks to form rock cuttings. As the cutter advances, the cutting force fluctuates with the formation of chips. Due to the randomness of crack propagation and chips, the peak value and period of cutting force in each cycle are not consistent. In the whole cutting process, the main cutting force Fx along the tool speed direction is dominant, which is also the main source of energy consumption. Thus, the subsequent detailed analysis related to the cutting force is only completed for the main cutting force Fx.

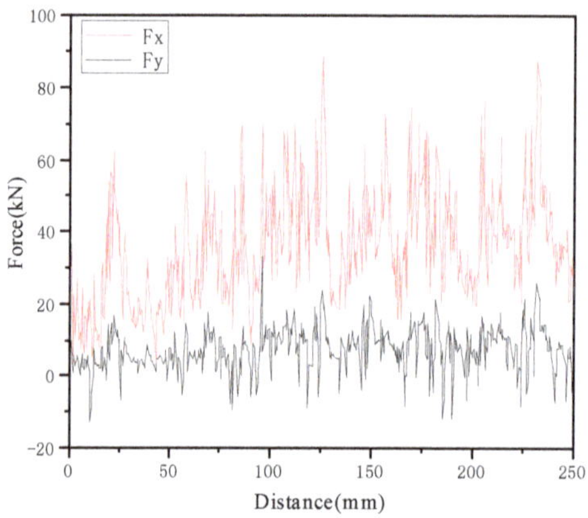

Figure 11. A typical cutting force history (ambient pressure of 10 MPa).

6. Results and Analysis

6.1. Crack Evolution

In order to further analyze the influence of fluid pressure on fragmentation characteristics of the SMS specimen from the mesoscopic point of view, the cumulative changes of microcracks were recorded by the program. As the simulation starts, the tool gradually contacts the sample, and then an initial crack is generated beneath the pick tip, which is called crack initiation. As the tool continues to advance, the crack propagates further until the fragment spalls. In PFC2D, the microcracks can be divided into tensile cracks and shear cracks according to fracture forms, which are represented by red and green segments, respectively, in this work. Figure 12 shows the distribution of cracks with displacement at the end of the simulation of SMS samples under different ambient pressures. Cutting rock under different fluid pressures is mechanically a comprehensive failure mode, accompanied by tensile and shear cracks. Although there is no obvious signal that tensile cracks or shear cracks are dominant, we can observe that the crack propagation is deeper under low confining pressure. At the same time, the internal damage of the model is more likely to occur under higher pressure, such as 20 MPa and 30 MPa. Regardless of the pore pressure, some chips collide with the high-speed moving tool in the simulation, accompanied by secondary breakage.

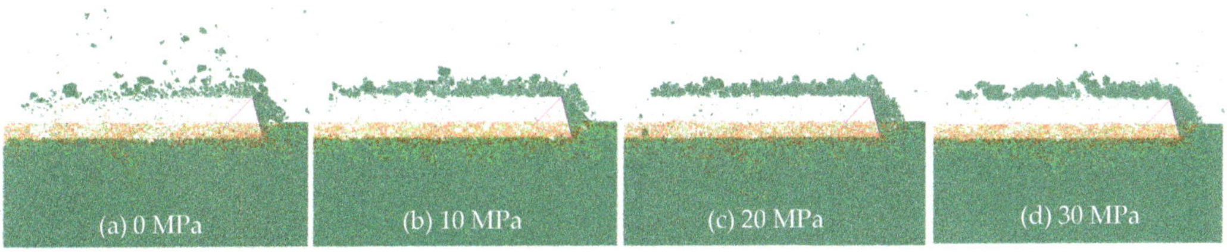

Figure 12. Crack distribution diagram.

The cumulative number history of micro-cracks under different confining pressures was further recorded by the program. As illustrated in Figure 13, the cumulative number of cracks under no confining pressure is similar to a step-by-step change. Thus, the fracture process can be divided into two types of stages: one is the crack accumulation and propagation stage, which is the main cutting process. The other stage is when the chips fall off, the tool pushes the chip forward, accompanied by a certain degree of secondary crushing, and the number of cracks is less in this stage. However, the alternating frequency of the two stages accelerates, and the cumulative crack growth rate is obviously accelerated with the increase in pressure. Figure 14 shows the relationship between the ratio of tensile crack number to shear crack number and hydrostatic pressure. The ratio has no obvious regularity with the increase of pressure, which is different from the usual simulation in which pressure is favorable for shear failure. This is mainly determined by the shear strength and tensile strength between particles in the parameter calibration. Grima et al. [14] interpret that the cutting process under high-pressure changes into an apparent ductile (cataclastic) mode and the failure of the rock will be predominately shear. When ductile behavior prevails, the crack formation is mostly along shear planes, and more force will be required to create a chip.

6.2. Load Characteristics

The horizontal cutting force Fx with respect to cutting distance for different rock models is shown in Figure 15. We can see that the overall horizontal cutting force is positively correlated with the ambient pressure, which is true for both black and white sample models. Moreover, pressure seems to have a greater influence on the black model. It is worth noting that for the two samples, the tool force will not drop to 0 when it drops from the peak value. This is due to the high-speed movement of the cutter, which causes the cutter to continuously contact a large amount of debris and cause secondary crushing.

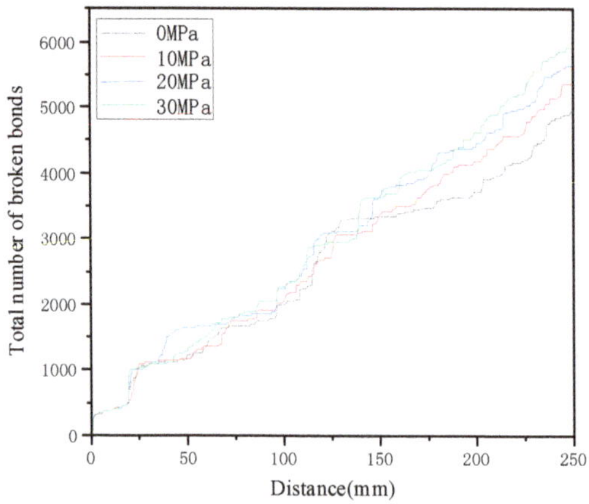

Figure 13. Cumulative number of broken bonds.

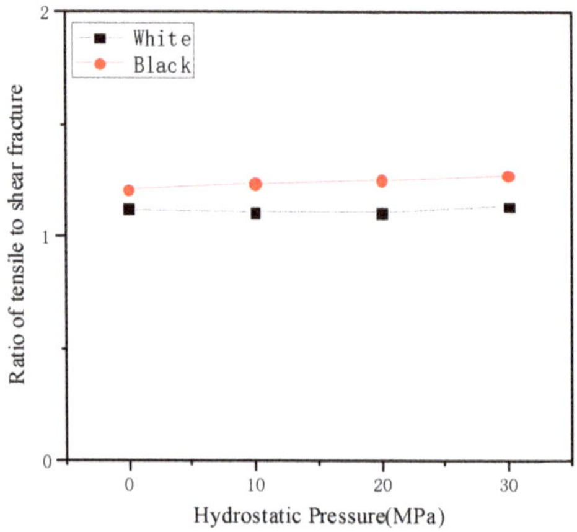

Figure 14. Variation of ratio of tensile crack to shear crack.

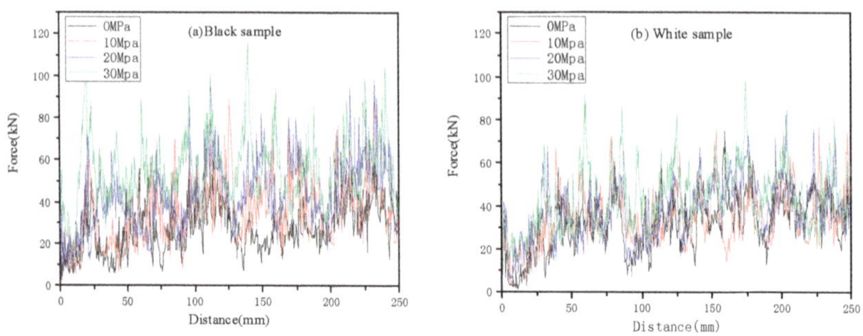

Figure 15. Cutting force variation under different ambient pressure.

Figure 16 indicates the relationship between the ambient pressure and the average cutting force for the black and white rock models. The average force for both the two rock models increases with the increases in ambient pressure. For instance, the average cutting force of the black sample under the pressure of 10 MPa, 20 Mpa, and 30 MPa is 30.8%, 55.2%, and 97.8% higher than those in the atmosphere, respectively. The average peak cutting force also increased from 57.6 MPa to 76 MPa, 85.2 MPa, and 102 MPa, respectively. While the average cutting force and the peak cutting force of the white sample increase with ambient pressure at a lower speed than those of the black sample. Although the strength of the black sample is about 1.5 times that of the white sample, the force changes of the two different materials do not show significant differences. In other words, there is no evidence that the difference in mechanical properties plays a dominant role in high-pressure and high-speed rock cutting. Under the combined action of high pressure and high speed, serious secondary crushing occurs in the cutting process. In contrast, the influence of rock strength on cutting force is relatively weak. As the average peak cutting force is significantly greater than the average cutting force, it is a more reasonable parameter for evaluating the working state of the pick and designing the excavation equipment. The fitting curve of the black specimens was $y = 1.43x + 58.7$, $R^2 = 0.985$. The fitting curve of the white specimens was $y = 0.72x + 62.99$, $R^2 = 0.984$. The correlation coefficient R^2 of the two models was greater than 0.98, which shows a good correlation between the ambient pressure and average peak cutting force for the two SMS specimens.

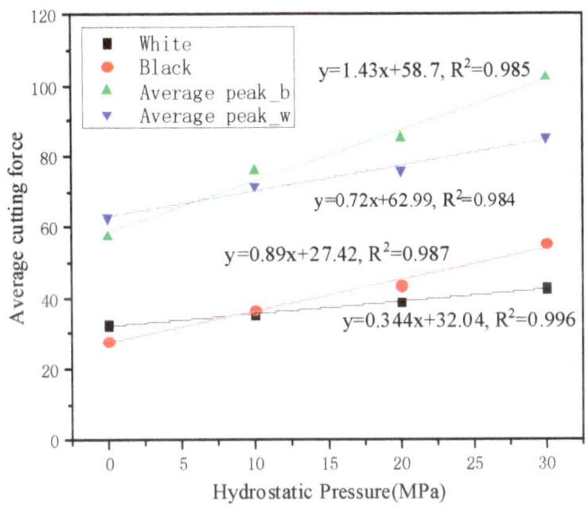

Figure 16. Average cutting force under different ambient pressures.

6.3. Fragment Morphology

Figure 17 shows the shape of debris at the end of the simulation under different hydrostatic pressure, where different color blocks represent chips generated in the cutting process. It can be seen that chips of different sizes are dispersed above the model, and the randomness of size is very strong, showing strong brittleness in atmospheric conditions. In contrast, in high-pressure conditions, the chip is finer and uniform in general. This is because the increase of confining pressure leads to more severe fragmentation in the crush zone and more difficult crack propagation, thus resulting in finer fragments. According to the research of Grima [14], lumps with the appearance of clay rather than chips are formed in the high-pressure cutting process, and a more ductile process will occur. This is consistent with the chip transformation shown in Figure 17, which verifies the correctness of the simulation in this study. We can also observe that the hydrostatic pressure also has a significant effect on the cutting surface of the model. Since the high pressure inhibits the crack growth, a finer cutting surface is generated, and the size of the crushed zone decreases. In atmospheric conditions, chips are scattered over the model due to the instantaneous breaking of the bonds, while with the increase of the pressure, the particle flow property of debris is stronger.

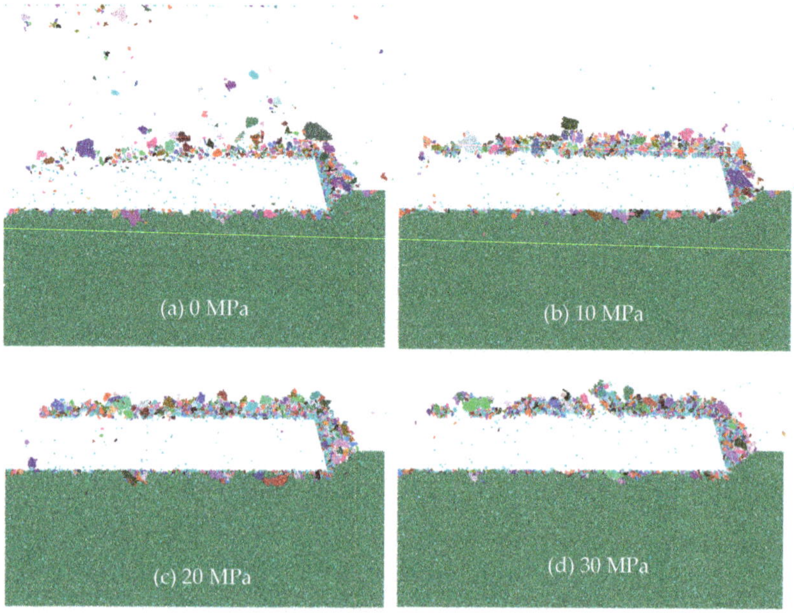

Figure 17. Chip morphology under different ambient pressures.

We also make statistics on the volume of fragments generated in the cutting process (it is actually the area in 2D), in which the volume of each fragment is the sum of the volumes of all particles in the fragment. The statistical data under different pressures is shown in Figure 18. As depicted, the chip size distribution of the two samples with different pressures is similar, and the small size accounts for the majority. Meanwhile, the variation trend of chip size produced by the two samples is basically the same, and there is more small debris and less large debris with the increase of pressure. In other words, the fluid pressure will make the generated debris more uniform, which is consistent with the chip morphology diagram in Figure 17. The existence of fluid pressure will inhibit the extension of cracks and make the size of debris smaller.

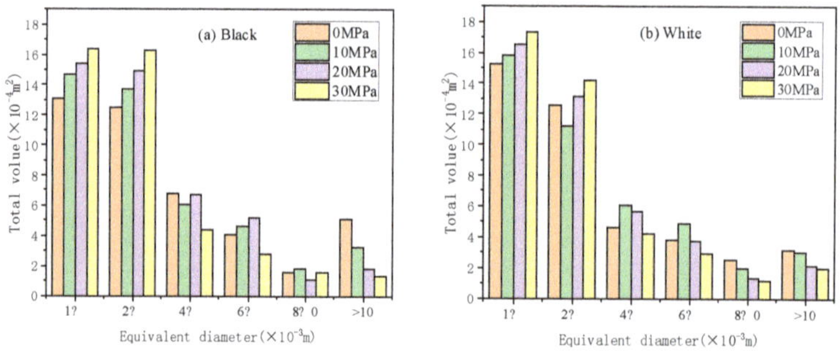

Figure 18. Fragment size distribution.

6.4. Specific Energy

Specific energy (SE)is an important and valid parameter to evaluate cutting efficiency, and it is widely used to evaluate the performance of mining and excavation machinery. SE is calculated by:

$$SE = \frac{F_R L}{V} \tag{14}$$

where F_R refers to the average horizontal cutting force, L is the cutting distance, and V denotes the total volume of chips. A large amount of powder or fine particles is produced during cutting, but some rocks are still relatively complete. Note that a large amount of powder or fine particles is produced in the cutting process, so it is inaccurate to calculate the volume directly by summing the volumes of all fragments. In fact, the calculation error can be reduced by subtracting the remaining volume from the original volume of the sample model.

Figure 19 shows the relationships between ambient pressure and specific energy for different rocks. For each rock sample, specific energy is positively correlated with the pressure. Moreover, the growth rate of specific energy of black samples with higher strength is obviously higher than that of white samples. This is due to a higher cutting force and a lower total amount of chips with the increase in pressure. Moreover, the slope of the "SE vs. ambient pressure" curve of the black sample is bigger than that of the white sample. This means that the cutting efficiency of black rocks with high strength is more sensitive to ambient pressure than that of white rocks with lower strength.

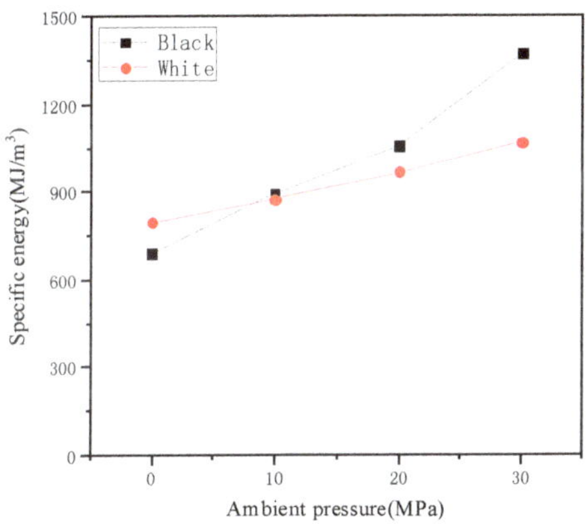

Figure 19. Relationships between ambient pressure and SE.

7. Conclusions

The purpose of this work is to study the effect of high hydrostatic confining pressure on the cutting performance of two SMS seabed minerals. The analysis results draw the following conclusions:

(1) The cutting force is significantly improved with the combination of high cutting speed and high pressure. Furthermore, the cutting forces increase with increasing ambient pressure.

(2) The increase in ambient pressure leads to the variation of chip morphology. With the increase of pressure, fragmentation in the crush zone becomes more severe, and crack propagation is more difficult. For these reasons, the fragments are finer and more uniform, and the cutting surface is flatter.

(3) Compared with the white sample, the cutting force of the black sample with higher strength increases faster with the pressure. At the same time, the specific energy of the black sample is more sensitive to the ambient pressure than that of white rocks with lower strength.

The combination of DEM and SP provides an effective method for the analysis of underwater fluid-solid coupling cutting processes, especially a valuable method for the interaction between materials and tools under hydrostatic pressure. The study of rock fragmentation mechanisms provides important information for the exploitation of SMS. It should be noted that only the influence of pore pressure is considered in the discussion, and the influence of fluid viscous resistance on the trajectory of debris is not considered. In addition, geotechnical characteristics may also have a strong impact on seabed excavation tools. For example, the silica content in seabed minerals can have a strong negative impact on the wear resistance of mining tools. Therefore, in future deep-sea excavation research, a more in-depth and large-scale geotechnical investigation is also necessary.

Author Contributions: The paper emerged from the long-term cooperation of the authors. Conceptualization, Y.L. and H.D.; methodology, H.D.; test and analysis, H.D. and H.L.; writing—original draft preparation, H.D.; writing—review and editing, Y.L. and H.L. All authors have read and agreed to the published version of the manuscript.

Funding: This research was supported by the National Key Research and Development Program of China, grant number 2021YFC2801701-02 and Major Science and Technology Projects of Hunan Province, China, grant number 2020GK1022.

Institutional Review Board Statement: Not applicable.

Informed Consent Statement: Not applicable.

Conflicts of Interest: The authors declare that they have no known competing financial interests or personal relationships that could have appeared to influence the work reported in this paper.

References

1. Miller, K.A.; Thompson, K.F. An overview of seabed mining including the current state of development, environmental impacts, and knowledge gaps. *Front. Mar. Sci.* **2018**, *4*, 418. [CrossRef]
2. Levin, L.A.; Amon, D.J. Challenges to the sustainability of deep-seabed mining. *Nat. Sustain.* **2020**, *3*, 784–794. [CrossRef]
3. Leng, D.X.; Shao, S. A brief review of recent progress on deep sea mining vehicle. *Ocean Eng.* **2021**, *228*, 108565. [CrossRef]
4. Kato, Y.; Fujinaga, K. Deep-Sea Mud in the Pacific Ocean as a Potential Resource for Rare-Earth Elements. *Nat. Geosci.* **2011**, *4*, 535–539. [CrossRef]
5. Boschen, R.E.; Rowden, A.A. Mining of deep-sea seafloor massive sulfides: A review of the deposits, their benthic communities, impacts from mining, regulatory framework and management strategies. *Ocean Coast. Manag.* **2013**, *84*, 54–67. [CrossRef]
6. Interridge. InterRidge Vents Database Ver.3.4. Available online: https://vents-data.interridge.org (accessed on 15 July 2022).
7. Yamazaki, T.; Park, S.H. Relationships between geotechnical engineering properties and assay of seafloor massive sulfides. In Proceedings of the 13th International Offshore and Polar Engineering Conference, Honolulu, HI, USA, 25–30 May 2003.
8. Spagnoli, G.; Jahn, A. First results regarding the influence of mineralogy on the mechanical properties of seafloor massive sulfide samples. *Eng. Geol.* **2016**, *214*, 127–135. [CrossRef]
9. Hu, J.H.; Liu, S.J. Deformation characteristics and constitutive model of seafloor massive sulfides. *J. Cent. South Univ.* **2017**, *24*, 1986–1991. [CrossRef]
10. Dai, Y.; Ma, F.Y. Mechanical tests and numerical simulations for mining seafloor massive sulfides. *J. Marine Sci. Eng.* **2019**, *7*, 252. [CrossRef]
11. Hunter, R.J. Investigation of the Application of Mechanical Mining of Ocean Floor Polymetallic Sulphide Deposits. Master's Thesis, University of British Columbia, Vancouver, BC, Canada, 2007.
12. Yutaro, S.; Jizhao, L. First successful mining of submarine hydrothermal deposits in the world. *Constr. Project.* **2020**, *72*, 83–87.
13. JOGMEC Conducts World's First Successful Excavation of Cobalt-Rich Seabed in the Deep Ocean. Available online: http://tokiox.com/wp/ogmec-conducts-successful-excavation-of-cobalt-rich/?lang=en (accessed on 15 July 2022).
14. Kaitkay, P.; Lei, S. Experimental study of rock cutting under external hydrostatic pressure. *J. Mater. Process. Tech.* **2005**, *159*, 206–213. [CrossRef]
15. Grima, M.A.; Miedema, S.A. Effect of high hyperbaric pressure on rock cutting process. *Eng. Geol.* **2015**, *196*, 24–36. [CrossRef]
16. Detournay, E.; Atkinson, C. Influence of pore pressure on the drilling response in low-permeability shear-dilatant rocks. *Int. J. Rock Mech. Min.* **2000**, *37*, 1091–1101. [CrossRef]
17. Li, X.F.; Wang, S.B. Investigation on the influence mechanism of rock brittleness on rock fragmentation and cutting performance by discrete element method. *Measurement.* **2018**, *113*, 120–130.
18. Xue, Y.D.; Zhou, J. Rock fragmentation induced by a TBM disc-cutter considering the effects of joints: A numerical simulation by DEM. *Comput. Geotech.* **2021**, *136*, 104230. [CrossRef]
19. Shimizu, H.; Murata, S. The distinct element analysis for hydraulic fracturing in hard rock considering fluid viscosity and particle size distribution. *Int. J. Rock Mech. Min.* **2011**, *48*, 712–727. [CrossRef]
20. Zeng, W.; Yang, S.Q. Numerical investigation on permeability evolution behavior of rock by an improved flow-coupling algorithm in particle flow code. *J. Cent. South Univ.* **2018**, *25*, 1367–1385. [CrossRef]
21. Zhang, Y.L.; Shao, J.F. An improved hydromechanical model for particle flow simulation of fractures in fluid-saturated rocks. *Int. J. Rock Mech. Min.* **2021**, *147*, 104870. [CrossRef]

22. Lei, S.T.; Kaitkay, P. Distinct element modeling of rock cutting under hydrostatic pressure. *Key Eng. Mat.* **2003**, *250*, 110–117.
23. Rizo, M.A.J. Considerations for Discrete Element Modeling of Rock Cutting. Ph.D. Thesis, University of Pittsburgh, Pittsburgh, PA, USA, 2013.
24. Tropin, N.; Manakov, V. Implementation of boundary conditions in discrete element modeling of rock cutting under pressure. *Appl. Mech. Mat.* **2014**, *598*, 114–118.
25. Lv, Y.; Li, H. Bonded-cluster simulation of rock-cutting using PFC2D. *Clust. Comput.* **2017**, *20*, 1289–1301.
26. Helmons, R.L.J.; Miedema, S.A. Simulating hydro mechanical effects in rock deformation by combination of the discrete element method and the smoothed particle method. *Int. J. Rock Mech. Min. Sci.* **2016**, *86*, 224–234. [CrossRef]
27. Helmons, R.L.J.; Miedema, S.A. Modeling fluid pressure effects when cutting saturated rock. *Eng. Geol.* **2016**, *211*, 50–60. [CrossRef]
28. Potyondy, D.O. The bonded-particle model as a tool for rock mechanics research and application: Current trends and future directions. *Geosys. Eng.* **2015**, *18*, 1–28. [CrossRef]
29. Helmons, R.L.J. Excavation of Hard Deposits and Rocks on the Cutting of Saturated Rock. Ph.D Thesis, Delft University of Technology, Delft, The Netherlands, 2017.
30. Cleary, P.W.; Monaghan, J.J. Conduction modelling using smoothed particle hydrodynamics. *J. Comput. Phys.* **1999**, *148*, 227–264. [CrossRef]
31. Muhammad, N.; Rogers, B.D. Understanding the behaviour of pulsed laser dry and wet micromachining processes by multi-phase smoothed particle hydrodynamics (SPH) modelling. *J. Phys. D Appl. Phys.* **2013**, *46*, 95101. [CrossRef]

Journal of
Marine Science and Engineering

MDPI

Article

Organic Geochemical Signatures of the Upper Miocene (Tortonian—Messinian) Sedimentary Succession Onshore Crete Island, Greece: Implications for Hydrocarbon Prospectivity

Angelos G. Maravelis [1,*], George Kontakiotis [2,*], Spyridon Bellas [3], Assimina Antonarakou [2], Chrysanthos Botziolis [4], Hammad Tariq Janjuhah [5], Panayota Makri [2], Pierre Moissette [2,6], Jean-Jacques Cornée [7], Nikolaos Pasadakis [3], Emmanouil Manoutsoglou [8], Avraam Zelilidis [4] and Vasileios Karakitsios [2]

1 Department of Geology, Aristotle University of Thessaloniki, 54124 Thessaloniki, Greece
2 Department of Historical Geology-Paleontology, Faculty of Geology and Geoenvironment, School of Earth Sciences, National and Kapodistrian University of Athens, Panepistimiopolis, Zografou, 15784 Athens, Greece
3 Institute of Geoenergy—Foundation for Research and Technology—Hellas (FORTH/IG), Building M1, University Campus, Akrotiri, 73100 Chania, Greece
4 Laboratory of Sedimentology, Department of Geology, University of Patras, 26504 Rion, Greece
5 Department of Geology, Shaheed Benazir Bhutto University, Sheringal 18050, KPK, Pakistan
6 Département Origines et Evolution, UMR7207 CR2P, Muséum National d'Histoire Naturelle, 8 rue Buffon, 75005 Paris, France
7 Géosciences Montpellier, Université des Antilles-Université de Montpellier-CNRS, Pointe à Pitre (FWI), 34095 Montpellier, France
8 Laboratory of Geology, School of Mineral Resources Engineering, Technical University of Crete, 73100 Chania, Greece
* Correspondence: angmar@geo.auth.gr (A.G.M.); gkontak@geol.uoa.gr (G.K.)

Citation: Maravelis, A.G.; Kontakiotis, G.; Bellas, S.; Antonarakou, A.; Botziolis, C.; Janjuhah, H.T.; Makri, P.; Moissette, P.; Cornée, J.-J.; Pasadakis, N.; et al. Organic Geochemical Signatures of the Upper Miocene (Tortonian—Messinian) Sedimentary Succession Onshore Crete Island, Greece: Implications for Hydrocarbon Prospectivity. *J. Mar. Sci. Eng.* **2022**, *10*, 1323. https://doi.org/10.3390/jmse10091323

Academic Editor: Luca Cavallaro

Received: 25 July 2022
Accepted: 12 September 2022
Published: 18 September 2022

Abstract: The definition of pre-Messinian source rocks in the eastern Mediterranean is of paramount importance for hydrocarbon exploration because of the ability of salt to act as a high-quality seal rock. This research evaluates the organic geochemical features of the Upper Miocene (Tortonian—Messinian) sedimentary succession onshore Crete Island, Greece. The study employs original (Messinian, Agios Myron Fm) and published (Tortonian, Viannos Fm, Skinias Fm, Moulia Fm, and Messinian Ploutis section) results from organic geochemical analyses of mudstone samples. One hundred and one samples were examined using standard organic geochemistry methodology (Rock-Eval II and VI-TOC) to define the origin, type, and degree of organic matter maturity. The data indicate that the studied samples have poor to fair gas-prone source rock potential. These possible source rock units have not experienced great temperatures during burial, and, thus, their organic matter is thermally immature. The sub-salt (Tortonian—Messinian) source rock units are likely to be of higher thermal maturity in the western and eastern south Cretan trenches because of tectonic subsidence and a thicker sedimentary overburden. Several traps can grow in these regions, associated with normal faults, rotated blocks and unconformities (both below and above the unconformities). This research provides a basis for the further evaluation of the hydrocarbon potential in Crete Island. It is an area that shares geological similarities with the surrounding regions that contain proven reserves and is of crucial economic and strategic importance.

Keywords: Rock Eval analysis; pre-evaporitic source rocks; hydrocarbon potential; sedimentary basin dynamics; thermal maturity; kerogen type; hydrocarbon exploration; evaporites sealing quality; Messinian salinity crisis; eastern Mediterranean

1. Introduction

Technological advances (including modeling and 3D mapping) in geophysical research have been the driving force for hydrocarbon exploration in tectonically complicated settings, such as the eastern Mediterranean [1–3]. The petroleum industry has paid more attention to the eastern sector of the Mediterranean Sea because of recent discoveries of significant gas resources and the existence of promising frontier regions [3–5]. These promising results stimulated interest in further and more detailed exploration activities in the region. An important parameter that enhances the prospectivity of the eastern Mediterranean is the widespread accumulation of thick Messinian evaporitic successions. Such rock types are related to first-class seal rocks in numerous regions across the world, such as the Middle East, Russia, the Gulf of Mexico, and offshore Brazil [6–9]. Thus, the documentation of suitable source and reservoir rocks below the Messinian evaporites in the eastern Mediterranean could be very beneficial for future exploration activities.

There is active exploration in the region, such as in the Levantine Basin, Egypt, Israel, and, more recently, offshore Cyprus, while additional promising regions (e.g., Herodotus Basin and offshore Crete) have also been suggested [10–14]. The Levantine Basin contains significant conventional and unconventional resources, thus, there is a high likelihood for future significant discoveries offshore Cyprus [12]. Exploration in Egypt and Israel has proven very successful, resulting in a number of discoveries, which in turn has prompted extensive regional exploration [15]. The Hellenic part of the Herodotus Basin is largely unexplored, despite the active mud volcanoes that have been documented on seismic profiles in several locations along the Mediterranean Ridge [16–20]. The areas west of Crete Island exhibit comparable geotectonic histories to those of nearby regions, such as Cyprus, Egypt, and Israel, where major discoveries have been reported (e.g., Zohr, Calypso, Tamar, Leviathan, and Glafkos). This information is of paramount importance for Greece and has stimulated the petroleum industry to explore the potential for hydrocarbon resources located outside the already established petroleum province of western Greece [14,21–26]. As a result, exclusive right licenses were granted for west and south Crete. The potential for onshore Crete and nearby regions to contain Tortonian to Messinian (sub-salt) source and reservoir rocks has been suggested based on studies from Gavdos Island south of Crete [27], the Messara and Heraklion Basin in central Crete [28–31], and the Levantine Basin in the south-eastern Mediterranean [4,5,10–12].

Rock Eval (RE) pyrolysis is viewed as an economic and reliable methodology to evaluate the hydrocarbon generative potential and maturation level of potential source rocks in sedimentary basins [32–34]. This study integrates both original and already published data derived from outcrop-based organic geochemical studies to assess the hydrocarbon generative potential of the Upper Miocene sub-salt formations in onshore Crete. This research examines available data from the Tortonian deposits that belong to Viannos Fm, Skinias Fm, and Moulia Fm (at Faneromeni) and from the Messinian Ploutis section (central Crete). Additionally, it presents new numerical constraints to the organic geochemical signatures of the Messinian Agios Myron Fm (new data), providing information about the type, quantity, quality, and maturation level of the organic material. Our results offer a comprehensive insight into the existence and nature of Late Miocene source rocks in the region. The results obtained from an inland frontier basin could enhance current exploration activities in the region and stimulate future exploration activity, such as the drilling of new exploration wells and the acquisition of new seismic data.

2. Geological Setting

The Aegean Sea has attracted both scientific and economic attention. The Aegean region is positioned on continental crust that comprises several stacked upper crustal nappes [35,36] (Figure 1). These nappes were stacked because of the northward dipping subduction of the African plate under the European plate that has occurred since the Cretaceous [35–37]. These plate tectonics were responsible for: (1) the orogenic processes, (2) the disruption of the marine junctions to the Indo-Pacific and Paratethys, and (3) the progressive restriction of the marine passage to the Atlantic Ocean [38]. The southeast Hellenides form part of a synorogenic orocline that illustrates modifications regarding the internal deformational pattern and the nature of the margin, as evidenced by geotectonic analysis [39]. Although orthogonal collision prevails in the western edge of the orocline, the central and eastern part is developed because of oblique subduction that is influenced by strike-slip tectonics (NE-directed transpressional faults). These strike-slip faults developed transverse zones that split apart some of the structural elements, such as salients, recesses and linear segments [40,41]. Structural and stratigraphic analyses have documented that the style of the compressional tectonic activity and the total thickness of Mesozoic sedimentary successions [41–43] are highly variable across strike, suggesting that these transverse zones had an impact on the Mesozoic sedimentary basins. From the Late Miocene to more recently, the regional post-orogenic extension occurred along these transverse zones that are oriented parallel to the thrust belt in eastern Crete [39,44,45].

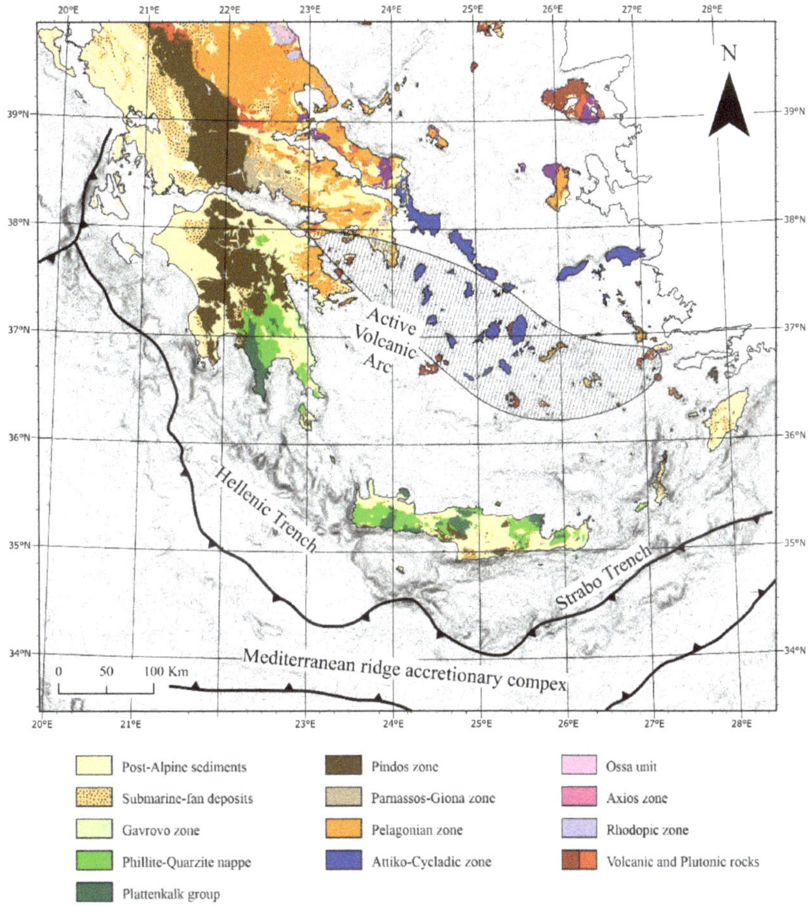

Figure 1. Scheme illustrating the tectonic zonation of Greece based on the main tectonostratigraphic zones (modified after [14]).

The Island of Crete includes several nappes of Triassic to Quaternary rock units that belong to different geotectonic zones (Figure 2). The nappes are subdivided into a lower member that contains high pressure and low temperature (HP-LT) metamorphic rocks and an upper member with no metamorphic rocks [46–48]. These nappes are thought to have been placed together through south-directed subduction and/or collision and accretion to the northern Gondwanan edge [49,50]. The pre-Miocene basement rocks were tectonically fragmented into numerous parts because of uplift and exhumation of the nappe pile and were responsible for the development of Lower to Middle Miocene sedimentary basins.

Several mechanisms have been proposed to explain this uplift and exhumation and involve both extensional [51–53] and contractional tectonics [50,54,55]. Younger (Late Miocene to Early Pleistocene) thrust-associated depocenters are ascribed to the migration of thrust activity to the southwest [56,57]. Comprehensive studies in central Crete have revealed two phases of compression [54]: An early phase of ductile exhumation of high-pressure metamorphic rocks that is related to NNW–SSE compression, responsible for the formation of low-angle reverse faults (thrusts) that contributed to nappe emplacement. A subsequent phase of brittle exhumation ascribed to NNE–SSW compression and dominated by large fault-associated folds and tectonic imbrication. The older, underlying nappes are commonly intersected by younger, large-scale thrust faults [54].

Figure 2. Scheme illustrating the tectonic zonation and main structural elements and depocenters of Crete Island and surrounding regions. Red cycles correspond to the sampling locations.

3. Stratigraphic and Environmental Constraints

Updating former lithostratigraphic models [58], the Tortonian-upper Messinian stratigraphic succession of central Crete has been divided into five formations, namely Viannos, Skinias, Kasteliana, Moulia, and Agios Myron formations that are overlain by the Marine Upper Messinian (MUM) and Lago Mare deposits [48] (Figure 3).

The pre-Neogene basement underlies Viannos Fm, and its boundary contains abundant coarse, angular debris [48]. Viannos Fm is approximately 400 m thick and includes sandstone and mudstone alternations, with some rare conglomeratic beds. Additionally, mud-rich (silty) limestone occurs, whereas fine-grained material (mud) with an increased level of coalification is rarely present. Viannos Fm includes a broad array of depositional elements and environments that include channel-belts, overbanks, and lakes. The clasts within the conglomeratic deposits are mainly sub-angular, pointing to an alluvial fan with sheet-flood deposition as the most suitable depositional environment [48]. Paleocurrent directions suggest a westward transport direction within this formation. The contact between Viannos Fm and the overlying Skinias Fm is conformable. Skinias Fm is 150–250 m thick and is represented by marine fine-grained (muddy) deposits that are intercalated with sandstone beds. Coarser-grained deposits are represented by scarce pebbly beds [48]. The depositional environments for Skinias Fm are regarded as shallow to deep marine (from the bottom to the middle of the formation) and display an upward trend towards shallower environments [48]. Paleocurrent measurements in this area suggest an east to southeast transport direction. The overlying Kasteliana Fm is 300–350 m thick and consists of both fine- and coarse-grained deposits (mudstone, siltstone, sandstone, and conglomerate). Deposits, such as travertine and lignite, along with oyster beds and coral bioherms, are scarcely present. Kasteliana Fm is interpreted to include several depositional environments, including fluvial-lacustrine, lagoonal, and inner neritic. The contact between Kasteliana Fm and the underlying Skinias Fm is unconformable [48]. Moulia Fm overlies Kasteliana Fm and is 70–80 m thick. It is represented by both siliciclastic and carbonate deposits and is divided into two members. The lower part contains sediments that have been deposited in an inner neritic environment, and the upper part illustrates an upward increase in water depth towards deeper marine environments [48]. In most of the cases, Moulia Fm conformably overlies Kasteliana Fm.

The samples from Moulia Fm have been collected from the Faneromeni section that comprises a sedimentary succession (up to 37-m-thick) and can be subdivided into two members. Coarse-grained deposits (conglomerate) separate these members, with the lower one containing thick-bedded sandstone and mudstone couplets (12 m thick) and the upper one including thin-bedded sandstone and mudstone alterations [59,60]. Sapropels occur in the upper parts of the succession. Agios Myron Fm is approximately 120 m thick and consists of bioturbated and fossiliferous sandstone, along with calcarenite and calcrete palaeosols. The upper parts, Messinian in age, designated here by the analyzed samples from the eponymous section (7.2–6.5 Ma) contain cyclically-bedded alternations of deep marine bluish-grey homogeneous and laminated brownish marls (sapropels) along with three distinct ash beds [48,61]. Paleobathymetric calculations of the samples from this location have shown that this sedimentary succession was deposited under upper bathyal conditions of 350–650 m [62]. Overall, Agios Myron Fm displays an upward increase in water depth, similar to the trend observed in Moulia Fm, and unconformably overlies Viannos Fm [48]. The overlying formation consists of Upper Messinian alternations of turbidites, sapropels and marls (up to 40 m thick, MUM, Marine Upper Messinian in Zachariasse et al. [63]) that are overlain by evaporites, where exposed (e.g., Ploutis section) [30], and Lago Mare deposits (generally up to 60 m thick) [63,64] or Lower Pliocene gravity failure deposits (mass wasting deposits of Zachariasse et al. [63]).

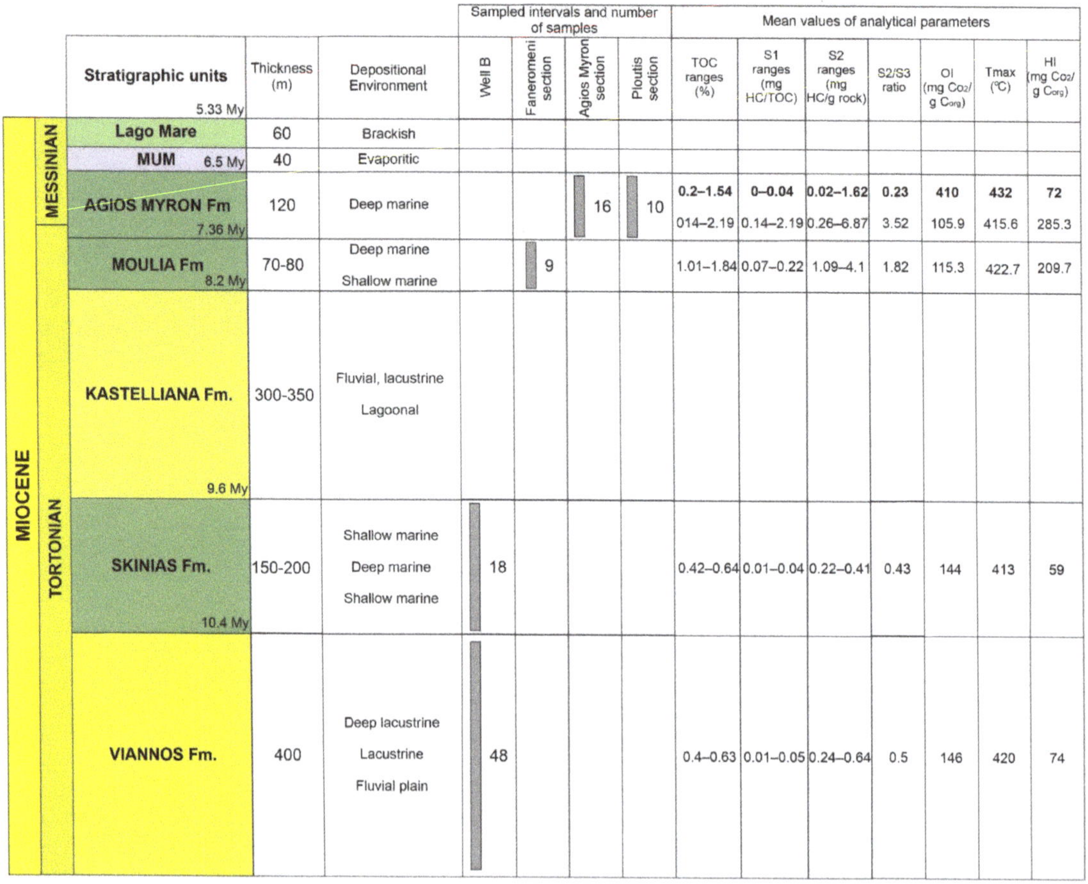

Figure 3. Stratigraphic column of the study region that illustrates the evolution of the Upper Miocene (Tortonian—Messinian) depositional environments. The age constraints of the study formations come from Zachariasse et al. [48]. To the left upper corner, the stratigraphic interval that the study sections and formations cover is displayed. Abbreviation, MUM: Marine Upper Messinian.

4. Material and Methods

One hundred and one (101) samples in total were selected for organic geochemical analysis. Forty-eight (48) samples from Vianos Fm, eighteen (18) samples from Skinias Fm, nine (9) samples from Moulia Fm (at Faneromeni), sixteen (16) samples from Agios Myron Fm, and ten (10) samples from Ploutis section, respectively. The samples from Agios Myron are new data (16 samples) and the data from the other formations come from previous studies (85 samples). Most of the samples that belong to Vianos and Skinias Fms come from an exploration well (well D, Figure 2). The samples that belong to the other Fms were collected from outcrops. The outer surface of the sections was removed prior to sampling in order to avoid contamination from weathering. The fresh samples were immediately placed, and protected from oxidization, in plastic sample bags.

The organic geochemical features of the sedimentary rocks studied were assessed using RE analysis. After overnight drying, the rock samples were crushed and sieved using a 60 mesh (250 μm) sieve. The analysis was performed using RE II and VI (Delsi Inc., Edison, NJ, USA) analyzers under standard conditions utilizing ~100 mg of pulverized rock. The samples were then heated in a helium atmosphere using a suitable oven [65–67].

The RE analysis was performed at the Technical University of Crete, Greece (Hydrocarbons Chemistry and Technology Research Unit). Analytical descriptions of the RE VI methodology and interpretations of the obtained results have been offered by several authors [66–73]. Principal parameters, including total organic carbon content (TOC, wt%), free (S1, mg HC/g rock) and pyrolysable (S2, mg HC/g rock) hydrocarbons, total hydrocarbon generative potential (SP, S1 + S2, mg HC/g rock), hydrogen index (HI, mg HC/g Corg), oxygen index (OI, mg CO_2/g Corg), Tmax (°C), and production index (PI, S1/S1 + S2) were documented. The data assessment was based on the works of Tissot and Welte [74], Peters [70], Burwood et al. [75] and Dymann et al. [76].

5. Results

The experimental data of the RE VI-TOC analysis for the 101 samples are illustrated in Supplementary Materials, Table S1.

The TOC contents fluctuated within the examined samples, from 0.4% (D304–307) to 0.63% (D415–420) in Viannos Fm, 0.42% (D128–132) to 0.64% (D14–17) in Skinias Fm, 1.01% (DT15) to 1.84% (DT16) in Moulia Fm (at Faneromeni), 0.14% (PL04) to 2.19% (PL07) in Ploutis section and 0.2% (AM18C) to 1.54% (AM14) in Agios Myron Fm. Based on the TOC values and the criteria established by Peters [70] and Dembicki [77], the studied samples could possess poor to excellent hydrocarbon generation potential. The studied sections included samples that could serve as source rocks (0.5% > TOC > 1%, Table S1) and samples that deserve a more thorough evaluation (TOC >1%, Table S1). Nevertheless, TOC values alone are not sufficient to characterize the samples as potential source rocks. The determination of the hydrocarbon generative potential is a function of the TOC and the hydrogen that is related to the organic material (Dembicki, 2009). Therefore, additional parameters (e.g., S2 values) need to be taken into consideration. The samples exhibited S2 values that fluctuated between 0.24 (D234–238 and D254–260) to 0.64 mg HC/g rock (D420–424) in Viannos Fm, 0.22 (D128–132) to 0.41 mg HC/g rock (D14–17) in Skinias Fm, 1.09 (DT15) to 4.1 mg HC/g rock (DT18) in Moulia Fm (at Faneromeni), 0.26 (PL04) to 6.87 mg HC/g rock (PL02) in Ploutis section, and 0.02 (AM4) to 2.62 mg HC/g rock (AM14) in Agios Myron Fm.

Several parameters from RE analysis were utilized to define the type of organic matter in the sedimentary rocks, including S1, HI and OI values, and the S2/S3 ratio [73]. The examined samples exhibited S1 values that ranged between 0.01 (D402–406 and D406–409) and 0.05 mg HC/g TOC (D415–420) in Viannos Fm, 0.01 (D210–214) and 0.04 mg HC/g TOC (D6–14) in Skinias Fm, 0.07 (DT5) and 0.22 mg HC/g TOC (DT16) in Moulia Fm (at Faneromeni), 0 (PL04) and 0.7 mg HC/g TOC (PL02) in Ploutis section and 0 and 0.04 mg HC/g TOC (AM14) in Agios Myron Fm (Table S1). The mean HI values in the studied succession were 74 mg HC/g Corg (Viannos Fm), 59 mg HC/g Corg (Skinias Fm), 209.7 mg HC/g Corg (Moulia Fm at Faneromeni), 285.3 mg HC/g Corg (Ploutis section) and 72 mg HC/g Corg (Agios Myron Fm). The average S2/S3 ratios were 0.5 (Viannos Fm), 0.43 (Skinias Fm), 1.82 (Moulia Fm at Faneromeni), 3.52 (Ploutis section) and 0.23 (Agios Myron Fm). The average OI values were 146 mg CO_2/g Corg (Viannos Fm), 144 mg CO_2/g Corg (Skinias Fm), 115.3 mg CO_2/g Corg (Moulia Fm at Faneromeni), 105.9 mg CO_2/g Corg (Ploutis section) and 410 mg CO_2/g Corg (Agios Myron Fm). The level of thermal maturity of the examined samples can be assessed using Tmax values. The mean Tmax values were 420 °C (Viannos Fm), 413 °C (Skinias Formation), 422.7 °C (Moulia Fm at Faneromeni), 415.6 °C (Ploutis section), and 432 °C (Agios Myron Fm). Most of the samples exhibited values of S2 > 0.2 mg HC/g rock, and, therefore, the obtained Tmax values were thought trustworthy [70]. Seven samples from the Agios Myron Fm exhibited S2 values that were below 0.2 mg HC/g rock and were not included in the interpretations relating to thermal maturity [73].

6. Discussion

6.1. Amount, Type and Maturation Level of the Organic Material

The diagrams of S2 vs. TOC [77] (Figure 4A) and SP vs. TOC plot [74,76] (Figure 4B) indicate the potential of the sedimentary rocks to generate hydrocarbons. The examined samples exhibited poor to good hydrocarbon generative potential. The most promising samples came from Moulia Fm (at Faneromeni) and the Ploutis section, whereas the Viannos, Skinias and Agios Myron samples exhibited little hydrocarbon generative potential. Analogous information was obtained from the HI vs. TOC [78] (Figure 4C), where, additionally, the samples clustered in the fields of both oil and gas generative potential. However, the determination of oil prone source rocks requires the integration of additional information (i.e., gas chromatography–mass spectrometry data; GC-MS). Except for Moulia Fm (at Faneromeni), such data were not available for the other formations and sections. The GC-MS data in Moulia Fm illustrate the prevalence of type III kerogen and a gas-prone source rock [59]. It needs to be borne in mind that processes such as surface weathering and oxidization need to be taken into consideration when evaluating organic geochemical data because it is possible to reduce the S1, S2 and TOC values obtained by the RE analysis [66,70,79]. Although the sampling was performed with the purpose of reducing the impact of weathering and oxidization in the samples (see methodology section), it is highly possible that the studied samples may exhibit more promising organic geochemical signatures.

The integration of S2/S3 ratios and HI values suggest that the organic matter contained is of kerogen types II, III and IV. A similar interpretation of the kerogen type is provided by the HI vs OI, S2 vs. TOC and the S2/S3 vs. TOC plots that illustrate a well-oxygenated environment of deposition and a terrestrial origin for the contained organic matter (Figure 5A–C). The samples that contain type IV kerogen organic material suggest severe alteration and/or oxidation of organic material in the depositional site [74]. The diagrams suggest that Moulia Fm (at Faneromeni) and Ploutis section could represent source rocks with the potential to generate gas. In contrast, Viannos Fm, Skinias Fm and Agios Myron Fm principally contain kerogen IV organic matter, with no or little hydrocarbon generative potential. The distinction between migrated and autochthonous hydrocarbons, as well as evaluation of sample contamination and kerogen mixing are important when defining the organic geochemical features of sedimentary rocks [80]. The application of S1 vs. TOC cross-plot [81] can shed light on the determination of hydrocarbons and suggests that autochthonous hydrocarbons predominate in all the studied samples and have the potential to generate gas (Figure 5D). The kerogen mixture can obscure the nature of the generated hydrocarbons [77,82]. However, the studied samples contain kerogen of type III and IV only and provide a correct representation of the source rocks.

A low maturation status resulting from low temperatures during burial is suggested for the studied succession by the HI vs. Tmax diagram [73,74] (Figure 6). The PI index is also considered as a parameter to evaluate thermal maturity, with mature rocks displaying values above 0.1 and post-mature rocks displaying values over 0.4 [83]. PI values below 0.1 point to low levels of thermal maturity. The studied succession has PI values that fall below the bottom threshold of 0.1, supporting the interpretation of a thermally immature succession.

In sum, the studied sections exhibit a similar degree of thermal maturity, but a different quantity and quality of organic material. This could be partly related to the nature of the depositional environments and sub-environments that encompass the different sections, and partly to the shoreline trajectories that are associated with their accumulation. The stratigraphic framework proposed by Zachariasse et al. [48] implies that, in contrast to Vianos and Skinias Fms, Moulia Fm (at Faneromeni) and the Ploutis section exhibit a deepening upward trend and associated transgression during their deposition. Relative sea level rise increases the available accommodation space, favoring the accumulation of organic material [2]. Furthermore, shoreline transgression decreases the sediment supply in the deep parts of the sedimentary basins because the terrigenous clastic material is accumulated in the nearshore environments. This depositional trend develops condensed sections that are often rich in organic matter [84]. Additionally, enriched organic productivity, eutrophication, and water stratification often lead to anoxia [85]. The difference in the quantity and quality of organic material between the different Fms and sections could also be associated with the nature of the depositional environments, with the continental-in-origin Vianos Fm being less favorable for accumulation and preservation of organic material. Furthermore, even though shallow marine settings contain sufficient sources that can deliver organic material, their high oxygenation levels can prevent the formation of suitable source rocks and can explain the unpromising geochemical signatures of the marine Skinias and Agios Myron Fms.

6.2. Exploration Opportunities

With worldwide energy demand rising and stricter environmental regulations being introduced, natural gas has become an increasingly important substitute for fuel oil. In this context, prospective regions that deserve further investigation have been the focus of the petroleum industry, despite existing technical challenges (stemming from the complex geology and deep water depths). The eastern Mediterranean is a region with proven hydrocarbon reserves and is the subject of extensive exploration and exploitation activity [4,10–12]. Despite the promising exploration results in western Greece [14,21–26,86] and the Thrace Basin [87], several regions in the Hellenic Domain are still underexplored. In Crete, this study enriches existing knowledge about the potential regional source rocks by including new organic geochemical results on the pre-evaporitic (Tortonian–Messinian) sedimentary succession. This study defines the organic geochemical signatures of the Agios Myron Fm (Messinian, weak source-rock potential and low maturation levels) and builds on the hydrocarbon generative potential of the pre-evaporitic deposits that contain units with fair to good source-rock attributes (Tortonian Moulia Fm at Faneromeni [59], and Mesinian Ploutis [30] section). A common feature of all existing onshore rock units, with merit for further exploration, is their low maturity levels that indicate low thermal evolution. Nevertheless, the offshore counterparts of these units may have experienced higher burial temperatures because of a thicker sedimentary overburden. Indeed, the sedimentary succession that crops out in Gavdos Island reflects the deepest parts of the Crete Basin and contains units with more promising organic geochemical signatures. The organic material in these units (Tortonian—Messinian, Metochia Fm) is of kerogen types II, III and IV, with fair to very good potential for hydrocarbon generation [27]. Despite the promising results, the organic material in Metochia Fm is thermally immature. The regional structural framework suggests tectonic subsidence and the accumulation of thick Pliocene—Pleistocene deposits in the offshore central parts of the region (western and eastern south Cretan trenches) that cover the Tortonian—Messinian potential source rocks [14,88]. This thicker sedimentary cover could likely facilitate the maturation of organic material and the development of source rocks that have reached the oil and/or gas window [29,31,59]. Pre-Messinian mature source rocks have been reported in the eastern Mediterranean. In the adjacent Levantine Basin, the occurrence of asphalt in the Upper Cretaceous sedimentary succession onshore Lebanon suggests the existence of a thermogenic petroleum system (Nader [89]). As in the region covered by the present study, the source rocks in the Levantine Basin are likely pre-Messinian in age (Jurassic—Cretaceous and Cretaceous—Miocene) and occur at the margins and deeper parts of the basin, where they have probably reached the oil/gas window [3,90].

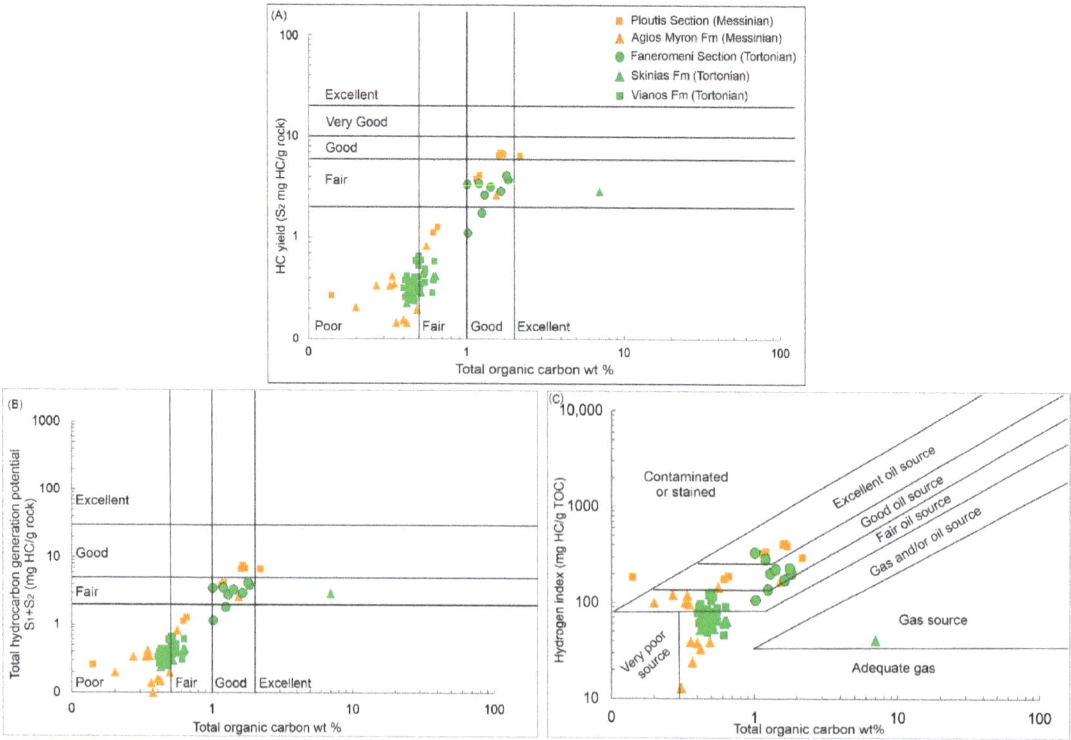

Figure 4. Evaluation of hydrocarbon generative potential of the examined samples from Crete Island. (**A**) Cross-plot of S2 vs. total organic carbon, (**B**) Classification diagram of total hydrocarbon generating potential vs. total organic carbon and, (**C**) Cross-plot of hydrogen index vs. total organic carbon. These diagrams suggest that the most promising samples come from Moulia Fm (at Faneromeni) and Ploutis section, whereas Viannos, Skinias Fm and Agios Myron Fm exhibit little hydrocarbon generative potential. Green color refers to samples that are Tortonian in age, whereas orange color refers to samples that are Messinian in age.

6.3. *Eastern Mediterranean Petroleum Plays Related to the Messinian Evaporites*

The seismic profiles illustrate the occurrence of numerous normal faults that crosscut the Tortonian—Pleistocene succession and develop smaller-scale graben, horst, and domino structures [88]. These tectonic configurations are highly prone to form migration paths terminating in structural traps constrained by evaporitic cap rocks and sealing faults. Similar trap styles are recognized in the Levantine and Herodotus Basins, where a combination of normal and reverse faults developed several large four-way closure structures, as well as Oligocene—Miocene tilted fault blocks [12]. These active basin tectonics promoted the development of regional-scale unconformities [88] that can form stratigraphic traps [14,91]. In this situation, hydrocarbon plays could exist both beneath and above the unconformity. The regional deepening that followed the accumulation of the Messinian evaporites likely provided the conditions for the accumulation of coastal, sand-dominated deposits. These deposits are often covered by finer-grained transgressive facies and can trap oil and/or gas above the unconformities. Apart from these mud-rich deposits that can prevent the upward movement of fluids (oil and/or gas), Messinian salt deposits constitute the principal regional seal rocks. Despite the mode of formation (e.g., shallow-water deep-basin vs. deep-water, non-desiccated scenario; [92,93]), these deposits are thick, potentially offering sufficient seal capacity. Further, they are laterally continuous, influencing the size of the potential plays. It should be noted that the thick evaporitic layer can hold fluids within the underlying sedimentary successions and cause increased overpressures, stimulated by sediment unloading ascribed to the relative sea level fall that occurred in the Mediterranean, and can increase overpressures and deteriorate the quality of the seal rock [94].

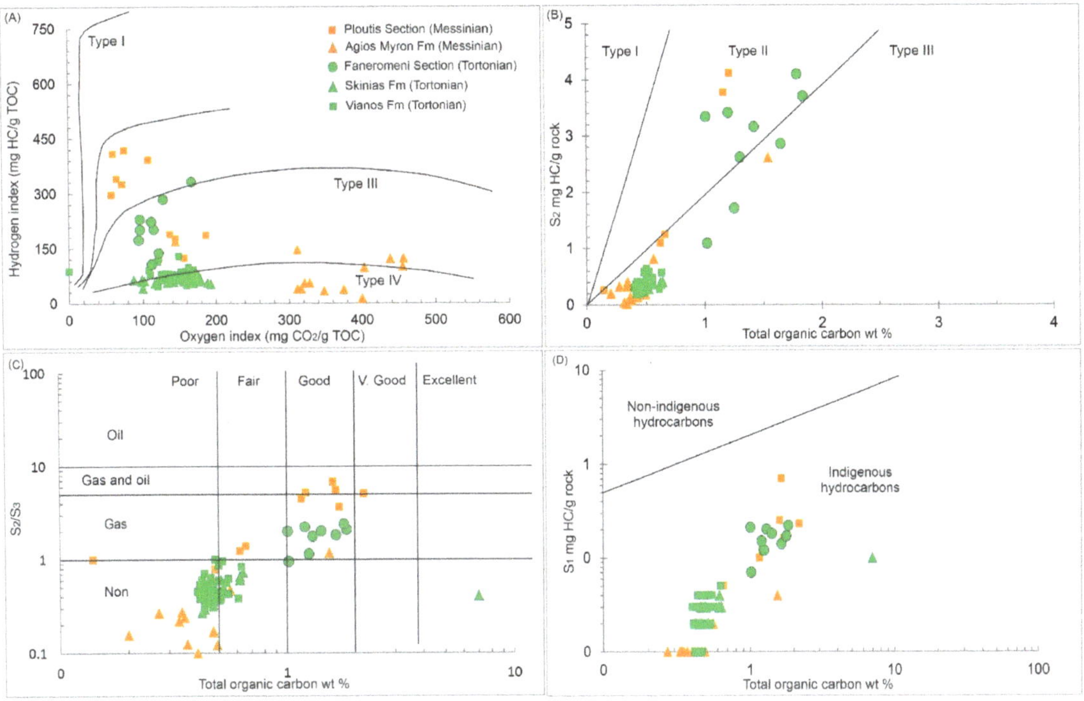

Figure 5. Evaluation of the kerogen type in the examined sedimentary rocks: (**A**) Classification diagram of hydrogen index vs. oxygen index kerogen, (**B**) Bi-plot of S2 vs. total organic carbon, (**C**) S2/S3 vs. total organic carbon, and (**D**) S1 vs. total organic carbon cross-plots. These plots suggest that Moulia Fm (at Faneromeni) and Ploutis section could represent source rocks with potential to generate gas. In contrast, Viannos Fm, Skinias Fm and Agios Myron Fm contain principally organic matter of kerogen IV, with no or little hydrocarbon generative potential. Green color refers to samples that are Tortonian in age, whereas orange color refers to samples that are Messinian in age.

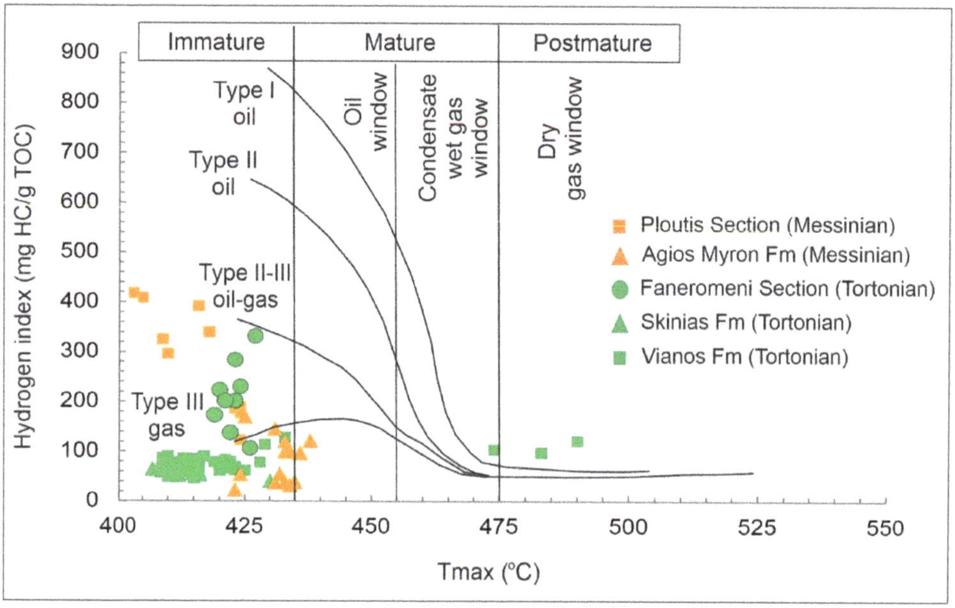

Figure 6. Assessment of maturation status of the examined samples, based on hydrogen index vs. Tmax cross-plot. The samples derived from all formations are thermally immature. Green color refers to samples that are Tortonian in age, whereas orange color refers to samples that are Messinian in age.

7. Conclusions

The Tortonian—Messinian sedimentary succession onshore Crete Island (eastern Mediterranean, Greece) was examined to assess the quantity, quality, and maturation level of organic matter as a source rock. Several samples yielded TOC values above 0.5 wt%, suggesting some hydrocarbon generating potential. The studied succession included samples that are likely to be interesting (TOC between 0.5 and 1.0%), and samples that are very promising (TOC above 1.0%). The organic geochemical analysis illustrates the presence of source rocks that exhibit poor to good potential to generate hydrocarbons. Rock units with fair to good geochemical values include the Moulia Fm (at Faneromeni, Tortonian) and Ploutis section (Messinian), whereas Viannos Fm, Skinias Fm, and Agios Myron Fm exhibit little hydrocarbon generative potential. Most of the samples contain organic matter that is of type III and IV kerogen, suggesting gas-prone source rocks. Maturity parameters (Tmax and PI) suggest that the analyzed samples have not reached the oil window. Despite the low levels of maturation, the offshore equivalents of the studied succession (in the western and eastern south Cretan trenches) could possibly have reached the oil and/or gas window because of tectonic subsidence and thicker sedimentary overburden. The hydrocarbons, possibly expelled by these mature source rock units, could be trapped in the hanging walls of normal faults, in the margins or the corners of tilted fault blocks, and below (and/or above) unconformities. The study region is a promising frontier basin with potential future exploration opportunities. As technological advances in the petroleum industry allow for the more precise evaluation of vintage geological, geochemical, and geophysical data, as well as the acquisition of high-quality new data, an updated re-evaluation of the regional geotectonic models is suggested. A better understanding of the regional geology can be achieved by refining the data from vintage seismic profiles and wells, along with the surveying of additional seismic campaigns and the drilling of new exploration wells.

Supplementary Materials: The following supporting information can be downloaded at: https://www.mdpi.com/article/10.3390/jmse10091323/s1, Table S1: Experimental data of Rock Eval-TOC analysis.

Author Contributions: Conceptualization, A.G.M. and G.K.; methodology, A.G.M., N.P. and G.K.; software, A.G.M. and G.K.; validation, A.G.M., N.P. and G.K.; formal analysis, N.P. and S.B.; investigation, A.G.M., G.K., S.B., A.A., C.B., P.M. (Panayota Makri) and H.T.J.; resources, A.G.M., G.K., A.A., E.M. and A.Z.; data curation, A.G.M., G.K., S.B., E.M. and N.P.; writing—original draft preparation, A.G.M. and G.K.; writing—review and editing, A.G.M., G.K., S.B., P.M. (Pierre Moissette), J.-J.C., H.T.J., V.K.; visualization, A.G.M. and G.K.; supervision, A.A., A.Z. and V.K.; project administration, A.A., A.Z. and V.K.; funding acquisition, A.G.M. and G.K. All authors have read and agreed to the published version of the manuscript.

Funding: This research received no external funding.

Institutional Review Board Statement: Not applicable.

Informed Consent Statement: Not applicable.

Data Availability Statement: The data used in this work is available on request.

Conflicts of Interest: The authors declare no conflict of interest.

References

1. Bruneton, A.; Konofagos, E.; Foscolos, A. Economic and geopolitical importance of Eastern Mediterranean gas fields for Greece and the E. U. Emphasis on the probable natural gas deposits occurring in the Libyan Sea within the Exclusive Economic Zone of Greece. *Miner. Wealth* **2011**, *160*, 7–30.
2. Bertoni, C.; Cartwright, J.A. Major erosion at the end of the Messinian Salinity Crisis: Evidence from the Levant Basin, Eastern Mediterranean. *Basin Res.* **2007**, *19*, 1–18. [CrossRef]
3. Bou Daher, S.; Ducros, M.; Michel, P.; Hawie, N.; Nader, F.H.; Littke, R. 3D thermal history and maturity modelling of the Levant Basin and its eastern margin, offshore–onshore Lebanon. *Arab. J. Geosci.* **2016**, *9*, 440. [CrossRef]
4. Grohmann, S.; Romero-Sarmiento, M.-F.; Nader, F.H.; Baudin, F.; Littke, R. Geochemical and petrographic investigation of Triassic and Late Miocene organic-rich intervals from onshore Cyprus, Eastern Mediterranean. *Int. J. Coal Geol.* **2019**, *209*, 94–116. [CrossRef]
5. Esestime, P.; Hewitt, A.; Hodgson, N. Zohr—A newborn carbonate play in the Levantine Basin, East-Mediterranean. *First Break* **2016**, *34*, 87–93. [CrossRef]
6. Warren, J.K. *Evaporites: Sediments, Resources and Hydrocarbons*; Springer: Berlin/Heidelberg, Germany, 2006. [CrossRef]
7. Morley, C.K.; King, R.; Hillis, R.; Tingay, M.; Backe, G. Deepwater fold and thrust belt classification, tectonics, structure and hydrocarbon prospectivity: A review. *Earth-Sci. Rev.* **2011**, *104*, 41–91. [CrossRef]
8. Basso, M.; Belila, A.M.P.; Chinelatto, G.F.; Souza, J.P.d.P.; Vidal, A.C. Sedimentology and petrophysical analysis of pre-salt lacustrine carbonate reservoir from the Santos Basin, southeast Brazil. *Int. J. Earth Sci.* **2021**, *110*, 2573–2595. [CrossRef]
9. De Freitas, V.A.; Vital, J.C.d.S.; Rodrigues, B.R.; Rodrigues, R. Source rock potential, main depocenters, and CO_2 occurrence in the pre-salt section of Santos Basin, southeast Brazil. *J. S. Am. Earth Sci.* **2022**, *115*, 103760. [CrossRef]
10. Gardosh, M.A.; Druckman, Y. Seismic stratigraphy, structure and tectonic evolution of the Levantine Basin, offshore Israel. *Geol. Soc. Lond. Spec. Publ.* **2006**, *260*, 201–227. [CrossRef]
11. Roberts, G.; Peace, D. Hydrocarbon plays and prospectivity of the Levantine Basin, offshore Lebanon and Syria from modern seismic data. *GeoArabia* **2007**, *12*, 99–124. [CrossRef]
12. Semb, P.H. Possible seismic hydrocarbon indicators in offshore Cyprus and Lebanon. *GeoArabia* **2009**, *14*, 49–66. [CrossRef]
13. Elia, C.; Konstantopoulos, P.; Maravelis, A.G.; Zelilidis, A. The tectono-stratigraphic evolution of SE Mediterranean with emphasis on Herodotus Basin prospectivity for the development of hydrocarbon fields. *Bull. Geol. Soc. Greece* **2013**, *47*, 1970–1979. [CrossRef]
14. Maravelis, A.G.; Koukounya, A.; Tserolas, P.; Pasadakis, N.; Zelilidis, A. Geochemistry of Upper Miocene–Lower Pliocene source rocks in the Hellenic Fold and Thrust Belt, Zakynthos Island, Ionian Sea, western Greece. *Mar. Pet. Geol.* **2015**, *66*, 217–230. [CrossRef]
15. Ford, J. Eastern Mediterranean Hydrocarbon Hotspot. *NVentures*. 2017. Available online: https://www.geoexpro.com/articles/2018/07/hydrocarbon-developments-in-the-eastern-mediterranean (accessed on 24 July 2022).
16. Montadert, L.; Nikolaides, S. The geological structure of the Eratosthenes continental block and its margins with the Levantine and Herodotus Basins (Eastern Mediterranean) from new seismic reflection data. In Proceedings of the AAPG European Region Conference, Athens, Greece, 18–21 November 2007.
17. Krois, P.; Hannke, K.; Novotny, B.; Bayoumi, T.; Hussein, H.; Tari, G. The emerging deepwater province of Northwest Egypt. In Proceedings of the AAPG International Conference and Exhibition, Rio de Janeiro, Brazil, 15–18 November 2009.

18. Panagiotopoulos, I.P.; Paraschos, F.; Rousakis, G.; Hatzianestis, I.; Parinos, C.; Morfis, I.; Gogou, A. Assessment of the eruptive activity and identification of the mud breccia's source in the Olimpi mud volcano field, Eastern Mediterranean. *Deep Sea Res. Part II Top. Stud. Oceanogr.* **2020**, *171*, 104701. [CrossRef]
19. Maravelis, A.; Manutsoglu, E.; Konstantopoulos, P.; Pantopoulos, G.; Makrodimitras, G.; Zoumpouli, E.; Zelilidis, A. Hydrocarbon Plays and Prospectivity of the Mediterranean Ridge. *Energy Sources Part A Recovery Util. Environ. Eff.* **2015**, *37*, 347–355. [CrossRef]
20. Bertoni, C.; Kirkham, C.; Cartwright, J.; Hodgson, N.; Rodriguez, K. Seismic indicators of focused fluid flow and cross-evaporitic seepage in the Eastern Mediterranean. *Mar. Pet. Geol.* **2017**, *88*, 472–488. [CrossRef]
21. Zelilidis, A.; Piper, D.J.W.; Vakalas, I.; Avramidis, P.; Getsos, K. Oil and gas plays in Albania: Do equivalent plays exist in Greece? *J. Pet. Geol.* **2003**, *26*, 29–48. [CrossRef]
22. Karakitsios, V. Western Greece and Ionian petroleum systems. *AAPG Bull.* **2013**, *97*, 1567–1595. [CrossRef]
23. Maravelis, A.G.; Makrodimitras, G.; Zelilidis, A. Hydrocarbon prospectivity in western Greece. *Oil Gas Eur. Mag.* **2012**, *38*, 84–89.
24. Makri, V.I.; Panagopoulos, G.; Nikolaou, K.; Bellas, S.; Pasadakis, N. Evaluation of Gas Generation Potential Using Thermal Maturity Modelling—The Katakolo Case: A Probable Pathway to Energy Transition. *Mater. Proc.* **2021**, *5*, 70. [CrossRef]
25. Kontakiotis, G.; Karakitsios, V.; Maravelis, A.G.; Zarkogiannis, S.D.; Agiadi, K.; Antonarakou, A.; Pasadakis, N.; Zelilidis, A. Integrated isotopic and organic geochemical constraints on the depositional controls and source rock quality of the Neogene Kalamaki sedimentary successions (Zakynthos Island, Ionian Sea). *Mediterr. Geosci. Rev.* **2021**, *3*, 193–217. [CrossRef]
26. Mavromatidis, A. Review of Hydrocarbon Prospectivity in the Ionian Basin, Western Greece. *Energy Sources Part A Recovery Util. Environ. Eff.* **2009**, *31*, 619–632. [CrossRef]
27. Pyliotis, I.; Zelilidis, A.; Pasadakis, N.; Panagopoulos, G.; Manoutsoglou, E. Source rock potential of the late Miocene Metochia formation of Gavdos island, Greece. *Bull. Geol. Soc. Greece* **2013**, *43*, 871–879. [CrossRef]
28. Pasadakis, N.; Dagounaki, V.; Chamilaki, E.; Vafidis, A.; Zelilidis, A.; Piliotis, I.; Panagopoulos, G.; Manoutsoglou, E. Organic geochemical evaluation of Neogene formations in Messara (Heraklion, Crete) basin as source rocks of biogenetic methane. *Miner. Wealth* **2012**, *166*, 8–26.
29. Maravelis, A.G.; Panagopoulos, G.; Piliotis, J.; Pasadakis, N.; Manoutsoglou, E.; Zelilidis, A. Pre-Messinian (sub-salt) source-rock potential on back-stop basins of the Hellenic Trench System (Messara Basin, Central Crete, Greece). *Oil Gas Sci. Technol. Rev. IFP Energ. Nouv.* **2016**, *71*, 1–19. [CrossRef]
30. Kontakiotis, G.; Karakitsios, V.; Cornée, J.-J.; Moissette, P.; Zarkogiannis, S.D.; Pasadakis, N.; Koskeridou, E.; Manoutsoglou, E.; Drinia, H.; Antonarakou, A. Preliminary results based on geochemical sedimentary constraints on the hydrocarbon potential and depositional environment of a Messinian sub-salt mixed siliciclastic-carbonate succession onshore Crete (Plouti section, eastern Mediterranean). *Mediterr. Geosci. Rev.* **2020**, *2*, 247–265. [CrossRef]
31. Panagopoulos, G.; Vafidis, A.; Soupios, P.; Manoutsoglou, E. A study on the Gas-bearing Miocene Sediments of MESSARA Basin in Crete (Greece) by Using Seismic Reflection, Geochemical and Petrophysical Data. *Arab. J. Sci. Eng.* **2022**, *47*, 7449–7465. [CrossRef]
32. Safaei-Farouji, M.; Kamali, M.; Hakimi, M.H. Hydrocarbon source rocks in Kazhdumi and Pabdeh formations—a quick outlook in Gachsaran oilfield, SW Iran. *J. Pet. Explor. Prod. Technol.* **2021**, *12*, 1489–1507. [CrossRef]
33. Ahmed, A.; Jahandad, S.; Hakimi, M.H.; Gharib, A.F.; Mehmood, S.; Kahal, A.Y.; Khan, M.A.; Munir, M.N.; Lashin, A. Organic matter characteristics and conventional oil potentials of shales from the Early Jurassic Datta Formation in the Upper Indus Basin, Northern Pakistan. *J. Asian Earth Sci.* **2022**, *224*, 104975. [CrossRef]
34. Chan, S.A.; Hassan, A.M.; Usman, M.; Humphrey, J.D.; Alzayer, Y.; Duque, F. Total organic carbon (TOC) quantification using artificial neural networks: Improved prediction by leveraging XRF data. *J. Pet. Sci. Eng.* **2022**, *208*, 109302. [CrossRef]
35. Jacobshagen, V. *Geologie von Griechenland*; Borntraeger: Berlin-Stuttgart, Germany, 1986; pp. 257–269.
36. Faccenna, C.; Jolivet, L.; Piromallo, C.; Morelli, A. Subduction and the depth of convection in the Mediterranean mantle. *J. Geophys. Res. Solid Earth* **2003**, *108*, 2099. [CrossRef]
37. Van Hinsbergen, D.J.J.; Kouwenhoven, T.J.; van der Zwaan, G.J. Paleobathymetry in the backstripping procedure: Correction for oxygenation effects on depth estimates. *Palaeogeogr. Palaeoclimatol. Palaeoecol.* **2005**, *221*, 245–265. [CrossRef]
38. Rögl, F.; Steininger, F.-F. *Neogene Paratethys, Mediterranean and Indo-Pacific Seaways*; J. Wiley & Sons: London, UK, 1984; pp. 171–200.
39. Kokkalas, S.; Xypolias, P.; Koukouvelas, I.; Doutsos, T. Postcollisional contractional and extensional deformation in the Aegean region. In *Postcollisional Tectonics and Magmatism in the Mediterranean Region and Asia*; Dilek, Y., Pavlides, S., Eds.; Geological Society of America: Boulder, CO, USA, 2006; Volume 409, pp. 97–123.
40. Kokkalas, S.; Doutsos, T. Kinematics and strain partitioning in the southeast Hellenides (Greece). *Geol. J.* **2004**, *39*, 121–140. [CrossRef]
41. Skourlis, K.; Doutsos, T. The Pindos Fold-and-thrust belt (Greece): Inversion kinematics of a passive continental margin. *Int. J. Earth Sci.* **2003**, *92*, 891–903. [CrossRef]
42. Robertson, A.H.F.; Clift, P.D.; Degnan, P.J.; Jones, G. Palaeogeographic and palaeotectonic evolution of the Eastern Mediterranean Neotethys. *Palaeogeogr. Palaeoclimatol. Palaeoecol.* **1991**, *87*, 289–343. [CrossRef]
43. Doutsos, T.; Koukouvelas, I.K.; Xypolias, P. A new orogenic model for the External Hellenides. *Geol. Soc. Lond. Spec. Publ.* **2006**, *260*, 507. [CrossRef]
44. Ten Veen, J.H.; Postma, G. Neogene tectonics and basin fill patterns in the Hellenic outer-arc (Crete, Greece). *Basin Res.* **1999**, *11*, 223–241. [CrossRef]

45. Caputo, R.; Catalano, S.; Monaco, C.; Romagnoli, G.; Tortorici, G.; Tortorici, L. Active faulting on the island of Crete (Greece). *Geophys. J. Int.* **2010**, *183*, 111–126. [CrossRef]
46. Van Hinsbergen, D.J.J.; Meulenkamp, J.E. Neogene supradetachment basin development on Crete (Greece) during exhumation of the South Aegean core complex. *Basin Res.* **2006**, *18*, 103–124. [CrossRef]
47. Papanikolaou, D.; Vassilakis, E. Thrust faults and extensional detachment faults in Cretan tectono-stratigraphy: Implications for Middle Miocene extension. *Tectonophysics* **2010**, *488*, 233–247. [CrossRef]
48. Zachariasse, W.J.; van Hinsbergen, D.J.J.; Fortuin, A.R. Formation and fragmentation of a late Miocene supradetachment basin in central Crete: Implications for exhumation mechanisms of high-pressure rocks in the Aegean forearc. *Basin Res.* **2011**, *23*, 678–701. [CrossRef]
49. Xypolias, P.; Dörr, W.; Zulauf, G. Late Carboniferous plutonism within the pre-Alpine basement of the External Hellenides (Kithira, Greece): Evidence from U–Pb zircon dating. *J. Geol. Soc.* **2006**, *163*, 539. [CrossRef]
50. Zulauf, G.; Dörr, W.; Fisher-Spurlock, S.C.; Gerdes, A.; Chatzaras, V.; Xypolias, P. Closure of the Paleotethys in the External Hellenides: Constraints from U–Pb ages of magmatic and detrital zircons (Crete). *Gondwana Res.* **2015**, *28*, 642–667. [CrossRef]
51. Meulenkamp, J.E.; Wortel, M.J.R.; van Wamel, W.A.; Spakman, W.; Hoogerduyn Strating, E. On the Hellenic subduction zone and the geodynamic evolution of Crete since the late Middle Miocene. *Tectonophysics* **1988**, *146*, 203–215. [CrossRef]
52. Fassoulas, C. The tectonic development of a Neogene basin at the leading edge of the active European margin: The Heraklion basin, Crete, Greece. *J. Geodyn.* **2001**, *31*, 49–70. [CrossRef]
53. Meulenkamp, J.E.; Sissingh, W. Tertiary palaeogeography and tectonostratigraphic evolution of the Northern and Southern Peri-Tethys platforms and the intermediate domains of the African–Eurasian convergent plate boundary zone. *Palaeogeogr. Palaeoclimatol. Palaeoecol.* **2003**, *196*, 209–228. [CrossRef]
54. Chatzaras, V.; Xypolias, P.; Doutsos, T. Exhumation of high-pressure rocks under continuous compression: A working hypothesis for the southern Hellenides (central Crete, Greece). *Geol. Mag.* **2006**, *143*, 859–876. [CrossRef]
55. Chatzaras, V.; Dörr, W.; Finger, F.; Xypolias, P.; Zulauf, G. U–Pb single zircon ages and geochemistry of metagranitoid rocks in the Cycladic Blueschists (Evia Island): Implications for the Triassic tectonic setting of Greece. *Tectonophysics* **2013**, *595–596*, 125–139. [CrossRef]
56. Tortorici, L.; Caputo, R.; Monaco, C. Late Neogene to Quaternary contractional structures in Crete (Greece). *Tectonophysics* **2010**, *483*, 203–213. [CrossRef]
57. Vafidis, A.; Andronikidis, N.; Economou, N.; Panagopoulos, G.; Zelilidis, A.; Manoutsoglou, E. Reprocessing and interpretation of seismic reflection data at Messara Basin, Crete, Greece. *J. Balk. Geophys. Soc.* **2012**, *15*, 31–40.
58. Meulenkamp, J.E.; Dermitzakis, M.; Georgiadou-Dikeoulia, E.; Jonkers, H.A.; Boger, H. *Field Guide to the Neogene of Crete*; University of Athens: Athens, Greece, 1979.
59. Zelilidis, A.; Tserolas, P.; Chamilaki, E.; Pasadakis, N.; Kostopoulou, S.; Maravelis, A.G. Hydrocarbon prospectivity in the Hellenic trench system: Organic geochemistry and source rock potential of upper Miocene–lower Pliocene successions in the eastern Crete Island, Greece. *Int. J. Earth Sci.* **2016**, *105*, 1859–1878. [CrossRef]
60. Moissette, P.; Cornée, J.J.; Antonarakou, A.; Kontakiotis, G.; Drinia, H.; Koskeridou, E.; Tsourou, T.; Agiadi, K.; Karakitsios, V. Palaeoenvironmental changes at the Tortonian/Messinian boundary: A deep-sea sedimentary record of the eastern Mediterranean Sea. *Palaeogeogr. Palaeoclimatol. Palaeoecol.* **2018**, *505*, 217–233. [CrossRef]
61. Kontakiotis, G.; Butiseacă, G.A.; Antonarakou, A.; Agiadi, K.; Zarkogiannis, S.D.; Krsnik, E.; Besiou, E.; Zachariasse, W.J.; Lourens, L.; Thivaiou, D.; et al. Hypersalinity accompanies tectonic restriction in the eastern Mediterranean prior to the Messinian Salinity Crisis. *Palaeogeogr. Palaeoclimatol. Palaeoecol.* **2022**, *592*, 110903. [CrossRef]
62. Zachariasse, W.J.; Kontakiotis, G.; Lourens, L.J.; Antonarakou, A. The Messinian of Agios Myron (Crete, Greece): A key to better understanding of diatomite formation on Gavdos (south of Crete). *Palaeogeogr. Palaeoclimatol. Palaeoecol.* **2021**, *581*, 110633. [CrossRef]
63. Zachariasse, W.J.; van Hinsbergen, D.J.J.; Fortuin, A.R. Mass wasting and uplift on Crete and Karpathos during the early Pliocene related to initiation of south Aegean left-lateral, strike-slip tectonics. *Geol. Soc. Am. Bull.* **2008**, *120*, 976–993. [CrossRef]
64. Cosentino, D.; Gliozzi, E.; Pipponzi, G. The late Messinian Lago-Mare episode in the Mediterranean Basin: Preliminary report on the occurrence of Paratethyan ostracod fauna from central Crete (Greece). *Geobios* **2007**, *40*, 339–349. [CrossRef]
65. Espitalie, J.; Deroo, G.; Marquis, F. La pyrolyse Rock-Eval et ses applications. Troisième partie. *Rev. Inst. Fr. Pét.* **1986**, *41*, 73–89. [CrossRef]
66. Lafargue, E.; Marquis, F.; Pillot, D. Rock-Eval 6 Applications in Hydrocarbon Exploration, Production, and Soil Contamination Studies. *Oil Gas Sci. Technol. Rev. IFP Energ. Nouv.* **1998**, *53*, 421–437. [CrossRef]
67. Behar, F.; Beaumont, V.; Penteado, H.L.D.B. Technologie Rock-Eval 6: Performances et développements. *Oil Gas Sci. Technol. Rev. IFP* **2001**, *56*, 111–134. [CrossRef]
68. Espitalié, J.; Laporte, J.L.; Madec, M.; Marquis, F.; Leplat, P.; Paulet, J.; Boutefeu, A. Méthode rapide de caractérisation des roches mères, de leur potentiel pétrolier et de leur degré d'évolution. *Rev. L'institut Français Pétrole* **1977**, *32*, 23–42. [CrossRef]
69. Espitalie, J.; Marquis, F.; Sage, L.; Barsony, I. Géochimie organique du bassin de Paris. *Rev. Inst. Fr. Pét.* **1987**, *42*, 271–302. [CrossRef]
70. Peters, K.E. Guidelines for evaluating petroleum source rock using programmed pyrolysis. *Am. Assoc. Pet. Geol. Bull.* **1986**, *70*, 318–329.

71. Jarvie, D.M. Total organic carbon (TOC) analysis. In *Treatise of Petroleum Geology: Handbook of Petroleum Geology, Source and Migration Processes and Evaluation Techniques*; Merril, R.K., Ed.; American Association of Petroleum Geologists: Tulsa, OK, USA, 1991; pp. 113–118.
72. Bostick, N.H.; Daws, T.A. Relationships between data from Rock-Eval pyrolysis and proximate, ultimate, petrographic, and physical analyses of 142 diverse U.S. coal samples. *Org. Geochem.* **1994**, *21*, 35–49. [CrossRef]
73. Peters, K.E.; Cassa, M.R. Applied source rock geochemistry. In *The Petroleum System—From Source to Trap*; Memoir 60; Magoon, L.B., Dow, W.G., Eds.; American Association of Petroleum Geologists: Tulsa, OK, USA, 1994; pp. 93–120.
74. Tissot, B.P.; Welte, D.H. *Petroleum Formation and Occurrence*, 2nd ed.; Springer: Berlin/Heidelberg, Germany, 1984.
75. Burwood, R.; De Witte, S.; Mycke, B.; Paulet, J. Petroleum geochemical characterization of the Lower Congo coastal basin Bucomazi Formation. In *Petroleum Source Rocks*; Katz, B.J., Ed.; Springer: Berlin/Heidelberg, Germany, 1995; pp. 235–263.
76. Dymann, T.S.; Palacas, J.G.; Tysdal, R.G.; Perry, W.J.; Pawlewicz, M.J. Source rock potential of Middle Cretaceous rocks in southwestern Montana. *AAPG Bull.* **1996**, *80*, 1177–1184.
77. Dembicki, H., Jr. Three common source rock evaluation errors made by geologists during prospect or play appraisals. *Am. Assoc. Pet. Geol. Bull.* **2009**, *93*, 341–356. [CrossRef]
78. Jackson, K.S.; Hawkins, P.J.; Bennett, A.J.R. Regional facies and geochemical evaluation of southern Denison Trough, Queensland. *APPEA J.* **1985**, *20*, 143–158. [CrossRef]
79. Maravelis, A.G.; Makrodimitras, G.; Pasadakis, N.; Zelilidis, A. Stratigraphic evolution and source rock potential of a Lower Oligocene to Lower-Middle Miocene continental slope system, Hellenic Fold and Thrust Belt, Ionian Sea, northwest Greece. *Geol. Mag.* **2014**, *151*, 394–413. [CrossRef]
80. Maravelis, A.G.; Boutelier, D.; Catuneanu, O.; Seymour, K.S.; Zelilidis, A. A review of tectonics and sedimentation in a forearc setting: Hellenic Thrace Basin, North Aegean Sea and Northern Greece. *Tectonophysics* **2017**, *674*, 1–19. [CrossRef]
81. Hunt, J.M. (Ed.) *Petroleum Geochemistry and Geology*, 2nd ed.; W.H. Freeman and Company: New York, NY, USA, 1996.
82. Katz, B.J. Limitations of 'Rock-Eval' pyrolysis for typing organic matter. *Org. Geochem.* **1983**, *4*, 195–199. [CrossRef]
83. Peters, K.E.; Moldowan, J.M. *The Biomarker Guide: Interpreting Molecular Fossils in Petroleum and Ancient Sediments*; Prentice-Hall, Inc.: Englewood Cliffs, NJ, USA, 1993.
84. Catuneanu, O. *Principles of Sequence Stratigraphy*; Elsevier: Amsterdam, The Netherlands, 2006; p. 375.
85. Vasiliev, I.; Karakitsios, V.; Bouloubassi, I.; Agiadi, K.; Kontakiotis, G.; Antonarakou, A.; Triantaphyllou, M.; Gogou, A.; Kafousia, N.; de Rafélis, M.; et al. Large Sea Surface Temperature, Salinity, and Productivity-Preservation Changes Preceding the Onset of the Messinian Salinity Crisis in the Eastern Mediterranean Sea. *Paleoceanogr. Paleoclimatol.* **2019**, *34*, 182–202. [CrossRef]
86. Kontakiotis, G.; Moforis, L.; Karakitsios, V.; Antonarakou, A. Sedimentary Facies Analysis, Reservoir Characteristics and Paleogeography Significance of the Early Jurassic to Eocene Carbonates in Epirus (Ionian Zone, Western Greece). *J. Mar. Sci. Eng.* **2020**, *8*, 706. [CrossRef]
87. Maravelis, A.; Zelilidis, A. Organic geochemical characteristics of the late Eocene–early Oligocene submarine fans and shelf deposits on Lemnos Island, NE Greece. *J. Pet. Sci. Eng.* **2010**, *71*, 160–168. [CrossRef]
88. Mascle, J.; Le Quellec, P.; Leité, O.; Jongsma, D. Structural sketch of the Hellenic continental margin between the western Peloponnesus and eastern Crete. *Geology* **1982**, *10*, 113–116. [CrossRef]
89. Nader, F.H. Insights into the petroleum prospectivity of Lebanon. In *Petroleum Systems of the Tethyan Region*; AAPG Special volumes memoir; Marlow, L., Kendall, C.C.G., Yose, L.A., Eds.; American Association of Petroleum Geologists: Tulsa, OK, USA, 2014; Volume 160, pp. 241–278.
90. Nader, F.H.; Inati, L.; Ghalayini, R.; Hawie, N.; Bou Daher, S. Key geological characteristics of the Saida-Tyr Platform along the eastern margin of the Levant Basin, offshore Lebanon: Implications for hydrocarbon exploration. *Oil Gas Sci. Technol. Rev. IFP Energ. Nouv.* **2018**, *73*, 50. [CrossRef]
91. Karakitsios, V.; Rigakis, N. Evolution and petroleum potential of Western Greece. *J. Pet. Geol.* **2007**, *30*, 197–218. [CrossRef]
92. Roveri, M.; Flecker, R.; Krijgsman, W.; Lofi, J.; Lugli, S.; Manzi, V.; Sierro, F.J.; Bertini, A.; Camerlenghi, A.; De Lange, G.; et al. The Messinian Salinity Crisis: Past and future of a great challenge for marine sciences. *Mar. Geol.* **2014**, *352*, 25–58. [CrossRef]
93. Roveri, M.; Gennari, R.; Lugli, S.; Manzi, V.; Minelli, N.; Reghizzi, M.; Riva, A.; Rossi, M.E.; Schreiber, B.C. The Messinian salinity crisis: Open problems and possible implications for Mediterranean petroleum systems. *Pet. Geosci.* **2016**, *22*, 283. [CrossRef]
94. Iadanza, A.; Sampalmieri, G.; Cipollari, P. Deep-seated hydrocarbons in the seep "Brecciated Limestones" of the Maiella area (Adriatic foreland basin): Evaporitic sealing and oil re-mobilization effects linked to the drawdown of the Messinian Salinity Crisis. *Mar. Pet. Geol.* **2015**, *66*, 177–191. [CrossRef]

Journal of
Marine Science and Engineering

MDPI

Article

New Model of Coastal Evolution in the Ria de Vigo (NW Spain) from MIS2 to Present Day Based on the Aeolian Sedimentary Record

Carlos Arce-Chamorro *, Juan Ramón Vidal-Romaní and Jorge Sanjurjo-Sánchez

Grupo Interdisciplinar de Patrimonio Cultural y Geológico 'CULXEO', Instituto Universitario de Xeoloxía "Isidro Parga Pondal", ESCI, Campus de Elviña, Universidade da Coruña, 15071 A Coruña, Spain
* Correspondence: carlos.arce@udc.es

Abstract: Galician Rias are fluvial valleys that were flooded during the last marine transgression in the Atlantic margin. The study of fossil dunes in the Cies Islands, a small archipelago in the mouth of the one of the rias (Ria de Vigo), allowed us to reconstruct the coastal evolution from the end of the Late Pleistocene to the present day. During this period, sea-level was 100 metres below the present one and the shoreline located about 5–10 kilometres away. About 15,000 years ago, sea-level rise began, radically modifying the coastline. This started with a gradual advance of large dune fields on both sides of the valley. The aeolian accretion continued until the Late Holocene, finishing when the sea reached its present level.

Keywords: aeolian accretion; coastal dunes; OSL-dating; MIS2; Holocene; sea-level; submerged forest; Ria de Vigo; NW Iberian Atlantic coast

Citation: Arce-Chamorro, C.; Vidal-Romaní, J.R.; Sanjurjo-Sánchez, J. New Model of Coastal Evolution in the Ria de Vigo (NW Spain) from MIS2 to Present Day Based on the Aeolian Sedimentary Record. *J. Mar. Sci. Eng.* **2022**, *10*, 1350. https://doi.org/10.3390/jmse10101350

Academic Editors: Dmitry A. Ruban, Assimina Antonarakou and George Kontakiotis

Received: 19 August 2022
Accepted: 17 September 2022
Published: 22 September 2022

Publisher's Note: MDPI stays neutral with regard to jurisdictional claims in published maps and institutional affiliations.

1. Introduction

The main landforms along the Atlantic coast of Galicia (inset in Figure 1) are represented by tectonic cliffs and rias affected by neotectonic uplift [1]. As to Galician rias (the western estuaries are indicated in the inset of Figure 1), these peculiar formations have a specifically continental origin [2] as fluvial valleys flooded during the Holocene transgression. With the exception of fluvial sediments, the sedimentary record during the Quaternary on this rocky coastal border is essentially represented by pebble, gravel and coarse sand deposits on rocky platforms at different elevations [3,4], which correspond to old marine terraces. These coarse-grained sediments are also represented on the present-day coastline by shingle beaches, although their formation and analysis are not discussed in this paper. The other type of sediments relates to aeolian sands that correspond to small but numerous outcrops distributed along the NW Iberian coast [5] (see some examples in Figure 2 and the aeolianite outcrop studied here (Figure 3)). Initially, these aeolian sands were considered erroneously as beach deposits [6] and were used to define Quaternary sea-levels along the Galician coast. However, an aeolian origin is now assumed [5] for this type of sediment.

Until now, the knowledge of sediments both inside the rias and on the continental shelf were based on high-resolution seismic-reflection profiles, complemented with deep-core or vibrocorer drillings [7–12], in which micropalaeontological studies were not performed. On the other hand, the chronology of the proposed sedimentary units was based on scarce radiocarbon dating of organic remains assuming only a few suspected Pleistocene fluvial sediments. Regarding the Upper Pleistocene fluvial deposits in the Ria de Vigo, these authors describe upper sedimentary units at the limit of the radiocarbon dating by accelerator mass spectrometry (40–45 ky), overlying suspected Tertiary deposits [13,14], with no other absolute ages beyond ^{14}C limit. In this sense, these infills in the Ria de Vigo would be positively related to the nearest fluvial deposits in the basin of the lower Miño

River (Pontevedra) (inset in Figure 1) dated by optically stimulated luminescence (OSL) and ascribed to the Upper Quaternary [1], as similar sediments of continental origin dated in the coast of Galicia and northern Portugal [15–19], although none of these data have been considered in previous assumptions.

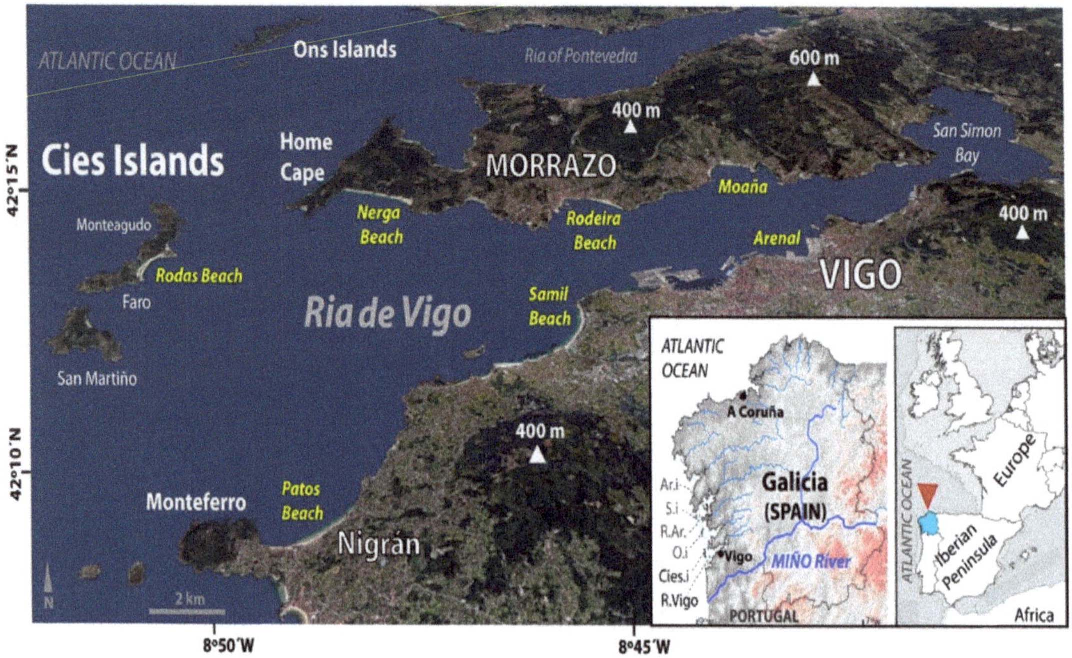

Figure 1. Present day configuration of the Cies Islands and Ria de Vigo showing the main sand beaches (Rodas, Patos, Nerga, Samil, Rodeira and Moaña) related to dune fields and climbing dunes, as examples of aeolian deposits. R.Vigo: Ria de Vigo; Cies.i: Cies Islands; O.i: Ons Islands; R.Ar.: Ria de Arousa; S.i: Sálvora Island; Ar.i: Areoso islet.

Regarding the sand deposits, the remaining sedimentary cover is mostly represented by sands of less than 15 ky, which have been interpreted only as deposits of marine origin, although neither saltwater diatoms nor foraminiferal associations are observed. Conversely, the data presented here suggest that they are of aeolian origin, similar to the current transgressive aeolian sand-sheets [20] described on other coasts of the world today, but mobilised by wind since the end of the Last Glacial period from a lower sea-level than in the present. Under this new hypothesis, the different levels reached by the sea in this Atlantic coast during this time period have not been satisfactorily understood, thus misunderstanding when the Galician rias were emerged or when the ocean waters flooded them again.

These aeolianite outcrops preserved in the coast of Galicia (Figure 2) are well-sorted azoic deposits that are mainly represented by a thin siliciclastic wedge less than 5 m thick [5], and they have been interpreted as climbing dunes correlated to a "regressive marine stage" [5]. Sedimentary structures, such as cross-stratification, have not been clearly preserved in these aeolian formations (Figure 3), something frequent, but not generalised, to all climbing dunes [5]. This is also the case of the aeolianite outcrop of Cies Island studied here (Figure 3c). However, these sedimentological issues are not discussed in the present paper.

Figure 2. Some examples of fossil dunes preserved along the Galician coast and interpreted as climbing dunes [6]. (**a**) Climbing dune from Trece (43°11′15″ N, 9°08′45″ W; WGS84); (**b**) 30 ky old aeolianite from Tal (42°47′16″ N; 9°00′55″ W) [16]; (**c**) Baldaio aeolianite (43°18′11″ N; 8°39′24″ W); (**d**) Penaboa aeolianite (43°22′50″ N; 8°26′18″ W), dating back ≈ 300 ky [4]; ae: aeolian sands.

The age of these sands (Figure 3), estimated by optically stimulated luminescence (OSL) [15,16] or infrared stimulated luminescence (IRSL) [21], can be established for all equivalent deposits located on the entire coast between 300 ky (aeolianite of Penaboa, A Coruña, Galicia, Spain) [4] and 12 ky [22], which are roughly related to regressive marine stages. Regarding the fossil dunes formed on the Galician coast during marine isotope stage 2 (MIS2), it is important to highlight that the sea-level rise of about 120 m during the last 15,000 years in the study area meant the disconnection due to marine flooding from their source area (of sand). This is confirmed by the presence of aeolian sands of 20 ky more than 100 m deep as described [12] in the continental platform near the Cies Islands (Figure 4; Table 1). Currently, under the wet and temperate climate of the North Atlantic peninsular coast [23], the most recent aeolian formations (<5 ky) are being affected by wind and sea waves even when stabilised by land vegetation [5,24,25], due to the breakage of the vegetal cover during the storms. In other cases, however, the preservation of the climbing dunes is due to the active slope dynamics developed in some steep cliff sections of the Galician coast, fossilizing the aeolianites by slope deposits [3,5] (Figure 3c,d).

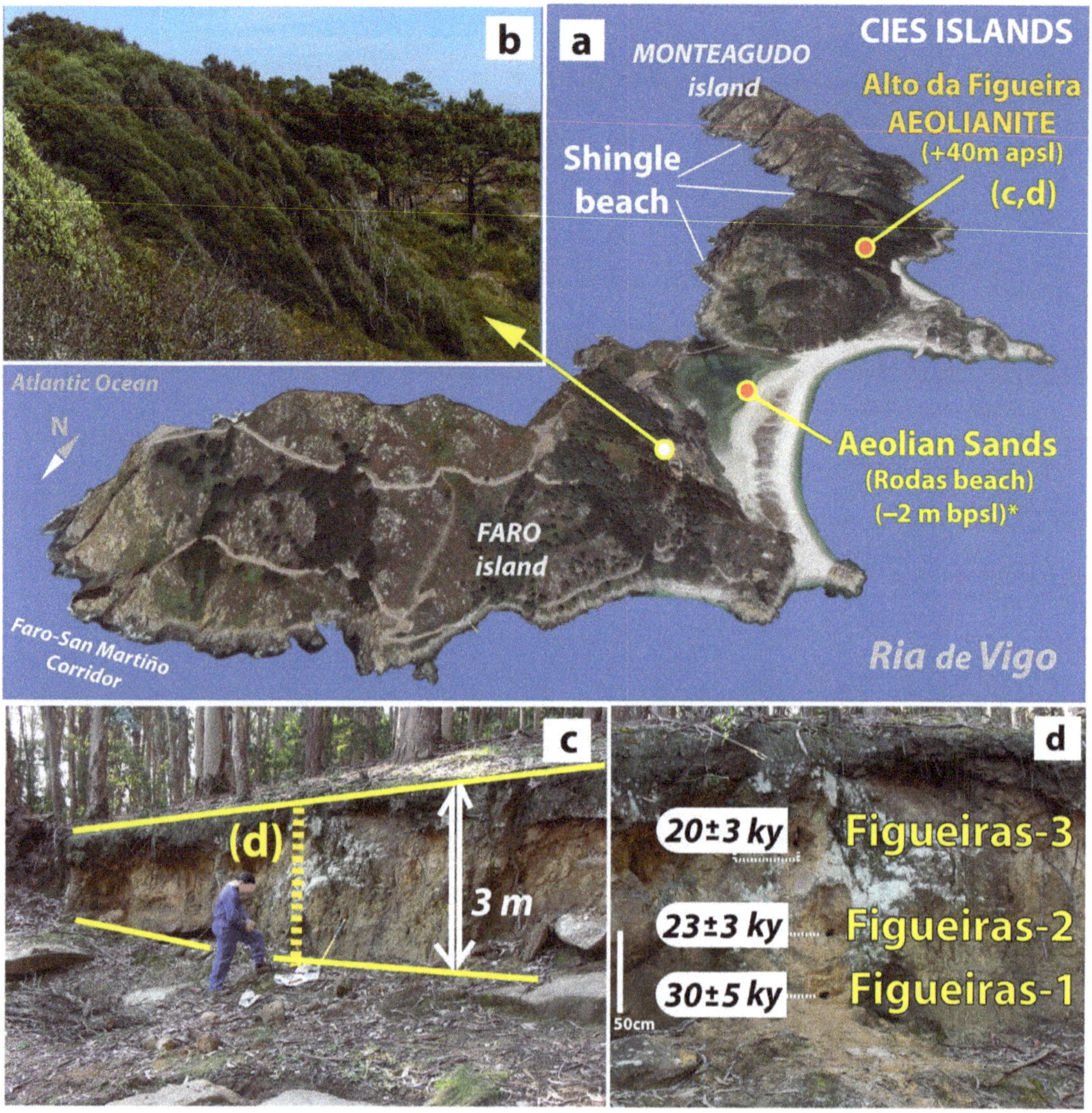

Figure 3. Monteagudo and Faro Islands (from the Cies Islands). (**a**) Aeolianite outcrop from Monteagudo Island (Alto da Figueira; 42°13′45″ N, 8°54′15″ W; WGS84) and aeolian sands (* [8]) location, as mainly aeolian sediments developed on the eastern slope at the end of the Upper Pleistocene. Shingle beaches are only observed in the western side. (**b**) Wind effect on vegetation on the eastern side of Faro Island, showing W–SW wind direction. (**c**,**d**) Fossilised climbing dune at +40 m above present sea-level (modified from [15]).

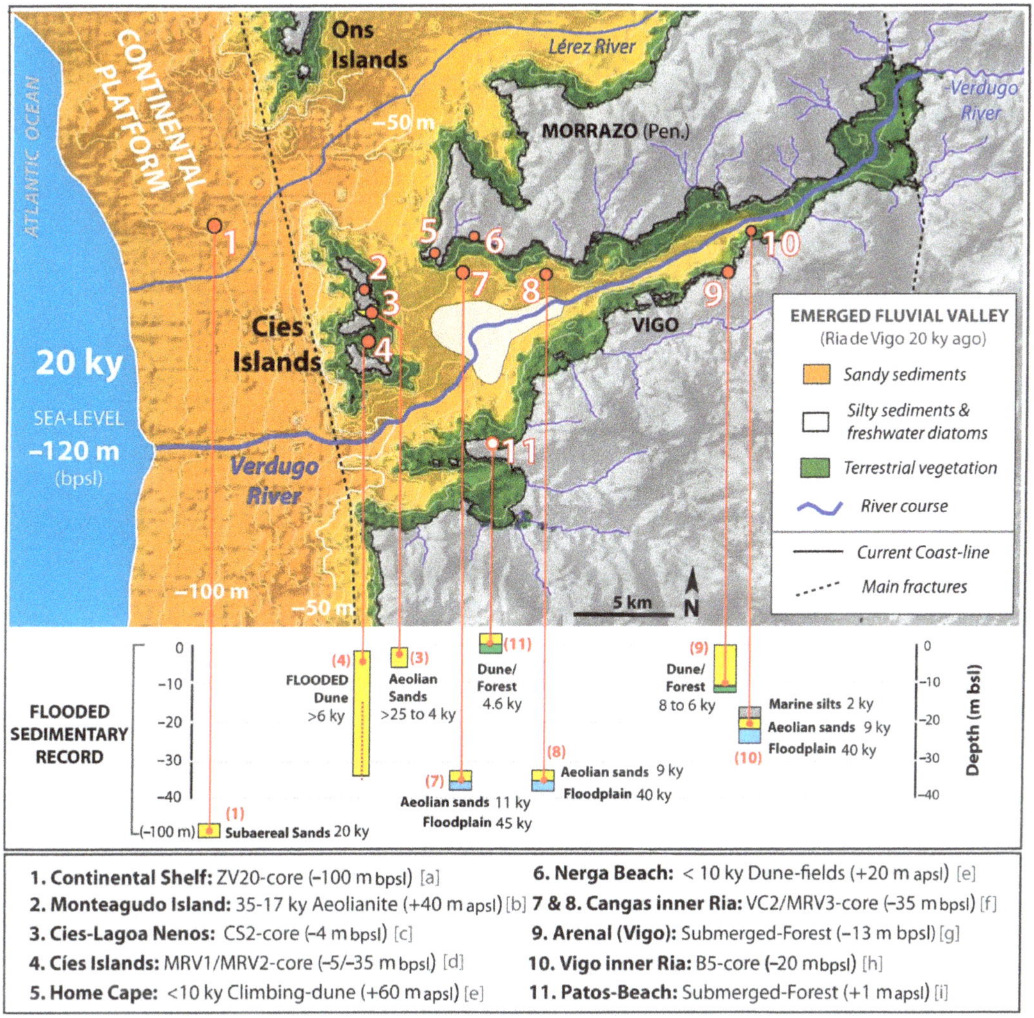

Figure 4. Geomorphology of the Ria de Vigo with sea-level −120 m below present sea-level (bpsl) at the end of the Last Glacial episode, showing the hypothetical position of the Verdugo River through the emerged fluvial valley; bathymetry from [26]. Green strip shows the advance of terrestrial vegetation and yellow area shows the predominance of sands at the bottom surface [7]. The white area represents silty sediments with freshwater diatoms [27]. Red dots show the main sand deposits related to aeolian sediments—some of them described as marine sediments (7–10) and reinterpreted in this paper as aeolian sands based absolute chronology and depth—[a]: [12]; [b]: [15]; [c]: [8]; [d]: [28]; [e]: [5]; [f]: [10]; [g]: [29]; [h]: [10]; [i]: [30]. Digital elevation database from [31]; apsl: above present sea-level.

Table 1. Conventional ^{14}C ages from sedimentary deposits related to flooded aeolian sands in the Cies Islands and Ria de Vigo. All these data have been recalibrated in this paper using the INTCAL20 curve [32] from the on-line OxCall program (v4.4), except for the calibrated age of the Patos forest (*) from [30]. The table specifies the dated material and the type of sediment in which it was found.

Sequence Location	Depth (m)	14C Uncal. Age (BP)	INTCAL20 cal yr (BP)	Material	Reference
Platform shore	100	18,010 ± 90	22,186–21,466	Mixed forams (subaerial sands)	[12]
Barrier lagoon (Rodas beach/Cies)	2	21,680 ± 60	26,030–25,835	Organic matter (sand dunes)	[8]
	2	17,240 ± 50	20,933–20,604	Organic matter (sand dunes)	
	2	3750 ± 30	4233–3916	Organic matter (sand dunes)	
Flooded dune (Cies)	5	5850 ± 30	6744–6561	Shell fragment (bioclast in sands)	[28]
	35	3220 ± 30	3482–3376	Shell fragment (bioclast in sands)	
Inner ria	35	39,890 ± 440	44,086–42,628	Bryophyte (floodplain)	[9,10]
	35	9960 ± 40	11,682–11,250	Organic matter (sands)	
	20	8010 ± 40	9010–8655	Shell fragment (bioclast in sands)	
	20	1490 ± 30	1055–1001	Shell fragment (bioclast in silts)	
Patos forest (mouth of the ria)	0	(*)	4600 *	Organic matter (fossil forest soil)	[30]
Vigo–Arenal forest (bottom of the ria)	13	7070 ± 30	7967–7836	Wood (whole trunks and roots)	[29]

In previous studies, the referred aeolian deposits on the coast of Galicia were mistakenly identified as marine sands although recently reworked by wind [33,34]. Conversely, sedimentological analysis of old climbing dunes [5] and chronological data [16] suggest that these sand deposits were formed when sea-level was much lower than today, being transported inland by the wind. Subsequently, as sea-level rose, the dunes were flooded or breached by waves, as it happens with sediments reworked by the sea. This is also suggested by both the grain size and impact marks on quartz grains observed for such deposits from a nearby ria at depths of 45 to 75 m [35]. However, some papers assumed such sand as old marine interglacial levels [36,37], while they were clearly identified as aeolian sands in some others [38,39]. These latter authors interpreted such deposits as remains of aeolian formations alternately deposited and eroded during the glacioeustatic oscillations of the Pleistocene. A complete sedimentological study of six outcrops of coastal aeolianites in Galicia carried out by [5] included, for the first time, grain size, morphoscopic, mineralogical and textural analysis by SEM (Scanning Electron Microscopy). All of these homogeneous sandy outcrops are constituted by fine–medium (98%) and subrounded (<95%) quartz grains (>90%), showing both V-shaped, arc-shaped and conchoidal or Hertzian fractures on grains, which support a true aeolian origin [5].

Regarding to the coastal dunes in the Cies Islands, previous studies show that they were formed from 25 ky to 4 ky [8] (Table 1). Such dune fields are located on the present coastline linked to sand beaches (as Rodas, Nerga, Patos and Samil beaches) (Figures 1, 3 and 4). The main objective of this work is the review of the aeolian sands series (Figure 4) based on their chronological, bathymetric and micropalaeontological data in the Ria de Vigo (Table 1). This would allow us to reinterpret the evolution of the ria from the end of MIS2 to the present day, which could be extrapolated to the Galician Rias based of the latest model of coastal evolution during the Holocene [4].

2. Study Area

The Ria de Vigo (Figures 1 and 4), as the Galician Rias, is a recent landform defined as a primary coast type [2:98]. During the present interglacial, marine waters flooded the lowermost areas of the Verdugo–Oitavén rivers drainage basin (of 710 km^2). The Ria de Vigo is the most southern ria of the coast because the Miño River, further south, is not a ria but an estuary [1] (see inset in Figure 1). This ria develops longitudinally in a WNW–ESE direction for 30 km, with the Cies Islands located at its mouth (Figure 1). Previous authors [7,40] identified shallow sedimentation (probably Holocene) overlying Pleistocene units of greater thickness. In the inner section of the ria there is a greater presence of silts and silty sands, while in the central and outer sections there is a predominance of sand-sized sediments [7,12,40] (Figure 4).

Regarding to the nearest shore sedimentation sequences (Figure 4), previous authors [12,41] reported (i) silt and very fine sand hypothetically carried by the marine currents from south to north [41] and (ii) fine, medium and coarse sands from the break-up of coastal sandy beaches during the storm events [41]. They also describe the occasional presence of gravels hypothetically transported by the main drainage networks (e.g., rivers Douro (Porto, Portugal), Miño-Sil (Galicia–Portugal) (Figure 1), Verdugo–Oitavén in the Ria de Vigo or Lérez in the Ria de Pontevedra (Figure 4), among others). These coarse-size materials were most likely reworked during the glacioeustatic Quaternary cycles [4]. For about 40 km, this platform has a slope of about 1% down to a depth of 200 m. Thereafter, it abruptly drops to depths of 1000 m through the continental slope.

The Cies Islands are located between the Ria de Vigo and the continental platform, belonging a tectonic alignment (Figure 4) of Cenozoic age [42]. This archipelago of Cies are constituted by three islands (Monteagudo, Faro and San Martiño) that show an N–S alignment about 4.5 km^2, reaching an altitude of up to 197 m above present sea-level (apsl). These islands are located in the northern half of the mouth of the Ria de Vigo. The northern foothills of the archipelago are 2.5 km away from the cape Cabo Home (Peninsula de Morrazo). These bottoms (Figure 4), up to 25 m deep, are covered by sands [7] describing a flat relief [26]. The southern foothills are 5 km from the coastline (Monteferro) (Figures 1 and 4), describing an irregular submarine relief up to 50 m depth [7,26] between granitic outcrops (Figure 4). At the base, a sand-dominated sedimentary cover of up to 20 m was described [7,12]. The topography of the Cies Islands offers a clear asymmetry, with a western face of steep slope due Cenozoic faulting (see main fractures in Figure 4). On the western side, shingle beaches are now exposed below the sand cover (Figure 3). On the east side of the Cies Islands, more sprawling and protected, the massive outcrop of aeolianites of Alto da Figueira (Monteagudo Island) (Figures 1, 3 and 4) is preserved at +40 m (apsl). The extent of this aeolian formation cannot be accurately detailed [15] and the only evidence available is the 3 m aeolianite outcrop shown in Figure 3.

3. Materials and Methods

For this study, three samples were taken of the aeolianite outcrop from Alto da Figueira at Montefaro Island (Cies Islands) (42°13′45″ N, 8°54′15″ W; WGS84) (Figure 3). Grain size was assessed by dry sieving or the raw sample (10 g). Morphological analysis was assessed with stereoscopic microscopy (50 quartz grains per sample). This information is complemented with an analysis of the surface of the most representative quartz grains of the samples and in better preservation conditions by Scanning Electron Microscopy (SEM) (JEOL mod. JAM-6400) by secondary electrons detection (25 Kv). For this, vacuum Au metallisation (0.05 mBa) of the cleaned and dehydrated samples is carried out by cathodic electrospray (Sputter Coater) (BAL-TEC SCD004).

OSL sampling was done by hammering three cylindrical steel-cores extracted along the sediment profile at depths of 200, 140 and 80 cm (distance from the top to the bottom of the outcrop). The outer part of the cores was removed under subdued red light in the Luminescence Laboratory of the University of A Coruña, and the central part dried and sieved. Coarse sand grains (180–250 μm) were treated with HCl and H_2O_2. Feldspars and heavy minerals were removed by density separation with sodium polytungstate solutions (densities of 2.62 and 2.70 g/cm^3) and the obtained quartz was etched in concentrated hydrofluoric acid (40%) to remove any remaining feldspars. This etching step removed approximately 10% of the beta dose rate (D_r) [43]. The quartz grains were checked with infrared (IR) stimulation to ensure the absence of minerals other than quartz. Luminescence measurements were performed on multigrain aliquots (2 mm in diameter) mounted on stainless steel discs in a Riso-DA15 automated TL/OSL reader equipped with blue light emitting diodes (LEDs) (470 ± 30 nm) for stimulation and a 9235QA photomultiplier. A Hoya U-340 filter was placed between the photomultiplier and the samples. To irradiate the samples, beta doses were used, using a $^{90}Sr/^{90}Y$ source that provided a dose rate of 0.120 ± 0.003 Gy/s. To estimate the equivalent dose (D_e), the Single-Aliquot Regenerative dose (SAR) protocol was used after performing preheat tests, and recovery tests were carried out [44]. The OSL signal was stimulated during 40 s, with the last 4 s of decay-curve used for subtraction of the fast component (first 0.4 s of stimulation) to determine the OSL-signal [45], known as Late Background (LBG). The Early Background (EBG) was also used to subtract the background signal from 0.5 to 4 s of decay-curve [46], with both signals compared.

The D_r was estimated using low-background gamma spectrometry on bulk samples. Marinelli beakers were used and measurements were taken in a coaxial Camberra-XTRA gamma detector (Ge-Intrinsic) model GR6022 within a 10 cm thick lead shield. A gamma cocktail solution (manufactured by LMRI, CIEMAT, Madrid, Spain) was used for calibration, with the spectrometer yearly intercalibrated with CSN (National Nuclear Security Council of Spain) and IAEA. Gamma spectrometry reports have assumed the activities of some isotopes to be those of their parents considering the equilibrium groups [47,48]. Guérin's conversion factors were used [49]. The alpha contribution was neglected for quartz dose rates, the beta dose-rate being corrected due to the HF etching step. An internal dose rate contribution from U and Th in the quartz grains of 0.02 ± 0.01 Gy/ka was assumed [50]. Water content and water saturation values were assessed in the laboratory for all samples to estimate average water content, and the cosmic dose rates were calculated in accordance with Prescott and Hutton [51].

4. Results

4.1. Grain Size, Morphology and Microscopic Analyses

The grain size shows greater than 95% medium and fine sand [52,53] in all three samples (Figure 5a). The symmetrical trend of the grain size distribution and good sorting fits well with most aeolian sands [5]. Morphological analysis mostly shows subrounded grains, following the criteria of [54]. Microscopic analysis by SEM clearly shows aeolian impact marks (Figure 5b).

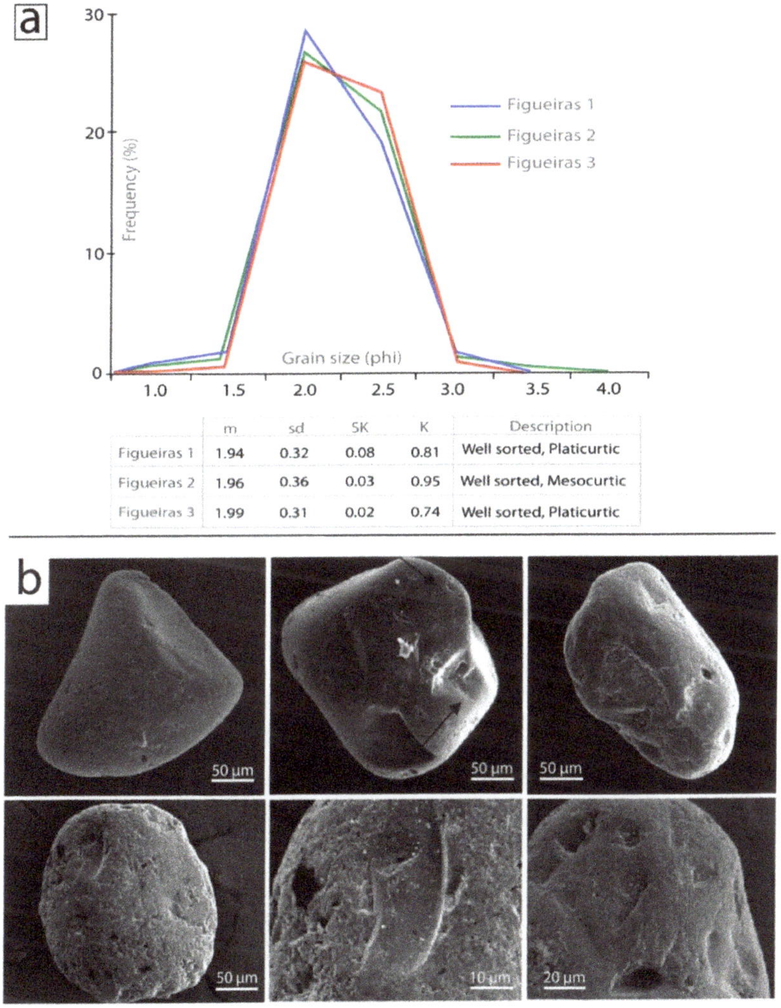

	m	sd	SK	K	Description
Figueiras 1	1.94	0.32	0.08	0.81	Well sorted, Platicurtic
Figueiras 2	1.96	0.36	0.03	0.95	Well sorted, Mesocurtic
Figueiras 3	1.99	0.31	0.02	0.74	Well sorted, Platicurtic

Figure 5. (**a**) Dry-sieving particle size and distribution of the samples from Figueiras aeolianite (Cies Islands): (m) mean, (sd) standard deviation, (SK) skewness coefficient, (K) kurtosis from the method described in [55]. (**b**) Scanning Electron Microscope (SEM) images of quartz grains showing aeolian impact marks.

4.2. OSL-Dating: Sampling, Equivalent Dose (D_e) and Dose Rate (D_r)

To estimate the dose rate (D_r) (Table 2), a saturation percentage of 20 ± 2% was assumed for all the samples. In this material, there is an average proportion of fine sand over 90%, so the drainage conditions are favourable, reducing the attenuation of the radiation for content in interstitial water [56]. The high grade of homogeneity minimises the variations related to beta-dosimetry [57]. For these samples, no disequilibrium is observed in the uranium and thorium decay chains, as the values of ^{226}Ra and ^{228} Ra are the same or very similar to those of ^{238}U and ^{232}Th (Table 2). Total D_r of the Figueiras samples is very similar, ranging between 2.7 and 2.1 Gy/ka (Table 2) [15]. These rates are also similar to those estimated for other aeolianite outcrops [16] and quartz-rich fluvial deposits in the region [1,16–18].

Table 2. Dose-rate estimation of the samples from aeolianite outcrops of the Cies Islands. Radioisotopic activity of the series of ^{238}U and ^{232}Th (^{226}Ra and ^{228}Ra values are also shown), and ^{40}K (Becquerel/kg); Dose rate (Total-D_r) (Grays/ka). Sampling distance (cm) from the top to the base of the outcrop.

Samples	^{238}U (Bq/kg)	^{226}Ra (Bq/kg)	^{232}Th (Bq/kg)	^{228}Ra (Bq/kg)	^{40}K (Bq/kg)	TOTAL-D_r (Gy/ka)
Figueiras-1 (200 cm depth)	18 ± 2.4	18 ± 2.4	20 ± 2.6	20 ± 2.6	660 ± 38	2.7 ± 0.4
Figueiras-2 (140 cm depth)	27 ± 1.3	19 ± 2.1	19 ± 2.8	19 ± 2.8	570 ± 35	2.4 ± 0.3
Figueiras-3 (80 cm depth)	19 ± 1.2	16 ± 2.2	15 ± 4.8	15 ± 4.8	485 ± 31	2.1 ± 0.3

The measured aliquots show bright OSL-signals and suitable OSL-growth curves to interpolate [58] (Figure 6), and the D_es show symmetrical distributions that fit a normal distribution, although scattered (Figure 6b–d), so the Central Age Model (CAM) [59] was used to estimate the D_es, summarised in Table 3. For these aliquots, D_e uncertainty is lower than 10%. The Overdispersion (*OD*)—with respect to the CAM average for these samples—falls within an acceptable limit (20–30%) for samples with complete bleaching [60], with the exception of Alto da Figueira2 that has an *OD* of 50% (Table 3). Nevertheless, despite this high overdispersion, the LBG-CAM burial-age estimation is stratigraphically consistent. If we compare the CAM D_es estimates using the LBG with these obtained using the EBG [46], for sample Figueiras 3 the obtained *OD* and D_es are very similar considering the uncertainty (see Table 3). For Figueiras 2, the comparison is not possible as the EBG provides just five acceptable aliquots, therefore not statistically significant. However, the *OD* of the EBG for Figueiras 1 (FIGR-1 in Table 3) is very small (2 ± 1%) when compared with the *OD* of the LBG (20 ± 4), although both are small. The D_e obtained with the EBG is around 23% smaller than these obtained with the LBG, but while the age obtained with the EBG is 23 ± 4 ka, these obtained with the LBG is 30 ± 5 ka. Similar ages are obtained for Figueiras 3 (FIGR-3 in Table 3) between EBG and LBG.

Table 3. Dose rate (D_r) (Grays/ka), equivalent dose (D_e) (Grays) and ages of the samples from the Cies Islands (FIGR: Figueiras 1, 2 and 3) (42°13′45″ N, 8°54′15″ W; WGS84), from EBG and LBG integration methods. Sampling distance (cm) from top to the base. (N) Number of aliquots accepted/analysed; (*OD*) Overdispersion percentage. (*) CAM-LBG ages (ka = kiloannum before dating) from [15]. Sampling distance (cm) measured from the top to the base of the outcrop.

Samples	D_r (Gy/ka)	N EBG	D_e EBG (Gy)	OD-EBG (%)	Age (ka) (CAM-EBG)	N LBG	D_e LBG (Gy)	OD-LBG (%)	Age (ka) (CAM-LBG)
FIGR-1 (200 cm)	2.7 ± 0.4	24/96	62 ± 2	2 ± 1	23.3 ± 4.2	31/96	80 ± 3	20 ± 4	30.6 ± 4.8 *
FIGR-2 (140 cm)	2.4 ± 0.3	5/54	–	–	–	36/54	56 ± 5	50 ± 6	23.3 ± 3.9
FIGR-3 (80 cm)	2.1 ± 0.3	26/51	38 ± 2	30 ± 5	18.9 ± 3.8	31/51	42 ± 3	33 ± 5	20.1 ± 3.3 *

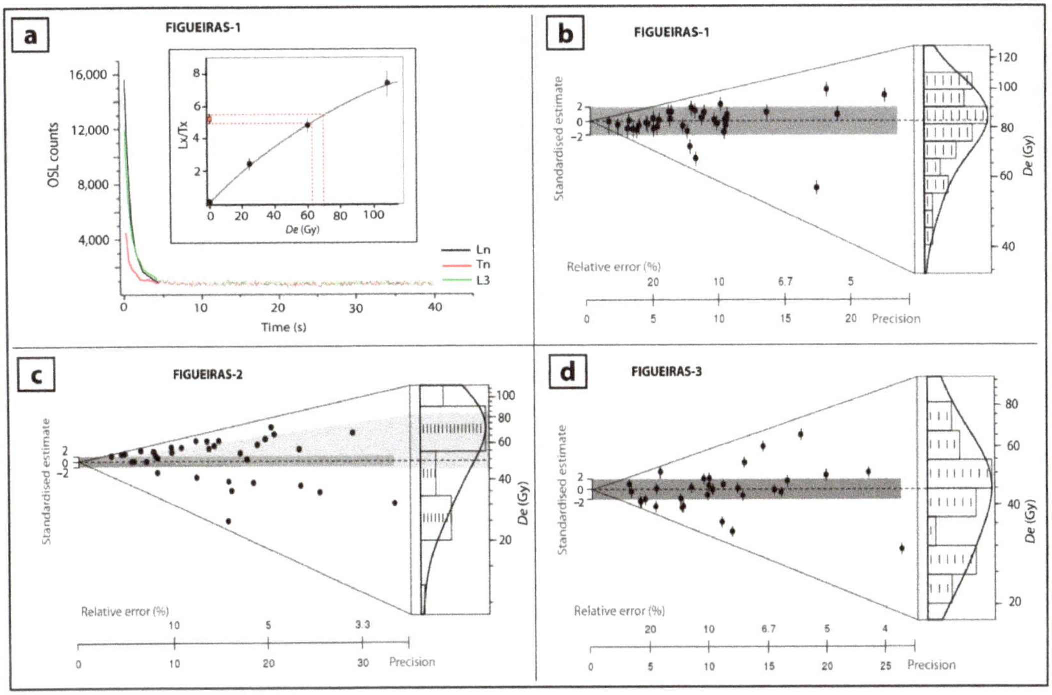

Figure 6. Quartz OSL signal and D_es distribution. (**a**) OSL-signal decay-curve from one aliquot of Figueiras 1 sample. Lines show the intensity (counts per second) from natural OSL-signal (Ln), natural OSL-signal from test-dose (Tn) and OSL-signal from third regenerative cycle from SAR protocol [44]. (Inset) Exp + lineal adjust for OSL growth-curve from normalised OSL-signal from SAR (Lx/Tx) versus equivalent dose (D_e), including the linear interpolation [58] from natural normalised OSL-signal. (**b–d**). Distribution of data from Figueiras 1, 2 and 3 samples, through Abanico plots [61]. Black dots represent D_e estimates from accepted single aliquots, showing error bars. Grey bar in the radial plots represents the standard error ($\pm 2\sigma$) from the standardised estimate (weighted mean) by the Central-Age Model (CAM) (dotted line). Kernel-density estimate plot is also represented. Light-grey area in (**c**) represents the standard error ($\pm 2\sigma$) from the kernel density estimation.

Thus, there are not significant differences in the obtained age when both methods are used. Final ages from Figueiras 1, 2 and 3 (Table 3) are considered using the LBG method. These results are coherent with the ages obtained for other aeolian sands (aeolianites) along the Galician coast [16], which show similar behaviour of the OSL signals. All these ages are also consistent with other absolute ages calculated for fossil dunes in the coast of SW France, Portugal, SW Iberia, Gibraltar and the Canary Islands [22,62–67]. As discussed later, this clearly indicates a major episode of coastal aeolian accretion since the end of MIS2 in this Atlantic region.

5. Discussion

The evolution of the Atlantic coast of Galicia during the last regression (Upper Pleistocene) is poorly understood. The same happened with the stages of sea-level rising and flooding the Galician fluvial valleys (Rias) during the Holocene transgression. This is due to sediments of aeolian origin having been misinterpreted as marine deposits (see Section 1). Concerning Ria de Vigo, isolation of the old river valley from the sea waters 9 ky ago [30] and an incipient connection with marine waters from less than 3 ky was proposed [68]. In the innermost areas of this ria, a fossil forest 7.9 ky old was buried at −13 m (bpsl) (Table 1) by a sand dune more than 10 m thick, indicating a lower sea-level at that time. The same occurs in the outermost areas of the ria where partially flooded dunes appear between 25–4 ky (Table 1)—along with 20 m thick dunes that accreted from more than 6 ky (Table 1), now flooded up to 35 m in depth. All of these sands indicate a lower sea-level that fits very well with the aeolianite outcrop preserved in the Cies Islands [15], older than 20 ky, correlated with a coastline lower than 100 m below present sea-level (bpsl) (Figure 7). Such data have allowed us to contextualise many of the isolated results available in references for the Ria de Vigo (Figure 4; Table 1) and the Galician coast, with the base of the reconstructed geomorphologic history for the coastal evolution from the end of the Upper Pleistocene to the present day (Figure 7).

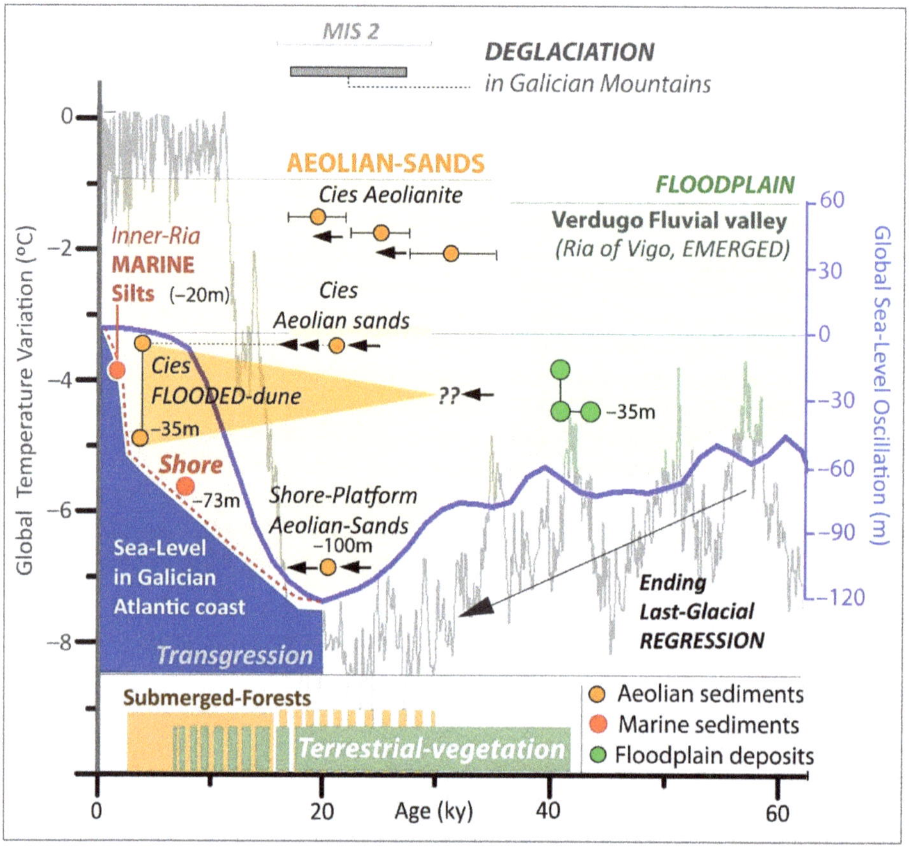

Figure 7. Local sedimentary record of the last 45 ky (continental and marine) in the Cies Islands and Ria de Vigo, compared to trends in the global climate record [69] indicated by temperature variations from Vostok ice cores [70] (grey line) and sea-level oscillations [71] (blue line), including the deglaciation processes in the mountains of Galicia and northern Portugal [72,73] at MIS2 [74]. Sedimentary record at the end of the Last Glacial period are represented by (i) floodplain deposits (green areas) and (ii) aeolian sands (yellow area) ([8,10,12,15] and this paper). Submerged forests buried by dunes from 30–3.5 ky along the coasts of Galicia and the northern Portugal [4]. Dark-blue area under the red dashed line shows the sea-level rise in the Ria de Vigo during post-glacial transgression, from the local record.

5.1. Aeolian Sediments in the Cies Islands and Ria de Vigo

The age of the aeolianites outcrop on Alto da Figueira (Isla de Monteagudo, Cies Islands) allowed us to establish an important aeolian accretion stage in the Atlantic margin of Galicia at the end of the Last Glacial period (MIS2) (Figure 7), coinciding with the maximum regressive levels in the northern hemisphere [71]. Additionally, within this aeolian accretion event there are other aeolianites preserved at various points along the Galician coast (Figure 2a,b) with age ranges of 20.9 ± 6.5 ky, 30.9 ± 3.6 ky and 29.7 ± 4.8 ky [16]. In the continental platform close to the Cies Islands, a 20 ky old deposit of aeolian sands [12] (Table 1) has been reported at −100 m (bpsl) (see record 1 in Figure 4; Figures 7 and 8a), which implies a lower sea-level at the end of MIS2 (Figures 7 and 8).

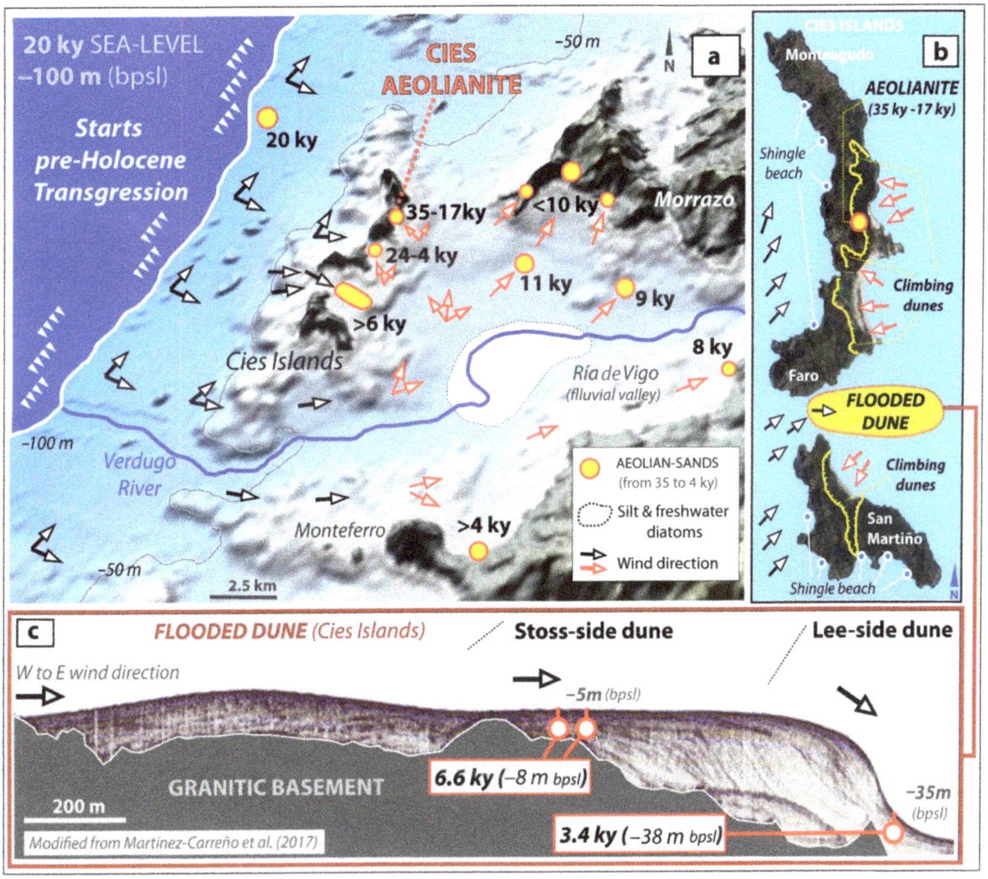

Figure 8. Aeolian sands in the Cies Islands, Ria de Vigo and continental platform. (**a**) Sea-level in the study area, showing the prevailing coastal winds from SW to NE (black arrows) in the Atlantic margin of Iberian Peninsula at the end of the Last Glacial period (modified from [15]). Red arrows show local wind variations, as suggested by the aeolian sands outcrops (yellow dots). The white area represents silty sediments with freshwater diatoms [27] in the uppermost levels, most probably related to freshwater diatoms in silty sediments described in [75] between 45 to 8 ky. Relief from [26]. (**b**) Prevailing wind direction in the Cies Islands at present sea-level and local variations suggests the climbing dunes advance (yellow stripes) as yet preserved in the archipelago [76]. (**c**) Flooded dune between Faro and San Martiño Islands, showing its internal structure [77] and chronology [28]. Black arrows show wind direction. (bpsl): below present sea-level.

At that time, glacial ablation processes were already taking place in the mountainous systems of the interior of Galicia (Figure 7) and northern Portugal [72,73]. This means, from a glacioeustatic perspective, that the Last Glacial maximum in the NW Iberian Peninsula must have been earlier, even reaching a regressive sea-level below −100 m (bpsl); similar features are also present in other part of the world [78,79]. Therefore, the beginning of this aeolisation began at the end of the last regressive stage, confirming the existence of an emerged sand strip more than 5 km wide in the continental platform of Galicia (Figure 8a), as the source of aeolian sand [15,21]. Considering that no other sources of aeolian sands are observed on this coast, the location of the fossil dune fields around Cies were built by sands transported by prevailing winds from SW to NE (Figure 8a,b) [15]. This regional wind dynamic has also been described to explain the evolution of the aeolian sands in the Atlantic coast of Portugal (Iberian Peninsula) at the end of MIS2, based on the sedimentary structures of dunes, OSL chronology and long-term regional climate models [22]. Furthermore, the bedding structures from the stoss-side and the lee-side of the now flooded dune in Cies (Figure 8c) agree with this regional palaeowind direction.

5.1.1. Aeolianite Outcrop in the Eastern Side of the Cies Islands at MIS2

Grain size, morphology and SEM analysis show that the sand deposit at Alto da Figueira (Monteagudo Island; Cies Islands; Figure 3) is of aeolian origin [5]. During the aeolian accretion that formed the aeolianite outcrop dated here, this fossil climbing dune advanced on the eastern side of Monteagudo Island (Figure 8a,b), reaching 3 m thickness at +40 m (apsl) [15]. This clearly indicates that the aeolian sands transported from the emerged shelf (Figures 4 and 8a) reached the interior of the Ria de Vigo, thus overcoming the topographic obstacle of the Cies Islands. Assuming a sea-level of −120 m (bpsl) at that time [71] (Figures 7 and 8), the three islands of Cies were connected to each other, and to the mainland (Figures 4 and 8a), reaching heights over 300 m above sea-level. This vigorous relief of Cies would be presumably a sedimentary trap for the aeolian sands mobilised from west to east [15] (Figure 8a,b), stopping the dunes from advancing and causing their concentration at the lowest levels. Nevertheless, similar to other relict climbing dunes in the cliff coast of Galicia [5] (Figure 9), also present on the northwest coast of the Ria de Vigo (records 5 and 6 in Figure 4; Figures 8a and 9c,d), the topography of the Cies Islands would not cause a major problem for the movement of aeolian sands that, in any case, could be channelled through the corridor between the islands of Faro and San Martiño (Figures 1, 3a, 4 and 8a,b), as explained below (Section 5.1.2). The growth of climbing dunes on the eastern slope of Cies (Figure 8b) implies a twist in the wind direction to drag the sands westwards [15] (Figure 8b). These local variations in wind movement are a common phenomenon even today as proven by the wind-faceted vegetation present on the islands (Figure 3b). All of these circumstances are also observed on nearby islands such as Ons Island [76] (Figure 1), located less than 30 km to the north (Figures 1 and 4). In this sense, more comprehensive studies of the ventifacts present on these islands are highly recommended.

Figure 9. Relict climbing dunes still preserved in the Galician coast, reaching heights of more than 100 m. (**a**) Ponteceso (43°14′00″ N, 8°56′11″ W); (**b**) Trece (43°11′15″ N, 9°08′45″ W). (**c**,**d**) Relict climbing dunes near the Cíes Islands (northwest coast of the Ria de Vigo). (**c**) Home cape (42°15′07″ N, 8°51′57″ W); the yellow dotted line shows the climbing dune advance. (**d**) Nerga Beach and adjacent climbing dune (42°16′09″ N, 8°51′15″ W) (WGS84); cl-d: climbing dunes.

5.1.2. Flooded Dunes in the Cies Islands and Sea-Level

The sedimentary sands reported in the corridor between Faro and San Martiño Islands (Cies) (Figure 8c) were recognised by submarine geophysical campaigns [77], although they were misunderstood as an underwater sedimentary construction (never before described) formed by bottom currents during strong storms and where aeolian sands were confused with marine sands. Nevertheless, sedimentary features, absolute chronology and hypsometric data confirm that it is not a submarine formation but a subaerial one, as other studies described on other coasts of the world (see seismic profile of Figure 8 in [80]). Firstly, this sandy formation at 5 m depth has physical continuity with the dune fields of Rodas Beach, located at less than 1500 m between the islands of Monteagudo and Faro of Cies (Figure 1; aeolian sands in Figure 3a; record 3 in Figure 4; Figure 8a,b), which were formed between 20 and 4 ky [8] (Table 1). In this sense, another flooded sand deposit more than 13 ky old [14] and up to 20 m thickness was also described on the Areoso islet of the Ría de Arousa (Figure 1), close to the Ría de Vigo. The upper levels (2 m) of this deposit were accreted from 13 to 4 ky (see core A14-VC4 in [14]), as dune fields between the Cies Islands [8], abovementioned. In addition, the prolongation of these currently flooded sands culminates in a dune field whose upper level was dated by OSL at 2.5 ky [81]—covering both a 6 ky old megalithic tomb and a 4.6 ky old fossil forest. This reiteration of submerged sand bodies that are physically connected to emergent dunes would indicate that these are large partially flooded dune fields, although only the upper levels that remain emerged are currently visible.

Secondly, the sedimentary structures of this flooded formation (Figure 8c) are related to a topographic-impeded dune close to lee-dune type [82,83], composed of fine to medium sands about 30 m thick. It clearly shows aeolian structure [84], with an extreme located stoss-side at −5 m (bpsl) (Figure 8c) whose formation culminated 6.6 ky ago (Table 1). Furthermore, these sands are extended in this corridor for over 1500 m in an easterly direction (towards the interior of the ria), where it is presently a well-developed lee-side reaching a depth of more than −35 m (bpsl) (Figure 8c). There, dune vegetation has been identified at 3 m below the deposit (−38 m bpsl) [28], dating back to 3.4 ky cal BP (Table 1). According to this, dune vegetation from 6 ky is also described in the nearby and now submerged dune fields of Rodas Beach [8] (Figure 3), reinforcing the idea of a continuum formed by aeolian sands. It should be noted that in this sand body, only the occasional presence of test is described in some of the few cores studied [28,77], without determining the species and community associations that define a marine environment—dissimilar from other studies carried out on the Galician coast [39,85]. Studies carried out in the Ria de Vigo have also failed to consider the wind mobilisation of the scarce test fragments found (e.g., from aeolian polishing marks and sorting by size), as recognised in previous studies on the Galician coast [39,85], thus dismissing the marine origin of these sands in favour of an aeolian formation.

Although this impeded dune is now completely flooded (Figure 8c), the reinterpretation proposed here is consistent with the data that places a transgressive sea-level at −73 m (bpsl) on the inner platform of Cies 9 ky ago cal BP [30]. Even more, its aeolian origin is really consistent with the data that place sea-level well below −13 m (bpsl), 7.9 ky ago cal BP (Table 1), as proven by the aforementioned submerged fossil forest of the Arenal (city of Vigo) in the inner section of the Ria de Vigo (record 9 in Figure 4), where whole fossil trunks and roots buried by a sand dune thicker than 10 m can be observed [29]. Later deposits indicate a well-developed continental ecosystem that was buried by those dune fields that reached the bottom of the ria and, at the same time, the aeolian accretion of the sand dunes with dune vegetation at the mouth of the ria (Cies Islands) was coming to an end. That was the case for the dune fields of Rodas Beach between the islands of Monteagudo and Faro [8] (Figure 3), and the now-flooded impeded-dune between the islands of Faro and San Martiño (Figure 8a,c). This clearly indicates that all of these areas were under subaerial conditions, in agreement with the disconnection from ocean waters suggested for this fluvial valley (the emerged ria) until 4 ky [30], as also suggested by test studies carried out in the Ría de Vigo at −40 m (bpsl) less than 3 ky ago [68]. Thus, the bathymetry of this flooded dune between the Faro and San Martiño Islands and its chronology should undoubtedly indicate a sea-level position lower than the stoss-side dune (−5 m bpsl) 6 ky ago (Figure 8c). Moreover, the sea-level would be lower than −38 m (bpsl) 3.4 ky ago on the basis of the chronological data [28], both the age and the depth of the uppermost level at lee-side dune (Figure 8c). This type of flooded aeolian formation can also be observed on the nearby Ons Islands (see the same transect PT7 in [86–88]) and Sálvora Island (see Figure 13c in [11]), developing a similar topographic and sedimentary structure—although they have not been mentioned until now. These sands would later be transported by the wind, covering the eastern leeward side of these islands (more sheltered from the strong coastal winds) and also reaching the current coastline of the Ria de Vigo (Figure 10), as will be explained later. According to the model proposed here, these data would appropriately fit the sea-level rise curve for the western Galician Rias during the Early–Middle Holocene, as represented in Figures 7 and 10.

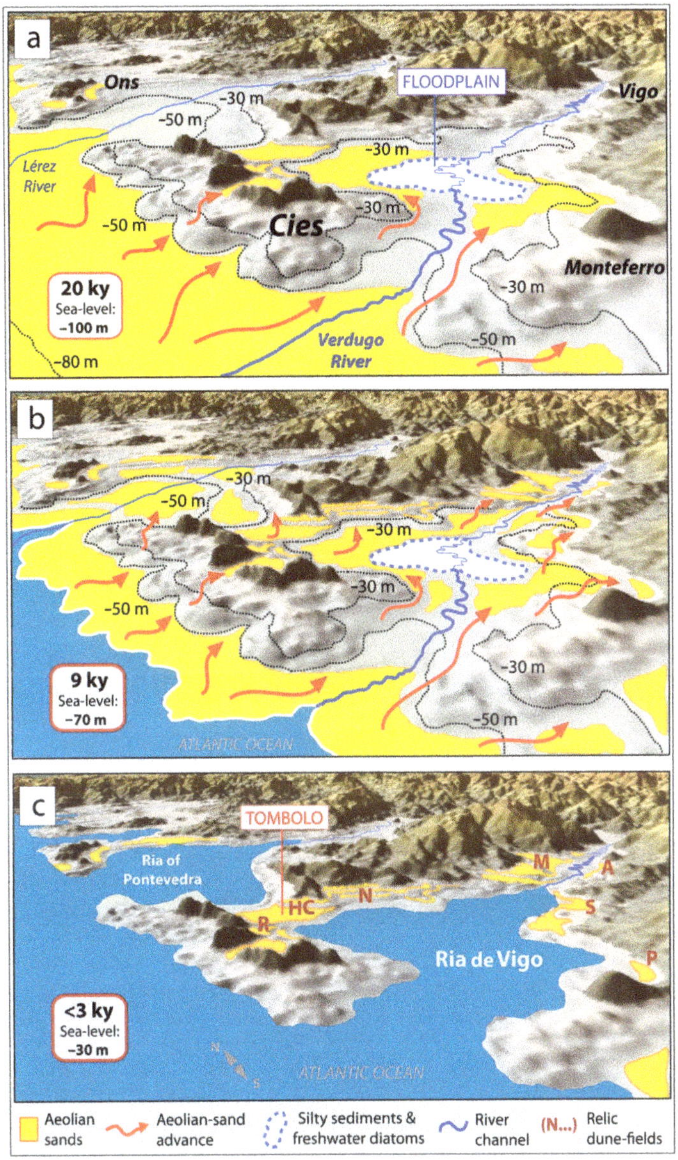

Figure 10. Sequence of the aeolian accretion process in the fluvial valley of Verdugo River at the end of the Last Glacial period, showing the hypothetical sand dunes advance towards the continent (red arrows) and subsequently sea-level rise in the last post-glacial transgression: (**a**) 20 ky aeolian sands advanced through the emerged continental platform and the Cies Islands from a sea-level below −100 m (bpsl); (**b**) 9 ky sea-level about −70 m (bpsl), with the aeolian sands upwards within the fluvial valley; (**c**) incipient configuration of the Ria de Vigo with a sea-level about −30 m (bpsl) from less than 3 ky, flooding the aeolian sands and the river plain. Capital letters indicate aeolian sands that are related to the relict aeolian formations (as dune fields and climbing dunes) currently preserved around the ria: Rodas (R), Home Cape (HC), Nerga (N), Moaña (M), Arenal-Vigo (A), Samil (S) and Patos (P).

5.1.3. Dunes Filling the River Valley and Environmental Collapse

Mobilisation of dune fields towards the continent remains active as the sea-level rises during the Holocene transgression [4], although the aeolian supply was reduced as the source area was flooded by the rising sea-level. Considering a sea-level at −73 m (bpsl) 9.4 ky ago [30] and below −38 m (bpsl) 3.5 ky ago (Figure 10), as suggested by the chronology and bathymetry of the flooded dune of Figure 6c, the relief of Cies at that stage was still emerged and joined the Peninsula de Morrazo through another sand [7] tombolo (Figures 8a and 10). In general, a marine origin has been assigned to the sand deposits around the tombolo, as shown by the models of coastal evolution previously proposed [8,10,86,88]. Nevertheless, in the cores sampled at −35 m (bpsl) in the same area (see cores VC2 and MRV3 in [10] and records 7 an 8 in Figure 4), a deposit with a 2 m thickness of fine sand and silty sand (upper levels) are reported having an age of 11 ky cal BP at the base (Table 1). These authors assigned a marine origin to these sands based on the sporadic presence of possibly marine bioclasts, without considering the possibility of artefacts typical induced by the same vibrocorer sampling, quite frequent and also mentioned in other studies in Galicia [5,38,39,85]. However, the chronological and bathymetric data analysed and discussed here show that these 11 ky sands located at −35 m (bpsl) (Figures 8a and 10) can only be subaerial and therefore aeolian sands. It is only under this context that we could favourably justify their physical connection with the Holocene aeolian deposits [8] located on the northwestern coast of the Ria de Vigo (Peninsula de Morrazo), such as the relict dune fields and climbing dunes of Nerga beach and the cape Cabo Home (Figures 1, 4, 8a, 9c,d and 10). Once again, a continuum is observed, as transgressive aeolian sand sheets accreted during the Holocene from a lower sea-level (Figure 8).

The coastal aeolian accretion described here also triggered the ecological collapse of the continental environments produced by all those aeolian sands that were able to reach the innermost areas of this valley (now the present-day Ria de Vigo) (Figures 4 and 10). For instance, in the sedimentary record described at −35 m (bpsl) in cores VC2 and MRV3 [10], lower strata of silty layers with woody remains from 45 to 40 ky cal BP (Table 1) were identified (see floodplain deposits in records 7 and 8 in Figure 4; Figures 8a and 10). These materials, similar to other silty sediments described in the inner ria with the presence of freshwater diatoms [75], are related to the fluvial valley (Verdugo River) when sea-level was below −100 m, and later covered by aeolian sands dated back 11 ky. It should be noted that the descriptions of these cores (MRV3 or VC2) [9,10] also differ, with no presence of foraminifera in [9]; in contrast, [10] mentions a sporadic presence of foraminiferal tests that are unable to establish a marine origin for these sediments, contrary to what the authors suggest. Thus, these currently flooded sands were surely accreted by wind, although the advance of dunes was stopped by the river course, as proven by the contact between the aeolian sands and the finest sediments [12] of fluvial origin (Figures 4, 8a and 10). Additionally, in this muddy area with silts and clay silts, an association of freshwater diatoms were also identified [27] along the path followed by the river (Oitavén–Verdugo River), as suggested by other authors based on hypsometric data (see Figure 3 in [10]). This silty area with a low slope [26] is interpreted here as a floodplain with a meandering course (Figures 5 and 10) and the lack of aeolian sands on it would be positively justified by the fluvial dynamics that prevented the accretion of the windblown sands.

As to the southern coast of the Ria de Vigo, the accretion of aeolian sands up the valley (Figures 8a and 10) is also represented by a 10 m thick sand dune that buried the forest of Vigo–Arenal (see record 9 in Figures 4 and 11a–c), located at −13 m (bpsl) and dated back 7.9 ky [29] (Table 1). This dune preserved in the bottom of the ria is spatially and temporally related to 2 m thick flooded sandy sediments, described at −20 m (bpsl) (see MRV4 and B5 cores in [9,10]), which dated back 8.8 ky cal BP (Table 1) (see sequences 9 and 10 in Figure 4). Although, once again, a marine origin was assumed by these authors (cores MRV3 and VC2) based mainly in global correlations; these sand sediments are also reinterpreted in our work as aeolian sands due to their absolute ages and depth, considering their physical connection with nearby sand deposits from aeolian origin (Figures 4, 5 and 10). These aeolian sands entered the ria on the southern side of the mouth of the Oitavén–Verdugo River and Monteferro promontory (Figure 10). This would be the case for the dune fields related to Samil Beach (Vigo) or Patos Beach (Nigrán) (Figures 1 and 10), where remains of fossil wood (in living position) have now appeared within the intertidal zone, dated back 4.6 ky [30] (sequence 11 in Figure 4; Figure 11d,e). Other forest soil with roots and trunks in living position (Figure 11d,e), were described along the Galician coast between 8 and 3.5 ky [4]. Thus, the woody remains and arboreal pollen under the 20 ky old dune fields in the Cies Islands [8] and all these ancient forests were buried by aeolian sands. This meant the ecological collapse of coastal environments by dunes from 25 to 4.5 ky, which preceded the marine flooding of the valley, as a major environmental crisis caused by climate change and global warming during the Holocene.

All chronological and sedimentological evidence allow us to relate these continental deposits to the remaining submerged forests buried by sand dunes that have been described along the Galician and northern Portuguese coastline [89–91]. These were well-developed forest formations with deciduous species growing away from marine influence [4]. In turn, all of these continental formations affected by coastal aeolian sands, as proposed here, are related to submerged forests that appeared in the west of France and all around the British coasts at the same time [92–95]. Therefore, this aeolisation stage could be extended to the whole southern half of the European Atlantic coast, also suggesting that the absolute ages of the aeolianites preserved in SW France [67,96,97] together with SW Iberia, Gibraltar and Canary Islands [22,62,68,69] are from MIS2 to Late Holocene. Based on the chronology established for other coastal aeolianite outcrops in Galicia [21], the model can also be applied to the penultimate glacial episode and the Eemian transgression, and could also be extended to other coastal areas worldwide where coastal aeolianites have been dated [80,98–105].

5.2. Present-Day Sea-Level and Lack of Aeolian Supply

As mentioned, the fluvial valley of Ria de Vigo was disconnected from ocean waters until less than 4 ky, at which time benthic test associations are described in silty surface sediments in the Ria de Vigo at −40 m (bpsl) [68]. These data correspond to the final stage of the lowermost areas of this valley as the ocean waters reached the present-day levels, leading to the configuration of this wide estuary as it is today. This agrees with the marine silty sediments described at −20 m (bpsl), in the inner section of the Ria de Vigo (see MRV4 and B5 cores in [9,10]) dated back to 2 ky cal BP (Table 1), overlying older aeolian sands (record 10 in Figure 4). From that moment onwards, sea-level rise completely stopped the aeolian supply described above and initiated an erosive aeolian phase [5] that affected the emerged dune located on the present-day shoreline. Paradoxically, sea-level rise has favoured the preservation of a large part of the old aeolian formations within the Ria de Vigo (such as the flooded dune between the Cies Islands), probably due to the Cies relief hindering erosion by wave action at the mouth of the estuary, unlike the processes of wave washing and destruction of Early and Middle Holocene sand-dunes suggested in other types of coasts characterised by the presence of large islands as a natural barrier [106]. Subsequent flooding and/or marine reworking of aeolian sands would favourably explain the presence of salt crusts and polished surfaces on the sand grains [35], along with the presence of neoformation marine minerals, such as glauconite, associated with shallow environments [77].

Today, as sea-level rises, the waves and tides sweep away the coastal dunes forming the current sand beaches, as in the case for the Rodas, Nerga, Moaña, Arenal-Vigo, Samil and Patos beaches (Figures 1 and 10). As mentioned, this suggests that the sand beaches in this area are aeolian sands reworked by the sea. It means that the relict dunes in this coast are not simply an extension of sand beaches inland [107] (p. 11). On the contrary, they are sediments formed from coastal aeolian accretion from a lower sea-level that began at the end of MIS2 and continued into the Late Holocene.

Figure 11. Submerged forests buried by Holocene dunes in the Ria de Vigo. (**a–c**) Submerged forest of Arenal (Vigo, Pontevedra) located at −13 m (bpsl) under a dune of over 10 m. (**b**) W: woody remains and whole trunks; bd: base of dune. (**c**) Detail of the aeolian sands from the dune of Arenal; S: aeolian sands (modified from [29]). (**d**,**e**) Submerged forest in Patos Beach (Nigran, Pontevedra); f: forest soil with roots in living position; gr: gravels; s: sand beach; w: wood remains.

6. Conclusions

The chronology assessed for the aeolianite outcrop of Monteagudo Island (Cies Islands) allows us to define a prolonged sequence of aeolian sedimentation between 35,000 and 17,000 years before present. Considering the age of this and other climbing dunes developed in the Atlantic margin of Galicia, our data suggest an episode of aeolian accretion coinciding with the end of the Last Glacial period (MIS2) with a massive mobilisation of aeolian sands towards the Cies Islands and the current coast of the Ria de Vigo (emerged during the Last Glacial period). Based on the correlation with the global palaeoclimatic curves, the dating of subaerial sand sediments in the continental platform of Galicia and the age of the aeolian deposits, we propose the existence of large areas of the continental platform (>5 km) completely emerged during the Last Glacial period.

The thickness of the Alto da Figueira aeolianite in Monteagudo Island (Cies Islands) and its location on the eastern slope (the most protected) suggests that during the last stage of the Upper Pleistocene the archipelago was emerged, connected with the continent and partially covered by aeolian sands. During this period (MIS2), the presence of active fluvial courses would prevent the advance of the dunes through the southern extent of Cies. Along the western slope, the aeolian processes would be constrained against abrupt vertical relief, being channelled by corridors such as those existing between the actual islands of the Cies archipelago (Monteagudo, Faro and San Martiño). In all of these areas, dunes that are totally or partially flooded are still preserved today. As to the north extreme of Cies, the connection with the continent would also be constituted by aeolian sands (as a tombolo), facilitating the development of Holocene dune fields and climbing dunes, such as those currently preserved on the northwesternmost coast of the Ria de Vigo (Peninsula de Morrazo).

During the Last Glacial period, oceanic waters completely abandoned the Ria de Vigo, reactivating the fluvial dynamics of the Oitavén–Verdugo fluvial system in this emerged valley. This main riverbed was developed over a wide floodplain for 50 km (approx.) covered by marine and continental sediments. On these emerged materials and throughout the Upper Pleistocene, the expansion of forests occurred along the emerged valley, present even in Cies. These well-developed forests were dramatically affected by the advance of the dune fields into the Ria de Vigo (then emerged) from 30 ky to less than 5 ky. This progressive aeolisation can be extrapolated to the entire coast of Galicia and northern Portugal, and most probably the southern half of the European Atlantic margin.

Based on chronological and hypsometric data available in the local sedimentary record, a sea-level rise curve is fitted for the Ria de Vigo, reaching less than −100 m (bpsl) at the end of the Last Glacial period (20 ky), −70 m (bpsl) in the Early Holocene (9 ky), −50 m to −40 m (bpsl) in the Late Holocene (3 ky) and −20 m (bpsl) to 0 m from 2 ky. Thus, the sea-level rise was fast enough (from 7 to 12 mm/y) to preserve all flooded aeolian formations (dunes) around the Cies Islands. This aeolian accretion stopped completely as the sea reached its present level and was replaced by wind and marine erosion.

Author Contributions: Conceptualisation, C.A.-C. and J.R.V.-R.; funding acquisition, J.R.V.-R.; methodology, C.A.-C. and J.S.-S.; investigation and resources, C.A.-C. and J.R.V.-R.; supervision, J.R.V.-R.; writing and reviewing, C.A.-C. and J.R.V.-R. All authors have read and agreed to the published version of the manuscript.

Funding: This research was funded and supported by Xunta de Galicia (programmes ED431B 2018/47 and ED431B 2021/17) through the Grupo Interdisciplinar de Patrimonio Cultural e Xeolóxico (CULXEO).

Institutional Review Board Statement: Not applicable.

Informed Consent Statement: Informed consent was obtained from all subjects involved in the study.

Data Availability Statement: The data used in this study were obtained from the doctoral thesis of the first author (C.A.-C.) (https://ruc.udc.es/dspace/handle/2183/19810) (accessed on 7 August 2022).

Acknowledgments: We would like to thank José Antonio Fernández Bouzas, director of the PNMTIIAA-Islas Cíes, for his permission to visit the park and take samples. We would like to thank the Xunta de Galicia for its support through the Grupo Interdisciplinar de Patrimonio Cultural e Xeolóxico (CULXEO) (programmes ED431B 2018/47 and ED431B 2021/17). We would also like to thank most sincerely the Reviewers and the Editors for their suggestions and comments.

Conflicts of Interest: The authors declare no conflict of interest. The funders had no role in the design of the study, in the collection, analyses or interpretation of data, in the writing of the manuscript or in the decision to publish the results.

References

1. Viveen, W.; Braucher, R.; Bourlès, D.; Schoorl, J.M.; Veldkamp, A.; van Balen, R.T.; Wallinga, J.; Fernandez-Mosquera, D.; Vidal-Romani, J.R.; Sanjurjo-Sanchez, J. A 0.65 Ma chronology and incision rate assessment of the NW Iberian Miño River terraces based on 10Be and luminescence dating. *Glob. Planet. Chang.* **2012**, *94–95*, 82–100. [CrossRef]
2. Ottman, F.C. *Introducción a la Geología Marina y Litoral*; Editorial EUDEBA: Buenos Aires, Argentina, 1967; 287p.
3. Trindade, M.J.; Prudêncio, M.I.; Sanjurjo-Sánchez, J.; Vidal-Romaní, J.R.; Ferraz, T.; Fernández-Mosquera, D.; Dias, M.I. Post-depositional processes of elemental enrichment inside dark nodular masses of an ancient aeolian dune from A Coruña, Northwest Spain. *Geol. Acta* **2013**, *11*, 231–244. [CrossRef]
4. Vidal-Romaní, J.R.; Grandal-d´Anglade, A. Nota sobre la última transgresión marina en la costa de Galicia. *Cad. Lab. Xeol. Laxe* **2018**, *40*, 229–246. [CrossRef]
5. Gutiérrez-Becker, L. Caracterización de los Sistemas Dunares Costeros del NW Ibérico y su Evolución Durante el Cuaternario. Ph.D. Thesis, Universidad de Coruña, A Coruña, Spain, 2008. Available online: https://dialnet.unirioja.es/servlet/tesis?codigo=45245 (accessed on 7 August 2022).
6. Nonn, H. *Les Régiones Cotières de la Galice (Espagne). Etude Géomorpholigique*; Tomo III; Publications de la Faculté des Lettres de L´Université de Strasbourg, Foundation Baulig: Paris, France, 1966; 592p.
7. Rey-Salgado, J. *Relación Morfosedimentaria entre la Plataforma Continental de Galicia y las Rías Bajas y su Evolución durante el Cuaternario*; Instituto Español de Oceanografía: Madrid, Spain, 1993; 233p.
8. Costas, S.; Muñoz-Sobrino, C.; Alejo, I.; Pérez-Arlucea, M. Holocene evolution of a rock-bounded barrier-lagoon system, Cies Islands, northwest Iberia. *Earth Surf. Proc. Land.* **2009**, *34*, 1575–1586. [CrossRef]
9. Martínez-Carreño, M.; García-Gil, S. The Holocene gas system of the Ría de Vigo (NW Spain): Factors controlling the location of gas accumulations, seeps and pockmarks. *Mar. Geol.* **2013**, *344*, 82–100. [CrossRef]
10. Martínez-Carreño, M.; García-Gil, S. Reinterpretation of the Quaternary sedimentary infill of the Ría de Vigo, NW Iberian Peninsula, as a compound incised valley. *Quat. Sci. Rev.* **2017**, *173*, 124–144. [CrossRef]
11. Cartelle, V.; García-Moreiras, I.; Martínez-Carreño, N.; Muñoz Sobrino, C.; García-Gil, S. The role of antecedent morphology and changing sediment sources in the postglacial palaeogeographical evolution of an incised valley: The sedimentary record of the Ría de Arousa (NW Iberia). *Glob. Planet. Chang.* **2022**, *208*, 103727. [CrossRef]
12. Mohamed, K.J.; Rey, D.; Rubio, B.; Rubio, B.; Vilas, F.; Frederichs, T. Interplay between detrital and diagenetic processes since the last glacial maximum on the northwest Iberian continental shelf. *Quat. Res.* **2010**, *73*, 507–520. [CrossRef]
13. García-Gil, S. A natural laboratory for shallow gas: The Rías Baixas (NW Spain). *Geo-Mar. Lett.* **2003**, *23*, 215–229. [CrossRef]
14. Cartelle, V. Stratigraphy, Depositional Environments, Shallow Gas and Seismic-Sequential Analysis of the Sedimentary Record of the Rias of Arousa and Ferrol. Ph.D. Thesis, Universidade de Vigo, Vigo, Spain, 2018; 351p.
15. Arce-Chamorro, C.; Vidal-Romaní, J.R.; Sanjurjo-Sánchez, J. Islas Cíes: Una trampa eólica en la Ría de Vigo (NO de la Península Ibérica) al final del último glacial. Estudio del afloramiento de eolianitas de la Isla de Monteagudo (Cíes, Pontevedra, Galicia). *Geogaceta* **2021**, *69*, 3–6. Available online: https://sge.usal.es/archivos/geogacetas/geo70/Geo70_p_03_06.pdf (accessed on 7 July 2022).
16. Arce-Chamorro, C. Datación por Luminiscencia de Depósitos Fluviales y Eólicos en el Margen Occidental de Galicia. Ph.D. Thesis, Universidade da Coruña, A Coruña, Spain, 2017. Available online: http://hdl.handle.net/2183/19810 (accessed on 7 August 2022).
17. Arce-Chamorro, C.; Vidal-Romaní, J.R. Late-Pleistocene chronology of coastal fluvial deposits at the mouth of the Ulla River in the Ría de Arousa (Galicia, NW Spain) by OSL dating. *Cad. Lab. Xeol. Laxe* **2021**, *43*, 61–88. [CrossRef]
18. Ribeiro, H.; Pinto de Jesus, A.; Mosquera, D.F.; Abreu, I.; Vidal-Romaní, J.R.; Noronha, F. Estudo de un Terraço de Lavadores. Contribuiçao Para a Deduçao das Condiçoes Paleoclimáticas No Plistocénico Médio. E-Terra: Revista Electrónica de Ciências da Terra. VIII Congreso Nacional de Geologia, Sociedade Geológica de Portugal, Portugal. 2010. Available online: https://repositorio-aberto.up.pt/bitstream/10216/53062/2/48388.pdf (accessed on 7 August 2022).
19. Ribeiro, H.; Pinto de Jesus, A.; Sanjurjo-Sánchez, J.; Abreu, I.; Romani, J.R.V.; Noronha, F. Multidisciplinary study of the quaternary deposits of the Vila Nova de Gaia, NW Portugal, and its climate significance. *J. Iber. Geol.* **2019**, *45*, 553–563. [CrossRef]
20. Hesp, P.A.; DaSilva, M.; Miot da Silva, G.; Bruce, D.; Keane, R. Review and direct evidence of transgressive aeolian sand sheet and dunefield initiation. *Earth Surf. Process. Landf.* **2022**, *47*, 2660–2675. [CrossRef]

21. Arce-Chamorro, C.; Vidal-Romaní, J.R.; Sanjurjo-Sánchez, J. Chronology of four climbing-dunes outcrops on the Atlantic coast of Galicia (NW Spain) linked to the sea-level fall during the MIS6 and MIS4 isotopic stages. *J. Mar. Sci. Eng.* **2022**, *10*, 312. [CrossRef]
22. Costas, S.; Jerez, S.; Trigo, R.M.; Bogle, R.; Rebêlo, L. Sand invasion along the Portuguese coast forced by westerly shifts during cold climate events. *Quat. Sci. Rev.* **2012**, *42*, 15–28. [CrossRef]
23. Fick, S.E.; Hijmans, R.J. WorldClim 2: New 1-km spatial resolution climate surfaces for global land areas. *Int. J. Climatol.* **2017**, *37*, 4302–4315. [CrossRef]
24. Jackson, D.W.T.; Costas, S.; González-Villanueva, R.; Cooper, A. A global 'greening' of coastal dunes: An integrated consequence of climate change? *Glob. Planet. Chang.* **2019**, *182*, 103026. [CrossRef]
25. Flor-Blanco, G.; Alcántara-Carrió, J.; Jackson, D.W.T.; Flor, G.; Flores-Soriano, C. Coastal erosion in NW Spain: Recent patterns under extreme storm wave events. *Geomorphology* **2021**, *387*, 107767. [CrossRef]
26. EMODnet Bathymetry Consortium. EMODnet Digital Bathymetry (DTM). 2018. Available online: https://sextant.ifremer.fr/record/18ff0d48-b203-4a65-94a9-5fd8b0ec35f6 (accessed on 7 August 2022).
27. Bao-Casal, R. Las Tanatocenosis de Diatomeas (*Bacillariophyta*) en Sedimentos Superficiales de las Rías y Plataforma Continental de Galicia (NW de la Península Lbérica) y su Relación con las Características Oceanográficas de la Columna de Agua; Implicaciones Paleoceanográficas. Ph.D. Thesis, Universidade da Coruña, A Coruña, Spain, 1991. Available online: https://dialnet.unirioja.es/servlet/tesis?codigo=49163 (accessed on 7 August 2022).
28. Muñoz-Sobrino, C.; García-Moreiras, I.; Martínez-Carreño, N.; Cartelle, V.; Insua, T.L.; Da Costa, J.F.; Ramil-Rego, P.; Fernández-Rodríguez, C.; Alejo, I.; García-Gil, S. Reconstruction of the environmental history of a coastal insular system using shallow marine records: The last three millennia of the Cíes Islands (Ría de Vigo, NW Iberia). *Boreas* **2016**, *45*, 729–753. [CrossRef]
29. César-Vila, M. *Documentación, Levantamiento de Estructuras y Control Arqueológico de los Movimientos de Tierras en la Parcela 4 de la U.A. I-06 Rosalía de Castro 2 de Vigo (Pontevedra)*; Memoria técnica ED 102A 2016/072-0; Xunta de Galicia: Santiago, Spain, 2018; 177p, Available online: https://www.residenciallassalinas.com/wp-content/uploads/2019/01/informe-arqueologico-las-salinas.pdf (accessed on 7 August 2022).
30. Nombela, M.A.; Alejo, I.; Bernárdez, F.; Clemente, F.; Costas, S.; Diz, P.; Fernández-Bastero, S.; Francés, G.; Gago-Duport, L.; García, T.; et al. Evolución sedimentaria desde el último máximo glacial en la costa y plataforma continental de las Rías Baixas (Galicia, NW de la Península Ibérica). In Proceedings of the Iberian Coastal Holocene Paleoenvironmental Evolution: Coastal Hope Conference Proceedings, Lisbon, Portugal, 24–29 July 2005; pp. 99–100. Available online: http://hdl.handle.net/10400.9/2804 (accessed on 7 July 2022).
31. N.A.S.A. *SRTM90m Digital Elevation Data from the CGIAR-CSI Consortium for Spatial Information*; University of Maryland: College Park, MD, USA, 2000. Available online: https://srtm.csi.cgiar.org/srtmdata/ (accessed on 7 August 2022).
32. Reimer, P.; Austin, W.; Bard, E.; Bayliss, A.; Blackwell, P.G.; Bronk-Ramsey, C.; Butzin, M.; Cheng, H.; Edwards, R.L.; Friedrich, M.; et al. The IntCal20 Northern Hemisphere Radiocarbon Age Calibration Curve (0–55 cal kBP). *Radiocarbon* **2020**, *62*, 725–757. [CrossRef]
33. Devoy, R.J.N.; Delaney, C.; Carter, R.W.G.; Jennings, S.C. Coastal stratigraphies as indicators of environmental changes upon european coasts in the Late Holocene. *J. Coast. Res.* **1996**, *12*, 564–588.
34. Sáez, A.; Carballeira, R.; Pueyo, J.J.; Vázquez-Loureiro, D.; Leira, M.; Hernández, A.; Valero-Garcés, B.L.; Bao, R. Formation and evolution of back-barrier perched lakes in rocky coasts: An example of a Holocene system in north-west Spain. *Sedimentology* **2018**, *65*, 1891–1917. [CrossRef]
35. Pazos, O.; Vila, A.; Alejo, I.; Vilas, F.; Nombela, M.A. Sedimentos costeros relictos frente a las Rías Bajas. *Thalassas* **1997**, *13*, 135–141.
36. Caraballo-Muziotti, L.F. Estudio Fisiográfico-Sedimentológico de las Rías y Frente Costero Comprendidos Entre la Estaca de Vares y el Cabo Prior (Provincia de La Coruña). Ph.D. Dissertation, Universidad Complutense, Madrid, Spain, 1969. Available online: https://eprints.ucm.es/id/eprint/51850/ (accessed on 7 July 2022).
37. Alonso, A.; Pagés, J.L. Evolución del Nivel del Mar Durante el Holoceno en el Noroeste de la Península Ibérica. *Rev. Soc. Geol. Esp.* **2010**, *23*, 157–167. Available online: https://sge.usal.es/archivos/REV/23(3-4)/art05.pdf (accessed on 7 August 2022).
38. Santos, M.L.; Vidal-Romaní, J.R. La transgresión holocena en la Ria de Ares (A Coruña, Galicia, España). Datos cronológicos, sedimentarios y geomorfológicos. In Proceedings of the 3ª Reunião do Quaternario Ibérico, Coimbra, Portugal, 27 September–1 October 1993; pp. 339–345.
39. López-Cancelo, L. Cambios Paleoambientales en el NW Peninsular, Durante el Holoceno, Determinados a Partir del Estudio de Foraminíferos Bentónicos. Ph.D. Dissertation, Universidad de A Coruña, A Coruña, Spain, 2004. Available online: https://dialnet.unirioja.es/servlet/tesis?codigo=47918 (accessed on 7 August 2022).
40. Vilas, F.; Nombela, M.A.; García-Gil, E. *Cartografía de Sedimentos Submarinos: Ría de Vigo*; Xunta de Galicia: A Coruña, Spain; Consellería de Pesca, Marisqueo e Acuicultura: Galicia, Spain, 1995; 40p.
41. Dias, J.M.A.; González, R.; García, C.; Díaz-del Río, V. Sediment distribution patterns on the Galicia-Minho continental shelf. *Prog. Oceanogr.* **2002**, *52*, 215–231. [CrossRef]
42. De Vicente, G.; Vegas, R. Large-scale distributed deformation controlled topography along the western Africa –Eurasia limit: Tectonic constrains. *Tectonophysics* **2009**, *474*, 124–143. [CrossRef]
43. Brennan, B.J. Beta doses to spherical grains. *Radiat. Meas.* **2003**, *37*, 299–303. [CrossRef]

44. Murray, A.S.; Wintle, A.G. Luminescence dating of quartz using an improved single- aliquot regenerative-dose protocol. *Radiat. Meas.* **2000**, *32*, 57–73. [CrossRef]
45. Banerjee, D.; Bøtter-Jensen, L.; Murray, A.S. Retrospective dosimetry: Estimation of the dose to quartz using the single-aliquot regenerative-dose protocol. *Appl. Radiat. Isot.* **2000**, *52*, 831–844. [CrossRef]
46. Cunningham, A.; Wallinga, J. Selection of integration time intervals for quartz OSL decay curves. *Quat. Geochronol.* **2010**, *5*, 657–666. [CrossRef]
47. Oczkowski, H.L. Calibration standard for use in gamma spectrometry and luminescence dating. *Geochronometria* **2001**, *20*, 31–38.
48. Murray, A.S.; Aitken, M.J. Analysis of low-level natural radioactivity in small mineral samples for use in thermoluminescence dating, using high-resolution gamma spectrometry. *Appl. Radiat. Isot.* **1988**, *39*, 145–158. [CrossRef]
49. Guérin, G.; Mercier, N.; Adamiec, G. Dose rate conversion factors. *Anc. TL* **2011**, *29*, 5–8.
50. Vandenberghe, D.; De Corte, F.; Buylaert, J.; Kucera, J.; Van den Haute, P. On the internal radioactivity in quartz. *Radiat. Meas.* **2008**, *47*, 771–775. [CrossRef]
51. Prescott, J.R.; Hutton, J.T. Cosmic ray contribution to dose rates for luminescence and ESR dating: Large depths and long-term time variations. *Radiat. Meas.* **1994**, *23*, 497–500. [CrossRef]
52. Udden, J.A. Mechanical composition of clastic sediments. *Bull. Geol. Soc. Am.* **1914**, *25*, 655–744. [CrossRef]
53. Wentworth, C.K. A scale of grade and class terms for clastic sediments. *J. Geol.* **1922**, *30*, 377–392. [CrossRef]
54. Krumbrein, W.C.; Sloss, L.L. *Stratigraphy and Sedimentation*, 2nd ed.; W.H. Freeman and Co.: San Francisco, CA, USA, 1963.
55. Folk, R.L.; Ward, W.C. Brazos River bar: A study in the significance of grain size parameters. *J. Sediment. Petrol.* **1957**, *27*, 3–26. [CrossRef]
56. Guérin, G.; Mercier, N. Preliminary insight into dose deposition processes in sedimentary media on a scale of single grains: Monte Carlo modelling of the effect of water on the gamma dose rate. *Radiat. Meas.* **2012**, *47*, 541–547. [CrossRef]
57. Nathan, R.; Thomas, P.J.; Murray, A.S.; Rhodes, E.J. Environmental dose rate heterogeneity of beta radiation and its implications for luminescence dating: Monte Carlo modelling and experimental validation. *Radiat. Meas.* **2003**, *37*, 305–313. [CrossRef]
58. Thomsen, K.J.; Murray, A.S.; Bøtter-Jensen, L. Sources of variability in OSL dose measurements using single grains of quartz. *Radiat. Meas.* **2005**, *39*, 47–61. [CrossRef]
59. Galbraith, R.F.; Roberts, R.G.; Laslett, G.M.; Yoshida, H.; Olley, J.M. Optical dating of single and multiple grains of quartz from Jinmium rock shelter, northern Australia: Part, I. Experimental design and statistical models. *Archaeometry* **1999**, *41*, 339–364. [CrossRef]
60. Galbraith, R.F.; Roberts, R.G. Statistical aspects of equivalent dose and error calculation and display in OSL dating: An overview and some recommendations. *Quat. Geochronol.* **2012**, *11*, 1–27. [CrossRef]
61. Burow, C.; Kreutzer, S.; Dietze, M.; Fuchs, M.C.; Fischer, M.; Schmidt, C.; Brückner, H. RLumShiny—A graphical user interface for the R Package 'Luminescence'. *Anc. TL* **2016**, *34*, 22–32.
62. Rodríguez-Vidal, J.; Cáceres, L.M.; Finlayson, J.C.; Gracia, F.J.; Martínez-Aguirre, A. Neotectonics and shoreline history of the Rock of Gibraltar, southern Iberia. *Quat. Sci. Rev.* **2004**, *23*, 2017–2029. [CrossRef]
63. Zazo, C.; Mercier, N.; Silva, P.G.; Dabrio, C.J.; Goy, J.L.; Roquero, E.; Soler, V.; Borja, F.; Lario, J.; Polo, D.; et al. Landscape evolution and geodynamic controls in the Gulf of Cadiz (Huelva coast, SW Spain) during the Late Quaternary. *Geomorphology* **2005**, *68*, 269–290. [CrossRef]
64. Salvany, J.M.; Larrasoaña, J.C.; Mediavilla, C.; Rebollo, A. Chronology and tectono-sedimentary evolution of the Upper Pliocene to Quaternary deposits of the lower Guadalquivir foreland basin, SW Spain. *Sediment. Geol.* **2011**, *241*, 22–39. [CrossRef]
65. Roettig, C.B.; Kolb, T.; Wolf, D.; Baumgart, P.; Richter, C.; Schleicher, A.; Zöller, L.; Faust, D. Complexity of Quaternary aeolian dynamics (Canary Islands). *Palaeogeogr. Palaeoclimatol. Palaeoecol.* **2017**, *472*, 146–162. [CrossRef]
66. Leira, M.; Freitas, M.C.; Ferreira, T.; Cruces, A.; Connor, S.; Andrade, C.; Lopes, V.; Bao, R. Holocene sea level and climate interactions on wet dune slack evolution in SW Portugal: A model for future scenarios? *Holocene* **2019**, *29*, 26–44. [CrossRef]
67. Sitzia, L.; Bertran, P.; Bahain, J.J.; Bateman, M.D.; Hernandez, M.; Garon, H.; de Lafontaine, G.; Mercier, N.; Leroyer, C.; Queffelec, A.; et al. The Quaternary coversands of southwest France. *Quat. Sci. Rev.* **2015**, *124*, 84–105. [CrossRef]
68. Diz, P.; Francés, G.; Pelejero, C.; Grimalt, J.O.; Vilas, F. The last 3000 years in the Ria de Vigo (NW Iberian Margin): Climatic and hydrographic signals. *Holocene* **2002**, *12*, 459–468. [CrossRef]
69. Cohen, K.M.; Finney, S.C.; Gibbard, P.L.; Fan, J.X. The ICS International Chronostratigraphic Chart. *J. Int. Geosci.* **2013**, *36*, 199–204. [CrossRef] [PubMed]
70. Petit, J.R.; Jouzel, J.; Raynaud, D.; Barkov, N.I.; Barnola, J.-M.; Basile, I.; Bender, M.; Chappellaz, J.; Davis, M.; Delaygue, G.; et al. Climate and atmospheric history of the past 420,000 years from the Vostok ice core, Antarctica. *Nature* **1999**, *399*, 429–436. [CrossRef]
71. Waelbroeck, C.; Labeyrie, L.; Michel, E.; Duplessy, J.C.; McManus, J.F.; Lambeck, K.; Balbon, E.; Labracherie, M. Sea-level and deep water temperature changes derived from benthic foraminifera isotopic records. *Quat. Sci. Rev.* **2002**, *21*, 295–305. [CrossRef]
72. Vidal-Romaní, J.R.; Fernández-Mosquera, D.; Marti, K.; De Brum-Ferreira, A. Nuevos datos para la cronología glaciar pleistocena en el NW de la Península Ibérica. *Cad. Lab. Xeol. Laxe* **1999**, *24*, 7–29.
73. Vidal-Romaní, J.R.; Fernández-Mosquera, D.; Marti, K. The glaciation of Serra de Queixa-Invernadoiro and Serra do Geres-Xurés, NW Iberia. A critical review and a cosmogenic nuclide (10-Be and 21-Ne) chronology. *Cad. Lab. Xeol. Laxe* **2015**, *38*, 25–44. [CrossRef]

74. Lisiecki, L.E.; Raymo, M.E. A Pliocene-Pleistocene stack of 57 globally distributed benthic d18O records. *Paleoceanography* **2005**, *20*, PA1003. [CrossRef]
75. García-Moreiras, I.; Delgado, C.; Martínez-Carreño, N.; García-Gil, S.; Muñoz-Sobrino, C. Climate and vegetation changes in coastal ecosystems during the Middle Pleniglacial and the early Holocene: Two multi-proxy, high-resolution records from Ria de Vigo (NW Iberia). *Glob. Planet. Chang.* **2019**, *176*, 100–122. [CrossRef]
76. Rodríguez-Fernández, R. *Guía Geológica del Parque Nacional de las Islas Atlánticas de Galicia;* Organismo autónomo de Parques Nacionales, Instituto Geológico y Minero de España: Madrid, Spain, 2013; 202p.
77. Martínez-Carreño, M.; García-Gil, S.; Cartelle, V. An unusual Holocene fan-shaped subaqueous prograding body at the back of the Cíes Islands ridge (Ría de Vigo, NW Spain): Geomorphology, facies and stratigraphic architecture. *Mar. Geol.* **2017**, *385*, 13–26. [CrossRef]
78. Valenzano, E.; Scardino, G.; Cipriano, G.; Fago, P.; Capolongo, D.; De Giosa, F.; Lisco, S.; Mele, D.; Moretti, M.; Mastronuzzi, G. Holocene Morpho-sedimentary Evolution of the Mar Piccolo Basin (Taranto, Southern Italy). *Geogr. Fis. Dinam. Quat.* **2018**, *41*, 119–135. [CrossRef]
79. De Santis, V.; Caldara, M.; Pennetta, L. "Continuous" Backstepping of Holocene Coastal Barrier Systems into Incised Valleys: Insights from the Ofanto and Carapelle-Cervaro Valleys. *Water* **2020**, *12*, 1799. [CrossRef]
80. Cawthra, H.C.; Jacobs, Z.; Wadley, L. Winds of change: Climate variability in a mild glacial on the east coast of South Africa, inferred from submerged aeolianites and the archaeological record of Sibudu. *Quat. Int.* 2022; *in press*. [CrossRef]
81. López-Romero, E.; Bóveda-Fernández, M.J.; Güimil-Fariña, A.; Mañana-Borrazás, P.; Sanjurjo-Sánchez, J.; Vázquez-Collazo, S.; Vilaseco-Vázquez, X.I. Prehistoric human occupation in the Western Rias of Galicia (Norwest Iberia): A review of methods and prospects. In *1st International Webinar: Investigate the Shore, Sound the Past. New Methods and Practices of Maritime Prehistory;* IRN PrehCOAST-French Prehistoric Society: Brest, France, 2020; Available online: https://www.inrap.fr/sites/inrap.fr/files/atoms/files/investigate_the_shore-sound_the_past_booklet.pdf (accessed on 7 August 2022).
82. Greeley, R.; Iversen, J.D. *Wind as a Geological Process on Earth, Mars, Venus and Titan;* Cambridge Planetary Science Old; Cambridge University Press: Cambridge, UK, 1985; 333p. [CrossRef]
83. Pye, K.; Tsoar, H. *Aeolian Sand and Sand Dunes;* Unwin Hyman Ltd.: London, UK, 1990; 396p.
84. Reineck, H.E.; Singh, I.B. *Depositional Sedimentary Environments. With Reference to Terrigenous Clastics;* Springer: Berlin/Heidelberg, Germany; New York, NY, USA, 1973; 240p. [CrossRef]
85. Mosquera-Santé, M.J. Evolución Post-Glaciar del Nivel del Mar en el NO de la Península Ibérica: El caso del Golfo Ártabro. Ph.D. Thesis, Universidad de A Coruña, A Coruña, Spain, 2000. Available online: http://hdl.handle.net/2183/1167 (accessed on 7 August 2022).
86. Méndez, G.; Vilas, F. Geological antecedents of the Rias Baixas (Galicia, northwest Iberian Peninsula). *J. Mar. Syst.* **2005**, *54*, 195–207. [CrossRef]
87. García-Gil, S.; García-García, A.; Durán, R.; Vilas, F. Estratigrafía Sísmica de Alta Resolución en las Rías Baixas: Pontevedra y Vigo (NO España). *J. Iber. Geol.* **2000**, *26*, 217–231.
88. Durán, R. Estratigrafía Sísmica Desde el Último Máximo Glacial en la Ría de Pontevedra (NO de España). Ph.D. Thesis, Universidade de Vigo, Vigo, Spain, 2005. Available online: https://dialnet.unirioja.es/servlet/tesis?codigo=222310 (accessed on 7 August 2022).
89. García-Amorena, I.; Gómez-Manzaneque, F.; Rubiales, J.M.; Granja, H.M.; Soares de Carvalho, G.; Morla, C. The Late Quaternary coastal forests of western Iberia: A study of their macroremains. *Palaeogeog. Plaeoclimatol. Palaeoecol.* **2007**, *254*, 448–461. [CrossRef]
90. Ribeiro, H.; Bernal, A.; Flores, D.; Pissarra, J.; Abreu, I.; Vidal-Romaní, J.R.; Noronha, F. A multidisciplinary study of an organic-rich mudstone in the Middle Holocene on the Northern coast of Portugal. *Comun. Geol.* **2011**, *98*, 93–98.
91. Ribeiro, H.; Pinto de Jesus, A.; Oliveira, F.; Vidal-Romaní, J.R.; Abreu, I.; Noronha, F. Estudo da "Formação Areno-Pelítica" na zona Litoral Porto—Vila Nova de Gaia. Contribuição do conteúdo polínico. *Comun. Geol.* **2014**, *101*, 631–634.
92. French, C.N. The submerged forest palaeosols of Conrwall. *Geosci. South-West Engl.* **1999**, *93*, 65–366.
93. Williams, D.M.; Doyle, E. Dates from drowned mid-Holocene landscapes on the central western Irish seaboard. *Int. J. Earth Sci.* **2014**, *32*, 23–27. [CrossRef]
94. Bicket, A.; Tizzard, L. A review of the submerged prehistory and palaeolandscapes of the Brithish Isles. *Proc. Geol. Assoc.* **2015**, *126*, 643–663. [CrossRef]
95. Westley, K.; Woodman, P. Ireland: Submerged Prehistoric Sites and Landscapes. In *The Archaeology of Europe's Drowned Landscapes;* Bailey, G., Galanidou, N., Peeters, H., Jöns, H., Mennenga, M., Eds.; Coastal Research Library; Springer: Cham, Switzerland, 2020; Volume 35, pp. 221–248. [CrossRef]
96. Bosq, M.; Bertran, P.; Beauval, C.; Kreutzer, S.; Duval, M.; Bartz, M.; Mercier, N.; Sitzia, L.; Stéphan, P. Stratigraphy and chronology of Pleistocene coastal deposits in northern Aquitaine, France: A reinvestigation. *Quaternaire* **2019**, *30*, 275–303. [CrossRef]
97. Bertran, P.; Andrieux, E.; Bateman, M.D.; Fuchs, M.; Klinge, M.; Marembert, F. Mapping and chronology of coversands and dunes from the Aquitaine basin, southwest France. *Aeolian Res.* **2020**, *47*, 100628. [CrossRef]
98. Andreucci, S.; Clemmensen, L.B.; Pascucci, V. Transgressive dune formation along a cliffed coast at 75 ka in Sardinia, Western Mediterranean: A record of sea-level fall and increased windiness. *Terra Nova* **2010**, *22*, 424–433. [CrossRef]

99. Derese, C.; Vandenberghe, D.; Zwertvaegher, A.; Court-Picon, M.; Crombé, P.; Verniers, J.; Van den Haute, P. The timing of aeolian events near archaeological settlements around Heidebos (Moervaart area, N Belgium). *Neth. J. Geosci.-Geol. Mijnb.* **2010**, *89*, 173–186. [CrossRef]
100. Bateman, M.D.; Carr, A.S.; Dunajko, A.C.; Holmes, P.J.; Roberts, D.L.; McLaren, S.J.; Bryant, R.G.; Marker, M.E.; Murray-Wallace, C.V. The evolution of coastal barrier systems: A case study of the Middle-Late Pleistocene Wilderness barriers, South Africa. *Quat. Sci. Rev.* **2011**, *30*, 63–81. [CrossRef]
101. Cooper, J.A.G.; Green, A.N. Geomorphology and preservation potential of coastal and submerged aeolianite: Examples from KwaZulu-Natal, South Africa. *Geomorphology* **2016**, *271*, 1–12. [CrossRef]
102. Lipar, M.; Webb, J.A.; Cupper, M.L.; Wang, N. Aeolianite, calcrete/microbialite and karst in southwestern Australia as indicators of Middle to Late Quaternary palaeoclimates. *Palaeogeog. Palaeoclimatol. Palaeoecol.* **2017**, *470*, 11–29. [CrossRef]
103. Ho, L.-D.; Lüthgens, C.; Wong, Y.-C.; Yen, J.-Y.; Chyi, S.-J. Late Holocene cliff-top dune evolution in the Hengchun Peninsula of Taiwan: Implications for palaeoenvironmental reconstruction. *J. Asian Earth Sci.* **2017**, *148*, 13–30. [CrossRef]
104. Jin, J.H.; Ling, Z.; Li, Z.; Liu, X. Chronology of coastal aeolian deposition and its paleoenvironmental implications on the Liuao Peninsula of South China. *J. Mt. Sci.* **2019**, *16*, 2754–2769. [CrossRef]
105. Decker, V.; Falkenroth, M.; Lindauer, S.; Landgraf, J.; Al-Lawati, Z.; Al-Rahbi, H.; Hoffmann, G. Collapse of Holocene mangrove ecosystems along the coastline of Oman—Corrigendum. *Quat. Res.* **2021**, *100*, 260–262. [CrossRef]
106. Lü, T.; Sun, J.; Feathers, J.K.; Sun, D.; Shen, X. OSL dating of coastal sand dunes in southeastern China provides new insights into the relationship between aeolian activity and eustatic sea-level fluctuations. *Palaeogeog. Palaeoclimat. Palaeoecol.* **2022**, *600*, 111082. [CrossRef]
107. Ley, C.; Gallego, J.B.; Vidal, C. *Manual de Restauración de Dunas Costeras*; Dirección General de Costas, Ministerio de Medio Ambiente: Madrid, Spain, 2007; 248p. Available online: https://www.miteco.gob.es/es/costas/publicaciones/cap01_introduccion_tcm30-161387.pdf (accessed on 7 August 2022).

Journal of
Marine Science and Engineering

MDPI

Article

Seasonal Variability in Present-Day Coccolithophore Fluxes in Deep Eastern Mediterranean Sea: A Multi-Year Study (2015–2017) of Coccolithophore Export in SE Ionian Sea at 4300 m Depth

Sikandar Hayat [1,2], Elisavet Skampa [1], Alexandra Gogou [3], Spyros Stavrakakis [3], Constantine Parinos [3] and Maria Triantaphyllou [1,*]

1 Faculty of Geology and Geoenvironment, National and Kapodistrian University of Athens, Panepistimioupolis, 15784 Athens, Greece
2 Department of Earth Sciences, University of Lille 1, Cité Scientifique, CEDEX, 59655 Villeneuve d'Ascq, France
3 Institute of Oceanography, Hellenic Centre for Marine Research, P.O. Box 712, 19013 Anavyssos, Greece
* Correspondence: mtriant@geol.uoa.gr

Abstract: This study is the first attempt to understand the coccolith flux and its seasonal variability at the deepest part of the Mediterranean Sea. Samples were obtained from the deepest Mediterranean time-series sediment trap (4300 m) moored in the SE Ionian Sea (Nestor site) from January 2015 to November 2017. Throughout the study period, the coccolith fluxes displayed a seasonality signal with high values during the late winter–early spring convective mixing period (February to April) and low flux values during summer except for some solitary peaks in June. The maximum coccolith flux was observed in March 2015 while the minimum value was recorded in November 2017. Among the nineteen identified species of heterococcoliths, the dominant species in all the samples was *Emiliania huxleyi* reaching up to 79%, followed by *Florisphaera profunda* that comprised up to 33% of the total coccolith count. For the annual cycle of 2015, the average coccolith flux for the Nestor Site at a relatively shallower depth (2000 m) was comparable and for some time intervals was lower than the coccolith flux recorded in the present study at 4300 m, while coccolith flux peaks appeared simultaneously in both traps indicating a fast sinking rate. The higher *E. huxleyi, F. profunda, Gladiolithus flabellatus,* and *Calciosolenia brasiliensis* coccolith flux at 4300 m compared with their corresponding fluxes at 2000 m can be attributed to lateral advection, resuspension, and/or the influence of Eastern Mediterranean Deep Waters (EMDWs).

Keywords: coccolith flux; Eastern Mediterranean; biogenic flux; Ionian Sea; sediment trap; *Emiliania huxleyi; Florisphaera profunda*

Citation: Hayat, S.; Skampa, E.; Gogou, A.; Stavrakakis, S.; Parinos, C.; Triantaphyllou, M. Seasonal Variability in Present-Day Coccolithophore Fluxes in Deep Eastern Mediterranean Sea: A Multi-Year Study (2015–2017) of Coccolithophore Export in SE Ionian Sea at 4300 m Depth. *J. Mar. Sci. Eng.* **2022**, *10*, 1761. https://doi.org/10.3390/jmse10111761

Academic Editors: Antoni Calafat and Weidong Zhai

Received: 14 September 2022
Accepted: 12 November 2022
Published: 16 November 2022

1. Introduction

Coccolithophores are a group of unicellular calcifying marine algae that dwell in the photic zone and cover their cells with an exoskeleton (coccosphere) comprised of tiny calcite platelets (1–20 μm across) called coccoliths [1]. They account only for 1–10% of marine primary productivity (PP); however, they exert a key influence on the ocean chemistry as they utilize CO_2 via photosynthesis to produce particulate organic carbon (POC) (carbon pump), as well as release CO_2 during the production of coccoliths/particulate inorganic carbon (PIC) (counter carbon pump) [2]. The flux of CO_2 between the photic zone of the ocean and the atmosphere is mainly dependent on PIC to POC ratio [2,3]; thus, the production of coccolithophore blooms results in an increase in the CO_2 content of the ocean on a short timescale. Due to the high density of calcite, coccoliths act as ballasts and expedite the removal of POC out of the surface waters [4,5]. Coccolithophores are expected to adapt better than other phytoplankton groups, e.g., diatoms, to the nutrient limitation that is expected due to enhanced stratification of the water column because of rising global temperatures [6,7]. Their assemblage composition is sensitive to seasonal changes and mirrors the oceanographic and environmental conditions of the photic zone. Hence, coccolithophore fluxes transfer the ecological signals of the photic zone to deep water and seafloor sediments [8].

Coccolithophores are the dominant phytoplankton group in the Eastern Mediterranean constituting a major portion of settling biogenic particles with pronounced seasonal variations [9–13]. Earlier sediment trap studies in the area have shown seasonal and interannual fluctuations in coccolithophore fluxes in response to changing sea surface temperature (SST), nutrient availability, and precipitation; maxima were observed during the winter–spring mixing period [8,13–16]. Occasional high fluxes have been recorded during summer, driven by short-lived events, e.g., atmospheric input, or upward movement of nutrient-rich intermediate waters [13]. The species *Emiliania huxleyi* dominates the upper photic zone, while *Florisphaera profunda* thrives in the deeper layers [17,18].

Nestor site is located in the SE Ionian Sea, the deepest basin of the Mediterranean (between 4600 and 5264 m depth) [12]. Hence, the Nestor sediment trap deployment at five successive depths, with the deepest sediment trap moored at 4300 m [12], offers a unique opportunity to study downward fluxes and document coccolithophore export fluxes at the deepest part of the Eastern Mediterranean water column. Previous time series studies at the Nestor site have focused on the downward fluxes of the major geochemical components investigating the mechanisms of vertical transport from 700 m down to 4300 m depth during 2006–2010 [12], whereas the analysis of coccolithophore fluxes has been restricted only to 2000 m depth for the time interval 2010–2015 [13]. This study presents, for the first time, the analysis of coccolithophore export fluxes from the deepest basin of the Mediterranean Sea. Thus, the aim of the present study is to analyze the deepest sediment trap samples acquired from Nestor site, in order to (a) quantify the coccolithophore primary productivity fluxes at 4300 m during the 2015–2017 time interval, (b) investigate the assemblage composition and detect the signal of surface seasonality signal in the deep water masses, and (c) compare the obtained coccolithophore fluxes with data of previous studies from different depths at Nestor site to determine potential alterations in the sinking coccolithophore assemblages.

Oceanographic Setting

Even though the semi-enclosed water body of the Mediterranean Sea covers only 0.8% of the global ocean, it has comparable features, particularly thermohaline circulation, complex bathymetry, and deep water convection [19]. Nestor site (Figure 1) is located in the Eastern Mediterranean near the SW Peloponnese peninsula in the SE Ionian Sea. The SE Ionian Sea, which contains the deepest basins of the Mediterranean Sea, has complex morphology with steep slopes, plateaus, and valleys [12]. Terrestrial input is negligible as there is no riverine inflow in the area [20]. The slopes are almost bare and the valley/basin floor contains thin layers of pelagic sediments indicating the absence of turbidity flow [12]. Sea surface temperature (SST) ranges from 15.1 °C to 29.3 °C where May–October are warm and November–April are cold months with the highest precipitation in the colder period [13]. The surface chlorophyll concentrations are highest during the late winter–early spring convective mixing period, lowest during the summer months [21], generally remaining below 0.5 mg L^{-1} in the open sea [22,23]. During winter and spring, the combined effect of the two westerly upper air jet streams namely the polar front jet stream and the subtropical jet stream, result in the east–southeast propagation of subtropical cyclones which outcomes in the fertilization of Mediterranean water by Saharan dust [24], while most of the atmospheric deposition in the Mediterranean Sea occurs as wet deposition [25].

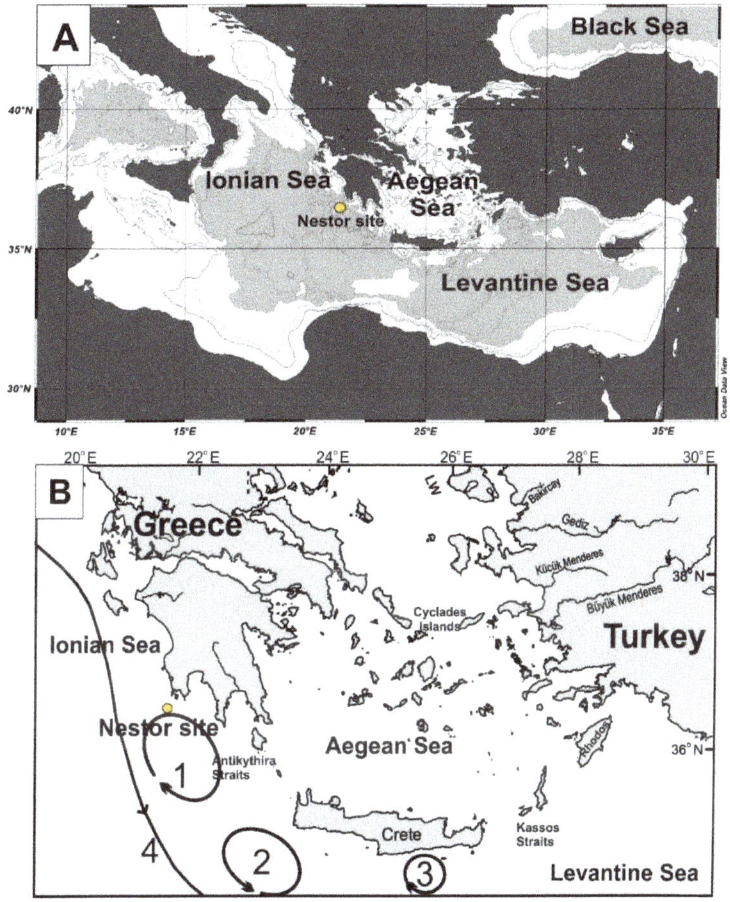

Figure 1. (**A**). Location of sampling site (Nestor site) in the SE Ionian Sea (Eastern Mediterranean Sea), (**B**). General circulation pattern in top 500 m (1. Pelops anticyclone, 2. Cretan cyclone, 3. Ierapetra anticyclone, 4. Atlantic–Ionian stream).

Eastern Mediterranean surface waters are P-starved due to the anti-estuarine circulation, atmospheric input from Sahara with high N/P (100:1), and limited denitrification [26,27]. The oligotrophic signature of the Ionian Sea is evident from the 80–100 m Deep Chlorophyll Maximum (DCM), which shows low seasonality [28].

Modified Atlantic Water (MAW), Levantine Intermediate Water (LIW), the Transitional Waters, and Eastern Mediterranean Deep Water (EMDW) are the four major water masses of the Ionian Sea from the surface downwards.

With temperature and salinity of 17.4 °C, 38.72 psu (25–100 m water depth); 15 °C, 39 psu (100–500 m water depth); 13.8 °C, 38.76 psu (500–1200 m water depth), and 13 °C, 38.6 psu (>1200 m water depth), respectively [29,30]. Nestor site is located at the edge of Pelops anticyclone, which disperses the intermediate waters that enter the Ionian Sea through southern Crete to the south, north, and west [31]. Pelops anticyclone was observed to turn off or weaken during summer, which results in the upward movement of intermediate waters [32].

2. Materials and Methods

2.1. Satellite Environmental Data

The satellite data (chlorophyll-a(Chl-a), Precipitation, and SST) were obtained from EarthData-Giovanni open access database (http://disc.sci.gsfc.nasa. gov/giovanni/ accessed on 11 October 2022) [33]. An average of 8 days of data was obtained for satellite Chl-a concentration from MODIS-Aqua (MODerate resolution imaging spectroradiometer) (4 Km resolution) through the National Aeronautic and Space Administration (NASA) Giovanni website (http://disc.sci.gsfc.nasa.gov/techlab/giovanni/ accessed on 11 October 2022) for the following boxes. The coordinates of satellite data for Chl-a and SST are 36.2292–36.5625° N, 21.2292–21.5625° E and for precipitation: 36.375–36.375° N, 21.375–21.375° E.

2.2. Deployment and Recovery of the Sediment Trap Mooring

PPS3/3 Technicap sediment trap (0.125 m^2 collecting area) was deployed at 4300 m depth from January 2015 to November 2017, at the Nestor site (36°2.96′ N, 21°28.93′ E) by the Hellenic Centre for Marine Research (HCMR; R/V Aegaeo, project KRIPIS), with a number of time gaps (see Table 1). After recovery, samples were stored in the dark at 4 °C until further processing. To remove large organisms, samples were sieved through 1 mm nylon sieve whereas swimmers smaller than 1 mm were removed under the light microscope using tweezers. Afterward, each sample was divided into subsamples using a high-precision peristaltic dispensing pump.

Table 1. Time intervals and sample codes of the analyzed sediment trap samples.

2015		2016		2017	
Sample Code	**Sampling Date**	**Sample Code**	**Sampling Date**	**Sample Code**	**Sampling Date**
KMS XV 7	16–31 January 2015	KMS XVII 8	16–31 March 2016	KMS XVIII 2	1–31 May 2017
KMS XV 8	1–15 February 2015	KMS XVII 9	1–15 April 2016	KMS XVIII 3	1–30 June 2017
KMS XV 9	16–28 February 2015	KMS XVII 10	16–30 April 2016	KMS XVIII 4	1–31 July 2017
KMS XV 10	1–15 March 2015	KMSXVII 11	1–15 May 2016	KMS XVIII 5	1–31 August 2017
KMS XV 11	16–31 March 2015	KMS XVII 12	16–31 May 2016	KMSXVIII 6	1–30 September 2017
KMS XV 12	1–15 April 2015	No data (1 June 2016 to 30 April 2017)		KMSXVIII 7	1–31 October 2017
No data (16 April to 31 May 2015)				KMS XVIII 8	1–30 November 2017
KMS XVIII 8	1–30 November 2017				
KMS XVI 1	1–15 June 2015				
KMS XVI 2	16–30 June 2015				
KMS XVI 3	1–15 July 2015				
No data (16 July to 15 August 2015)					
KMS XVI 6	16–31 August 2015				
KMS XVI 7	1–15 September 2015				
KMS XVI 8	16–30 September 2015				
KMS XVI 9	1–15 October 2015				
KMS XVI 10	16–31 October 2015				
No data (1 November 2015 to 15 March 2016)					

2.3. Sample Preparation

Each subsample was first split into 10 equal fractions by McLane rotary wet splitter (<4% deviation between aliquots). One fraction of each subsample was processed in ultrasonic for two minutes to disintegrate pellets and aggregates, then further subdivided with the rotary splitter, and necessary sub-fractions depending upon the coccolith density of samples were filtered onto Millipore cellulose acetate filter (diameter: 47 mm; pore size: 0.45 μm) using a vacuum pump. Afterward, the filter was put in the oven (40 °C) for 30 min to dry and then stored in petri dishes. The solution used in the process of splitting was pre-distilled water that was buffered with sodium carbonate (pH > 8). A small random triangular portion of the filter (approx. 30 mm^2) was mounted on a glass slide for the microscopic analysis.

2.4. Quantitative Microscopic Analysis

For the coccolithophore analysis, we followed standard sample preparation and light microscopy techniques of previous sediment trap studies in the Eastern Mediterranean, e.g., [8,13–16] to ensure the accuracy of data comparison.

The quantitative analysis of coccolith assemblages was performed using Zeiss Axioscope 5 polarizing light microscope at the magnification of 1000×. Coccolith counting for each sample was performed in a filter area of 2 mm^2, except for *E. huxleyi* which was counted in the first 1 mm^2 area. The list of identified species is shown in Table 2. The total coccolith flux and each species' coccolith flux were calculated using the equation of Ziveri et al. [34]: F = (N × Af × S)/(af × Ast × T), where F = coccolith flux (coccolith m^{-2} day^{-1}), N = number of counted coccoliths, Af = effective filtration area (mm^2), S = split factor, af = investigated filtration area (mm^2), Ast = sediment trap aperture area (m^2), T = duration of sampling (days).

Table 2. List of coccolithophore species recorded in the studied sediment trap samples.

Algirosphaera robusta (Lohmann, 1902) Norris, 1984
Calcidiscus leptoporus (Murray and Blackman, 1898) Loeblich and Tappan, 1978
Calciosolenia brasiliensis (Lohmann, 1919) Young, 2003
Ceratolithus cristatus Kamptner, 1950
Coccolithus pelagicus (Wallich, 1877) Schiller, 1930 subsp. *pelagicus*
Discosphaera tubifera (Murray and Blackman, 1898) Ostenfeld 1900
Emiliania huxleyi (Lohmann, 1902) Hay and Mohler in Hay et al. 1967
Florisphaera profunda Okada and Honjo, 1973
Gephyrocapsa oceanica Kamptner, 1943
Gladiolithus flabellatus (Halldal and Markali, 1955) Jordan and Chamberlain, 1993
Helicosphaera carteri (Wallich, 1877) Kamptner, 1954
Pontosphaera spp.
Rhabdosphaera clavigera Murray and Blackman, 1898
Scyphosphaera spp.
Syracosphaera mediterranea Lohmann 1902
Syracosphaera pulchra Lohmann, 1902
Umbellosphaera tenuis (Kamptner, 1954) Paasche, 1955
Umbilicosphaera sibogae (Weber-van Bosse, 1901) Gaarder, 1970

3. Results

3.1. Satellite Measurements of Surface Chl-A Concentration, SST, and Precipitation

Satellite measurements of surface Chl-a concentration (Figure 2) displayed a seasonal pattern with the highest values occurring during late winter–spring (max.: February 2015; 0.2 μg L^{-1}; Figure 2). The annual cycle can be subdivided into a warm (May–October) and a cold season (November–April) according to the SST with values ranging from 15.3 to 29.5 °C, whereas the precipitation followed an anticorrelating pattern with the rainy season maxima during late winter–spring and increased precipitation interval beginning with distinct events within the fall of each annual cycle (Figure 2).

3.2. Total Coccolith Flux

The total coccolith flux at Nestor site at 4300 m ranged from 3.63 × 10^8 coccoliths m^{-2} day^{-1} (late February 2015) to 0.06 × 10^8 coccoliths m^{-2} day^{-1} (November 2017) (Figure 3). Despite the existence of several sampling gaps, the total coccolithophore fluxes displayed an overall uniform seasonal pattern with higher values from February to April and lower values for the rest of the year, except for some solitary peaks during summer (Figure 3). In particular, high coccolith flux was observed from February to March 2015, with an additional single peak in June (3.5×10^8 coccoliths m^{-2} day^{-1}). For 2016, the highest coccolith flux was recorded in early April (1.37×10^8 coccoliths m^{-2} day^{-1}) in contrast to 2017, when the highest flux was observed in June (3.57×10^8 coccoliths m^{-2} day^{-1}).

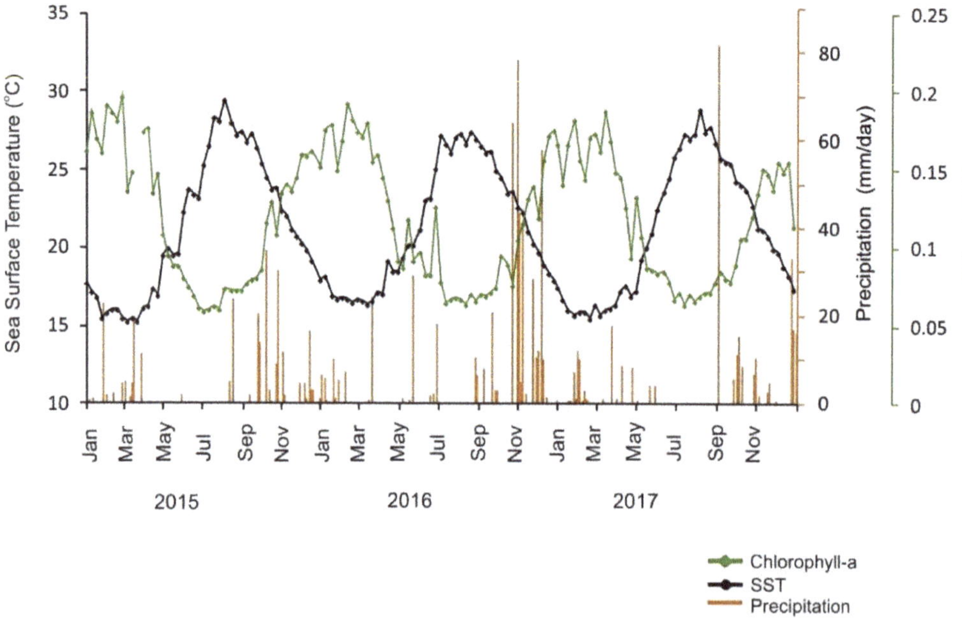

Figure 2. Satellite-derived data of the study area's SST, chlorophyll-a, and precipitation variations during the studied time interval. (Obtained from http://disc.sci.gsfc.nasa.gov/techlab/giovanni/ accessed on 11 October 2022) (all analyses of this study were produced with the Giovanni online data system that is developed and maintained by the NASA GES DISC).

3.3. Coccolithophore Assemblage Composition

In total 19 species of heterococcoliths were identified in the studied samples (see Table 2). *Emiliania huxleyi* was the dominant species (Figure 3), reaching maxima in coccolith flux and relative abundance during early March 2015 (2.4×10^8 coccoliths m^{-2} day^{-1}, 77%) and June 2017 (2.39×10^8 coccoliths m^{-2} day^{-1}, 67 %). The second-most abundant species was *F. profunda* (9–33 %) with maximum coccolith flux recorded in June 2017 (0.9×10^8 coccoliths m^{-2} day^{-1}), late February 2015 (0.73×10^8 coccoliths m^{-2} day^{-1}), and early June 2015 (0.59×10^8 coccoliths m^{-2} day^{-1}), while the highest relative abundance of the species was recorded in July 2017 (33 %). Other species that contributed significantly (max 5–8%) to total coccolith flux included *S. pulchra* (max. late February 2015: 0.2×10^8 coccoliths m^{-2} day^{-1}), *U. sibogae* (max. late June 2015: 0.28×10^8 coccoliths m^{-2} day^{-1}), *R. clavigera* (max. late February 2015: 0.2×10^8 coccoliths m^{-2} day^{-1}), *G. flabellatus* (max. late February 2015: 0.2×10^8 coccoliths m^{-2} day^{-1}), and *C. brasiliensis* (max. late February 2015: 0.16×10^8 coccoliths m^{-2} day^{-1}). Finally, *Calcidiscus leptoporus* and *Helicosphaera carteri* (both up to 2%), displayed their maxima in June 2017 (0.06×10^8 coccoliths m^{-2} day^{-1}) and February 2015 (0.03×10^8 coccoliths m^{-2} day^{-1}), respectively.

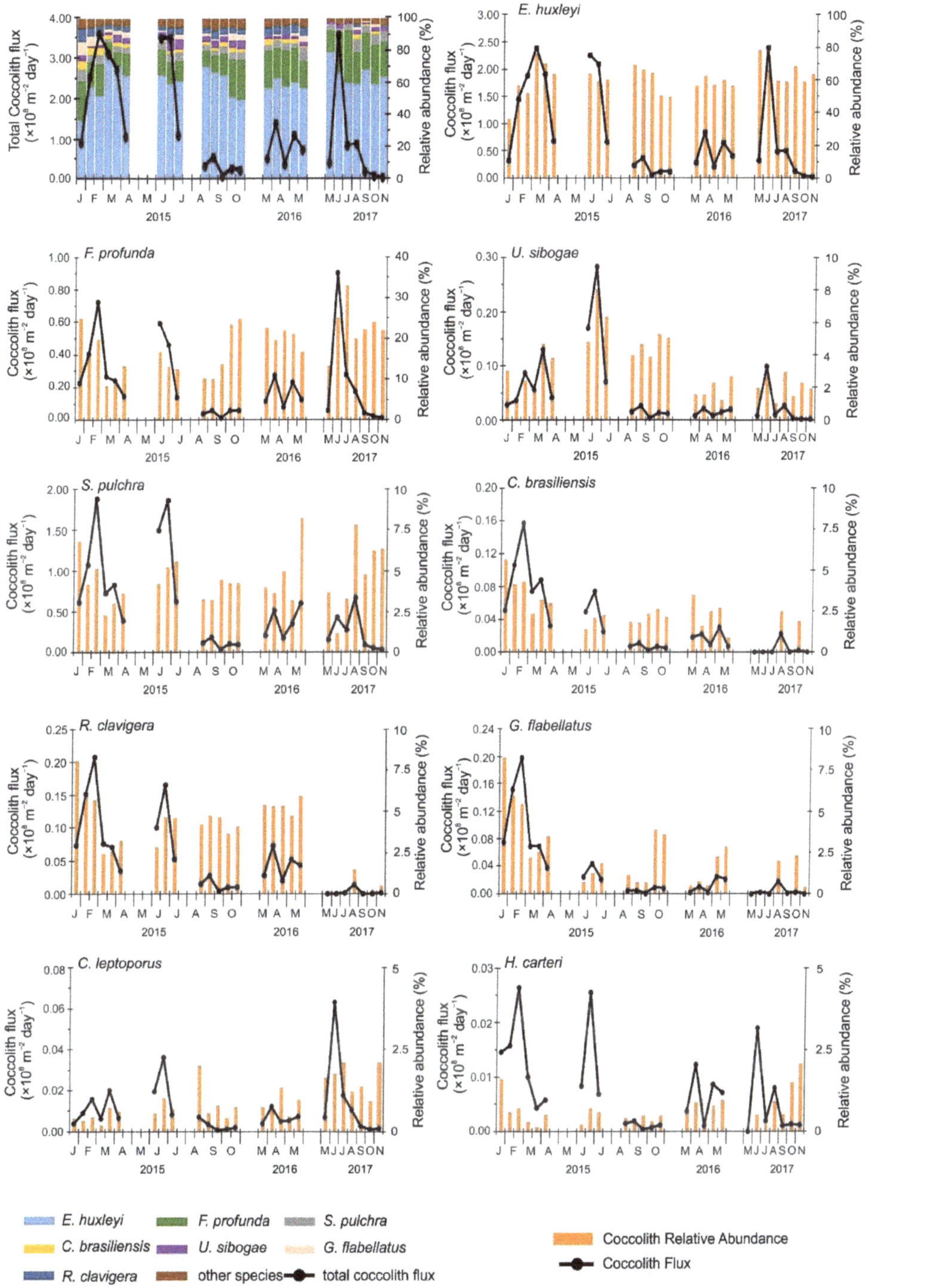

Figure 3. Total coccolith flux and major species coccolith fluxes and relative abundances recorded at 4300 m at Nestor site, SE Ionian Sea.

4. Discussion

4.1. Total Coccolith Flux and Seasonal Trends of Major Coccolithophore Species

Throughout the study period, the coccolith flux at 4300 m displayed a seasonality signal following sea surface productivity, with high values during the late winter–early spring converting mixing period (February to March; Figure 3), when the water column stratification breaks down providing nutrients to the photic zone and enhancing PP [8,12–15,35,36]. Similarly, Stavrakakis et al. [12] detected a peak in total mass fluxes in the late winter/early spring interval, coinciding with relative increases in organic carbon (OC) and opal fluxes at all investigated depths. These seasonal flux maxima were attributed by Stavrakakis et al. [12] to the development of siliceous and nannophytoplankton blooms in the euphotic zone, as also verified by our findings. The coccolith flux peaks at 4300 m depth are well-associated with increased Chl-a values, precipitation events (Figure 2), and increase in atmospheric dust input [24] that contributes significantly to fertilize the Mediterranean surface waters [13,14]. Secondary high isolated peaks were observed during summer and fall time intervals (June 2015, September 2015, June-August 2017; Figure 3), reflecting occasional high productivity in the summer months that is linked to nutrient intrusion due to the upwelling of high salinity CIW and/or LIW water masses [12,13]. This upwelling can be attributed to the seasonal pattern of Pelops anticyclone (Nestor Site is located at the edge of Pelops anticyclone, Figure 1) that repeatedly weakens or totally disappears in the summer months [32]. Likewise, Gogou et al. [37] attributed a similar isolated PP peak during June–July 2012 to the intrusion of LIW/CIW intermediate waters. In agreement with the summer peaks recorded in the present study, Stavrakakis et al. [12] noticed a pronounced flux maximum in all biogenic (OC, carbonate, and opal) fluxes in late spring/summer (May–July) of each year. During this interval, the lithogenic fluxes were also considerably high, indicating enhanced lithogenic inputs in the marine environment as a result of aeolian transport, as supported by Stavrakakis et al. [12].

The coccolith flux signal of all the major species followed the same pattern as of the total coccolith flux; with the highest values documented during high productivity seasons, as also revealed by the increased Chl-a values (Figures 2 and 3). In addition, during the late winter–early spring high productivity season, the conditions are favorable for zooplankton, thus production of pellets increases, which expedites the sinking of coccolithophores [24,38]. Alternatively, high productivity is linked to high input of Saharan dust that fertilizes the surface waters [12,13] but also acts as a ballasting mechanism [39] and can explain the simultaneous increase in fluxes. Skampa et al. [13] reported a significant increase in lithogenic fluxes during the winter–spring high coccolithophore productivity period in the Ionian Sea associated with prominent dust events. Dust is high over the Mediterranean Sea throughout the year but most of the dust deposition (65–80%) occurs as wet deposition during colder months [40] triggering the formation of large aggregates and expediting the sinking process [41]. Hence, it seems rather possible that pellets and dust are both acting as ballast during the spring peak in the SE Ionian, facilitating fluxes down to 4300 m depth, while the high abundance of coccoliths due to the productivity peak is also the reason to expedite the overall flux of both coccoliths and ballast [42]. As shown in previous studies [13–15], the total mass, total coccolith, and total $CaCO_3$ fluxes were reported to increase simultaneously, implying the above-described driving mechanisms. In contrast, during summer the high water column stratification in the Eastern Mediterranean hampers vertical transport, while the upwelling of intermediate waters mostly causes the observed productivity and the subsequent increase in grazer's pellets is primarily responsible for the ballasting. However, isolated precipitation events coupled with slight productivity increase (e.g., June 2016, June 2017; Figure 2) may also have expedited the sinking process, resulting in sporadic flux peaks in summer.

The major species observed at the 4300 m depth are comparable to those detected in other shallower Eastern Mediterranean sites [8,10,13–18]. The coccolith assemblage was dominated by *E. huxleyi* with species' relative abundance ranging from 38 to 79% (practically above 50 %, except for January 2015; Figure 3). The species dominance was also observed in other deep eastern Mediterranean sites, e.g., at 500 m (up to 85%) and 1700 m depth in the Cretan Basin [15] and at 3000 m depth in the Bannock Basin (up to 95%) [14], linked to the phosphate-limited conditions of the eastern basin [15], under which *E. huxleyi* thrives [43].

The relative abundance of the intermediate and deep photic zone (DPZ) species *C. brasiliensis* (0–5.6%), *F. profunda* (9–33%), and *G. flabellatus* (0–8.3%) followed the same seasonal pattern (Figure 3). The year-round presence of DPZ species supports the constant presence of Deep Chlorophyll Maximum (DCM) in the SE Ionian Sea [8]. *Florisphaera profunda,* which represented the second-most abundant species in the assemblages (9–33%), has been widely used to understand the changes in nutricline depth in stratified, oligotrophic waters [40,44,45]. The species' constant presence has been previously observed in other sites of the SE Ionian Sea at 500 m and 2800 m (average 23 and 35%, respectively) [8] and at 500 m and 1700 m depth in the Cretan Basin (average 9%) [15]. Except for the high PP interval, the DPZ species were observed to maintain more or less the same relative abundances annually (Figure 3), which indicates their capacity to endure to the summer stratified water column and deep nutricline. However, high relative abundance for all three species was observed during the pre-high PP period in January 2015. In addition, the high relative abundance of *F. profunda* and *C. brasiliensis* was recorded during the pre-high PP interval of March 2016. The prominent pre-high PP abundance of DPZ species can be explained alternatively by two processes. During December–February, the mixed layer is deep but nutrients slowly start to penetrate the lower photic zone, which facilitates the increase in DPZ species, while as the mixing of water column increases during spring, *E. huxleyi* starts to dominate the assemblages [16]. The alternative explanation for high DPZ species relative abundance in January–February can be the influence of atmospheric triggers and surface water circulation patterns that creates turbidity, reducing the penetration of light and forcing DPZ species to move upwards to shallower depths [46], where they can benefit from nutrient availability, e.g., via wet dust deposition.

The highest peak of *S. pulchra* coccolith flux was observed during February and June 2015 (high productivity seasons). Similar to the findings of the Cretan basin (average *S. pulchra* coccolith relative abundance 3.32%, [15]), the species high relative abundance was recorded in our study for the high SST months (average 5%, max 8%) (Figures 2 and 3), showing its affinity to stratified and oligotrophic waters as well as to higher nutrient availability [14,16]. *Rabdosphaera clavigera* followed the same trend as *S. pulchra*, with the highest flux in February 2015; however, for the sampling period of 2017 it is completely absent, except for August 2017 where it contributes to the assemblages but much less compared with *S. pulchra*. *Umbilicosphaera sibogae*, which prefers warm, stratified, oligotrophic waters and displays a negative correlation to upwelling [47], and showed the highest relative abundance during the summer (June to October 2015), coinciding with the high SST-low precipitation period (Figures 2 and 3).

Helicosphaera carteri flux was overall low and makes up less than 1% of total coccolith flux for most of the year, which confirms the oligotrophic character of the study site [10,48]. The species' coccolith flux was highest during the high PP interval of early February 2015, while it also showed high values during the rest of the high productivity intervals, confirming its correlation with an increase in nutrient availability and PP [15,18]. Likewise, previous studies at southwestern margin of Crete [16] have shown that *H. carteri* (average 2.6%) dominated over the *Helicosphaera* species, with highest values recorded during the late winter to spring high flux period. At the present study, its' high relative abundance was observed during the pre-mixing period (January 2015) and in the post-high PP interval (September to November 2017), which either indicates its tolerance for a variety of trophic conditions or suggests high input of freshwater via precipitation [49–51].

4.2. *Comparison of Coccolith Fluxes at 2000 m and 4300 m for the 2015 Sampling Interval*

For the better understanding of the coccolithophore fluxes in the deep Ionian Sea we performed comparisons with previous studies in the area [13], concerning a single annual cycle (2015) when there is a time overlap of the available data. Skampa et al. [13] studied the coccolithophore export flux at the Nestor site at 2000 m depth from 2010 to 2015. The sampling interval of 2015 overlaps with the time period of the present study; therefore, the comparison of coccolith fluxes at 2000 m and 4300 m would contribute to understanding the coccolith flux dynamics in the Ionian Sea water column (Figure 4). Skampa et al. [13] documented the peak season of coccolithophore flux at 2000 m in between February to March 2015, reflecting the sea surface productivity. The second peak in coccolith flux at 2000 m was observed in June 2015 (Figure 4), while the major contributors were *E. huxleyi* (30–87%) and *F. profunda* with the secondary presence of *S. pulchra*, *R. clavigera*, *A. robusta*, *S. mediterranea*, and *U. sibogae* with the peak fluxes of all major species occurring simultaneously with the total coccolith flux. The first coccolithophore flux maxima at 2000 m coincided with the time of convective mixing during the late winter–spring transition [13], while for the summer peak the intrusion of LIW/CIW intermediate water triggered by dry and cold northerly winds was considered responsible based on a similar observation in 2012 by Gogou et al. [37].

The comparison of the total coccolith and the major contributing species' fluxes at 2000 m obtained by Skampa et al. [13] with their fluxes at 4300 m recorded in the present study (Figures 4 and 5), reveals that the coccolith flux is comparable between the two depths with flux peaks appearing almost instantaneously. In particular, the total coccolith flux correlation between 2000 m and 4300 m during 2015, showed a statistically significant common monthly pattern between the two studied depths ($R^2 = 0.92$ and $\rho < 0.0005$; Figure 5). The period of peak productivity at 4300 m was from February to March, which coincided with the peak productivity at 2000 m (Figure 4), with both traps deployed within the EMDW water mass, thus suggesting fast sinking particle velocities in the deep Ionian Sea water column.

Interestingly, the peak fluxes of *E. huxleyi*, *R. clavigera*, and the DPZ species *F. profunda*, *G. flabellatus*, and *C. brasiliensis* were higher at 4300 m (Figure 4). This may be attributed to their increased contribution to the total coccolith flux during the recorded peaks, as well as to their dissolution tolerance. In contrast, *U. sibogae*, *S. pulchra*, *H. carteri*, *C. leptoporus*, and other minor species presented higher fluxes at 2000 m [13], possibly implying that these taxa are dissolution prone and cannot fully survive the sinking process up to 4300 m. Overall, the total coccolith flux was higher at 4300 m during February and June 2015, i.e., the highest flux periods; in addition to the increased PP, the vertical sinking mechanism and aggregate formation are triggered during these intervals, adding to fast transfer of the coccoliths to deeper layers of the Ionian water column.

The comparison between the export fluxes at 2000 m [13] and 4300 m (present study), shows that apart from export production from the photic zone resuspension or lateral inputs occasionally enhanced the coccolith fluxes at 4300 m, as the sediment trap is located near the sea bottom (4500 m). Similarly, Stavrakakis et al. [12] observed enhanced total fluxes at 4300 m depth and ascribed them to the influence of resuspension and lateral advection from the slope margin. Higher coccolith fluxes of *E. huxleyi* and *F. profunda* (Figure 4) at 4300 m, supports the impact of the above-mentioned bottom processes that alters the coccolithophore export flux signal of the photic zone in the Ionian Sea.

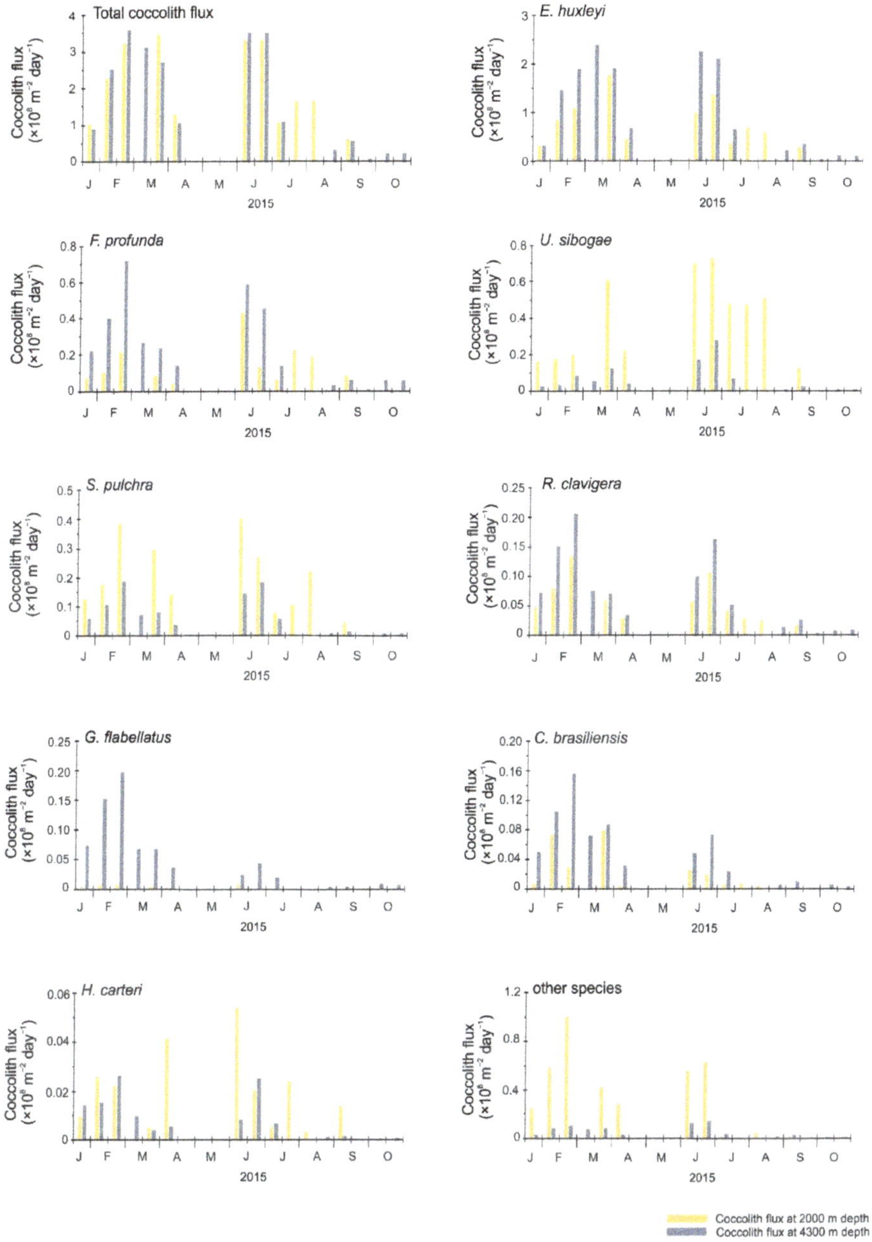

Figure 4. Comparison of the total and major species coccolith fluxes recorded at 2000 m [13] with their fluxes at 4300 m during the 2015 annual cycle sampling interval at Nestor site (2000 m [13] and 4300 m, present study). The total and major species coccosphere units from 2000 m depth (data from Skampa et al. [13]) were converted to coccoliths following the method by Boeckel and Baumann (2008) [52] and were incorporated to the coccolith fluxes at 2000 m depth.

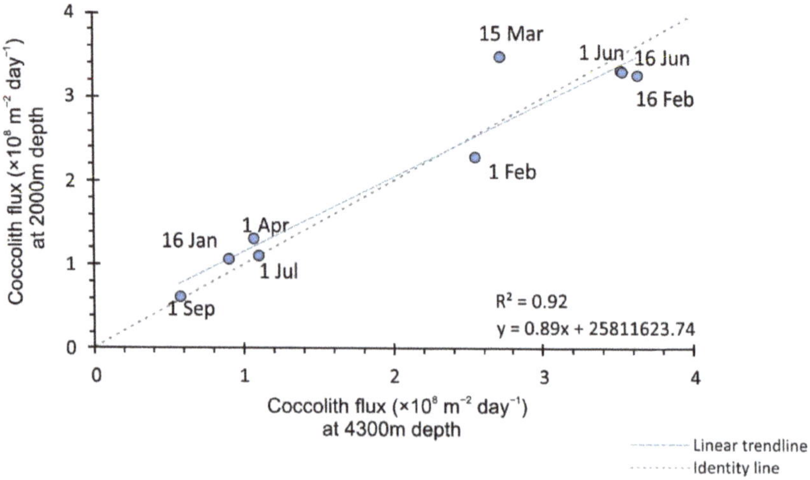

Figure 5. Correlation of the total coccolith fluxes recorded at 2000 m by Skampa et al. [13] with their fluxes at 4300 m during 2015 recorded in the present study.

5. Conclusions

The current study provides important insight into the seasonal variability, environmental control, and quantification of coccolithophore export fluxes at the Nestor site, located in the deepest basin of the Mediterranean. The coccolith flux in the studied sediment trap (4300 m), the deepest of the deployment, was documented and analyzed for the time interval of 2015–2017. In all the studied samples, a high abundance and diversity of coccoliths were observed. The results of the present study were compared with the coccolithophore flux data from 2000 m water depth, reported by Skampa et al. [13]. The main conclusions are described below:

Coccolith fluxes followed a seasonal pattern with the highest values during February–March 2015 (max 3.63×10^8 coccoliths m^{-2} day^{-1}) followed by June 2017 (max 3.57×10^8 coccoliths m^{-2} day^{-1}). The coccolith flux peak during the winter–spring transition period was attributed to convective mixing, while summer peaks were potentially linked to the weakening of Pelops anticyclone and the simultaneous nutrients increase due to regional upwelling. Our findings suggest fast downward particle velocities and preservation of the sea surface seasonal pattern within the Eastern Mediterranean/Ionian Sea deep water column.

In total 19 species of heterococcolithophores were identified. *Emiliania huxleyi* dominated the assemblage (38–79%), indicating the P-limited conditions of the Ionian Sea. The second-most abundant species was *F. profunda* (9–33%) and its significant presence points to the year-round presence of DCM in the study area. Other species that contributed significantly to the coccolith flux were *G. flabellatus, S. pulchra, C. brasiliensis, R. clavigera,* and *U. sibogae.* The assemblage composition was comparable to other shallower E. Mediterranean sediment trap sites, further verifying the preservation of the euphotic zone productivity signal at the deep Nestor site.

The comparison with coccolith flux at 2000 m for the 2015 sampling interval showed that flux at 4300 m was similar and in some cases even higher, possibly due to the influence of resuspension and lateral advection. Nevertheless, the high flux peaks were observed almost instantaneously at both depths pointing to the fast transfer within the deeper layers of the Ionian Sea water column, further enhanced by wet dust deposition and the associated aggregate formation that accelerated the sinking processes.

Supplementary Materials: The following supporting information can be downloaded at: https://www.mdpi.com/article/10.3390/jmse10111761/s1, File S1: Major and minor species coccolith relative abundance in Nestor Site sediment trap (4300 m), File S2: Satellite Measurements of Surface Chl-A Concentration, SST, and Precipitation.

Author Contributions: Conceptualization, S.H., E.S., and M.T.; methodology, S.H., and E.S.; validation, A.G., S.S., and C.P.; writing—original draft preparation, S.H.; writing—review and editing, M.T., E.S., and A.G.; visualization, S.H.; supervision, M.T.; project administration, M.T. All authors have read and agreed to the published version of the manuscript.

Funding: This research is partly supported by Erasmus+ programme of European Union through the scholarship awarded to Sikandar Hayat by PANGEA consortium (https://master-pangea.eu/; "European Joint Master in Paleontology, Geoconservation, Applications", 2019-2025; Erasmus Mundus Joint Master Degree Project Number—610506-EPP-1-2019-1-FR-EPPKA1-JMD-MOB) for his joint master's degree education at University of Lille and National and Kapodistrian University of Athens. This research was funded by the Greek National Project CLIMPACT: Flagship Initiative for Climate Change and its Impact by the Hellenic Network of Agencies for Climate Impact Mitigation and Adaptation.

Institutional Review Board Statement: Not applicable.

Informed Consent Statement: Not applicable.

Data Availability Statement: Data supporting reported results can be found in the Supplementary Materials (Files S1 and S2).

Conflicts of Interest: The authors declare no conflict of interest.

References

1. Pienaar, R.N. Ultrastructure and calcification of coccolithophores. In *Coccolithophores*; Winter, A., Siesser, W.G., Eds.; Cambridge University Press: Cambridge, UK, 1994; pp. 13–38.
2. Rost, B.; Riebesell, U. Coccolithophores and the biological pump: Responses to environmental changes. In *Coccolithophores*; Thierstein, H.R., Young, J.R., Eds.; Springer: Berlin/Heidelberg, Germany, 2004; pp. 99–125. [CrossRef]
3. Cermeño, P.; Dutkiewicz, S.; Harris, R.P.; Follows, M.; Schofield, O.; Falkowski, P.G. The role of nutricline depth in regulating the ocean carbon cycle. *Proc. Natl. Acad. Sci. USA* **2008**, *105*, 20344–20349. [CrossRef] [PubMed]
4. Armstrong, R.A.; Lee, C.; Hedges, J.I.; Honjo, S.; Wakeham, S.G. A new, mechanistic model for organic carbon fluxes in the ocean based on the quantitative association of POC with ballast minerals. *Deep Sea Res. Part II Top. Stud. Oceanogr.* **2001**, *49*, 219–236. [CrossRef]
5. Ziveri, P.; de Bernardi, B.; Baumann, K.H.; Stoll, H.M.; Mortyn, P.G. Sinking of coccolith carbonate and potential contribution to organic carbon ballasting in the deep ocean. *Deep Sea Res. Part II Top. Stud. Oceanogr.* **2007**, *54*, 659–675. [CrossRef]
6. Šupraha, L.; Gerecht, A.C.; Probert, I.; Henderiks, J. Eco-physiological adaptation shapes the response of calcifying algae to nutrient limitation. *Sci. Rep.* **2015**, *5*, 16499. [CrossRef]
7. Lazarus, D.; Barron, J.; Renaudie, J.; Diver, P.; Türke, A. Cenozoic planktonic marine diatom diversity and correlation to climate change. *PLoS ONE* **2014**, *9*, e84857. [CrossRef] [PubMed]
8. Malinverno, E.; Maffioli, P.; Corselli, C.; De Lange, G.J. Present-day fluxes of coccolithophores and diatoms in the pelagic Ionian Sea. *J. Mar. Syst.* **2014**, *132*, 13–27. [CrossRef]
9. Ziveri, P.; Grandi, C.; Stefanetti, A.; Cita, M.B. Biogenic fluxes in Bannock Basin: First results from a sediment trap study (November 1991–May 1992). *Rend. Lincei* **1995**, *6*, 131–145. [CrossRef]
10. Malinverno, E.; Ziveri, P.; Corselli, C. Coccolithophorid distribution in the Ionian Sea and its relationship to eastern Mediterranean circulation during late fall to early winter 1997. *J. Geophys. Res. Ocean.* **2003**, *108*, 8115. [CrossRef]
11. Stavrakakis, S.; Chronis, G.; Tselepides, A.; Heussner, S.; Monaco, A.; Abassi, A. Downward fluxes of settling particles in the deep Cretan Sea (NE Mediterranean). *Prog. Oceanogr.* **2000**, *46*, 217–240. [CrossRef]
12. Stavrakakis, S.; Gogou, A.; Krasakopoulou, E.; Karageorgis, A.P.; Kontoyiannis, H.; Rousakis, G.; Velaoras, D.; Perivoliotis, L.; Kambouri, G.; Stavrakaki, I.; et al. Downward fluxes of sinking particulate matter in the deep Ionian Sea (NESTOR site), eastern Mediterranean: Seasonal and interannual variability. *Biogeosciences* **2013**, *10*, 7235–7254. [CrossRef]
13. Skampa, E.; Triantaphyllou, M.V.; Dimiza, M.D.; Gogou, A.; Malinverno, E.; Stavrakakis, S.; Parinos, C.; Panagiotopoulos, I.P.; Tselenti, D.; Archontikis, O.; et al. Coccolithophore export in three deep-sea sites of the Aegean and Ionian Seas (Eastern Mediterranean): Biogeographical patterns and biogenic carbonate fluxes. *Deep Sea Res. Part II Top. Stud. Oceanogr.* **2020**, *171*, 104690. [CrossRef]
14. Ziveri, P.; Rutten, A.; De Lange, G.J.; Thomson, J.; Corselli, C. Present-day coccolith fluxes recorded in central eastern Mediterranean sediment traps and surface sediments. *Palaeogeogr. Palaeoclimatol. Palaeoecol.* **2000**, *158*, 175–195. [CrossRef]

15. Triantaphyllou, M.V.; Ziveri, P.; Tselepides, A. Coccolithophore export production and response to seasonal surface water variability in the oligotrophic Cretan Sea (NE Mediterranean). *Micropaleontology* **2004**, *50* (Suppl. S1), 127–144. [CrossRef]
16. Malinverno, E.; Triantaphyllou, M.V.; Stavrakakis, S.; Ziveri, P.; Lykousis, V. Seasonal and spatial variability of coccolithophore export production at the South-Western margin of Crete (Eastern Mediterranean). *Mar. Micropaleontol.* **2009**, *71*, 131–147. [CrossRef]
17. Dimiza, M.D.; Triantaphyllou, M.V.; Dermitzakis, M.D. Seasonality and ecology of living coccolithophores in Eastern Mediterranean coastal environment (Andros Island, Middle Aegean Sea). *Micropaleontology* **2008**, *54*, 159–175. [CrossRef]
18. Dimiza, M.D.; Triantaphyllou, M.V.; Malinverno, E.; Psarra, S.; Karatsolis, B.T.; Mara, P.; Lagaria, A.; Gogou, A. The composition and distribution of living coccolithophores in the Aegean Sea (NE Mediterranean). *Micropaleontology* **2015**, *61*, 521–540. [CrossRef]
19. El Hourany, R.; Abboud-abi Saab, M.; Faour, G.; Mejia, C.; Crépon, M.; Thiria, S. Phytoplankton diversity in the Mediterranean Sea from satellite data using self-organizing maps. *J. Geophys. Res. Ocean.* **2019**, *124*, 5827–5843. [CrossRef]
20. Trimonis, E.; Rudenko, M. Geomorphology and bottom sediments of the Pylos area. In Proceedings of the 2nd NESTOR International Workshop; Resvanis, L.K., Ed.; Fortress of Niokastro: Pylos, Greece, 1992; p. 321.
21. D'Ortenzio, F.; Ragni, M.; Marullo, S.; Ribera d'Alcalà, M. Did biological activity in the Ionian Sea change after the Eastern Mediterranean Transient? Results from the analysis of remote sensing observations. *J. Geophys. Res. Ocean.* **2003**, *108*, 8113. [CrossRef]
22. Crombet, Y.; Leblanc, K.; Queguiner, B.; Moutin, T.; Rimmelin, P.; Ras, J.; Claustre, H.; Leblond, N.; Oriol, L.; Pujo-Pay, M. Deep silicon maxima in the stratified oligotrophic Mediterranean Sea. *Biogeosciences* **2011**, *8*, 459–475. [CrossRef]
23. Karageorgis, A.P.; Georgopoulos, D.; Kanellopoulos, T.D.; Mikkelsen, O.A.; Pagou, K.; Kontoyiannis, H.; Pavlidou, A.; Anagnostou, C. Spatial and seasonal variability of particulate matter optical and size properties in the Eastern Mediterranean Sea. *J. Mar. Syst.* **2012**, *105*, 123–134. [CrossRef]
24. Theodosi, C.; Markaki, Z.; Pantazoglou, F.; Tselepides, A.; Mihalopoulos, N. Chemical composition of downward fluxes in the Cretan Sea (Eastern Mediterranean) and possible link to atmospheric deposition: A 7 year survey. *Deep Sea Res. Part II Top. Stud. Oceanogr.* **2019**, *164*, 89–99. [CrossRef]
25. Rutten, A.; De Lange, G.J.; Ziveri, P.; Thomson, J.; Van Santvoort, P.J.M.; Colley, S.; Corselli, C. Recent terrestrial and carbonate fluxes in the pelagic eastern Mediterranean; a comparison between sediment trap and surface sediment. *Palaeogeogr. Palaeoclimatol. Palaeoecol.* **2000**, *158*, 197–213. [CrossRef]
26. Tanaka, T.; Zohary, T.; Krom, M.D.; Law, C.S.; Pitta, P.; Psarra, S.; Rassoulzadegan, F.; Thingstad, T.F.; Tselepides, A.; Woodward, E.M.S.; et al. Microbial community structure and function in the Levantine Basin of the eastern Mediterranean. *Deep Sea Res. Part I Oceanogr. Res. Pap.* **2007**, *54*, 1721–1743. [CrossRef]
27. Powley, H.R.; Van Cappellen, P.; Krom, M.D. Nutrient cycling in the Mediterranean Sea: The key to understanding how the unique marine ecosystem functions and responds to anthropogenic pressures. In *Mediterranean Identities—Environment, Society, Culture*; Soc Cult. InTech: London, UK, 2017; pp. 47–77. [CrossRef]
28. Casotti, R.; Landolfi, A.; Brunet, C.; d'Ortenzio, F.; Mangoni, O.; Ribera d'Alcalà, M.; Denis, M. Composition and dynamics of the phytoplankton of the Ionian Sea (eastern Mediterranean). *J. Geophys. Res. Oceans* **2003**, *108*. [CrossRef]
29. Nittis, K.; Pinardi, N.; Lascaratos, A. Characteristics of the summer 1987 flow field in the Ionian Sea. *J. Geophys. Res. Ocean.* **1993**, *98*, 10171–10184. [CrossRef]
30. Malanotte-Rizzoli, P.; Manca, B.B.; d'Alcalà, M.R.; Theocharis, A.; Bergamasco, A.; Bregant, D.; Budillon, G.; Civitarese, G.; Georgopoulos, D.; Michelato, A.; et al. A synthesis of the Ionian Sea hydrography, circulation and water mass pathways during POEM-Phase I. *Prog. Oceanogr.* **1997**, *39*, 153–204. [CrossRef]
31. Estournel, C.; Marsaleix, P.; Ulses, C. A new assessment of the circulation of Atlantic and Intermediate Waters in the Eastern Mediterranean. *Prog. Oceanogr.* **2021**, *198*, 102673. [CrossRef]
32. Larnicol, G.; Ayoub, N.; Le Traon, P.Y. Major changes in Mediterranean Sea level variability from 7 years of TOPEX/Poseidon and ERS-1/2 data. *J Mar Syst.* **2002**, *33*, 63–89. [CrossRef]
33. Acker, J.G.; Leptoukh, G. Online analysis enhances use of NASA earth science data. *Eos Trans. AGU* **2007**, *88*, 14–17. [CrossRef]
34. Ziveri, P.; Young, J.R.; Van Hinte, J.E. Coccolithophore export production and accumulation rates. *GeoResearch Forum* **1999**, *5*, 41–56.
35. D'Ortenzio, F.; Ribera d'Alcalà, M. On the trophic regimes of the Mediterranean Sea: A satellite analysis. *Biogeosciences* **2009**, *6*, 139–148. [CrossRef]
36. Patara, L.; Pinardi, N.; Corselli, C.; Malinverno, E.; Tonani, M.; Santoleri, R.; Masina, S. Particle fluxes in the deep Eastern Mediterranean basins: The role of ocean vertical velocities. *Biogeosciences* **2009**, *6*, 333–348. [CrossRef]
37. Gogou, A.; Stavrakakis, S.; Triantaphyllou, M.; Paraskos, F.; Parinos, C.; Dimiza, M.; Kambouri, G.; Lykousis, V. Seasonal and interannual variability of sinking particulate matter in the deep Ionian Sea: Ecological and biogeochemical perspectives. *Rapp. Comm. Int. Mer Medit.* **2016**, *41*. Available online: https://ciesm.org/online/archives/abstracts/pdf/41/CIESM_Congress_2016_Kiel_article_0083.pdf (accessed on 11 October 2022).
38. Turner, J.T. Zooplankton fecal pellets, marine snow and sinking phytoplankton blooms. *Aquat. Microb. Ecol.* **2002**, *27*, 57–102. [CrossRef]

39. Schulz, M.; Prospero, J.M.; Baker, A.R.; Dentener, F.; Ickes, L.; Liss, P.S.; Mahowald, N.M.; Nickovic, S.; García-Pando, C.P.; Rodríguez, S.; et al. Atmospheric transport and deposition of mineral dust to the ocean: Implications for research needs. *Environ. Sci. Technol.* **2012**, *46*, 10390–10404. [CrossRef]
40. Cascella, A.; Bonomo, S.; Jalali, B.; Sicre, M.A.; Pelosi, N.; Schmidt, S.; Lirer, F. Climate variability of the last ~2700 years in the Southern Adriatic Sea: Coccolithophore evidence. *Holocene* **2020**, *30*, 53–64. [CrossRef]
41. Molinaroli, E.; Guerzoni, S.; Rampazzo, G. Contribution of Saharan dust to the Central Mediterranean Basin. *Geol. Soc. Am. Spec. Pap.* **1993**, *284*, 303–312. [CrossRef]
42. Klaas, C.; Archer, D.E. Association of sinking organic matter with various types of mineral ballast in the deep sea: Implications for the rain ratio. *Glob. Biogeochem. Cycles* **2002**, *16*, 63-1. [CrossRef]
43. Young, J.R. Variation in *Emiliania huxleyi* coccolith morphology in samples from the Norwegian EHUX experiment, 1992. *Sarsia* **1994**, *79*, 417–425. [CrossRef]
44. Molfino, B.; McIntyre, A. Precessional forcing of nutricline dynamics in the equatorial Atlantic. *Science* **1990**, *249*, 766–769. [CrossRef]
45. Triantaphyllou, M.V.; Ziveri, P.; Gogou, A.; Marino, G.; Lykousis, V.; Bouloubassi, I.; Emeis, K.-V.; Kouli, K.; Dimiza, M.; Rosell-Melé, A.; et al. Late Glacial–Holocene climate variability at the south-eastern margin of the Aegean Sea. *Mar. Geol.* **2009**, *266*, 182–197. [CrossRef]
46. Bonomo, S.; Schroeder, K.; Cascella, A.; Alberico, I.; Lirer, F. Living coccolithophore communities in the central Mediterranean Sea (Summer 2016): Relations between ecology and oceanography. *Mar. Micropaleontol.* **2021**, *165*, 101995. [CrossRef]
47. Kleijne, A. Morphology, taxonomy and distribution of extant coccolithophorids (calcareous nannoplankiton). *Neth. J. Sea Res.* **1993**. Available online: https://www.amazon.co.uk/Morphology-Distribution-Coccolithophorids-Calcareous-Nannoplankton/dp/9090061614 (accessed on 11 October 2022).
48. Knappertsbusch, M. Geographic distribution of living and Holocene coccolithophores in the Mediterranean Sea. *Mar. Micropaleontol.* **1993**, *21*, 219–247. [CrossRef]
49. Flores, J.A.; Sierro, F.J.; Francés, G.; Vázquez, A.; Zamarren, I. The last 100,000 years in the western Mediterranean: Sea surface water and frontal dynamics as revealed by coccolithophores. *Mar. Micropaleontol.* **1997**, *29*, 351–366. [CrossRef]
50. Colmenero-Hidalgo, E.; Flores, J.A.; Sierro, F.J.; Bárcena, M.Á.; Löwemark, L.; Schönfeld, J.; Grimalt, J.O. Ocean surface water response to short-term climate changes revealed by coccolithophores from the Gulf of Cadiz (NE Atlantic) and Alboran Sea (W Mediterranean). *Palaeogeogr. Palaeoclimatol. Palaeoecol.* **2004**, *205*, 317–336. [CrossRef]
51. Dimiza, M.D.; Triantaphyllou, M.V.; Malinverno, E. New Evidence for The Ecology of *Helicosphaera Carteri* In Polluted Coastal Environments (Elefsis Bay, Saronikos Gulf, Greece). *JNR* **2014**, *34*, 37–43.
52. Boeckel, B.; Baumann, K.-H. Vertical and lateral variations in coccolithophore community structure across the subtropical frontal zone in the South Atlantic Ocean. *Mar. Micropaleontol.* **2008**, *67*, 255–273. [CrossRef]

Journal of
Marine Science and Engineering

MDPI

Correction

Correction: Fazal et al. Geochemical Analysis of Cretaceous Shales from the Hazara Basin, Pakistan: Provenance Signatures and Paleo-Weathering Conditions. *J. Mar. Sci. Eng.* 2022, *10*, 800

Abdul Ghaffar Fazal [1,*], Muhammad Umar [2], Faisal Shah [1,3], Muhammad Armaghan Faisal Miraj [4], Hammad Tariq Janjuhah [5,*], George Kontakiotis [6] and Abdul Khaliq Jan [7]

1 Department of Earth Sciences, Abbottabad University of Science and Technology, Havelian 22500, Pakistan
2 Department of Earth Sciences, The University of Haripur, Haripur 22620, Pakistan
3 National Centre of Excellence in Geology, University of Peshawar, Peshawar 25000, Pakistan
4 Institute of Geology, University of the Punjab, Lahore 54590, Pakistan
5 Department of Geology, Shaheed Benazir Bhutto University, Sheringal 18050, Pakistan
6 Department of Historical Geology-Paleontology, Faculty of Geology and Geoenvironment, School of Earth Sciences, National and Kapodistrian University of Athens, Panepistimiopolis, 15784 Athens, Greece
7 Department of Chemistry, Shaheed Benazir Bhutto University, Sheringal 18050, Pakistan
* Correspondence: ghaffargeology@aust.edu.pk (A.G.F.); hammad@sbbu.edu.pk (H.T.J.)

Citation: Fazal, A.G.; Umar, M.; Shah, F.; Miraj, M.A.F.; Janjuhah, H.T.; Kontakiotis, G.; Jan, A.K. Correction: Fazal et al. Geochemical Analysis of Cretaceous Shales from the Hazara Basin, Pakistan: Provenance Signatures and Paleo-Weathering Conditions. *J. Mar. Sci. Eng.* 2022, *10*, 800. *J. Mar. Sci. Eng.* **2022**, *10*, 1654. https://doi.org/10.3390/jmse10111654

Received: 12 September 2022
Accepted: 12 October 2022
Published: 4 November 2022

Addition of an Author

In the original publication [1], there was a mistake in the authorship. Muhammad Umar should be added as the co-author. His affiliation is the Department of Earth Sciences, The University of Haripur, Haripur 22620, Pakistan.

New Author Contributions Statement

Conceptualization, A.G.F. and M.U.; methodology, A.G.F.; software, A.G.F. and H.T.J.; validation, A.G.F., M.U. and H.T.J.; formal analysis, A.G.F., M.U. and H.T.J.; investigation, A.G.F., M.U. and H.T.J.; resources, A.G.F. and M.U.; data curation, A.G.F. and M.U.; writing—original draft preparation, A.G.F. and M.U.; writing—review and editing, H.T.J., F.S., M.A.F.M., G.K. and A.K.J.; visualization, M.U., H.T.J., F.S. and M.A.F.M.; supervision, M.U.; project administration, M.U.; funding acquisition, G.K. and H.T.J. All authors have read and agreed to the published version of the manuscript.

Reference

1. Fazal, A.G.; Umar, M.; Shah, F.; Miraj, M.A.F.; Janjuhah, H.T.; Kontakiotis, G.; Jan, A.K. Geochemical Analysis of Cretaceous Shales from the Hazara Basin, Pakistan: Provenance Signatures and Paleo-Weathering Conditions. *J. Mar. Sci. Eng.* **2022**, *10*, 800. [CrossRef]

MDPI
St. Alban-Anlage 66
4052 Basel
Switzerland
Tel. +41 61 683 77 34
Fax +41 61 302 89 18
www.mdpi.com

Journal of Marine Science and Engineering Editorial Office
E-mail: jmse@mdpi.com
www.mdpi.com/journal/jmse

www.ingramcontent.com/pod-product-compliance
Lightning Source LLC
Chambersburg PA
CBHW061510220926
51850CB00004B/221